L 111

L 112

L 112

L 110

W5

L 101

W1

L 109

L 102

W2

WHITE FLOOR

RED NUBIAN SANDSTONE

LIMESTONE

GRANITE

WHITE SANDSTONE

0   1m.

Metal in History: Two

Edited by Beno Rothenberg

Editorial Committee: *A. Arribas-Palau,*
H. G. Bachmann, P. A. Clayton, J. D. Evans,
M. Gichon, C. T. Shaw, R. F. Tylecote,
P. Wincierz

Researches in the Arabah
1959–1984

The Arabah Project sponsored by Stiftung Volkswagenwerk

Volume I

Institute for Archaeo-Metallurgical Studies
Institute of Archaeology, University College London

The Egyptian Mining Temple
at Timna

by

Beno Rothenberg

with contributions by
H. G. Bachmann, I. L. Barnes, R. H. Brill, P. Craddock, A. Fahan,
Ch. Frydmann, M. Gichon, J. Glass, T. Kertesz, M. E. Kislev,
G. Lehrer-Jacobsen, H. Lernau, A. Lupu, E. Minoff,
W. A. Oddy, D. Price, D. S. Reese, A. Sheffer,
A. Schulman, R. F. Tylecote, E. Werker, A. E. Werner,
M. Woudhuysen, F. Woodward, E. Zamski
and the archaeological team of the Arabah Expedition

1988
Institute for Archaeo-Metallurgical Studies
Institute of Archaeology, University College London

To
DICK ALTHAM
in friendship

Distributed for IAMS by

Thames and Hudson (Distributors) Ltd.
44 Clockhouse Road,
Farnborough, Hants. GU14 7QZ.

to whom all orders should be addressed.

© 1988, The Institute for Archaeo-Metallurgical Studies, London

British Library Cataloguing in Publication Data
The Egyptian mining temple at Timna.—
    (Researches in the Arabah 1959–84; v. 1).
    1. Temples, Egyptian—Israel   2. Timna
    Site (Israel)   3. Israel—Antiquities
    I. Institute for Archaeo-Metallurgical
    Studies   II. Series
    932   DS110.T6/

ISBN 0 906183 02 2

Published by The Institute for Archaeo-Metallurgical Studies (IAMS)
The Institute of Archaeology, University College London, 31–34 Gordon Square, London WC1H 0PY.

Text and illustration origination by Pardy & Son (Printers) Limited.

Printed in Great Britain by Pardy & Son (Printers) Limited, Ringwood, Hampshire.

Contents

Foreword

During the twenty-five years of our fieldwork in the Arabah and, parallel to it, the intensive archaeo-metallurgical investigations, a huge body of finds and information was accumulated. The definitive publication of this material will be presented in the four volumes of *Researches in the Arabah 1959–1984*. Contrary to the natural sequence of the Arabah research programme – field investigations of the site, finds and work, archaeo-metallurgical research, experiments and theory – the first volume of *Researches in the Arabah* contains the excavation report on the Egyptian Mining Temple dedicated to the goddess Hathor which provided much of the basic chronological and culture-historical evidence for the southern Arabah. The second volume, *The Ancient Extractive Metallurgy of Copper*, presents the factual, experimental and theoretical investigations into the archaeo-metallurgy of the area. It was felt that the publication of the basic archaeological and metallurgical evidence prior to the detailed reports on the explorations and excavations in the Arabah would be appropriate. Volume 3 will report on the extensive explorations in the Arabah and the archaeological and archaeo-metallurgical finds. Volume 4 will contain the comprehensive report on the excavations of the copper mining and smelting sites at Timna and in the Arabah.

Twenty-five years of almost continuous research often produced new ideas, facts or interpretations which naturally meant reconsidering our own previously held and published views. Even one single fact which could not be accommodated by our current archaeological or metallurgical concepts often led to renewed intensive investigations and, consequently, revised culture-historical concepts and archaeo-metallurgical process-models. To bring these developments into perspective, and also to emphasise the significance of the discovery of the Egyptian Mining Temple at Timna, it was thought useful to include in the present volume a concise review of the history of the earlier investigations of the Arabah.

Chapters I, II and IV of the present volume contain the final report on the methodology and excavations of the Temple, prepared by the excavator. Although the detailed field notes by Reinhard Maag (1969) and Morag Woudhuysen (1974), together with the find records of the Arabah Expedition team, were of decisive importance for the preparation of this report, the writer accepts sole and full responsibility for its contents and conclusions.

Most of the specialist studies of the finds from the Temple are signed by their authors. It should be mentioned, however, that some of these reports were written and submitted in the early 1970s, soon after the conclusion of the first major season of excavation at Site 200. In some cases it has not been possible to

update them and wherever it seemed necessary, the date of submission of the respective contributions is recorded in the Notes.

In the preparation of Chapter III, section 13, 'Debris from metallurgical activities at Site 200', I was greatly assisted by analytical studies of the material made by the late Alexandru Lupu, formerly of the Haifa Technion, and I would like to acknowledge here the very valuable contributions he made during the early stages of our Arabah research. Chapter III, sections 1, 2, 3 and 7 are unsigned and represent a collective effort over a number of years by the Arabah Expedition team. They have been edited, partly re-written and updated by myself and I therefore accept the responsibility for their present content. I would especially like to thank five former members of the Arabah Expedition for their important contributions: Gloria London and Ivan Ordentlich (pottery), and Benett Kozloff and Ilana Mozel (flints). Benett Kozloff also prepared the original catalogue of the metal finds (revised by myself in 1984). Craig Meredith was responsible for the detailed recording, classification and photography of the huge number of small finds from the Temple.

In 1969, when I joined the academic staff of Tel Aviv University as a Senior Lecturer for a number of years, the excavation of Site 200 was carried out under the auspices of the Institute of Archaeology, Tel Aviv University, and its Director, the late Professor Johanan Aharoni. I would like to acknowledge here, with deep gratitude, the decisive role he played as my teacher and friend in the most difficult years of my struggle for the right to free and unbiased archaeological research in Israel, as indicated in the Introduction to this volume. He provided a proper academic basis for my independent research work and the moral support so badly needed in those days. It was also Johanan Aharoni who initiated the setting up of the independent Institute of Mining and Metallurgy in the Biblical World, Tel Aviv, which, under my direction and in collaboration with the Institute for Archaeo-Metallurgical Studies, London University, brought the Arabah Project to its conclusion in 1984.

It was mainly the discovery of the Egyptian Mining Temple at Timna which created considerable professional and public interest in our work in the Arabah and its significance. Largely due to the intellectual vision of the late Sir Mortimer Wheeler (British Academy) and the late Dr Richard Barnett (British Museum), our work was exhibited at the British Museum in 1971. It was due to this exhibition and the initiatives of Sir Mortimer Wheeler, the late Sir Val Duncan (chairman of Rio Tinto Zinc Corporation, London) and Sir Sigmund Sternberg, that the Institute for Archaeo-Metallurgical Studies (IAMS), based on the Institute of Archaeology, University College London, was founded. Since 1973 IAMS has provided the major base for my research work and that of several of my research associates.

I would like to take this opportunity to express my sincere gratitude to the current Trustees of IAMS for their vital support, past and present, which makes it possible for me and my colleagues to apply the lessons learned at Timna to many other important areas of significance in the history of metal: R. J. L. Altham; D. Rafael Benjumea Cabeza de Vaca; Professor J. D. Evans; Sir Alistair Frame; Tom Kennedy; Nigel Lyon; Sir Ronald Prain OBE; Robert Rice; Sir Sigmund Sternberg KCSG, JP; Simon D. Strauss; Professor R. F. Tylecote, Patricia Walker and Casimir Prinz Wittgenstein.

Many thanks are due to the expedition's team at Site 200, listed in the following report, and I wish to add here my appreciation of the devoted professional work on the Temple material by the following: Drawing of finds – Ruth Halfon, Naomi Schechter, Ora Semmer and, especially, Ana Hasson who prepared the final versions and the plates for this volume; Photography – Craig Meredith, Jacob Marcus and Abe Hai; Draughtsmen (plans and maps) – (late) S. Moscovits, Judith Gavish and Ora Paran (for the final, revised version of all the plans and sections of the excavations). We also had much support from local people and institutions; (late) Yoske Levy, the first Mayor of the City of Eilat; Avinoam Finkelmann, formerly head of the Eilat Regional Council, and the Timna Mining Company.

The detailed investigation of the finds at Site 200 and the preparation of the present excavation report would not have been possible without the generous support of The Volkswagen Foundation, Hannover, and a group of friends of Timna in London, New York and Geneva: Ernest Fraenkel, Walter Griessmann, Ludwig Jesselson, Ralph Kestenbaum, Nigel Lion, Sir Leslie Porter, Felix Posen, Lionel Schalit and Leo Schiff. To all of them we would like to record here the most sincere thanks of all the members of the Arabah team.

Grateful thanks are expressed to Judy Allen in Madrid, who turned a bewildering variety of texts into a tidy, reasoned and ordered typescript ready for the editors and printer, thereby making an immense contribution by her efficient assistance.

A last word of thanks and appreciation goes to Peter Clayton who edited and saw through the press 'Rothenberg 1971' and 'Rothenberg 1972' – he has carried out the similar and more onerous task here for this first volume of the Arabah series and has always been ready in the intervening years with help and advice.

Beno Rothenberg
Tel Aviv and London,
September 1987

Introduction to Researches in the Arabah

1 Wadi Arabah

The Arabah, as part of the huge African rift valley, extends from the south end of the Dead Sea to the shores of the Red Sea (*Illus.* 1). It is a 175 km. long valley, which separates the high plateau of Trans-Jordan and, further south, adjacent Hejaz (N.W. Arabia) from the South Negev and the Sinai Peninsula. An almost eternal nomans land at the crossroad of two continents – Asia and Africa – it borders on four geographically and culture-historically different but in many ways interconnected territories: Israel, Jordan (Edom), Hejaz (Midian–N.W.Arabia) and Sinai. Yet the rather arid and inhospitable nature of this long stretch of land, hemmed in on both sides by steep mountain ranges with very few pathways, did not favour sedentary human habitation. Although geographically the shortest route between south and north, it only became a main thoroughfare in modern times (Rothenberg, 1971a). With the exception of several kilometres at its north and south end, occupation of the Arabah was limited by the meagre water resources, saline soil and high temperatures to camping sites, road stations and several Roman-Byzantine castles, mainly found along roads crossing the Arabah from west to east. This picture of an inhospitable and alien desert area, shunned even by the Bedouin, drastically changed wherever copper mineral deposits attracted miners and metallurgists, small groups of primitive tribal artisans or large scale Pharaonic expeditions, to mine and smelt copper.

Clusters of settlements and copper smelting camps were found in the wadis and side valleys of the Arabah wherever copper ores could be extracted from the depth of the sandstone layers and also in the wider surroundings of the mineral deposits along the fringes of the Arabah itself, where primitive settlements and smelting camps were located on isolated hilltops and rocky mountain slopes. It is here that small heaps and concentrations of blackish slag indicated intensive copper smelting activities, connected with clusters of stone enclosures and building foundations, datable by archaeological evidence from prehistoric to early medieval times.

The unfortunate political realities of the area – the border between Israel and Jordan runs down exactly in the middle of the Arabah, from the Dead Sea to the Red Sea – limited our research activities to the west side of the Arabah Valley and this terrain was first surveyed archaeologically by the Arabah Expedition.[1] The main emphasis of this work was the location and investigation of ancient copper mines and smelters against the culture-historical background of the Arabah as a whole.[2] Although minor copper mineralization and mine workings occur in several wadis of the southern Arabah, the major mining site of the western Arabah was located in the Timna Valley (W. Meneijeh),[3] with a similar, but much smaller copper mining and smelting area in the Nahal Amram (W. Amrani) *c.* 10 km. further south.[4] Most of the archaeological excavations of our research group were located in the Timna Valley and its close surroundings, although minor excavations were also undertaken at mining-related sites in the Arabah and even as far as southern Sinai.[5]

2 Previous research in the south-western Arabah and the Timna Valley

Of the long list of explorers and travellers who visited the Arabah during the 19th and early 20th century, only very few mentioned copper mining or smelting remains in the south-western Arabah.

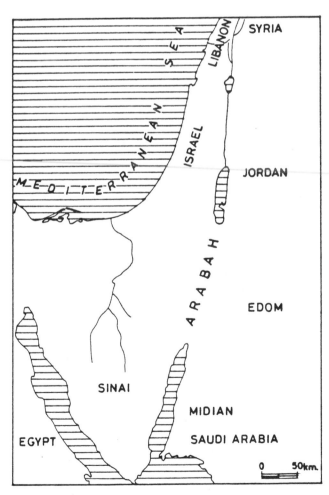

Illustration 1. The Arabah Valley location map.

1

(1) *1845: John Petherick* The British mining engineer, John Petherick (British Consul at Khartoum), who for 16 years undertook wide exploratory tours of Egypt, the Sudan and Central Africa, undertook the investigation the southern Arabah on behalf of Mehemet Ali, Viceroy of Egypt (Petherick, 1861). He reported a number of slag heaps and copper smelting camps, including copper slag heaps at 'Riguel Hadid' (Site 4 on the Arabah Survey Map) and extensive copper smelting installations in 'Wadi-il-Muhait' (now Site 30, Timna Valley). 'At Riguel Hadid and Wadi-il-Muhait, on the west side of Wadi Arabah, are two very interesting spots, where copper ores were formerly smelted; the slag still remained, which contained a large proportion of copper. The latter of the two must have been the most considerable smelting locality, judging by the quantity of slag lying there, the whole of which, comprising a large area, is enclosed within a dry-stone wall; the greater number of stones, being lime-stone, were probably brought there as a flux for the reduction of the ores. But from whence those ores, or the fuel with which to smelt them, were derived, or who were the operators, were questions which the Arabs could not answer, nor myself divine.' (Petherick, 1861, 37). According to this description, Petherick was the first European to visit the walled campsite Site 30 in the Timna Valley[6] and recognise its function as a copper smelting installation. For unexplained reasons, Petherick did not look for the ancient copper mining sites of the Arabah.

(2) *1902: Alois Musil* Alois Musil, the Austrian topographer and theology professor, visited the Timna Valley ('al Meneijje') on 10 September 1902, between 11.32 and 14.35 hours, and observed 'a few remains of human settlements'. Musil mentioned the Arab guide's story about a city in the Meneijjeh Valley, the inhabitants of which possessed numerous ships. As these inhabitants offended Allah, a very prolonged rainfall ('Platzregen') caused great floods and the sea was lowered down to al-Akaba, and the city of Meneijje totally destroyed. Musil summed up: 'Zweifellos liegt dieser Sage eine dunkle Erinnerung an eine noch in historischer Zeit durch das Zurückweichen des Meeres erfolgte Veränderung der Terrainverhältnisse im N. von el-'Akaba zugrunde' (Musil, 1908, 186–8).

(3) *1932/33: Fritz Frank* During the winters of 1932/33 and 1933/34, the German Templer Fritz Frank, Civil Engineer, hunter, planter and explorer, undertook an extensive exploration of the Arabah, lasting many weeks. Travelling mostly on foot and accompanied by Bedouin guides, Fritz Frank visited all the major wadis running into the southern Arabah and published the first detailed description of numerous habitation, mining and smelting sites in the Arabah and north-east Sinai. Frank was, in fact, the first modern explorer of the Arabah and his report is still a valuable source of reliable information (Frank, 1934).

One of Frank's most important discoveries was Tell el-Kheleifeh, located about 500m. from the shore line of the Red Sea, about halfway between modern Aqaba and Eilat (Frank, 1980, 243f., Tab. 41B, 42). Observing that the remains at this site seemed to be pre-Roman, he suggested the identification of the site with the Biblical port of Ezion-Geber, a suggestion later taken up by Nelson Glueck (see below).

In the Timna Valley, Frank described seven copper smelting camps, including a 'Brandstätte' surrounded by a semi-circular wall with a diameter of 70–80m., undoubtedly our Site 30 excavated in 1974/76. (Frank, 233f, 241f, Tabl 39f, Plan 32). He also described 'copper and iron ores in round "Gänge" with a diameter of up to 50cm. in the yellow-white sandstone of the side valleys, underneath the limestone',[7] and assumed that copper was produced in installations similar to 'Palestinian limekilns'. According to Frank, there were two different types of slag: small and thin (1–2cm.) as well as large and thick (5–10cm.) lumps, an observation confirmed by our excavations at Site 30 in 1974/76.[8]

At most of the ancient sites discovered by Frank in the Arabah he found pottery but, with the one exception of Tell el-Kheleifeh, he refrained from dating them or speculating on their historical significance.[9]

(4) *1934/40: Nelson Glueck and 'King Solomon's Copper Mines'* In 1934 the American Biblical scholar and archaeologist Nelson Glueck undertook a trip, mainly on camelback, through the Arabah and, after visiting some of the copper mining areas of the north-east Arabah,[10] also visited for several hours (from the afternoon of the 30 March to the morning of the 31 March, 1934) the Timna Valley (Wadi Mene'iyyeh) (Glueck, 1935, 42).[11] In his reports (Glueck, 1935, 42–5; 1940, 77–9) he describes seven smelting camps (our sites 2, 12, 13, 15, 30, 34, 35) with stone-built copper smelting furnaces, habitations and copper slag heaps. According to Glueck (in 1935) the Wadi Mene'iyyeh was 'the largest and richest copper smelting site in the entire 'Arabah' (Glueck, 1935, 42) dated by Edomite pottery to the 13th–8th centuries BC (Glueck, 1935, 138).

In his later publications, Glueck dated the Arabah mines to the 'Iron Age, that is to the time of the Kings of Israel and Judah and particularly to the time of Solomon' (Glueck, 1940, 56, 77), and the chapter on the Arabah mines is now simply captioned 'King Solomon's Copper Mines' (Glueck, 1940, 50; 1959, 153).

According to Glueck, mining of copper ore in the Timna Valley (he specified cuprite and malachite) 'was a very simple task . . . because it protruded all over the surface of the entire wadi' (Glueck, 1970, 77). He also described details of a 'copper smelting furnace' measuring 3 × 3m., built of roughly hewn blocks in 'two compartments one above the other' (Glueck 1940, 60).[12] However, in these furnaces the ore was only 'partially treated', i.e. 'roasted' (Glueck, 1939, 10; 1959, 164) '. . . to be further smelted and refined and worked up partly into finished metal products at the smelters and foundries and factories of Ezion-Geber . . .' (= Tell el-Kheleifeh); 'Solomon's port and industrial city' on the shores of the Red Sea (Glueck, 1940, 64; 1959, 164).

In 1938–39, Glueck excavated Tell el-Kheleifeh, following Frank's suggestion that this tell contained the remains of Solomon's famous Red Sea port: Ezion-Geber.[13] A large brick building, 13 × 13m., heavily scorched and fire-blackened, was interpreted by him to be a complicated and sophisticated copper smelting furnace where the copper ores, roasted at the mining sites of the Arabah, were smelted to refined copper. In fact, according to Glueck, Solomon's copper smelters had already invented the Bessemer process for copper smelting (Glueck, 1959, 165). A number of large, hand-made and fire-blackened pots found in a square of buildings surrounding this central 'smelter', was taken as evidence for crucible smelting of copper in further smelter buildings. Glueck summed up his discoveries (Glueck, 1940, 94): 'The entire town, in its first and second periods, was a phenomenal industrial site. A forced draft system for the furnaces was employed, and later abandoned and forgotten, to be rediscovered only in modern times. Ezion-Geber was the Pittsburg of Palestine, in addition to being its most important port'.

According to Glueck (1940, 99, 104) Ezion-Geber was built by King Solomon. The 'roasted ores' from the Arabah mines were here smelted to metallic copper to provide King Solomon with copper for the Temple of Jerusalem and refined and cast into finished copper implements to provide export goods for Solomon's Tarshish metal trading ships to legendary Ophir.[14] There, the copper metal and/or implements were exchanged for gold, silver, ivory and spices.[15]

The archaeological and historical picture of the Arabah and the Red Sea, based on the new discoveries of 'King Solomon's Copper Mines' and the 'Smelter and Port of Ezion-Geber', immediately became an archaeological sensation[16] and was unreservedly integrated into the professional literature, as well as into all current standard text books of Biblical archaeology and history.[17] To demonstrate the spirit of this unique chapter in the history of archaeological research – perhaps an offspring of the ideological and theological overtones of Biblical archaeology – we quote here the great American scholar W.F. Albright (Albright, 1949, 128) without any further comments: 'There can be no doubt whatsoever that Tell el-Kheleifeh was a great smelting plant, but just how the reduction of copper was accomplished remains a mystery to specialists in metallurgy who have studied the problem'.

3 The Arabah Expedition

(1) *The Arabah Survey 1959-61* Although the main objective of the Arabah survey in 1959 (*Illus.* 2) was to systematically search the whole of the western Arabah for remains of ancient habitations, roads and activities, special attention was paid to the investigation of its copper mining and smelting sites. It had

been felt for a long time that the information published so far by scientists and archaeologists interested in the history of technology – mainly by Frank and Glueck – did not allow the formation of a coherent, technologically acceptable picture of the ancient copper industries in the Arabah. It was mainly for this reason that, since the completion of the first overall survey in 1961, our expedition concentrated its major efforts on the investigation of the Timna Valley which, already at this early stage of our investigations, presented the picture – and challenge – of a totally unexplored copper mining and smelting area, covering about 80 square kms.

Our detailed mining-related survey was extended into the side valleys of the Arabah, where geologists had reported copper mineralisation, including the area of Beer Ora, Nahal Amram and Nahal Jehoshaphat (south-west of Eilat), as well as the area around the estuaries of these wadis where we had previously recorded clusters of settlements, many with clear evidence of copper smelting, like slag heaps and furnace fragments, especially on hill tops and slopes near the habitations.[18]

In the Timna Valley, which during the first stage of our investigations (up to 1967) was only partially investigated in detail, we recorded ten camps, in each of which we found slags, tuyère fragments and intensive signs of burning. There were also workshop and habitation buildings, many in a surprisingly good state of preservation and containing stone tools, crushers, grinders and potsherds. We were obviously the first archaeologists at those sites.

Although the metallurgical debris, including many fragments and pieces of metallic copper, were clear evidence of copper smelting at Timna, the rock outcrops of greenish colour in different parts of the valley, published by Glueck as the source of ore for Solomon's smelters, turned out to be mainly very low-grade chrysocolla, or just tinted sandstone, of no possible use to an ancient metallurgist.[19] For this reason the present author, assisted by the geologist Y. Bartura, undertook an intensive search for the ancient copper mines in the central part of the Timna Valley and these were finally located at the foot of the c. 300m. high Timna Cliff, on the western side of the Timna Valley (Sites 7, 9, 16, 19, 20, 21, 124, etc.).[20] Here a whitish sandstone formation – the Middle-White Nubian Sandstone horizon – sandwiched between brightly coloured variegated sandstone layers, showed pockets and streaks of green copper ore nodules and banks of finely diseminated flakes of copper ore (Y. Bartura *et al.*, 1980, 41–56, Slatkin, 1961, 292–301). First analyses of these ores identified mainly malachite but many of the nodules also contained a core of high-grade chalcocite. The Middle-White rockface showed clear evidence of mining, further strengthened by the find, nearby, of numerous stone tools and pottery fragments. These archaeological finds apparently represented different mining methods, and led to the identification of three periods of mining in Timna[21] – periods which at this stage of our investigations (1961) we had also identi-

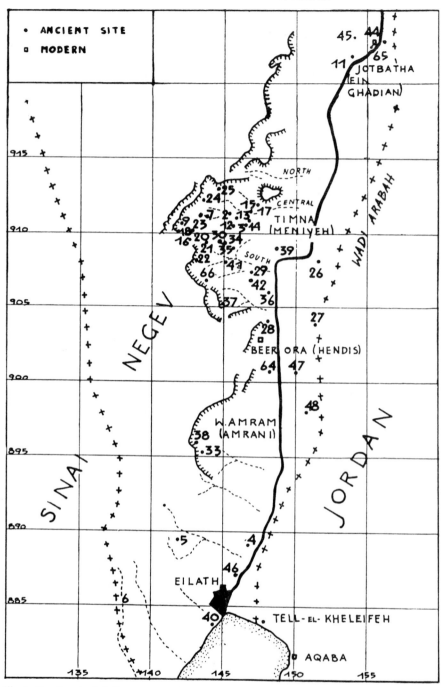

Illustration 2.   Arabah Survey map 1962.

fied at the smelting sites in the Timna area: Chalcolithicum (4th millennium BC); Early Iron Age I (11th–10th centuries BC) and Roman-Byzantine.[22]

A few open and many filled-in mining shafts, observed in the mining area (investigated up to 1961) were understood to be water cisterns and ore dressing 'plates', saucer-like flat areas on the slopes of the mining area, clear of the usual surface rocks and stone litter (Rothenberg, 1962, 11; McLeod, 1962, 68). Only in 1976 did the Arabah Expedition reinvestigate the problem of the 'plates' and established them as mine workings.[23]

(2) *First Publication and its Aftermath*   The first comprehensive archaeological field report of the Arabah and Timna survey, including proposals for a metallurgical process model, was published in 1962 (Rothenberg, Aharoni, McLeod, 1962). Although at first (1962) we had reluctantly adopted a Solomonic date for the copper workings in the Timna Valley – with the exception, of course, of the Chalcolithic and Roman sites – the archaeo-metallurgical finds in Timna posed numerous pertinent questions regarding Glueck's publications of 'King Solomon's Mines and Smelters' in the Arabah and Timna and, es-

pecially, concerning Tell el-Kheleifeh as a huge copper smelter. Besides the fact that Glueck, according to his published field diary (1934), had never even been near the actual ancient copper mines of the Timna Valley, which were discovered by our expedition in 1959, many of his descriptions and theories turned out to be in gross discrepancy with the bare facts in the field (Rothenberg, 1962 and 1967). One of these discrepancies, of basic significance for Glueck's reconstructions of the Solomonic copper industry in the Arabah, were connected with the nature of the slags found at the Timna sites. Even a casual investigation showed that these slags contained quite a lot of metallic copper pellets and further metallurgical studies proved the same to be proper (fayalite) copper smelting slag – *not* by-products of 'roasting' copper ore, as published by Glueck. The copper ores of the Arabah would not have required any roasting prior to their reduction to copper metal in a one-step smelting process, and there could not be any doubt that in the smelting camps of Timna, and the other mining sites of the western Arabah, copper ore had been smelted to the final product: metallic copper.[24]

Furthermore, when investigating Glueck's stone-built smelting furnaces (Glueck, 1935, 79 and also figs. 26, 28), these turned out to be human burials or workshops with stone grinders *in situ*. Also, nowhere could we find Glueck's 'large smelting crucibles', though we found sherds of fire-blackened handmade cooking pots, known as 'Negev-ware', because they were found in the Iron Age settlements of the Negev mountains (Aharoni, 1960, 98–102, Pl. 13).

In the wake of these rather surprising implications of our first survey results, we scrutinized in detail the excavation reports of Tell el-Kheleifeh (Glueck, 1938, 3–17; 1939, 8–22; 1940, 2–18; 1940, chap. 4; 1959, 157–68; Pinkerfeld, 1942),[25] which led to the proposals for a fundamental revision of Glueck's facts and their interpretation, his dates and historical concepts. Tell el-Kheleifeh could not possibly be accepted as a metallurgical, industrial city, nor could it be accepted as the Solomonic port of Ezion-Geber: Eilat (Rothenberg, 1962, 44–56; 1967a, 191–213).

Our critical review of 'King Solomon's Mines – Tell el-Kheleifeh = Ezion Geber: Eilat', was immediately and incessantly met by strong opposition, especially by Israeli and American Biblical archaeologists and historians. (Yadin, 1961; 1965; Avigad, 1963; Albright, 1964, 67; Wright, 1961).[26] However, in 1965, Glueck himself accepted many of our proposals for a re-evaluation of the excavations of Tell el-Kheleifeh (Glueck, 1965, 2–18) including the reinterpretation of the huge smelting furnace, the centre of his 'industrial city', as a burned down grain store (see also Glueck, 1977, 713). Glueck also followed our arguments against the identification of Tell el-Kheleifeh as the port of Ezion Geber. Yet Glueck continued to insist on Israelite dates and stratigraphy at Tell el-Kheleifeh and saw King Solomon as the founder of this city and of the copper mining enterprises in the Arabah. Neither Glueck, nor any of his fellow archaeologists (see Amiran, 1969, 300–301) gave an answer

to our basic argument that all the burnished pottery from Tell el-Kheleifeh did not belong to the Israelite ceramic corpus; in fact it was a stranger to the pottery of western Palestine and indicated a foreign culture, perhaps 'Amon, Edom, Ashur or Arabia' (Rothenberg, 1967a, 200 and lately Rothenberg and Glass, 1983).

Glueck's theories of Ezion Geber and King Solomon's Mines lingered on even after he himself had revoked many of his interpretations and, strangely enough, even after the discovery of the Egyptian Mining Temple in Timna (Meshel, 1975, see also Glueck, 1967, 1969).[27] The chapter Tell el-Kheleifeh = Ezion Geber–Eilat, has been finally closed by the reinvestigation of Glueck's excavations by G. Pratico, who studied all available material and documentation from Tell el-Kheleifeh and established that the ceramic finds from these excavations are in fact Edomite and belong to the 8th–6th centuries BC. Consequently, he proposed to reject the identification of Tell el-Kheleifeh with Biblical Ezion Geber (Pratico, 1982, 6–11; 1985, 1–32).

4 Excavations 1964-1970[28] (*Illus. 3*)

Let us briefly return to 1962 and our first survey report. The archaeological and metallurgical facts, established by our surveys in the Arabah, necessitated not only a fundamental reappraisal of Glueck's metallurgical interpretations of Tell el-Kheleifeh, but also created considerable apprehension concerning the dating and culture-historical and ethnical identifications of the copper mines and smelters of the Arabah, which at that phase of our work were still dated to the 10th century BC, i.e. to the period of King Solomon (Rothenberg–Aharoni, 1962, 40, 66). Adding to these queries was the fact that nowhere does the Bible even as much as indicate the existence of any Solomonic copper mines. On the contrary, according to Chronicles, much copper for the Solomonic Temple was captured from Hadadezer, King of Zobah (in the Lebanon) (I Chronicles, 18:8) and more was collected from the population (I Chronicles, 29). As a matter of fact, the very concept of 'King Solomon's Mines' was not mentioned anywhere before Glueck's publications, except by Henry Ryder Haggard in his famous adventure novel published in 1885 and set in East Africa.

Moreover, the dating problems of the Timna sites showing chronologically obviously differing mining techniques, as well as different copper smelting processes, could not be resolved by surface exploration and sampling because we had found on the surface of most sites three totally different kinds of pottery: 1, normal, wheel-made 'kitchenware'; 2, primitive, handmade vessels, mainly fire-blackened cooking pots, well known from the Iron Age II settlements in the Negev; and 3, hand-made vessels of simple shapes, covered with painted bi- and tri-chrome decorations of unknown origin. We therefore had to assume that we were, in fact, dealing with quite different periods or even cultures. Since no single

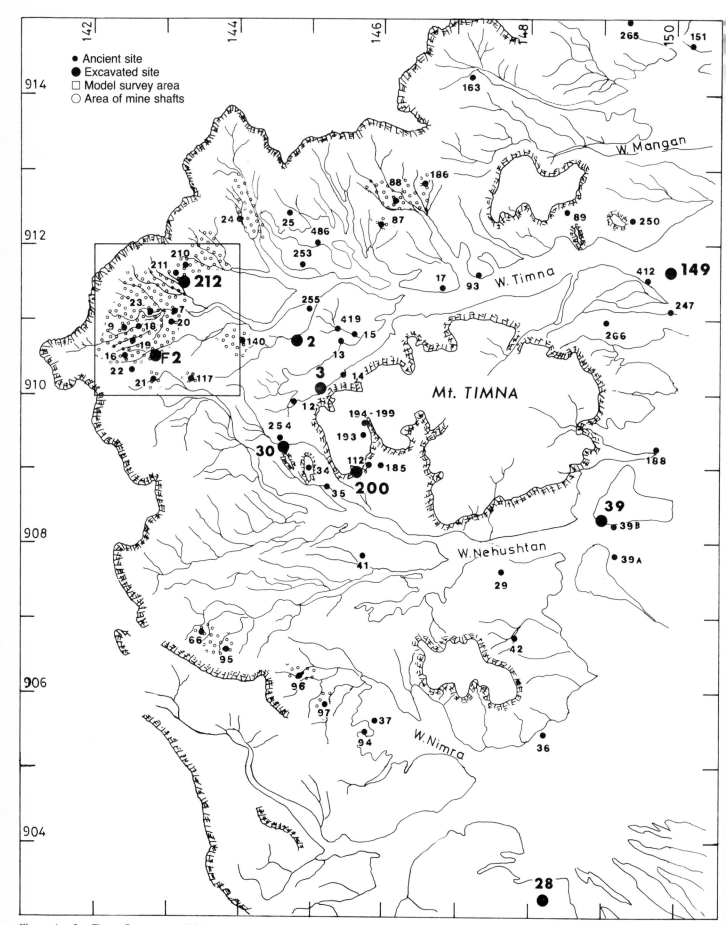

Illustration 3.   Timna Survey map 1984.

ancient copper smelting furnace had been published at the time to serve as a model for the explanation of the different furnace fragments littering the surface of the smelting camps of Timna, the extractive metallurgy of Timna could not be effectively reconstructed without additional information which could only be obtained by systematic excavations.

From 1964 to 1970, the Arabah Expedition, led by Beno Rothenberg and with the occasional participation or advice of Y. Aharoni, A. Lupu, B.H. McLeod, H.H. Coghlan, R.F. Tylecote and H.G. Bachmann, undertook a series of problem-related excavations in smelting camps of Timna and Beer Ora, which had been tentatively dated by the survey finds to the 4th millennium BC, Iron Age I, and Roman-Byzantine (Rothenberg, 1972).[29] These excavations culminated in the discovery of the Ramesside Hathor Mining Temple in 1969. In order to fully appreciate the significance of this discovery and its impact on the archaeological scene, it will be helpful to briefly unfold the development of the archaeological and archaeo-metallurgical picture during these years of systematic excavations.

(1) *Site 39*[30]   Site 39 was a metallurgical work and smelting site dated to the late Chalcolithic Period, the 4th millennium BC – the 'Sinai-Arabah Copper Age-Early Phase'.[31] The site consisted of a small, circular working area and several stone-built round houses, situated at the foot of a steep hill on top of which remains of a hole-in-the-ground copper smelting furnace were excavated in 1965. This primitive smelting installation, together with peculiar slags, potsherds, stone and flint tools, indicated a very early phase of metal production. In fact Site 39 shows the earliest type of a copper smelting furnace found so far.

The excavations at Site 39, as well as a series of experiments, made it possible to reconstruct for the first time the earliest phase of the copper smelting process (Rothenberg, Tylecote and Boydell, 1978).

(2) *Site 2*[32]   Site 2 is a large smelting camp located in a side arm of Nahal Timna, where in 1964 and 1966 extensive excavations uncovered a series of copper smelting, as well as melting-casting, furnaces. These installations, which stood next to fairly large slag heaps outside the built-up workshop area of the camp, were in most cases stone-built,[33] lined with clay mortar, and had facilities for proper ventilation (tuyères) and slag tapping (slag pit) – evidence for a very advanced metallurgical technology.

The pottery, already found in the first season of this excavation in 1964, came as a great surprise. The three, basically different, kinds of pottery,[34] found on the surface of all Arabah smelting sites and previously dated to the 10th century BC, which had created so many problems and discussions, were now stratigraphically proven to be contemporary and typologically had to be dated to a pre-Solomonic phase of the Early Iron Age: the 12th–11th century BC at the latest (Rothenberg, 1967b, 53–70).

The normal wheel-made 'kitchenware' was related to a transitory phase between the Late Bronze Age and Iron Age I of Palestine, although it seemed difficult to find really fitting comparisons in the relevant corpus of Palestinian pottery. The Egyptian origin of most of this pottery was neither recognised nor even suspected by us at the time. The hand-made Negev-type pottery showed slag inclusion as temper and was therefore obviously locally made – but by itself quite unsuitable for dating.[35] For the beautifully decorated pottery, found in abundance in all layers of Site 2, we could find no comparisons in the relevant literature but, because of some similarities in its decorative motifs with Edomite pottery decorations, published in Glueck's survey (Glueck, 1935), we related the same – and the mining operations of this period in the Arabah – to the Edomites of Eastern Palestine (today the Kingdom of Jordan). This idea was made rather plausible by the 'Edomite pottery' found at the mining and smelting sites in the northeastern Arabah, related to the Kingdom of Edom (Glueck, 1935, 123–37).[36]

(3) *Site 28 – Beer Ora*[37]   In 1969 we continued the chronological series of excavations at the smelting site of Beer Ora, at the south end of the Timna Valley, where a very large slag heap was located near the ancient well and modern settlement of the same name. The excavations uncovered a series of copper smelting furnaces in a fair state of preservation, as well as a crucible furnace for casting. Pottery finds dated this site to the 2nd century AD, the time when the Legio III Cyrenaica was stationed in the southern Arabah.[38]

The work at Beer Ora and its copper smelting furnaces completed the series of excavations projected to establish the history of copper smelting installations and processes in the Arabah (Rothenberg, 1972, 212–23, 237; 1983, 14–15; 1985, 131–2, Figs. 15, 16, Pl. 15, 16).[39]

(4) *Site 200 – The Hathor Mining Temple*[40]   Glueck's reconsideration of his metallurgical interpretations (Glueck, 1965) did not greatly weaken the opposition to our overall conclusions concerning the Arabah, its mines and smelters and, especially, their dating. The newly proposed date of the 12th–11th century BC and the Edomite connection with the copper industry in the southern Arabah, were bitterly attacked by Biblical archaeologists.[41] These arguments were mainly directed against our dating and ethnical identification of the Timna pottery. Not one of the archaeologists concerned took the trouble to study this in our workshops or, to the best of our knowledge, in the field – and because much of this pottery could not be found in any existing ceramic corpus, 'historical logic' was called in as a decisive argument.

Y. Yadin emphasised the view widely propagated in Biblical archaeology text books, that the main reason for the repeated wars between the Kings of Israel and Judah and the Edomites, must be seen in the struggle for the possession of the rich copper deposits of the Arabah – and therefore the Timna mines had to be dated to King Solomon and the Judean Kings after him (Yadin, 1961, 109). Under these circumstances it had become extremely frustrating to try to discuss the basic problems of the history of the Arabah in relation to the findings in our excavations and, in fact, we refrained for several years from engaging in such arguments; instead we continued with our systematic fieldwork.

In March 1969, following a season of work at Beer Ora, we turned to an unusual site at the foot of 'King Solomon's Pillars', in the centre of the Timna Valley, which we had discovered during our surveys in 1967. Here, a few pieces of decorated pottery and many finely dressed white building stones, scattered on the surface of a low hillock, as well as three strange niches cut into the rock face of the Pillar, indicated a very special cult site (Rothenberg, 1967, 313).

The subsequent discovery of the Ramesside Hathor Temple in Timna, published here, was a decisive turning point in the history of the Arabah research, and brought with it the end of the repeated arguments about the date and ethnical connections of the Arabah sites.[42] As the three different kinds of pottery, present at all the Timna sites, were found in the Hathor Temple together with hieroglyphic inscriptions of the Egyptian New Kingdom, from Seti I to Ramesses V (Rothenberg, 1972, 163–6), an absolute date from the end of the 14th century to the middle of the 12th century BC became obligatory for the evidently Egyptian copper industry of the southwestern Arabah.

This identification of the large scale copper mining and smelting operations in the Arabah as Egyptian industries, where Egyptian expeditions worked together with local inhabitants from neighbouring territories – 'Midianites' from N.W. Arabia, (Hejaz = the Biblical Midian) (Rothenberg, 1972, 182–4; 1983 and below) and Amalakites from the Negev (Rothenberg, 1967a, 92–101; 1972, 180–82; and below) – necessitated fundamental changes in the culture-historical concepts concerning the areas adjacent to the Gulf of Eilat-Aqaba. Instead of King Solomon's mines and the copper smelters on the shores of the Red Sea, as part of the Solomonic Empire, we are now faced with the historic-geographical integration of the south-western Arabah into the sphere of activities of the Egyptian Pharaohs of the New Kingdom.

5 The New Timna Project 1974–1976

(1) *Publications in 1972 and open questions*

The discovery of the Hathor Mining Temple in the Timna Valley, created considerable international interest. Due to the initiative of Sir Mortimer Wheeler, Secretary of the British Academy, and Dr Richard Barnett, Keeper of the Department of Western Asiatic Antiquities in the British Museum, the Timna finds and findings were exhibited in the British Museum, London, in 1971 and subsequently also in the archaeological museums of Manchester, Birmingham and Newcastle. From England the Timna exhibition travelled to the German Mining Museum, Bochum, the German Museum in Munich and the Kestner Museum in Hannover (Rothenberg, 1971;

Rothenberg and Bachmann, 1973). It was mainly the exhibition of Timna in Bochum which brought about the New Timna Project 1974–76, as a collective research undertaking by the Arabah Expedition and a team from the German Mining Museum, Bochum, sponsored by the Volkswagen Foundation (see reports: Conrad and Rothenberg (ed.), 1980).

In 1972, the first comprehensive report on the Timna excavations was published in the series 'New Aspects of Archaeology', edited by Sir Mortimer Wheeler (Rothenberg, 1972) which included a chapter on the first season of excavations of the Timna Temple.[43] We conceived this summary of our surveys and excavations in the Arabah, in 1959–70, as a first techno-historical synthesis of our work, fully aware that this could only be tentative, since many questions were still open, even after the discovery of the Egyptian Mining Temple and its absolute dates.

Most of the unsolved problems arose out of the slow processing of the huge number of archaeological as well as metallurgical finds and, furthermore, our subsequent, more comprehensive surveys of the Timna and Amram areas. Many mining relics found during these surveys could not be explained by purely archaeological criteria and it had become imperative to involve experienced mining experts in our research. If, at an early stage of our work, we had observed only a limited number of 'plates', and their explanation as ore dressing installations (McLeod, 1962, 69; Rothenberg, 1972, 63) seemed plausible, this was no longer acceptable after our subsequent discovery of thousands of such 'installations'.

There was also the very basic question of the geomorphological structure of the ancient landscape. Certain phenomena in the field seemed to indicate that floods, wind erosion and tectonic disturbances had caused substantial changes in the morphology of the terrain; the early mine workings visible today became therefore unexplainable without a geomorphological 'correction'. For example, a great many shafts found at quite 'illogical' locations, i.e. on the top of hills or in the rock faces of steep wadi banks or high mountain cliffs, could not be explained as water cisterns, but were equally difficult to understand as mining shafts. There was also the often observed fact that, wherever in the wall of a wadi a gallery opening appeared, a similar opening appeared also on the opposite side of the wadi, making it obvious that we were dealing with one gallery, which originally ran underground and was later cut into two parts by the deep erosional cutting of the wadi bed. Taking these observations into account, it seemed plausible to assume that many such workings lay hidden in the depth of the hills and wadis and the real mine entrances had not yet been discovered.

Furthermore, many small and shallow workings, visible in the variegated and white rockfaces at the foot of the high Timna cliffs, which had been understood as mine workings for copper ores (Rothenberg, 1962, 9–10, Pl. 1)[44] were subsequently found to contain only iron ore. These were evidently mined in ancient times, but the ores found there could only have been used as iron flux in the copper smelting process and not as copper ore. Consequently, it had become imperative to look for the real copper mines of the Timna Valley. Further explorations in the narrow and deep wadi beds at the foot of the Timna Cliffs, in 1966–67, resulted in the discovery of many mining galleries and shaft openings, some of which showed different tool marks of apparently different technological horizons. Also the find of typical prehistoric mining picks in some of the galleries previously considered as Roman workings (Site 23 and Site 212 – Rothenberg, 1972, 208–10, figs. 115–18) caused considerable difficulties.

Although quite a lot of work on the extractive metallurgy of Timna had been done by the Arabah Expedition (Lupu and Rothenberg, 1970; Rothenberg, 1972; Tylecote, Lupu and Rothenberg, 1967) the new discovery of additional different slag types as well as quite different furnace fragments in the smelting camps of the Arabah, indicated much more complex – and historically more intricate – developments not yet properly understood.

These were some of the basic problems of mining and extractive metallurgy investigated in 1974–76 by the New Timna Project.[45]

(2) *First season of fieldwork (1974)*

(a) *Mining* The first excavations in the mine workings were undertaken in an area (Site 212)[46] where exposed gallery openings had been observed in walls of two adjacent wadi beds, as well as a large shaft (S 14),[47] located on top of a high hill, which was found only partly silted up and showed rock-cut footholds and regular, diagonal chisel marks. In this first season of work, large parts of two widespread gallery systems could be cleared and the large shaft S 14 excavated to a depth of 17m., although the intrinsic connection between the shafts and the galleries was not recognised by the mining team at that time.

A group of archaeologists from Bochum, directed by G. Weisgerber, also investigated the problems of the enigmatic 'plates'. They excavated a number of the 'plates' in the area of Site 212, and produced a theory of ore dressing by flotation (Weisgerber, 1975).[48] As this theory seemed to us for many reasons quite unacceptable, a team of the Arabah Expedition undertook in 1976 a series of new excavations in the 'plates' of the same area which led to a completely new understanding of the mining systems of Timna (see below).

(b) *Site 30 – A walled copper smelting camp* Site 30 was chosen for excavations for several reasons: it was the only well-preserved walled smelting camp in Timna; various different pottery and slag types, furnace and tuyère fragments, were found on its surface and some of the latter had not been seen in any other smelting camps at Timna. We therefore expected to clarify, through systematic excavations at Site 30, not only the stratigraphy of the New Kingdom smelting camps of Timna, but also the stratigraphic and chronological context of its different types of smelting remains. As all our previous, rather small-scale excavations in the smelting camps of Timna had been directed towards the discovery of smelting installations, the excavation of Site 30 represented a new approach, which also took into consideration the logistic and topographic organisation of Egyptian copper smelting operations on a real industrial scale. It was therefore planned to excavate two wide transverse trenches with their meeting point at the centre of the metallurgical operations, identified by a large slag heap and dark, scorched ground all around it.

In the first season of work, the stratigraphy of Site 30 could already be established, showing three phases of copper smelting activities, all belonging to phases of the Egyptian New Kingdom. However, the detailed chronology of these different phases and the archaeo-metallurgical investigations of the many newly found smelting installations and slag formations, had to be left to subsequent typological, archaeo-metallurgical and archaeometrical investigations.[49]

(3) *Second season of fieldwork (April–July, 1976)*

(a) *Mining* The second season of fieldwork carried out by a small Israeli team and a few volunteers from England, led by Beno Rothenberg,[50] was dedicated to the investigation of problems which had arisen out of the work of the first season: (i) the enigma of the 'plates' which, to our mind, had become more obscure after Weisgerber's excavations; (ii) the relation between the geomorphological location of the mine workings and their technology and dates; (iii) the relationship between shafts and galleries which, though found in close proximity, seemed so far (1974) to be without any physical connection.

In early spring 1976, the 'plates' excavated in 1974 were investigated by digging a simple trial trench through the 'clay lining' of the bottom of the 'flotation bowls' (Weisgerber, 1975); they turned out to be funnel-shaped, silted-up openings of mining shafts and not, as previously assumed, ore dressing sites (Rothenberg, 1962; McLeod, 1962) nor bowl-shaped ore flotation installations (Weisgerber, 1979). This discovery demanded a total revision of our previous concepts of mining in Timna. During the past years, several thousand 'plates' had been observed by us all along the slopes, terraces and even in shallow wadi beds at the foot of the Timna Cliffs and their identification as mining shafts implied their major significance – as thousands of mine shafts of different periods – in the mining operations of Timna.

In June–July 1976, the Arabah Expedition therefore followed up this first, rather small scale control of Weisgerber's excavations by proper excavations, choosing the 'plates' to be investigated in accordance with their geomorphological and topographical location, but also taking into consideration related archaeological surface finds, such as mining tools, pottery and flint implements.[51]

In Model Area Section S, four 'plates' were excavated and found to be mining shafts, funnel-shaped on top and tube-shaped below. These shafts penetrated through a very hard top layer of conglomerate into the copper ore-bearing middle-white sandstone horizon. At the bottom of one of these shafts (S 11), a horizontal gallery was found, but this could not be excavated at the time.

The final and definitive proof that we have in Timna of proper shaft and gallery mining, was produced in subsequent excavations (T 1 and T 31) in the Model Area Section T. Although instead of the New Kingdom shafts and galleries in Section S, of almost standard dimensions, the earlier proto-historic mine shafts in Section T[52] were just large circular holes, cut into the white sandstone (almost without conglomerate cover) and these became at a depth of *c.* 3m. a system of large and irregular, but still gallery-like cavities (Ordentlich *et al.,* in Conrad and Rothenberg, 1980). The basic shaft and gallery mining technology proved to be common to both.

(b) *The Model Area*[53] In addition to the investigation of the 'plates' and the excavations in Model Area Section T,[54] the Arabah Expedition undertook a detailed geological, geomorphological and archaeological survey of a model area of *c.* 4 km.² (maps in Conrad and Rothenberg, 1980), which contained the characteristic elements of the ancient landscape and all types of mine workings observed in the Timna Valley. Although copper ores and 'plates' had been found by our surveys along an entire strip of terrain, *c.* 3 km. wide and *c.* 15 km. long, at the foot of the Timna Cliffs, the Model Area appeared to be the centre of the ancient mining activities. It showed the greatest concentration of mining relics, including all hitherto discovered gallery openings and thousands of 'plates' of varying forms and topographical density and at quite different geomorphological locations. This choice of a restricted area for detailed research proved itself during subsequent fieldwork and excavations.

However, to undertake such detailed investigations of four square kilometres of extremely rugged countryside, and also to enable a proper cartographic recording of the ancient remains in the Timna Valley, new small-scale maps of the entire Timna region (1:5000) and of the Model Area (1:2000) had to be prepared, based on new aerial photography.[55]

In the summer of 1976, the detailed geological and geomorphological mapping of the Model Area was completed and produced irrevocable evidence for widespread and rather intensive, deep-going geomorphological changes in the Timna landscape within the time-span between the ancient mining activities and the present. This new understanding enabled us to clarify for the first time a great number of topographical and mining-technological problems related to the mining relics of the Timna Valley and other mining sites in the southern Arabah.

Parallel to the geological and geomorphological mapping, a very minute archaeological survey of the

Model Area was undertaken which produced a detailed record and inventory of all types of mine-workings, such as 'plates' (silted-up mine shafts), shafts sunk straight into outcropping sandstone bedrock and often found partly empty, heaps of copper ore and rock dumps next to silted-up mine shafts and primitive stone structures used as windshields in the mining region.

The archaeological finds, mainly pottery and flint implements, indicated four periods of mining activities in Timna:[56] the earliest finds, belonging to the Eilatian Period (5th–4th millennium BC) were mainly in Model Sections F and G; finds of the subsequent (Timnian) Period (4th–3rd millennium BC) appeared mainly in Model Area Section T, but also in Model Area Section S and in other places. The Egyptian New Kingdom (14th–12th centuries BC), which was evidently the period of most intensive and extensive mining activities in the southern Arabah, left behind very numerous mine workings and archaeological remains spread throughout the mining area of Timna. Although a number of Roman sherds found in the Model Area indicated Roman activities, no Roman mine workings could be identified within this area.[57]

(4) *Third season of fieldwork (September–November 1976)*

(a) *Mining* The excavation of several 'plates' in Model Area Sections S and T during the previous season, established for the first time that the rockcut shafts, the 'plates' and the galleries were, in fact, parts of the underground mining systems of almost all periods of mining in the Timna Valley. This new understanding and the systematic geomorphological and archaeological investigations in the Model Area, made it possible to plan properly the third season of fieldwork as a strictly problem-related undertaking.[58] The mine workings investigated during the third season were selected from numerous possible objectives, according to their geomorphological and geological situation and the archaeological context; it became possible thereby to comprehend all types of mining in Timna.

Several extensive underground mining systems in Model Area Sections S and T were almost completely excavated (Conrad and Rothenberg, 1980, 69–167), as well as a number of other, individual shafts, found in different geomorphological settings. These latter shafts created a highly interesting picture of the systematic 'geological' surveying methods employed by the miners of the New Kingdom, who explored the underground ore deposits by the use of exploratory shafts as 'geological boreholes'.

Based on the 'geological' information obtained by a series of prospection shafts radially sunk through the semi-circular ore-bearing region, sophisticated shaft and gallery systems – many also in several superimposed levels – could be systematically developed by the Egyptian miners.

Following the lead of the obvious significance of the geomorphological location of the mine workings, a number of trial trenches were also excavated in

Model Area Sections F and G. Here, in wide and flat wadi beds, very dense concentrations of 'plates' had been observed, which surprisingly were located underneath the previous, by now completely eroded, ore-bearing sandstone horizon. As a result of these excavations, we discovered the earliest 'mine workings' of Timna: primitive pit-mining into conglomerate to extract the heavy copper ore nodules held in considerable quantity from the eroded sandstone horizon. Related archaeological finds made it possible to date these primitive pitting operations to the Early Chalcolithic (Eilatian) Period.[59]

The investigations of the mines of Timna brought about an essential revision of mining history: After an incipient prehistoric phase of pit-surface collecting and mining, proper – though still primitive – shaft-and-gallery mining methods were already used by apparently indigenous miners in the 4th millennium BC. More than a 1500 years later,[60] at the end of the 14th century BC, Egyptian mining expeditions arrived in the southern Arabah – especially in the Timna Valley and Nahal Amram – set up and operated sophisticated shaft-and-gallery mines and large scale copper smelting plants.

The major phase of Pharaonic, Ramsesside copper mining in the southern Arabah came to its end in the middle of the 12th century BC with the collapse of the Egyptian Asian Empire. Evidence for a short revival of Egyptian activities during the 10th century BC was found in the excavation of Site 30 in the Timna Valley (Rothenberg, 1980a, 198–201), but the related mine workings could not be identified.

(b) *Site 30 - New Kingdom smelting camp* The excavations at Site 30 were continued (Rothenberg, 1980a, 187) and were extended to contain all of the actual smelting area. At least two different types of smelting furnaces were uncovered as well as a workshop for refractory ceramics, storage pits for ores and clay, and extensive slag heaps. A 10m. wide east-west trench was excavated to converge with the south-north trench of 1974, in order to explore parts of the camp not directly concerned with the actual smelting operations and the wall surrounding the site.

Site 30 is the first ancient copper smelting camp to be systematically excavated, with the emphasis on the stratigraphy of the metallurgical installations and workshops, and the application of strict archaeometric research methods. These excavations produced for the first time a clear picture of the technological developments of the smelting furnaces and processes during this culture-historically decisive transitional period between the Bronze Age and the Iron Age.[61]

In Site 30, three main phases of copper smelting activities, each with its own installations and specific furnace technology, could be stratigraphically distinguished and this led to more detailed and improved process models than had hitherto been possible. These models became the starting point for the subsequent archaeo-metallurgical research programme of the years 1978–83.

The beginning of Pharaonic copper production in Layer III of Site 30, could be dated to the 14th century BC. Under Seti I and Ramesses II (1290–1224 BC) the Egyptian workings attained the proportions of a large-scale industry (Site 30, Layer II). It was at that time that extensive building activities took place in Timna, including the erection of a central Hathor Mining Temple (Site 200) in the middle of Egyptian Timna. The Pharaonic copper workings of Timna operated apparently without much interruption throughout most of the 19th and 20th Dynasties, until the reign of Ramesses V (1156–1152 BC), when all activities in Timna ceased. However, identifiable in Site 30, Layer I, Egyptian activities were renewed for a very short time and on a small scale, during the 22nd Dynasty, perhaps under the pharaoh Sheshonk I (946–925 BC) – the Shishak of the Biblical story (I Kings 14: 25–26). The 22nd Dynasty smelters used a much improved furnace and process technology, which seems to have spread over the entire region and was used there in the later Iron Age II (Hauptmann et al, 1985, 8).[62] This metallurgical technology already stood at the threshold of modern extractive metallurgy.

6 The New Arabah Project 1978–1983

The New Arabah Project, planned as the conclusion of almost 20 years of field research in the Arabah,[63] was carried out in three parallel, interdisciplinary research undertakings:

(1) *Fieldwork* A thorough exploration of the southern Arabah was carried out with much improved survey methods and aerial photography, including the re-investigation of all sites discovered during our previous surveys (*Illus.* 4). This major fieldwork, often accompanied by trial trenching, had become necessary because the excavations of 1964–78 produced stratigraphic evidence for several periods of occupation not previously identified in the Arabah. We also looked for undetected sites in some areas not thoroughly explored previously.

Important revisions of a number of previously published dates were made, such as the identification of Early Islamic (7th century AD) activities in a Ramesside smelting camp (Site 2) (Rothenberg, 1985, and IAMS Newsletter No. 9, 1986) and an Early Islamic phase in Roman mines and smelting sites (Sites 4, 64, 28, 37). Proto-historic smelting sites were newly discovered on top of many of the hills along the southwestern Arabah and, most important, a smelting site of the Sinai-Arabah Copper Age – Late Phase[64] (Rothenberg in IAMS Newsletter No. 5, 1983). The latter period is the beginning of the appearance of tin bronze in this part of the Near East and the identification of a copper smelter of this period, the first ever found, fills a significant gap not only in the archaeological history of the Arabah, but also in the history of metallurgy.

(2) *Archaeometric investigation of the Arabah finds* The systematic application of scientific research methods,

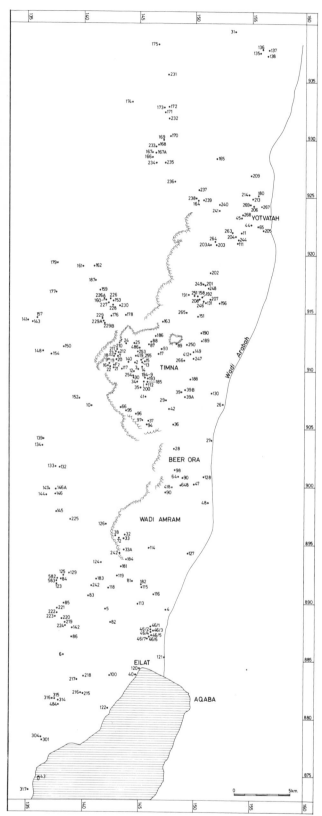

Illustration 4.   Map of Southern Arabah Survey 1984.

in addition to the traditional typological investigation of the finds from the excavations and surveys, produced a comprehensive chronological and culture-historical picture of the southern Arabah and its copper industries and the synchronization of most of the ancient sites of the western Arabah.

(a) *Flint and pottery*   The most significant contributions to this comprehensive picture of the history of mining, of metal production, work-related workers' habitations and other settlements in the Arabah, especially during the proto-historical periods, was the detailed typological classification of all flint implements and flint working debitage collected at hundreds of sites in the Arabah and Sinai[65] and, later on, the petrographic investigation of every sherd found by us in the Arabah (and Sinai) surveys and excavations.[66] These investigations provided new criteria for the dating of the often rather sparse and difficult archaeological material found at the Chalcolithic-to-Bronze Age sites in the Arabah and Sinai.

(b) *Metal objects*   All the metal objects and fragments from the Arabah sites, especially the very numerous votive gifts of metal jewellery and metal fragments from the Timna Temple, were analysed and a selected group was also metallographically investigated in order to establish their manufacturing methods and casting techniques. This was by far the largest group of metal finds from a single site to be completely analytically investigated.[67]

(c) *Iron in copper*   One of the basic problems encountered during the analytical research of the Arabah metal objects and the metal produced in the smelting experiments, was the presence of substantial quantities of metallic iron in the copper. Although the occurrence of metallic iron inclusions in copper smelting was by itself not, of course, a new discovery, the fact that in the Late Bronze Age (early Ramesside period) obviously locally made iron jewellery was found amongst the votive offerings in the Hathor Mining Temple of Timna, drew our attention once more to the possibilities of finding evidence here that iron smelting originated from copper smelting.[68]

A detailed study was therefore made of the conditions most favourable for the reduction of the flux from a copper smelting charge to metallic iron and the nature of the iron thus produced. The final link in the chain of evidence for the origin of the iron objects found in the Temple from local ores and out of a copper smelting process, was provided by metallographic and lead isotope studies. It may now be proposed, based on our factual evidence, that iron smelting was first discovered in a copper smelting furnace (see Vol. II of this publication).

(3) *Theoretical and experimental extractive metallurgy*
The main emphasis of the archaeo-metallurgical research programme of the New Arabah Project was on extractive metallurgy. As the previous process models of copper smelting at Timna (Tylecote, Lupu and Rothenberg, 1967; Rothenberg, 1972; Milton *et al*, 1976; Tylecote and Boydell, 1978; Bachmann and

Rothenberg, 1980) had left many questions un-answered, it was felt that the smelting installations and debris found *in situ* at the Arabah sites – showing the development from a primitive hole-in-the-ground bowl furnace of the Chalcolithic period to the low shaft furnace of the fully industrialized smelting industries of the Bronze Age – offered unique research opportunities for theoretical as well as experimental extractive archaeo-metallurgy.

This research project was carried out as two parallel working programmes:[69]

(a) M. Bamberger, working in collaboration with P. Wincierz and H.G. Bachmann, began his work with a series of laboratory tests and the calculation of a mathematical process model of copper smelting, based on a reconstruction of furnaces at Timna, Site 30 Layer I (Bachmann and Rothenberg, 1980, 218–23). This work established that the previously proposed reconstructions of the Timna furnaces, based on purely archaeological considerations, i.e. mainly the shape and size of fragmentary installations and furnace debris found at Timna, require substantial alterations to represent workable smelting furnaces. Based on those calculations, a series of smelting experiments[70] in full scale furnaces (several with the use of copper ores from Timna) were carried out in order to establish the various process parameters and the conformity of the proposed process model with the data – furnace shape and smelting products – from Timna.

(b) In his research programme,[71] John Merkel com-menced with smelting experiments in full scale fur-naces,[72] basing his work mainly on the Timna Site 2 New Kingdom furnaces (Rothenberg, 1972; 1985). He systematically investigated the various parameters of the copper smelting process and their interdepen-dence, including the problems relating to the nature of the ore and fuel charge, and the charging pro-cedures as well as changes in furnace dimensions caused by the furnace operation and the problems of proper furnace ventilation, bellows and tuyères. Mer-kel extensively used copper ores and iron fluxes from Timna in order to achieve a realistic reconstruction of the Bronze Age smelting operations and their pro-ducts – metal and slag. Merkel's concluding smelting experiment was carried out at Timna Site 2, using a furnace built of local clay and (reconstructed) Egyp-tian-type pot bellows with a charge of locally col-lected copper ore and iron flux.

Both archaeo-metallurgical programmes, although based on quite different archaeological furnace models and carried out by quite different methods, achieved similar, in many ways complementary re-sults, which substantially advanced our knowledge of ancient copper smelting (see Vol. II of this publica-tion).

7 Chronology

(1) *Problems of metology* The protracted discussions around the dating of the Ramesside sites in the Arabah, related above, were symptomatic of the archaeological dating problems of the Arabah and the surrounding territories. This inhospitable, arid rift valley with few good water sources and generally rather saline soil, did not attract much sedentary habitation and therefore almost no proper stratified sites came into existence. Even the areas of rich copper mineralisation, which during different periods were indeed a considerable attraction for the ancients, produced only temporary, often rather shortlived centres of activities, mainly of intrusive groups of miners from more habitable areas nearby, or foreign, well-organised expeditions from afar. Each of these groups brought into the Arabah their own style of architecture and their own types of artifacts and different techniques – most of which were not archaeologically datable by themselves. This problematic archaeological situation was magnified by the fact that only very few stratigraphic excava-tions were undertaken in the territories adjacent to the Arabah and only very little, tentatively dated, comparative material was available. The sherds from the surveys in the Arabah were few and not of typologically known types and therefore rather diffi-cult to date. The three kinds of pottery of totally different character and manufacturing techniques reported from each of the smelting sites of Timna characterized the chronological dating problems dur-ing the early years of our research in the Arabah.

(2) *The early sites and their dating* Whilst for the Ramesside occupation of the southern Arabah the decisive chronological evidence was finally produced by the hieroglyphic inscriptions from the Hathor Temple, which abruptly concluded the protracted arguments against our dating proposals, the exis-tence of numerous, rather flimsy sites of obviously early periods, containing flint industries with distinct local characteristics and rough, hand-made pottery belonging to a Chalcolithic-to-Early Bronze Age horizon, again created continued chronological disputes.[73] These difficulties were increased by the fact that some of the surveys in Sinai were made by Biblical archaeologists from Israel, who simply extended their chronological criteria into the newly-investigated territories. A typical case was the dis-covery in southern Sinai of a group of settlements containing typical EBII pottery imported from south-ern Canaan (probably from EBII Arad). Because at several sites this Arad-type EBII pottery was found together with local hand made vessels (Amiran, Glass, Beit Arieh, 1973), all sites in Sinai and the Arabah which contained such pottery were sum-marily dated EBII (Beit Arieh, 1981; 1982, map Fig. 1) without taking into consideration the possibility that this local pottery, and the related sites, might have had a much longer life, i.e. might be earlier and/or later.

(3) *Local cultures in the Arabah and Sinai* At an early stage of our surveys in the Arabah and Sinai we had already reached the conclusion that many of the early settlements contained evidence for the existence in this area of local population groups which, during

Chronological and petrographic correlation table (rotated)

RELATIVE ISRAEL CHRONOLOGICAL TABLE	ABSOLUTE ISRAEL CHRONOLOGICAL TABLE	RELATIVE AND ABSOLUTE EGYPTIAN CHRONOLOGICAL TABLE	ARABAH, TIMNA AND UVDA VALLEY POTTERY TRANSITIONS	PETROGRAPHIC CHARACTERISTICS AND CULTURAL AFFINITIES	SITES
IRON AGE II	—1000—	NEW KINGDOM — DYN.XXII	MINING AND SMELTING IN THE SOUTHERN ARABAH (in Dyn.XIX,-XX and XXII)	Characteristic pottery: Midianite (for the 1st and 2nd Phase); Negev-ware (for all phases); Egyptian red polish type (for 1st Phase); Egyptian red polish type B (for 3rd Phase); Egyptian white slipped type A (for 2nd Phase); Egyptian white slipped type B (for 3rd Phase); Egyptian regular pottery (for all Phases); rough hand made pottery (for 1st and 2nd Phases).	2, 30, 200, 3; 12, 13, 14, 15, 19; 24, 25, 32, 33, 34; 35, 43, 52, 68, 86; 102, 103, 104, 112; 185, 193, 198, 25A; Mines S
IRON AGE I	—1200—	DYN.XXI 1080; DYN.XX 1196			
LB IIB	—1300—	DYN.XIX 1305			
LB IIA	—1400—	DYN.XVIII	NO SITES IDENTIFIED		
LB I	—1550—	1554			
MB II (MB IIB)	—1750—	INTERMED. II — DYN.XIII-XVII 1785			
MB I (MB IIA)	—2000—	MIDDLE KINGDOM — DYN.XII 1991; DYN.XI 2040	II LATE PHASE I	Typical combed, thin-walled high grade, compact wares of EB IV and related wares.	31, 149, 153, 166; 167, 167A, 168, 170; 172, 226, 227, 231; 233, 234, 236, 237; 239, 248, 249, 263
EB IV (MB I or EB-MB)	—2200—	INTER. I — DYN.VII-X 2155	MINING AND COPPER SMELTING IN THE SOUTHERN ARABAH	Degenerated Holemouth rims of arkosic and calcareous wares (mainly of Uvdah Valley)	167A, 226, 227, 231, 237
EB III		OLD KINGDOM — DYN.IV-VI 2570; DYN.III 2635	MIDDLE PHASE	Mixed arkosic wares Site 201A	111, 131; 144, 167, 168, 177, 191, 201A, 234, 236, 239
EB II	—2650—	PROTO DYNASTIC — DYN.II 2780; DYN.I 2955	EARLY PHASE	EGYPTIAN NILE WARES; ARADIAN WARES (A.III-I) (ABSENT IN ARABAH); HIGH GRADE, HARD BLACK VARIANT TO P.C. WARE; OXIDED WARES; VARIOUS HIGH GRADE; MUSCOVITE BEARING THIN-WALLED ARKOSIC WARES OF T.EBI SALAH TYPE; NORMAL BUFF TO GREY P.C. WARES	
ARAD I-III / ARAD IV (Hor-Aha, Narmer)	—2780— Qa'a; —2925— Djer				
EB I	—2970—		SINAI – ARABAH COPPER AGE	Characterised by absence of thin-walled slipped arkosic wares and thickened holemouth rims, Egyptian Nile-wares. Aradian wares. Dominated by rough hand made thick-walled wares of arkosic, calcareous and other composition. Arkosic wares un-slipped and use of chaff is common. The typical thin-walled calcareous p.c. wares are already present.	F2, 39, 44, 112, 119B, 152, 137, 189A, 200, 201A, 203, 209, 213, 250, 250B, 668, Mines G and T
ARAD V	—3150—	PRE DYNASTIC — Lower / Upper; El-Omari, Maadi, Mer-tan; Badar-Amrat-tan (Nagada II); Gerzian; Tasian; Fayum A			
CHALCOLITHIC					
EARLY	—4500—				

14

many hundreds of years, had developed their own indigenous flint and pottery industries. This conclusion was based on the facts that, besides very distinct local flint industries, the early inhabitants had their own specific ways of constructing their habitation sites and burials, their hand-made, rough pottery was different from the early pottery of Egypt and Palestine – although here and there some common formative tendencies could be discerned – and the extractive metallurgy found at many sites in the mining areas showed specific local, mostly rather primitive characteristics (Rothenberg, Tylecote, Boydell, 1978; Rothenberg, 1979; 137–91).

For our first attempt to establish a chronology suitable for the local cultures of the Arabah and Sinai we turned to the very large collection of flint objects and debitage from several hundred sites, because the typology of the relatively small number of diagnostic sherds from the surveys and the few stratigraphic excavations did not produce reliable chronological criteria (Rothenberg and Ordentlich, 1979).

(4) *First attempts: flint sequences* Two distinct flint industries in the Arabah and the Sinai peninsula[74] were identified and related to the chronological range between the Neolithic and EBIV. These were called, after key sites in the Arabah, 'Eilatian' and 'Timnian' (Ronen, 1970: 30–41; Kozloff, 1974: 35–49), a nomenclature henceforth used to define two distinct local cultures (Rothenberg and Ordentlich, 1979; Rothenberg, 1979), roughly contemporary with the Pre- and Proto-Dynastic and Chalcolithic to EB II cultures of Egypt and Palestine (see comparative chronological tables in Rothenberg and Ordentlich, 1979, 234 and Rothenberg, 1979, 238; in Conrad and Rothenberg, 1980, 26).

However, because of the intrinsic problems involved in the dating of flint industries, the reliance on a flint sequence, even if occasionally backed by imported diagnostic sherds of known date, did not allow for an entirely satisfactory definition of the chronological 'boundaries' between the 'Eilatian' and 'Timnian' cultural phases which often seemed to be contemporary or at least overlapping.[75] We therefore returned to pottery, hoping that systematic petrographic investigations[76] backed by archaeological typology would produce a methodologically as well as factually more satisfactory basis for a chronology of the indigenous, pre- and proto-historic cultures of the Arabah and Sinai.

(5) *Petrography: The Sinai–Arabah Copper Age Phases* The petrographic investigation of the Arabah and Sinai pottery was, of course, not expected to produce absolute dates, but to 'create order' in our fairly large collection of sherds, mostly worn body fragments of simple, hand-made 'kitchen ware' vessels.[77] At the conclusion of these studies the investigated pottery had been arranged into several distinct mineralogical and technological groups and their interfaces. It also became possible to distinguish between local and imported wares and point to their likely origin.[78]

As an overall result of this study, it became poss-

ible to define distinct and sequential pottery phases and transitions of major chronological significance, which we call 'The Sinai-Arabah Copper Age – Early, Middle and Late Phase'.[79] It is now possible to demonstrate that these Copper Age Phases are in fact the major phases of the local cultural developments in the Arabah and Sinai, clearly reflected in the archaeological, archaeo-metallurgical and political history of these territories. We propose, therefore, to replace the 'foreign' chronologies of the neighbouring regions – African Egypt on the one side and Palestine and the Levant on the other – by the Sinai-Arabah Copper Age chronology (*Illus.* 5).[80]

Notes

1. The Arabah Expedition was founded in 1959 by Beno Rothenberg, to investigate the Arabah rift valley and, especially, the copper mining areas at its southern end. First affiliated to the Haarez Museum, Tel Aviv and later (1970–75) to Tel Aviv University, by 1975 the Arabah Expedition had developed into an international research group and was incorporated into the newly formed Institute of Mining and Metals in the Biblical World, Tel Aviv, later affiliated to the Institute for Archaeo-Metallurgical Studies (IAMS), Institute of Archaeology, University of London.

2. The final report of the Arabah survey will be published as Vol. III of this publication.

3. 'Timna' on the Israel map is called Wadi Meneijeh on the map of the Survey of Palestine, 1939 (1:250.000). In the following, the old Arabic names, used generally in the literature prior to the publication of the new map of Israel, will be given in parenthesis, together with the first mention of any new names of sites and locations.

4. See the detailed report on the survey of the Nahal Amram mines and smelters in Vol. III of this publication. No Nahal Amram sites have yet been excavated.

5. See preliminary report by Rothenberg, B., *Sinai*, 1979.

6. See Rothenberg, B. in Conrad and Rothenberg, 1980 and final excavation report in Vol. IV of this publication.

7. Since Frank did not mention the numerous plate-like features (which turned out to be filled-in mining shafts) most conspicuous in the actual mining areas of the Timna Valley (Conrad and Rothenberg, 1980), we must assume that he never visited the main mine area of Timna, located on the west side of the valley, and saw in the natural round holes in the sandstone layers, caused by decomposing petrified trees 'Gänge' of mining. Incidentally, Frank did observe such 'Plates' in Wadi Taba and identified same as 'sleeping places of the miners and their supervisors', (Frank, 1934, 247, Tab. 45B).

8. Bachmann and Rothenberg, in Conrad and Rothenberg, 1980.

9. Contrary to Fritz Frank, the German Biblical scholar A. Alt tried an historical interpretation of Frank's discovery (A. Alt, 1935), including the reconstruction of a north-south main road from the Dead Sea to the Red Sea, based on numerous 'Roman strongpoints'. Alt's reconstruction was generally accepted by historians and archaeologists (see M. Avi-Yona's Map of Roman Palestine, 1940). However, our Arabah survey proved that most of these 'Roman strongholds' were, in fact, much earlier sites and that no major thoroughfare ever ran along the Arabah (Rothenberg, 1971).

10. Those important mining areas at the north-east side of the Arabah now located in the Kingdom of Jordan, were first explored by Glueck in 1934 (Glueck, 1935), but this did not result in a clear archaeological picture and nothing became known about the mining and smelting operations of this area until 1984, when H.G. Bachmann and a team from the German Mining Museum, Bochum, started systematic explorations of the area. From preliminary reports (Bachmann and Hauptmann, 1984; Hauptmann, Weisgerber and Knauf, 1985), it appeared that after intensive early prehistoric smelting, the main period of activities was in the later first millennium BC which, according to pottery finds, seems to be mainly connected with the Kingdom of Edom (8th–6th centuries BC). So far, no second millennium BC activities were reported from

the north-east Arabah, which during most periods of its history presents a picture quite different from the southern Arabah.

11. In the early fifties, the present author was field director, archaeological chief assistant and keeper of records of Glueck's Negev Survey. During a visit of the survey team to Timna in 1953, he had the opportunity to discuss with Glueck his work at Timna in the early thirties. Glueck apparently did not return to Timna after his first and rather short visit in 1934, and the plans of sites in Timna published by him (Glueck, 1935; idem, 1940) were drawn from R.A.F. aerial photographs which he obtained after his survey campaigns. These plans were never confirmed on the ground, a fact which explains the total difference between his plans and the site plans of the Arabah Expedition (Rothenberg, 1962; idem, 1967a).

Glueck, during his Negev survey, returned to the Arabah in 1953–54 but no additional work was done in the Timna Valley. However, Glueck later investigated several sites in Nahal Amram, convinced that he was dealing with additional sites of King Solomon's Mines (Glueck, 1959, 1960).

12. This particular structure in Site 2 was subsequently excavated by the present author and turned out to be a human burial.

13. I Kings, 9: 26; 22: 48. For the proposed identification of the port of Ezion-Geber with the island anchorage Jezireth Fara'un ('Pharaoh's Island') c. 10km. south of Eilat, see Rothenberg 1967a, 207–13; idem, 1972, 202–7; Flinder, 1985, 63–81) .

14. The Biblical Tarshish-ships have long been considered by Biblical archaeologists as nautical metallurgical workshops, a 'refinery fleet' (Albright, 1941, 21–22) often connected with the metal trade between Phoenicia and Iberia (see in this connection, Rothenberg and Blanco, 1981, 171–3).

15. Glueck, 1940, 104.

16. See frontispiece to *The Story of Man*, Time-Life Publications, New York; also Keller, W. *Und die Bibel hat doch recht*, Düsseldorf 1955, T.V. Kap. 2.

17. See *Cambridge Ancient History*, 1975, Vol. 2, 594; Noth, M. *Die Welt des Alten Testaments*, Berlin 1962, 152; Bright, J. *A History of Israel*, London 1960, 195; Wright, G.E. *Biblical Archaeology*, London 1955, 132–6.

18. At the time of our first survey, most of the geological information on the Arabah was still unpublished and we received most of our information verbally from the geologists of the Timna Mines Ltd., Y. Bartura, M. Preiss (see now Bartura, et al., 1980; Preiss, M. 1967).

19. The first analyses of our survey were made gratuitously by the laboratory of Timna Mines Ltd. Later, after A. Lupu joined our group, and until our New Arabah Project in 1978, most analyses were made by wet chemistry at the Technion, Haifa, Department of Material Science.

20. At the same time, we also investigated the ancient mining areas of Nahal Amram which showed a rather different topography but a very similar mineralisation (see Rothenberg, 1962, 33–8; idem, 1967a, 41–9).

21. Rothenberg, 1962, 5–65.

22. See for first dating of the Timna pottery, Y. Aharoni, 1962, 66f. Although in Rothenberg, 1960, we first dated the Timna pottery as pre-Solomonic, the Solomonic date subsequently proposed by Aharoni after his study of our survey collection, was adopted by us up to 1964. Subsequently the Timna excavations produced stratified as well as new types of pottery not found during the surface survey: these caused a first shift of the date (by Aharoni and the present author) to the 12th–11th centuries BC and later, after the discovery of the Hathor Temple, to be revised to the 14th–12th centuries BC.

Dates of smelting sites in 1962: Site 39 = Chalcolithic; Sites 3, 12, 13, 14, 15, 30, 34, 35 – Early Iron Age; Site 28 (Beer Ora) = Roman-Byzantine.

23. Already at the time of their first discovery, the 'plates' represented an enigma, discussed with geologists, mineralogists and the experts of the Timna Mines Ltd., but no other explanation was offered. Reluctantly (because of the ever-increasing number of 'plates' which we continued to find on the slopes of Timna) we adopted the ore dressing installation explanation of these 'mysterious plates' (Rothenberg, 1962, 38).

24. The metallurgical processes of the Arabah were systematically investigated in the early eighties. cf. Vol. II of this publication.

25. Pinkerfeld was the field architect of the Tell el-Kheleifeh excavation. It is rather unfortunate that the excavation of Tell el-Kheleifeh was only published in preliminary reports in *BASOR*, and neither the pottery nor any records of the proposed stratigraphy were ever published (but see now Pratico, 1983).

26. We quote here some of the publications which rejected our evaluation of Glueck's theories, as these are representative of the rationale of this protracted, rather one-sided argument. After my detailed critical review (Rothenberg, 1967a, 194–206), I refrained from taking part in any further discussions of the subject and, instead, continued our systematic excavations in Timna, culminating in the discovery of the Hathor Mining Temple and the identification of the Ramesside copper industries in the Arabah.

27. Glueck quotes in 1969 from an 'unpublished paper' by W.F. Albright, strongly rejecting our pre-Solomonic dating of the Timna pottery: 'Every new discovery of pottery convinces me that Nelson Glueck is right in his chronology and that Aharoni and Rothenberg are wrong'.

28. Although beginning in 1964, our main emphasis was placed on systematic excavations in the Timna Valley, we nevertheless continued our surveys in the Arabah. In 1966–67, right up to the Six Days War, we undertook very intensive explorations in the southernmost Arabah (the 'Eloth Survey' sponsored by the Eloth Regional Council, and its Chairman, Avinoam Finkelmann of Kibutz Yotvata), discovering many previously unknown sites and reinvestigating sites we, or others, had found on earlier surveys (Rothenberg, 1967, 283–331: cf. also *Illus.* 4 above, and Vol. III of this publication).

29. In 1964–70, the Arabah Expedition only excavated in the smelting camps and not in the mines of the Arabah, but returned to Timna in 1974 together with the German Mining Museum, Bochum, to explore the mine workings, (Conrad and Rothenberg, 1980).

30. See the final report on the excavation of Site 39 in Vol. IV of this publication. For preliminary reports, see: Rothenberg, 1966; idem, 1972; idem *et al.*, 1978.

31. On the new chronological terminology used in this publication, see above, pp. 13–15.

32. See final report in Vol. IV of this publication. For preliminary reports see: Rothenberg, 1967b; idem, 1972; Lupu and Rothenberg, 1970.

33. One of these furnaces (No. IV) was, in fact, only a clay lined hole-in-the-ground bowl type furnace, but in every other respect identical with the stone-built furnaces of Site 2 (see Rothenberg, 1985, and Vol. II of this publication).

34. Y. Aharoni, who had dealt with the survey pottery of 1959–60 and also undertook the first study of the pottery from the excavations, was the first to propose an earlier date for the Timna pottery, based on excavated material.

35. The Negev-ware of Timna, predating (by its archaeological context) the Solomonic period, is considerably earlier than any of its kind found previously (Aharoni, 1958; idem, 1960) or since (see Cohen, 1980) in the Negev, where it is dated Iron Age II. This is the main reason for our view that this pottery represents a local non-Israelite tradition, which we propose to identify with the first settlements in the Negev mountains of the previously itinerant Biblical Amalekites (Rothenberg, 1967a).

36. Only in the last few years was this area properly investigated (Bachmann *et al.*, 1984; Hauptmann *et al.*, 1985) and large scale Edomite smelting sites identified at Feinan and Kh. en-Nahas, and other sites in the area. However, these sites must be dated Iron Age II-III, and no reliable traces of Late Bronze Age (New Kingdom) smelting has yet been reported from the northern Arabah.

37. The original Arabic name of the well and site was Wadi (Bir) Hindis. Preliminary report on Beer Ora: Rothenberg, 1972, 212–23.

For the final excavation report, which after further intensive studies shows quite important alterations compared with the preliminary report, see Vol. IV of this publication.

38. Further studies established that part of the Beer Ora pottery, as well as some of its slag-built installations, should be dated to the Early Islamic Period, the 7th century AD.

On the presence of the Legio III Cyrenaica in the Arabah (see Rothenberg 1967a, 167).

39. Some of the preliminary publications: Rothenberg, 1972; idem, 1983; idem, 1985.

40. Preliminary publications: Rothenberg, 1969a; idem, 1970; idem, 1970a; idem, 1971; idem, 1972.

41. It is quite significant for the 'spirit' of these renewed 'discussions', that the most outspoken and bitter criticism of our dates was published by Professor Y. Yadin in an open letter to the press: Yadin, Y., 'King Solomon's Mines – how did they disappear?' in *Haaretz* (Hebrew Daily), Tel Aviv, 03.12.65. For our reply see Rothenberg, B.,'This is how King Solomon's Mines disappeared' in *Haaretz*, Tel Aviv, 10.12.65; Aharoni, Y., 'King Solomon's Mines – a reply to Yadin's question', in *Haaretz*, Tel Aviv, 10.04.66.

42. It was rather curious to read the solitary argument by S. Meshel (Meshel, 1975, 51) in support of Glueck's old theories, even after Glueck himself had expressed 'second thoughts' about the same (Glueck, 1965) and several years after the discovery of the Egyptian Mining Temple.

43. Most of the subsequent publications about the Timna Temple were mainly based on this report (e.g. Rothenberg, 1978), as the final stratigraphic and chronological data of the Temple site, published here for the first time, became available only after further excavations at the site and the processing of the large body of finds.

44. During the first years of our work, we had no analytical facilities nor funds to pay for analyses of our numerous metallurgical samples. This situation was greatly improved after A. Lupu of the Department of Material Science, of the Haifa Technion, joined our group (Lupu, A. and Rothenberg, 1970).

45. The New Timna Project 1974–76 was directed in the field by Beno Rothenberg, with H. G. Conrad (1974), Director of the German Mining Museum, Bochum, and I. Ordentlich (1976) of the Institute of Mining and Metals in the Biblical World – the Araba Expedition, Tel Aviv, in charge of the excavations in the mines.

46. The Timna site numbering in use in 1974 was in accordance with the overall system of running site numbers in the original Arabah Survey (*Illus.* 2). This is still being used, with the exception of the 'Model Area' in the Timna Valley, which has a different numbering system (see Conrad and Rothenberg, 1980, and also Vol. III of this publication).

47. 'S 14' belongs to the new numbering system of the 'Model Area' in the Timna Valley.

48. Weisgerber's overhasty publication of his 'theory of flotation' was not authorised by the director of the expedition, besides being unacceptable for an area where the annual rainfall is approximately 5mm.

49. Only rather restricted archaeo-metallurgical investigations could be carried out in the framework of the New Timna Project, since this was originally planned as a purely mining research programme. The processing of all the Arabah finds and, especially, the archaeo-metallurgy of Timna, were the main objectives of the final New Arabah Research Project 1978–83, published in the present series of publications.

50. Besides a small team of the Arabah Expedition (Ivan Ordentlich, Abraham Bercovici, Judith Gavish) the geologist Aharon Horowitz, Tel Aviv University, and Y. Bartura, formerly of Timna Mines Ltd. also took part in the second and third seasons of work in Timna. The former concluded a detailed geomorphological survey, the latter produced detailed geological maps which provided the groundwork for the later, additional investigations (third season), together with members of the Bochum group (see A. Horowitz *et al.*, 1980; Y. Bartura *et al.*, 1980).

51. Many of the finds had already been made during our earlier surveys in Timna (Rothenberg, 1962; idem, 1967; idem, 1967a), but many more finds were recorded during our intensive, detailed survey of a 'Model Area' in Timna (see Conrad and Rothenberg, 1980).

52. One early system in Area T was almost completely excavated during the third season of work (September 1976), and this excavation produced the archaeological evidence for its dating to the Sinai-Arabah Copper Age-Early Phase (Chalcolithic-Early Bronze I).

53. The 'Model Area' was partitioned into geographical units according to topographical criteria (for example, plateaus between wadis) and a new nomenclature was introduced: capital letters for the geographical units (Model Area Sections) and numbers for individual mining relics (see Rothenberg, 1980, Abb. 6, Beil. 2–3).

54. I.Ordentlich, staff member of the Institute of Mining and Metals in the Biblical World – the Arabah Expedition, Tel Aviv, was responsible for the excavations in the mine workings of Area T.

55. The cartographic work in Timna and adjacent areas was made possible by a special grant from the Volkswagen Foundation. The aerial photography also proved to be of great value for the actual field survey, making it possible to map the survey findings and sites on enlarged aerial photographs.

56. Although we now propose a new chronological terminology for the Sinai-Arabah area (see below), we are still using here the terminology in use at the time (Rothenberg and Ordentlich, 1979). In 1984 an additional period of mining and copper smelting in Timna could be established: Early Bronze Age IV (see below, note 66).

57. The mine workings identified as Roman mines in the early surveys of the 60's (Rothenberg, 1972, 208–11) turned out to belong to the New Kingdom mining systems, and apparently had been partially reworked during later periods. However, Roman mineworkings, identifiable by unique mining dumps of finely crushed sandstone debris, mixed with minute copper ore particles and Roman sherds, were located during subsequent Timna explorations, carried out as part of the New Arabah Project in 1978–83 (see below). These workings are located at the extreme north as well as the south ends of the Timna Valley (Sites 87, 88, 95). Copper mines worked during the Early Islamic Period (7th century AD) could also be identified in the south of Timna (Site 37).

58. In the third season, Ivan Ordentlich remained as the archaeologist in charge of the excavations in the mines. (B. Rothenberg carried on his excavations at Site 30.)

We wish to take this opportunity to emphasise once more the magnificent job done by the technical team of the Bochum Mining Museum, especially its miners and mine surveyors.

59. Pottery and flint objects collected on the surface of Model Area Sections F and G, indicated an 'Eilatian' date, which could be confirmed by subsequent excavations of the small smelting site F2 (see Chronological Table in Rothenberg 1980, 26, and the excavation report of F2, to be published in Vol IV of this publication).

In our new chronological terminology, F2 belongs to an early stage of the 'Sinai-Arabah Copper Age – Early Phase' (see below, *Illus.* 5).

60. Additional fieldwork within the framework of the New Arabah Project 1978–83 produced some evidence for mining and smelting during this 'interregnum' – Sites 201 and 149. See forthcoming reports in Vol. III-IV of this publication (see also Rothenberg, 1980, 31; and *IAMS Newsletter* No. 5, 1983, 2).

61. Rothenberg 1972, was mainly a preliminary summary of the work done at the Arabah furnace sites up to 1970. Conrad and Rothenberg 1980, is the preliminary report on the New Timna Project up to 1976. For the furnaces in Timna, see now Rothenberg 1985, and see also Bamberger, 1985, and Vol. II of this publication.

62. Hauptmann's proposal to consider a 'technological transfer' between Timna and Feinan (North Arabah) seemed at first not quite acceptable because of the totally different history of both ends of the Arabah (Rothenberg 1971a). However, it is of course possible that the Egyptians of the 22nd Dynasty, or at least their technology, also reached the Feinan and Kh. en-Nahas area, although during this period the area must have been a stronghold of the Edomite Kingdom. Further work in the northern Arabah may reveal new facts.

63. The New Arabah Project was carried out by the permanent team of the Institute of Mining and Metals in the Biblical World, Tel Aviv (see above, Note 1), lead by Beno Rothenberg, in collaboration with an international group of specialists. A list of the participants in our fieldwork and the processing of finds, will be published in the relevant reports in the volumes of the present publication.

The New Arabah Project was made possible through the generosity of the Volkswagen Foundation, Hannover, and the Institute for Archaeo-Metallurgical Studies (IAMS), London.

64. 'The Sinai-Arabah Copper Age – Late Phase' is approximately equivalent to Early Bronze Age IV (MBI or EB-MB) in the Chronology of Palestine. See below, para. 7.

65. For preliminary reports see: Ronen, 1970; Kozloff, 1974; Bercovici, 1978. See also Rothenberg and Ordentlich, 1979. A

report on the Arabah and Sinai flints will be published in Vol. III. (See below, note 74).

66. The petrographic report on specific pottery groups will be published in the relevant final excavation reports in Vol. IV. The final, comprehensive petrographic report on the early Arabah pottery will be included in Vol. III (The Arabah Survey 1959–84) of this publication. The Sinai pottery report will be published in Rothenberg, B. (ed.), *New Researches in Sinai* (in preparation).

67. Because most of the Arabah metal finds came from the Timna Temple it was decided to publish all Arabah metal analyses in the present volume and to include the same in the comprehensive metallurgical discussion below (Chap. III, 8).

68. The first initiative for these studies came many years ago from H.G. Bachmann (in an internal, unpublished IAMS research review) and considerable work on these problems was subsequently done by R.F. Tylecote (Tylecote and Boydell, 1978, 45–8).

69. The definitive report on the extractive research programmes will be published in Vol. II of this publication. See preliminary reports: Merkel, 1983; idem, 1983a; Bamberger, 1984; idem, 1985; Bamberger, Wincierz, Bachmann and Rothenberg, 1986 and 1987. See also Rothenberg, 1985.

70. Bamberger's smelting experiments were carried out at the Haifa Technion with the technical assistance of A. Amram.

71. John Merkel's research programme was carried out at the Institute of Archaeology, University of London, as the central part of his doctorate (supervised by R.T. Tylecote). This fact put certain restrictions on the exchange of information between Bamberger and Merkel, at least up to the time of the completion of Merkel's Ph.D. thesis (Merkel, 1983).

72. The smelting experiments and the extensive analytical research were carried out at the premises of Geomet Services-Borax Consolidated Ltd., Chessington, England.

73. See the characteristic argumentation by J. Muhli, 1976, 92–3, who opposed our Chalcolithic date for the flint implements and pottery of Site 39 (Rothenberg, Tylecote, Boydell, 1978), without ever having seen the material which, at the time of his publication, was mostly still unpublished.

74. Although the Arabah, because of its geographical situation and specific characteristics, has to be considered a historio-geogra-

phical unit of its own, in many aspects of its early developments it must be seen as part of the large semi-arid region between Egypt, Palestine and Arabia; in fact in predynastic (Chalcolithic-EB I) times it was essentially part of Sinai. It is for this reason that our detailed studies of the material cultures of these territories dealt with the finds from the Arabah and Sinai in conjunction. (See above, note 65.)

75. Radiocarbon analyses obtained by B. Kozloff for charcoal samples collected at 'Timnian' habitation sites during his excavations in eastern Sinai (Themed area) produced pronounced dating clusters in the late 5th millennium and again in the late 4th millennium BC. Although the hiatus may be the result of the sampling and/or excavation methods, it contributed to our decision to look for further chronological criteria.

76. The petrographic investigation of all sherds from our Arabah and Sinai surveys (by J. Glass) was already started in 1973 but could only be concluded in 1985.

77. A detailed description of the methodological strategy of these investigations, will be published as part of the definitive report on the Arabah survey, in Vol. III of this publication.

78. J. Glass took part in one of our survey seasons in Sinai and spent some time in the Arabah, investigating on the spot the available clays and tempers. For the necessary comparisons he also had at his disposal a large number of ceramic samples from the territories adjacent to the Arabah and Sinai, including collections of sherds from Egypt, Arabia and the Negev.

79. A change in the outdated 'metallurgical' chronological nomenclature – which starts the 'Bronze Age' long before Bronze actually appeared – is long overdue. For this reason we are using the term 'Copper Age' instead of 'Bronze Age'. The Late Phase of the Sinai-Arabah Copper Age saw in some parts of the Near East the transition to the use of tin-bronze, but in the Arabah and Sinai no traces of such early use of tin-bronze were ever found.

80. For the relative and absolute chronology of Israel used by us, see Avi-Jonah, M. and Stern, E. (ed.), *Encyclopedia of Archaeological Excavations in the Holy Land*, Vol. IV, Jerusalem, 1978, 1226–28. For the Egyptian chronology we refer to Beckerath, J. von, *Abriss der Geschichte des Alten Aegyptens*, München, 1971, 63–68.

I. The Site and Method of Excavation

1 Location and Discovery of Site 200

During the Arabah Survey in 1966, attention was paid to the vague outlines of what appeared to be a small structure built against one of 'King Solomon's Pillars' (hereinafter: 'The Pillar') in Nahal (Wadi) Nehushtan. These 'Pillars' are huge, picturesque, palaeozoic sandstone formations at the south-western end of Har (Mount) Timna (453m.) (*Pl.* 1) which is the central massif of the Timna Valley, formed mainly of Pre-Cambrian granites of strong red, brown and black colour, capped by huge tabular dolomitic rocks. The Pillars are located almost in the centre of the ancient mining and smelting area of Timna.

The site (*Pl.* 2) No. 200 on the Arabah Survey map, at Isr. Gr. R. 14579090 – was a low mound of fairly loose red sand, measuring about 15 × 15m. and *c.* 1.10m. in height, piled up against the face of one of the 'Pillars'. On its surface building debris was dispersed, including some well dressed, large stone blocks, seemingly the collapsed upper part of a rectangular structure (*c.* 9 × 10m.) vaguely discernible on the present surface of the mound. It seemed highly interesting that much of the building debris was of white sandstone, which had to be carried to the site from quite a distance, instead of the easily available local red sandstone. At the west end of the mound, several large, red sandstone boulders were lying on top of the building debris and it was obvious that a huge stonefall had come down from the top of the 'Pillar' and smashed into the structural remains of the site. Luckily, this stonefall only involved the very edge of the structure, and took place after the site had already been abandoned for some time.

Among the debris, a small number of bichrome and other sherds, as well as a small copper arrowhead (or spatula) were collected. There were no traces of any metallurgical activities at the site, which was rather unusual for an ancient site in the Timna Valley. Furthermore, three niches cut into the rockface of the 'Pillar' immediately above the mount (*Pl.* 3), in the shade of a huge overhanging rock ledge, added interest to this peculiar site. In the first published description of Site 200 (Rothenberg, 1967, 313) it had already been suggested that '. . . we are probably dealing with a very unusual cultic site which should be excavated. . .'. However, the excavation of Site 200 had to be postponed until the end of 1969, at the conclusion of the first series of our systematic archaeo-metallurgical excavations in the smelting camps of the Timna Valley (Rothenberg, 1972).

Plate 1.   'King Solomon's Pillars' with fenced-in Site 200.

Plate 2.   Site 200 before excavation.

Plate 3.   The three niches cut in the rockface behind the site.

2 The Excavations, Organisation and Team Members

The first and main season of excavations at Site 200, began on March 24, 1969 and continued until late May 1969.[1] It was directed by Beno Rothenberg (Arabah Expedition, Tel Aviv University); supervisor and surveyor: Reinhard Maag (Zürich); final plan of the excavation: Shmuel Moskovitz (Tel Aviv University); site photography: Beno Rothenberg; find records and photography: Craig Meredith (USA); R.F. Tylecote, A. Lupu and Johanan Aharoni spent some time at the excavations and advised on metallurgical and stratigraphical aspects.

The following volunteers from the USA, Denmark, Switzerland, Britain, Australia, Holland and Israel, took part in the excavations: Richard Aron, Arthur Bankoff, Andrea Bankoff, William Beren, W. Brewer, Tony Elitclier, Gene Friedlander, Dianne Grossman, Hans-Peter Gruenfelder, Christian Herz-Petersen, Awyneth L. Jones, Philippe Lachenal, Marlene Lang, Bob Leiteras, Brian G. O'Dea, Chester J. Pavlovski, John Pavlovski, Frederick S. Rosenfeld, Nancy Rubel, Jan van der Laak, Maya Zahavi, Rochele Zaltzman.

A second season of excavation took place in September 1974 and was carried out by Morag Woudhuysen, with the participation of Gloria London, Paul Woudhuysen and several team members of the New Timna Project.[2] Further work at Site 200 carried out during 1976–84, in connection with the development of Timna as an 'Archaeological Park', and the need to consolidate the Temple structure, enabled us to reconfirm some details of architecture and stratigraphy.[3]

3 Method of Excavation

Grid Squares, Walls and Trial Trenches: In order to facilitate three-dimensional recording, a one metre square grid was laid over the whole site, extending for several metres outside the area of the actual mound, onto the adjacent level ground. The whole area of the mound was segmented quadrant-wise (Harris, 1979) into four 'Areas' (I–IV), separated by baulks which served as the main stratigraphic sections of the excavation (*Pl. 4*). As it turned out, one of these baulks ran over the entrance to the Temple courtyard, as well as through the centre of the *naos*. For this reason, and because it became more convenient to change over to a proper open area excavation combined with cumulative sections (Barket, 1977), the baulks were taken down as soon as the excavation had reached a clear man-made interface, i.e. a solid floor made of crushed white sandstone.

Walls: Walls were the only feature which needed a separate numbering system (W1 to 5) as all other, much smaller features could be identified by their grid reference (e.g. Basin D–E 13).

Trial Trenches: Whilst most of the excavation was area clearing over one or a group of grid units, in conformity with the bed-lines of the layers, the cutting of trial trenches (TT) became necessary occasionally in order to determine the relationship and integration between features and obtain detailed sections of particular stratigraphic situations.

The first season of excavations ended when a solid white sandstone floor had been reached under most of the Temple area. However, the second season, planned to remove most of the White Floor and excavate to bedrock, took the form of a system of adjacent trial trenches which allowed a close scrutiny of the stratigraphic sequences and relationships.

Heights: A convenient 0–benchmark was fixed on the face of the Pillar, above the mound, and all heights recorded are therefore negative, i.e. height given indicates the depth below the 0–benchmark.

4 Method of Recording

Supervisor's Field Notebook: The Supervisor's Notebook consisted of a daily graphic record – sketches, plans and sections – of features, layers and interfaces and their stratigraphic relationship, exposed during the day's work, as well as a detailed description of the material aspects of the individual units of stratification, for example, the kind of debris (sand, humus, gravel, etc.), colour, consistency, and the heights reached at the end of the day's work.

As a separate findbox contained the finds from each layer, the Supervisor marked the box number into his daily graphic record, to ensure a reliable record of the stratigraphic provenance of the finds. The box numbers of the first season started with 1, those of the second season started with 500. Each findbox also contained its own 'passport': grid and layer number, a short description of the nature of the same and the relevant heights. Three-dimensional grid references were recorded for especially interesting individual finds. The Supervisor's Notebook concerned itself exclusively with the stratigraphic sequences and did not deal with the phasing of the stratification or its chronology (see below in Director's Field Journal and Post-Excavation Analyses and Chronology).

Findbook: A Register of Finds was kept for the whole of the excavation. The Recorder was also responsible for the supply of properly labelled findboxes to the trench supervisors whenever a new layer was reached, they also had to make sure that at the end of the day all necessary data was properly recorded on the labels.

Each findbox occupied a separate page in the Register of Finds, which contained all the data recorded on the box label, including a short description of the relevant layer, a brief description of the finds and also, in the case of important objects, their three-dimensional location, often with a small sketch.

Plans and Sections: The Daily Field Notebook contained single-layer plans and detailed local sections (sketches relating to the specific areas excavated during the day). Wherever necessary, such as before the

Plate 4. Site 200 under excavation, note main baulks.

removal of structures, debris or other features, detailed small-scale plans (1:10) were prepared. Similarly, plans were drawn of architectural details of special significance.

At the end of the first, major season of excavations, a composite plan, 1:25, was made including all man-made and natural layers,[4] i.e. architectural elements, 'standing strata', man-made and natural interfaces exposed at this stage (*Illus.* 6).

As the second season was mainly a series of trial trenches concerned with stratigraphic problems in the layers below the White Floor level, the White Floor, which was of fundamental stratigraphic significance and also the main floor of the Temple structure to be preserved as a National Monument, was retained on this plan. This composite plan remained the Work Plan of the excavations.[5] A number of details, exposed during trial trenches in 1974, (e.g. some Chalcolithic pits inside the *naos* structure) were added to the Plan.

Two main sections criss-crossing the whole site (the baulks) and a number of smaller, partial sections of special stratigraphic significance were drawn as the excavation progressed (*Illus.* 7). The main W–E profile (8a–8b) was drawn as a cumulative section down to bedrock, during both seasons of excavation (1969 and 1974) (*Illus.* 8, 9).

Director's Journal: The somewhat disturbed stratification in the loose sand over part of the site, plus the spotty character of some of the late occupation layers and interfaces, demanded the immediate correlation of layers and their phases during excavation, instead of during post-excavation investigations.[6] The Director's Journal therefore correlated the layers of the various areas and trenches of the excavation to form an overall stratigraphic model of the site. This model was, of course, a dynamic process until the very end of the excavation and was daily summed up in the Director's Journal, which included a review of the problems solved or still to be investigated.

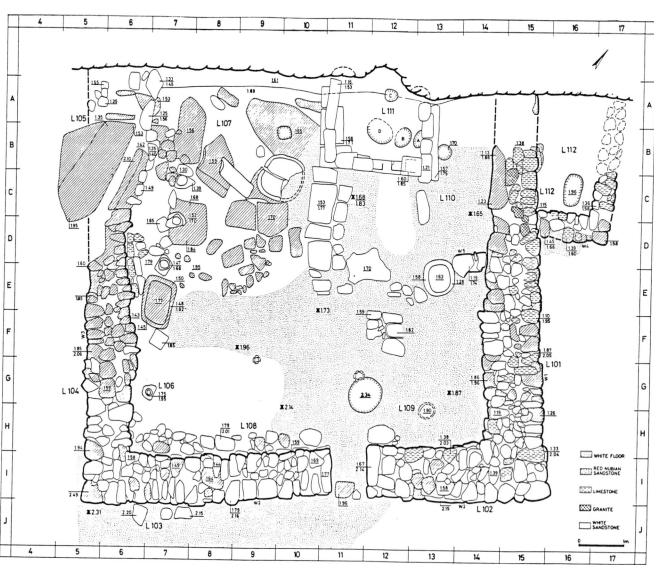

Illustration 6.   Plan of excavation of Site 200.

22

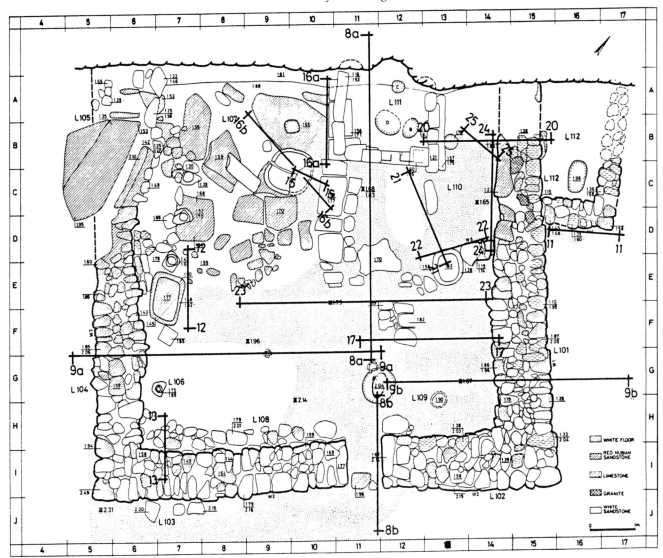

Illustration 7.   Plan of loci and sections.

Although during the excavation itself the main emphasis was on stratigraphic sequences, separately established for each grid unit or group of grid units, the Director's Journal also recorded the overall stratification of the whole site and the appearance and chronological significance of features and finds. Thus, the primary recognition of the chronological strata and periods represented at the site, could be achieved as the excavation drew to its close.

During the last few days in the field, the Director recorded systematically in this Journal, and by photography, a summary of the different areas, their stratigraphic sequence, phasing and tentative chronology; these notes proved to be especially useful during the preparation of the present final excavation report. It was at this concluding stage of the fieldwork that locus numbers were introduced. Each locus number marked a specific area of the excavated site, previously recorded in terms of grid-units which, according to the character, and topography of the archaeological remains, was more conveniently treated as a separate unit. However, it should be remembered that such loci had no separate archaeological significance, but are to be understood only as a matter of convenience to enable cohesive descriptions and discussions in the excavation report.

Photographic Record: The photographic record of the excavation, in black and white and colour, was kept by the Director and was linked into the Director's Journal as an extension of its written and drawn records. At the end of the excavation, a systematic and very detailed photographic coverage of the final findings was made, including the numerous stray architectural elements found at the site.

5 Post-Excavation Analyses and Chronological Periods

The very large number of votive gifts and other small finds, plus the numerous architectural elements found throughout the often disturbed layers of the site, many found in secondary use, and the often

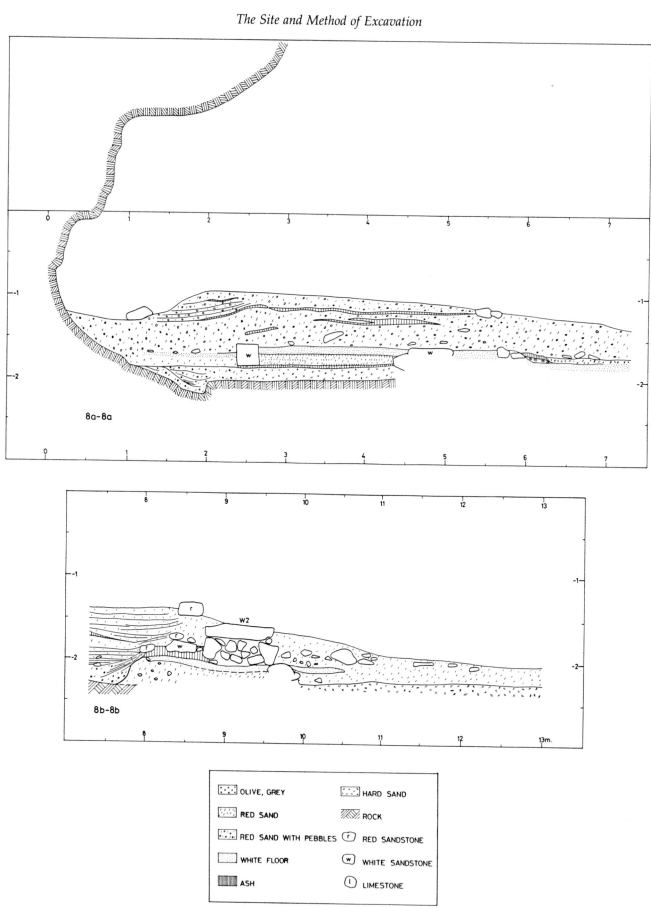

8a-8a

8b-8b

OLIVE, GREY		HARD SAND	
RED SAND		ROCK	
RED SAND WITH PEBBLES		RED SANDSTONE (r)	
WHITE FLOOR		WHITE SANDSTONE (w)	
ASH		LIMESTONE (l)	

Illustration 8. Main section E-W.

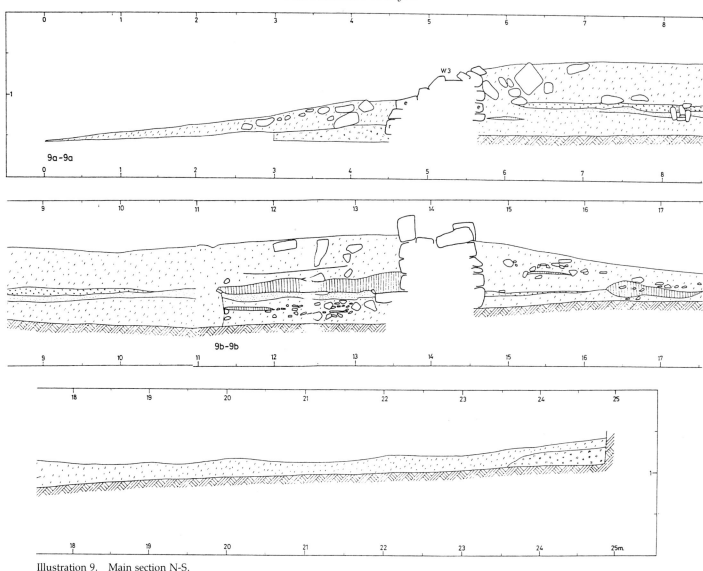

Illustration 9.   Main section N-S.

fragmentary and spotty nature of the stratigraphic evidence, demanded very painstaking, systematic post-excavation investigations.

Processing of the Finds, their Investigation and Analysis: During the excavation a primary, albeit tentative chronological model of the site could be established and recorded in the Director's Journal. The first task after the excavation was the processing of the find-boxes and the artifactual analyses, by a group of scientists and specialized archaeologists. The contents of each findbox was recorded on a box index card, which listed all its identifying topographic and stratigraphic data, as well as the nature of the deposit and composition of the layer. Out of this group of finds a number of objects were chosen for further processing and analysis and given an individual card. This find index card carried a short description or definition, often also an identifying sketch of the object, and served as its 'passport'. It also recorded the movements of the objects to and from laboratories, specialist workers, drawing office, photographer and a brief review of the results. If archaeometric work was done on the object, its Sample Numbers were also listed. As a separate archaeometric card index was kept, the sample card contained characterisation of the sample, the method used for its analysis and a table of results.

The different groups of finds were separated for specialist work and it was the specialists' decision which of the objects should be selected for inclusion in the final excavation report. The only non-selective approach concerned the very large number of metal objects. If an object was too fragmentary to be drawn, photographed and included in the final catalogue, it was sent to the British Museum Research Laboratory for metallurgical investigation (Chapter III, 8–10).

Comparative investigation of the finds was only possible to a limited degree because of the unfortunate lack of relevant comparative excavations,

25

material and publications. However, this did not affect the chronology of the Temple because this could be based almost entirely on hieroglyphic inscriptions and cartouches.

Correlation of stratigraphic sequences, artifacts and habitation phases: Post-excavation correlation of stratigraphy and artifactual analyses is, of course, the backbone of every excavation, as only thereby can chronological and cultural significance be obtained for the stratification of the site. Because of the complexity of the stratigraphic problems of the Temple, this work had to be done in several stages:

(1) *Reconstruction and phasing of the stratigraphic sequences:*
The primary definition of the stratigraphic phases and chronological strata of the site in the Director's Journal, was rechecked and finalized, prior to writing the present excavation report. The stratigraphic features of each grid unit or group of units, if excavated and recorded as such, were separately tabulated, and then correlated and combined with its neighbouring grid units, until a whole locus was covered. At this stage it became possible to establish phases of habitation, disturbances, destruction and rebuilding. The successive tying together in diagrams[7] of the neighbouring loci, resulted in the recognition of overall stratigraphic phenomena, in contrast to locally restricted stratigraphic features. Although the resulting cohesive, overall stratigraphic picture of the whole site clearly indicated its successive habitation phases (*Illus.* 8, 9), the local sequences, mostly around special features and recorded in large-scale sections, often retained their decisive significance for the understanding of the site's occupational history.

(2) *Habitation phases and chronological strata:*[8]
Once the stratigraphic sequences and habitation phases, wherever preserved, had been clarified, all data produced by the artifact analyses were systematically correlated with the same – using similar diagrams as produced for the phase analyses – in order to establish the chronological strata and culture-history of the site. The primary, naturally tentative,

concept of the Temple's chronological and culture-historical strata, as suggested in the Director's Journal during the excavation and published in several preliminary reports (Rothenberg, 1969a; 1970; 1970a; 1972; 1978) was tested against all information obtained by the post-excavation stratigraphical and artifactual analyses. As a result, it could be shown that this fundamental concept remained valid, though in some details and their interpretation a number of changes became necessary.

Several special diagrammatic studies were carried out to test special stratigraphic problems of the correlation of phases with artifacts and chronological strata:[9]
(a) A series of diagrams showed the finds (pottery groups, metal, faience and textiles, as well as inscribed objects) at their find levels above, inside and below the White Floor, which was the interface assumed to be of major stratigraphical significance (separating the first and second Egyptian phases of the Temple).
(b) Another series of diagrams recorded the find-boxes from each numerical grid segment, i.e. only horizontal grid units 1–22 were considered, ignoring the vertical grids A-K, and showed the pottery finds (New Kingdom Egyptian, Negev-ware, Midianite, local, as well as Roman and Chalcolithic) at their levels against the different deposits and interfaces of each such segment (early hard packed floor, White Floor, olive-grey or green sand, red sand, Roman, etc.).
(c) A special diagrammatic study was made in order to relate the find levels of the Pharaonic cartouches to the White Floor and thereby establish their stratigraphic significance.

Although the uncorrected levels (see footnote 9) in some of these diagrams distorted the real distribution pattern in height of the finds, their position relative to the interfaces, especially the stratigraphically dominant White Floor, had important stratigraphic and chronological implications and in this sense the diagrammatic study, although a big job without the help of a computer, proved worthwhile.

II. The Excavations 1969 and 1974

In the concluding stage of the first season of excavation, it had already become necessary to cluster groups of the original 1×1 metre grid units ('Squares') into loci, in order to overcome the rather fragmentary nature of the information from each of the one metre grid squares, and to facilitate a more meaningful grouping of the findings and finds, and their stratigraphic and chronological correlation. As the post-excavation investigations proceeded, the boundaries of these loci were finalized as shown on *Illus.* 10a.

In the following description of the excavation, we describe the stratigraphic findings and chronologically significant finds of each locus. In the concluding Chapter IV, the overall stratigraphy of Site 200 will be discussed.

In the first season, the whole site was systematically excavated down to the base of the courtyard walls, which was over most of the site connected with a major interface – the White Floor – of basic stratigraphic significance.

Because of the importance of the Timna Mining

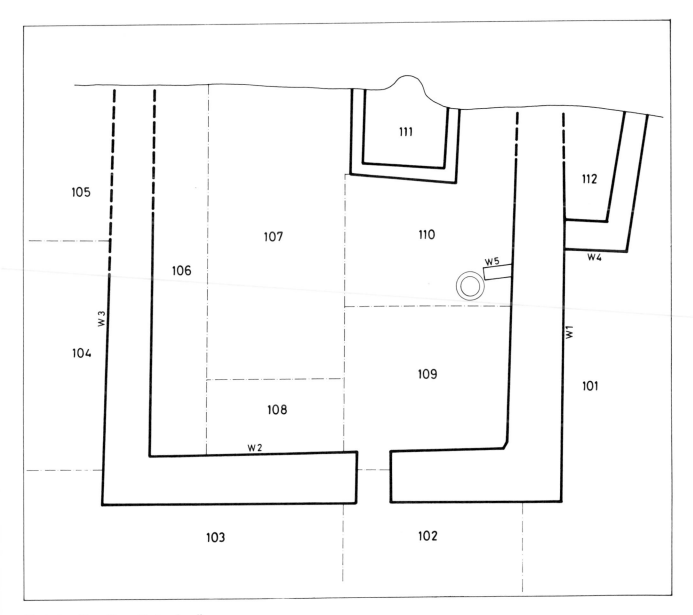

Illustration 10a. Plan of loci and walls.

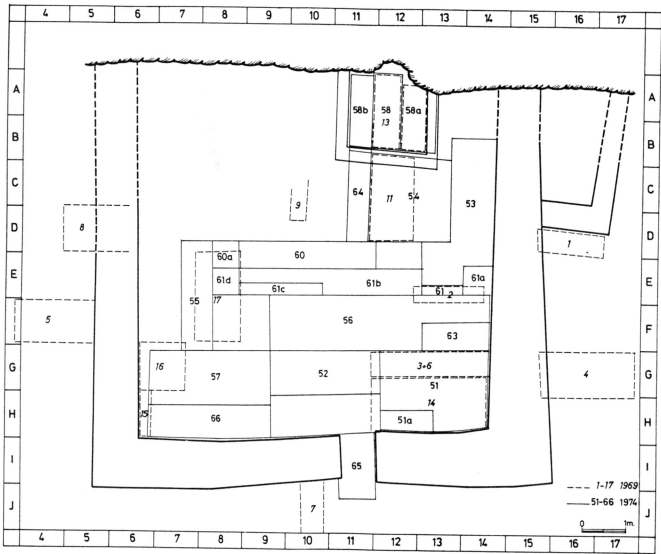

Illustration 10b.   Trial trenches 1969 (1–17) and 1974 (50–66).

Temple, it was decided to preserve at this stage as much as possible of the original architecture; we therefore refrained from cutting through the walls or removing the White Floor or the rather fragile stone basins in loci 106, 107 and 110 (see footnote 3). However, in order to establish the overall stratigraphy, we also investigated the bottom layers of the site by digging a number of small trial trenches, going through the White Floor or other interfaces (inside the *naos* or outside the Courtyard where no White Floor existed) down to virgin subsoil or bedrock. At the end of the excavations in 1969, in the light of the results of the trial trenches and, especially, the subsequent investigation of the finds, it had already become clear that more of the lower levels underneath the White Floor had to be investigated, in order to clarify some of the basic stratigraphic problems of the site.

During the second season of excavation, in 1974, as we were then mainly concerned with detailed stratigraphic problems, the excavation was carried out by a very small team and by systematic trial trenching. In 1974 most of the White Floor was removed and the excavation taken down to virgin subsoil (disintegrating sandstone) or bedrock (*Illus.* 10b).

1 Locus 101

Locus 101 comprised Squares D-L 15–25, including sections of Wall 1 and Wall 4, and the junction Wall 1–Wall 2. Locus 101 is the north-eastern end of the mound, gradually sloping down towards the east in D-L 15–18, and tilting downwards steeply in D-L 19–20, to reach the ground level of the surrounding area. At the end of the grid area, at grid line 25, windblown red sand piled up against the face of the 'Solomon's Pillar' formation.

In Squares D-K 15–18, alongside Wall 1, intensive signs of occupation were uncovered at several different levels, containing very numerous finds, whilst from grid line 19 onwards only relatively few finds came to light.

D-L 15-18: The surface layer of *c.* 15cm. consisted of fine, windblown sand. At the depth of *c.* 8cm., a Roman sherd was found – the first find at Locus 101. In D-I 15, the top of a wall appeared after the removal of *c.* 10cm. of the surface layer (*Pl.* 5). Underneath this loose surface layer, a layer of more compact red sand appeared.

Along Wall 1 in D-I 15–16, and in K 12–15, rubble of medium sized stones appeared right under the surface layer. This debris was found to be lying on top of a greenish-grey sand layer containing numerous finds and was obviously debris from the uppermost part of Wall 1.

In E 15–16 at level 147,[10] i.e. only 27cm. below the surface (at l20) a very rich hoard was recovered from under some rubble, lying right against the outside of Wall 1 (Box 51, Level 147) (*Pl.* 6). It consisted of a pile of copper and bronze objects, mainly jewellery, mixed with numerous copper ore nodules, several iron bracelets,[11] and a large number of beads of many shapes, made of faience and shells.[12] In the same layer, at a distance of 30–50cm. from the hoard (Box 173, Level 161), were Midianite, local wheelmade, Egyptian and Negev-ware sherds,[13] together with many faience fragments.

At this level, and at the equivalent sloping levels of the whole excavated area of Locus 101 up to gridline 18, an olive green-grey sand interface appeared underneath the red sand. The olive green-grey interface, although not really a continuous deposit, was spread over the excavated area up to about 3 metres from Wall 1, but it was not everywhere the same. At some spots it disappeared or was just a greenish tint of the sand, and was replaced by red or grey with

Plate 6.   The hoard in E 15–16.

small stones. However, wherever we met the olive green 'layer', there was a revealingly large number of copper objects, copper prills and many copper ore fragments.

Regarding the finds related to the olive green-grey interface, it must be remembered that, although most of the time we were dealing with a very thin interface, there was often a proper greenish layer which continued down to another interface below. It was therefore often extremely difficult to separate the finds of this interface from the many finds in the soft fill right underneath and even further below in the fill. (See below for further discussion of this aspect.)

E15–16 had a very distinct stratification throughout, typical for the stratigraphical situation of Squares D-G 15–18. The olive green-grey sand layer started at level 147 and continued down to level 167 where ashes, charcoal and bone fragments indicated an active interface, i.e. an occupation surface (although not a proper floor) with many finds of various character: many copper objects, Midianite, Negev and Egyptian pottery, faience objects, Egyptian glass fragments, Egyptian amulets and numerous beads, including complete strings; and also bits of slag, copper ores, animal bones, charcoal pieces. The slag was casting waste, not smelting slag (see below, Chap. III, 13). The majority of the sherds found at this level was Midianite, with only a few Negev-ware sherds present.

In several squares along Squareline 15–16, mainly underneath the rubble from Wall 1 and often lying right against the face of Wall 1, were packed masses of rough woollen cloth (Boxes 201, 303, 173), sometimes with beads attached to them.[14] Lying mainly on the ashy interface, on the bottom of the olive green-grey deposit, the cloth was mostly in thick lumps, apparently folded over many times, as a large hanging textile sheet would fold up when falling straight down on itself (*Pl.* 7).

Plate 5.   Top of wall in D-I 15.

Plate 7.   Textile find next to Wall 1.

There were relatively few stone tools at the Temple site (see Chap. III, 29), but in this layer, i.e. the uppermost 'occupation' layer of the Temple, two saddle-backed querns of fine-grained red sandstone were found which are typical of the New Kingdom smelting sites in the Arabah (J 16: Box 66, Level 175; G 16: Box 209, Level 169) (*Pl.* 8).

Underneath the olive green-grey deposit appeared again a red sand layer, spreading throughout the area D-I 15–18. If we return to E 15–16 as the typical situation for the area up to 3 metres from Wall 1, this red sand layer began at level 167 and ended at level 180, where it meets another interface indicated by fireplaces in and on greyish sand with miscellaneous debris,[15] i.e. concentrations of ashes and charcoal which seemed to be of stratigraphic significance.

The red sand layer down to the ashy interface contained many finds similar to those in the olive green-grey layer above. The finds of this layer were more evenly dispersed throughout the sand deposit. However, on the ashy interface itself (see below) a

Plate 8.   Saddle-backed quern *in situ*.

denser find distribution was recorded, indicating a second occupation-related active interface. It is important to note that in areas of greater concentration of copper ores and objects in this layer, the sand took on a greenish tint, though much less noticeable than in the olive green-grey layer higher up.

Amongst the finds at this (second) ashy level (Boxes 224, Level 180–204; 234, Level 185–201; 302, Level 170–191; 311, Level 167–180) were many faience objects including faience amulets bearing cartouches of Seti I (?) (Egyptian Cat. No. 195) and Ramesses IV (Eg. Cat. No. 34),[16] building stone fragments bearing hieroglyphs, Midianite, Negev-ware, Egyptian and local wheelmade pottery, many beads, Egyptian glass fragments, copper objects, a very fine Midianite votive cup (241/6, see *Pl.* 10) and a sistrum fragment.

The ashy (with charcoal) interface with much miscellanea, though occasionally rather spotty, lay on top of a reddish grey sand layer which was about 20–30cm. thick. Below this red layer (in E 15–16 at Level 196) was another ashy (in spots) charcoal interface on top of a pebble-rich layer of mainly undisturbed disintegrating red sandstone – the original, virgin surface. The ashy (and charcoal) spots on top of the undisturbed subsoil were obviously occupation-related, probably fireplaces. All along Squarelines 15–18, this layer, which was only a few centimetres thick, contained besides miscellaneous debris, some finds partly similar to the finds in the layers above and partly of a different character, and obviously of a much earlier period (Boxes 232, Level 201–211; 254, Level 187–218; 268, Level to 196; 290, Level 200–215; 315, Level 183–203): Midianite, Egyptian, local wheelmade sherds; some beads and copper objects, slag pieces, glass, a gold jewel fragment, faience fragments; flint tools and debitage, together with many miscellaneous occupation remains.

In Squares J-K 15, on the north-east corner of the Temple, was a particularly rich, though 'mixed', bottom layer, starting with olive green at Level 190 and ending with an ashy (with charcoal) layer of only 2–3cm. on bedrock (at Level 215). This layer extended for about 2 metres from Wall 2. In the uppermost part of this layer, c. 25cm. above bedrock, were recorded (Boxes 241, Level 189–214; 243, Level 203–210; 250, Level 186–213): very many Midianite sherds, faience objects, including a leopard figurine (Eg. Cat. No. 87), copper rings, many beads, Negev-ware and Egyptian sherds, a faience sistrum handle with the cartouche of Seti (?) (Eg. Cat. No. 19),[17] menat fragments with cartouches of Queen Twosert (Eg. Cat. No. 33) and Ramesses IV (Eg. Cat. No. 27). In the lowest part (2–3cm. thick) occupation remains included flint objects and debitage, in greyish sand mixed with ashes and charcoal. In J-K 15 there was no identifiable interface at Level 190, just a level of many finds, some of which were found also in the sand layer below, in fact down to the ashy interface at the very bottom of the sandlayer, mixed with Chalcolithic debris.[18] This Square, part of the area in front of Wall 2, was rather exceptional and should be considered as functionally belonging to Locus 101.

D-L 15-18 was excavated to bedrock or to the level of typical subsoil of disintegrating coarse-grained sandstone. Everywhere there was a thick pre-Temple layer with spotty ash and charcoal concentrations and miscellaneous debris and flint, and handmade early pottery, but no architectural remains.

D-L 19-25: In D-L 19 the surface had already been flattened out almost to the level of the surrounding area, but we cleared most of it down to bedrock. It was essentially a surface layer of 10–25cm. of red sand, lying on bedrock with very few stray finds (Boxes 87, Level 181; 136, Level 172; 297, Level 160–169; 307, Level 124–140; 308, Level 138–170; 312, Level 159–174): Midianite, Egyptian, local wheelmade, Negev-ware, also Roman and black Gaza-ware sherds, a few beads, and metal (Cu and Pb) fragments, and bits of slag. In some of the Squares (L 16–18; G 19; E 19–20) a flimsy ashey-grey occupation layer of 2–3cm. was found under red sand, containing flint and miscellaneous occupation debris.

E 15-24: Squares E 15–24 were excavated to bedrock. In E 15 a thin layer of grey-ashey with charcoal occupation layer was found which in E 15 (Level 190–196) also went underneath Wall 1. This layer of pre-Temple occupation was found in several squares of E. In E 24, a pile of building debris of red and white sandstone (a piece of the latter showing fine mason's toolmarks), was an indication for post-Temple activities at the site, which must have been the cause of much of its destruction.

Plate 10.   Close up of Wall 1.

Wall 1 (in D-I 15):[19] (Pl. 9)

Wall 1, as it appears in Locus 101, was built without a foundation trench, using large stones for the outer face and with an inner core of smaller stones and sandfill. The maximum width: 1.25m. (*Pl. 10*).

As became apparent when clearing E 15–22 to bedrock (see below TT 1) this part of Wall 1 was laid on top of a thin layer of hard and undisturbed, coarse pebbly sand. A very flimsy and pebbly ash interface, with some charcoal bits, which was on top of this subsoil, testified to an occupation phase prior to the erection of Wall 1.

From D 15 to about the middle of G 15, Wall 1 was relatively well-built with large stones laid stretcher-like as foundations. It was a dry-built wall of

medium-sized undressed field stones, mainly red Nubian Sandstone, but also some dark coloured limestone and granite boulders. Some white sandstone boulders were also used but these, lying on top of the wall seemed to be a later addition. At its best preserved part, in D-E 15, Wall 1 had five courses *in situ* and was about 90cm. high. Taking into consideration the quantity of building debris found along the wall, it must have been about 1.5m. in height.

Whilst at the junction with Wall 4, Wall 1 seemed to have been damaged and rather roughly rebuilt (see *Pl. 103, Locus 112, page 000*), it ended in G 15 at a proper construction corner, built in almost header and stretcher fashion (*Pl. 11*).

The east end of Wall 1, from the middle of G 15 to the corner, was built in a quite different and rather coarse fashion, by just piling small- and medium-sized undressed field stones on top of each other. This part of the wall was built mainly of white

Plate 9.   Wall 1 in D-I 15.

Plate 11. 'Corner' in Wall 4.

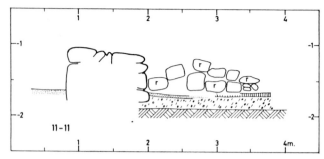

Illustration 11. Section 11–11.

C-D 15 and ran in a curved line up to the rock face of 'Solomon's Pillar' at A 17 (*Illus.* 6). In order to clarify the stratigraphic relation between Wall 4 and Wall 1, a 2 × 0.5m. trial trench, TT1, was excavated along Wall 4 in D 15.

Wall 4 was built against Wall 1, but the walls were not bonded together. The base of Wall 4 was about 20cm. above the base level of Wall 1. The stratigraphy of the walls and the layers, as seen in the section (*Illus.* 11), proves that Wall 4 was built later than Wall 1. At the bottom, above bedrock, under both Wall 4 and Wall 1, was a layer of hard sand with a flimsy ashy interface at its top. Wall 1, at this spot, was built directly on this interface, but underneath Wall 4 there were several more stratified deposits: above the bottom layer of red sand and the ashy interface layer, lay a 3–4cm. thick, loose layer of crushed white sandstone (see 'White Floor' further on) which, however, was not met with in D 15–17 on the other side of Wall 4. This white sandstone layer ran against the bottom course of Wall 1, and petered out after about 70cm. where, in the same level, a sandy ash layer appeared. This seems to have been another interface, laid down after Wall 1 was built, but obviously before the erection of Wall 4. Above this interface was a layer of grey-red sand with bits of charcoal, on top of which stood Wall 4.

The Stratigraphy of Locus 101

In Squares D-G 15–16, we found three interfaces with many typical habitation-related finds, of which the bottom interface contained mixed New Kingdom and Chalcolithic finds. In I-J 15 (in front of Wall 2) and around the corner in H-I 15–17 (in front and east of Wall 1) we found two interfaces only: the uppermost was an olive green-grey layer with exceptionally numerous finds, as well as fireplaces and occupation-related miscellaneous debris; the bottom one with the same mixed New Kingdom and Chalcolithic finds as in D-G 15–16. Finds of the New Kingdom were also made in the sand layers between the interfaces, but no habitation-related clusters of finds nor any kind of recognisable surface were encountered in these 'fill' layers.

In the light of the above, we must assume that the part of Wall 1 in D-G 14–15 is earlier than Wall 1 in H-I 15, which was obviously built against the finely constructed 'corner' in the middle of G 15.

sandstone, some of which, bearing traces of mason's tools, appeared to be in secondary use.

At the corner of Wall 1 with Wall 2, as well as at the frontal face of Wall 2 (see below Loci 102–103) an effort was made to create a very solid looking wallface by using more regular shaped and larger rocks, mainly white sandstone (*Pl.* 12).

Wall 4 (in D 15–17): (*Pl.* 103)

Wall 4 was a very roughly built, low stone fence, today only one or two courses high, which started in

Plate 12. Wall 2 in I 11–14.

2 Locus 102

Locus 102, part of the eastern slope of the mound, is comprised of Squares I-L 11–14, including a part of Wall 2 in I 11–14. It was not excavated to bedrock, though in several Squares (e.g. I-K 12) undisturbed subsoil was reached. After removal of a windblown surface layer of fine-grained sand (approximately Level 140 to 149–159), the outline of Wall 2 appeared.

Squares I-K 14 showed very much the same layer sequence and find distribution as the adjacent I-K 15: In the red sand top layer, under some windblown surface cover, a few finds appeared (Box 208): Midianite, Negev-ware and local wheelmade and Egyptian sherds, some faience fragments, glass, beads, a copper object and pieces of stray miscellanea.

At approximately Level 202 an olive green-grey interface appeared, with ashy and charcoal patches. Many finds were made in and above this interface (Boxes 80, 81, 82, 89, 151, 176): many Midianite, Negev-ware, Egyptian and local sherds, especially large storage jars, faience fragments, glass, copper objects, a gold band and a faience sistrum fragment with a Hathor inscription (Eg. Cat. No. 220).

Underneath the olive green-grey interface, the red sand deposit continued to the foundation level of Wall 2 (in I 14 212–215). At this level and also in the sand layer above, were a number of finds (Box 253, Level 192–215): Midianite, Negev-ware and local pottery, a crucible fragment, many beads, a menat fragment (Eg. Cat. No. 36) and a bracelet with a Ramesses II cartouche (Eg. Cat. No. 47) and miscellaneous debris. At the foundation level of Wall 2, flint debitage and bones were found, but we did not excavate down to the level of a possible pre-Wall interface. However, in adjacent I 13 the excavation reached Level 225 and here (Box 257) many more flint artifacts and miscellaneous occupation debris appeared, typical of the pre-Wall Chalcolithic occupation interface found in other loci. This may be taken as an indication for the stratigraphic situation also in Square-Line 14.

South-west of Square-Line 14, the stratigraphic sequence remained fairly unchanged up to the south-western half of Square-Line 12, where a different kind of deposit, the 'White Floor', made its first appearance.

Squares I-J 12 were typical for this area: In the red sand layer underneath the windblown surface sand layer, approximately at Level 190, a few finds appeared (Box 207): Midianite, Negev-ware and local sherds, ore and slag bits, a copper object, faience fragments, a faience bracelet with Queen Twosert's cartouche (Eg. Cat. No. 41) and some miscellanea.

An olive green-grey interface, without ashy spots of fireplaces, appeared at Level 207 with numerous finds of a kind typical for the find clusters of the green-grey interfaces in many loci of the excavation (Box 260, Level 206–213): Midianite, Negev, local and Egyptian sherds, many beads, many copper objects, some glass, faience fragments, copper ore bits and copper prills, slag pieces. However, it should be noted that in I-J 12 there were noticeably fewer finds than in the slope further east and in Locus 101.

Underneath the green-grey interface, the red sand deposit continued with occasional grey sand patches and a few finds. In the western half of Squares I-L 12 (in J at Level 215, in L at Level 223) a thin layer of crushed white sandstone showed up. In I-J it was a loose but very obvious interface, in K-L it became just a very flimsy scatter. Compared with the white sandstone interface underneath part of Wall 2 and inside the Temple courtyard, where we found a very substantial, hard cemented and well-laid White Floor, the white sandstone interface on this slope was probably only a spread of wastage from the laying of the White Floor inside the Temple area. However, although on the slope outside the Temple it was not a proper floor, the white interface was of considerable stratigraphic significance. The finds (Box 270) related to this interface and also to the red sand layer to the east, included: Midianite, local and Egyptian, Negev-ware sherds, faience and copper objects, some small slag pieces and ore bits, glass, many beads, a saddle-backed quern, an inscribed sistrum fragment (Eg. Cat. No. 31) and miscellanea.

Continuing in J 12 down to Level 233, signs of occupation were found including miscellaneous debris and worked flint (in Box 270). This material came from a slightly dusty-grey area on a pebbly red sand layer, probably the pre-Wall Chalcolithic interface or close to it. There were no architectural debris connected with this layer in the limited area of Square J 12.

At the end of the excavation in Locus 102, Baulk J-L 11 was removed down to the White Floor without close attention to its different layers. We mention here some of the finds from this excavation (Box 337, Level 213–231) which can only be marked as 'coming from I-L 11 above the White Floor': Roman sherds (from right on top of the baulk), faience bowl with the cartouche of Merenptah (Eg. Cat. No. 102), faience fragment with a cartouche of Ramesses IV (Eg. Cat. No. 216), a fine Midianite bird-juglet (*Fig.* 8:2) and a faience fish bowl (*Fig.* 42).

About three metres from Wall 2, in L 11–14, almost at the bottom of the slope, the surface layer was hard packed red sand (surface at level 205–209) and underneath it an olive green-grey horizon was discernible (at Level 209–218), approximately 4cm. thick, but not a continuous layer. A few stray finds were recovered in this green-grey interface: Midianite and local wheelmade pottery, some faience fragments, glass, beads, and some miscellaneous debris. In the western half of L 12, at Level 223, a scatter of crushed white sandstone in red sand (not further excavated) indicated the White Floor horizon.

Wall 2 in I 11-14[20]

Wall 2 in I 11–14 was built of roughly dressed, rectangular white sandstone boulders. Similar to Wall 1, it was constructed by a double line of stones forming the faces of the wall, with smaller stones and

Plate 13.   Close up of Wall 2.

sand as an inner fill. Two or three courses of stones were found *in situ*, the foundations were mostly laid in stretcher-like order; the wall above in regular courses of headers (*Pls*. 12–13).

We excavated the Squares next to the wall, to several centimetres below the foundation level of Wall 2.[21] The hard red sand layer, cleared in Square-Line I down to the level of the foundations of Wall 2, continued below this level. As far as we could establish without excavating or cutting through the wall, it also went below Wall 2. However, about 1.3m. from the entrance, i.e. in part of I 11–12, a layer of crushed white sandstone was visible below the foundation of Wall 2, which indicated that Wall 2 was built above the White Floor. Whilst the White Floor underneath and right next to Wall 2 was several centimetres thick, in places approximately 10 cm., and fairly solid, the white sandstone layer further away from the wall gradually petered out down the slope.

The Stratigraphy of Locus 102

In the squares excavated, we found two habitation-related interfaces. The uppermost interface running against Wall 2 at considerable height, was the olive green-grey layer with fireplaces and very many finds and habitation-related miscellanea. The second, lower interface was only clearly defined in the squares where the White Floor or the related white sandstone interface appeared. However, in Square-Line 12–14 at the same level, a noticeable increase in finds and grey-ashy patches indicated a habitation-related interface.

The foundations of Wall 2 lay at this level and the related interface – though not everywhere a 'floor' – was formed by activities in the Temple at the phase of the White Floor. Only in J-K did the excavation in Locus 102 penetrate below the White Floor horizon and here, after several centimetres of dusty red sand, Chalcolithic flint and miscellanea were found – obviously signs of the pre-Temple habitation at the site.

3 Locus 103

Locus 103, Squares I-M 1–10, is the southern part of the eastern slope of the mound, including part of Wall 2 (in I 5–10). It was only excavated down to the interface connected with the foundations of Wall 2,[21] except trial trench TT7 in I-M 9, which penetrated down to bedrock.

Locus 103 was a relatively steep slope with the very top of Wall 2 slightly protruding out of the sandy Present Surface of the mound. Excavating along grid-line I, i.e. the front of Wall 2, there appeared a solid mass of rubble, mainly medium-sized building stones, which had slipped down from the wall (*Pl*. 14). Underneath this rubble was a layer of wind-blown red sand containing a few stray finds. Clearing

Plate 14.   Debris in front of Wall 2.

this layer we reached a hard interface of crushed white sandstone, i.e. the White Floor (Levels in I 5: 231; I 7: 227; I 9: 216).

Although the same simple stratigraphy was found over most of Locus 103, the frequency of artifacts in the layers was noticeably different south and north of gridline 7. Whilst in I-K 7–10 quite a number of finds were made, almost nothing was found in the layers of I-K 4–6.

I-K 8 was typical of the northern half of Locus 103. Right underneath the stonefall in I 8, at Level 190, appeared the hard surface of a red sand layer. At Level 222 the White Floor appeared. On this floor a fair number of artifacts were uncovered (Box 257, Level 190–225): many sherds of local wheelmade, Egyptian, Negev-ware and decorated Midianite pottery, part of a leopard figurine (Eg. Cat. No. 94) and of a menat. In J 8 a small clay tuyère was found. In K-L 8, where the White Floor was very thin and appeared to be intermingled with red sand from a lower layer, some miscellanea were found and also flint flakes.

To check the stratigraphy of the northern half of Locus 103, we excavated trial trench TT7 in I-L 10–11 down to bedrock.[22] Starting at Level 178 by removing the solid mass of rubble, we reached a layer of red sand, sloping down along Squares J-L (to level 200). At level 221 the White Floor appeared accompanied by a number of typical finds (Box 261, Level 178–221): Negev-ware, Midianite, local wheelmade and Egyptian pottery, a copper chainlink, faience fragments, bones and miscellanea. Underneath the White Floor was a layer of red sand, some of it obviously disintegrated bedrock, and in this layer only one bead and some flint debitage were found (Box 321, Level 224–249. At Level 249 (in J 10) bedrock was reached) (*Pl.* 15).

Plate 15.   Trial trench TT7 showing White Floor and layer below.

Plate 16.   The White Floor outside Wall 2 I-L 5.

There was no olive-grey layer in Locus 103, and no sign of any fireplaces; by comparison with Locus 102 and, especially Locus 101, there were only a small number of finds. It was obvious that we were here only at the fringes of the find-rich area further north, along Wall 2 and Wall 1. Amongst the miscellanea found on the White Floor was a quantity of metallurgical waste, bits of slag, crucibles and tuyère fragments which indicated the proximity of a casting-workshop (see below Locus 109).

The White Floor extended from the corner of Wall 2 in I 5 in a fairly straight line to approximately the middle of L 5 (*Pl.* 16). The eastern border of the White Floor ran through the middle of L 5–10. Near Wall 2 the White Floor was a solid, well-laid interface of several centimetres, but sloping down towards the east it gradually petered out into a thin and spotty scatter of crushed white sandstone pieces. It was at this eastern end that most of the miscellanea of an earlier occupation (beneath the White Floor) were found. The only Roman sherd of Locus 103 was found in Square L 7. The White Floor was clearly traceable in Section 8 of Baulk G-M 11, and ran everywhere beneath the foundation of Wall 2 (in Locus 103), including its gateway.

In I 6–7, three flat stones were found on the White Floor, placed right against Wall 2, perhaps as a sort of bench (*Pls.* 16 and 17). One of these stones carried masonry tool-marks and may have been in secondary use. Further evidence of stratigraphic significance for the secondary use of stones in Wall 2, was a fragment of an offering stand[23] found in I 6 under *c.* 40cm. of rubble fallen from the wall.

Wall 2 in Locus 103 (Pl. 17)

Wall 2 in Locus 103 was approximately 93cm. wide. Its lowest course consisted of small rocks piled onto the White Floor. However, the Wall itself was carefully drybuilt with roughly-dressed, medium-sized, rectangular flat white sandstone boulders, laid in header fashion, similar to Wall 2 in Locus 102. As in Locus 102, the stones were arranged in two parallel rows with smaller stones as fill in between. The White Floor, which was clearly discernible underneath the foundations all along the outer side of the

Plate 17. Wall 2 in Locus 103.

wall, was approximately 10cm. thick. Only 2–3 courses of Wall 2 were found *in situ*. The southern corner (with Wall 3) was badly damaged, its heavy corner stones slipped down. Almost all of Wall 2, in Locus 103 as in Locus 102, was built of white sandstone.

The Stratigraphy of Locus 103

In Squares I-M 1–4, no proper stratification was found in the red sand and only very few stray finds. Squares I-M 4–10 showed only one habitation-related interface, the White Floor, onto which Wall 2 was built. Compared with the adjoining Locus 102, Locus 103 was poor in finds, most of which were found either directly on the White Floor or just above it, at the bottom of the red sand layer lying on top of the White Floor.

The extremely find-rich olive green interface, most conspicuous in Locus 101 and Locus 102, was missing in Locus 103, although at the extreme north end of Locus 103 and in Baulk G-M 11, a small group of finds appeared in the upper red sand layer at a level equal to the level of the olive green interface in Locus 102. These finds could indicate a habitation phase of the Temple which left only rather flimsy evidence in the area in front of the Temple (Locus 103), although it was clearly represented in Loci 101–102 and inside the Temple.

Only very little evidence for a pre-Temple occupation of the site was found in Locus 103, although the miscellanea, especially under the very thin White Floor, in the Squares (L-M 5–10), contained quite a quantity of flint debitage and other items, indicative for the early occupation phase found in Loci 101–102.

4 Locus 104

Locus 104, Squares D-I 1–5, is the eastern part of the southern slope of the mound, including a section of Wall 3 in D-I 5.

The surface of Locus 104 was littered with rubble but no structural traces were discernible before excavation. This situation was complicated by a huge rockfall, weighing many tons,[24] which lay on top of the south-western part of the mound. Although the centre of this rockfall was located in Locus 105–106, many broken boulders were spread all over the south side of the mound and it was obvious, even before excavations, that great damage must have been done to any structures which stood at this side of the site (*Pl.* 18).

Locus 104 was the southern slope of the mound, but the surface of the mound and the ground surrounding it sloped eastwards, a fact which clearly

Plate 18. Rockfall on top of mound in Locus 104.

36

showed, after excavation in the levels of the interfaces and foundations of the walls and other features.

The whole area of Locus 104 was excavated to a hard interface at the base of Wall 3 (in D-I 5), except TT5 in F 2–5, which penetrated the hard virgin soil (disintegrating red sandstone). Squares G 1–5 were first left as a baulk and were excavated at the end of the excavations in Locus 104.

After removing loose debris, rubble and wind-carried red sand from the surface of D-I 4–5, it became obvious that we were dealing with a solid mass of building rubble, which in D-F extended also into 3–2 (*Pl.* 19). There were almost no finds in this mass of wall debris but, on reaching the bottom – at the level of the base of Wall 3 – a hard interface, obviously an old surface, was found and here some finds could be recorded: (Box 275, Level 140–203): in D 5 a Negev-ware sherd and a copper ring; in E 4–5 Midianite, Egyptian and local sherds, a fine goldleaf jewellery fragment (*Fig.* 84:131), faience fragments, beads, a faience bracelet fragment with the cartouche of Ramesses V (Eg. Cat. No. 46) and miscellanea, including some flint fragments. In I 4–5 there were no finds.

This picture of only a few finds on the old surface underneath the rubble of the fallen wall (see also in trial trenches in Square-Lines F and H), continued in the Squares further away from Wall 3, but here almost no finds were made. In D 2–3 (Box 226, Level 183–210) under about 20cm. of rubble, Roman sherds, few beads, fragments of copper objects, glass and miscellanea, including some pieces of plaster, shells and bits of slag. In G 2–3 (Box 333, Level 195–225): Egyptian sherds and a few bits of miscellanea, including flints.

A trial trench (TT5) was excavated in F 3–5 (*Pl.* 20). From Present Surface down to the hard interface was

Plate 20.   Trial trench TT5 in F 3–5.

a solid mass (45cm.) of fallen and broken stones, with some wind-carried, finely straticulated red sand fill lying on an interface of hard, red sand. At F 5, in the 20cm. deep sand layer overlying the wall, was a Roman sherd (Box 244, Level 158). On top of the hard sand layer under the rubble, were a few artifacts (Box 266, Level 208): Midianite and Egyptian sherds, some beads, copper ore pieces. There were no finds in the hard bottom layer (Level 232), which appeared to be undisturbed virgin ground. To check on the stratigraphic results of TT5, the excavation of H-I 4–5 was continued for *c.* 20–25cm. into the bottom layer of hard red sand (*Illus.* 9).

Plate 19.   Solid mass of rubble from Wall 3.

Plate 21.   Traces of the White Floor below Wall 3 in H-I 5.

The wall rubble in H-I 4–5 consisted entirely of white sandstone, much of it crumbled into small pieces. Some of the white stones showed toolmarks. In H 5, right next to Wall 3, under the mass of rubble, a fireplace appeared on a sandy interface (Level 237). At this level the interface was littered with many pieces of crushed white sandstone, and in the 20cm. thick section underneath the foundation course of Wall 3 (in H-I 5) traces of the White Floor were clearly visible (*Pl.* 21). In this old surface layer the following finds were recorded (Boxes 248, Level 203; 272, Level 190–237): Midianite, Egyptian and local sherds, beads, a faience bracelet fragment with the cartouche of Merenptah (Eg. Cat. 90). In the hard red sand layer underneath the old surface, no finds were made.

Wall 3 in D-I 5[25]

After lightly brushing away the rather thin Present Surface cover of wind-carried sand, the slope of the mound (D-I 2–5) appeared as a solid mass of rubble which 'somewhere' contained the southern wall of the Temple (*Pl.* 22). After removing some of the uppermost rubble in D-I 5, a rough outline of a wall top became barely discernible in F-G 5, but no traces of the wall could be found in D-E 5, nor in H-I 5. We therefore decided to approach the problem of tracing the wall by excavating Locus 104 in strips of squares, starting from the south (D-I 1). It was by this method that we were finally able to trace the outside of Wall 3, discussed below.

D-E 5: Squares A-D 5 were under a huge mass of fallen rocks (*Pl.* 23). When lifted, most of the original Wall 3 structure was found completely smashed and its outlines distorted. In D 5 the wall's courses had been pushed out of their original position, but by clearing a trial trench (TT8) across the rubble mass in D 5–6 (see Locus 106), it could be established that Wall 3 was definitely built onto the top of a huge

Plate 23. Wall 3 smashed by rockfall (A-D 5).

rock, extending into Square D 5 from further west (*Pls.* 24–25 and *Illus.* 16). At the early stage of the excavation, this rock appeared to be a native outcrop, but later on it became obvious that it was part of a huge rockfall predating not only the construction of the Hathor Temple, but even the Chalcolithic occupation of the site.[26]

Whilst carefully clearing away the rubble, the former line of the wall's outside face could be distinguished but it just fell apart during excavation.[27]

D-F 5 was entirely built of red sandstone, whilst further east, in G 5, the wall was built of white sandstone mixed, especially in the uppermost course, with red sandstone boulders. H 5 to the corner in I 5 was built entirely of white sandstone (with traces of the White Floor underneath (*Pl.* 26; also *Pl.* 21).

Plate 22. Solid mass of rubble of Wall 3 in D-I 5.

Plate 24. Wall 3 build on top of huge stone of earlier rockfall.

Plate 25.   Huge rock underneath Wall 3 in D 5.

Plate 26.   Wall 3 in H5 to I 5 built of white sandstone.

The building methods of Wall 3 in E-F 5, as compared with the same in H-I 5, were quite different. In E-F 5 the two lowest courses were apparently intended to function as solid foundations for the wall structure above and were therefore carefully laid, contrary to the rest of the wall, which was built of undressed field stones of various sizes, picked up in the vicinity. H-I 5 was throughout carefully built, mainly in header fashion, and the building stones brought to the site from some distance (the mining area ?) were simply dressed to a rectangular shape most convenient for wall building. G 5 seems to have been disturbed, perhaps prior to the building of H-I 5, and was rather rough and irregular in its structure.

There was no foundation trench; Wall 3 in E-F 5 was built on a hard sand surface. In H-I 5 it was laid on a fairly thick White Floor.

The Stratigraphy of Locus 104

Because of the rather disturbed state of Wall 3 in D-I 5 and its immediate surroundings, it is difficult to relate to the relative heights of this Locus any basic stratigraphic meaning. All along Wall 3 one hard red sand interface was found on which Wall 3 was built, at least up to Square G 5. In H-I 5, traces of White Floor material were found in the red sand, but we did not find a distinct White Floor interface. The few finds found in Locus 104 came almost exclusively from the interface on the hard sand bottom layer, including some miscellanea and some flint debitage. With the exception of the few Roman sherds and glass fragments found in the uppermost fill on top of Wall 3, all the other finds belong to the New Kingdom, but some of the miscellanea may belong to the Chalcolithic, pre-Temple occupation found in most of the other Loci of the site.

5 Locus 105

Locus 105, Squares A-B 1–6 and C 1–5, is the westernmost part of the southern slope of the mound and gridline A runs along the rockface of the Pillar.[28]

Most of this area was covered by a huge rockfall (*Pl.* 18) of red sandstone, outcropping about three metres above the surface of the mound, which by available evidence took place some time after the final abandonment of the site. Amongst this evidence should be mentioned the find of Roman sherds right under the (removed) huge mass of fallen rock in A 6 and also further north. The stratigraphic sequence of the layers in the south-western part of Site 200 is a history of repeated rockfalls from the rock-shelves overhanging the site and the consequent destruction and reconstruction of its features. This sequence starts with a rockfall pre-dating even the earliest occupation of the site, and ends with a huge rock-collapse on top of the finally abandoned Temple site. The Loci most affected by this last rockfall were Locus 105 and Locus 106.

The excavation of Locus 105 was made possible by the removal of the large fallen rock mass but, as could be expected, a great deal of damage had been done to the Temple structure and archaeological layers underneath, a fact which made the stratigraphic excavation of this Locus rather difficult.

A-C 1-3: After removing the surface rubble (Level 180–190), we cleared the surface layer of still rather stoney red sand which contained only very few stray finds (Box 274, Level 190–200; 320, Level 140–180): single Egyptian, local handmade, Midianite and Negev-ware sherds, a few bits of slag, some plaster pieces and miscellanea. Under the surface layer appeared an undisturbed hard layer of red sand.

A-B 4-6 and C 4-5: After removal of the rockfall 'outcrop', the area was found covered by a thick, coarse-grained mass of disintegrated red sandstone and small rock fragments, apparently still part of fallen rock which was completely shattered by the subsequent impact of huge rocks on top. In C 4–5 this sand layer was about 70cm. thick and underneath it lay the crushed remains of a red sandstone feature (Wall 3). In fact, the whole area of A-B 4–6 and C 4–5 (also partly extending further south into A-C 2–3) was one solid mass of red sandstone rubble with some red sand between the stones and it was extremely difficult to distinguish any features. However, it was obvious, especially by the very existence of such a mass of rubble of building stones (similar to the stones used in the better-preserved parts of Wall 3 and in Wall 1), that Wall 3, fairly well-preserved up to E 5–6, extended westwards into Locus 105 (see below).

After the removal of the mass of wall rubble, a huge red sandstone boulder showed up in B-C 4–6 (*Pl.* 27), with a second boulder apparently broken away from the first in C 6–7 (Locus 106). First thought to be an outcrop connected with the rock-face of the Pillar to the west, further excavations proved it to be one of several large and many smaller red sandstone boulders which belonged to a very early pre-Temple, and even pre-Chalcolithic rockfall. Rather deformed outlines of Wall 3 could be traced, albeit somewhat tentatively, in the massive body of crushed red sand-

Plate 27.   Huge red sandstone boulder in B-C 4–6, Locus 106.

stone building stones, overlaying the large rockfall sandstone boulder see (*Pls.* 24 and 25).

Very few finds were recorded in Locus 105 and because of the rather disturbed layers and features, especially in the upper layers, these did not contribute a great deal to the stratigraphy of Locus 105. In A-B 4–6, the stony red sand layer under the rockfall produced a group of finds (Box 310, Level 80–144): Midianite, Negev-ware, Egyptian sherds, a faience amulet of an unidentified Pharaoh (Eg. Cat. No. 206), a copper sistrum part, faience fragments, a string of beads, glass, a small tuyère and miscellanea. Level 144, in A-B 4–5, must have been very close to the base level of Wall 3 and this find group could well belong to its occupation surface, although no very clear interface was discernible in the much disturbed layers.

In A 5–6, a row of stones (in A 5) and a squarish, basin-like installation (in A 6) were found *in situ*, underneath the badly distorted foundations of Wall 3 (*Pl.* 28). The builders of this 'installation' placed the same into the narrow space between the edge of the large fallen boulder in B 6 and the rock-face of the Pillar, i.e. the stones of the installation were carefully put parallel to the big boulder. Inside the installation was some light-grey material, perhaps vitrified wood-ash, and in the narrow space between the boulder and the installation (at Level 168) a Chalcolithic sherd, two flint pestles and miscellanea were found.

Wall 3 in Locus 105

Little can be said about Wall 3 in Locus 105, except that its badly smashed remains were found as a solid body of crushed and broken red sandstone pieces, more or less as a continuation of a similar situation in

40

Plate 28. A Chalcolithic installation under Wall 3.

D-E 5–6. However, contrary to C 6, where some of the lowest courses were still found in fairly good condition, nothing remained of the original structure of Wall 3 in Locus 105. Under the impact of the huge rockfall, it was squashed to a formless body of rubble.

The Stratigraphy of Locus 105

Although the area was very disturbed by the impact of large rockfalls and the red sandy masses of disintegrated sandstone boulders, we found indications of several habitation phases. In a sandfill between the layers, huge rockfall boulders and the top of the wall rubble in A 6, Roman pottery indicated a Roman secondary occupation of the abandoned Temple site. Under the large mass of wall rubble, which covered almost all of Locus 105, an interface could be traced, at least at some spots, and here New Kingdom finds, including Midianite, Negev-ware and Egyptian pottery, helped to fit this phase – habitation surface and Wall 3 – into the general stratigraphy of the Hathor Temple.

Underneath the New Kingdom remains, traces of pre-Temple occupation, including installations, with one Chalcolithic sherd and some flint tools, could be identified, relating to a still earlier rockfall found also in other Loci at the south-west corner of the mound. Many stratigraphic questions concerning Squares A-C 5–6 could better be handled together with the remains in Locus 106, and will be discussed there.

6 Locus 106

Locus 106, Squares A-B 7, C-I 6–7, contains the area, and a row of special features, along the inside of Wall 3, from the rockface of the Pillar to the junction with Wall 2. The western side of Locus 106 was covered by

the huge rockfall mentioned above which extended as far as Gridline E, though in C-D 7 the mass of fallen rock was much less substantial than in A-B 6–7.

All along Squareline 7, with the exception of the eastern part, which was badly disturbed by rockfall, the mound was covered by a layer of wind-carried sand, its Present Surface, sloping from Level 80 in A 7 to 130 in H 7. It is stratigraphically important that under this surface sand layer quite a lot of rubble from Wall 3 was found, lying close to the wall all along A-G 6–7, and some of it also on top of a number of features (Standing Stones, etc.) which stood near and along the wall.

Standing Stones in B-H 7

In A-B 7, underneath the uppermost rockfall (at Level 80), appeared a mass of rubble, white sandstone masonry and sand. Whilst clearing this mass, an Offering Stand and several Standing Stones were uncovered which, to the excavators' great surprise, were *in situ* (*Pl.* 29). In the rubble and sand heap on which this Offering Stand was found, lay the head of a sphinx (Eg. Cat. No. 13), but this find belongs to another context.[29]

At Level 156 in A 7 and 151 in B 7, underneath the rubble and sand deposit, an olive green-grey interface, the surface of a thin olive green-grey slightly dusty layer, became visible over most of A-B 7.

Along the Gridline A-C 7 a row of white stones was found lying across several red sandstone boulders, which were found dispersed over this part of Locus 106 and must have been part (together with the large red boulders in B-D 5–7 and similar boulders in neighbouring Locus 107) of a pre-Temple red sandstone rockfall. As this row of white stones was found

Plate 29. Offering Stand and Standing Stones *in situ*.

in situ, after the excavators cleared the red sand and sandstone rubble all around, it seemed most likely that it served as a 'demarcation line', marking the Temple boundaries at this phase of its existence. The 'line' was related to the olive green-grey interface and some of its stones were lying on top of, or next to rubble from Wall 3. As there was a thin layer of sandfill between these red boulders and some of the white 'demarcation' stones, it must be assumed that the latter were put there some time after the devastation of Wall 3.

Starting in B 7, the row of Standing Stones (*Pl. 30*) found standing all along, and almost parallel to Wall 3, consisted of a mixture of pillar-like, rectangular white upright stones and typical Egyptian offering stands,[23] also of white sandstone. The latter were standing on a base of one or two flat, white stones, except for one which stood on red stones. There was a definite pattern to the order of this row of Standing Stones, as the simple Standing Stones alternated with the composite offering stand-on-stone-base type, at least up to the red sandstone basin in E-F 7. Most of the Standing Stones stood in a red sand layer showing clear traces of an olive green-grey interface. One of the stela tapered to a sharp end and to ensure its stability the builders cut a socket-like pit into an underlying stone. It was found *in situ*. (*Pl. 31*).

In E-F 6–7 a basin of rather brittle red sandstone (rectangular on the outside but with an oval inside), stood in the line of the Standing Stones, with a heavy granite boulder on top of it (*Pl. 32*). The basin and granite boulder were found *in situ*, under a mass of

Plate 30.   Row of Standing Stones in Locus 106.

Plate 31.   Standing Stone set in socket-like pit.

Plate 32.   Basin with granite boulder *in situ*.

rubble and sand, including parts of Wall 3, which just slipped onto the basin and was found in this position during the excavation. This part of Wall 3 must have been higher by several courses at the time the basin was integrated into the row of Standing Stones.

The heavy granite boulder (*Fig.* 23.6, *Pl.* 33)[30] so far remained an enigma. It was carefully flattened on its narrowest side, perhaps to create a more solid base for its use as a Standing Stone – but in the basin it was lying on its side. It was also not put right into the middle of the basin, but was found lying on top of the rather brittle sidewalls at the west half of the basin, which seemed to have cracked under its weight or the impact of its deposition. Underneath the (removed) boulder the basin had a fill of stones and sand and,

right on top of this fill lay a bundle of textiles (*Pl.* 34). In the fill a surprising assembly of finds turned up (besides the textiles) (Box 369): Midianite, Egyptian and local sherds, faience fragments, a bone pendant, 27 beads, some bits of slag and miscellanea.

The inner surface of the basin was found covered by a thin, hard layer of whitish plaster, obviously an attempt to make the highly permeable red sandstone watertight (*Pl.* 35). This indicated that as a member of the row of Standing Stones the basin was in secondary use and that originally it may have served in the Hathor Temple as an ablution basin.[31] A similar basin with plaster repairs on the outside, originally belonging to an earlier stratum, was found *in situ* in the adjoining Locus 107, close to the *naos*.

Plate 33.   Enigmatic granite boulder.

Plate 35.   White plaster inside red sandstone basin.

Plate 34.   Textiles and other finds inside basin (under boulder).

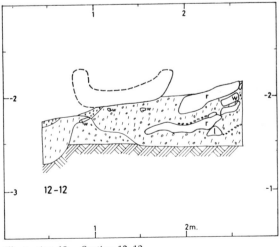

Illustration 12.   Section 12–12.

43

Plate 36.   Trial trench in front of basin (TT 55).

The area in front of the basin was disturbed. However, a trial trench (TT55), right in front of and along the basin (see below, p. 47) was of considerable help in working out the stratigraphical situation in E 6–7 and of the basin therein (*Illus.* 12; *Pl.* 36). Underneath the basin itself (Level 192), a thin interface of the White Floor was discernible.[32] However, a thick layer of olive green-green ashy sand was piled up against the north-west side of the basin and the olive green-grey interface ran against the outside of the basin (Level 183) in front and to the east of it. It would therefore be reasonable to assume that the basin in E-F 6–7 was actually in its original position and was a feature of the White Floor phase of the Hathor Temple, re-used again in the later olive green-grey interface-related phase of the site. This conclusion fits very well with the findings during the clearing of the inside of the basin.

Plate 37.   Square Pillar as found.

In F 7, after clearing some rubble from the nearby wall, and a thick layer of fine-grained red sand fill, a Square Pillar of white sandstone was found, its base lying right next to the large basin and its head in rubble in G 7 (inside the baulk) (*Pl.* 37). The pillar base was at Level 178, where clear traces of the olive green-grey interface (Level 178–180) could be discerned. Stratigraphically, the collapse of Wall 3, falling onto the basin, and the knocking over of the Square Pillar, appeared to be contemporaneous and it is clear that the square pillar stood next to the basin, in line with the row of Standing Stones.

The Square Pillar (Eg. Cat. No. 2, *Fig.* 23:1, *Pls.* 38–39) showed excellent masonry and on two of its opposite sides was a Hathor head in relief, both unfortunately thoroughly defaced.[33] This probably occurred when the pillar was taken from the debris of the Hathor Temple to be re-used as one of the Standing Stones (*Pl.* 40).

The last of the row of Standing Stones was found in G 6 under a pile of rubble from Wall 3: an Offering Stand which had fallen off its base,[34] both lying on an ashy olive green-grey interface (Level 190). The stand was fragmentary and was here obviously in secondary use (*Pl.* 41).

Finds from the olive green-grey interface

The stratigraphy in Locus 106 was disturbed by rockfall and the first clear stratigraphic feature was the olive green-grey interface, often only fragmentarily preserved under heavy rubble, at other times difficult to identify in the turmoil of broken masonry and local repairs, outcropping rockfall debris and replacements.

The olive green-grey interface in Locus 106 appeared at the following levels (of A-H 6–7): A 7:156; B 7:151–156; C 6–7:144; D 6–7:165; E 6–7:183; F 6–7:175; G 6–7:184–187; H 6–7:184–190. To convey the varying depth of the mostly sterile fill above the olive green-grey interface, we list here some of the levels of the Present Surface (wherever the final huge rockfall left some of the original surface of the mound intact): A 7:80; C 7:84; D 7:100; E 7: 128; F 7:136;

Plate 39.   Close up of Hathor head.

Plate 38.   Hathor head in relief on Square Pillar.

Plate 40.   Square Pillar in restored position.

45

Plate 41.   Offering Stand fallen off its base, next to Wall 2.

Plate 42.   Textiles lying on olive-grey interface next to animal bones.

G 7:130. In this olive green-grey interface (which was often only a shade of colour on top of a basically red sandstone layer, but at other times a proper layer of olive green-grey, dusty and ashy sand), a fair number of finds were recorded.

In A-B 7 the olive green-grey interface was only partly preserved and only a few finds were made (Box 305, Level 80–156): Midianite and Egyptian sherds, some beads, Egyptian glass fragments and miscellanea.

A most striking group of finds of special significance for the understanding of the Temple (related to the olive green-grey interface), was the large quantity of textiles (see below, Chapter III, 19) found behind the row of Standing Stones, close to Wall 3. These textiles lay all the way from the first Standing Stone in B 7 to the basin in E 6–7, and appeared to have been large pieces of rather rough cloth, which fell down from above and formed folded-up bundles.

In B 7, behind the Standing Stones, many animal bones were found, together with textiles lying on olive green-grey under rubble from Wall 3 (*Pl.* 42). On both sides of the 'demarcation line' (in B 6–7) a number of finds were made, together with a quantity of textiles (Box 323, Level 144–165): Negev-ware, Midianite, Egyptian and local pottery, and beads; (Box 330, Level 149–174): Egyptian sherds, faience fragments, beads and miscellanea. In C 6–7, on both sides of the 'demarcation line' (Box 342, Level 169, under wall rubble – olive green-grey interface): Midianite, Egyptian and local sherds, copper objects, some charcoal. On top of and between the two large, early rockfall boulders (Box 357, Level 150): textiles, Midianite and Egyptian sherds, beads, an iron ring.

In D-E 6–7, between the Standing Stones and Wall 3, a row of flat stones on edge, was carefully put

against Wall 3 as a kind of bench (*Pl.* 43) and on this we found a considerable quantity of textiles and many other artifacts in olive green-grey sand (Box 280, Level 159–174): textiles all along the 'bench', Midianite, Negev-ware and Egyptian pottery, a sistrum fragment bearing cartouche (*Fig.* 29: 2), faience, copper objects, a large votive bronze earring (*Fig.* 55: 16), bronze chain links, many rings, beads, a crucible part, miscellanea; (Box 213, Level 148–179, behind the

Plate 43.   Bench put against inside of Wall 3.

46

Standing Stone next to the basin): faience fragments, beads, Egyptian sherds.

A trial trench dug below one of the stones of the 'bench' exposed a layer of fine red sand, but no finds. In E-F 6–7, in front of the basin and on the olive green-grey interface, a number of finds were recorded (Box 221, Level 183–185):[35] a faience amulet (Eg. Cat. No. 208), a copper amulet, many copper-based small objects, gold leaf, one iron ring, beads, Egyptian pottery and miscellanea.

Baulk G 1–10 was excavated at the end of the excavation in 1969, after its profile was drawn as part of the main south-north Section of the site (*Illus.* 9). The Present Surface level was 130. In G 6 a solid mass of rubble lay along the wall and under it and, in G 7, an olive green-grey interface (at Level 184–187). On this interface lay a white rock on which the upper end of the Square Pillar rested.

On the strength of this evidence, it can be assumed that the Square Pillar was knocked down at the same time as Wall 3, in D-H (which, as mentioned above, also covered under its rubble an Offering Stand *in situ*)[36] and that these events are related to the olive green-grey horizon.

The olive green-grey interface, here with ashes and some charcoal, was also found in H 6–7 and there was a fair number of finds (Box 258, Level 153–184): Midianite, Negev-ware, Egyptian and local pottery, a small gold ring, some copper-based small objects, faience fragments, an Egyptian seal (Eg. Cat. No. 186), beads, an alabaster bowl, glass, a tuyère, bones and miscellanea.

At the level of this interface (c. 184–190), a bench-like row of flat stones was found *in situ*, built against the whole length of Wall 2 (see below) with its southernmost end also touching Wall 3. Although the clearly visible connection between the olive green-grey interface and the stone bench, as well as the equally obvious difference in height between the level of the stone bench and the base of Wall 2 and Wall 3, were sufficient stratigraphic evidence for the building phases involved, two trial trenches (TT 15–16) were dug in G-H 6–7 and both verified the olive green-grey interface relation of the bench (see below, p. 48 and *Illus.* 13).

Locus 106 below the olive green-grey horizon

In contrast to the olive green grey interface (which was in a fair state of preservation in most parts of Locus 106 and provided a reliable stratigraphic horizon, except where the huge final rockfall had caused havoc) the layers underneath were quite complex, mainly due to earlier, apparently repeated disturbances, rockfalls, repair operations and architectural alterations.

In A-B 7, no further interface was found under the spotty, olive green-grey interface until bedrock was reached. This, however, should not be taken as evidence that no such interface existed here originally, as it is quite possible that any interfaces, e.g. a scatter of crushed white sandstone pieces (White Floor), badly disturbed by rockfall, may not have been recog-

nizable by the excavators. In A 7, the face of the Pillar bulged out and here bedrock was very soon reached. A quantity of miscellanea, copper fragments and flint debris found close to bedrock indicated that the pre-Temple occupation established in most loci described above, continued also in Locus 106 as it was, indeed, found in all the loci of the site. In B 7, under the olive green-grey, was a layer of medium-grain red sand fill amongst red sandstone boulders belonging to an earlier rockfall. In this sand deposit, at Level 165, the same early occupation miscellanea with flint debitage were found (in Box 323), but no interface until bedrock. The same situation was found in C 6 in the sand fill between the two large, red sandstone boulders – on top were typical finds of the main Temple period and further down (Level 215) miscellanea and flint tools and debitage (in Box 280).

In Square C-D 7, below the olive green-grey interface (Level 144–165) was a layer of fine-grained sand which lay on a hard layer of medium-grained red sand, packed around large and medium sized flat, red sandstone boulders. These boulders lay in sand fill and on bedrock. Scattered over this hard sand and boulder mass were numerous crushed white sandstone fragments (at Level 180), which appeared to be remnants of a former White Floor (badly damaged by the huge rockfall on top). In this fragmentary interface a number of finds were recorded (Box 283, Level 179–185):[37] an Egyptian seal, copper and iron objects, some gold leaf.

Behind the Standing Stones and Offering Stands in D 7, a Chalcolithic mace-head and flint debitage were found (Box 214) in the bottom of a sand fill layer. Early type miscellanea with flint objects also appeared in the lowest sand fill (Level 179), behind the Standing Stones and Offering Stand, next to the basin (west side).

About halfway through E 7 the hard-packed red sand and sandstone layer of C-D 7 continued (Level 187) underneath the olive green-grey interface; here it gave the impression of an intentionally laid pavement.[38] As in C-D 7 and, in fact, also in F 6–7, some White Floor remnants were found in the sand fill right above the hard sand and stone mass (the 'pavement') but, as in C-D 7, it was very difficult to definitely ascertain its nature. However, the very fact that a White Floor of much greater certainty – though still rather flimsy compared with the solid, 10cm. thick proper floor construction in the northern parts of the Temple – was found right at the eastern edge of the 'red pavement' (in E 7) and, in fact, at many spots all around it (e.g. in D-E 10), made it most probable that here too we are dealing with remnants of a White Floor, i.e. the 'red pavement' was below the White Floor horizon. In 1974 we therefore re-investigated the stratigraphic significance of the 'red pavement' in its relation to the White Floor, by excavating Trial Trench 55 (in E-F 7–8),[39] including the 'red pavement' in E 7.

TT55 (2.50 × 0.70m.) ran along the front of the basin in E-F 7 (*Pl.* 36), starting in D 7–8 at the large, flat red sandstone boulder and ending at Gridline

G.7–8. It was excavated down to bedrock. The western part of TT55 was covered by a solid sand and stone 'pavement'; the eastern part showed clear traces of the White Floor interface in red sand. Cutting into the 'pavement', only very few finds were recorded (Box 526, Level 216–220): two faience beads, a copper ring fragment, bits of copper ore; (Box 532, Level 220): a faience fragment and faience bead. The 'pavement' in TT55 (*Illus.* 12) turned out to be red sandstone tumble with fine, windblown red sand as fill in between. The clear tumble lines in the fill sloped eastwards and a typical picture of a tumbled mass of rocks could be seen in the section. Below the rock tumble, a single Egyptian glass bead was found in the fill. Underneath and down to bedrock (Level 238), was a layer of sand fill with straticulated pebble patches and some miscellanea, including flint, ostrich egg-shell, charcoal, animal bones; the latter indicated there had been some early occupation below the rock and sand tumble.

The eastern part of TT55 produced quite a different picture: below the olive green-grey interface (mostly disturbed during the first season of excavation in 1969) traces of the White Floor were discernible in the soft sand (Level 193) which followed westward the slight slope onto the top of the 'red pavement', i.e. originally the White Floor interface must have also extended over the 'red pavement'. Below this White Floor level, sand fill of about 40cm., with a few bits of miscellanea and very decayed sandstone on bedrock (Level 250) completed the stratigraphic picture of TT55. At the eastern end of TT55 a pit was found in the section which appeared to have been dug down from the olive green-grey interface to bedrock. It contained greenish sand fill.

In G-H 6–7 and TT16,[40] under the olive green-grey interface (Level 184–195) a layer of red sand fill went down to a hard, red sand layer (its interface at Level 218) which lay on undisturbed subsoil (at Level 227). A few rocks, which could be seen below the stone bench along Wall 2 (see below, p. 58 and *Pl.* 44), lay on this subsoil, but these were tumbled rocks and not part of any structure.

A few finds were recorded (Box 378) at Level 222: one Midianite and one Egyptian sherd, a Chalcolithic rim sherd with rope decoration, miscellanea, including flint.

There were no signs of any White Floor interface in G-H 6–7 (in fact, not up to G-H 9) but this area seemed disturbed, perhaps by the builders of the bench along Wall 2.[41]

Trial Trench 15: To clarify the stratigraphic relation between Wall 2 and the stone bench built against it and Wall 3, as well as the stratigraphic significance of the White Floor, present and missing, we excavated TT15 in the inside corner of Wall 2 and 3, adjacent to TT16, which ran further along Wall 3 up to Gridline G.6–7[42] (*Illus.* 13, *Pl.* 44).

TT15 was only *c.* 40cm. wide, the width of the end member of the stone bench and a metre long (along Wall 3). The top of Wall 3, behind TT15, was at Level

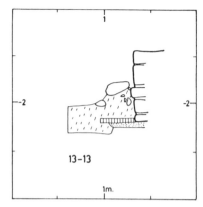

Illustration 13.   Section 13–13.

Plate 44.   Trial trench 15.

144. When the bench stone (Level 180–190) was removed, it was found to be sitting in an olive green-grey layer. Here a number of finds were recorded (Box 374): faience fragments, one with a cartouche of Ramesses IV (Eg. Cat. No. 43, Level 198), a small copper earring, beads and miscellanea, including numerous grape pips (see Chaper III, 22).

Under this olive green-grey layer was red fill, down to a thin ash layer at Level 215, which ran against the base of Wall 3. Right below the ashy interface, at Level 218, was a White Floor interface *c.* 6cm. thick, which clearly continued under Wall 2. At Level 220 an Egyptian sherd was found. Underneath the White Floor was a layer of red gravelly sand down to a hard interface (Level 227) (not excavated).

The Walls in G-I 6-7

Wall 3: As shown above, Wall 3 in A-C 6 ran originally over large red sandstone boulders of an early, pre-Temple rockfall, up to the face of the Pillar. This part of Wall 3 was found completely smashed, all of it red sandstone. The cause of the damage at this stage – apparently accompanied by the devastation of the whole Hathor Temple – could not be definitively established.[43] The heavy and often repeated rockfall at this corner of the site made proper stratigraphic excavations of Wall 3 in Squares A-C 6, rather difficult.

In D-F 6 the wall was found better preserved, although its uppermost courses had slipped onto the Standing Stones and basin alongside the wall (see above, p. 42 and *Pl.* 32). However, the inner side of Wall 3 showed clear evidence of repeated repairs and rebuilding down to its very base, which is also obvious from the crooked line of this part of the wall (*Illus.* 6). For this reason, and also because of the rather disturbed nature of the sandlayers along the inner side of the wall in D-E 6 (man-made disturbances, most of them apparently in relation to the olive green-grey interface activities, but mainly rock-falls and rubble), the detailed stratigraphy of this part of Wall 3 remained unclear, though we have a much clearer picture at its outer side (in Locus 105).

Wall 3 in G-J (*Illus.* 14) consists of three architecturally and stratigraphically quite different parts, which well reflect the complex archaeological problems of the site. Up to and including F 6, Wall 3 (including the repairs) was built entirely of undressed flat red sandstone boulders. The base of the wall was built on top of hard red and gravelly fine sand. At Gridline G-F a large, flat white stone on edge was used as a kind of pinning stone, obviously for wall repair, as was also indicated by a few more stones on edge (red sandstone, one white) right next to it (*Pl.* 45). The use of stones on edge was not a common building technique in the Hathor Temple, and was not actually met with before the olive green-grey interface (see the thin skin of stones on edge at the inner face of Wall 3 in D-E 6). The fairly well-constructed continuation eastwards (in G) of the wall was still of flat, medium-sized red sandstone boulders with a few white stones put on the very top (perhaps a later addition), which came to an abrupt end near Gridline H. This end was noticeably well-constructed and had a straight face towards H-G (*Pl.*

Plate 45. White pinning stone in Wall 3.

46). Underneath the base of this part of the wall (Levels 209–216), which because of its repairs was very irregular, was a layer of very gravelly sand, also with some miscellanea. At Level 200–215 a few very small white sandstone pieces were visible between the red stones of the wall (right next to H 6), but this may have had some connection with the White Floor interface found (at approximately the same level) underneath the continuation of Wall 3 in H 6.

Wall 3 in H 6, up to Wall 2, was of a completely different construction (*Illus.* 14, *Pl.* 46): it was a roughly built link, mainly of irregular white sandstone pieces, between the noticeably well-built red sandstone endpiece at Gridline G-H and the well-built face of Wall 2 at Gridline H-I. This part of the wall appeared to have been a doorway between Wall 2 and Wall 3 which was blocked by roughly piling up white sandstone pieces of any available size. Further evidence for such an operation may be seen also in the differences in the heights of the base levels of Wall 2 (in H 6) and Wall 3 (in I 6) which were noticeably lower on both sides of the doorway.

The very gravelly red sand layer below the base of G 6 continued also into H 6, but here a thick White Floor interface was found on top of it (at Level 206–210), underneath the base of the wall (in H 6). This White Floor layer, slightly sloping eastwards, ran against Wall 2 as well as Wall 3 (in G 6) and was therefore evidently a later addition, although it

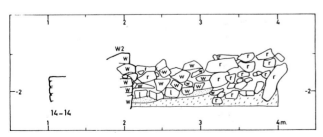

Illustration 14.   Section 14–14.

Plate 46.   Close up of Wall 3 in HG.

obviously belonged to the White Floor phase of the Temple.

Wall 2 (Pl. 59): The southern end of Wall 2, in I 6–7 (Pl. 47), was in every way a continuation of this wall as described in Locus 103. Its inner face was much less impressive than its outer face and it was obvious that the builder laboured harder for a fine appearance of the Temple's façade. Amongst the white building stones of Wall 2 a good number seemed to be in secondary use, a phenomenum met with already in Loci 103 and 104.

TT15 (above, p. 48) and TT66 (below, p. 58), established that the stone bench, built against the inner face of Wall 2 (see below, Locus 108, p. 59, *Pl.* 59) and (in the corner) against Wall 3, was built in an olive green-grey layer which ran against Wall 2 and Wall 3, and that the White Floor interface found in the fine sand fill underneath the bench continued under Wall 2 (Level 220) and the corner of Wall 3.

The white sandstone part of Wall 3 (in H 6) was built against the inner face of Wall 2, but some effort was made to bind the two walls together at the top courses. This does not contradict our assumption that this section of Wall 3 (in H 6) was built later than Wall 2 (and the other parts of Wall 3) probably in order to block a doorway between Wall 2 and Wall 3 or, alternatively, as a later repair.

The Stratigraphy of Locus 106

At the western end of Locus 106, close to the rockface of the Pillar, a mass of red sandstone boulders towered above the Present Surface of the mound, and this was obviously a heavy rockfall after the site was finally abandoned.[44] Locus 106, as in all of the other parts of the site, was covered by a layer of wind-carried, fine-grained yellowish-red sand (the Present Surface) which, however, was not discernible in Squares A-D under the mass of fallen rocks.

Underneath the wind-carried uppermost sand layer, structural features in semi-disturbed condition

Plate 47.   Wall 2 in I 6–7.

– such as the knocked-over stela and Offering Stands, part of Wall 3 (E-F 6) which had fallen onto the basin and quite some rubble in the upper fill layer along the inside of Wall 3 – indicated deterioration due to abandonment of the site and, perhaps, short-lived, temporary occupation during the Roman period and/ or by Bedouins. There was no indication in Locus 106 of any intentional destruction during or after the final abandonment of the site.

The uppermost stratigraphic horizontal feature[45] was an olive green-grey interface found *in situ* over most of Locus 106, though often fragmentary due to disturbances by rockfall, etc. It was also found running against Wall 2 (in I 6–7) and Wall 3 (in D-I 6). The vertical features belonging to this interface were the row of Standing Stones along Wall 3 (except the red sandstone basin in E-F 6–7), the 'demarcation line' of white sandstone (in A-C 6–7) and the stone bench built along the inner face of Wall 2.

Most of Wall 3 in Locus 106 was already in ruins before the olive green-grey interface came into being. Furthermore, sections of this wall, in Squares D-F 6, behind the basin and some of the Standing Stones, were roughly repaired or even partly rebuilt during the olive green-grey interface-related phase of the site. This part of the wall was badly out of line (*Illus.* 6) and partly rebuilt on a thick layer of sand and rubble, at a noticeably higher level than the original base of Wall 3. It was stratigraphically quite certain that these repair/rebuilding activities were connected with the setting up of the row of Standing Stones on the olive green-grey layer.

Amongst the numerous finds on the olive green-grey interface, was a conspicuously large number of copper-based metal objects, mainly of votive character. A number of finds made in the fine-grained to medium-grained sand fill beneath the olive green-grey interface, occurred either at the upper or lower levels of the fill and, considering the rather soft nature of the sand deposit, must be considered as belonging either to the olive green-grey interface above or to another interface further below.

Underneath the olive green-grey interface and the vertical features belonging to this phase, Locus 106 had quite a complex stratigraphy, dominated by areas of the White Floor interface and, as mentioned above, heavy damage to Wall 3.

In Squares A-B 7, a number of architectural elements of white sandstone were found under red sandstone rockfall. As Squareline 7 was here only at the very edge of a peculiar and most significant deposit of debris in the adjacent area, its context will be discussed in the chapter on Locus 107. However, already here we should take into account that the whole of A-C 7–10, i.e. the area between the *naos* and (destroyed) Wall 3, was covered by a large pile of white sandstone building fragments, sculptures, stelae, etc. – fragments of the destroyed Hathor Temple[46] – which had been still further damaged by the red sandstone rockfall on top of the same. It was evident that this destruction predated the olive green-grey interface-related phase of the site and was related to the White Floor interface. Although there was no chance to separate this white sandstone interface from the crushed and broken mass of white sandstone building debris, traces of the White Floor were discernible in the nearby Squares C-D 7 and adjacent Squares in Locus 107 and we may assume that the pile of building debris was located on top of the 'White Floor', or related, interface.

The White Floor, or clear traces of the same, were found at several other locations in Locus 106 and as far as the stratigraphy of Locus 106 is concerned, we must consider it the dominating interface for this Temple phase, connected, as we have shown above, with the building of Wall 2 and the eastern end of Wall 3. The basin in E-F 6–7 was part of the Temple furnishings standing on the White Floor.

A special stratigraphic feature of the south-western part of the Temple courtyard was the 'red pavement' but in Locus 106 (TT55) this 'pavement' turned out to be a mass of rock tumble with the White Floor originally going over its top. The 'red pavement' will be discussed in more detail with the adjacent Locus 107, where most of it was located; but as far as the White Floor horizon is concerned, the 'red pavement' obviously pre-dated the White Floor interface – at least in Locus 106.

Wall 3 in Locus 106 implied a stratigraphic problem which could not be finally solved. Whilst Wall 2 and the earlier part (H 6) of Wall 3 were clearly built on the White Floor and almost exclusively built of white sandstone, Wall 3 in A-G 6 was built of red sandstone and the White Floor did not continue under its foundation. Because of repeated destructions and repairs of Wall 3 and the rockfall disturbances alongside it, there was no clear stratigraphic connection between this part of the wall and the interfaces in the Temple courtyard. We shall come back to this problem – which touches in fact on the basic problem of the existance of one or two main phases of the Hathor Temple – in Chapter IV.

The pre-Temple phase at the western part of Locus 106 was dominated by the large red sandstone rockfall and large red boulders, rubble, disintegrated red sandstone and sand fill which were found on bedrock or virgin subsoil at the very bottom of the site (A-E 5–10).

All over Locus 106, as well as at the other loci, quite a quantity of miscellanea of an early type was found in the bottom sand layer, including Chalcolithic flints, sherds and a fine mace-head. These remains were also found around and between the big rockfall boulders (C 6, B 7, D 7, etc.), and there can be no doubt that these boulders pre-dated the first, pre-historic habitation at the site.[47]

7 Locus 107

Locus 107 is comprised of Squares A-F 8–10, from the rockface of the Pillar to the baulk in Squareline G, and from the row of Standing Stones (Locus 106) to the southern part of the *naos* and *pro-naos* structure.[48]

The Present Surface of Locus 107 was almost completely flat (Levels at A:108; B:118; C:110; D:128;

E:128), with a fairly dense scattering of white and red rubble (*Pl.* 2).

A-C 8-10:[49] Under the Present Surface at A-C 8–10, a red sand layer of 20–30cm. covered a mass of white sandstone building debris (*Pl.* 48). A small fireplace and Roman sherds were found in this uppermost red sand layer of A 9–10 (Level 130) and more Roman sherds appeared, at similar levels and with a locally restricted ash layer, in B-C 9–10. In A-B 9–10 some of the Roman sherds were found mixed with Egyptian artifacts (in Box 245) amongst the white building debris.

The rather chaotic mass of white sandstone building debris was found from the rockface of the Pillar in A 8–10 to the edge of D 8–10. It appeared to have been thrown there without any order (*Pl.* 49).[50] Amongst the architectural elements, many of which were rectangular, finely dressed slabs of white sandstone, as well as relief decorated cornices and lintels (*Figs.* 22–26, *Pls.* 48–52, 110–15), Square Pillars with hieroglyphs (Eg. Cat. No. 5), and small flat basins and Offering Table fragments (Eg. Cat. Nos. 245, 243) (*Pls.* 48, 49, 52), we mention here in particular several decorated and inscribed building elements of stratigraphic significance: In A 10, right next to the *naos* wall, a large white building stone carried a fragmentary cartouche of Ramesses II (Eg. Cat. No. 1, *Fig.* 22:10; *Pl.* 50); in B 9 a stone statuette of a queen-goddess (Eg. Cat. No. 14, *Pls.* 51, 117); in B 9–10 a fragmentary, Ramesside sphinx (Eg. Cat. No. 12, *Pl.* 115); in C 8 an Offering Stand (Eg. Cat. No. 256, *Pl.* 49) and the base of a large statuette (*Pl.* 116:2).

In A 8 an olive green-grey interface appeared (at Level 154) under white building debris, continued north from Locus 106, but this interface could not be traced in A 9–10 or in B-C 8–10.

Plate 48.   Mass of white sandstone debris in Locus 107, A-C 8–10.

Plate 49.   Chaotic mass of debris in A 8–10.

Plate 51.   Statuette of a queen-goddess as found.

Plate 50.   White stone with fragmentary cartouche of Ramesses II.

In A-B 8–10, underneath the building debris, remains of the White Floor were found. Most of the very numerous small finds at this part of Locus 107 were found in sand and stone fill above the White Floor interface or on the interface itself (Level 153–165)[51] (Box 245, Level 126–153): Negev-ware, Midianite, Egyptian and local sherds, very numerous

copper-based objects (many of same in A 10 on the White Floor, underneath the large white boulder with a Ramesses II cartouche), a faience ring (Eg. Cat. No. 98), a faience bracelet with a cartouche of Ramesses V (Eg. Cat. No. 44), a Midianite sherd with the drawing of a magic human figure (*Fig.* 7:2), many beads; (Box 269, Level 153–178): a faience bowl with the cartouche of Ramesses IV (Eg. Cat. No. 102), a small Egyptian seal (Eg. Cat. No. 185), menat fragment, many beads, Egyptian glass, miscellanea.

In C 9–10 and D 10, amongst white debris and sand fill, a hard pebbly sand, crust-like, interface appeared in the red fill (Levels 157–165), which seemed to be wind and water laid. In this fill and debris layer above the White Floor interface, numerous finds were recorded:[52] (Box 225, Level 130–158): (at level 138: Roman sherds and ashy lenses), a faience wand (Eg. Cat. No. 176), Midianite, Egyptian and local sherds, glass, beads, copper objects, faience fragments, copper ore and slag pieces; (Box 239: Level 155): menat fragment, Midianite and local sherds, faience fragments, copper objects; (Box 242, also C 11, Level 150–154): part of a Hathor faience mask (Eg. Cat. No. 25),[53] a bracelet with Hathor inscription (Eg. Cat. No. 55); (Box 251: Level 150–165): Egyptian pottery, part of a Hathor mask (Eg. Cat. No. 15), miscellanea; (Box 277: 155–176):[54] an inscribed faience bracelet (Eg. Cat. No. 51), inscribed sistra handles (Eg. Cat. Nos. 20, 21), many copper objects, Midianite, Negev-ware, Egyptian pottery, plaster fragments, copper ores, slag and metallic copper prills.

The thick layer of white debris and sandfill in D 10 reached the White Floor interface at Level 183. Here a finely dressed flat and rectangular Egyptian Offering Table (Eg. Cat. No. 244, Pl. 52) was found lying on its face, perhaps put there as an extension of the white pavement of the *pro-naos*, or fallen down from it. Underneath the offering table, on the White Floor, a number of artifacts were found (in Box 248): Midianite and Egyptian sherds, an inscribed faience frag-

Plate 52.   Offering Table, D 10.

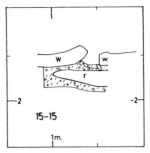

Illustration 15.   Section 15–15.

ment (Eg. Cat. No. 217), several copper objects and beads.

In B-C 9–10, a round white sandstone basin was found standing partly on the edge of a large red sandstone boulder (see below), and partly on smaller red boulders, with the White Floor interface running underneath the basin and over the red sandstone boulders (*Illus.* 15).[55] The basin must have been badly damaged and broken into several parts at an early stage of its use; it was repaired by a layer of lime plaster[56] around its outer circumference, still found *in situ* (*Pl.* 53).

In A-C 9–10, a large oval-shaped boulder of coarse-grained red sandstone was found underneath the White Floor interface. The latter was especially well preserved in B-C 10, on top of the large boulder along the wall of the *naos*, and between the white sandstone basin in B-C 9–10 and the *naos* (*Pl.* 53). In B-C 10, the red sandstone boulder ran underneath the south wall of the *naos*, where a small step was cut into its surface to create a level base for the *naos* wall.

A large, shallow cup-hole or pit was cut into the face of this red sandstone boulder. It was found filled with crushed white sandstone of the White Floor interface which ran over it. The cup-hole seemed to have been made by a rather coarse grinding tech-

Plate 53.   White sandstone basin in B-C 9–10 standing on large boulder.

nique, similar to other large cup-holes or pits found cut into bedrock, underneath the *naos* structure.

The trench-like space between the large red sandstone boulder and the natural rock-bench at the foot of the Pillar was intentionally packed with longish red rocks (*Pl.* 53), probably to create a level ground for the White Floor, which was also found to run over the top of the rock-bench. During the excavation of this 'trench' (to Level 196) only a few stray beads and miscellanea were found.

In area B-C 8–10 (and further east – see below, D-E 8–10), a compact layer of red sandstone boulders packed in coarse-grained red sand appeared under the White Floor horizon. These were rock boulders of various sizes, lying side by side or, occasionally, partly on top of each other, forming a kind of solid surface (*Pls.* 30, 54). This layer was first investigated[57] by a small trial trench (TT9) in C 10 (1.10 × 0.50m.) between the southern naos wall and the white sandstone basin in B-C 9–10, which was dug through the red sandstone layer next to and underneath the basin (*Illus.* 15). Underneath the basin was a large rock boulder of irregular shape. It had a flat top on which stood part of the white basin (see above, p. 53). More, but smaller, red sandstone boulders, adjacent to this large boulder, formed a compact layer of stone and sand. Underneath it was fill of medium- to coarse-grained sand with fine-grained red sand around the large boulder and basin. There were no finds in the sandfill.

In C 10, at least, the sandstone layer of B-C 8–10 appeared to be a tumble of stones – not a properly laid pavement though its upper surface could have been used as a 'floor' or may have served as a levelled base for the White Floor, remains of which were found running over the top of this 'red sandstone surface' (p. 57). This conclusion was strengthened by a small trial trench excavated in 1984 (Note 3, *Illus.* 16), along the large red sandstone boulder with cup-hole (in A-L 9–10) which showed a mass of finely stratified fill (by no means a habitation layer), mixed with tumbled rocks underneath the boulder and in the adjacent sandlayer. Although some stray finds were recorded in the uppermost layer of this sandfill, these may have slipped between the red boulders during the early stages of the Temple.[58]

TT 62: TT62, in A-D 8–9 (3.00 × 1.00m.), was excavated in 1974 to further investigate the nature and stratigraphy of the 'red sandstone layer' underneath the White Floor horizon in the south-western quarter of the Temple courtyard. Below the upper surface of the red sandstone layer (Level 170) – on which traces of the White Floor interface had been found pre-

Plate 54.   Compact layer of redstone boulders (B-C 8–10).

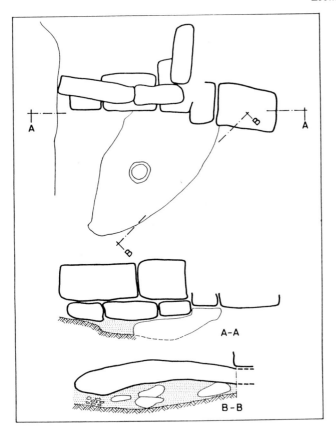

Illustration 16.   Plan and section TT 1984 (Locus 7).

Plate 55.   Strange red sandstone feature in D-E 8–9.

viously – the compact red sandstone tumble of various sizes continued. At about Level 200 a small Midianite sherd was found between the tumble stones. Down to Level 210 a few more finds were made (Box 567): several beads, a copper ring fragment, bits of copper ore, ostrich egg-shell. The entire trench down to bedrock was cut through a mass of sandstone tumble without any apparent stratigraphy.

D-F 8-10: The western half of Locus 107 was dominated by a strange, mushroom-like red sandstone feature which protruded out of the Present Surface of D-E 8–9 (*Pl.* 55). Clearing the uppermost sand layer around this feature, it became apparent that it was a large rock that had fallen from the top of the Pillar and disintegrated into relatively soft sand on impact, after the mound had already been covered by a thick layer of wind-and-water-carried sand.

In the thick uppermost sand layer of wind-and-water-carried fill, a few Roman sherds (Levels: D 10:134; E 9:168; F 9:138; E 10: 131) were found.

In E 9–11 a 'crusty interface' with several Roman sherds appeared under up to 40cm. of red sandfill. Apparently the inside of the Temple was still not completely covered by wind-carried fill at the time of the Roman re-occupation of the site. Judging by the levels of the Roman sherds and the 'crusty interface' (where found), the 'Roman occupation' was at a much deeper level in the middle as compared with

the edges of the mound (near the Temple walls) and evidently there was still a considerable depression in the middle of the mound in Roman times. This would explain the occasional Roman sherd in the soft sand layers much below the actual Roman occupation horizon, often together with Egyptian New Kingdom material (e.g. in E 9 at Level 168).[59]

Underneath 25–30cm. of the red fill top layer, the olive green-grey interface was found over most of the area. (Levels: D 8–9:165; E 8–10:180; F 8–10:183). Numerous finds were recorded related to this inter-, face (D 8–9 – Box 283, Level 165–185): textile remains; (E-F 7–8 – Box 221, Level 180–186): very many copper objects, Midianite and Egyptian sherds, etc.; (E-F 8–9 – Box 279, Level 180–196): a faience jar stand with a Ramesside cartouche (Eg. Cat. No. 97), a faience bracelet with Ramesside cartouche (Eg. Cat. No. 48), very many copper objects, many beads, Midianite, Negev-ware and a few Egyptian and local sherds, some gold jewellery, a gilded iron earring (Fig. 54:16), one iron bracelet.

In F 9, at the eastern edge of Locus 107 (on Gridline G) a small stone-lined pit was found (24cm. in diameter including the stone lining), dug from the olive green-grey interface (Level 184) into the layers beneath, including also the White Floor interface. This appeared to be a socket-like posthole (*Pl.* 56).

Underneath the olive green-grey interface and a thin layer of red sand fill (with some stray finds) appeared a White Floor interface: D 8–9 (continuation from D 7): 180; D-E 8–10: 185–200; F 8–10: 195–214. It was often a hard, white crushed stone mass (in E-F 7–10), but sometimes (in D-E 8–10, west half) only traces of it were found.[60] In D-E 8–10, the White Floor interface was only several centimetres above the 'red sandstone interface', but the stratigraphic relations were quite evident (*Pl.* 57).

55

Plate 56.   Socket-like posthole in F 9.

Plate 57.   White Floor in D-E 8–10.

We list here some of the artifacts found in the White Floor horizon (D 8–9 – in Box 283 at Level 185): menat fragment with a Ramesses II cartouche (Eg. Cat. No. 30), a faience wand (Eg. Cat. No. 179), many copper objects, gold leaf, many beads, an Egyptian seal, an iron bracelet, a lead object, a few Egyptian and Midianite sherds, miscellanea; (E-F 10 – in Box 278, Level 184): a faience Hathor mask (Eg. Cat. No. 17), a faience sistrum handle, an inscribed faience bracelet (Eg. Cat. No. 107), a faience leopard (Eg. Cat. No. 92), a scarab (Eg. Cat. No. 182), a

faience wand (Eg. Cat. No. 177), and very many copper objects and beads, few Midianite, Negev-ware, Egyptian and local sherds, miscellanea; (E-F 9–10 – in Box 236, Level 157–177): Midianite sherds, copper objects, faience leopard figurine (Eg. Cat. No. 88).

In B-E 10, the White Floor interface ran against the white stone wall and pavement of the *naos* and *pronaos* and this interface was especially solid in the squares near those features, and also in Squares E-F 9–10.

In D-E 8–9, White Floor material was found embedded in a compact red sandstone interface (at Level *c.* 187). Here the stratigraphic problems, caused by the flimsy appearance of the White Floor interface on top of the red sandstone mass, were increased by the very shallow layer of fill separating the White Floor on Red Sandstone interface from the olive green-grey horizon. It is quite possible that here the White Floor interface was already disturbed during the olive green-grey phase of the Temple (as apparently happened in adjacent G-H 8–10 of Locus 109).

The compact red sandstone and sand 'interface' in D-E 8–9, which gave the impression of having served as a 'red pavement' during an early phase of the Temple, or even prior to the erection of the Egyptian sanctuary[61] presented a major stratigraphic problem, dealt with by a series of trial trenches (TT60, 60a and 61c, d).

TT 60-60a in D-F 8-10 (2.50 × 0.75m.):[62] In TT60 the compact top layer of coarse red sandstone boulders and fill (Level 185–196) was rock tumble without any order or features. The same rock tumble continued down to bedrock (Level 226). At the bottom of the trench a flint blade, one bead and some ostrich egg-shell fragments were found. At the east end of TT60 a layer of fine red sand was found, *c.* 10cm. above bedrock, but nothing was found in this layer.

TT60a (in D-E 8) presented a different picture (similar to TT61d, see below): a hard red sandstone layer – 'red pavement' – was embedded in the top of a thick layer of brownish-red, fine to medium sand, which continued as one layer down to decayed bedrock (Level 220). Near the top of this layer (Level 211) a few finds were recorded (Box 590): two faience beads, four carnelian beads, copper ore bits and ostrich egg-shells. There were no finds further down in the red sand layer.

TT 61c-61d in E 8-10 (2.50 × 0.75m.): TT 61c (E 9–10) started at Level 200 where traces of White Floor material could be seen.[63] Decayed large and small boulders of coarse red sandstone were embedded in the top layer of the trench, especially at the west side of the TT (in continuation of the red sandstone tumble to the west and south). Right at the top (Level 201–216) was a brownish-red dusty layer and here a few finds were recorded (Box 584): one cowrie-shell bead, iron fragment, ostrich egg-shell, bones; (Box 585): a tiny fragment of Midianite pottery, one faience bead, ostrich egg-shell.

Below this dusty layer appeared a medium-grained

pebbly sand layer (Level 220), containing many large and small flat rocks on edge or at acute angles. At the bottom of this layer some finds were made (Box 586): a Chalcolithic sherd, some flint, ostrich egg-shell, and copper ore bits. A layer of coarse red sand over bedrock had no finds.

TT 61d in E 8: The top layer of red sandstone contained many much-decayed boulders, which formed a compact mass – the 'red pavement'. It was embedded in a fine to medium grained brownish-red dusty layer (Level 212–225) which also contained a few finds (Box 588–589): two beads, a copper ring fragment, a small faience fragment, an Egyptian sherd, two bits of slag, copper ore, ostrich egg-shell – all from the very top of this layer.

The medium-grained sand continued down to a course, red, hard layer of decayed sandstone, without any artifacts, overlying bedrock.

The Stratigraphy of Locus 107

The western and eastern halves of Locus 107 were stratigraphically rather different. Contrary to A-C 8–10, which was dominated and disturbed by repeated rockfalls from the top of the Pillar and by a mass of destruction rubble from the Temple structure and furnishings, D-F 8–10 was typical for the overall stratigraphy of the site, although it also had a very special feature in its 'red pavement'.

A-C 8-10: Square-line 8 ran along the huge rockfall of Loci 105–106 some of which was also scattered over A-B 8–9. The Present Surface of the mound was a wind and water carried red-grey sand layer. In its upper level a few Roman sherds and ashy charcoal lenses (in the sections) indicated temporary Roman occupation of the mound a considerable time after the Temple's ultimate abandonment. Under the upper sand layer a compact mass of white sandstone building debris, including also decorated lintels and even sculptures, lay directly on a White Floor interface, with the exception of A 8, where an intermediate olive green-grey interface was found (not found in A 9–10 and B-C 8–10). Very many artifacts, mainly small and fragmentary objects, were found amongst this debris and on the White Floor interface. A white sandstone basin was the sole feature found *in situ* in the White Floor interface.

D-F 8-10: A mushroom-like red sandstone feature on top of the Present Surface in D-E 8–9 was proof of repeated rockfalls even in more recent times. Below the Present Surface was a thick layer of red sand, containing in Square-line 10 and north/south of it, a thin crusty, pebbly sand layer with some Roman sherds.

Underneath the red sand layer an olive green-grey interface occurred over most of the area accompanied by numerous finds, especially very many metal objects. It was from this interface that a small stone-lined post-hole was dug into the ground (a second was found in Locus 109). Underneath the olive green-grey interface a layer of red sand, containing some

stray finds lay on top of the White Floor interface. Many finds were made on this interface, which appeared to be a solid floor of the Temple. Along Square-line 10 the White Floor ran against the *naos* wall and the *pro-naos* pavement (see discussion in Locus 110–111).

Underneath the White Floor interface a compact layer of coarse-grained red sandstone boulders, many completely disintegrated, spread over most of B-E 8–10 and formed a distinct interface – the 'red pavement'. These boulders, several of which were very large and heavy, originated from the upper regions of the Pillar, overhanging the Temple mound, and are the remains of a huge, ancient, pre-Temple rockfall. One of these large boulders (in A-C 9–10) had a shallow cup-hole in its upper face, which when found was full of White Floor material, with part of the White Floor running over the large boulder. This cup-hole is related to similar cup-holes found under the adjacent *naos*, together with evidence for Chalcolithic activities.

Investigating the 'red pavement', it appeared that its core was a mass of rock tumble lying in a thick layer down to bedrock (in TT9, TT60) where some Chalcolithic debris, pottery and flint, were found (underneath this rock tumble).

In TT60a and 61 (c and d), at the edge of the solid rock-fall tumble, the stratigraphy was different. The rock tumble was found as a solid layer of red stones and sand, embedded into the top of the sand layer underneath, i.e. underneath the top layer of sandstones. In TT60a there was only a fine to medium grained sand layer underneath the sandstone layer, with some Chalcolithic remains almost on bedrock. In TT61 the 'red pavement' top layer was embedded in a thin fine-grained dusty layer which contained a few small finds. In the medium-grained sand layer below the dusty layer, a Chalcolithic sherd, some flints and intrusive flat stones (remains of 'installations'?) indicated pre-Temple occupation of this area. A very coarse-grained sand layer underneath the medium-grained layer, was close to bedrock.

Underneath the White Floor and/or the 'red pavement' interface, there was no structural evidence of any kind indicating a pre-White Floor Temple phase and the mainly very tiny New Kingdom finds from beneath this, may well be stray finds which had slipped down below the interface (see further discussion of this important problem in Chapter IV). The Chalcolithic habitation debris and artifacts were not related to a distinct interface, but all were found close to bedrock and well below all Temple-related layers.

8 Locus 108

Locus 108 comprised Squares G-I 8–10. Square-line G was left as a baulk in Locus 108 and removed at the end of the excavation of 1969 down to the level of the base of Wall 2, respectively to the White Floor in H 9–10. In 1974 this strip was cleared to bedrock in TT57, TT52 and TT51 (see Locus 109). The baulks served as part of the main section of the excavation.

Present Surface was at Level 130–140. Removing the top layer of wind-carried red-grey sand in H-I 8–10, the top of Wall 2 appeared at approximately Level 150. During the later removal of baulk G 8–10, a harder surface was noticed at approximately Level 135 and here some Roman sherds and glass fragments (Box 304) were found. More Roman sherds were found in G 10 at Level 170 (and also in G 11 at Level 167).

Going down in Squares H 8–10, between baulk G 8–10 and Wall 2, red sand fill and some rubble were cleared down to Level 177, where a 'pavement' of flat stones appeared (*Pl.* 58). This was a single line of flat stones against the face of Wall 2 in H 8–10 (also in H 6–7 in Locus 106 and in H 11 and 12 in Locus 109) apparently as a bench (see below). This bench was lying in an olive green-grey sand layer, which was found at a similar level (180–185) all over H 6–10 and also in baulk G 8–10 (Level 184–187).

Finds from the olive green-grey layer

Whilst almost no finds were made in the red-grey sand and rubble top layer, the olive green-grey layer, found well preserved all over Locus 108, was very rich in finds, especially copper-based metal objects, but also very much miscellanea typical for this occupation phase of the site:

H 8–10 on the olive green-grey interface (Box 244, Level 140–180):[64] Midianite, Negev-ware and local wheelmade pottery, a quantity of copper pellets, a tuyère fragment, beads.

G 8–10 (baulk) (Box 304, Level 135–184): Midianite, Negev-ware, Egyptian and local wheelmade sherds, textile fragments; (Box 319, Level 184–187):[65] Midianite, Negev-ware, Egyptian and local sherds, a broken scarab, a faience seal, faience fragments,

Plate 58. Bench put against inside of Wall 2 in H 8–10 showing up in excavation.

many metal objects including a piece of gold leaf, an iron ring, a lead fragment, very numerous copper-based metal objects and fragments (chains, amulets, wire, rods, etc.), many beads and a quantity of textiles and very much miscellanea.

Locus 108 below the olive green-grey horizon – the White Floor

G-H 8–9: Under the olive green-grey interface was a layer of red sand. At Level 200 a harder interface was discernible containing some stray finds (Box 347): several beads, an Egyptian amulet, a copper ring, a faience fragment, miscellanea including some flint. This interface seemed to indicate a further occupational surface, but (as in adjacent G-H 6–7) no White Floor was found in this area, though such White Floor did appear at a similar level in the adjacent western half of G 8–9 (Level 195), the northern half of G 9 (Level 195), and also over G-H 10 (Level 198–214). The red sand layer continued down to Level 218, containing further stray finds (Box 362 – Level 209–218): Midianite, Negev-ware, local wheelmade and Egyptian sherds, a faience wand fragment, one faience sherd, some beads and a flint implement. At Level 218–227, a layer of hard, coarse red sand was found to overlie sterile subsoil, containing a mixed lot of finds (Box 363): Midianite, Negev-ware and local wheelmade sherds, some beads and miscellanea including one Chalcolithic sherd and some flint fragments.

In G-H 10 the White Floor (Level 198) was a very solid mass of crushed white sandstone, approximately 8cm. thick, and its edge in G-H 9 looked broken away. A few finds were made in this interface (Box 510):[66] Midianite sherds, copper ore, one copper object, some beads, animal bones.

Immediately below the White Floor, in a dusty reddish-brown layer (Level 203–209), were few stray finds (Box 511): faience and shell beads, animal bones. Below this layer a coarse-grained, stony sand layer showing an orangy tint[67] overlaid red bedrock and contained a few finds (Box 515): two faience beads, some copper ore bits, ostrich egg-shell, animal bones and several Chalcolithic sherds.

Bench H 8–10 – TT66 (Pl. 59):[68] Below the olive green-grey interface, onto and into which the row of flat bench stones were laid, was a layer of red sand containing a number of medium-sized stones (white and red sandstone) which were not arranged in any particular order and appeared to be rubble of a previous occupation phase. Most of the stones seemed to belong to the White Floor horizon underneath.

At Level 215–225 the White Floor interface was located running towards and underneath the base of Wall 2. On this interface a thin dusty, probably ashy, sand layer indicated the occupation on top of the White Floor. In the layer from the olive green-grey interface to the White Floor, a small number of finds were made (Boxes 598, 599 – Level 195–225): some Midianite, Negev-ware, local and Egyptian sherds, faience fragments, one scarab (Eg. Cat. No. 193), some beads and Egyptian glass.

Plate 59.   Offering bench in H 8–10, built against Wall 2.

The White Floor underneath the bench was not a very solid layer, but one spot where a more solid White Floor was found (its continuation towards Square H 8–9 obviously broken away), clearly indicated that the White Floor had existed there but was obliterated at some stage.[69] TT66 was excavated to a level just below the White Floor.

Further cleaning of Wall 2 led to the finding of a number of Midianite sherds inside the wall itself, especially towards the south end of the bench. We concluded that these sherds were deliberately put into the wall because of the depth of the sherds in the wall; there was often a non-structural stone put in front of the sherd; no plain sherds were found in the wall in a similar position, which might be expected if this was the result of sherds filtering down into the wall, and no such sherds-in-the-wall were found anywhere else along the walls of the site. Bones were also found in wall 2, some tightly wedged between rocks, and should also be considered as deliberately inserted in the wall.

Wall 2 in Locus 108: Little can be added to the description of Wall 2 given above (in Locus 103). When the wall was finally exposed to its full height, there were only white sandstone boulders of various sizes, mainly built in header fashion with smaller stones as fillers. Only the bottom course showed some stones arranged in stretcher fashion.

The Stratigraphy of Locus 108

In the top layer of wind-carried red-grey sand a harder interface was noticeable in some places, which seemed to be just a trampled surface. A few Roman sherds indicated a secondary Roman occupation in the area of Locus 108.

Under approximately 50 cm. of upper sandfill (including the Roman 'stratum'), the olive green-grey layer was the uppermost major occupation stratum, found all over Locus 108 and running against Wall 2. This stratum produced a very large quantity of finds, especially copper-based metal objects and copper ore particles.

Underneath the olive green-grey layer was red sand fill with occasional small-to-medium sized white or red sandstone rocks and a few stray finds. At the bottom of this sandfill appeared a thin layer of fine, dusty sand including some finds and occupational miscellanea, which were in fact lying right on the White Floor interface (wherever it appeared in Locus 108).

The White Floor appeared originally to have spread over the entire south-east corner of the Temple court-yard, although apparently at its south-eastern end as a rather thin and flimsy interface. It was found as a fairly solid floor in G-H 10 going about halfway into G-H 9 and was well-preserved at the western side of G 8–9. In the eastern part of G 8–9 and in most of H 8–9 the White Floor was not found, but here a hard interface, at the level of the White Floor (in the adjacent squares), was found, containing a number of New Kingdom finds.

Wall 2 in Locus 108, like the eastern end of Wall 3, was built onto the White Floor, which was found well-preserved underneath its entire length (in Locus 108). Underneath the White Floor was a thin dusty layer changing into more coarse-grained sand. This dusty layer was not found where the White Floor was missing in Locus 108.

In part of Locus 108 (mainly G 8–10), where the excavation reached bedrock or sterile subsoil, an orangy, fine-grained sand layer[70] contained early Chalcolithic occupation miscellanea, pottery and flint, though even here some New Kingdom objects were found.[71]

The only special feature in Locus 108 was the stone bench, built against the inner face of Wall 2. It was laid into the olive green-grey layer and must be considered as functionally belonging to the olive green-grey stratum.

Although the stratigraphical connection between the Midianite pottery and the animal bones found inserted into Wall 2, perhaps as a deliberate act of votive offering, was not strictly determinable, their position in the wall in relation to the stone bench make their connection with the olive green-grey phase of the Temple most probable.

9 Locus 109

Locus 109 comprised Squares F-I 11–15. Square-lines G-I 11 and F 12–15 were left as baulks to serve as part of the main sections of the excavation (*Illus.* 8 and 9). In 1974, most of Locus 109, with the exception of baulk F 11–14[72] was excavated down to bedrock

The Present Surface of Locus 109 (at Level 124–140) was almost completely flat, with a scatter of sand-stone rubble including some well-dressed white building stones.

G-H 12-14 (to White Floor): Removing the fairly loose surface layer of red-grey sand, a few Roman sherds were collected from the uppermost 10cm. (Box 73 Level 136–150). Below Level 140 the red sand became more compact. In G 13–14 it was mixed with rubble from Wall 1. Below, at Level 178, an olive green-grey interface with patches of charcoal, occurred all over the Square-line, also near Wall 1, where it appeared under wall rubble at Level 174.

In the olive green-grey layer a small number of finds were made (Box 240, Level 178–186): Negev-ware and local wheelmade sherds, some beads and animal bones; (Boxes 220 and 235, Level 174): Midia-nite, Egyptian, local plain sherds, copper-based rings, Egyptian glass fragments, many beads, some copper ore and metal prills and some miscellanea.

At Level 187 the olive green-grey was found mixed with fine woodash which continued downwards until the White Floor interface appeared. This fine grey woodash was a very special feature of Locus 109. It was found on top of the White Floor and next to Wall 1 it was a thick layer of *c.* 30cm. It was piled up against the wall, becoming gradually more shallow towards G 11 where it was *c.* 8cm. thick. As it turned out, this layer of woodash, which was found all over Locus 109, was connected with another special feature of Locus 109: metallurgical furnaces on and in the White Floor.

The White Floor interface in G-H 12–14, was at Level 188–190. On it a number of finds were recorded (Box 256, at Level 190): Midianite and Negev-ware sherds, faience fragments, beads, one iron ring, plaster fragments, slagged soil, and some copper ore; (Box 247, Level 187–197 in fine woodash): Midianite, Negev-ware and Egyptian pottery, a copper ring, beads, crucible fragments and crucible slag.[73] The White Floor was a solidly cemented, *c.* 12cm. thick layer containing very few, probably trodden-in artifacts (Box 299, Level 188–200): Negev-ware and Egyptian sherds and miscellenea.

Baulk G-I 11 showed the same strata as G-H 12–14, though a number of special features demanded more detailed investigations (see below). In G 11, under the Present Surface layer (Level 130–142), a layer of hard, very stony red sand contained some Roman sherds. At Level 167 (Box 331): Roman sherds, copper

ore and miscellanea, belonged to a flimsy but clearly discernible interface which we assumed to represent the secondary Roman occupation of the abandoned Temple site.

The red, stony sand layer continued below the Roman interface down to a packed olive green-grey layer at Level 186. In the olive green-grey layer a small, stone-lined, socket-like pit was found in G 11. The White Floor interface appeared only *c.* 15cm. below the olive green-grey interface. Between the two interfaces some finds were recorded (Box 338, Level 186–201): Midianite and local wheelmade sherds, a faience Bes amulet, several copper-based objects, beads and miscellanea.

At the north side of Square G 11 (and going into G 12) a round pit appeared which will be dealt with below.

Because the area of the baulk right next to the doorway was disturbed during the excavation[74] H 11 was cleared from Present Surface to the White Floor level without stratigraphic records. The finds from H 11 were therefore unstratified (Box 339, Level 130–190): a faience bowl fragment with the cartouche of Ramesses II, a phallic bronze figurine (*Pl.* 126:5–6), very many metal objects, Egyptian, Midianite and Negev-ware pottery, beads.

G-H 11–14 below the White Floor: A small trial trench (TT6) was excavated in 1969[75] from the White Floor at Level 191 to bedrock at Level 225. In the fill layer below the White Floor a distinct hard interface was encountered at Level 210 and, connected to this interface – a possible floor – a number of finds (Box 368): several Midianite, Negev-ware, Egyptian sherds, many fragments of faience bowls, a hammer stone, three beads. Near the bedrock a further group of finds was recorded (Boxes 300 and 368, Level 225): miscellanea including several flint implements.

TT 6, as well as TT51 (1974) in G-H 12–14, showed a more complex stratification underneath the White Floor, as had previously been noticed in other loci. (*Illus.* 17; *Pl.* 60). In G 14, right underneath the White Floor, a red sandstone boulder in very coarse red sand (decayed sandstone), appeared to be rubble from Wall 1, i.e. rubble from a part of the wall which might have pre-dated the White Floor, because the White Floor, here 10cm. thick, was found running above the rubble and against the base of the wall (and not underneath it as in H 14, etc.) (*Illus.* 8).

The layer of rubble in coarse-grained sand was separated from a layer of fine-grained sand by a clear, distinctive interface. Moreover, some of the red boulders, lying flat right under the White Floor could well indicate deliberate leveling of this disturbed area before the White Floor was laid. In the fine-grained sand layer was a quantity of small red rocks, but one white sandstone piece (not native at Site 200) found amongst it could serve as an indication for the artificial nature of this layer, i.e. evidence for pre-White Floor occupation of this site. This occupational phase seemed to be connected with the finds at Level 210 in TT6 (see above, Box 368).

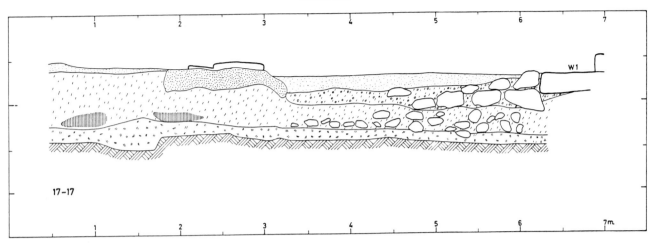

Illustration 17.   Section 17–17.

Plate 60.   Section in TT 6, showing pre-White Floor debris of Wall 1.

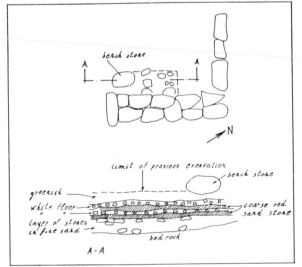

Illustration 18.   Sketch plan and section, Locus 109 (1974).

Square-line H 11-14, below the White Floor (*Illus.* 18) was investigated in 1974 by TT51 and TT51a. In H 11–13 (TT51a) along Wall 2, there appeared to be a continuation of the bench in Locus 108, though only one flat stone was found *in situ*, right at the corner of the doorway (in Square H 12). This stone was lying

on an olive green-grey interface. Under this layer the White Floor appeared at Level 185. The White Floor in TT 15a showed a rather peculiar stratification, not encountered anywhere else in the Temple: below the White Floor interface was a thin layer of coarse-grained red sand and below this layer appeared a second White Floor layer. In this formation very few finds were made (Box 506, Level 185–215): a Midianite sherd, Egyptian glass, two beads.

Below this White Floor/red sand/White Floor in H 11 (baulk) and H 12,[76] appeared another thin layer of coarse-grained red sand which contained a few stray finds (Boxes 509, 527, Level 215–223): three faience beads, a faience fragment (leopard) an inscribed stone, miscellanea. Underneath this coarse-grained red sand an interface of very fine-grained sand appeared at Level 223 and on its very top lay a pavement-like layer of flat stones, which appeared to be deliberate. In the fine-grained red sand layer, which reached bedrock at Level 232–236, a few finds were recorded (Boxes 509 and 530, Level 223–232): three Negev-ware and one wheelmade sherd, five beads, miscellanea, including some flint. In H 13 a very dark patch of ash and charcoal, several heat-crazed small rocks and one single piece of slag were found, at Level 236, close to bedrock.

Baulk F 12-14: On the Present Surface and in the uppermost red sand fill of baulk F 12–14, were many medium-sized stones which must have been rubble from Wall 1 (*Pl.* 60). However, there were also a few rectangular, finely dressed white sandstone blocks, which seemed to have been originally part of a well-built structure.[77]

The upper red fill and rubble layer extended from the Present Surface at Level 124 down to a dark, ashy interface at Level 174. In F 12 an olive green-grey interface was noticeable just on top of this ash layer. In the red sand layer a few finds were recorded, most of which came from close to the olive green-grey and ash interface (Box 334, Level 124–174): local wheelmade and Egyptian sherds, several copper-based objects, a red saddle-backed quern, miscellanea, including small pieces of slag, some copper ore and copper prills; (Box 301, at Level 152): a glass fragment with cartouche of probably Ramesses II (Eg. Cat. No. 180).

The ash layer, as in G-H 12–14, was piled up against Wall 1 and (*c.* 30cm. high) gradually thinning out towards F 12. In this ashy layer were a number of finds (Box 314 in F 14, Level 175–185): Midianite, Negev-ware and local pottery, copper-based metal objects, Egyptian glass fragments, two saddle-backed quern parts, some gold leaf, beads, much miscellenea, including slag, copper prills and ores; (Box 366, Level 174–185): a fine scarab (Eg. Cat. No. 133). The ash layer was found lying on the White Floor (at Level 189). In F 12 a small, heavily charred structure – Fu I – was found set into the White Floor (see below, p. 64), and around it was a thin layer of yellowish burnt sand.

Clearing the White Floor in F 14, it was found running against the base of Wall 1 – not underneath the wall, which was built on a thick stony layer of red sand (*Illus.* 9). The White Floor (as in G-H 14) also ran over a pile of rubble which lay next to the wall.

In F 12–13, during the removal of some of the White Floor, a few finds only were made (Box 367, Level 199–209): solitary Midianite, Negev-ware and Egyptian sherds, some copper fragments, beads, miscellanea, including some slag pieces and flint debitage.

TT63 in F 14 produced a very accurate picture of the strata under the White Floor next to Wall 1. Below the White Floor layer (Level 200) lay a layer of medium-sized rocks in red, coarse-grained pebbly sand, apparently rubble from Wall 1. After removing the wall rubble, a layer of fine sand appeared which contained many small rocks and pebbles. The finds (Boxes 517, 518, Level 200–224): a faience fragment, few beads, miscellanea; (Box 561, Level 204–216): a solitary rough handmade sherd;[78] (Boxes 562 and 569, Level 224): five beads, one faience fragment, miscellanea, including flint. The fine sand layer went down to a thin layer of coarse, decayed sandstone overlying red bedrock (Level 224).

Wall 1 in Locus 109

Although Wall 1 has already been dealt with above (Locus 101), further stratigraphic evidence was obtained by the clearance down to bedrock of Square F-H 14 (TT63 and 51) (*Pl.* 61; *Illus.* 9).

The inner face of Wall 1, up to the middle of Square G, like its outer face in Locus 101, showed quite a different building technique on its west side compared with its east side, from Square G to the junction with Wall 2 (*Illus.* 19). In Square F to half way along G, Wall 1 was fairly well-built, mainly of red sandstone and red-grey limestone, though on top of the wall, in F-G, a few white sandstones seemed to be a later addition or repair. The east end of Wall 1 was very roughly built, with the use of many more white sandstones. Moreover, Wall 1 in Square F to half G was built on a thick, stony layer of red sand (see TT63) whilst under its eastern end, up to its meeting with Wall 2, a White Floor layer was found below the wall, overlying the red sandstone bottom layer (*Pl.* 62).

Wall 2 in I 11-14 and the doorway

The north end of Wall 2 and its corner junction with Wall 1 was fairly well built, mostly of white sandstone boulders of various sizes. At their junction, Wall 1 and Wall 2 seem to have been built together also because the White Floor layer (ca. 10cm. thick)

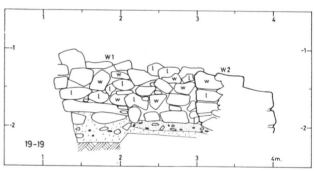

Illustration 19.   Elevation of Wall 1 – Section 19–19.

Plate 61.   Different building techniques in Wall 1 TT 63 and 51.

Plate 62.   White Floor below eastern end of Wall 1 continuing also below Wall 2.

62

below Wall 1 (in G-H 14) continued at the same level underneath Wall 2 (*Pl.* 62).

During the removal of baulk G-I 11 the doorway to the Temple courtyard was uncovered in I 11–12 (*Pl.* 63). It consisted of an opening in Wall 2, 1.28m. wide, flanked on each side by a white sandstone boulder laid across the width of the wall as a capstone.

Another long sandstone boulder (*c.* 1.28m. long) was found on top of the baulk in Square I 11 and partly in I 12, and was found to lie right across the doorway. Although it gave the impression of a lintel, we could not establish any secure structural relationship with the doorway itself.

Going down from the level of the Roman interface (in G 11), the olive green-grey interface was found to continue into I 11 but it stopped near the entrance of the doorway. The White Floor, which was found all over Locus 109, was also found underneath the doorway (*Pl.* 64). It was only missing at the south-west

Plate 63. Doorway to Temple showing up, with long boulder across.

corner of the doorway which apparently was damaged and repaired at some stage (see Wall 2 in *Illus.* 9). It is quite possible that this side of the doorway was almost completely rebuilt during these repairs. Here, as further along Wall 2 (see TT15, Locus 106, *Illus.* 13), a darker, ashy layer several centimetres thick, was lying right on the White Floor but it seemed to have stopped at the base of the wall, whilst the White Floor ran on underneath the wall. This dark layer appeared to be occupational debris on top of the White Floor interface.

The stratification below the White Floor in I 11, underneath the doorway, was investigated by a trial trench through the entrance (TT65 in 1974; *Pl.* 65). The White Floor in the doorway was extremely hard, presumably trampled by the visitors to the Temple, but this paving stopped in line with the outer face of Wall 2, and on the slope outside (in Locus 103) it became only a loose scatter of small white stones. Embedded in this White Floor layer (Level 216 to 230) was a small hammer stone and one faience fragment (Box 595). Right at the line where the White Floor stopped, at the outer side of the doorway, a large stone appeared under the White Floor layer (*Pl.* 65), but this stone was later on found to be lying on bedrock and belonged to the pre-Temple phase of the Site (see below).

Below the White Floor layer was pebbly brown sand and right on this brown interface (at Level 230), i.e. right under the White Floor layer, was another faience fragment (Box 596). The brown layer went down to bedrock (at Level 251). A large boulder found in this layer, at a level close to bedrock, may have belonged to a pre-Temple installation, because leaning against it were several large Chalcolithic sherds (*Fig.* 1:3) and next to it was a broken red sandstone grinding stone.

Plate 64. White Floor underneath doorway to Temple.

Plate 65. TT 65 through the doorway, through White Floor.

Special Features in Locus 109

1. *Posthole in olive green-grey layer in G 11 (Pl. 66).* In the olive green-grey interface at approximately Level 170, was a small pit (25cm. diameter), lined with small flat stones on edge, which seemed to be a socket-like posthole. Around this posthole in the olive green-grey layer, a number of finds were recorded, including a few Midianite sherds, beads and a copper fragment.

2. *Pit in G 11-12 (Pl. 67).* The posthole described above, related to the olive green-grey layer, was located above and at the very edge of a round pit in G 11–12. This pit, already mentioned briefly above, was 75cm. in diameter and its bottom was at Level 204. It was cut into the White Floor. The upper fill of the pit consisted of wind-carried sand, interspersed by waterlayed thin crusts, i.e. it had been filling-in during a period of abandonment. The pit contained (Box 349, Level 182–204): very many copper-based metal objects, Midianite sherds, one faience leopard figurine (fragment), beads and some crucible slag (casting debris). The thick ash layer, overlying the White Floor interface, stopped at the edge of the pit. It appeared that a row of small stones had been laid around the edge of the pit as a marker or upper rim (*Illus.* 8). Stratigraphically the pit was dug through the White Floor interface and clearly seems to belong to the functional horizon of the ash layer, which was related to the metallurgical installations nearby. The crucible slag is in this connection of clear stratigraphic significance.

3. *Fu I – Casting Installation in F 12 (Illus. 6; Pl. 68).*[79] When clearing the sandfill and ash layer of F 12, a roughly circular patch of red/yellow burned ground appeared on the White Floor interface and in its centre a fragmentary structure of heavily charred red sandstone, approximately 80 × 80cm. It was a small, stone-built compartment, open at one side (north-east), with a hard-baked, slaggy floor (approximately 30 × 30cm.) found covered with woodash. The high temperatures obviously involved in its operation, the dark ashy material in front of it, much woodash and charcoal, crucible fragments and crucible-casting slag found in Locus 109, indicated that the installation was a crucible furnace (Fu I) for the casting of small objects (see below, Chapter III, 13).

In the burnt and ashy layer close to Fu I a number of finds were recorded (Box 343, Level 174–184): Midianite and Egyptian pottery, many copper-based metal objects, iron rings, glass fragments, some copper ore nodules and copper prills, many beads, miscellenea.

Fu I was built into a white mortar setting, apparently specially provided at the time of laying

Plate 66.   Posthole in olive-grey layer (G 11).

Plate 67.   Pit in G 11–12, cut into the White Floor.

Plate 68.   Casting installation in F 12.

down the White Floor. Stratigraphically, and according to the finds close to it, Furnace I clearly belonged to the White Floor stratum (*Illus.* 17).

4. *Fu II – a Casting Installation in G-H 12-13* (Pl. 69). Fu II was found as a dark red, burned, slaggy mass *c.* 25cm. high, surrounded by a yellowish-red charred patch on the White Floor interface. There was only a vague outline of a hearth discernible, but otherwise it was rather similar to Fu I. Close to Fu II a layer (*c.* 8cm. deep) of dark ashy material contained slagged bits of stone and a crucible fragment with slag incrustation. The thick layer of woodash found spread all over Locus 109 stopped right next to Fu II and already for this reason Fu II must be functionally related to the ash layer, both belonging to the White Floor stratum.

The Stratigraphy of Locus 109

In some parts of the red-grey Present Surface layer, a hard interface was noticed which was related to a few Roman sherds found at the same level, obviously traces of a secondary Roman occupation in Locus 109.

The olive green-grey interface and layer were clearly discernible over most of the area of Locus 109. In G-H 14 it ran against Walls 1 and 2. A flat stone, seemingly a continuation of the votive bench in Locus 108, also put against Wall 2, was lying in the olive green-grey layer. Connected with the olive green-grey layer were very many metal finds and Midianite sherds, but also Negev-ware, Egyptian and locally wheelmade pottery. The olive green-grey interface was also noticed on top of the thick ash layer found almost all over Locus 109.

The stone-lined posthole in G 11 belonged to the olive green-grey layer; it penetrated into the White Floor beneath it. The White Floor layer, in most of Locus 109, was a hard cemented floor on which many finds were recorded.

A thick wood-ash layer was found spread all over Locus 109 and formed an especially thick pile (*c.* 30cm.) against Wall 1. It sat right on the White Floor interface and must be considered the wastage of metallurgical activities on this Floor. Two melting-for-casting installations – Fu I and Fu II – belonged to the White Floor phase. Fu I was built onto a hard mortar layer of *c.* 10cm. thickness, especially inserted into the White Floor for this purpose at the time the White Floor was originally laid. The pit in G 11–12 also belonged to the metallurgical operation in the White Floor phase.

The relation of the White Floor to Walls 1 and 2 (in Locus 109) is of considerable stratigraphic significance: in Squares F 14 and the western half of G 14, Wall 1 was well-built of mainly limestone boulders, its base sat on a stony layer of red sand. The White Floor ran against the bottom course of the wall, but not underneath it. In the eastern half of G 14 and H 14, Wall 1 was very roughly built, mostly of white sandstone, and the White Floor ran on underneath it. Under this White Floor, the layer of stony, hard red sand, found under Wall 1 in F and part of G 14, was also found. Stratigraphically and according to the building technique used, the two parts of Wall 1 were quite different and the western part appeared obviously earlier than its eastern part.

Wall 2, though more regularly built as the eastern part of Wall 1, was built together with this part of Wall 2 and, like it, on to the White Floor. The doorway in Wall 2, which at this phase was the main entrance to the Temple, also sat on the White Floor but its southern half seemed a later repair or reconstruction of the original doorway.

Below the White Floor phase, Locus 109 was of particular stratigraphic importance: next to Wall 1, in F-G 14, the White Floor that ran against the base of the wall, lay on top of a pile of rubble, apparently from Wall 1. In H 14, a row of sandstone boulders lay along Wall 1 with the White Floor running over it against the wall. Here, a hard interface with related finds (Box 368) was found underneath the row of stones and this interface continued also in G-H 14 below Wall 1, and below the White Floor under the wall. The trial trenches in Locus 109 (1974), revealed wall rubble and other stone assemblies which indicate pre-White Floor activities, though no related architectural features could be discovered. Although some of the flat stone arrangements under the White Floor could be the result of levelling as preparation for the White Floor laying, others, including non-native white sandstone fragments, indicated a pre-White Floor phase of the Temple.

In Locus 109 a number of Chalcolithic artifacts, found close to bedrock, indicated a pre-Temple occupation phase at the site. This was particularly evident underneath the White Floor inside the doorway in Wall 2. It must be emphasised that some of the pre-White Floor disturbances and features in Locus 109, where mixed Chalcolithic and New Kingdom finds

Plate 69. Casting installation in G-H 12–13.

were uncovered, could belong to this Chalcolithic phase, but a number of features, especially those located right below the White Floor layer and in the related coarser sand layer, could not belong to this early phase.[80]

10 Locus 110

Locus 110 comprised Squares B (partly)-E 11–13 (partly) and A-E 13–14.[81] Squareline B-E 12 was left as a baulk to serve as part of the main E-W section of the excavation (*Illus.* 8) and towards the end of the first season it was excavated down to the White Floor level (found in part of B-E 12 only). In 1974 it was cleared down to bedrock, together with most of the area of Locus 110.

The Present Surface of Locus 110 (Levels 104–120) was littered, like the surface of the adjacent loci, with rubble some of white sandstone (see *Pl.* 2).

A-E 13–14 (to the olive green-grey interface): In the red-grey surface sand layer a few Roman sherds were found in the uppermost 10 cm. related, locally, to two distinct ashy layers. The upper, wind-carried red-grey sand layer was particularly deep in the western half of Locus 110 and seemed more disturbed than the surface layer of the adjacent loci. In Square-Line 14 (and 15 of Locus 112) which was originally slightly protruding, the outlines of a solid wall (Wall 1) appeared after the removal of several centimetres of sand cover (at Levels 103–123).[82]

In the upper sand layer of Squares C-D 13–14 some Roman sherds were found, together with New Kingdom finds (Box 138, Level 150): Roman, Egyptian, Midianite sherds, a copper ring, an iron fragment and beads. Similarly, in Squares D-E 13–14 Roman sherds appeared in the red sand layer from Level 138 to 155 (Box 55, Level 138): one Roman sherd; (Box 135, Level 155): one Roman sherd, Negev-ware, beads, copper parts from sistra, faience amulets, an Egyptian seal (Eg. Cat. 187), hieroglyphs of Hathor on faience (Eg. Cat. 222), miscellanea, including a flint object. This situation, i.e. the presence of solitary Roman sherds in obviously pre-Roman contexts, was especially noticeable in the north-western corner of the Site, including Locus 111.

In Squares A-B 13–14 no interfaces were found under the very thick layer (c. 50cm.) of upper sand-fill.[83] Below this red sand layer, which also extended into Squares A-B 15, a number of finds were recorded (Box 116, A-C 14–15, Level 118–156): Midianite, Negev-ware, local handmade sherds, few copper objects, beads, plaster fragments and miscellanea. At Level 158–183 a hard red sand layer was found, containing a few finds (Box 210, Level 158–183): one Midianite sherd and some miscellanea; (Box 215, Level 170–183): one local rough sherd, one bead. Bedrock was reached at Level 195.

In Squares B (the undisturbed half)-D 13–14, under the red-grey upper sand layer, the olive green-grey interface appeared (first at Level 147) on top of an ashy layer which spread over most of Squares B-D

13–14 (Levels 151–160). In the olive green-grey and ashy layer, the following finds were recorded (Box 98, Level 155): a copper snake with gilded head (Metal Cat. 3; *Fig. 53:3*; *Pl.* 125:5);[84] (Box 206, Level 147–155): Midianite, Egyptian, Negev-ware pottery, including Midianite bird-decorated juglet, beads, faience menat fragment and bracelet with cartouches of Ramesses IV (Eg. Cat. 28 and Eg. Cat. 42), glass, slag bits, miscellana; (Box 228, Level 154–160, in red sand, olive green-grey, ash and charcoal): Egyptian, local wheel-made, Midianite sherds, one inscribed faience menat (Eg. Cat. 34), copper rings, beads, copper and iron fragments, copper prills and ores; (Box 228A, Level 144–154): a group of metallurgical remains, including some ore fragments, casting crucible fragments, casting slag, and miscellanea;[85] (Box 138, in C-D, Level 150): one Roman sherd, Egyptian, Midianite sherds, one copper ring, one iron fragment, beads, miscellanea including many small slag pieces, some plaster fragments.[86]

Area E 13–14, along the east side of Wall 5, was investigated (TT 2) before the baulk in Squareline F was cleared. Under the thick upper layer of red grey sand, an olive green-grey-brownish layer appeared (Level 157–180) which was, in fact, an almost solid mass of heavy copper ore nodules, mixed with artifacts (*Pl.* 70). The ore nodules were heaped against Wall 5 and the basin attached to it, and also against Wall 1. The finds in the heap of ore nodules (Box 204, Level 157–180): Midianite and local wheelmade sherds, very many copper rings, one iron ring, beads, much miscellanea including slag pieces, plaster fragments. At Level 166–178 and 180, a mass of textile was found (*Pl.* 71) lying close to Wall 1 (in the olive green-grey layer).

The large number of metal artifacts was obviously intentionally collected and stored in this corner. This

Plate 70. Mass of ore nodules heaped against Wall 5.

Plate 71. Mass of textile found close to Wall 1.

unusual store clearly belonged to the olive green-grey horizon and was perhaps connected with the function of Wall 5.

In E 13–14, TT 2 reached the White Floor interface almost directly below the ore storage heap (at Level 180), but the thin reddish-grey interface, found directly above the White Floor interface over most of B-E 14, would not have been discernible in this colourful storage area.

A-E 13–14 (White Floor to bedrock): Below the olive green-grey interface and related ash layer, a thin interface (2–3 cm.) of wind-carried reddish-grey fill lay over the White Floor interface, which was found all over the area of B-E 13–14 (at Levels B 13–14: 158; C 13–14: 163; D 14: 164; E 14: 179). This White Floor was a *c.* 20cm. thick, solidly cemented white mass. Its stratigraphic relation to the special features of Locus 110 and 111 (*naos, pro-naos* and Walls 1 and 5) was the main objective of several sections excavated in B-C 13–14.

Most important was the Section (*Illus.* 20) which showed the connection between the foundation stone

at the eastern end of the *naos* structure and Wall l, indicating that the White Floor clearly ran against both the red slab on edge in Wall 1 (see below, also *Illus.* 21) and the foundation stone of the *naos*. Here the White Floor was laid to exactly the same height as the top of the foundation stone, a fact of considerable stratigraphic significance (see below).

This Section also showed that the thick White Floor layer in B 12–14 lay solidly on top of a red-grey sand layer. Below this sand layer a shallow, *c.* 10cm. deep pit (*Pl.* 72) was found cut into bedrock (Level 170), one of several such pits which were found underneath the Temple strata and seemed everywhere related to Chalcolithic occupational miscellanea, pottery and flints.

Across B 13–14, the White Floor stopped almost in a straight line with the end of Wall 1 (B 14–15) (see *Pl.* 72). As already explained above, no interfaces were recorded in the area between this line and the rockface of the Pillar. We assume that this gap in the wall was intentionally made as a passage to the structure (Wall 4) attached to the outside of Wall 1, belonging to a later phase of the Temple (see below, Locus 112) and it appears that the White Floor was removed at the same time.

On the White Floor only a few finds were recorded (Box 281, Level 158): one Roman sherd,[87] Midianite and Egyptian sherds, two beads and Miscellanea; (Box 237, Level 160–163): one iron ring, copper objects, miscellanea, including ores, slag, shells.

Along the western half of Square-Line C-E 13, going into C-E 12, a row of large and flat white sandstone boulders was found lying on top of the White Floor interface (*Pl.* 73). As there was no order discernible in this pile of rocks, it was obviously not a structure but gave the impression of structural elements removed from a dismantled building nearby (see below, our description of adjacent B-E 12). Underneath these stones, the White Floor interface appeared in B-E 13, but along about a third of the west side of this Square-Line, almost in a straight line, the White Floor was missing apparently intentionally dug away. As some of these rocks were

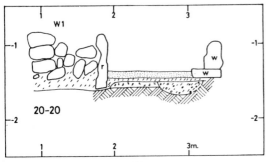

Illustration 20. Section 20–20.

Plate 72. Shallow pit underneath the Temple structure.

Plate 73. White boulders on top of White Floor, probably Roman disturbance.

Plate 75. Same boulder as Plate 74, perhaps podium of Square Pillar.

found lying also on top of the basin in E 13, this disturbance must have been fairly late in the history of the site, perhaps during its Roman occupation. On the White Floor interface below the row of flat rocks, a number of finds were made, including strings of beads (in the find boxes of B-E 13–14 listed above).[88]

Whilst clearing the heavy mass of flat boulders, one large, square, finely dressed boulder was found standing on edge, leaning over under the weight of the heavy rocks put against it (see *Pl.* 74). The upper edge of this boulder, in fact, protruded through the Present Surface before the excavations. After these

Plate 74. Finely dressed boulder *in situ* (secondary use).

rocks were removed, the upright boulder was found to be of considerable stratigraphic significance. Standing on edge, it was found deeply imbedded in the White Floor layer (which stopped abruptly a short distance further west)[89] and it must have been a free standing feature at the time the White Floor interface was the active surface of the Temple courtyard, i.e. it was clearly at that time already in secondary use.

The boulder, *c.* 0.90 × 0.90 × 0.20m. (*Pl.* 75), was apparently originally a podium for a Square Pillar. It was diagonally dressed around the upper half of its sides (see *Pl.* 74) and had a square boss in the middle of its well dressed upper surface. The bottom of the Pillar-Podium was roughly dressed. There was no indication of the original location of this Pillar-Podium and little to suggest its original function in the Temple.

Wall 5

The function of the feature 'Wall 5' (*Pl.* 76, *Illus.* 6) remained somewhat obscure,[90] mainly because of the inbuilt basin of very soft white sandstone, which was probably in secondary use and intentionally cut in front and therefore of little practical use (*Pl.* 77). Wall 5 was more in the nature of an attachment to Wall 1, built against its face (not bonded). The short link connecting the basin to the wall was only *c.* 85cm. long and *c.* 60cm. high; it was built of three courses of white sandstone boulders, one of them finely dressed and obviously also in secondary use (*Pl.* 76). At a later stage of the excavation (1974), this boulder was extracted from Wall 1 and was apparently an Offering Table (*Pl.* 78).[91] It had been inserted into the structure upside down.

Wall 5, including the basin, was built onto the White Floor interface. To secure the stability of the

Plate 76.   'Wall 5' south side.

basin, wedges of white sandstone and a large (local wheel-made ware) sherd were put underneath it (*Pl.* 76). Although there was only a very thin sand layer between the solid White Floor interface and the foundations of Wall 5, it was obvious that the latter was an addition built after the White Floor had been in use for some time.

To try to establish the stratigraphic relations between Wall 5, the White Floor and the *naos* (in Locus 111), a connecting section was excavated (*Illus.* 21). This section showed clearly that the White Floor ran underneath the basin in Wall 5, it was missing in

Plate 78.   Offering Table found in Wall 5.

Plate 77.   Intentionally cut white sandstone basin in Wall 5.

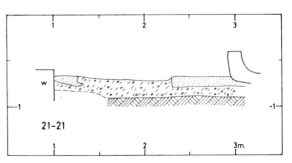

Illustration 21.   Section 21–21.

69

the middle of the section, obviously disturbed at some time, and it appeared again at the west end of the section, put against the face of the eastern wall of the *naos*.

The stratigraphic context of Wall 5 was further ascertained in the excavation of the Squares D 13–14 and E 13–14, from the White Floor interface (reached during the first season) to bedrock (TT 53 and TT 61A).

TT 53: TT 53 was first excavated only in D 13–14, but later extended over the whole area of B–D 13–14. The White Floor interface in D 13–14 at Level 164, sloped down towards Wall 5[92] and clearly continued below it. The White Floor layer was here 22 cm. thick, it contained only very few finds (Box 501, Level 164–186): one copper ring, some beads. Below the White Floor was a brownish sand layer (Level 186–192) with ashy patches which also contained some finds (Box 505): one Egyptian sherd, one faience fragment, two beads.[93] Bedrock was reached at Level 202. There were no finds in the bottom layer of D 13–14, but at the edge of C 13–14 one Chalcolithic sherd was found in the hard sand layer, close to bedrock (Box 375, Level 186).

The eastern section of TT 53, dug along Wall 5, confirmed our previous stratigraphic conclusion concerning Wall 5 and added important stratigraphic evidence for Wall 1 (*Illus.* 22, *Pl.* 79). The solid White Floor layer ran below Wall 5, and was also clearly

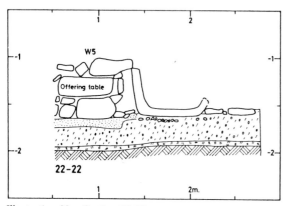

Illustration 22. Section 22–22.

Plate 79. Eastern section of TT 53, showing White Floor below Wall 5.

visible in D 14 under the foundations of Wall 1 (see below). The White Floor below the basin appeared to be somewhat disturbed, probably by the stone wedges pushed underneath it.

TT 61A: In TT 61A, E 13–14, the White Floor layer (at Level 179–180) was removed. There was only one Midianite sherd (Box 563) within this layer. Below the White Floor a thin (1–2cm.) interface of fine, dusty sand[94] lay on top of a dark red, medium coarse, pebbly sand layer, which contained a number of finds (Box 564, Level 190–192): one Midianite and one Egyptian sherd, three beads, two saddle-backed quern fragments and one Chalcolithic sherd; (Box 570, Level 192–203): one Midianite and one Egyptian sherd, one Chalcolithic sherd, flint tools and miscellanea. Underneath this medium coarse sand layer, the texture of the sand became very coarse and hard, seemingly decayed bedrock. In this layer, directly above bedrock, one Chalcolithic sherd was recorded (Box 571, Level 206–216).

The stratification of the layers below the White Floor, described here, was very typical and was frequently found during the excavation of Site 200 (*Illus.* 23). However, a relatively large number of Chalcolithic artifacts was conspicuous in the layers below the White Floor of TT 61A.

The western section of TT 61A ran along Wall 5 and showed the same stratigraphical sequence as established by TT 53 (*Pl.* 76). However, the section along Wall 1 in E 14 proved that, contrary to D 14, the White Floor in E 14 did not continue below Wall 1 but ran against its foundation course.

Wall 1 in Locus 110[95]

Wall 1 in Locus 110, which showed up under a few centimetres of Present Surface, turned out to be a rather complex structure. In E 14 the wall was fairly well built of red sandstone and limestone, with a few smaller, white boulders, an obvious addition, on top. In E 14 and part of D 14, as in adjacent F–G (part) 14, the White Floor did not run underneath this part of the wall. Here, a few red boulders which had apparently fallen from the top of the wall before the White Floor was laid, were left *in situ*, in front of the foundations of Wall 1 and the White Floor ran against this rubble (*Illus.* 23).

In D 14, the thick (10 cm.) White Floor layer ran underneath the foundations of Wall 1[96] (at Level 163), but stopped at the eastern end of a very large red sandstone boulder (*Illus.* 24, *Pl.* 80) which was inserted into, and was the continuation of Wall 1 in B–C 14. In B–C 14 the White Floor was clearly laid against this large flat rock (see *Illus.* 20, above).

A section along Gridline C 13–14 (*Illus.* 25) revealed an interesting stratigraphic situation: To prevent this huge flat rock standing on its edge from falling over, it was set into a foundation trench, the only proper foundation trench recorded at the Temple site. This trench was cut from a layer of greyish-red sand (with some charcoal and ash

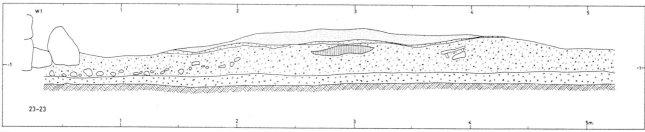

Illustration 23.   Section 23–23.

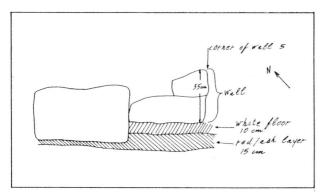

Illustration 24.   Sketch section (partial) of Wall 1.

Plate 80.   Wall 1, section along inside shows large boulder on edge as pre-White Floor structure.

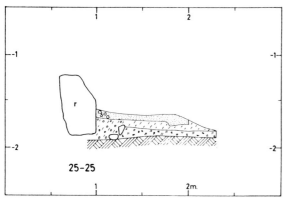

Illustration 25.   Section 25–25.

patches), which contained (see above) some Chalcolithic artifacts, into the hard, coarse-grained layer of decayed bedrock. It was found filled with greyish-red sand. This section proved that the huge flat rock was put into this position in Wall 1 during a pre-White Floor occupation phase.[97]

Wall 1 ended in B 14 at the end of the huge rock on-edge.[98]

B-E 11–13, to the White Floor: Present Surface in B-E 11–13 was at Level 104–110. In the upper levels of this red-grey fill layer, some Roman sherds were found (Box 317, Level 110 (Present Surface) – 137,[99] related to some localised trampled and ashy interfaces. In the red sand layer below, the following finds were recorded (Box 238, Level 130–155): Roman sherds, two beads, small slag pieces, plaster fragments; (Box 63 and 72, Level 120–142): Roman sherds, Midianite, Negev-ware and local wheelmade pottery, a faience bracelet, copper objects, one iron fragment, glass, beads, miscellanea including plaster fragments.

In B-D 13–14, no olive green-grey interface appeared, but this interface was found in E 11–12, outside the area of the *pro-naos* (see below). Here Present Surface was at Level 120. In the red sand top layer some ashy patches contained Roman sherds (Box 318, Level 154). The olive green-grey interface appeared at Level 174, related to some finds (Box 332, Level 154–174): Egyptian sherds, faience, a bead, some copper rings. At Level 185 the White Floor interface appeared (in E 11–12).

In B-D 11 the White Floor interface appeared (Levels 164–168). In part of B-D 12 the White Floor was missing, apparently intentionally removed, although here and there along the edge of this area fragments of the White Floor were still found *in situ* (*Pl.* 81).

After clearing the whole area of B-D 11–13 down to the White Floor interface level, the cause of this missing White Floor became evident. The heap of large, flat white sandstone boulders found lying on the edge of the White Floor along the west side of B-D 13 (*Pl.* 73), had evidently been forcibly removed from the squares in front of the *naos* (Locus 111)[100] and we may assume that the obvious disturbance of the White Floor underneath the (now missing) 'pavement' was also caused by the same operation.

On the (still preserved) White Floor interface in B-D 11–12, some finds were recorded (Box 344, Level 167):

Plate 81. Intentionally removed White Floor in B-D 12 (Roman disturbance).

a Roman sherd, Midianite, Negev-ware and Egyptian pottery, faience fragments, amulet of Sekhmet (?) (Eg. Cat. 205), beads, copper objects, glass, miscellanea including plaster fragments; (Box 325, Level 137–164): Roman sherd (at Level 138), Egyptian, Midianite, Negev-ware pottery, inscribed faience

bracelet (Eg. Cat. 53), beads, glass, miscellanea including plaster, ore bits.

We first only excavated the area of B-D 12–13, where the White Floor was found to be missing, down to the level of the White Floor interface in B-D 11–12. The layers below this level were subsequently investigated together with the layers below the White Floor in the adjacent squares (see below, p. 73).

The Pro-Naos

Along the Squareline 11 (at the south side of Locus 110 and going partly into Locus 107) a 'pavement' of two rows of flat, white, roughly dressed sandstone stones appeared at Level 153–159 and a further row of similar white stones continued in E 11–12, at a right angle to the former row of stones (Level 153–155). It was quite apparent that these stones represented only part of the outer border of a solid pavement which had originally covered the area in front of the *naos* (B-E 11–13), and appeared to have served as a kind of *pro-naos* (Pl. 82). The flat stones were not bonded together, no mortar was used, and gaps of several centimetres were left open between the

Plate 82. *Pro-naos* in front of *naos*.

Plate 83.   Thick mass of plaster sticking to corner of *naos*.

stones. From these gaps, a number of small finds were collected.

At the north-east corner of the *naos*, traces of the White Floor, which had previously also run against the front of the *naos* (under and between the 'pavement' of the *pro-naos*) were found still sticking to the well-dressed cornerstone (*Pl.* 81).

At the south-east corner of the *naos* the 'pavement' touched its foundation stone and here both this foundation stone and the end of the pavement (at the same level: 154) had a thick mass of pinkish-coloured lime plaster[101] sticking onto their surfaces (*Pl.* 83). A number of similar fragments of plaster, some of which had obviously been shaped, were also found dispersed in the immediate vicinity (in B 11) and also in the disturbed layers further away, all over the area of the *pro-naos*. The plaster mass lay over the narrow gap between the *naos* and the 'pavement' stone and was obviously part of a plaster-made structure, fragments of which were still *in situ* on top of the *naos* foundation (see below, Locus 112). As will be shown below (Locus 112) this lime plaster was part of a later addition to the *naos* structure, or part of its total rebuilding, and was also found on top of the foundation stones, at the opposite, north-east corner of the *naos*.

B-E 11–13 below the White Floor (TT 64, 54, 53): To investigate the stratigraphy below the White Floor and the stratigraphic relation between the *pro-naos* pavement and the White Floor interface, the inner row of stones (in B-D 11) was removed. These stones, perhaps only fragments of originally much larger, dressed boulders, had previously been knocked about (probably by the ancient removers of the stones from the adjacent part of the *naos*), and simply disintegrated on trying to remove them. The stones appeared to have been set into a thin layer of White Floor, which also continued under the adjacent solid pavement. The heavy white sandstone boulders of the pavement seemed to be embedded in White Floor material. This also became obvious by comparing the carefully taken levels of the White Floor right next to

the big stones with their bottom levels: the flat stones in D 11 were at Level 157 (top) and 170 (bottom) respectively 159 (top) and 178 (bottom) whilst the White Floor interface next to the stones was at 164.

The same situation was found at the row of stones in E 11–12, where a thin layer of White Floor material was found below the stones, but the White Floor interface ran against their lower parts at a higher level. E.g. Level of stone bottom 167, Level of White Floor interface against its base: 164; resp. stone bottom Level 183, the White Floor interface at Level 164 (running against the stone).

We conclude from these facts that the *pro-naos* pavement was laid together with the White Floor. To emphasise the stratigraphic significance of these facts, it should be pointed out that the White Floor all around the *naos* was found to run against its foundation, and both at nearly the same level, but no traces of such White Floor layer have been found underneath any part of the *naos* structure (see Locus 111).

In TT 64 (B-D 11 southern half) and adjacent TT 54 (C-D 11–12) (which together covered the whole area of White Floor south of the disturbed squares of the *pro-naos*) the White Floor was removed and the area excavated to bedrock (*Pl.* 84). Below the White Floor layer a dusty brown layer appeared which contained a few finds (Box 507 – mainly from C 11 – at Level 185): Midianite, Negev-ware, Egyptian sherds, faience beads, faience fragments, copper objects, glass, miscellanea. Below this relatively find-rich dusty layer, a red-greyish sand layer contained ashes, crazed stones and a flint tool (Box 592). This layer lay on bedrock (Level 205).

In E 12, where remnants of the White Floor had been found on removal of the large pavement stone, the dusty brown layer below the White Floor layer contained several finds (Box 503, Level 186–194): a scarab, a faience rim sherd, beads, glass, a copper ring, some ore nodules. Above bedrock (Level 206) some crazed stones and ashes appeared in coarse-grained sand.

TT 11 in B-D 12–13: TT 11 investigated the area in B-D 12 where the pavement appeared to have been removed in ancient times. From Level 164 (the level of the White Floor interface in the adjacent squares),

Plate 84.   TT 64, section underneath *pro-naos*.

Plate 85.   TT 11 underneath *pro-naos* structure.

where bits of White Floor material could be discerned (*Pl.* 85), the red-grey, pebbly sand fill continued until about Level 172, where a hard interface was found near the *naos* foundation. It was also discerned locally at some other areas along the edges of TT 11. As this was the approximate level of the bottom of the pavement stones in B-D 111, we assumed that this harder interface was not indicative of an occupation surface but was caused by a long period of pressure from the flat stones of the removed pavement.[102] When TT 11 was subsequently extended westwards into B-C 11, clear traces of the White Floor were found at this level (with the typical thin layer of dusty, fine sand underneath). On the hard interface close to the *naos* a few finds were recorded (Box 353, Level 172–174): Egyptian sherds, a faience ring-stand with the cartouche of Ramesses III (Eg. Cat. 96), faience fragments, some beads. In the red sand fill below, over the whole area of TT 11, more finds turned up (Box 370, Level 174–180): faience bracelet, Negev-ware, Egyptian sherds, faience bracelet, stone jar lid, copper ore bits.

The centre part of TT 11, especially C-D 12, clearly showed a major disturbance of the layers below the White Floor level. At level 183, a hard interface on coarse-grained red sand was reached, showing several ashy patches. Here a few finds were made (Box 371, Level 183): Midianite and local wheelmade sherds, two beads and a copper pin (see similar findings in adjacent TT 64 and *Illus.* 9).

Bedrock was reached at approximately Level 208.

E 11–12 below the White Floor (TT 61B): Squares E 11–12 had previously been excavated down to the White Floor interface (Level 184). In 1974 this area was cleared in TT 61B.

Several of the smaller flat white pavement stones in E 11–12 were found to be very fragile and were removed. They were embedded in White Floor material. In the gaps between these, as well as other pavement stones, some copper rings and a sistrum fragment were found (included in Box 575).

The stratigraphic situation here was the same as in adjacent area E 13–14. In the dusty fine sand layer below the White Floor layer, a few beads were found (Box 575, Level 192). Further down, but still in fine, dusty red sand, some ashy patches were discernible, containing crazed stones. Here a Chalcolithic sherd was found (Box 576, Level 194). Below this, at Level 201, in a layer of medium coarse sand, more Chalcolithic artifacts were found (Box 578, Level 201–205): two Chalcolithic sherds; (Box 579, from ashy patches at same level): miscellanea, including flints; (Box 577, Level 205–219): one Chalcolithic sherd. The medium coarse sand layer lay on the coarse-grained, pebbly sand layer of decomposed bedrock.

The Stratigraphy of Locus 110[103]

The Roman secondary occupation of the Temple mound was much in evidence in the centre of the courtyard and in particular in C-D 12–13 where there were two distinct trampled interfaces with some ashy patches in the upper levels of the top layer. A deeply penetrating disturbance was evident in most parts of the *pro-naos*, which practically destroyed its architecture and stratification. It was apparent that the Romans intentionally removed most of the pavement of the *pro-naos* and dug through the layers underneath, perhaps to hunt for treasure. A number of finds were recorded in all levels of the disturbed fill, but these were unconnected with any features of stratigraphic significance.

The bottom layer of the site, containing a number of fireplaces with Chalcolithic artifacts, was found undisturbed.

An olive green-grey interface was found in all Squares of Locus 110, except in the area of the *pro-naos*. It was often related to an ashy layer right below it. Very many finds, especially metal objects and metallurgical debris, came from this horizon. As the olive green-grey interface was not found in the adjacent *naos*, we may assume that there was probably a functional cause for the absence of an olive green-grey interface, not indicative for the architectural stratigraphy of those features (see discussion in Locus 111).

Wall 5, with its white basin and the store of ore nodules and metal objects – obviously considered as scrap – found piled against it, appeared to belong to the olive green-grey phase. Only a very thin layer of sand, which could have been occupational debris belonging to the White Floor phase, separated the White Floor (below Wall 5) from the foundations of Wall 5; but the fact that Wall 5 was not bonded into Wall 1 behind it, but roughly put against it, as well as the secondary use of two architectural elements of white sandstone (Basin and Offering Table) in this wall, seemed to provide some evidence for the stratigraphic integration of Wall 5 into the olive green-grey horizon. Stratigraphically it is important to emphasize that this White Basin in Wall 5 was obviously in secondary use.

One of the building stones found upside down within Wall 5 was a large, rectangular Offering Table, obviously also in secondary use, perhaps from the White Floor phase or, alternatively, from a pre-White Floor phase of the Temple.

The White Floor interface was found in all squares of Locus 110 except in the disturbed areas of B-D 12–13. The layer of this White Floor material was mostly very solid and up to 20cm. thick. A fair number of artifacts, mainly votive objects, came from this White Floor layer.

As almost everywhere, a very thin, dusty fine sand layer lay immediately below the White Floor mass. We did not consider this layer as a stratum of its own (see footnote No. 94), although the findings and finds in this layer need further scrutiny. In this connection, it should be mentioned that many relatively small finds were collected between the stones of the white pavement and in other cracks in the White Floor interface. Considering the enormous number of sometimes very small votive artifacts which had been brought to the Temple, it should not be surprising to find a few of those small artifacts somehow infiltrating (almost seeping) through the interfaces (cf. above, note 71).

The pavement of the *pro-naos* was laid together with the White Floor, obviously as an integral part of the White Floor phase of the Temple. Sometimes – presumably during the Roman occupation and treasure hunting at the site[104] – a great part of this pavement was intentionally removed and the White Floor, and the sand layers below it, were also dug up.

A mass of lime-plaster, part of a plaster-made structure found still partly preserved on top of the *naos* foundation, covered part of the pavement and many plaster fragments were found in the disturbed area of Locus 110. These facts on their own did not provide clear-cut stratigraphic evidence for the relation of this plaster structure to the *pro-naos* and *naos*. However, taking into consideration also the evidence from Locus 111, we consider the plaster debris in Locus 110 as belonging to the last pre-Roman phase of the Temple.

The Pillar-Podium, set into the White Floor interface at the time the White Floor was laid, was obviously in secondary use in this phase of the Temple's history. This is one of the few architectural elements which provided some evidence for a pre-White Floor phase of the Egyptian Temple.

Wall 1 in Locus 110

The large red sandstone boulder on edge in Wall 1 was set into a foundation trench which was cut from the medium-grained red-grey fill layer below the White Floor layer and penetrated into the decayed coarse-grained sand layer close to bedrock. This foundation trench was filled-in (around the inserted boulder) with medium-grained sand material which then also formed a layer above it running against the face of the boulder. The White Floor overlay this medium-grained sand layer and also ran against the face of the large boulder. The two sections (*Illus. 20* and 24) show that the large boulder was set prior to the laying of the White Floor, perhaps during the earliest occupation phase of Site 200; it may have been connected with the Chalcolithic pits found in the area around it. It was obviously a 'stranger' in the whole architectural context of the Temple.

The two large boulders of Wall 1 in D 14 (and 15), were standing on a White Floor interface not found in adjacent C 14 and E 14. The wall here was very roughly built, perhaps during the major repair or even rebuilding of Wall 1 in the White Floor phase or sometime afterwards (see also Locus 101 and Locus 112).

The stones of Wall 1 in part of D 14 (and 15) and in E 14 (and 15) lay on red sand fill, perhaps the Chalcolithic interface, with the White Floor layer running against them at some height. This part of Wall 1, already discussed in Locus 101, seemed clearly to indicate a pre-White Floor phase of the Temple courtyard.

The excavations of the trial trenches below the White Floor layers of Locus 110 provided very little stratigraphic evidence for a pre-White Floor phase of the Temple, although the evidence from parts of Wall 1, the Pillar-Podium, and perhaps the Offering Table, do clearly point in this direction.[105]

The red-greyish sand layer below the White Floor layer, topped by the fine, dusty red sand layer, showed many signs of occupation and at its bottom, on top of the coarser sand level close to bedrock, was a distinct hard interface (*Illus. 23*). However, no structural evidence was forthcoming to help identify the occupants of this layer. The fact that in Locus 110, as in some of the other loci, the finds from this layer were mostly mixed (New Kingdom and Chalcolithic artifacts) added to the problem. It should however be pointed out that the finds on the hard interface at the bottom of this layer, i.e. on top of the coarse-grained sand layer, consisting of fireplaces, crazed stones, shallow pits, Chalcolithic pottery and flint tools, provided evidence for an exclusively Chalcolithic occupation.

11 Locus 111

Locus 111 comprised Squares A-B 11–12 and about half of A-B 13,[106] i.e. it contains the *naos*,[107] the central feature of the Temple.

Locus 111, before the excavations (*Pl. 2*), was the central sector of the low Temple mound which lay right against the face of the Pillar (see also *Illus. 6* and 8). When excavated, the side walls of the *naos* structure were found to have been built right up against this rock face (*Pl. 86*). Three niches were cut into the rock face behind the *naos*, visible before the excavations (*Illus. 26*): in the centre of the *naos* was a man-

Plate 86.   *Naos* and niches in rockface behind, Locus 111.

high niche, originally of rectangular shape and about 1.50 × 0.50m., which may have housed a statuette or a stele. Its outlines were badly distorted by wear and erosion. At the height of approximately 1.80m. from the top of the foundation stones of the *naos* walls and immediately above them, two smaller niches (0.65 × 0.45m.) with round tops had been cut into the rock face. As those niches were approximately in line with the apparent location of two square pillars at the front of the *naos*, we presumed at the time that they served to hold the end of large architraves (Rothenberg, 1972, 130).

The Present Surface (Level 95–100) was somewhat disturbed by recent campers who removed the rubble they had found there to make room for sleeping bags and campfires.

In the upper levels, the red-grey fill layer contained ashy patches and local hard trodden interfaces and Roman pottery, complementing the stratigraphic picture of the adjacent loci, of a secondary occupation of the Temple site during the Roman period. However, in the *naos* area mixed groups of artifacts, i.e. Roman and New Kingdom objects, started to appear high up in the top sand layer (Box 42, Level 110–120): Roman

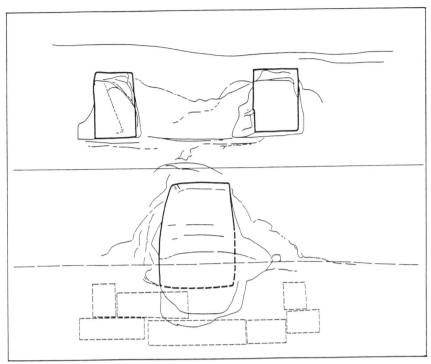

Illustration 26.   Elevation of niches in rockface behind the *naos*.

and Midianite sherds; (Box 316 in A-B 12, Level 100): Roman, Negev-ware, local wheelmade sherds, faience fragments, beads, glass, miscellanea. Going further down into the thick red-grey fill layer, the Roman presence continued to be surprisingly conspicuous. At Level 159 in B 11 a Roman lamp was found (Box 276), but Roman sherds also appeared much further down in the layers, together with artifacts from the Egyptian strata of the site. From the findings listed below, it became apparent that during the Roman occupation of the Temple site, part of the *naos*, similar to the adjacent area of the *pro-naos* in Locus 110, was dug into and part of the archaeological stratification of the *naos* was disturbed.

Clearing the red-grey top layer down to about Level 170, a hard interface of scattered White Floor material was uncovered over most of the *naos* area. Here some finds were recorded (Box 354, Level 170): one faience fragment, one Negev-ware sherd, beads; (Box 372, Level 171): a faience cartouche fragment, one ring, a copper pendant, glass, beads. At the north-east corner of Square B 12, a door lintel of red sandstone was found on the White Floor interface, this was the only architectural element of red sandstone found in the Temple (*Pl. 87*, also *Pl. 113:1*).

In most of the front part of the *naos*, especially in B 12, close to the *naos* entrance, the White Floor interface was missing and the area appeared disturbed. This disturbance in the stratigraphy was enhanced by the fact that some material containing white stones, sand and plaster fragments, and also artifacts, was apparently backfilled into the dug-up hole. The direction of this dump (found sloping down from the *naos* entrance to the middle of A-B 12, where it was touching an interface at Level 185), suggested

Plate 87.   Red sandstone lintel found on White Floor of *naos*.

strongly that this operation was contemporary with the similar operation – the 'Roman treasure hunt' – in the *pro-naos*. During the subsequent excavations it became obvious that this disturbance also affected the other adjacent squares inside the *naos* (cf. TT 12 and TT 58).

Going down in the centre of the *naos* (TT 13) (*Pl. 88*) where traces of White Floor material were only found closer to the Pillar face (Level 170), medium-coarse sand was encountered which lay on a hard interface (Level 185). This interface contained a few finds (Box 376, Level 185, dug down to below Level 218): 6 beads, glass, alabaster bowl fragment, miscellanea; (Box 373, Level 185): one Roman and one Egyptian sherd, beads, glass.

Below this interface a layer of coarse-grained sand lay on bedrock (Level 206). Cut into the bedrock was a shallow pit (Pit C) but only half of it became uncovered in our trench excavation. It was subsequently fully exposed, together with several similar pits inside the *naos*.

Plate 88.   TT 13 in the centre of the *naos*.

Plate 89.   TT 58 inside *naos*, shallow pits in bedrock underneath.

The above description refers to some remains of orderly stratification left *in situ* in A-B 12. In A-B 11 the stratification was found somewhat better preserved (in TT 58 A-B)[108] and will be reported in the following.

TT 58 B: Underneath the previously excavated White Floor (scattered) interface in TT 58 B (to the left of TT 13) was red sand and below the same (Level 190) lay a thin interface of fine red sand which contained some finds (Box 551, Level 190–194): one faience fragment, 7 Egyptian beads, glass, a copper fragment and a Roman sherd.[109]

In the layers of medium-coarse red sand, below the fine red sand interface, a few finds were recorded (Box 552, Level 194–204): one copper fragment, one bead and part of a Roman flask. At level 205, on a coarse-grained hard sand interface, a few finds were recorded (Box 555, Level 205–218): one faience bowl fragment, one bead.

Reaching bedrock (Level 218), a shallow pit appeared. Around and above this pit the fill was brownish and soft, and contained flints and Miscellanea (Box 549).

TT 58 to bedrock: The whole of the inside of the *naos* was cleared down to bedrock – TT 58 (1974).[110] The bottom rock surface was very rough and sloped eastwards from the rockface of the Pillar. Whilst clearing the bottom layers in B-C 12–13, a brownish sandfill was noticed which lay on top of several shallow pits (A-D), cut into bedrock (*Pl.* 89; *Illus.* 6). One pit (A) was found to be partly underneath the northern foundations of the *naos*, separated by about 15cm. of brownish fill. In the brownish layer close to Pit B, a Chalcolithic sherd was found (Box 545, Level 228), but in the higher levels of this brownish fill New

Kingdom artifacts were recorded (Box 545 in B 12–13, Level 195–204): one Egyptian sherd, 2 beads, 13 faience fragments, glass; (Box 541, in A 12, Level 204): one faience rim, 3 beads; (Box 548, in A 12–13, Level 190–200): one faience fragment, 2 beads.

The brownish layer which overlay the rock-cut pits, was not found in A-B 11.

The *naos* structure[111]

Only two stone courses of the *naos* structure were found *in situ* (*Illus.* 27; *Pls.* 85, 86). Because of the different character of these two building courses, we shall describe each of them separately.

(1) *The upper course of building elements found in situ*
Going down in the upper red-grey sand layer a rectangular stone block appeared in A 11 (Level 116). When fully exposed, it turned out to be a finely dressed, stele-type, white sandstone block on edge, the end of which lay against the face of the Pillar (*Pl.* 90). Although the face of the stone block facing the *naos* was badly damaged, clear traces of carvings could be distinguished on it. Two lines, one on each side of the block, ran along the edge and may have been the frame of a hieroglyphic inscription, traces of which could be discerned. As shown by the toolmarks made by a sharp, chisel-like instrument, this inscription appeared to have been intentionally obliterated. There could be no doubt that this fine piece of masonry was originally an inscribed stele and in the present location was in secondary use as a building stone in the wall of the *naos*. Next to it stood another white stone block on edge, which was partly covered by pink plaster (*Pl.* 91). It was bonded by a plaster mass to the corner and the first 'building stone' around the south-east corner of the *naos*. This corner was made almost entirely of distinctly pink

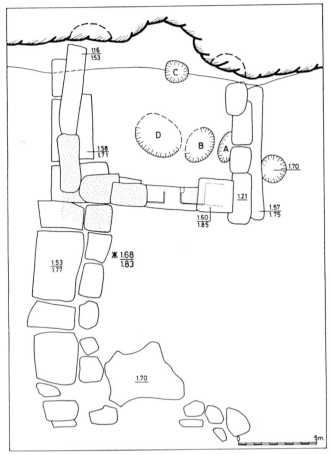

Illustration 27.   Plan of *naos* and *pro-naos*.

and white plaster. When excavated, this part of the east wall of the *naos* was an almost formless mass of plaster which also extended onto the adjacent pavement of the *pro-naos* (*Pl.* 83). Apparently attached to this course stood a structural element, but only its formless plaster 'bed' remained *in situ*.

Further along the east wall a gap in the upper course of building stones must have been the entrance to the *naos* (*Illus.* 27; *Pl.* 86), with a long foundation stone of the *naos* structure as a doorstep.

At the north end of the east wall, where the red-grey sand fill reached the White Floor interface (Level 170), lay several Roman sherds right on top of a finely dressed, square white stone, built into the foundations of the *naos* (*Pl.* 92). The presence of several Roman sherds at this level provided further evidence for the deep-reaching disturbance of this area of the Temple site in Roman times.

A formless mass of pink plaster was stuck to the surface of the square stone and also extended onto the adjacent building stone (*Pl.* 81). This plaster mass stopped in a straight line near the edge of the square stone; here a thin line of plaster (*Pl.* 93) indicated that

Plate 92.   Roman sherds on square base of pillar, part of foundation structure of *naos*.

Plate 90.   Stele with traces of hieroglyphs in secondary use in the *naos* structure.

Plate 91.   South half of *naos*. Note plaster on corner stone.

Plate 93.   Base of Square Pillar in the foundation structure of *naos*.

originally this square stone must have been the base of a Square Pillar. It was also apparent that such a Square Pillar had been held in position by a bed of pinkish plaster and at the time of its removal – i.e. before the Roman disturbance which left Roman sherds in a fill layer overlaying the already damaged pillar base – its plaster bed was ripped apart. According to the plaster outlines on the Pillar base, the Square Hathor Pillar, found in secondary use as Standing Stone in F 7, could well have originally stood here.

The upper course of the north wall (*Pl.* 94) consisted of two white stone blocks, of which the one nearest to the Pillar face was badly damaged. The other, close to the north-east corner, was also partly damaged, but on its face towards the inside of the *naos* were clear traces of thin plaster lining (*Pl.* 89).

(2) *The foundations of the* naos *structure*
In contrast to the not very accurate building technique of the upper course of the *naos* wall, the lower foundation course of the *naos* structure was very meticulously layed. Whilst the upper course building blocks standing on edge were not exactly in line and not accurately laid along the middle of the foundations, the stone blocks of the foundation lay on their flat sides and accurately in line, thus providing a solid base for the *naos* structure.

The lower course of the southern *naos* wall (*Pls.* 91, 95) consisted of three rectangular stone slabs. The slab which touched the Pillar face lay on a protruding rock ledge. This rock ledge was intentionally flattened to provide a good base for the stone slab. The

gap between the rock ledge and the large red sandstone boulder below the second foundation stone (see above, p. 54), was found closed by a fill of small rocks and sand with traces of White Floor material spread above. The White Floor also lay over part of the large red sandstone boulder and against the outside of the foundation stones of the *naos* (see already in Locus 107, above p. 53).

A small step was cut into the northern part of the large red boulder (which continued below the southern *naos* foundations) in order to provide space and a better hold for the foundation slab which lay on top of it. The stone blocks of the foundation were wider than the upper stone course and extended on both sides of the upper course.

The third stone in the row of foundation stones was the corner stone of the south-east *naos* corner. Because it was almost completely covered by a stone and by plaster,[112] only a small part of it could be investigated. However, it appeared to be a well-dressed square block, slightly larger than the similar inner corner stone at the opposite, north-east *naos* corner.

The foundation of the north wall of the *naos* structure was in fact one long and roughly dressed stone slab (*Pl.* 94). Similar to the south wall foundation, this slab rested with its head on a rock-ledge, protruding from the Pillar, but here was no need to artificially cut a flat, supporting surface. Partly underneath this long slab were two shallow pits – one mainly outside and the second mainly inside the *naos* (*Pl.* 89) – cut into the here rather high rising surface of the bedrock (Level 170). These pits were covered by brownish sand which was also found below the long foundation slab, in contrast to the more complex stratification below the foundations of the southern *naos* wall (described above, in TT 58).

Plate 94.   North wall of *naos*.

Plate 95.   South wall of *naos* structure (outside).

Plate 96.   Inner cornerstone on north-east corner of *naos*.

The inner corner stone[113] on the north-east corner (*Pl.* 96) is a good example of the fine Egyptian masonry met in the Temple. It was perfectly square (*c.* 40 × 40cm.) and flat on top. This impressive building block rested on the hard interface at Level 185 inside the *naos* and also seemed clearly related to a similar hard interface under the pavement of the *pro-naos*. A solid White Floor layer (*c.* 10cm. thick), as well as a layer of medium coarse sand (in Locus 110), lay against the outside of the square corner stone (*Illus.* 8). It is important to notice here that inside the *naos* a hard interface, with a scatter of White Floor material, was found at almost the same level as the hard interface below the missing White Floor, outside and close to the *naos* (at approximately Levels 170–172).

Although the area outside the *naos* has already been described above (in Locus 110), it should be emphasised here that the thick White Floor layer ran everywhere (except in B 12, where it was disturbed) against the foundation stones of the *naos*; in fact, its upper interface was obviously intentionally laid at the same level as the top of the *naos* foundations (*Pls.* 81, 82, 86).

The two corner stones in the eastern wall of the *naos* were connected by a long, rectangular white stone slab, which appeared damaged on the side turned towards the inside of the *naos* (*Pl.* 97). Except for an area of 35 × 30cm. in the centre of its upper face (see below), it was very smoothly dressed on all of its sides and by its overall appearance could well have been in secondary use in its present location in the *naos* wall. This long slab was about 28 cm. wide and 10 cm. thick and lay on the interface on top of the

medium-coarse sand layer, with remains of the White Floor found inside and outside the *naos* structure.

The foundation stones at the front of the *naos* showed additional features of considerable stratigraphic significance (*Illus.* 27; *Pl.* 97). Although the stone slabs were of various heights, they were laid to form one perfectly flat and level surface (159–160). Subsequently, a series of lines was cut into this surface which showed typical chisel-and-hammer tool-marks. It appeared possible that these lines were builder's markers for the structure above, but the plaster-structure found above *in situ* did not appear to have been actually built along these lines. At the south-east corner the plaster covered the whole of the corner stone (and also the adjacent stone of the *pro-naos* pavement) without any obvious reference to the 'border line'.[114] There was also the very roughly dressed, small area on the surface of the long front slab (*Pl.* 98). On one side of it a line was roughly cut out with a hammer-and-chisel, whilst on its opposite side a finely dressed, boss-like surface had a small, square notch cut into the inside edge of the slab. This area and its different marks were difficult to explain as builder's marks, and appeared to have already been on the slab before its removal from its previous location. It was also obvious that there could not have been an entrance to the *naos* with this slab as a doorstep, had there really been an upper structure related to these 'markers'. However, the outer borderline along the long central slab did continue onto the north-east corner stone and stopped at the plaster outline of the Square Pillar.

The Stratigraphy of Locus 111

As the stratigraphical interpretation of Locus 111 is necessarily a decisive part of the overall stratigraphical interpretation and phase analysis of the Temple site, we shall discuss here only the stratigraphical sequence of its Standing Strata (the *naos* structure, the niches in the Pillar face, etc.) in relation to the interfaces and layers of Locus 111. As mentioned above, we shall include here also some stratigraphic considerations concerning the *naos–pro-naos* relation.

As in the adjacent loci, Roman occupation of the site was indicated by local trampled surfaces, ashy patches and Roman sherds in the upper levels of the red-grey sand fill layer. However, the encountering in certain areas of Locus 111 (and Locus 110) of relatively many mixed groups of finds (Roman sherds and Egyptian New Kingdom related artifacts) already in these high levels and, especially, of Roman sherds in very low levels of the Temple stratification, indicated a major disturbance during the Roman period. This disturbance became evident as a sort of treasure hunt in the front parts of the *naos* (and in the *pro-naos* close to the *naos* foundations).

No traces of the olive green-grey interface were found inside the *naos*, the same as in the area of the *pro-naos*. We assumed that there was a functional reason for the absence of this interface inside the

Plate 97. Long connecting slab between corner-stones of east wall of *naos*.

Plate 98.   Markings on surface of eastern foundation stone.

central Temple region, especially in view of the fact that the olive-green grey interface was found in the squares around the *naos* and also lay against the outside of the northern *naos* wall. Although there was no solid White Floor interface inside the *naos*, a dense scatter of White Floor material, at a level (170) equivalent to the base-level of the White Floor layer adjacent to the *naos* and *pro-naos*, clearly indicated the White Floor horizon also inside the *naos*.

A thin layer of fine, dusty sand lay a on medium-coarse sand layer, which ended at a hard interface on a coarse-grained layer of decayed red bedrock. This sequence of sand layers, typical for the stratification of most of the Temple site, and also the vicinity of the rock-cut pits, was partly disturbed by Roman intruders. The shallow pits were filled and covered by brownish, dusty sand which also contained ashes, fireplaces, (Chalcolithic) pottery and flints. In the top levels of this brownish fill there were also New Kingdom artifacts, seemingly individual and mainly fragmentary stray finds out of context.

The niches in the face of the Pillar could not be stratigraphically related to the Temple. However, this relationship was strongly indicated by the location of the large niche exactly in the centre between the walls of the *naos*. It probably originally housed a standing votive feature. The same applied to the two niches higher up in the rock face, which were in line with the two pillar-bases at the east corners of the *naos* and appeared to have been sockets for the architraves of the roof.

Of the *naos* structure itself, only two building courses of distinctively different character were found, and it was evident that the *naos*, like most of the other features of the Temple, underwent a series of partial, seasonal repairs and even major rebuildings. For these reasons it was difficult to reconstruct the details of the sequential building phases of the *naos*. However, a number of stratigraphic relations and the following sequence, could be established.

The foundation course of white sandstone blocks as found *in situ* contained several elements in secondary use, i.e. the *naos* structure as found was not the first Egyptian structure at the site, though we have very little information about the shape, nature and exact location of a previous structure. The two pillar-bases at the east corners of the *naos* rested solidly on the lowest interface found in Locus 111, and no

foundation trenches were discovered to indicate their 'intrusion' into this low level. We assume that these pillar-bases were *in situ* and, together with the niches in the Pillar face, belonged to the first building phase of the *naos*.

The long flagstone between those two pillar-bases was clearly also in secondary use and not in its original location.

The rather different masons' techniques discernible on the foundation stones of the *naos*, clearly indicated that these building elements were not specially prepared for the *naos* structure as found *in situ*. Furthermore, the outside faces of those building blocks had been finely dressed – but lay hidden by the White Floor layer, which was put against the foundation course of the *naos* to its full height. Comparing those building stones with the stones of the *pro-naos* pavement, and also with the upper course of the *naos* itself, it seemed evident that this fine masonry was originally meant to be seen – in another structure and context. This stratigraphic conclusion should also be evaluated in the light of the fact that a large number of finely carved and decorated building elements were found dumped next to the *naos* (in Locus 107) and these could by no means be debris from the *naos* structure as found *in situ*.

The stratigraphic relation of the *naos* foundation to the White Floor is also revealing. The section under the south wall of the *naos* showed that its foundation stones lay on top of a red sandstone boulder (from a previous rock-fall in Chalcolithic context) and on a rock ledge protruding from the Pillar. The medium-coarse sand layer below these foundation stones was covered by fine, wind-carried sand which had penetrated into the empty space under and between the stones, after the foundations of the *naos* had been laid (on the slightly higher rocks). At this level, inside the *naos*, the White Floor related scatter indicated the White Floor horizon.

Outside the *naos*, the White Floor layer ran against the *naos* foundation and both appeared to have been laid during one and the same construction phase – most of the *naos* foundation stones were in secondary use and the White Floor material itself consisted evidently of crushed fragments of previously shaped white sandstone building elements. Stratigraphically, the *naos* foundations as found *in situ*, the *pro-naos* pavement and the White Floor, should be considered contemporary and belonging to a second building phase of the *naos* structure.

The upper, second course of building stones must be considered as belonging to a later, post-White Floor phase of the naos structure. The reasons for this reconstruction are: Beside the distinctive difference in the quality of the building technique and the building material between the lower and the upper course, there was the obvious change of function of the southern corner stone, connected with the use of lime plaster, which during this later phase could no longer be used as a pillar-base. It was during this phase that some of the roughly chiseled lines across the face of the flagstones at the front of the *naos* must have been cut, perhaps as builders' markers. However, some of those lines must have originated from a previous use in a different context. The stela in secondary use as a building block in the south wall of the *naos* does not really 'fit' the rough building stones of the upper course, but by itself it has to be considered out of its original context and, apart from being additional evidence for the existence of another, previous Temple structure or layout, was only of restricted stratigraphic significance.

On the inside of the *naos*, the upper course of the *naos* structure was contained in a thick layer of red sand fill without any interfaces (down to the White Floor level which was already at the base-level of its foundations) and there was therefore no indication of its Phase-relation or even purely stratigraphical situation.

Considering the repeated repair and rebuilding activities in evidence at many of the other standing strata of the Temple site, this late restructuring of the *naos* need not necessarily indicate a major changeover to a new chronological Phase or Stratum. However, the olive green-grey interface which lay against the outside of some of the stones of the upper course (in B 13) was a clear indication of the use of this structure during the olive green-grey Phase of the Temple, though this was not enough evidence for the actual rebuilding of the *naos* at this time. It should again be emphasised here that during the olive green-grey Phase of the Temple many architectural and votive elements of the original Temple – such as the Square Pillars which appeared to have stood on the corner stones of the *naos* – were removed from their original location, or from the places they had fallen, and re-used for different purposes, sometimes just as building stones (altars and offering tables built into a wall, etc.). As almost no building debris was found around the *naos* walls during the excavations, we assumed that the last users of the Temple – belonging to the olive green-grey horizon – had cleared away the fallen building elements of the original (or previously rebuilt) *naos* and used the remaining structure as a kind of 'sacred' enclosure without any major changes or repairs. This would be much in line with the very primitive secondary use of other structural elements during the olive green-grey phase of the Temple.

Below the *naos* structure a cluster of shallow pits, accompanied by Chalcolithic sherds, flints and

miscellanea cut into bedrock, indicated the pre-Temple occupation of the site by prehistoric people. Whether these Chalcolithic remains indicate the former existence of a prehistoric shrine at the site of the Egyptian New Kingdom Mining Temple, is a question beyond the frame of our stratigraphic deliberations.

12 Locus 112

Locus 112 comprised Squares A-C 15–17, i.e. it contains Wall 4, and about half the width of Wall 1 (in B-C 15)[115] (*Illus. 6*). As part of Wall 4 extended into Locus 101, the stratigraphic relation of Wall 4 to Wall 1 has already been dealt with above (p. 32). However, a number of problems, mainly related to building techniques, will be discussed in the following.

The Present Surface of Locus 112 was at approximately Level 118, sloping gradually towards the north. Locus 112 was excavated to bedrock.

In the uppermost 10cm. of the red-grey sand layer in A-B 15–16, i.e. almost on the Present Surface, a few finds were already recorded (Box 8, Level 118–128): Midianite, Egyptian, Negev-ware sherds, a faience fragment, beads. At approximately Level 138 the outlines of Walls 1 and 4 were visible in the red sand fill and it became apparent that there was a cell-like attachment to the outside of the Temple courtyard (*Pl. 99*). Going down in the inside of this cell, more finds showed up in the red sand (Box 95, Level 145): fragment of a faience leopard figurine; (Box 115, Level 138–158): menat with cartouche of Seti II (Eg. Cat. 26), a few Egyptian, Midianite, Negev-ware sherds, copper rings, beads. There were no interfaces discernible in the red sand fill until Level 171, where ashy patches indicated occupation in a brownish sand layer close to bedrock. Here some finds were made (Box 218, Level 164–171): Midianite, local wheelmade sherds and miscellanea.

Clearing the red sand fill along Wall 4 (in B 17) down to bedrock, some finds were made (Box 285,

Plate 99. Cell-like attachment to Temple wall becomes visible during excavation.

Level 164–170): faience fragment, 2 beads, Negev-ware (?), local handmade sherds, miscellanea including some flints. Again along Wall 4 in C 15–16, the red fill went down to bedrock, with a few finds close to the latter (Box 335, Level 168–180): 2 beads, miscellanea including flints and bits of slag. The same red sand fill above an ashy occupation layer, close to bedrock, was encountered in C-D 17 outside Wall 4. Here the following finds were recorded (Box 218, Level 164–171): Midianite and local wheelmade sherds and miscellanea including animal bones.

Reaching bedrock in B-C 16, a large oval pit, *c.* 18cm. deep (*Pl. 100*), was found cut into bedrock (Level 180). There was a distinctive brownish tint to the sand layer over the pit and its surroundings, and a few finds were recorded (above and inside the pit) (Box 336, Level 180–198): one sherd, some bits of copper ore, bones, charcoal.

With the exception of a few artifacts in the ashy patches on bedrock, which seemed to belong to the Chalcolithic horizon, all the other finds in the red fill layer at Locus 112 were unconnected with any feature or interface.

Wall 1 (in B-C 15)

When Wall 1 (in B-C 15) was uncovered (Level 123–175), it gave the impression of a very sloppy repair job made by just piling rough fieldstones against the back of the large, red rock on edge (in B-C 14) (*Pl. 101*). As we removed a number of boulders from the top of the wall, which seemed to be just rubble on sand, quite a few beads and bones were found in the sand fill. Apparently this red sand was taken from nearby and intentionally used as filler between the building stones. Almost the whole of the upper part of Wall 1 had to be removed before we reached a better built stone course and could establish the outlines of the wall.

Wall 1 stopped in B 15, at the west end of the large rock on edge. It could not be stratigraphically established that the wall originally continued up to the

Plate 101.   Top of Wall 1 in Locus 112..

rock face of the Pillar, but a single rock, which stuck to the rock face in a direct line with the wall, could indicate that such was the case (*Illus. 6*). Apparently, this part of the wall was removed in order to open a doorway into the 'cell' attached to Wall 1. Stratigraphically, the part of Wall 1 behind the large rock on edge lay on the same thick layer of fill found also underneath Wall 4 and appeared to have been built (or, better, repaired) together with Wall 4. It was therefore not surprising that at the junction an Egyptian Altar Stand was found built into the uppermost course of the wall (*Pl. 102*).

Plate 100.   Large pit cut into bedrock in B-C 15, Locus 112..

Plate 102.   Altar Stand built into Wall.

At the junction of Wall 4 and Wall 1, already described in Locus 101, the rough walling continued; in fact, the stone courses in this part of Wall 1 were just as coarse as rubble-wall 4.

Wall 4

Wall 4 was just a rubble fence, roughly built by piling up red fieldstones (*Pl.* 103). It lay against Wall 1 (not bonded) and went in a semi-circle to the rock face of the Pillar[116] (*Illus.* 6). Below the foundations of Wall 4 (in TT 1) lay three layers (from the bottom up): a hard, coarse-grained sand layer on bedrock with ashy patches, which was also found below Wall 1; above it was the *c.* 3–4cm. thick White Floor interface, which ran against the outer face of Wall 1; and above this interface lay a thicker layer of greyish-red sand. Wall 4 lay on top of this greyish-red fill layer (*Illus.* 11).

The Stratigraphy of Locus 112

Stratigraphically Wall 4 belonged to a post-White Floor rebuilding activity of the Temple compound. Although we did not find the olive green-grey interface in the 'cell', we relate these rebuilding activities, including the repair and/or rebuilding of Wall 1 in B-C 15 to this last phase of the Temple architecture.

The White Floor interface, which was observed below the repair to Wall 1 in D 14, seemed to have continued to D 15, below the eastern outer half of Wall 4, but it was not found anywhere inside the 'cell'. Perhaps this interface was disturbed by the builders of Wall 4 at the time of the removal of the western end of Wall 1, to clear a doorway from the Temple courtyard to the 'cell'.

The prehistoric debris and pit below the 'cell' were a continuation of very similar features below the *naos* and represent a pre-Temple Chalcolithic occupation at the site.[117]

A Stela on the Pillar's rock face

The Pillar formation, against which the Temple had been built, formed a small enclave around the Temple area which was partly excavated (*Illus.* 6 and 9). Right at the northern end of the excavation the Pillar formation showed a fissure through which it was possible to climb up to the top of the cliff (*Pl.* 104). In 1972, Alfonso Nussbaumer, a volunteer member of the Arabah Expedition, discovered a rock stela carved on a towering, smooth rock face (*Pl.* 105), *c.* 20m. above the Temple. A rough rock ledge, outcropping right below the stela, must have served as a platform for the ancient sculptor, as it served our team for the preparation of a silica copy of the stela.

Plate 104. The rock formation bearing the Ramesses III stela. The Temple is inside the fence.

Plate 103. Wall 4 built against Wall 1, just a rubble fence.

Plate 105. The rock stela (slightly deformed as photographed from below).

The Timna Temple at end of excavation (some of the Standing Stones were replaced in their assumed original position).

The rock stela was *c.* 90cm. high and *c.* 55cm. wide. It showed rather primitive artistic skills, though this may have been the result of the precarious position of the artist, who had to stand on the narrow rock ledge.

Details of the rock stela, its investigation and reading, will be given in Chapter III, 6.

Whilst there is of course no stratigraphic connection between the Temple and the rock stela 20m. above it, the stela must have been related to the Temple as the cult centre of the Egyptian activities in the Timna Valley. It also indicated the importance of the period of Ramesses III in the history of these activities and of the copper industry at Timna ('Atika'; Rothenberg, 1972, 201) during the New Kingdom.

Notes

1. Preliminary reports on the 1969 season were published by Rothenberg (1969; 1969a; 1970; 1970a; 1971; 1972; 1973; 1978).

2. On the 'New Timna Project', see above p. 8; Conrad-Rothenberg, 1980: 27–31; including the list of team members.

3. In August 1984, when this report was already in its final stage, Uzzi Avner, on behalf of the Eilot Regional Council and the Israeli Department of Antiquities, undertook consolidation work at the Temple and found in its lower layers a few faience fragments and small beads. Two objects deserve special mention: a fragmentary inscription with the hieroglyph of Hathor and a cartouche of Seti I on a faience bracelet. As this cartouche was found in an apparently clear stratigraphic context, it had to be taken into account in the discussion of the stratigraphic sequence of the Temple. (See below, Chapter IV; also Schulman, below, Chapter III, 6:19).

4. On the differentiation between natural, man-made and upstanding layers, see Harris, 1979: 36–42.

5. The plan of Site 200, prepared by the late Shmuel Moscovitz, Tel Aviv University, records the situation in 1970. The second season, in 1974, did not essentially alter the architectural plan, and the figures inside the courtyard refer to the level of the White Floor (1970) and not to the levels reached in 1974 in our Trial Trenches.

6. On correlation and phasing during or after excavation, see Kenyon, 1971: 272, and Alexander, 1970: 71–72.

7. The tabulation in two dimensions of stratigraphic sequences over a large area originally recorded in three dimensions must, of course, take into consideration the relative nature of the measured layer levels in cases of irregular topographic conditions. Hence at a site with sloping Present Surface the top 10cm. over the whole area will appear on the diagrams as a layer of great depth. Since this was the case at our site, the necessary corrections had to be calculated for each level measurement, based on the chart of the heights of the surface of the site before the commencement of the excavation.

8. K. Kenyon (1961: 129; see also 1971: 274) suggested the phasing of layer sequences as 'Working Periods' in the order of their excavation, i.e. from the top downwards, but then to reverse their order, i.e. from the earliest upwards, and relate to 'Final Periods'. Dealing with the excavation, irrespective of the historical

development of the site, we retained throughout the natural order of the excavation, i.e. from the top downwards. See Harris (1979) for discussion of the methodology of phasing archaeological sequences.

9. In the early seventies, when this work was done by Gloria London, we did not have computer facilities to introduce the very numerous corrections needed because of the horizontal inclination of the sloping interfaces (see Note 3). The tabulation of the distribution heights in these diagrams was therefore significant only in reference to the recorded relative height of the floors, which there were drawn as thick 'floors', i.e. indicating the whole range of floor levels between the uppermost and the lowest level of the particular floor, and the whole A-K coordinate in the numerical grid segment (e.g. along the numerical Grid 10 over the area A to K).

10. As explained above, all levels are given in centimetres, as negative heights relating to a 0-benchmark on the rock face behind the site.

11. The iron bracelets of hoard E 15–16 were found completely corroded and disintegrated into powder when touched. The bracelets had a diameter of 60mm. and the metal itself was almost 10mm. thick (see *Pl.* 6). The appearance of iron jewellery at the 14th–12th century BC Temple, i.e. in an early New Kingdom (Late Bronze Age) context is, of course, of particular interest. See related chapter 'The adventitious production of iron in the smelting of copper in Timna' in Vol. II of this publication; and *IAMS Newsletter* 6 (1984) 6–7.

12. All finds from this hoard are marked Box 51. A phallic figurine (48/1), though found a few centimetres above the hoard, probably belonged to the same layer, also Boxes 110, 160 and 173.

13. The ceramic finds in Locus 101 comprised all the various groups of pottery typical for the New Kingdom sites of Timna. See below, Chapter III, 3.

14. See below the chapters on textiles from Timna Site 200.

15. As 'Miscellaneous occupation debris', we understand typical small debris as found in/on most occupational interfaces in the Temple: charcoal bits, seeds, twigs, bark and wood fragments, ostrich egg-shells, sea shells, fossils, copper ore, manganese ore, gypsum bits, carnelian, rocks, copper prills, tiny slag bits, cloth fragments – all these often mixed, either with New Kingdom objects or, in the occupation layer at the very bottom, flint tools and debitage, sherds, etc., related to the Chalcolithic Period.

16. On the stratigraphic significance of Pharaonic cartouches in the Timna Temple, see below, Chapter IV.

17. See below, Schulman's discussion of this cartouche.

18. These finds are part of the widespread distribution of Chalcolithic remains underneath the Temple and all along the face of 'Solomon's Pillars'. See below, Chapter IV.

19. The architecture of the Temple as a whole will be discussed in the concluding chapter of this volume, based on the analyses of the different architectual elements as they appear in the loci of the excavation.

20. The section of Wall 2 bordering Locus 102 also contained the doorway to the Temple courtyard in Square I 11. This will be dealt with together with Locus 103 and its stratigraphy related to the Temple architecture in Locus 109.

21. As the walls of the Temple were in many parts rather rough and partly disintegrating, we preferred not to excavate down to bedrock, in order not to weaken the foundations of the walls.

22. Because of the sloping layers, we shall indicate here the Square where the heights were actually measured.

23. This type of pillar-shaped Offering Stand was first described by Flinders Petrie in the Egyptian Temple of Serabit el Khadim in Sinai, and he identified them as 'incense altars' (Petrie, 1906: 133–4, Fig. 142, 143). See below, Chapter III, 6:16, Schulman's discussion of these stands.

24. To make it possible to excavate this part of the site, we had to remove the huge rocks. By courtesy of Timna Mines Co., we lifted the rocks from the top of the site with as little damage as possible to the remains beneath, with the help of a front end loader.

25. See further discussion of Wall 3 in Locus 105 and 106 below.

26. It seems that the repeated rockfalls were caused by the collapse of a part of the huge rock shelf overhanging Site 200. The gap in this rock shelf is today still clearly visible (see *Pl.* 2).

27. The final plan of the excavation was drawn at the end of the excavation (see Note 5) and does not contain the removed debris

of Wall 3 in A-D 5. However, the continuation of the wall, as established during the excavation, is indicated by broken lines.

28. In the following, 'The Pillar' means the part of the 'Solomon's Pillars' formation against which Site 200 was erected.

29. The whole area of A-B 7–10 turned out to be covered by a very thick deposit of mainly white sandstone building debris, including sculptures in the round and reliefs as well as decorated and inscribed building fragments. This destruction deposit, stratigraphically underneath the olive green-grey levels, was not apparently touched when the olive green-grey interface and Standing Stones-related occupiers of the Temple site (called by us 'Midianites', cf. Rothenberg, 1970, 1972, 1983) put up their obviously rather temporary features in, or better, on the abandoned and mostly wrecked Hathor Temple. The sphinx head from the rubble and sand heap under the olive green-grey horizon must therefore be considered as originating from an earlier phase of the site – the Egyptian Hathor Temple proper.

30. The granite boulder was the only one of its kind used in ancient times in a Timna site. It probably originates from the nearby Timna Massif (see Rothenberg, 1972, Fig. 3) and could have been picked up in one of the wadi-beds near the Temple.

31. On ablution basins in the Hathor Temple of Serabit el Khadem, see Petrie, 1906: 105–7.

32. The White Floor was a solid mass of white sandstone pieces and dust in other parts of the Temple, but it became thinner in its southern parts. In the light of this overall situation, the thin spread of crushed white sandstone underneath the basin fits the stratigraphic situation of Squares E 6–7 very well.

33. Several other Hathor pillar fragments were found at Site 200, unfortunately not in context, and all had the face of Hathor flattened. One of these pillars was found at some distance from the mound and was used as a convenient table stand by the excavation team – only at the end of the excavation, when all stones in the vicinity of the Temple were carefully scrutinized, was the table stand identified as an additional Hathor Pillar.

34. We replaced the Stand onto its base as shown on the final plan of the excavation.

35. Box 221 contained finds from E-F 7–8, i.e. also from the neighbouring Locus 107.

36. The sequence and nature of the destructions, especially of the southern part of the site, will be discussed at the end of the detailed excavation report, when an overall picture will become available (see Chapter IV).

37. Box 283, Level 179–185, also contained finds from the same interface in neighbouring Squares D 8–9.

38. This 'red pavement' will be further discussed in Locus 107 because only a small area of this feature is actually located in Locus 106. However, TT55 (see below) in Locus 106, provided important stratigraphic details (see *Illus.* 12).

39. The 'red pavement' was not excavated in the first season of the excavation in 1969 because it was then intended to preserve as much as possible of the Hathor Temple. However, in 1974 it was found necessary to remove most of the White Floor, and also some part of the 'red pavement' in order to clarify the stratigraphy of the site, especially of its early phases. TT55 was actually partly in Locus 107, but as the Section (12–12) is located in Locus 106, it will be discussed in the present chapter.

40. In 1969, G-H 6–9 was only partially excavated to virgin subsoil but a *c.* 120cm. wide trench, TT16, was dug to bedrock from Gridline G 6–7 up to TT15 in H 6, and right against the face of Wall 3. This detailed work helped a great deal towards the sorting out of the rather complex stratigraphical problems of Locus 106. TT57, excavated in 1974, confirmed the results of 1969.

41. As the White Floor was rather flimsy in some parts of the southern part of the Temple courtyard, often no more than a thin scatter of crushed white sandstone lumps over and in soft red sand, it is possible that in the early stages of the excavations in 1969 this was not recognised as being a feature on its own. These squares were particularly difficult to excavate because the baulk along G ran right across it and there was only a narrow trench between it and Wall 2.

42. TT15 and TT16 were excavated in 1969, but the bench itself in Squareline H-I 7–12, was left *in situ* until 1974, when most of it was excavated.

43. Clear evidence for earthquake damage in Timna was found at New Kingdom sites 2 and 30, and at mining site 9.

44. It would probably be more accurate to speak about several rockfalls because the history of the south-western part of the site was dominated by a series of small and large rockfalls, some of which (but by no means all of them) could be stratigraphically isolated. The extent of these rockfalls can be judged by the huge gaps in the rock ledge overhanging the site.

45. No Roman traces were found in Locus 106.

46. The cause of the destruction of the Hathor Temple – whether by earthquake or wilful destruction after its abandonment by the Egyptians in the middle of the 12th century B.C. – will be discussed below.

47. As rockfalls at this part of the site must have been a repeated phenomenon, it is of course possible that not all of the boulders found at the bottom of the site's strata fell at the same time. However, our trial trenches (see below) showed that most of the sand layers between and below the boulders consisted of finely straticulated wind-or-water-carried fill. Occasionally there was some differentiation of grain sizes in this fill layer (see below, p. 54) which gave the impression of different layers, but this was caused by sedimentary processes or decaying sandstone.

48. The *naos* structure as a whole will be dealt with in detail in Locus 110–111, except for the stratigraphic implications of the rockfall underneath the southern wall of the *pre-naos*, which will be discussed in the present chapter.

49. Because of the different stratigraphic conditions of major significance, caused mainly by rockfall of various dates, our excavation report on Locus 107 will present its western half (A-C) apart from its eastern half (D-F) .

50. See below the detailed record of architectural elements, stone sculptures and other artifacts from A-C 8–10, also *Figs. 22–26*. See Schulman in III, 6.

51. In this part of Locus 107, it was often very difficult to differentiate between the thick mass of broken and disintegrating white sandstone debris and the White Floor interface, especially since the White Floor interface in the southern part of the Temple courtyard was much less solid than in its northern parts. We therefore recorded the White Floor interface only where a continuous stretch of ground was covered by a clear layer of crushed white sandstone. Luckily, such an interface was usually accompanied by numerous finds of artifacts and debris, indicating a 'habitation' horizon.

52. The differences in the levels of the interfaces between Square-lines A-C and D-F were caused by the general tendency of the site layers to slope gradually towards the East.

53. The beautiful Hathor mask was found in two separate parts: one part at Level 150 on top of the white stone pavement of the *pronaos*, at the very edge of L 10, and the second at Level 153 in L 9 (Box 251) on the White Floor.

54. Box 277 contained finds from two horizons of find concentrations in C-D 10–11, the interfaces of which could not be differentiated because of the massive debris in this area. Comparison with adjacent D 9 and E 8–10, indicated the probable relation of these 'horizons' to the olive green-grey and White Floor interfaces.

55. In Section 15–15, a thin layer of fill seemed to separate the basin from the large red sandstone boulder, but this was only fine wind-blown sand which filled the cavities between the White Floor and the irregularly shaped bottom of the damaged basin.

56. See Gourdin, W.H. and Kingery, W.D., below, *J.F.A.*, 2, 1975, 133–50 and Note 101.

57. In 1969, the first season of work at Site 200 stopped at the level of the White Floor (see above, note 39) and the 'red sandstone interface' (the southern half of the site). During the second season of work in 1974, we investigated the stratification below these interfaces by a series of trial trenches in D-G 9–10.

58. See discussion of the stratigraphic significance of these and similar stray finds in the lowest fill layers of the Temple site in Chapter IV.

59. This situation led at the time to a mistaken identification of the casting installation in Locus 109 as Nabataean-Roman (Rothenberg, 1972: 177–179).

60. The White Floor interface was not everywhere a proper floor, or even a thick spread of White Floor material. In the southern part of the Temple courtyard the White Floor as found was sometimes no more than a scatter of typical White Floor crushed sandstone pieces. However, the overall appearance of this interface, its levels and nature, allowed the establishment of this interface over almost the whole Temple courtyard (see Notes 41, 51).

61. It was for this reason that the 'red pavement' was not removed in the first season of work, in order to preserve as much as possible of the Temple structure. After working through the documentation and records of the first season, it seemed likely that the 'red pavement' was not a 'floor' and in the second season (1974) this was very thoroughly investigated.

62. TT60a was an extension of TT60 (61d an extension of 61c) and both parts are here reported together. TT60 and 61 joined up with TT55 in Locus 106 and the findings are in fact complimentary.

63. Most of the White Floor interface (at Level 185) was removed in 1969 to approximately Level 200, where the excavation was stopped at the 'red pavement'.

64. Find Box 244 also contained finds from H 6–7.

65. Find Box 319 also contained some finds from G 7.

66. The excavation from the White Floor to bedrock (TT 52) was made in 1974 and the Find Boxes 510, 511 and 515 therefore also contain finds from G 10–11.

67. This orangey tint was typical of the bottom layer of many of our excavations in Timna. See Note 70.

68. In 1974, the entire length of the bench along Wall 2, from Wall 3 to the doorway in Wall 2 (see Locus 109) was excavated (TT66). The stratigraphy all along the bench was found to be the same.

69. See above, Note 41.

70. The orangey tint of this layer, not recorded previously in the Temple excavation (recorded in TT52 in 1974), was caused by the admixture of yellow loess with the disintegrated NSP bedrock of this area of Timna. It was common in the excavation of Site 30 and indicated the bottom layer of the stratification, i.e. the loess-covered virgin subsoil.

71. There is, of course, the simple explanation that these stray finds filtered down through the soft sand layer into the Chalcolithic bottom layer, though we must also consider the possibility that the red sand layer below the White Floor was deliberately put there in a levelling operation before the laying of the White Floor. Incidentally, the dusty layer right below the White Floor, found in most Squares of the excavation, could well be caused by the admixtures of fine white sandstone dust with fine loess and wind carried fine red sand below.

72. Squares F 11–13 were left intact at the level of the White Floor, in order to preserve the Crucible Furnace in F 12 for further investigation. A part of H 10–11, right inside the doorway, was also left unexcavated at the level of the White Floor.

73. Crucible slag is formed in a melting crucible during the melting of metal for casting, in contrast to smelting slag formed in a smelting furnace during the smelting of ore to metal.

74. The doorway appeared early in the excavations and was used during the excavations as the passage out of the Temple courtyard, causing the upper layers of Square 11 to be damaged.

75. To avoid repetitions, the detailed description of the various and varying layers underneath the White Floor, will be given with the description of the Trial Trenches of 1974. The special features in Square-lines F and G are described below in a separate paragraph.

76. The area of TT51 and TT51a, extending over G-H 11–14, was successively excavated in small parts but from the White Floor interface downwards it should be considered as one area of similar stratification and nature.

77. Quite a number of such finely dressed building stones, even part of a Hathor pillar, were found either on the Present Surface of the site or even just lying on the wadi surface at some distance from the Temple mound. It is obvious that long after the Temple was abandoned some of its building elements were removed from their original location, although from their secondary location it was impossible to tell the purpose of this operation.

78. This sherd, found close to bedrock, was too small and shapeless to be defined. However, its ware was very similar to some Chalcolithic sherds found in the bottom layer of the site.

79. The casting installations in Locus 109 were previously assumed to be Nabataean-Roman (Rothenberg, 1972: 177–179) because a few sherds of this period had been found in the ash layer related to these installations (at Level 184). Subsequent excavations (1974) and radiocarbon dating (BM 1117 – 2779 + 55 B.P., *Radio-*

carbon, Vol. 21, 349) provided evidence for their New Kingdom date.

80. The problem of the New Kingdom finds below the very compact White Floor in Locus 109 will be discussed in Chapter IV.

81. The extent of Locus 110 and 111 was fixed to include complete features as architectural units (*pro-naos, naos*, Wall 5, etc.). It is for this reason that two halves of some squares belong to different loci. Consequently some of the find boxes, especially from the top layers, contained finds from both sides of the loci border lines. However, the more important finds, recorded in detail, could be correctly placed in the final finds records.

In the following, we shall not constantly mention the fact that the locus borders ran through the middle of some of the squares, a fact obvious from the excavation plans (*Illus.* 6 and 10a-b).

82. At the east end of Locus 110, one or two stones of the top course of Wall 1 protruded out of the Present Surface (Level 123) and were visible among the surface scatter of rubble.

83. Squares A-B 13–15 presented problems which we were unable to clear up completely. Whilst we were setting up the excavation grid, some of the volunteer workers were asked to remove a thick mass of recent camping rubbish which had accumulated along the rock face of the Pillar. Unfortunately, this was done without proper control by the Square Supervisor and some of the occupation layers and the flimsy rest of Wall 4 were removed. Although the verbal explanations by the workers involved did answer some of the questions, such as the west end of Wall 1, our understanding of this area remains somewhat tentative.

84. The copper snake found by R.T. Tylecote, who participated for a short time in the Temple excavation in 1969, was originally reported to have come from inside the *naos* (Rothenberg, 1972, 173). Whilst investigating the problems referred to in footnote 83, we succeeded in accurately pinpointing the find spot of this interesting copper serpent: it was found outside the *naos* in B 14, *c.* 80cm. from the *naos* wall and definitely in the olive green-grey context.

85. The presence of casting remains in the ashy layer of B-D 13–14, related to the olive green-grey horizon, indicated casting activities in the Temple phase related to the olive green-grey interface. This conclusion, was backed up by the fact that the ashy layer was separated from the White Floor interface by a thin layer of reddish-grey, wind-carried sand (in B-D 13–14, see below), and also by other evidence for metallurgical activities in the olive green-grey related horizon of the Temple.

86. The plaster fragments which frequently turned up in the olive green-grey horizon in the area adjacent to the *naos*, provided clear additional stratigraphic connections to the last construction phase of the *naos* (see below, pp. 78–80).

87. We have already commented above on the Roman sherds in the deeper levels of Loci 110 and 111, and we shall return to this phenomenon in our summary of the stratigraphy of the Temple (see below, Chapter IV).

88. These finds were collected during the clearing of the building debris, before the White Floor was recognised and were therefore included in the Find Boxes containing the material from the higher levels (Boxes 228, 237, etc.).

89. When investigating the area around the base of this boulder and clearing the area to the west of it (TT 11) the boulder, which had become insecure because of the previous weight against it, fell into TT 11 (*Pl.* 75).

90. In Rothenberg, 1972, 178, Wall 5, together with the casting installations and ashes in the Temple courtyard and the store of iron and metal objects in the corner of Wall 5 and Wall 1, were understood as a metallurgical workshop unit and related to the Nabataean-Roman occupation (Phase I) of the site. In accordance with the change of date for the casting installation (see above, Chapter III, note 79) and its stratigraphic context, it is now related to Phase II.

91. The dressed boulder had two shallow, bowl-like 'cupmarks' in its upper surface, unlike the other Offering Tables found in the Temple. We had, however, no other suggestions to make for its possible original use in the Temple.

92. The White Floor interface generally sloped towards the south-east with Level 158 in B 15 and 190 in H 15, but in Square-Line C-D 13–15 it was particularly noticeable (Levels 164–165).

93. The particularly detailed record of even obvious stray finds

from below the White Floor is meant to provide the actual data for the stratigraphic considerations of the pre-White Floor horizon(s) of the Temple site.

94. As mentioned above, such a fine dusty interface was often noticed in the excavation, always directly below the White Floor. We suggested that it may have been formed by seepage of white dust out of the White Floor mass into the sand layer below, in which case it should stratigraphically be related to the medium-grained red sand layer below the White Floor. However, we did also find a similar fine, dusty layer of greater thickness, up to 8–10cm. (see main North-South Section, *Illus.* 9), but this was also directly below the White Floor.

95. Wall 1 was already discussed in Locus 101 and Locus 109. However, as the stratigraphy of Locus 110 is obviously also related to part of Wall 1, these aspects are discussed here.

96. In fact, the White Floor continued also in Locus 101, at the other side of Wall 1 (see above, p. 32).

97. This pre-White Floor phase interpretation is based on the fact that the foundation trench at the base of the huge rock was refilled with greyish-red sand and not with white crushed sand-stone. Had the trench been made together with the White Floor, this fill would obviously have been of White Floor material.

98. This wall structure behind the huge rock showed a different stratigraphy, to be discussed in Locus 112.

99. We excavated the baulk in Square-Line B-E 12 early in the excavation (after drawing the main North-West section) because it had run right through the centre of this most essential part of the excavation in Loci 110 and 112. We report the findings and finds from the baulk together with the adjacent squares.

100. As Roman sherds were found in and below this pile of stones, we concluded (above, p. 67) that part of the pavement in front of the *naos* – the structure of the *pre-naos* (see below) – was removed during the Roman occupation of the abandoned Temple structure and assumed that this happened in the course of Roman 'treasure hunting' in the Temple, clear traces of which were also found within the *naos* (Locus 111).

101. Samples of the plaster from the *naos* were investigated by W.H. Gourdin and W.D. Kingery (*J.F.A.*, 1975) who established that it was lime plaster. The authors pointed out that according to Lucas (1962, 78) lime plaster was not used in Egypt prior to the Ptolemaic period and they therefore proposed to revise this view in the light of the Timna find. However, the above authors were not aware of the fact that this plaster may have been used by other, non-Egyptian occupants of the Egyptian Temple (e.g. the Midianites from north-west Arabia).

102. In Rothenberg, 1972, 126, this hard interface was related to an interface at a similar level inside the *naos*. Whilst there exists, in fact, a similarity between both these interfaces and the interfaces below, the interpretation of the stratigraphy in 1972 has been somewhat changed after further excavations in 1974.

103. The stratigraphic relations between the *pro-naos* (Locus 110) and the *naos* (Locus 111) will be discussed below, at the end of the excavation report on Locus 111 (Chapter III, 11).

104. No Roman sherds were actually found in the disturbed sand layers below the White Floor level, but this fact is not of stratigraphic significance especially as in the similarly disturbed layers in the adjacent *naos* Roman pottery, including a large flask fragment, was indeed found almost at the bottom of the dug-up area (see below, Locus 111) .

105. We intentionally did not include in this list the finds of artifacts from levels below the White Floor interface, because the stratigraphic significance of these finds will have to be discussed together for all the Temple finds (see below, Chapter IV).

106. In the following report on Locus 111 we shall refer to the western half of A-B 13 (inside the *naos*, as well as to the eastern wall of the *naos*) as 'A-B 13'.

The baulk in A-B 12 was taken down early in the excavation of the *naos* and a trench (TT 13) dug alongside to make it possible to draw this part of the main N-S Section of the site (*Illus.* 9). Later on the *naos* was completely cleared (1974) .

107. We refer to the remains of the structure found in Locus 111, as the *naos*, the 'Holy of Holies' of the Temple, because of its central position in the Temple, the special characteristics of its remains as found *in situ*, and the niches in the rockface against which the *naos* was erected. As we do not have any archaeological

evidence for the reconstruction of the *naos*, nor any other central structure in the Temple, which would take into account the many architectural elements and cult objects found in a large heap next to the *naos* and astray all over the area of Site 200, we relate in the following only to the structural remains found *in situ*.

108. In the first season (1969), TT 58A was only excavated down to the White Floor interface. In 1974 this area was found badly damaged by visitors and only the very bottom layer was found intact.

109. The solitary Roman sherds found in the layers of TT 58B below the hard White Floor interface must have infiltrated from the Roman 'treasure hunting' hole in adjacent TT 13, and are not to be considered of any stratigraphic significance. No Roman finds were made in the equivalent, undisturbed layers anywhere else at the site.

The levels in TT 58B were sloping eastwards, the same as the bedrock underneath.

110. In 1969, some parts of the lower layers of B 12–13 were left unexcavated. In 1974 the *naos* was completely cleared down to bedrock (TT 58).

111. Because of its complexity, it was difficult to define the *naos* structure as found in the excavation and we refer to it in the following as the *naos* 'wall', though in its final phase it may have been just a two-course high enclosure.

112. This corner was not further cleared in order to preserve the typical elements of the *naos*.

113. Contrary to the corner stone at the south-east corner, which was the real corner of the foundation of the *naos*, the square block at the north-east corner lay against a long foundation stone belonging to the north wall of the *naos*. We therefore refer to it as an 'inner corner stone'.

114. On *Pl.* 97 the plastered building stone seems to stop in front, according to the borderline, but this is not as found *in situ*. The front part of this stone was cut back during the excavation to try to find the continuation of this line under the plaster.

115. This somewhat unusual partition was made in order to discuss together the relevant sections of interrelated architectural elements, i.e. the outer face of Wall 1 is the inside of the cell attached to the Temple courtyard.

116. At its end, close to the Pillar, Wall 4 was just rubble and was removed as such by the excavators. Only at a later stage was it realized that those loose stones in red sand were in fact a rubble wall. For this reason, on the plan of the excavation the end of the wall is only indicated.

117. Chalcolithic debris, flint and pottery, as well as votive installations, were found during our survey along the whole length of the King Solomon's Pillars' formation. As no other site in this area was excavated, we could not say whether there are other clusters of pits, fireplaces, etc. among those sites. (See Vol. III of this publication).

III. The Finds

A. STRATUM V

1. THE EARLIEST POTTERY

Only a small number of sherds were recovered from the earliest strata (V). There were two rim types, one with plastic cord decoration (*Fig.* 1:4) and the other with a prominent ridge on the body (*Fig.* 1:3).

The first type is known in Israel from the Chalcolithic Age to the Early Bronze Age (Koeppel *et al.*, 1940, Pl. 87). At Arad it is also common in the Early Bronze I period (Amiran 1978, Pl. 8:13–23). The second type is known in Sinai from the material of the Arabah Expedition. At Site 676 (Isr. Gr. R. 9857 8765) one sherd, found together with Elatian flint tools, belonged to the first rim type. At Site 567 (Isr. Gr. R. 9454 9324) another sherd was discovered which was related to Timnian flints (Kozloff, 1974, 47, Pl. VI).

In conclusion, the sherds found in Strata V can be dated to the Sinai-Arabah Copper Age – Early Phase (Chalcolithic – Early Bronze I) (see above, Introduction, par. 7).

Bibliography

Amiran, R. 1978. *Early Arad, the Chalcolithic and Early Bronze City,* Jerusalem.
Koeppel *et al.* 1940. *Teleilat Ghassul II,* Rome.
Rothenberg, B. (In preparation). *New Researches in Sinai.*

2. THE LITHIC OBJECTS

The lithic material from Stratum V belongs to the Timnian Lithic Industry of Sinai and the Arabah, which is a flake industry characterised by the diminutive size of flakes and tools, and the high ratio of waste to finished tools (B. Kozloff, 1974).

Table 1. Typology of the Lithic Finds

	No. of Tools	%
End-scraper, simple flake	1	2.27
End-scraper, on blade	1	2.27
End-scraper, on retouched flake	1	2.27
End-scraper, nosed, thin	1	2.27
End-scraper, core (Timnian)	3	6.81
Side-scraper, concave	1	2.27
Side-scraper, transversal convex	2	4.55
Side-scraper, on flake	1	2.27
Side-scraper, on blade	1	2.27
Notches	4	9.09
Denticulates	2	4.55
Burin, dihedral simple	1	2.27
Hump-backed bladelet	1	2.27
Borers	7	15.90
Retouched pieces	11	25.00
Utilised blade	1	2.27
Blades, retouched one side	3	6.81
Varia	2	4.55
	44	

Description of tools:

End-scraper, simple flake (1): This scraper is on a small thin flake from which the platform is broken off. The retouch is confined to the distal end and consists of small semi-steep retouch.

End-scraper, on blade (1) (*Fig.* 2:2): This scraper is on a broken natural backed blade. The retouch is abrupt. One portion of the scraping end has its cortex left almost intact, except for some very small nibbling which may be from use. The natural backing is partly retouched with abrupt retouch.

End-scraper, on retouched blade (1) (*Fig.* 3:2): The scraping end is retouched with semi-steep retouch and is partly broken. One side of the blade is naturally backed and the other is backed with abrupt retouch.

End-scraper, nosed, thin (1): The scraper is on a small flake with an unprepared platform. The nose is 6.1mm. wide. One side of the nose is formed by a single blow notch with the other formed by a semi-steep retouch notch. The sides are retouched with flat entral and dorsal retouch on alternate sides.

End-scraper, core (Timnian) (3) (*Fig.* 2:4, 6): In previous reports, this type of core scraper has been called a Timnian scraper (Kozloff, 1974), because of some unique features which it displays. They are usually on thick core-like scrapers or thick crest blades, have a definite curve to the scraping edge, are nibbled along the scraping edge (possibly from use) and the scarping edge is usually perpendicular to the ventral side of the tool. One unique example from this site (*Fig.* 2:4) is triangular in cross-section and is retouched along the junction of the sides, forming a straight scraping-like edge.

Side-scraper, concave (1): This scraper is on a broken flake and is retouched with a flat step-like retouch (*Fig.* 2:5).

Side-scraper, transversal (2) (*Fig.* 2:8): This scraper appears on one of the largest flakes from the collection. The retouch is flat.

Side-scraper, on blade (1): One of the largest blades in the collection.

Notches (3): Two of the notches are on flint, both of which have plain platforms, the third is on a quartz flake and has a faceted platform. All the notches are made by retouch technique.

Denticulates (2): One denticulate is on a small flake and the denticulation is formed by a series of single blow notches. The other denticulate is on a thick (23mm.) elongated flake with a plain platform. Both sides are retouched with irregular abrupt retouch.

Burin, dihedral simple (1): This burin is on a core tablet. The single burin spall is removed from a flat distally retouched side.

Hump-backed bladelet (1) (*Fig.* 2:1): A typical example of a broken bladelet. The side opposite the backing shows signs of utilisation.

Borers (7): Three of the borers are alternately retouched; one is on a blade (*Fig.* 2:3) and the other two are on broken flakes. Five borers are retouched with abrupt retouch. The two remaining borers are on flakes and have small boring points formed by notching on either side of the boring point (*Fig.* 2:7). One of these has notches made by single blow technique. The other example has notches ventrally retouched with semi-steep retouch.

Utilised blade (1): This blade is broken on the distal and proximal ends. One side shows signs of use and is quite worn.

Blades, retouched on one side (3): One of the blades has a preserved plain platform. It is retouched along one side with a semi-steep retouch.

Varia (2) (*Fig.* 3:1): The working edge of one item is bifacially retouched and the side opposite the working edge is also bifacially retouched but dulled for hafting (knife). Similar bifacially retouched items have been found at other Timnian sites. The second item is continuously retouched with alternate retouch. Three-quarters of the flake is retouched with flat dorsal retouch, while the remaining quarter is retouched with flat ventral retouch.

Flakes, Blades, Cores, Varia:

Flakes (322): 304 of the flakes tended to be small, seldom exceeding 35mm. in length. Of the flakes with preserved platforms, the majority have plain platforms. Most of the flakes are of flint with a few quartz examples.

Blades (11): All of the blades are irregular, i.e. with non-parallel sides. All are made on flint.

Chips and chunks (169): 119 of the chips and chunks were found in Stratum V. A flake which is less than 15mm. in length is defined as a chip.

Crested blade (1): One crested blade was also recovered.

Cores (27): All of the cores are quite small and well worked. Twelve of them do not exhibit any pattern of flake removal and are classed as undefined. Three of the cores are globular in shape. Four are single platform flake cores and the remaining core is a single platform blade core. One unique core is shaped like a pick, but does not exhibit any signs of use (*Fig.* 3:3).

Hammerstones (3): Three of the items are irregularly shaped chunks of flint which show utilisation as hammerstones.

Macehead (1) (*Fig.* 1:5): One macehead with a biconical hole was also recovered.

Conclusions

The lithic objects from Stratum V belong to the Timnian Lithic Industry (see Kozloff, 1974, and also Vol. 3 of this publication). However, because of the small number of lithic objects found at Site 200, comparison with other sites is somewhat difficult. Site 649 in Sinai (Isr. Gr. R. 1016 9009), 30 km. southwest of Site 200, produced the largest sample of any Timnian site in the area. Here 328 tools were systematically collected. Of these, end-scrapers (13.8%), notches and denticulates (20.8%), borers (13.1%) and scrapers on tabular flint (12.1%) dominate. Although no tabular flint scrapers were recovered from Site 200, a few fragments of tabular flint flakes were found. The lack of tabular flint scrapers at Site 200 may be due to the small number of lithic finds. The end-scrapers at Site 649 are dominated by small simple end-scrapers on flakes and shouldered end-scrapers. Two core scrapers (Timnian) from Site 200 (*Fig.* 2:4, 6) are very similar to those from Site 649.

Site 39 in Timna was also defined as Timnian (Bercovici, 1978). Unlike that site, Site 200 does not have an axe component which may, however, be explained by the metallurgical activities carried out at Site 39 (copper smelting and processing). The other lithic material from Site 39 is similar to that from Site 200. Both are dominated by a diminutive flake industry. The core-scrapers at each site are identical. The boring tools are also similar. Site 39 was dated to the late Chalcolithic era (Bercovici 1978, 18).

Bibliography

Bercovici, A. 1978. Flint Implements from Timna Site 39, in Rothenberg, B., Tylecote, R. and Boydell, P.J., 1978, 16–20.

Kozloff, B. 1974. A Brief Note on the Lithic Industries of Sinai, *YMH*, 15–16, 35–49.

B. STRATA IV–II

3. THE POTTERY GROUPS

The pottery found in Strata IV-II, can be divided into five groups:[1] 1) Midianite (25%); 2) Local rough handmade, including Negev-ware (10%); 3) Local wheel-made; 4) Egyptian; 5) Painted, Egyptian and local (3–5 c. 65%). These groups were found together in all strata of the site, except in Stratum V (the Chalcolithic habitation phase). With the exception of the painted pottery, these groups are also known from all contemporary sites in the Timna Valley, but at the Temple site a number of more refined vessels were found.

1 Midianite Pottery (*Figs.* 4–13)

The provenance and distribution of this pottery has previously been discussed (Rothenberg, 1972, 162–3; Rothenberg and Glass, 1983). Since Rothenberg (1970, 1971) pointed out its connection with the area of Hejaz = Midian, the term 'Midianite pottery' has been widely accepted (Parr, Harding and Dayton, 1970; Aharoni, 1978, 124–6). The present report covers the various types of Midianite pottery, their decorative motifs and the relation between types and motifs.

Vessel Types

Small Bowls (*Figs.* 4–6)
Found in a range of sizes, the small bowls share characteristics such as flat bases, almost vertical sides, and the presence of decorations.

(a) The vertical side meets the base at right angles; the rim is flat (*Fig.* 4:2).
(b) The side flares slightly and the joint at the flat base is rounded (*Figs.* 4:1, 3, 5, 7; 5:10, 16, 17).
(c) The side flares outward and its joint at the base is rounded (*Fig.* 4:4).
(d) A deeper bowl, its wall slightly flaring; the joint at the base is rounded (*Figs.* 4:6; 5:12, 14).
(e) Composite curved side; the joint at the flat base is rounded (*Figs.* 4:8–13; 5:1, 4, 13, 15). One vessel has loop handles (*Fig.* 4:9), another has a vertical line suggesting a handle (*Fig.* 4:12) since all specimens with handles have such a vertical line.
(f) The straight wall tapers, then straightens towards the rim (*Fig.* 5:11).
(g) A rounded rim turns outwards (*Figs.* 5:5, 18).
(h) Some broken bases with centre decorations may belong to the small bowls (*Figs.* 6:11, 16).

Goblets (*Figs.* 6:20, 21, 22)
An almost straight wall and a flat base (Fig. 6:1). Some bases as well as body sherds seem to belong to this type (Fig. 6:3–5, 9, 10, 12).

Jugs (*Figs.* 6:17–19)
The jugs have a spherical body and a narrow flat base. A handle was probably drawn from the rim to the shoulder (see Timna Site 2: 199/1).[2]

Juglets (*Figs.* 7:3, 4, 6; 8:1, 2; 9:1, 2, 3, 5, 6)
The base is flat and the body is either round or pyriform. The handle is drawn in a perfect curve from the shoulder to the cylindrical neck.

Mug (*Fig.* 10; *Pl.* 106:1)
The body is concave, with rim and base of equal diameter. There are protrusions from the base and rim connected by two parallel, tubular handles.

Varia
(a) A large vessel with a thick side and wide mount (Fig. 9:8). This is probably a large bowl (Kalsbeek and London, 1978, Fig. 2b, p. 52).

(b) A disc-base (*Figs.* 6:13–15) probably of a jug-type vessel.

The decorations (Figs. 11–13)

Table 1

A –	Lines	*Fig.* 11 A1, 5, 8
B –	Crosses	*Fig.* 11 B1–2
C –	Nets	*Fig.* 11 C1–2
D –	Chevrons	*Fig.* 11 B1
E –	Triangles	*Fig.* 12 E1
F –	Lozenges	*Fig.* 12 F1
G –	Zigzags	*Fig.* 12 G1–5
H –	Arches	*Fig.* 12 H1
I –	Joined semi-circles	*Fig.* 12 I1–2
J –	Wavy lines	*Fig.* 12 J1
K –	Simple dots	*Fig.* 13 K1–2
	Dot-centred circles	*Fig.* 13 K3–4
L –	Scrolls	*Fig.* 13 L1–2
M –	Independent motifs	*Fig.* 13 M1–6

All vessels are covered with a thick layer of cream-coloured slip which in some cases was also burnished. The decorations, in brown, black and red-brown, are usually in darker colours than the slip. The decorations were hand-made, some perhaps with the aid of a slow wheel, especially the bands around the vessels (Kalsbeek and London, 1978). With the exception of the small bowls, which are often decorated inside and outside, all Midianite vessels are decorated only on the outside.

The motifs fall into three categories: geometric designs, birds and human figures.

Geometric motifs are very common on the Midianite vessels. They may be divided into thirteen basic motifs which appear together in various combinations. Such a combination usually has a central motif and additions that serve as fill or frame. On *Table 1* (see *Figs.* 11–13) the motifs are listed in order of complexity.

The bird motif (*Fig.* 13:II), of which only one complete example was found (*Fig.* 7:4), appears on several vessels. It was drawn in brown (*Figs.* 7:4, 5; 8:1; 10) or in red (*Figs.* 7:6; 8:2; 9:8). On three vessels the bird's head is drawn as a dark circle with a dot in the middle and against a light background (*Figs.* 7:4, 6; 10). The beak is either long (*Fig.* 7:4) or short (*Fig.* 10). The body is painted solid, with an 'eye' in the middle. The 'eye' in the brown bird is small and rhomboidal (*Fig.* 7:4), in the red bird the 'eye' is large and oval, and fills almost the whole body (*Fig.* 7:6). The oval 'eye' consists of an oval red band with a small line in the centre, and has an outer frame made by a row of black dots. Another 'eye' is drawn in the same way but is rhomboidal (*Fig.* 8:2).

In the drawing of the wings and tail there is also a difference between the brown and the red birds, although both have long wings, and their tail feathers fan out from one point. The feathers of the brown bird are parallel bands with a row of dots along the top of the wings and body. In contrast, each of the feathers of the red bird, including its tail, ends in a brown dot. The legs, similar in both the brown and the red bird, are bent and the claw is cleft. The birds appear in metopes of geometric motifs and there is a cross with dots in front of each bird (e.g. *Fig.* 7:4).

A human figure appears on two sherds (*Figs. 7:1, 2*): the upper part of a human figure appears in *Fig. 7:2*, while only a head appears in *Fig. 7:1*. The head is schematic and drawn very similarly on both specimens, a fact that points to the two sherds probably being part of the same vessel. The head is constructed of three concentric circles and the hair indicated by short lines. There are two longer lines starting from the head and going down to the shoulders. The body and arms are drawn as black bands; the fingers are large in proportion to the body. The arms are bent at the elbow and the forearms are raised. The figure is drawn in black on a light background.

Motifs and vessel types

The geometric motifs appear on all the vessel types. The small bowls, goblets and jugs have only geometric motifs. Some common features can be distinguished:

(a) The geometric motif is arranged as a frieze around the bowl, outside or inside.
(b) All the small bowls have a red or brown band on the rim.
(c) Each of the small bowls has one independent motif of Group M (*Fig. 13: M1–6*) on the inside of its base.
(d) All the small bowls which are not decorated on the outside, are decorated on the inside with motifs from Group G and I, and
(e) most of them show a group of three to six parallel lines (*Fig. 11: A6*) on their inner side below the rim.
(f) Most of the vessels have two parallel lines below the rim on the outer side. All the vessels which have a central motif on the outer side have two parallel lines near the base as well.
(g) Except on the sample in *Fig. 5:5*, the small bowls which are decorated inside and outside have different motifs on each surface.

Except for the cross (B), triangle (E) and arches (H), all the geometric motifs appear on the small bowls.

On goblets a central motif – zigzag G3, 5 and cross B1 – appears in the centre part of the exterior. The jugs usually have several friezes of geometric motifs, in contrast to the small bowls and goblets which have only one. The friezes here are also more crowded and show a variety of motifs: zigzags G (*Fig. 6:19*), lines A (*Fig. 11:A1, 5, 8*) and semi-circles (*Fig. 12:1, 2*).

Only one juglet (*Fig. 9:1*) has geometric motifs, whilst the rest have a combination of birds and geometric motifs. The frieze decoration – with motifs G2; H1; A1, 7 – of this 'geometric' juglet covers the whole of its outside. On the other juglets (*Figs. 7:4, 6; 8:1–2*) there is a division into metopes, each containing the bird as central motif. Each juglet shows two geometric metopes, containing the bird motif as well as the cross with dots (B1). The vertical friezes contain a net motif (C2).

Like the juglets, the mug is decorated with metopes containing the bird as central motif (*Fig. 10*).

There is the same frieze at the top and the base, emphasizing the unique shape of the piece. The other unique vessel, probably a large bowl (*Fig. 9:8*), shows a combination of a geometric design and a bird. However, as the relation between its different fragments is not quite clear, it is difficult to know whether we are dealing here with friezes or metopes.

2 Local, Rough Hand-Made Pottery

This group consists of two kinds of ware: the Negev-ware and another ware of different character and shape.

a. Negev-Ware (Fig. 14:1–9)[3]

Negev-ware pottery was only found in relatively small quantities and mostly only as body sherds. However, a few rims and bases were recovered, several with textile impressions (see below, Chapter III, 19), and the following vessel types could be identified: bowls (*Fig. 14:9*), hole-mouth jars (*Fig. 14:5*), small bowls (*Fig. 14:6*). Most of the bases resemble the ring-base type (*Figs. 14:6–8*).

b. Rough hand-made pottery (not Negev-ware)

Small bowls (Figs. 14:10–11)
These are small bowls of different sizes, but all have flat bases and rounded sides.

Large bowls (Figs. 14:12–13)
One bowl-like cooking pot (*Fig. 14:12*) has a flat base, a low, vertical side but no handles. This pot differs from the small bowls also in the material, since its clay contains much temper. The other bowl (*Fig. 14:13*) has a thickened rim and a handle.

Goblet (Fig. 15:1)
The goblet has a wide mouth, rounded sides and a rounded base.

Juglets (Fig. 15:2–4)
(a) Juglet *Fig. 15:3* has a spherical body and a short neck. Its loop-handle is drawn in a perfect curve from the rim to the shoulder. There are perforations at the base arranged in a circle, punched from outside in.
(b) Juglet *Fig. 15:4* has a spherical body, a longer neck and a trefoil mouth. On the shoulder, there is an indication of a handle which may have been attached to the rim.
(c) *Fig. 15:2*, with an oval-to-cylindrical body, is perhaps a dipper juglet.

Pilgrim flasks (Fig. 15:5, 7)
Both pilgrim flasks are lentoid in section. Flask *Fig. 15:7* has two handles going from the shoulder to the slightly flaring neck, below the (missing) rim. The two halves of its body are joined in a decided angle. Flask *Fig. 15:5* is more spherical and has longer handles. The joint of its two halves is rounded.

Holemouth jar (Fig. 15:8)
The jar is pyriform, its side tapering towards its mouth. The base is flat. There is no handle.

Appliqué decorations (Fig. 16:1–3)
Two sherds show decorations in high relief, in the shape of a snake (*Fig.* 16:2–3). On one of them (*Fig.* 16:2) the snake curls between the handles, and on the other is a snake's head. The three protrusions from a bodysherd (*Fig.* 16:1) may have been triple handles.

3 Local, Wheel-Made Pottery

The following division into groups is mainly based on rims, bases and other fragments, since no complete vessels were found in the excavation.[4]

Jars
Figs. 16:4–5: In view of the size, we assume that these are jars with a short neck and simple, thickened rim.
Fig. 17:6: Storage jar, thickened inverted rim.
We may assume that these jars had shoulder handles, many of which were found in the excavation (*Fig.* 19).
Fig. 18:3: Probably a holemouth jar with sloping wall, tapering towards its wide mouth.
Figs. 19:2–5: Loop handles, protruding on the middle of the body, probably of large storage jars. Some have a round section (*Fig.* 19:3, 5), others an oval section (*Figs.* 19:2, 4).
Figs. 19:6–9: Cone-shaped and rounded (*Fig.* 19:9) bases, probably belonging to storage jars.
Figs. 20:1–3: Ring bases from jars or kraters (*Fig.* 20:3 perhaps a jug).
Figs. 20:4–8: Storage jars. There are different types of bases: 1. a simple flat base with an angular joint to the body (*Figs.* 20:4–6); 2. a flat base protruding at the joint with the body (*Fig.* 20:8); 3. a concave ring-base (*Fig.* 20:7).

Bowl
Fig. 17:4: Carinated bowl with rounded rim.

Jug
Fig. 19:1: Oval (section) loop handle drawn from the rim to the shoulder; probably a jug.
Fig. 18:5: A trefoil rim, probably of a large jug.

Cooking pots
The cooking pots do not appear to have had handles. There are some variants (*Figs.* 16:6–7) with a ridge around the thickened rim.
Fig. 16:8: Cooking pot with flat rim
Fig. 17:3: A carinated cooking pot, triangular rim (in section).

Kraters
Figs. 17:1–2: Krater with thickened, round rim.
Fig. 17:5: A krater with a thickened everted rim.
Figs. 18:1, 2: Krater with an S-like profile and loop handles drawn from the rim to the shoulder.

Goblet
Fig. 18:4: A goblet with a high ring base and a handle springing from the base.

Chalice
Fig. 18:6: A chalice or, most probably, a pedestalled incense burner.

4 Undecorated Egyptian Pottery[5] (*Figs.* 20:9–12, 21:1–3)

This group of undecorated Egyptian pottery contains only a few sherds of tentatively identifiable types: *Fig.* 20:9: Jug; *Fig.* 20:10: Bowl; *Figs.* 20:11–12: Kraters; *Fig.* 21:1: Storage jar; *Fig.* 21:3: Juglet; *Fig.* 21:2: Flat base, perhaps of a jar or krater.

5 Painted Egyptian (*Fig.* 21:4–10) and Local (*Fig.* 21:11–14) Pottery

A number of small vessels were painted red and decorated with black or brown bands. From the point of view of provenance, some of these vessels come from the Nile Valley (*Fig.* 21:4–10), others are local (*Fig.* 21:11–14) (see Chapter III, 4). Since only fragments were found in the excavation, we can only tentatively identify several types: bowls, juglets and jars. This kind of painted pottery was not found at any other site of the Timna Valley and it seems, therefore, that these vessels were connected with ritual or votive functions.

Remarks

The pottery groups dealt with in this chapter, although of different origins, belong chronologically to one and the same period (19th–20th Dynasties of the New Kingdom) and appeared together in all phases of the Hathor Temple. However, these groups did not appear in equal quantities. The Midianite and local wheel-made pottery represent the largest groups and showed more, and more differentiated, types. The local rough hand-made pottery appeared in small quantities and a restricted range of types. Negev-ware, as well as Egyptian and local painted pottery also appeared only in very small quantity.

Because of the cult nature of the site, a number of sophisticated vessels of obviously votive character, appear in the various groups: in the Midianite group – *Fig.* 10; in the rough hand-made pottery group – *Fig.* 15:2; in the local pottery group – *Figs.* 18:4, 6; and all of the painted pottery, *Figs.* 21:4–14, which is unknown at other sites.

Most of the vessels are of small or medium size. In contrast to the relatively large-size Midianite vessels found at the smelting sites of Timna (Sites 2 and 30),[6] most of the Midianite vessels found at the Temple are small and beautifully decorated bowls, jugs and juglets. It is important to emphasize that the decorations on the Midianite vessels are not only hand-painted, but also different on each vessel, i.e. the potter did not serialize his artistic creations, although these decorations were combinations of standard motifs.

Painted pottery was only found at the Temple and Negev-ware pottery was only represented by a few sherds, in strong contrast to the contemporaneous smelting camps of Timna, where a considerable quantity of Negev-ware vessels was found.

Egyptian pottery originating in the Nile Valley, appeared in small quantities in the Temple, as in the other Timna sites. It seems that this pottery was initially brought to the site by the Egyptian mining expeditions which eventually made their 'Egyptian vessels locally and of local material. This fact can be clearly seen in the painted pottery group, part of which is made in Egypt from nilotic clays and tempers, whilst the rest is made locally and of local materials. All of these painted vessels are similar in technique and decoration (see Chapter III, 4).

A very similàr situation can be found when comparing vessels of the locally-made group with the same types of vessels in the Egypt-made group: Egyptian *Fig.* 21:1 and Local *Fig.* 16:7; Egyptian *Fig.* 21:3 and Local *Fig.* 19:7; Egyptian *Fig.* 20:12 and Local *Fig.* 17:5; Egyptian *Fig.* 21:2 and Local *Fig.* 20:7.

The pilgrim flasks (*Figs.* 15:5, 7) also have Egyptian parallels from the 18th–20th Dynasties (although not in the Temple) (Keeley, 1976: 62.3/29, 30; 63.2/26; *Pl.* 68.5/98; *Pl.* 75.1/32; *Pl.* 76.2/13).

Similar to the Egyptian potter, the Midianite potter also made locally vessels of typical Midianite characteristics (see below, Chapter III, 4) although these are shown mainly in the pottery technique and type of vessel, rather than by the raw materials used.

Notes

1. Originally the excavator only distinguished between three groups of pottery, found together everywhere at the Timna smelting sites: Midianite, Negev-ware and Normal (Rothenberg, 1972, 153–162). After the petrographic investigations (see Chapter III, 4), the Normal group could be split up into three different ceramic groups.

2. Rothenberg, 1972, Fig. 32:9.

3. In previous publications (see Rothenberg, 1972, 70) all the coarse, hand-made pottery was called 'Negev-ware'. Petrographic studies made it possible to distinguish between several different kinds within this group, now called 'local, rough hand-made pottery'.

4. The descriptions and definitions of the wheel-made pottery of Site 200 also took into consideration the pottery finds at Site 30, which will be published in Vol. IV of this publication.

5. The Egyptian pottery was identified by petrographic investigation as having been made of nilotic material (see Chapter III, 4).

6. Rothenberg, 1972, Fig. 32; Conrad and Rothenberg, 1980, Fig. 210.

4. PETROGRAPHIC INVESTIGATIONS OF THE POTTERY

Introduction

Various types of pottery were found during the excavations of the Timna Temple, differing significantly in form, material, technology and decoration. Rothenberg (1972, 105, 153) proposed a threefold subdivision for the pottery of the New Kingdom period of the Timna Valley: (1) Negev-type pottery, which was primitive-looking, hand-made and poorly fired, and was considered by him to be locally made, reflecting a technological tradition of the local Amalekite tribal population;[1] (2) Midianite pottery, characerised by typical decorations, unusual materials and high firing grade, was considered to have been imported from the Hejaz (Midian)[2] in North-west Arabia; (3) Normal pottery, the largest group, was described as an unhomogeneous group of various wheel-made wares, its provenance was an open question. Typologically, the Normal pottery was considered to resemble

Palestinian pottery, but the possibility of connections with the north was rejected on the basis of general archaeological arguments.[3] At the time, none of the pottery groups were suspected of being Egyptian. Slatkine (1974), in an earlier petrographic investigation of the Timna pottery, reinforced Rothenberg's threefold subdivision, but no definite conclusion concerning the provenance of the Normal pottery could be reached. Slatkine concluded that 'it may have come from the Timna-Eilat area, but equally from much further away, e.g. Sinai or Arabia'.

However, the apparent absence of Egyptian pottery among the ceramic objects found in the Timna Valley, and especially in the Temple with its numerous Egyptian objects, was rather difficult to understand; it was therefore decided to devote major efforts to the investigation of this problem. Since it was expected that Egyptian wares would only be found among the Normal pottery, the main attention was directed to the detailed study of this group.

Methodologically, several steps were involved. The first was a detailed petrographic study of two collections of Egyptian pottery from the Nile Valley, one representing 18th Dynasty pottery from Malkata, Thebes,[4] and the other representing various periods from several sites along the Nile: Kom-Ombo, El-Kab, Thebes, Dendera, Abydos, Amarna, Meydum, Sakkara and Abusir.[5] This study was carried out to provide a basis for comparison.

The second step was a brief examination of the sherds from the Timna Temple by Colin Hope, in order to sort out suspected Egyptian sherds by unaided visual study of typology, surface treatment and other properties related to technology and materials.

The third step involved a binocular study of all the sherds of the Normal pottery group found in the Temple, including the samples suspected of being Egyptian. This made it possible to select pottery samples for thin-sectioning in a way that ensured the maximum petrographic variation needed to provide a representative picture of the Normal Pottery group. Special emphasis was therefore put on samples thought to be Egyptian.

In the last step, this representative collection of thin sections was compared with the reference collections of Egyptian pottery. As a result, 'Normal Pottery' is now given a new definition excluding the true Egyptian pottery and some other types of ware which deviate petrographically. Attempts to understand the petrographic variations within the Normal pottery led to the conclusion that two technological attitudes are represented in this pottery.

The present study also deals with the Midianite Pottery, its technology, provenance and distribution, as well as with the Local Hand-made Pottery and the so-called 'Negev-ware.

1 Egyptian Pottery (Col. Pl. 1)

Introduction

The Nile Valley, in terms of its young, fine-grained deposits, is a rather homogeneous environment and the materials available to the potter were very similar along the whole valley which could, therefore, be considered one petrographic province.

From a previous study dealing with the identification of Canaanite pottery imported into Egypt (Amiran and Glass, 1979), it became obvious that the petrographic properties of Egyptian ceramics, which have a major component of Nile silt-sand, are rather distinct and that in almost all the geological environments of Israel such materials would be considered foreign. This situation is, of course, of great advantage in provenance studies relating to Egyptian pottery imported into other countries. It has a disadvantage, however, when attempting to trace the movements of Egyptian pottery within Egypt.

These preliminary perceptions indicated that the question of the presence of Egyptian pottery in the Timna Valley could be answered with a satisfactory degree of certainty by petrographic analyses. However, the few samples studied by Amiran and Glass (1979) and the examination of several samples of Egyptian pottery from the thin-section collection of the Institute of Mining and Metals in the Biblical World, Tel Aviv, indicated that the petrographic picture of Egyptian pottery is not so simple as it first appeared, or as the general geological circumstances might suggest. In fact, although many petrographic properties were shared by most of these samples, it was difficult to find two precisely identical samples. Several variation trends could be observed in the mineralogical composition of the siliceous silt-sand fraction, in the maximum grain size and the grain-size distribution of the siliceous nonplastics, also in the presence or absence of granular carbonates, organic materials, clay-rich aggregates, microfossils, shell fragments, weathered micas and other mineralogical and textural attributes. Many of these differences can be explained by vertical variations of the young Nile sediments, lateral variations within the flood plain due to mixture with non-nilotic materials at the boundaries of the Nile Valley and at junctions of wadis running into the Nile Valley, or even due to artificial, man-made mixtures.

In order to place our provenance study on safer ground, we carried out a systematic sampling of Egyptian pottery so that the largest possible selection of fine-grained deposits used by the ancient Egyptian potters would emerge in our reference collection.

In recent years, questions have been dealt with concerning the clay types of Egyptian pottery; Hope, Blauer and Riederer (1981) presented chemical and petrographic data on 18th Dynasty pottery from Malkata and Amarna and compared these by visual examinations. However, as no attempt was made to define petrographic pottery groups on the basis of microscopic data, and each thin section was described separately, it is difficult to relate our microscopic data to the descriptions of Riederer. Other authors presented petrographic data on Egyptian pottery related to questions concerning the origin within Egypt of the materials used. Since our main interest in this study is the identification of true

Egyptian pottery in the Timna Temple, no attempt is made here to correlate our findings with the various groups of nilotic and other clay-rich materials proposed by the various investigators.

a. Reference collections of Egyptian pottery

The Malkata collection of 18th Dynasty pottery

This collection comprises 47 samples, selected to represent the different materials present among the 18th Dynasty pottery of Malkata, Thebes. The samples were first described archaeologically in terms of the clay type, surface properties and typology.[6]

Clay types were designated as common Nile clay, Nile ware, coarse Nile ware, marl, desert marl, pinkish desert marl, marl with numerous white inclusions, unusual type of marl and Qena ware.

Surfaces were designated as cream slipped, burnished, cream slipped and burnished, natural cream coating, thin white slip, thin cream wash, coated with green, yellow or red pigment and painted.

The 47 samples from Malkata were divided into 14 petrographic groups, five of which include about 80% of all samples. Of these five major groups, Groups I and II probably represent recent Nile sediment, Groups IV and V represent calcareous mixtures with nilotic components. Group III is an outstanding group that shows no nilotic affinity, i.e. a siliceous silt-sand fraction with abundant hornblende, pyroxene, epidote mica and feldspars is practically absent. The high carbonate component in the sand fractions of Group III, and the abundant fossil fragments in which Nummulite fragments are frequent, suggest that this group probably originated at the junction of the Nile Valley with one of the wadis draining the adjacent Eocene terrains.

The minor groups of Malkata represent various nilotic and non-nilotic provinces. The non-nilotic samples point to an igneous metamorphic province, a volcanic terrain of vitreous basalts with fine fossiliferous calcareous sediments, to the sandstone terrain of the Nubian facies, and to a desert sedimentary terrain governed by carbonate lithologies with flint and gypsum. The origin of these samples must be outside the Nile Valley and possibly outside Egypt; however it is impossible to determine their provenance by petrographic data alone because the same geological environments can be found, for instance, in Egypt, Sinai and Israel.

The Regional Reference Collection

This collection comprises 23 samples, two from Kom-Ombo, two from El-Kab, two from Thebes, five from Dendera, one from Abydos, four from Amarna, two from Meydum, three from Sakkara and two from Abusir. The samples represent different wares from different periods. Of the 23 samples, 18 exhibit very close petrographic affinities with samples from Malkata: 11 samples show close affinity with Groups I and II of Malkata; 4 with Malkata Group III, 3 with Malkata Group IV, one with Malkata Group V, and 2 with one of the minor Malkata groups showing nilotic affinity.

About 70% of the regional collection exhibits petrographic affinity with the five major Malkata groups. It is beyond the scope of this report to present a detailed comparison between these two collections. We may, however, conclude that the major Malkata groups also occur at other sites and over a range of periods, and that these are the main material groups in major parts of the Nile Valley.

b. Egyptian Pottery from Site 200

Seventeen thin sections of Normal pottery exhibit petrographic affinity with samples in our Egyptian reference collections, and are therefore considered to have been manufactured in the Nile Valley and imported into the Timna Valley.[7]

Close petrographic affinity with Malkata Group II

S.N. 1503	Body sherd	(Col. Pl. 1:6)
S.N. 1507	Base sherd	(Col. Pl. 1:8)
S.N. 1529	Cream coated body sherd	(Col. Pl. 1:1, 2)
S.N. 1139	Body sherd	
S.N. 1117	Body sherd	
S.N. 1558	Conical base	(Col. Pl. 1:9)
S.N. 1560	Handle	(Col. Pl. 1:12; Pl. 108:4, 5)

Of these seven samples, only S.N. 1503 and S.N. 1507 were selected by visual examination as probable Egyptian pottery.[8] There is nothing unique about the external appearance of these sherds. They are thin or medium-thick walled and belong to relatively small vessels. In this respect they differ from the non-Egyptian items in the Normal pottery group. Some, for example S.N. 1507, show a zoning pattern in the section, rather typical of Egyptian Nile wares: a wide core of dark grey colour is followed outwards by a symmetrical pair of reddish zones, followed again by a pair of external brown zones. The outer surface of these samples is usually somewhat lighter than the inner surface. In extreme cases, as in S.N. 1529, there is a cream-white outer surface, which appears to belong to a family of phenomena related to the concentration of soluble salts on the outer surface of the pottery. This occurs on some common Nile wares, but is certainly not a diagnostic feature of Egyptian pottery. This kind of surface coating is expected on ceramics from desert environments in which either the clays or the water, or both, contain considerable quantities of soluble salts.

Microscopically, the above listed samples share the following properties with Malkata Group II:

(a) Absence of optically visible carbonates.

(b) Relatively high volume proportion of a variegated siliceous non-plastic assemblage.

(c) A wide and almost continuous grain-size range covering the silt and parts of the sand sub-range.

(d) Presence of minerals such as micas, hornblende and colourless amphiboles, pyroxenes, epidotes, lath-like plagioclases, other feldspars, quartz and occasionally basalt fragments. Other minerals such as sphene, opaque iron oxides and garnet, occur in a much smaller quantity.

(e) As a rule, the upper portion of this grain-size range is occupied by quartz and feldspars (usually alkali feldspars), while the other minerals occupy the finer fractions.

(f) In some samples, rounded dark grey domains occur that have the appearance of grog, but under the microscope these domains prove to be silty clay islands of a somewhat different composition from the surrounding silty clay-rich groundmass. These 'islands' show a different (usually lower) volume proportion of non-plastics than their surroundings and the sand fraction is almost always absent. Quite commonly a crack develops at their boundary; this feature is best developed when the islands are poor in non-plastics; thus the cracks are explained as due to differential shrinkage at the island boundary. These phenomena occur in samples 1503, 1507 and 1119.

(g) In some samples occur either organic remnants or cavities formed by burning out of the organic matter.

(h) In strongly zoned samples, green hornblende occurs in unoxidised zones (usually the cores), and reddish brown hornblende in the relatively oxidised zones. Zoned hornblende crystals with reddish brown rims and green cores, occur both in external and internal zones of the sherd. In addition to the samples comprising Malkata Group II, these properties are shared by samples from Sakkara, Meydum, Dendera, Amarna and Abusir.

Some samples show transitional tendencies: S.N. 1117 is characterised by a high volume proportion of silt and almost complete absence of the siliceous sand fraction. This is a transitional sample between Malkata Group II and sand-free variants of Nile sediments that form one of the minor groups in Malkata. S.N. 1558 and S.N. 1560 are somewhat lighter and exhibit a uniform section, although under the microscope they show some differences. First, there is a break in the grain-size range between the large rounded sand quartzes and the silt and very fine sand fractions. The clay-rich groundmass is vitrified, which indicates a fine calcareous component, and explains the lighter colours of these two samples. This variant was not detected in the Malkata Group II.

Petrographic affinity with Malkata Group III

| S.N. 1505 | Cream slipped and burnished body sherd | (Col. Pl. 1:15, Pl. 108:6, 7) |
| S.N. 1506 | Cream slipped and burnished body sherd | (Col. Pl. 1:14) |

Besides these two samples, several other sherds with this cream slip were found in the Temple. To the unaided eye, the most conspicuous feature of these two sherds is their thick, burnished cream slip. Such cream slips occur in Malkata Groups II and III. Under the polarizing microscope, these two samples are different from all the other samples of Site 200, Egyptian and non-Egyptian. They closely resemble Malkata Group III and share with the sherds of this group the following properties:

(a) The cut section is light brown with a weakly defined light grey core.

(b) The non-plastic assemblage is governed by a medium-textured coarse sand, composed of almost equal quantities of well rounded quartz grains and various calcareous fragments.

(c) Many of the rounded calcareous grains are derived from broken microfossils with characteristic structures. Some appear to be derived from Nummulites.

(d) A fine siliceous silty fraction is almost absent but, when present, it has a strong nilotic mineralogical affinity (hornblende, mica, epidote, pyroxene).

(e) Longitudinal and oblique shrinkage cavities develop in this group to a high degree.

(f) The clay-rich groundmass is very fine and contains a fine calcareous component.

In addition to the samples comprising Malkata Group III, these properties are also shared by samples from Thebes, Dendera and Amarna.

Transitional between Malkata Group II and III
S.N. 1501   Burnished body sherd   (*Col. Pl.* 1:3)

To the unaided eye, the most conspicuous feature of this sherd is its burnished surface. Since burnishing is a very uncommon feature among the sherds from the Temple,[9] it is not surprising to find that its microscopic aspects are also unique. It is best described as a transition between Malkata Groups II and III because it contains the typical variegated siliceous silt and sand of Malkata Group II, and also a carbonate sand with rounded fragments of broken microfossils, common to Malkata Group III. This sample reinforces our previous conclusion that the Malkata Group III must have its origin in the Nile Valley.

Petrographic affinity with Malkata Group IV
S.N. 1570   Body sherd   (*Col. Pl.* 1:10 and
                                            Pl. 108:8)

This sample shares the following properties with Malkata Group IV:

(a) It contains typical grains of a micaceous habit, probably weathered micas.

(b) The non-plastic assemblage mainly covers the medium-textured subrange.

(c) Mineralogically the non-plastic assemblage contains both typical nilotic components such as horneblende, epidote and pyroxene, and also weathered micas and calcareous nodules, calcite crystals and microfossils all more or less in the medium-textured sand range.

(d) The clay-rich groundmass is calcareous.

(e) Zonal development is governed by decomposition of carbonates and reaction with clay minerals at the surfaces.

In addition to Malkata Group IV, these properties are shared by samples from Thebes, Dendera and El-Kab.

Other samples of Nilotic affinity
The five samples comprising this group cannot be compared with the major groups of Malkata, but they show a nilotic affinity in the mineralogical composition of their silt-sand range particles.

S.N. 1538 Sherd decorated with two nets of stripes: Light brown stripes parallel to the wheel marks were applied before the transversal dark brown stripes. This is the only sample of this kind of decoration found in the Timna Temple (*Col. Pl.* 1:7). The non-plastic assemblage is governed by a variegated silt-fine textured sand $(25–100_M)$. Mineralogically it shows strong nilotic affinity. The clay-rich groundmass is microfossil, ferrous and calcareous. A few inclusions in the granule range are composed of chalky-marly rock. Clay-pure domains occur also, exhibiting typical peripheral shrinkage cavities.

S.N. 1536 Neck fragment (*Col. Pl.* 1:5): The siliceous non-plastics mainly cover the fine and medium-textured sand fractions (most grains fall between 100 and 400_M). The composition of this sand has a strong nilotic affinity and contains various amphiboles, colourless, pale and common hornblende, epidote, zoisite, micas, various feldspars and a single garnet crystal (0.3mm.). The non-plastic assemblage exhibits close affinity with Malkata Group I. The clay-rich groundmass of this sample is very fine and calcareous and in its general optical appearance resembles Malkata Group V, the so-called Qena wares. This sample is therefore quite probably an artificial mixture of a fine marl with Nile sand.

S.N. 1563 Painted neck fragment of a juglet (*Col. Pl.* 1:11): There is a distinct nilotic element in the silt, very fine sand fractions defined by micas, epidotes and amphiboles. There is a grain-size gap between this nilotic element and the coarse quartz sand. The quartz sand could therefore indicate a different origin. The clay-rich groundmass is highly calcareous.

S.N. 1564 Painted handle with transversal dark brown stripes (*Col. Pl.* 1:13): As in sample 1563, the nilotic element occurs in the silt-very fine sand range. Again, there is a gap in the grain-size range up to the coarse sand fraction which, besides quartz, contains fine calcareous nodules, shell and microfossil fragments. The clay-rich groundmass is highly calcareous.

S.N. 1502 Base fragment of a cone shaped jar (*Col. Pl.* 1:4): As in samples 1564 and 1563, the nilotic element occurs in the silt-very fine sand fraction. Quartzes of the coarse sand range are very rare and carbonates of various types more abundant. The clay-rich groundmass is microfossiliferous and calcareous.

Conclusions
Petrographic analyses proved that true Egyptian pottery, i.e. manufactured in the Nile Valley, is present in the Timna Temple. Most of it is comparable to the major petrographic groups of 18th Dynasty pottery from Malkata. This does not mean, however, that this

pottery came from Malkata since this petrographic grouping is valid for a large region.

The Egyptian pottery of the Timna Temple is characterised by small- or medium-sized vessels with thin or medium-thick walls, whilst all the thick sherds of much larger vessels belong to other members of the Normal pottery group. All the unusual surfaces: burnished cream-slipped, commonly burnished, red slipped and painted surfaces, with a single exception, proved to be on true Egyptian pottery. This might suggest that most served in the Temple in a cult context.

2 Midianite Pottery (*Col. Pl. 4*)

Introduction
Midianite pottery, more than any other pottery type found in the New Kingdom context of the Timna Valley, can easily be distinguished due to its typical decorations and unusual combination of materials. With the exception of a few decorated Egyptian sherds, and one unidentifiable decorated sherd belonging to the Normal pottery group, Midianite pottery is the only decorated pottery found in Timna. Its unusual and diagnostic appearance, and the fortunate discovery of identical pottery (Parr, Harding and Dayton, 1970) in Qurrayyah, enabled Rothenberg to postulate, on the basis of unaided visual examination, the working hypothesis that the decorated pottery of the Timna Valley came from Qurrayyah in Midian (Rothenberg, 1972: 162). This hypothesis was first tested by A. Slatkine, who confirmed by petrographic analogy that the pottery from Timna and Qurrayyah must have originated at one and the same location and, by indirect argument, concluded that this place was probably Qurrayyah (Slatkine, 1974).

In a recent comprehensive study of the Midianite Pottery, based on a detailed petrographic analysis of numerous samples, Rothenberg and Glass (1983) reviewed the occurrences of Midianite pottery from Timna and the Arabah, Arabia and Jordan, the Gulf of Eilat and northern Sinai, the northern Negev and the Hebron mountains, and discussed in detail the technological, chronological and provenance aspects of this pottery. As the Midianite pottery of the Timna Temple was proved to be identical in all its petrographic properties with all its other occurrences, only a summarised description will be given here.

a. Unaided visual examination
Surface properties
Most outer surfaces of the Midianite pottery are smooth and slipped and their colour is light cream or yellowish, or varieties of whites. When a slip is absent, the surface shows a spotted texture due to numerous dark fragments. Slipping was therefore necessary in order to cover these dark fragments and to form a uniformly light surface as a background to the coloured decorations.

Light yellowish brown, red, dark brown and black, and transitional colours were used for the decoration. These decorations were applied in several steps on the light slipped background. When stripes of various colours transverse each other, the lighter colours were applied first and the darker colours later.

Turning marks which are mainly observed on inner surfaces, indicate the use of a turning device (see also Kalsbeek and London, 1978, on the Midianite potter's technique).

Section properties
The most conspicuous feature of the cut section is the occurrence of red and brown-black inclusions of usually elongate rectangular and platey shapes and lengths of up to a few millimetres. Their size and content varies from sample to sample. In thin-walled, small vessels the quantity of inclusions is low and their size usually relatively small. Some such fragments shown an internal splitting into several thin plates parallel to the length of the fragment. The rectangular-platey shape and the internal splitting suggest that they are shale fragments.

The colour of the shale fragments varies from sample to sample, but some variation is also observed in each one. In samples with a light grey body, brown-black fragments dominate, while the red fragments dominate in light reddish, yellowish and white bodies. This suggests that the variation in colour is not a primary feature of the fragment, but rather a phenomenon controlled by the kiln atmosphere. Black fragments show a tendency to lose their rectangular shape and to change gradually through eliptical into more spherical shapes. This change in shape is sometimes associated with the occurrence of fine spherical cavities (gas bubbles) and a glassy appearance. These changes suggest an advanced stage of melting of the fragments.

In addition to the dark fragments, some samples also show some light shale fragments that may be even lighter than the body. Only in a few samples do the lighter fragments predominate. A low content of some quartz sand is almost always present.

b. Microscopic examination
Coarse non-plastic ingredients
The shale fragments are the predominant coarse non-plastic ingredient and they are certainly an intentional addition to the plastic clay. Under the microscope the clay minerals in the fragments exhibit a strong lattice and dimensional preferred orientation, thus confirming the view that they are shale fragments rather than grog.

In composition the fragments represent various shales that differ in their content of the non-plastic, quartzo-feldspathic silt fraction and in the content and distribution of iron oxides. The different shale types appear to grade into one another and to form two main variation trends that cross each other. One trend begins with almost pure non-plastic shales, moves through various silty shales and ends with argillaceous silts and silts with almost no clay minerals. The other trend begins with white shales, moves through spotted shales with tiny iron oxide grains and concretions, and ends with dark red shales with a uniform distribution of iron oxides.

Some fragments show a banded structure parallel

to the shaley fabric, where each band represents a different shale type. This suggests close physical association between the different shale types in their natural occurrence. It appears that all the shale types can be found in one exposure of a banded silty shale formation. Although almost all the types of shale fragments encountered may be found in one sample, the predominant fragments are dark shales and dark silty shales. Light shales, argillaceous silts and silts occur in most samples in minor quantities. It is clear that the dark shales were deliberately selected to be used as temper and that they were never used for the plastic component.

It remains an open question whether the shale fragments were fired prior to mixing with the clay and then fired again together with the clay. However, it can be stated that if pre-firing took place it was most probably no more than a short, low-temperature baking, since major differential shrinkage between fragment and host must have occurred during the firing of the finished ceramic object, as indicated by the wide peripheral shrinkage cavities surrounding the less silty shale fragments. Several of the microscopic aspects of shale fragments are shown in *Pl.* 107: 7, 8, 9.

A well-sorted quartz sand is a minor component of the coarse non-plastic assemblage which occurs in almost all the samples. Carbonate impurities occur in a few samples, completely decomposed and showing yellowish isotropic reaction rims.

Clay-rich groundmass
The clay-rich groundmass is optically identical with the light variants of the shale fragments, only the preferred orientation of the clay minerals is absent. The shales used for the preparation of the plastic paste were mostly light silty shales, spotted shales and spotted silty shales. The dark shales were never used for the body. Typical of all the shale types in the Midianite pottery is the almost total absence of carbonates. The clay-rich groundmass of almost all the samples shows some degree of anisotropism, even in samples where the dark shale fragments were molten and the carbonate impurities are decomposed completely and show vitrified reaction rims.

A sample that was refired under oxidising conditions in an electrical kiln, to 900, 1000, 1100 and 1200°C became gradually less anisotropic. No other major changes were noticed, confirming that this material is rather refractory, resembling the so-called stone wares of modern potters. This is not surprising considering the absence of carbonates and the low content and concretional distribution of iron oxides. It seems possible that the light shales were selected for the body because of their refractory qualities.

c. Provenance
Microscopically, the Midianite pottery is a distinct group which can easily be distinguished from all other pottery types found in the New Kingdom context in the Timna Valley. Unlike the Normal and Rough Hand-made pottery, which do show some petrographic affinity, and also transitional types,

there are no transitional types between Midianite pottery and the other types of the Timna Temple pottery. The granitic sand temper, slag temper and grog temper of Normal pottery, features which are also common in the Rough Hand-made pottery of the Temple, are totally absent in Midianite pottery, as are the dolomitic and glauconitic shales, gypsum and various carbonate impurities that characterise the Normal pottery. On the other hand, the non-calcareous shales that characterise the Midianite pottery have not been encountered among Normal and Rough Hand-made pottery. Chemical analyses (H. Hatcher and R.E.M. Hedges, 1976) which showed the calcium-poor aspect of Midianite pottery, clearly confirm this picture, distinguishing Midianite pottery from the Normal and Rough Hand-made groups. If other chemical attributes, e.g. the ratio Al_2O_3/MgO, are plotted against the CaO content, a completely separate Midianite pottery field results. Thus it is extremely unlikely that the Midianite pottery could be manufactured in the Timna Valley.

Petrographically and chemically there is no difference between Midianite pottery from Qurrayyah and from the Timna Temple, which leads us to the conclusion that there must be a common source. Although similarities in the chemistry and petrography of pottery do not always mean a common source, in this case, due to the very unusual petrographic properties and unusual chemistry, this appears to be the only realistic conclusion for the Midianite pottery. From all the sites where Midianite pottery was found (Rothenberg and Glass, 1983) the only site where the local geological circumstances match the petrography of the pottery is Qurrayyah, situated in the heart of a Nubian sandstone terrain in North-west Arabia, where a formation of 100 metres of shales and sandy shales is exposed (Helal, 1965). Finally, Parr *et al* (1970: 220) described the Hill of Qurrayyah and stated that it is built of a 'grey green silt stone'. Thus, all the data available today serves to confirm Rothenberg's original hypothesis that the Midianite pottery of Timna originated in the region of Midian (Hejaz).

3 Normal Pottery (*Col. Pl.* 2)

Introduction
As already mentioned, the term 'Normal pottery' as it was used before this petrographic study, comprised a wide range of types, including everything that was neither Midianite nor Rough Hand-made pottery. The present petrographic analysis showed that part of what was earlier considered 'Normal pottery' is, in fact, imported Egyptian pottery and that several distinct types of Egyptian pottery occur in the Timna Temple.

The majority of the samples which were considered to be Normal pottery, after the Egyptian sherds were excluded, should in fact be considered as one group, and the term 'Normal pottery' could be applied to this group. The term must, however, be clearly defined. Because our deliberations are based on petrographic and not on typological analysis, the nature and significance of a petrographic group must first be explained: A homogeneous petrographic group comprises samples that share compositional and textural aspects or matrix and temper. However, if described in great detail, two thin sections of pottery are never exactly the same, and a certain degree of 'freedom' must therefore be allowed to avoid the formation of too many, often rather meaningless, groups. A homogeneous petrographic group represents a workshop, or a group of workshops, working in the same geological province and using the same materials in the same way. The 'freedom' of variations allowed in such a petrographic group must take into consideration the heterogeneous nature of clays, sands and various non-plastics used by the potters and the non-uniformity of the processes such as firing. At an earlier stage of this study, we attempted the subgrouping of the 41 samples that comprise the newly defined Normal pottery group. It soon became obvious that such a subgrouping would be meaningless and even confusing, because the number of different material combinations proved to be too large to form meaningful homogeneous groups, and because these material combinations in any case strongly relate to one another through several continuous variation trends.

It is therefore proposed to regard the 41 samples examined as one petrographic group, representing variants of pottery that were manufactured in the same area (quite probably in the Timna area), using various combinations of a range of closely related materials. This group is now given the name 'Normal pottery'. It will be shown later that within this group two technological attitudes can be detected which explain some of the variation tendencies observed, but which do not justify breaking up the Normal pottery into two distinct groups.

Description

The descriptions given here are based on observations made on 41 sherds and their polished thin sections. The order of this description is according to aspects of the pottery and not according to groups of variants or individual samples because, as stated above, no clear subgrouping emerged.

a. Unaided visual examination of the sherds
Surface properties

The surface texture changes from sample to sample. Some surfaces, e.g. S.N. 1571 and S.N. 1544, are extremely irregular and give the impression of primitive hand-made pottery. However, other samples, with rather rough outer surfaces, show well-developed turning marks on their inner surfaces. The rough hand-made appearance of S.N. 1544 could be due to the inclusion of a large quantity of coarse shale fragments coupled with a relatively low grade of firing, making this sherd more vulnerable to weathering.

No clear burnishing marks or any other marks indicating intentional surface treatment of any kind, were encountered. Only one sample – S.N. 1500 – shows a somewhat smooth surface with weakly

defined burnishing marks. The surface texture is governed also by other factors. Some surfaces are extremely pitted, such as the outer and inner surfaces of samples 1549 and 1537, and the inner surface of 1548. In many other samples a few such pits occur. Almost all these pits represent leached-out gypsum inclusions. The tendency for more pronounced leaching at inner surfaces can be either due to the protective nature of an external self slip, coupled with a higher degree of vitrification at the outer surface, and/or a leaching process that would take place in water containers mainly at the inner surface. Some samples, such as S.N. 1509 and S.N. 1559, exhibit a rough surface texture due to imprints of chaff temper.

Another type of surface roughness develops in variants carrying large quantities of a very coarse quartz sand temper. This surface texture is most clearly illustrated by S.M. 1547.

The inner surfaces of most samples show turning, wheel (?) marks, usually in the form of fine striations. Samples 1500 and 1532 show, in addition, wide ridges formed during the turning process. Some samples show turning marks on their outer surfaces also. These may be fine striations as on S.N. 1540 and S.N. 1545, or a few wider grooves as on S.N. 1548, or a combination of both as on S.N. 1562. The general impression from turning marks and shapes indicates a potter's wheel of a not very high standard or the use of some other, more primitive turning device.

The colour of the surfaces varies from white, cream-white to dark black through various browns, red browns, greys, buffs, etc. White and cream surfaces are most probably due to high temperature reactions on surfaces with high soluble salt concentrations. On almost all the samples, this white or cream surface developed on the outside of the vessel. The inner surface of these samples is always darker than the outer surface, but it may be lighter than the body underneath it. This is expected because during drying, especially in the case of closed vessels, humidity inside the vessel will rise, causing a sharper humidity gradient towards the outer surface. Thus evaporation at this surface will be more pronounced, leading to a higher concentration of the salts at this surface. In desert environments, where the clay or the water or both may carry soluble salts, the white surfaces are not necessarily a property intended by the potter.

Samples 1498, 1510, 1512, 1516, 1543, 1545 and 1552 show these white and cream surfaces well developed. On some other samples the surfaces are not cream or white, but tend to be lighter than the body underneath. A slip is clearly absent in association with the white cream surfaces. This is elegantly demonstrated in samples such as S.N. 1545 that show fine turning marks on the outside and a well developed cream surface. Had this surface been slipped, the rather tender turning marks would have certainly disappeared. Reversed colour relationships are observed in Samples 1513, 1514, 1533, 1534, 1537, 1540 and 1568. In all these samples the outer surface is darker than the body underneath. The surface may be grey or brown-grey as in S.N. 1513, or dark brown as in Samples 1333 and 1537. The colour of the body underneath these surfaces is generally red or light brown. A short reducing period at the final stage of firing could account for this phenomenon. This pottery was probably not decorated. Only one sample – S.N. 1498 – shows a dark brown band running down the rim, which could be a part of a decorative pattern.

Section properties
The sections cut perpendicular to the surfaces also show a great deal of variation. The thickness of the sherds ranges from 3mm. in S.N. 1541 to 12mm. in S.N. 1562, while the average thickness is about 8mm. Colour zoning patterns of the sections vary a lot. Some sherds are homogeneous, or almost homogeneous, and appear fully oxidised (Samples 1504, 1517, 1548, 1557 and 1555). Others show a homogeneous grey-black colour without any oxidised rims. These appear to have been fired under reducing conditions or subjected to an intensive smoking period after a normal firing procedure (Samples 1535, 1533, 1546 and 1549). The rest exhibit various colour zoning patterns. Wide black cores with narrow symmetrical oxidised rims occur in Samples 1498, 1540, 1552, 1553 and 1561. An asymmetrical zoning pattern with a wide black or grey core shifted towards or reaching the inner surface occurs in Samples 1500, 1510, 1512, 1513, 1516, 1532, 1534, 1545 and 1551. A simple correlation between the zoning patterns and the colours that develop at the surface, appears to be missing.

Cream surfaces may occur on homogeneous black or on narrow oxidised zones as in Samples 1535 and 1512 respectively and dark brown surfaces on fully oxidised section as in S.N. 1537.

Porosity also varies a great deal. Only a few samples show an almost compact section with few tiny cavities (Samples 1500 and 1543). The rest are porous, showing various shapes and sizes of cavities. The large cavities are due to shale fragments, burnt chaff and leached-out gypsum. Shale fragments split due to shrinkage, and develop elongated cavities parallel to the shale fragment fabric. Because the shale fragment in most cases, especially when coarse, are not aligned parallel to each other, the elongated cavities related to the shale fragment also lack parallel alignment. This type of porosity is well developed in Samples 1554 and 1547. Chaff develops a better dimensional preferred orientation, especially in samples lacking other coarse non-plastics. Thus the cavities formed after its burning are parallel or sub-parallel to each other, and parallel or oblique to the sherd wall. These cavities are easily recognised when carbonised coronas develop around them. Cavities due to leaching of gypsum inclusions are equidimensional or only slightly elongate. Leaching could have occurred during burial so that these cavities do not necessarily represent original porosity.

b. Microscopic examination
Clay-rich groundmass (Pl. 107:1–6)
A number of clays were used. Microscopically they can be distinguished by their non-plastic ingredients. Mineralogically they fell into four groups: (1) Silty clays with non-plastic silt fraction of mainly quartz; (2) dolomitic clays with fine zoned dolomites; (3) silty dolomitic clay with both a quartz silt and zoned dolomites; (4) pure clays with no optically visible non-plastic minerals or with very few such grains. These clays may occur with or without the mineral glauconite, and other minor constituents such as magnetite. A further subdivision of these clays, which is of meaning to pottery making, can be based on the quantity of the non-plastic minerals, and indeed the variation is quite large beginning with the almost pure clay variants and ending with non-plastic-rich dolomitic silty clays. Lastly, the clays can be characterised and divided according to the grain size of these non-plastics. Some of the silty clays have extremely fine quartz particles not exceeding 20 microns, while some dolomitic and dolomitic silty clays carry dolomites up to 200 microns. The presence of shale fragments of these clay types in the pottery indicates that the original clay-rich deposit was of a shaley nature. That the shale types are closely related to each other is indicated by the apparent continuous variation trends of the non-plastic mineral composition, quantity and size. This would mean that quite probably all the clay types derived from one heterogeneous shale formation.

Coarse non-plastic ingredients
Various coarse non-plastic ingredients occur in this pottery: rounded quartz grains, shale fragments, chaff, various calcareous inclusions, gypsum crystals and in a few cases flint and even some slag fragments. Of these, only the quartz, the shale fragments and the chaff appear to be added intentionally by the potters to serve as temper. The others are quite probably impurities.

Quartz (Pl. 107:4–6)
Quartz occurs as equidimensional grains, sometimes well rounded, and sometimes of a less spherical habit, while angular grains, such as those that occur in the granite tempered Rough Hand-made pottery, do not occur at all. The grain size varies between 1.3 and 1.5mm. However, in every individual sample the grain size range is narrow, suggesting that either sands of different coarseness were used or that unsorted sand was sifted to separate the grain size fractions. The first possibility seems more likely. There is always a pronounced difference in grain size between these quartzes and the silty quartzes that are original constituents of the shale, and thus the coarser fraction must be considered temper, intentionally added to the silty clay by the potter. The bimodal distribution of quartz grain size is a very typical property of the Normal pottery. Coarser sands were used together with large shale fragments in heavily tempered samples.

Although other non-plastics, such as carbonates, gypsum and flint occur as well, these are not part of the quartz sand. This sand is practically pure quartz and its most probable source would be one of the Nubian Sandstone formations.

The sandy quartz is an almost ubiquitous component of Normal pottery, but its volume proportion changes considerably. In some samples only a few quartz grains occur, while in others it may reach 20–30%. The sand deficient samples tend to belong to the apparently less plastic clays, characterised by a high content of silty quartz and dolomites, while the heavily tempered samples tend to belong to the variants of the clay types which are poor in non-plastics. This general principle also holds true for the other tempers, e.g. cooking pots usually contain a high quartz sand temper.

Carbonates (Pl. 107:1, 3)
Carbonate inclusions of various compositions, shapes and sizes, occur usually in minor quantities. Because of the high temperature of firing, most of them were strongly modified and their original texture obliterated. They are usually more or less rounded and may reach a diameter of 3mm. or more. Many of these inclusions are cryptocrystalline and appear to be derived from micritic limestones. Some contain a small quantity of silty or very fine sandy quartz. Others show a few large zoned dolomites, or both quartz and dolomite. Dolomite rock fragments also occur, showing well developed rhombohedral crystals. The dolomites are frequently silty or very fine sandy and some are glauconite-rich. Sandy (quartz) glauconitic dolomites and fragments with equal quantities of quartz glauconite and dolomite occur too. Some carbonate gypsum aggregates occur, but it is difficult to tell whether these represent an original or a secondary assemblage that was formed during burial by gypsum deposition in cracks and cavities. Decomposition of the carbonates during firing was almost complete. In most samples the carbonate inclusions are surrounded by a yellowish isotropic corona which very probably represents a vitrified reaction rim formed at the contact between the carbonate and the surrounding clay-rich groundmass. In some cases a more complex reaction rim was formed with an external yellowish isotropic zone, an internal zone of carbonate decomposition and a core of primary crystals or partial decomposition. In many cases the carbonate inclusion is missing, leaving a cavity, and only the yellow isotropic reaction rim that became physically unified with the matrix is preserved. The relatively low quantity of these inclusions suggests that they were not added as temper, but are impurities derived from rocks associated with the shale formation.

Gypsum
This mineral occurs in may of the Normal pottery samples, but does not exceed 5–10%. Various tempers were used by the potters of Normal pottery, but there appears to be no good reason why gypsum should be added intentionally, and it appears to be simply an impurity. The Ora shale formation which outcrops over vast areas around the Timna Valley, contains many gypsum beds, veins and veinlets, criss-crossing the shales. The microscopic analysis of the samples does not reveal any clear damage such as cracking at the boundaries of gypsum inclusions. Thus for primitive pottery the presence of gypsum is quite probably less critical than for modern industrial pottery.

Gypsum is also a highly soluble material compared with other constituents of pottery and many gypsum inclusions were leached out completely. Evidence for this pronounced leaching process is found in the parallelogramic shape of the emptied cavities, which resemble in shape and size those gypsum inclusions that survived dissolution. Leaching could have occurred during the use of pottery that functioned as water storage, or during burial.

Evidence for secondary deposition of gypsum in pottery is found in veinlets that follow shrinkage cracks in shale fragments and in the clay-rich groundmass and filling of cavities that were occupied by organic materials before firing.

Shale fragments (Pl. 107:4,5)
Shale fragments are defined as distinct domains of the clay-rich groundmass that possess perfect dimensional preferred orientation of their clay particles. These domains are fragments of the shale deposits that did not suffer internal strain or mechanical breakdown, and acted as rigid non-plastic particles during the various preparation stages of the clay and during the forming stage. Shale fragments frequently occur in the Normal pottery and in two distinct ways. In the first, their content is relatively low, their edges rounded and their grain size varies from large and well-defined fragments down to very small fragments that pass gradually into the fully broken down clay-rich groundmass. In this case, they reflect an incomplete breaking down of the original shale deposit. In the second, they may be very large, their content may be very high, their edges are quite commonly straight, their size more uniform and transition to fully broken down clay absent. In this case, they were most probably added intentionally as temper. This conclusion is best proved when their mineralogical composition differs from that of the matrix.

We dealt mainly with the intentional shale fragments, because they form a dominant ingredient of the non-plastic assemblage. In some samples, the shale fragments are mineralogically different from the matrix. Quite commonly they are less silty. Such a combination of materials tends to develop an open texture due to higher shrinkage of the shale fragments. In several of the cooking pots, extensive use of large shale fragments of this kind occurs, probably in order to give a high porosity. There is no way to prove whether such shale fragments were pre-fired before being added as temper to the plastic clay.

Organic temper (Pl. 107:1)
In several Normal pottery samples a considerable quantity of organic temper occurs. In most cases the plant temper was completely burnt out and only the elongated cavity and the slightly blackened periphery of the cavity bear evidence of its presence before firing. In a few cases a carbonised relict of the plant

temper is still present in the cavity. It is quite possible that, in fact, more of these relics are present and that during thin section preparation they were pulled out, leaving a clean cavity.

c. Provenance

Bartov *et al.* (1972) describe two columnar sections of the Turonian Ora shales, one from Be'er Ora, about 6km. south of the Timna Temple, and the second from Nahal Netafim, about 20km. SSE of the Timna Valley. This formation is composed of thick shale beds with calcareous intercalations and gypsum layers. The thickness of the formation at Be'er Ora is 88m. Three members are described by Bartov *et al.* In the lower member gypsum veins cut across the shales. Part of the shales carry silty quartz. In the upper part of the member calcareous shales occur. The limestone intercalations have clastic quartz and glauconite. Dolomitic shales occur in the middle member. The limestones, partly dolomitic, are micrites with minor amounts of glauconite and angular quartz grains. Glauconitic dolomites and glauconitic sandstones and gypsum occur with the shales of the upper member that contain autigenic dolomite and quartz grains.

From this summarised description it appears that all the shale types and coarse non-plastics that have been detected in the Normal pottery occur in the Be'er Ora section of the Ora shales. The gypsum comes most probably from the veins cutting the shales and the calcareous non-plastics from the limestone intercalations between the shales. The coarse quartz sand temper is probably the only constituent that derives from another part of the geological section. The pure nature of this quartz sand strongly indicates the Nubian Sandstone facies. The Ora shales are not exposed in the Timna Valley itself, but they occur in the wide outcrops around the Timna Valley at distances of only a few kilometres. A formation of 100 metres of various types of clays would be a paradise for potters, and it is unlikely that these clay deposits were overlooked by potters working in the area.

So is the Normal pottery local? This question cannot be answered definitely because the combination of materials that composes the Normal pottery is of wide geographic distribution. However, some boundaries to the potential source areas of Normal pottery can be placed. The vast areas to the north of Timna, the Galilee, the whole coastal plain, Samaria, the Judaean and Hebron mountains, the Jordan Valley and the north-western Negev cannot be regarded as potential source areas. On the other hand, the southern Negev and parts of Sinai could produce the combination of materials that compose the Normal pottery. For example, many ancient sites in Jordan (Edom) east of the Arabah, are situated geologically in a similar environment to the Timna Valley and, indeed, several polished thin sections of so-called Edomite pottery from Tawilan and other sites show close petrographic resemblance to the Normal pottery. S.N. 1076 from Tawilan is an almost perfect

analogy to some Normal pottery samples. Its clay contains a silty quartz fraction, glauconite and dolomite. Calcareous inclusions that occur among the coarse non-plastics include micritic types with quartz and dolomite, and glauconitic sandy dolomites. Typologically however, this sample is altogether different from the Normal pottery of Timna and it dates from the eighth century B.C.

An additional limit on the possible origin of Normal pottery is placed by the occurrence of slag fragments in two Normal pottery samples. This observation suggests closeness to a metallurgical site. From all this it follows that, although a definite proof cannot be given, the most logical conclusion would be that the Normal pottery is locally made. This conclusion leads to a quest for the exact location of the workshops that produced the Normal pottery. As the Ora shales are not exposed in the Timna Valley itself, either the clay was transported to Timna, or the workshops were situated near the shale outcrop. Had this pottery been produced in the heart of the Timna Valley, it would contain fragments of large angular quartzes, feldspars, micas and hornblende derived from the igneous complex that occupies the centre of the valley. The total absence of such fragments, even as infrequent accidental impurities, indicates that the Normal pottery was manufactured outside the Timna Valley.

d. Two technological attitudes

Our conclusion that the variants of the Normal pottery reflect the same geological province, and that they probably originated from one or several closely related workshops, requires a technological explanation for these petrographic variations.

It appeared promising to examine the variation trend which has at one end samples with an extremely high content of coarse temper, regardless of its mineralogical composition, and at its other end samples with a very low content of coarse non-plastics or none at all. The 41 samples of Normal pottery were thus divided into three types: (1) Normal Pottery Type I (N.P.T. I): a temper-rich type with a very high content of coarse non-plastics such as quartz sand and shale fragments; (2) N.P.T. II: a temper-poor type with a very low content of various non-plastics (quartz, shale fragments, chaff, carbonates and gypsum) or none at all; (3) N.P.T. III: an intermediate type with temper content transitional between the two end members. Each of these three types was subdivided according to four clay types: A: pure clays or clays with a negligible content of fine non-plastics; B: clays with a silty quartz fraction; C: calcareous clays with dolomite; D: calcareous clays with dolomite and silty quartz fraction (*Tables* 1–3).

Thus the 41 Normal pottery samples were divided into twelve classes. *Tables* 1, 2 and 3 show the distribution of the samples over these classes and tabulates additional petrographic properties such as the major coarse non-plastic ingredients that occur in each sample, the content of fine original non-plastics of the clay and the content of the mineral glauconite. The

Table 1. Normal Pottery Type I (N.P.T. I)
High content of coarse non-plastics

N.P.T. I A
Pure clays or clays with a negligible content of fine non-plastics

S.N.	Coarse Non-plastics Major	Minor	Glauconite Content	State of C.R.G.M.
1531	93>sh.fr.	carb.	xx	anisot.
1537	sh.fr>qz.	gyp, carb	–	inter.
1547	sh.fr>qz.	carb	–	inter.
1550	sh.fr>qz.	carb	–	isot.

N.P.T. I B
Clays with a silty quartz fraction

S.N.	Coarse Non-plastics Major	Minor	Silty Quartz Content	Glauconite Content	State of C.R.G.M.
1534	sh.fr,qz		xx	–	inter.
1555	sh.fr>qz		x	x	anisot.
1562	sh.fr>qz	carb, gyp	x	–	inter.

N.P.T. I C
Calcareous clays with fine-zoned dolomites

S.N.	Coarse Non-plastics Major	Minor	Fine Dolomite Content	Glauconite Content	State of C.R.G.M.
1554	sh.fr.	qz, glag. carb.	xx	xx	anisot.

N.P.T. I D
Calcareous clays with fine-zoned dolomites and a silty quartz fraction

S.N.	Coarse Non-plastics Major	Minor	Fine Dolomite & Silty Quartz Cont.	Glauconite Content	State of C.R.G.M.
1514	sh.fr>qz	–	xx	x	inter.
1530	sh.fr>qz.	gyp.	xx	x	isot.
1548	qz>sh.fr.	gyp.	xx	xxx	anisot.
1567	qz, pl.	gyp, carb	x	–	isot

Table 2. Normal Pottery Type II (N.P.T. II)
Low content of coarse non-plastics

N.P.T. II A
Pure clays or clays with a negligible content of fine non-plastics

N.P.T. II B
Clays with a silty quartz fraction

S.N.	Coarse Non-plastics Major	Minor	Silty Quartz Content	Glauconite Content	State of C.R.G.M.
1535	–	carb,pe,qz	xxx	–	inter.

N.P.T. II C
Calcareous clays with fine-zoned dolomites

S.N.	Coarse Non-plastics Major	Minor	Fine Dolomite Content	Glauconite Content	State of C.R.G.M.
1517		qz, carb. pl, gyp	xxx	xx	isot.

N.P.T. II D
Calcareous clays with fine-zoned dolomites and a silty quartz fraction

S.N.	Coarse Non-plastics Major	Minor	Fine Dolomite & Silty Quartz Content	Glauconite Content	State of C.R.G.M.
1510	–	qz,pl,carb	xxx	–	isot.
1513	–	pl,qz,carb	xxx	–	isot.
1516	–	pl,sh.fr.	xxx	x	inter.
1532	–	pl,carb, gyp,qz	xxx	–	isot.
1540	–	carb,gyp,pl, qz,sh.fr.	xxx	–	isot.
1543	–	pl,carb	xxx	xx	isot.
1545	–	pl,qz,carb	xxx	x	isot.
1551	–	pl,sh.fr,carb	xxx	xx	isot.
1552	–	pl,qz,carb	xxx	–	isot.
1553	–	pl,carb,gyp	xxx	xx	isot.

KEY TO TABLES 1–3

N.P.T.	–	Normal Pottery Type
S.N.	–	Sample Number
G.R.G.M.	–	Textural state of clay rich groundmass
qz	–	Quartz
sh.fr.	–	Shale fragments
pl.	–	Plants (chaff and other plant fragments)
carb.	–	Carbonates (various carbonate lithic fragments)
gyp.	–	Gypsum

isot.	–	Clay minerals optically isotropic, indicating a relatively high firing grade
anisot.	–	Clay minerals optically anisotropic, indicating a relatively low firing grade
inter.	–	Intermediate
x	–	Relatively low content of fine non-plastics
xx	–	Intermediate content of fine non-plastics
xxx	–	Relatively high content of fine non-plastics

Table 3. Normal pottery Type III (N.P.T. III)
Intermediate content of coarse non-plastics

N.P.T. III A
Pure clays or clays with a negligible content of fine non-plastics

N.P.T. III B
Clays with a silty quartz fraction

S.N.	Coarse Non-plastics Major	Minor	Silty Quartz Content	Glauconite Content	State of C.R.G.M.
1509	sh.fr>pl	carb	xx	–	isot.
1533	sh.fr,qz,pl	carb	x	–	isot.
1549	sh.fr>qz	carb,gyp	xx	–	isot.

N.P.T. III C
Calcareous clays with fine-zoned dolomites

N.	Coarse Non-plastics Major	Minor	Fine Dolomite Content	Glauconite Content	State of C.R.G.M.
498	qz>sh.fr	carb,gyp	x	x	inter.

N.P.T. III D
Calcareous clays with fine-zoned dolomites and a silty quartz fraction

S.N.	Coarse Non-plastics Major	Minor	Fine Dolomite & Silty Quartz Content	Glauconite Content	State of C.R.G.M.
1500	qz	carb.	xxx	xx	inter.
1504	sh.fr	carb,gyp,qz	xx	xx	anisot.
1512	sh.fr>qz	carb,gyp,flint	xx	–	isot.
1541	qz,sh.fr gyp	carb	x	–	anisot.
1546	qz>pl	carb,gyp,flint	xxx	–	isot.
1557	qz,sh.fr	carb	xxx	xxx	inter.
1559	qz,sh.fr carb	pl	xxx	–	isot.
1561	sh.fr,qz	pl,carb	xxx	xx	isot.
1565	qz	sh.fr,pl	xx	xx	inter.
1568	qz,carb,pl	flint	xxx	–	isot.
1571	qz,pl,carb		xxx	–	isot.
1528	sh.fr	gyp,flint, carb	xx	xx	isot.

content of the fine non-plastics (silty quartz and dolomite) were roughly estimated and graded into low, intermediate and high contents. Also for glauconite, three grades – low, intermediate and high content – were defined, but they stand for much lower absolute values compared with the same grades of the quartz and dolomite. As a result of this tabulation procedure, several correlation tendencies between the petrographic properties could be observed.

In N.P.T. I, a tendency is observed to prefer pure clays or clays with a negligible content of original fine non-plastics (N.P.T. I A) and towards the non-dolomitic silty clays with a low and intermediate content of fine non-plastics (N.P.T. I B). The calcareous clays that dominate in N.P.T. II occur in only about a third of the N.P.T. I samples, however, with a low and intermediate content of fine non-plastics. The dominance of samples with pure clays or with clays of a low and intermediate content of fine non-plastics, agrees well with the high content of the coarse temper.

In N.P.T. II there is a strong tendency to prefer one specific clay with a high content of silty quartz and dolomite (N.P.T. II D). The total absence or scarcity of a coarse non-plastic fraction is fully in agreement with the observed preference. Using the same logic, it is well understood why samples that belong to N.P.T. II A were not encountered at all. It is not certain why the highly calcareous clay with dolomite and silty quartz fraction was preferred to other highly

Table 4. Properties of two technological end members of Normal Pottery

	N.P.T. I A and N.P.T. I B 7 Samples	N.P.T. II D 10 Samples
Typology	Mainly cooking pots	Mainly big storage jars
Clay type	Plastic non-dolomitic, non-silty, somewhat silty clays	Non-plastic dolomitic silty clays
Firing colour of clay	Dark red or red brown	Buff
Surface colour	Weakly developed light cream outer surface in only some samples	Well developed light cream outer surface in some samples
Porosity	Open texture, highly porous. Elongate voids related to shale fragments	Dense texture. Slightly porous elongate voids with carbonised coronas (Burnt out chaff)
Temper	Heavily tempered with mainly coarse quartz sand and large shale fragments	Practically non-tempered
Content of coarse non-plastic impurities	Relatively low content or absolute absence of carbonate and gypsum impurities	Relatively high content of carbonate and gypsum impurities
Wheel marks	Mostly indistinct	Distinct

non-plastic clays. Either the other clays are only occasionally rich enough in non-plastics to be used without the need to add a coarse temper, or this clay was preferred because of another desirable quality, perhaps due to the finely dispersed carbonate phase that lowered the hardening temperature. Both the fine texture and the lowered hardening temperatures help to achieve density, low permeability and strength in the ceramic end product.

In N.P.T. III a tendency is observed to prefer calcareous clays with dolomite and silty quartz fraction (N.P.T. III D) and in this respect N.P.T. III shows close affinity with N.P.T. II. Mainly silty clays, with intermediate and low contents of fine non-plastic also occur in this group (N.P.T. III B), defining the affinity with N.P.T. I.

The most drastic differences in petrographic properties exist between N.P.T. II D on the one hand, and N.P.T. I A and N.P.T. I B on the other. Examining other properties of the samples which belong to these groups, more differences emerged (*Table* 4).

Although the number of typologically definable sherds is rather small, it appears that cooking pots belong to N.P.T. I A and N.P.T. I B, while large jars with handles belong to N.P.T. II D. We cannot, however, state definitely that there are no exceptions to this differentiation or that all the other types of vessels necessarily belong to other petrographic classes.

The difference in the firing colour noted in *Table* 4 is due to the difference in mineralogical composition of the clays. The highly calcareous clays of N.P.T. II D fire to a buff colour and the carbonate-poor clays of N.P.T. I A and N.P.T. I B fire to a dark red or brown colour. It was also observed that seven out of ten N.P.T. II D samples exhibit a light cream outer surface. The differences in porosity depend mainly on the content of coarse temper, especially of shale fragments. It appears also that the content of coarse non-plastic impurities is greater in N.P.T. II D samples. As these gypsum and carbonate impurities were considered to be derived from rocks associated with the clays in their geological setting, it would appear that the pure and silty clays are not so intimately associated with such rocks as the calcareous clays. Lastly, the wheel marks appear as distinct striations on N.P.T. II D samples, but as rather indistinct marks on N.P.T. I A and N.P.T. I B samples. Whether this difference between the two end members also reflects a difference in potter technique, or is merely an effect of textural differences, which result in better observed wheel marks on fine fabrics, cannot be determined. The two technological end members of the Normal pottery can be understood as a typological and functional differentiation between cooking pots and large jars. For cooking pots an open texture was attempted, hence a high content of shale fragments together with a quartz sand was used as temper, which in turn required a more plastic clay. For large jars which probably served as liquid containers and for the transport of products, both strength and impermeability were attempted. Thus a

highly non-plastic and calcareous clay was used so that coarse non-plastics would not be required, and the carbonates would decompose on firing and react with the clay, to give a strong and compact ceramic product.

Since the makers of Normal pottery could not be identified on the basis of pottery typology, we approached this problem from a different angle. The fabrics of N.P.T. II resemble the most common fabrics of Egyptian pottery. Coarse tempers are practically unknown in Egypt, where the Nile supplies silty clays of various types that can be used as such, or in combination with other fine-grained sediments, and therefore the whole tradition of pottery making is based on silty clays. Mixtures (natural and artificial) of a siliceous Nile mud and calcareous clays are also very common in ancient Egyptian pottery. Although mineralogically the silt fraction of the Nile mud is very different, the calcareous silty mixtures of Egypt are technically comparable with N.P.T. II D.

On the other hand, the large shale fragments and the quartz sand temper are basic components of Midianite pottery. Assuming that the Normal pottery was manufactured in the Arabah by the people who collaborated in the metallurgical industry, it is tempting to suggest that N.P.T. I and N.P.T. II reflect the meeting in the Arabah of two different technological traditions of pottery making, the Egyptian and the Midianite respectively. Working together in the same area and probably even using the same shale and sand outcrops, each chose the materials and treated them according to the principles of his technological traditions.

4 Rough Hand-made Pottery (*Col. Pl.* 3)

Introduction

The pottery treated under this heading, previously called 'Negev-ware pottery', comprises various types of rough hand-made pottery, containing coarse mineral and lithic tempers of various compositions which mostly agree with the local geology of the Timna Valley. A close examination of the rough hand-made pottery from the Timna Temple reveals the following facts:

(a) It is an unhomogeneous group and the samples differ from each other in the composition of clays and tempers, in their firing temperature and degree of oxidation.

(b) Some vessel types of the rough hand-made pottery group deviate significantly from the typical shapes of the so-called Negev-ware pottery.

(c) Some rough hand-made pottery exhibits petrographic affinity with members of the Normal pottery group.

These facts indicate that a variety of ceramic products share the general rough hand-made aspect and thus the term 'Negev-ware' pottery is too narrow for this group. It is therefore proposed to use the more descriptive term 'Rough Hand-made Pottery'. Some of the samples included in this group are genetically related to Negev-ware in its narrow sense.

Microscopic Examination
a. Rough Hand-made Pottery tempered with a coarse granitic sand
S.N. 1523 - Rim fragment of a bowl (*Col. Pl.* 3:4)
This sample contains a very coarse granitic sand in which most of the grains range between 1–2mm. and the largest reach 3mm. in diameter. Individual grains of quartz, alkali-feldspar of a fine vein perthitic habit and oligoclases occur together with lithic fragments of the same minerals, with or without hornblende and biotite. The sand is therefore derived from a hornblende biotite granite. This granatic sand is embedded in a silty clay-rich groundmass. The silt fraction is mainly quartzo-feldspathic with some amphiboles, biotite, pistacite and carbonates.

Some Sinai-Arabah Copper Age-Early Phase (Chalcolithic-Early Bronze) pottery from Timna Site 201A, resembles this material in its general composition. The difference is mainly textural. The volume proportion of the coarse non-plastics in Site 201A samples is greater. Another way to distinguish between coarse granitic pottery of this early period, and that of the New Kingdom period, is by the profile of the rim. All the granitic sherds of Site 201A, and other sites of this period so far analysed, prove to belong to closed hole-mouth cooking pots with thickening rims, while the rough hand-made pottery of the New Kingdom context from the Timna area belongs to relatively open shapes with uniformly thick or even thinning rims.

S.N. 1518, Rim fragment from a Negev-ware bowl (*Pl.* 108:3)
This fragment probably belongs to the typical Negev-ware bowls with a vertical wall. The coarse fractics of the non-plastics is not strictly granitic as it contains mainly large (up to 3mm. diameter), highly angular quartzes exhibiting internal growth zoning bands. This material could very well represent coarse quartz veins commonly found in igneous rocks. The potter could have collected this material from the veins themselves or from quartzitic pebbles that occur abundantly in the Timna Valley. In addition to these angular quartz grains, chaff was added to the clay in considerable quantities. Part of it is still present in a carbonised form. The clay is very fine, non-silty and completely anisotropic. This combination of materials was not observed among the Site 201A samples.

S.N. 1125 Body sherd (*Pl.* 108:2)
In this sample the granitic non-plastics occupy the whole range between silt and very coarse sand. The volume proportion of the coarse fraction is higher than in samples 1518 and 1523, approaching the large quantities of coarse non-plastics observed in the Site 201A samples. Compositionally this unsorted non-plastic assemblage belongs to a biotite granite with perthitic alkali feldspars and oligoclase. The clay-rich groundmass abounds in tiny biotites of various degrees of alteration.

Concluding remarks
The three samples representing this group are different from each other in the mineralogical composition of the non-plastics, the type of clay-rich groundmass and texture. They share the rough hand-made appearance and a low firing grade, indicated by the absence of oxyhornblende, absence of decomposed carbonates and the anisotropic nature of the clay-rich groundmass. They all indicate a geological environment of acid plutonic igneous rocks and were most probably manufactured in the Timna Valley.

b. Rough Hand-made Pottery tempered with crushed ceramic fragments (Grog)

S.N. 1124	Thick body sherd
S.N. 1126	Thick body sherd
S.N. 1520	Body sherd
S.N. 1522	Base (*Col. Pl* 3:2)

These four samples share the presence of abundant crushed pottery fragments, abundant chaff temper, black cores and patches and a relatively low firing grade. The sherds are thicker compared to those with a granitic temper and it would appear that they belong to larger vessels. The crushed pottery fragments used as temper in all four samples, belong to various types of high grade wheel-made Normal pottery. The optical contrast between grog and host is emphasised by the silty and vitrified nature of many grog fragments, versus the non-silty and anisotropic nature of the clay-rich host groundmass. Some grogs show the bimodal distribution of quartz grain size common to the Normal pottery, and others exhibit the typical zoned dolomites of the Normal pottery.

In S.N. 1124 some of the grogs have a granitic temper. In this case, granitic variants of the Rough Hand-made Pottery were crushed to be used as temper. Grogs of Midianite and Egyptian pottery are absent in these samples. In samples 1520 and 1522 some flint fragments occur and in sample 1522 some coarse quartzes also. The clay-rich groundmass of S.N. 1126 shows some tendency to form shale fragments with typical longitudinal shrinkage cracks. In these cracks, and in elongated voids associated with the chaff, gypsum occurs that was probably deposited there during burial.

Concluding remarks
The presence in the Rough Hand-made Pottery type of vitrified grog with close petrographic affinity to the wheel-made Normal pottery defines clearly the chronological relationships between these two groups. It is clear that this Rough Hand-made Pottery cannot be confused with earlier primitive pottery, as in the case of the granite tempered Rough Hand-made Pottery. The presence of Normal pottery grog strongly indicates its local origin.

c. Rough Hand-made Pottery tempered with slag fragments
S.N. 1527 Base fragment (*Col. Pl.* 3:1 and *Pl.* 108:1)
The outward appearance of this sample differs from the previously described rough hand-made pottery. The cut section shows oxidised rims of bright reddish

colours, indicating a relatively high firing grade, while the core is light grey. Chaff imprints are absent, but there are numerous black inclusions in a lighter groundmass. Under the polarizing microscope, these black inclusions prove to be slag fragments, ranging in size between 0.5 and 2.00mm. There are many different types of slag particles in this sample. Some are completely glassy with few or no gas bubbles, others are completely crystalline with radiating aggregates of fibrous and prismatic crystals. Most particles are partly glassy and partly crystalline. In some, relict, old minerals (usually quartz) occur together with the new phases. Some quartz aggregates exhibit an interstitial glassy phase. These may represent tiny chips of tuyères or other refractory elements. There can be no doubt that the slag particles in this sample were intentionally added to act as temper, since they occur in large quantity, while other coarse non-plastics, except for a few quartz and limestone inclusions, are absent. There can also be no doubt that these are metallurgical slags since the dense and homogeneous nature of these particles indicate very high temperatures. It also appears that they reflect a highly advanced metallurgical technology, because Chalcolithic slags usually show many gas bubbles and contain abundant relict minerals. It appears, therefore, that slag of the Ramesside industry was crushed and probably sifted before being added as temper to the pottery.

The clay-rich groundmass of this sample is a fine dolomitic shale with tiny magnetite crystals and it shows many shale fragments resembling in shape, size and composition the shale fragments that occur in some of the Normal pottery. The groundmass is still anisotropic but carbonates show various degrees of decomposition.

S.N. 1521 Body sherd (*Col. Pl.* 3:5 and *Pl.* 108:9)
This sample appears to be fired to even higher temperatures than S.N. 1527 and is oxidised all through. It is harder than S.N. 1527 and gives a metallic sound when knocked by a hard object.

Under the microscope this sample proves to be much poorer in slag fragments and the fragments are not so dense as those of S.N. 1527. They show a spongy structure due to a higher volume proportion and bigger size of the gas bubbles. The clay-rich groundmass is very fine silty (quartzes ranging in size from 10 to 20$_M$) and derived from a shaley sediment as the numerous shale fragments of identical composition indicate. The clay-rich groundmass and the shale fragments of this sample, show close petrographic resemblance to variants of the Normal pottery. A single carbonate inclusion is completely decomposed and it shows typical reaction rims at its contact with the surrounding clay, a clear indication of a high firing grade.

Concluding remarks
Several properties distinguish the slag-tempered Rough Hand-made Pottery:

(1) The very unusual procedure of using crushed slag as temper.

(2) The relatively high firing grade compared with the other rough hand-made pottery with granite and grog as temper.
(3) The use of shaley clays resembling those that occur among the Normal pottery.
(4) The occurrence of abundant coarse shale fragments, a technique that was used by Midianite potters and also by the potters that produced the Normal pottery.

We have as yet no explanation for these phenomena. The slag temper can be taken as an indication that this pottery was manufactured in the Timna Valley.

d. Other Rough Hand-made Pottery
S.N. 1525
To the unaided eye, this sample looks like the grog-tempered samples of Rough Hand-made Pottery. It is a thick body sherd with a wide dark grey core, shifted towards the outer surface, showing elongate cavities with blackened coronas which indicate the use of chaff temper. Under the microscope this sample proves to belong to none of the previous groups, although it shows some affinity with the granite sand and grog-tempered pottery. It contains a high volume proportion of a quartzo-feldspathic sand with a few zircon and mica grains. Mineralogically this sand resembles the coarse unsorted granitic temper that was found in other samples of Rought Hand-made Pottery, but it differs in its fine and better sorted grains (50–250$_M$). This sand probably derives from an igneous outcrop dominated by acid rocks (granite) and it could represent an original constituent of the clay-rich deposits that were exploited by the potters.

A coarses non-plastic fraction is represented by grog and chaff temper. Most of the crushed pottery fragments show petrographic resemblance to Midianite pottery. In small fragments, the 'Midianite' aspect is indicated by the light, noncalcareous nature of the silty clay, the fine spotty distribution of iron oxides and the presence of a quartz sand. In a large grog fragment, dark red shale fragments occur, identical with the shale fragments of Midianite pottery. A few crushed pottery fragments belong to the highly vitrified variants of Normal pottery. The provenance of this sample is the same as the provenance of the grog and granite sand tempered Rough Hand-made Pottery, i.e. the Timna Valley.

S.N. 1526
To the unaided eye, this sample looks like other Rough Hand-made Pottery, especially the samples tempered with grog. It shows an irregular grey core, buff rims and is rather friable, indicating a low firing grade. Under the microscope, however, it shows close petrographic resemblance to Normal pottery. The clay-rich groundmass is identical to some Normal pottery variants, it contains glauconite, zoned dolomites and a silty quartz fraction. Among the coarse non-plastics, shale fragments predominate, reaching a length of up to 2–3mm. There appears to exist a continuous transition from large shale fragments, through smaller ones with well-rounded edges,

down to a completely disintegrated clay-rich ground-mass.

Coarse carbonate inclusions are also the same as those that characterise Normal pottery: micritic carbonate inclusions, silty and fine sandy micrites, micritic dolomitic limestones, microfossiliferous carbonate inclusions, some large individual calcite crystals, re-crystallised shell fragments (fragments of fossilised shells), dolomitic and glauconitic fine quartz sand, dolomitic, calcitic and somewhat argillaceous fine quartz sand. As in most Normal pottery samples, coarse quartz sand is also present. Some slag fragments also occur in this sample.

Sample 1526 differs from Normal pottery only in the hand-made aspect and by its features related to the lower firing grade, i.e. the anisotropic nature of the clay-rich groundmass and shale fragments, and the almost unmodified state of the various calcareous non-plastics. The resemblance is so close that it is reasonable to assume that this rough hand-made pot was made of the typical materials of Normal pottery, probably even at the same site, and by the same potters. The provenance of this sample is the same as the provenance of the Normal pottery.

S.N. 1519

To the unaided eye the outstanding features of this sample are the dark red colour, the absence of a grey core and of chaff temper and a reduced and very irregular, hard outer surface. Microscopically it also shows some unusual features. The clay-rich ground-mass derives from a noncalcareous silty clay (silty quartzes not exceeding 25_M in diameter) with tiny iron oxide spots ($2–10_M$). Only shale fragments occur as coarse non-plastics. These shale fragments are only somewhat lighter than the clay-rich ground-mass, all their other properties being the same. Other coarse non-plastics, such as carbonates, gypsum and quartz sand, are totally absent. The clay-rich ground-mass and shale fragments show close resemblance to S.N. 1521 which contains abundant slag fragments.

Conclusions

This study of twelve polished thin sections of various types of Rough Hand-made Pottery showed that petrographically this is an unhomogeneous group. A variety of tempers were used: various sands derived from mainly acid igneous rocks, various grog assemblages, slag, chaff and shale fragments. Also a variety of clays were used: fine noncalcareous clay, calcareous clay with zoned dolomites with or without glauconite, and noncalcareous silty clay. There appears to be a correlation between temper and clay types used, but the number of samples analysed is too small to prove it. The samples vary also in properties reflecting firing conditions. The firing grade of most samples is very low, as evidenced by the wide black cores, narrow, partially oxidised, buff coloured rims, the undecomposed nature of carbonate inclusions, the presence of green hornblende, the aniso-tropic nature of the clay-rich groundmass and the soft friable nature of the sherd as a whole. The samples tempered with slag and shale fragments, however,

show evidence of a higher firing grade: the cores are fired to a bright red colour, their carbonates broken down, their groundmass only slightly anisotropic and they are much harder. Beyond all this diversity, the samples carry some evidence for a local origin: the samples tempered with crushed fragments of Normal and Midianite pottery are undoubtedly manufactured in the Timna Valley, and the same can be said regarding the slag-tempered samples. The samples tempered with granitic sand could theoretically originate also in other areas of granitic rocks, but this seems highly unlikely.

The realisation that the Rough Hand-made Pottery was locally manufactured leads to the question of who the potters were and in what way were they connected with the metallurgical activities in Timna. A partial answer can be obtained from the find of typical Negev-ware cooking pots, with a flat bottom and vertical walls (Rothenberg, 1972) among the Rough Hand-made Pottery. The fact that none of these samples contains materials foreign to the geology of Timna, strongly supports Rothenberg's view that the makers of these cooking pots, the local Negev population, were actually living together with the Egyptian mining expedition in the Arabah and produced, with the locally available materials, the pottery needed for their everyday life.

The presence of high grade Rough Hand-made Pottery, the presence of samples that deviate typologically from the typical Negev-ware cooking pots and the occurrence of samples showing close petrographic affinity to Normal pottery, suggest that some of the Rough Hand-made Pottery was made by other people, perhaps by the Egyptians or the Midianites.

5 Other types of pottery

To complete the petrographic analysis of the Timna Temple pottery, a brief description of five samples not comparable with any of the samples so far described, is given.

S.N. 1556 Body sherd

In this sample both clay and non-plastics are essentially different from the rest of the Timna Temple pottery. The clay-rich groundmass is rich in calcareous microfossils, while zoned dolomites, silty quartz and glauconite, so common among the Normal pottery, are totally absent. The coarser non-plastics occupy the $100–200_M$ range and only a few reach 400_M. Mineralogically this sand is composed of quartz, feldspar calcite and flint. One large (2mm.) aggregate of quartz feldspar and carbonate shows a coarse calcitic cement. The carbonates show no decomposition microstructures and the clay-rich groundmass is anisotropic, indicating a relatively low firing grade.

S.N. 1542 Rim fragment

The outstanding aspect of this sample is the occurrence of quartz feldspar sand aggregates with a coarse calcitic cement resembling the sandstone fragments of S.N. 1556. The clay-rich groundmass, however, shows zoned dolomites.

Table 5: List of petrographic samples

Pet. Sample No.	Field No.	Square	Locus	Pet. Sample No.	Field No.	Square	Locus
1105	306/5	B 20–21	101	1525	116	A-B-C 14–15	110, 112
1106	306/4B	20–21	101	1526	340	B-C 9–10	107
1107	306/2	B 20–21	101	1527	27/297	E-F 7–9	106, 107
1114	305/10	A-B 6%7	106	1528	277	C-D 10–11	107, 110
1115	305/9	A-B 6–7	106	1529	28	C-D 6–7	106
1116	305/8	A-B 6–7	106	1530	32	A-B 15–16	112
1117	305/7	A-B 6–7	106	1531	119	C-D 14–15	101, 110, 112
1118	305/6	A-B 6–7	106	1532	201	D-E 15–16	101
1119	305/5	A-B 6–7	106	1533	167	E-F 16–17	101
1120	305/3	A-B 6–7	106	1534	259	H 6–11	106, 108, 109
1121	305/4	A-B 6%7	106	1535	322	F 15–16	101
1122	305/2	A-B 6–7	106	1536	374	H-I 6	103, 106
1123	305/1	A-B 6–7	106	1537	220	G-H 12–14	109
1124	304/3	G 6–11	106, 108, 109	1538	235/3	G 12–14	109
1125	304/2	G 6–11	106, 108, 109	1539	–	–	–
1126	304/1	G 6–11	106, 108, 109	1540	170	I-J 16–17	101
1136	258/21	H 6–7	106	1541	302/10	F 17–18	101
1206	306/1	B 20–21	101	1542	233/11	I-J 18	101
1207	306/3	B 20–21	101	1543	245/73	A-B 8–10	107
1212	234/146	H 15–18	101	1544	67	I-J 16–17	101
1213	71/9	I-J 17–18	101	1545	67	I-J 16–17	101
1214	287/31	L 12–13	102	1546	127	I-J 16–17	101
1215	295/206	F 17–18	101	1547	–	G-H 16–17	101
1498	269	A-B 8–10	107	1548	364	I-J 12–13	101
1499	241	J-K 15	101	1549	259	H-G 11	106, 108, 109
1500	233	I-J 18	101	1550	293	Surface	Wall 3
1501	228	G 12–14	109	1551	211	G 15–17	101
1502	502	TT 53	110	1552	309	G-H 19	101
1503	228	G 12–14	109	1553	–	–	–
1504	319	G 7–12	106, 108	1554	295	F 17–18	101
1505	594	–	110	1555	311	F 15–16	101
1506	505	TT 53	110	1556	138	C-D 13–14	110
1507	502	TT 53	110	1557	134	A-C 13–14	110, 111
1508–1509	278	E-F 10–11	107, 109, 110	1558	303/646	F 15–16	101
1510	123	H-I 16–17	101	1559	253	J-K 14	102
1511	253	J-K 14	102	1560	–	I-J 16–17	101
1512	370	B-C-D 11–14	110	1561	172	G-H 16–17	101
1513	52	A-B 11–12	111	1562	220	G-H 12–14	109
1514	253	J-K 14	102	1563	269/78	A-B 8–10	107
1515	280	C-E 6–7	106	1564	232	I 15–17	101
1516	253	J-K 14	102	1565	231/149	H 15–17	101
1517	172	G-H 16–17	101	1566	234/149	H 15–18	101
1518	303	F 15–16	101	1567	16/50	–	–
1519	206	B-D 13–14	110, 111	1568	263/287	J-K 8	103
1520	325	C-D 12	110	1569	–	–	–
1521	Surface			1570	234	H 15–18	101
1522	258	H 6–7	106	1571	253	J-K 14	102
1523	237	B-C-D 13–14	110	1572	224/137	G 15–18	101
1524	323	A-B 6–7	105, 106				

S.N. 1566 Rim fragment of a juglet

The outstanding aspect of this sample is the highly microfossiliferous nature of the clay-rich groundmass.

S.M. 1569 Painted handle fragment

The outstanding features of this sample are the presence of small mica flakes, phyelite, mica schist, slag and gypsum fragments and a quartzo-feldspathic angular sand (100–300$_M$). Vitrified shale fragments are extremely abundant.

The presence of slag fragments points to a local origin, but this is contradicted by the phyelite and mica schist fragments that are foreign to the geology of the Timna Valley, where metamorphic rocks are not exposed.

S.N. 1508 Egyptian Offering Stand

There are no nilotic elements in this sample and it does not compare with any of the non-nilotic groups of the Malkata reference collection. It is essentially composed of a calcareous silty clay. The silt is quartzo-feldspathic and the grain reach up to 75$_M$. Coarse temper is dominated by chaff. Particles of calcareous silt occur in much lesser quantities.

Jonathan Glass

Notes

1. Rothenberg, 1972, 153–4.
2. Rothenberg, 1972, 162.
3. Rothenberg, 1972, 163.
4. This collection was kindly put at our disposal by Colin Hope, who also introduced us to the problems of Egyptian clay types.

Much of our awareness of the diversity of clay-rich materials used in Egypt is due to this carefully prepared collection from Malkata.

5. This collection was kindly prepared by V. Hankey.

6. The selection and description of the Malkata samples was carried out by Colin Hope. The detailed analysis of this collection will be published elsewhere.

7. The 17 samples with nilotic affinity represent about 23% of the Normal pottery sample, but this is quite probably an exaggerated figure since, during sampling, every sherd that looked distinctly different from others was sampled, even if it represented only a very small group of sherds. Some of the samples that later proved to be Egyptian actually represented a small number of sherds. Also emphasis was put on samples that, by unaided visual observations, appeared to resemble Egyptian pottery.

8. It became obvious during our study that visual examination of sherds alone was not sufficient to identify Egyptian materials with any certainty. Even the experience we gained by comparing thin-section data with aspects of general appearance only slightly improved our ability to make such identifications. A microscopic analysis is always needed for confirmation because most of the characteristic aspects of Egyptian materials are of a microscopic nature.

9. Slipped and burnished pottery of nilotic affinity occurs abundantly in the early phase (Layer IV) of Site 30, which most probably predates the construction of the Timna Temple.

Bibliography

Amiran, R. and Glass, J. 1979. An Archaeological-Petrographical Study of 15-W-Ware Pots in the Ashmolean Museum, *Tel Aviv*, Vol. 6, 54–9.

Bartov, J., *et al.* 1972. Late cretaceous and tertiary stratigraphy and paleogeography of southern Israel, *Israel Jour. of Earth Sci.* Vol.21, 69–97.

Hatcher, H. and Hedges, R.E.M. 1976. Elemental Compositions of sherds from Timna, and the relationship to possible clay sources (unpublished report).

Helal, A.H. 1965. Stratigraphy of outcropping Paleozoic rocks around the northern edge of the Arabian Shield, *Zeit. Deutsch. Geol. Ges.* 117, 506–43.

Hope, A.C., Blauer, H.M. and Riederer, J. 1981. Recent Analyses of 18th Dynasty Pottery in Arnold, D. *Studien zur altägyptischen Keramik*, Mainz, 139–66.

Parr, P.J., Harding, G.L. and Dayton, J.E. 1970. Preliminary Survey in N.W. Arabia 1968, *BIAL* 8 and 9, 193–241.

Rothenberg, B. 1972. *Timna*, London.

Rothenberg, B. and Glass, Y. 1983. The Midianite Pottery, in Sawyer, J.F.A. and Clines, J.A. (eds.), *Midian, Moab and Edom*, Sheffield, 65–124.

Slatkine, A. 1974. Comparative petrographic study of ancient pottery sherds from Israel, *YMH* 15/16, 110–11.

—— 1978. Etude Microscopic de Poteri Ancien de Negev e et Sinai, *Paleorient* 4.

5. COLOURED FRITS AND PIGMENTS ON EGYPTIAN AND MIDIANITE POTTERY

Among the numerous finds from the Timna Temple, a considerable quantity of sherds show decorations with pigments and coloured frits. A few characteristic specimens were selected for analysis. With the aid of the essentially non-destructive methods, an attempt was made to determine the nature of these decorations. The composition of the pigments was verified by semiquantitative X-ray flourescence (XRF) and X-ray diffraction (XRD). In addition, the surface appearance of frits and pigments on ceramic sherds was observed under a scanning electron microscope (SEM).

1 Egyptian faience

The concave-convex fragment investigated has a size of 25×30mm. and a thickness of 8mm. The convex side is coloured light green over a white slip. The concave side of the specimen is decorated with black lines and dots on a light green frit. On this side the ceramic body was first covered with a white slip before the green and black decorations were applied. The ceramic body itself is light reddish brown in colour (Munsell: 2.5 YR 6/4). The fine-grained ceramic is free from inclusions, but quartz (sand) was probably used as temper.

A semi-quantitative XRF of both sides of the sherd for elements of atomic order above 22, gave the following results:

	Main Constituents	Minor Constituents	Traces
Concave side (dark to light green)	Cu	Mn, Fe, Sn	Zn
Convex side (light green only)	Cu, Fe	Sn	Ba, Zr

The ceramic body was first coated on both sides with a slip or an engobe. The next step was the application of a copper-rich frit. The tin found by XRF is tin dioxide (SnO_2), either part of the slip or of the frit.[1] Irregular 'glazing' or inhomogeneities in the frit have resulted in patches of darker and lighter green. While the outside of the sherd is monochrome (light green), the concave inside shows additional ornaments black to brown in colour, composed of an iron-manganese-oxide pigment. It is not known whether engobe, copper frit and black decorations were applied to the fired or unfired vessel. Thus the question of one or two firings remains unanswered. The elements present as traces only (zinc, barium and zircon), are either impurities of the clay or the constituents of pigments and/or frit. Under the scanning electron microscope (*Pl.* 109:1, 2) pigment particles embedded in partly fused or molten frit are clearly visible.

Though the green colour of the frit was not identified by XRD, it may safely be assumed that it is composed of Egyptian blue, i.e. $CaCuSi_4O_{10}$. According to Noll (Noll and Hangst, 1975), Egyptian blue can occur in a variety of shades from light blue to green, depending on the method of manufacture. As stated by Denninger (1975) partial substitution of calcium by sodium can also result in colour changes from blue to green.[2]

2 Decorations on Midianite ware

Decorations on Midianite pottery are essentially composed of four colours: red, brown, black and white. The colour of the sherd bodies is buff to pinkish white (Munsell: 7.5 YR 8/2, 8/3, 7/4 or 6/4; pinkish white to light brown). The red decorations range from reddish-yellow to reddish-brown, the brown from medium to dark reddish-brown, and the blacks from dark brown to brownish-black. XRF of various red and brown decorations on sherds gave iron as the main colouring agent (with minor amounts of manganese). The black pigments are invariably rich in manganese. XRD of isolated flakes of black pigments, showed Mn_2O_3 as the main constituent (plus small admixtures of quartz, α-SiO_2). Though not analysed separately, the red and brown pigments probably contain α-Fe_2O_3 as colouring agent. They were probably applied in the form of clay suspensions enriched with iron oxide. The various shades of brown could have been made by diluting deeply coloured suspensions with light coloured or white clay slip. XRD of white decorations gave quartz reflections plus some faint lines due to admixed clay substances.

The colours of the ceramics bodies of the sherds indicated firing under oxidising conditions. Unlike black iron pigments, manganese black is unaffected by excessive air during firing.

The surface morphology (cf. *Pl.* 109: 3–5) does not reveal any sign of sintering or degassing. Clay and pigment particles still retain their original shapes. The firing process was thus carried out at relatively low temperatures (around or slightly above 900°C). Decorations were applied directly onto the body, prior to firing. Due to low firing temperatures, ceramic bodies as well as pigment decorations are porous, although the pigments adhere firmly to the body. The quality of the pink to cream-coloured Midianite ceramic, with its delicate decorations in various attractive shades of brown, red, black and white, is evidence of outstanding craftsmanship.

Hans-Gert Bachmann

Notes

1. The use of SnO_2 as opacifier in lead glazes, is well known from medieval and other glazes. It is a constituent of the glaze rather than of the engobe. Only very rarely was SnO_2 added to the slip or engobe because mixing tin dioxide with the glaze itself helped to save tin. Adding SnO_2 to lead glazes was common practice to make the glazes opaque. Its addition to alkaline glazes was unusual. To our knowledge, SnO_2 has not yet been reported either as a constituent or unintentional admixture to glazes, frits or slips in Egyptian ceramics.
[*Note by the editor:* The present chapter was written by H. G. Bachmann in 1976; see now Kaczmarczyk and Hedges, 1983, 78–94, a detailed study of tin in Egyptian faience; see also Tite, Freestone and Bimson, 1983, 17–27, on methods of faience production, and Kalsbeek and London, 1978, 47–56, on methods of production of Midianite pottery.]

2. XRD of a series of blue and green pigmented frits on Amarna pottery of very similar appearance, invariably gave the compound $CaCuSi_4O_{10}$ plus α-SiO_2 (quartz) as the main constitutents.

Bibliography

Denninger, E. 1975. (Personal communication quoted in Holl and Hangst, 1975.)

Holl, W. and Hangst, K. 1975. Grün- und Blau-pigmente der Antike, *N. Jahrbuch f. Mineralogie*, Monatshefte, 529–40.

Kaczmarczyk, A. and Hedges, R.E.M. 1983. *Ancient Egyptian Faience*, Warminster.

Kalsbeek, J. and London, G. 1978. A Late Second Millennium B.C. Potting Puzzle, *BASOR* 232, 47–56.

Munsell Color Company Inc. 197?. *Soil Color Charts*, Baltimore, Maryland.

Tite, M.S., Freeston, I.C. and Bimson, M. 1983. Egyptian faience: an investigation of the methods of production, *Archaeometry* 25, 17–27.

6. CATALOGUE OF THE EGYPTIAN FINDS

Introduction

The following catalogue of Egyptian finds, comprising 260 items, is divided into eighteen sections:

1. Displaced fragments of the shrine (Eg. Cat. 1–11).
2. Stone sculpture in the round (Eg. Cat. 12–15).
3. Sistra (Eg. Cat. 16–25).
4. Counterpoises of menat-necklaces (Eg. Cat. 26–40).
5. Bracelets (Eg. Cat. 41–83).
6. Ushabti (Eg. Cat. 84).
7. Cat figurines (Eg. Cat. 85–95).
8. Jar stands (Eg. Cat. 96–101).
9. Inscribed faience vessels (Eg. Cat. 102–110).
10. Decorated faience vessels (Eg. Cat. 111–174).
11. Amuletic Wands ('Magic Knives', Eg. Cat. 175–179).
12. Inscribed glass sherd (Eg. Cat. 180).
13. Scarabs, scaraboids, plaques (Eg. Cat. 181–194).
14. Amulets (Eg. Cat. 195–215).
15. Miscellaneous faience objects (Eg. Cat. 216–228).
16. Stone objects (Eg. Cat. 229–264).
17. Offering Tables 18)
18. The rock stela

The Egyptian glass objects are dealt with separately in Chapter III, 15–17; the beads and pendants are discussed in Chapter III, 14, and a number of apparently Egyptian metal objects are dealt with in Chapter III, 8–11.

With the exception of sections 1 and 2, and some objects in 13–14, 12 and 16, all the objects in the catalogue were made of faience. This fragile material was used for votive or funerary purposes, except for certain types of drinking vessels and jars, scarabs, amulets and beads. With the exception of the ushabti (Cat. 84), which is clearly funerary and difficult to explain in the Timna Temple, all the objects are either of a votive or ritual nature.

1 Displaced fragments of the shrine (Eg. Cat. 1–11)

A fair reconstruction of the shrine can be made, taking the displaced structural fragments and including several pillars, parts of a cornice, a fragment of relief and several pieces of inscriptions, in conjunction with the plan of walls and foundations uncovered during the excavation (*Illus.* 6). In a strict sense, it may be going too far to call this shrine a 'temple'. The Egyptian language, relatively rich in architectural terms, most frequently uses three separate expressions for this concept. These are *pr*, *hwt* and *r3-pr*. Pr, basically meaning 'house', when qualified by the specification 'of the deity NN' designated first 'temple' and was then extended to mean 'temple-estate', 'temple-domain' (Wb I, 515 no. 7; 514 no. 2; Badawy III, chap. III). *Hwt* 'castle', 'mansion' was frequently qualified with the explicit specification 'of the god'. *Hwt-ntr* was the usual designation for the actual temple as distinct from the temple-estate (Wb III, 2, nos. 2–8). *R3-pr* ⌐⌐, less frequently used in the 'temple' context although this appears to be the original meaning of the expression, was used to indicate a temple-precinct or chapel (Wb II, 397, nos. 6–7). In the spelling of these terms, the determinative ⌐⌐ clearly shows that a building was involved. The typical cult temple of the Egyptian New Kingdom, as defined by these terms, had three distinct elements: a) a forecourt; b) a pillared or columned hall; c) the sanctuary, around which subsidiary chambers might be clustered (Badawy, III, 176). To qualify as a 'temple' in the Egyptian sense of *hwt*, *pr* or *r3-pr*, the Timna structure should be a building containing both a sanctuary and a hypostyle hall, which was entered through an enclosed forecourt. A glance at the plan shows that it was a small structure, built flush against the base of one of 'Solomon's Pillars'. In front stood a *pronaos* of approximately the same size which, in turn, was fronted by a courtyard enclosed by three walls.

Egyptian has two common architectural terms for lesser religious structures: *'itrt* and *k3ri³*, both of which designated a chapel or shrine. The usual determinative ⌐⌐ of *'itrt* suggests simply that it was a walled structure and *'itrt* is the more general and vaguer of the two terms. The determinative of *k3ri³*, on the other hand, represents a specific structure, a large shrine. Usually this is the Upper Egyptian *naos*, with cavetto cornice and curved roof all seen from the side. When part of a large temple, it was frequently of stone and housed the cult statue of the god.

The main axis of the Timna inner sanctuary (*'naos'*) is in direct alignment with a niche carved into the cliff face, which was obviously intended to hold a cult-image. Flanking it, but higher up on the cliff-face, is an additional pair of niches, each apparently intended to receive one end of a pair of architraves, the other end of which would have been supported by a pillar seated on a square base close to each of the outer corners of the structure. Since fragments of four pillars, two square with no capitals (Cat. 4, 5) and two with Hathor-head capitals (Cat. 2, 3) were found, we may assume that two of them supported the side architraves, as well as the ends of the lintel in front of the Temple. To take the additional weight and thrust of the latter over the entrance, two additional pillars served as jambs. These were the Hathor-headed pillars which bore the goddess's face on two opposite sides, thus clearly intended to be seen from two sides. The square pillars, one of which bears a trace of an incised inscription (Cat. 5), were therefore located at the outer corners of the structure. The lintel and architraves which they supported had a cavetto cornice with torus moulding, two fragments of which were found (Cat. 6, 7). Although no recognisable fragments of the roof were found, it is clear that it must have had the characteristic curve normally found on the roofs of *naoi*, which the Timna inner sanctuary appears to have been: a large *naos*-shrine of the *k3ri³* type. That the side and front walls were solid from ground-level to roof, seems clear from the frag-

ment of inscribed relief (Cat. 1) where, although the scene itself is illegible, the presence of a pair of vertical cartouches shows that it came from high up on the wall, since they would have been written either in front of or behind the king's head. What may have been represented was a figure of the king, Ramesses II, who would then have been responsible for the construction of the *naos* (see below, par. 18), in a gesture of adoration, probably towards the cult image of the chapel.

Eg. Cat. 1*.[1] *Fig.* 22:10. *Pl.* 50; 110:3. Field No. 464/1, Locus 107. Fragment of relief and inscription, white sandstone. H. 39cm., W. 34cm., Th. 21.5cm.
Scene and inscription incised, but so badly weathered and damaged as to be virtually illegible. At the left, and apparently facing left, are traces of a scene which cannot be interpreted. All that can be made out is a slightly curved line at the centre, beneath which are small, squarish traces set above two vertical lines. To the right of these, a third vertical line seems to continue upwards, incorporating the lower part of the curved line and then continuing vertically almost to the top of the fragment. It is a borderline, setting off a pair of vertical cartouches at the right. Close examination of the supposed curved line reveals that it did not exist originally, but developed when part of the surface of the stone cracked and weathered. Of the first of the pair of cartouches to the right of the borderline, only the bottom stroke and the left half of the curve above it are visible ⋃ just below the upper edge of the stone. This is followed by a lacuna with illegible traces, but enough space for the restoration of the title *s3 R' nb h'.w*, 'the Son of Re, the Lord of Diadems', which normally stood before the second cartouche. The outline of the major part of the second cartouche is intact, and the hieroglyph *ms* is preserved in the lower left-hand corner, although the right hand edge is damaged by a lacuna which fills the lower right hand quarter of the cartouche. Unquestionably the hieroglyph *s* or *sw* is to be restored here. Clearly, the royal *nomen* in the cartouche ends in *-ms-s(w)* or *ms-sw* '-messes' and the name can only be restored as a '[Ra]messes', with some additional element in the cartouche to indicate which King Ramesses was meant. Immediately above the *ms*-sign are damaged but legible traces of a horizontally written *mry*. Between this and the top curve of the cartouche, the surface of the stone is badly damaged and traces of whatever hieroglyphs were there are completely illegible; but there is just enough room to restore the signs. The name would then be [Rs]-ms-[s(w)]-mry-['Imn], '[Ra]mes[ses]-mi[amun]', the *prenomen* of Ramesses II. Beneath and to the left of this cartouche is a trace of a hieroglyph which may be restored *dίw*, 'granted', the beginning of a stereotyped pious wish such as 'grant life like Ré forever and ever'.

Observation: Seemingly two objections may be made to the restoration of the *nomen* as 'Ramesses-miamun', but neither is decisive. The first could be that the sign read as *mry* is facing the wrong way, but this is not important and happens quite often. The second, and seemingly more serious, is that the *mry* and the traces appear on the photograph to be the hieroglyph 🜚 *Dhwty* (Kitchen, 1976a).[2] If this were the case, the name could only be read as *Dhwty-ms*, 'Thutmose', and this could only be the *nomen* of one of the Thutmoside kings of the 18th Dynasty. This would present serious chronological problems since not one of the dateable Egyptian finds is earlier than the 19th Dynasty. However, careful study of the stone at different times of day and under many different lighting conditions, showed that the apparent *Dhwty* on the photograph does not exist.

Eg. Cat. 2*. *Fig.* 23:2. *Pls.* 37–40. Field No. 433/1, Locus 106. Pillar, white sandstone. H. 11.3cm., W. 28cm., Th. 33cm.
Face of the goddess Hathor carved on two opposite sides of the capital in high relief. Due to weathering, virtually all detail of both faces is lost, but the cow-ears of the goddess can be distinguished. She wears the characteristic circlet-type head-dress, minus the sun-disk and horns. The tresses of her wig have pointed, rather than curved ends.

Eg. Cat. 3. *Fig.* 23:1. *Pl.* 116:1. Field No. 434/1, Surface find. Pillar, white sandstone. H. 48cm., W. 27cm., Th. 30cm.
Capital only. Head of the goddess Hathor carved on two opposite sides in high relief. On one side (*Fig.* 23:1b) only the right half of the goddess's head is preserved. On the other (*Fig.* 23:1a) it is intact. Otherwise the details are identical. The deity has bovine ears and is unmistakeably Hathor, with the characteristic head-dress (minus the horns) of the goddess on top of her wig. The lappets of the wig do not flare inwards, towards her chest, but hang straight down. The right lappet on each face is divided vertically into three tresses. The faces on the pillar are badly damaged by weathering, so that none of the facial features of the goddess are preserved.

Eg. Cat. 4. *Fig.* 23:3. Field No. 457/1. Locus 107. Pillar, white sandstone. H. 68cm., W. 41.5cm., Th. 35cm.
Fragment. No traces of inscription or decoration preserved.

Eg. Cat. 5. *Fig.* 23:4. *Pl.* 113:4. Field No. 470/1. Locus 107. Pillar, white sandstone. H. 93cm., W. 27cm., Th. 34cm.
Incised inscription. On one face, barely visible, are two vertical border lines and a few illegible traces of hieroglyphs, one of which may be the sign *mry* 🜚 'beloved'.

Eg. Cat. 6*. *Fig.* 24:5. *Pl.* 110:1. Field No. 424/1. Locus 107. Cornice, white sandstone. L 66cm., W. 30cm., Th. 66cm.
From the façade of the shrine, probably the front right side. Unquestionably a fragment of a cavetto cornice with torus moulding.

Eg. Cat. 7. *Fig.* 24:4. *Pl.* 110:2. Field No. 454/1. Locus 107. Cornice(?), white sandstone. L 46cm., W. 26cm., (Th. 30cm., broken).

Fragment, perhaps from a cavetto cornice with torus moulding. If so, the fragment would have come from a side end of the cornice.

Eg. Cat. 8. *Fig.* 22:8. Field No. 227/1. Locus 101. Fragment of inscription, white sandstone. H. 5cm., W. 5cm..
Incised with traces of a hieroglyphic inscription which may be restored ⬦≋ '[*ntr n*]*fr nb t3.wy*', 'the goo[d god], the lord of the Two Lands'.

Observation: This inscription containing the title and name (now lost) of a king, was originally a descriptive caption to a relief of the king, standing before or behind his head. As such, it probably came from one of the walls of the shrine.

Eg. Cat. 9. *Fig.* 22:9. Field No. 245/4. Locus 107. Fragment of inscription, white sandstone. L 14cm., W. 11.5cm., Th. 7cm.
Broken away from a much larger inscription, this small fragment preserved two damaged hieroglyphs written vertically within a cartouche, of which only the straight left side is preserved. The hieroglyphs appear to be 🜚 and ≋ (the latter is certain), indicating that the name should probably be restored [*Rs-ms-sw*] – *m[vy-'I] mn*, '[Ramesses] -m[ia]mun', the *nomen* of Ramesses II.

Eg. Cat. 10. *Fig.* 22:5. Field No. 276/1. Locus 111. Fragment of inscription, white sandstone. L 8cm., W. 3cm.
With traces of two badly damaged incised hieroglyphs, the upper of which may be the coil of rope 🜚 *w*, while the lower is a portion of a thin horizontal sign.

Eg. Cat. 11. *Fig.* 22:6. Field No. 234/2. Locus 101. Fragment of inscription, white sandstone. H. 9.3cm., W. 7.5cm., Th. 2.3cm.
Traces of an incised hieroglyph, possibly the sign ﹏ *n*.

2 Stone sculpture in the round (Eg. Cat. 12–15)
Of the four pieces of stone sculpture in the round dealt with here, the two sphinxes (Cat. 12 and 13) undoubtedly represent the customary pair which flanked the entrance to a shrine (Badawy, III, 479). Although in larger temples the number of sphinxes might be large enough to form processional avenues leading up to the façade, two would have been sufficient for Timna which was a small structure. Although the head of one (Cat. 12) and the body of the other (Cat. 13) are missing, they were probably identical for each sphinx.

Cat. 14 is a fragment of a female statue and it is tempting to see it as the cult-statue of the deity to whom the shrine was dedicated. However, this would be the goddess Hathor, but the statue lacks the characteristic attributes of the goddess, particularly the cow-ears. To discover the statue's identity, and assuming that Ramesses II was the actual builder of the shrine at Timna (see below, par. 19), the answer could be supplied by the façade of the small Hathor temple at Abu Simbel, where the façade features two colossi of the wife of Ramesses II, Ahmose-Nefertari, wearing a double-plumed horned sun-disk, the head-dress of Hathor and, as a queen, a uraeus on her forehead (Badawy, III, 314–318; Fig. 172). Since the Timna statue appears to have worn a uraeus, it is reasonable to suppose that the statue was of Nefertari and that, as at Abu Simbel, she is represented as Hathor. If this is correct, then it is quite possible that this was the cult-statue. It should be noted that Cat. 12–14 are of the same stone as the foundation of the *naos* itself, which is a good indication that they were contemporary with the building of the *naos* and not later additions. The same does not necessarily apply to Cat. 15, which is a fragment of a tiny alabaster statuette of an animal.

Temple statuary (as opposed to statues set up in temples) had a more than ornamental purpose. Ideologically it was of both a symbolic and religious nature, being simultaneously an expression of the king's power and a mediator between the worshippers at the shrine and the shrine's deity (Badawy, III, 473–478).

Eg. Cat. 12*. *Fig.* 25:2a-d. *Pl.* 114:1. Field No. 380/1. Locus 107. Sphinx, white sandstone. H. 30cm., W. 12cm., Fragmentary.

The head and upper portion of the body are lost, and there is some damage to the left upper buttock. The base is unusually high, almost as if it were to be set into something so that part of it was not to be seen, but this is uncertain. A groove running just beneath the upper edge on all four sides produces a cornice-like effect on the top of the base. The sphinx itself is in a semi-*couchant* pose. It sits on the haunches of its rear legs but is raised up on its front legs, the paws of which are missing. Even in this attitude, the body is quite shortened. No trace of the animal's tail is visible. The workmanship appears crude, but this may be due to the damaged and weathered condition of the stone.

Eg. Cat. 13. *Fig.* 26:1, 2. *Pl.* 115:1, 2. Field No. 380/1. Locus 107. Sphinx, white sandstone. H. 30cm., W. 12cm., Fragmentary.

Only the head and shoulders are preserved, although the facial features are much damaged and the nose lost. The head is of a king, as is clear from both the beard and the uraeus on his forehead. As a head-covering he wears the nemes-headcloth, the lappets of which are short and have a pronounced curve where they drape over the shoulders. The ears are slightly over large with no trace of any internal detail. The eyes are almond-shaped and deeply set over rather prominent cheekbones, so that they present an almost bulbous appearance. There is no trace of any hollows round the jowls and no apparent detail round the mouth. The mouth itself is no longer visible due to the weathering of the stone. It is difficult to generalise on the impression given by the face because of the lack of nose, but it does tend to present a remote and vacuous expression, so typical of much of Ramesside sculpture. On the left shoulder, faintly visible in a raking light, is the upper

portion of a cartouche containing the *prenomen usr-mz't* (?) – r' . . . 'user-ma'at-re' (?).

Seen in profile, the head juts up sharply at a right angle from the shoulders of the sphinx, without any apparent neck and on a straight axis with the shoulders, whose rear musculature is suggested rather than delineated by a fold or curve in the stone. Clearly the attitude of the sphinx when intact was one of crouching on all fours, with the front legs and paws stretched out forwards, while the rear of the animal rested on its haunches. While it is tempting to assume that this is the head and shoulders of the sphinx Cat. 12, it is clear from the different angles of the backs of each that this is not the case.

Eg. Cat. 14*. *Fig.* 25:1. *Pl.* 117:1. Field No. 245/3. Locus 107. Statuette, white sandstone. H. 20.5cm., W. 12cm., Fragment.

The head (minus the left side) and the right shoulder are preserved. The stone is weathered and pitted with resulting damage to the preserved portions of the face. Eyes, nose and lips are modelled simply with no detail. The lips are thick, the nose squat, the deep hollows below them emphasise the almond-shaped eyes. The right ear (the left one is not preserved) is set high up on the side of the head. The wig (or wig-cloth) flares out widely in front, beneath the chin. Traces of some sort of head-dress (crown ?) are visible on the top of the head and there are also traces of an ornament on the forehead. It is clearly a statuette of a deity, or a queen in the guise of a deity (see above, p. 116). If the former, the goddess may be Hathor, but this is not certain because the ear is human and not the characteristic cow's ear of Hathor. If it is a queen, the traces on the forehead are probably the remains of a uraeus. The head-dress might have been a simple round circlet. The presence of a what appears to be a thin back pillar (uninscribed), reaching as far as the head, suggests that the original complete statuette was a seated figure. The stylistic features, particularly the almond-shaped eyes and the serene, almost bland and vacuous stare, suggest a date in the Ramesside period.

Eg. Cat. 15. *Fig.* 22:2. Field No. 125/1. Locus 101. Statuette, alabaster. H. 4.7cm., Fragment.

All that is preserved is the left rear paw of an animal, possible a small sphinx, plus a small segment of the pedestal on which it rested.

3 Sistra (Eg. Cat. 16–25)

Sistra of two distinct forms were found, the first (Cat. 16, 17, 22, 24, 25) being Hathor-headed, the second (Cat. 18, 19, 23) having the shape of a pylon-tower with a cavetto cornice and torus moulding. The shape of the remaining two cannot be determined, since only fragments of their handles are preserved. The sistrum was a musical instrument, a type of rattle, which was used especially in cult ritual (Sachs, 1920; Hickmann, 1949; idem, 1954). The handle and head could be made of a number of materials, but the rattles themselves would have been of metal. The sistrum was particularly associated with Isis (although it was used in other cults as well, notably that of Hathor) and if no other evidence for the identity of the deity to whom the Timna shrine was dedicated had been available, the presence of the Hathor-headed sistra would have strongly pointed to her. Other evidence, however, is available in the form of her name on one sistrum handle (Cat. 21) as well as on other Timna finds (Cat. 33,55, 215, 218, 220, 221), where she is 'Hathor, Lady of Turquoise', the same title as she bears at Serabit el-Khadim where Petrie found a number of Hathor-headed sistra (Petrie, 1906, 146–7; Pl. 151:4–15). The three sistra fragments in the form of a pylon-tower may have been part of the Hathor-headed ones. The name Hathor means 'the house ⌷ of Horus' and the tower forms the base of her head-dress on which, in representations other than sistra heads, were mounted the cow's horns and plumed head-dress of the goddess. At least two of the Timna pieces (Cat. 24, 25) may not have been functional, since they are flat on one side. These may have been either model or funerary sistra, designed as an offering for a foundation deposit or ex-voto, or to be included among the funerary ornaments of a deceased worshipper.

Eg. Cat. 16*. *Fig.* 27:1. *Pl.* 118:1. Field No. 303/1. Locus 101. Sistrum, pottery, blue glaze, traces of black decoration. H. 8cm., W. 4cm., Th. 5 cm. Fragment.
Modelled in low relief. On each side is the head of the goddess Hathor with her characteristic wig and cow's ears. On one side, the left half of the face is preserved, with the corresponding right half of the face intact on the other. At the width of the neck, and penetrating inside to the height of the bridge of the nose, are traces of an oval hole which received the tang of the handle to which the sistrum head was originally fixed. The modelling of the facial features is simple. On the obverse (showing the left side of the face) the break comes just to the left of the corner of the right eye, continues perpendicularly down along the right side of the nose, curves around the right nostril, and ends with a sharp diagonal thrust from the right edge of the mouth to the inner edge of the left side of the head-dress. The cow's ear is modelled in detail. The ridge between the upper lid of the left eye and the brow is subtly modelled, as are the nostrils. The lips are thick and separated by a single straight line. The eye, which is represented by a narrow almond shape with an almost oriental cast, does not show the pupil, but this may have been painted on since the decoration of the wig is faintly indicated by a series of perpendicular striations in black paint beneath the ear. On the reverse (showing the left side of the goddess's face) much less of the facial features are preserved. Here the break has occurred immediately to the right of the nose and mouth, and the surface above the right temple has flaked off. Otherwise, the preserved details correspond to those of the face preserved on the obverse.

Eg. Cat. 17. *Fig.* 27:2. Field No. 278/1. Loci 107, 109. Sistrum, pottery, blue glaze, black decoration.

H. 4cm., W. 4.5cm., Th. 3 cm. Fragment of head. Decorated on each side with a head of the goddess Hathor modelled in low relief. The facial features of the goddess on both obverse and reverse have completely disappeared. The head is identified as being of Hathor because of the cow's ears, details of which have been reinforced and outlined in black. The tresses of the wig have also been delineated with black paint.

Eg. Cat. 18. *Fig.* 27:3. Field No. 289/1. Loci 101–102. Sistrum, faience, very pale brown frit, blue glaze, decorated with black paint. H. 4.5cm., W. 4.5cm., Th. 4.5cm.
Fragment consisting of the upper corner of a rectangular sectioned sistrum. On one face the corner of a doorway is preserved while above, just below the top, a cornice is indicated by a pair of vertical lines moulded in high relief.

Eg. Cat. 19*. *Fig.* 28:1. *Pl.* 119:1. Field No. 241/1 + 225/1. Locus 101. Sistrum, faience, white frit, traces of greenish glaze, decorated with black paint. H. 9cm., W. 4cm., Th. 2cm. Fragment.
Rectangular in section, this sistrum has the shape of a pylon tower with a cavetto cornice and torus moulding on all four faces. On the flat roof of the tower is a cartouche bearing traces of a *prenomen* which begins

Mn---R' which could be restored as the *prenomen* of Ramesses I, 'Men[pehty]re' (Gauthier, 1914, 2–9); Seti I, 'Men[ma]re' (ibid, 10–27), Amenmesse, 'Men-[mi]re-[setepenre]' (ibid, 127–9, especially 128 n.1) of the 19th Dynasty, or even Ramesses XI 'Men[ma]re-[setepenptah]' (ibid, 220–4) of the 20th, depending upon the width of the original, intact sistrum.

Observation: It is highly unlikely that the name was of either the first or last of these kings, and the choice is between Seti I and Amenmesse. The former is more probable, but the latter cannot be ruled out, cf. Cat. 97, 104, 180.

Eg. Cat. 20. *Fig.* 29:4. *Pl.* 120:3. Field No. 277/4. Locus 107. Sistrum, faience, white frit, green glaze, black ink. H. 2.7cm., Diam. 4.2cm.
Fragment from the lower part of a sistrum handle, inscribed on both sides. In each case the inscription is framed within a rectangular border and the preserved text concludes an inscription. On one side it reads: *d i.w 'nh '*[--- a royal name] granted life', and on the other: 'nh d.t '[------- who li]ves forever'.

Eg. Cat. 21*. *Fig.* 29:5. *Pl.* 120:2. Field No. 277/2. Locus 107. Sistrum, faience, brown frit, green glaze, black paint. H. 4.3cm., Diam. 2.7cm.
Fragment of the lower part of the handle. Between the two border lines, in a single vertical column of hieroglyphs, facing to the left, are the remains of an inscription reading [m]fkt mry '--- beloved of [Hathor, the Lady of] Turquoise'. This, in turn, indicates that a royal name was inscribed on the missing portion of the sistrum.

Eg. Cat. 22. *Fig.* 27:4. Field No. 276/18 B. Locus 111. Sistrum, faience, very pale brown frit, blue glaze, decoration in black paint. L. 2.4cm., W. 2.5cm., Th. 1.2cm. Fragment.

Decorated on one side only, this fragment is of the left ear of the goddess Hathor plus part of her head-dress, which consists of a pleated coiffure of which four tresses run across her forehead, while three hang down around her ear.

Eg. Cat. 23. Not illustrated. Field No. I-J/16–17 (B)/3. Locus 101. Sistrum, faience, white frit, green glaze. H. 3.8cm., W. 3.1cm., Th. 1.1cm.

Fragment representing a corner of the cornice-shaped top of a rectangular sectioned sistrum.

Eg. Cat. 24*. *Fig.* 30:2. *Pl.* 118:2. Field No. FG/11–12. Locus 109. Sistrum, faience, white frit, green glaze. H. 3cm., W. 5cm., Th. 1.6cm.

Fragment of the face of the goddess Hathor. Broken away at the jaws, the entire left and most of the right side of the visage is preserved. The details of the face, left ear, hair and head-dress are crudely modelled in high relief. Although the bottom half is missing, and the back of the piece is flat, there is no doubt that the fragment was part of a sistrum (cf. Petrie 1906, Pl. 151:12, 13 and 15 which has most of its handle intact).

Eg. Cat. 25*. *Fig.* 30:1. *Col. Pl.* 5. Field No. 242/1, 251/1. Locus 107. Sistrum(?), faience, brown frit, blue glaze, decorated with black paint. H. 5cm., W. 5.7cm., Th. 1.7cm. Fragmentary, reassembled from two pieces.

Although the left half of the face, from the cheekbone to the temple, is missing, as is the right ear and the forehead above the ridge formed by the eyebrows, the edges of the right and left jaws are intact enough to show that this mask was part of a composite piece and that the head-dress (now lost) was made separately. The face itself has large, well-shaped almond eyes outlined in black. The nose is straight and regular, the lips thin, the nostrils flaring; the two vertical lines between the top of the upper lip and the bottom of the nostrils are prominently marked. The chin is truncated and appears weak. Something was affixed to the chin at its lower edge and set in a little from the front. Whether this was a beard or a handle cannot be determined since it has been broken away cleanly and completely. However it is unlikely that it was a beard, considering that the chin, though seemingly truncated, presents a foreshortened pointed effect in conjunction with the angle formed by it and the sides of the jaws extending upwards and outwards to the ear-root. If a head-dress or coiffure can be visualised affixed to the head, it is clear that the inner edges of the lappets would angle inwards, following the line of the jaws and then curl outwards in a volute, a phenomenon is more common with a female head. Furthermore, the preserved right ear-root is on a axis exactly level with the eye. Although the ear is lost, the protruberance of its root is tiny, too tiny for a human ear. In view of its position and the angle of the jaw, what is suggested as a restoration is

a cow's ear. If this is correct, it is probable that the mask represented the goddess Hathor.

Observations: Petrie found a large number of plaques representing Hathor at Serabit el-Khadim (Petrie 1906; Pls. 152–3) but none of these were composites like the Timna mask. All depict the goddess with a coiffure. However, among the sistra was one (ibid: Pl. 151:10) which may have been a composite since it also was made without a wig, although it does have the base of the head-dress of the goddess intact. It further differs from Cat. 25 in that the back is not flat, but bears a second face of Hathor. If the projection – of which only the broken traces just below the chin are preserved – represents a handle, the Timna piece could have been a sistrum. That it is flat on the back would not argue against this (cf. Cat. 24). On the other hand, Randall-MacIver and Woolley found a number of masks at Buhen in Nubia similar in size to the Timna mask. These were placed on the face of the mummy (Randall-MacIver and Woolley 1911; Pl. 60–61). The Buhen masks superficially resemble the Timna mask in that they have neither ears nor head-coverings, but their material is different (plaster) and the appearance of the faces seems more like portraits and less like the stereotype narrow-jawed face of Hathor. However, if the projection, which is only a broken, unclear trace just below the chin, does in fact represent a beard, then it is possible that the Timna mask does not represent Hathor but, like the Buhen objects, might have been a mummy-mask.

4 Counterpoises of menat-necklaces (Eg. Cat. 26–41)
The menat-necklace ⟨𓏲⟩ consisted of a number of strands of tiny beads which were fastened at their ends to two single strands of large beads which, in turn, were each attached to a metal, stone or faience counterpoise of a long, narrow trapezoid which terminated in a flat disk. From the Middle Kingdom onwards it was associated especially with the Hathor cult. In art it is shown in the hands of her female worshippers and has been found among their funerary paraphenalia. Likewise it has been found at sites sacred to her, among the foundation deposits or as votive or cult objects. Only the counterpoises of such necklaces were found at Timna (although it is probable that many of the loose beads found came from them originally). These were probably votive in nature, rather than functional, since they were of faience, which is too fragile to survive a functional or cult use for very long (Barguet, 1953; 103–111; Aldred, 1971).[3]

Eg. Cat. 26*. *Fig.* 31:3. *Pl.* 121.1. Field No. 115/1. Locus 112. Counterpoise of necklace, faience, white frit, blue glaze. L. 2.6cm., W. 1.8cm., Th. 0.7cm.

Fragment, inscribed vertically on one face in black with a cartouche. What can be read is ⟨cartouche⟩ ------

hprw ------- *Rʿ*, to be restored ⟨cartouche⟩ [*Wsr-*] *hprw-[RʿStp-n-]Rʿ* 'User-khepru-re-sotep-en-re', the *prenomen* of Seti II, the sixth king of the 19th Dynasty (*c.* 1207–1202 BC).

Eg. Cat. 27*. *Fig.* 32:5. *Pl.* 121:6. Field No. 241/5, 248/1, 118/1. Loci 101, 104. Counterpoise of a *menat*-necklace, faience, white frit, blue glaze, decorated with black paint. H. 3.7cm., W. 4.1cm., Th. 0.5cm. Fragment reassembled from three pieces, decorated on one face only, pierced through the thickness at the upper corners for attachment to the collar. Framed between the two vertical border lines in the centre is a cartouche with the name ⬭ *Hk3-m3't-R'* 'Hika-mare', the *prenomen* of Ramesses IV of the 20th Dynasty. Beneath the cartouche, at the right, is a damaged solid sign, clearly an element in the title which preceded the king's *nomen* and is to be restored as ☉ 'the son [of Re]'.

Eg. Cat. 28*. *Fig.* 31:2. *Pl.* 120:6. Field No. 206/1. Locus 110. Counterpoise of a *menat*-necklace, faience, white frit, blue glaze, inscription in black paint. H. 4.6cm., W. 2.5cm., Th. 0.5cm. Fragment.
Inscribed on one side with a single, vertical band of hieratic signs facing to the right. What is preserved reads: ⬭⬭ *m3't? s3 R'R'-ms-s(w) di.w 'nh.x* where the sign ⬭, though written last in the first cartouche, is nevertheless to be read as the second element in the *prenomen* of Ramesses IV, [Hk3]-m3't-[R'], with the entire text reading '---- [Hika]-ma[re], the Son of Re, Ramesses, granted life'.

Eg. Cat. 29. *Fig.* 32:1. Field No. 258/20. Locus 106. Counterpoise of a *menat*-necklace, faience, brown frit, blue glaze, decoration in black paint. H. 1.8cm., W. 3cm., Th. 1cm.
Decorated on both sides, but the faint traces on one side (Fig. 32:1a) are illegible. On the other side (*Fig.* 32:1b) between two vertical border lines the signs ⬭ *stp n* may perhaps be discerned. If this reading is correct, it is an element in the names (*nomen* or *prenomen*) of several Ramesside kings.[4]

Eg. Cat. 30*. *Fig.* 32:6. *Pl.* 121:8. Field No. 283/203, 231/1, 263/4. Loci 101, 103, 106. Counterpoise of a *menat*-necklace, faience, white frit, traces of blue glaze, decoration in black paint. H. 10cm., W. 5.3cm. Lower portion only, reassembled from three fragments.
Decorated on one side only, the flat disk-like bottom bears a floral design consisting of eight rounded petals emanating from a central disk. Above, on the neck of the counterpoise, framed between two border lines, is a vertical cartouche inscribed with the pre-nomen ⬭ *User-m3't-R -ste-n-R'* 'Usermare-Setepenre', ostensibly the *prenomen* of Ramesses II. Above the cartouche are preserved the left halves of a pair of *t3*-signs ⬭, clearly to be restored as part of the title [*nb t*]3.*wy* '[the lord of the T]wo Lands'. The blue glaze is preserved only on the central fragment of the counterpoise. It is most probable that the name in the cartouche is that of Ramesses II, who used this *prenomen* more than any other.

Observation: It must be pointed out that the same *prenomen* in the same spelling was used by Sheshonk II of the 22nd Dynasty (cf. Gauthier 1914, III, 353, 361–7) and with a variant spelling by Pamai of the same dynasty (ibid. 372). Without an accompanying cartouche inscribed with the *nomen*, the *prenomen per se* cannot be unequivocaly assigned to Ramesses II. However, since nothing dating to the 22nd Dynasty has been found at Site 200, and Ramesses II is attested there, it is probable that the name here is his.

Eg. Cat. 31. *Fig.* 29:6. Field No. 270/105. Locus 102. Counterpoise of a *menat*-necklace(?), faience, white frit, green glaze, decorated with black paint. H. 2.2cm., W. 5.3cm., Th. 2.2cm. Fragment, inscribed on one side only.
It is uncertain whether this fragment actually comes from the counterpoise of a *menat*, although its shape and section suggest this. Only obscure traces remain of the inscription, in which only the head of the hieroglyph *stp* 'chosen' seems certain. The vertical line above it could be the hieroglyph ⬭ *n*. Since the element *stp*, if correctly read, forms part of a *prenomen*, the only Ramesside ruler possible here would be Siptah of the 19th Dynasty, whose *prenomen* was variously *3h-n-R'/stp-n-R'* 'Akhenre-setepenre' (the later form of his name) and *SH'-n-R'-stp-n-R'* 'Sekhaenre-setepenre' (the earlier form). However, as there is a very damaged but faint trace above and parallel to the centre of the '*n*', perhaps the hieroglyph should not be read *n*, but be restored ⬭ *mry* 'beloved'. In this case, the name in question could either be *Wsr-h'w-R'-mry-Imn-stp-n-R'* 'Userkhaure-miamun-setepenre', the *prenomen* of Setnakhte, or *Wsr-m3't-R'mry-Imn-stp-n-R'* 'Usermare-miamun-setepenre', the *prenomen* of Ramesses VII. However, the latter would be highly unlikely since no trace of Ramesses VII has ever been found outside Egypt. Because of the virtually illegible state of the hiero-glyphs, it is impossible to state with any certainty which of these four *prenomina* is to be read.

Eg. Cat. 32. *Fig.* 32:4. *Pl.* 121:5. Field No. 302/1, 301/1. Loci 101, 109. Counterpoise of a *menat*-necklace, faience, light brown frit, blue glaze, decorated with black paint. H. 4.7cm., W. 4.2cm., Th. 0.8cm. Fragment, assembled from three pieces.
Decorated on one face only, pierced through the thickness at the upper corners for attachment to the necklace. Inside a rectangular ornamental border is a rather large sun-disk, drawn in outline only, beneath which is the upper half of a vertical cartouche containing the hieroglyphic signs *W[sr-]m3't-]R'* 'U[ser]-m[a]re', clearly the beginning of the *prenomen* of a Ramesside king, although it is impossible to determine which. The ornamental border may represent bundles of reeds tied together to form a door frame.

Observation: Kertez (1972, 65, Pl. 13:6) reads this name as *Hk3-m3't-R'* 'Hikamare', the *prenomen* of Ramesses IV.

Eg. Cat. 33. *Fig.* 32:2. Field No. 241/3. Locus 101. Counterpoise of a *menat*-necklace, faience, brown frit, traces of blue glaze, decorated with black paint.

H. 3.8cm., W. 3cm., Th. 1cm. Fragment, inscribed on both faces.

On one face, which is very badly weathered, are traces of the word *mfk3t* 'turquoise', framed between two vertical lines. On the opposite face a badly preserved *nb.t* 'Lady' and the initial signs of *mfk3t* are likewise framed between two vertical lines. In each case, this is probably an epithet of the goddess Hathor, 'the Lady of Turquoise', which here probably qualified a royal name in the expression '[king NN beloved of Hathor] the Lady of Turquoise'. Since the position of this epithet normally comes towards the end of a titulary, this fragment clearly comes from the lower portion of the counterpoise, just a little above the disk-like knob at its bottom. The careful rendering of the individual hieroglyphs is reminiscent of the palaeography of the first half of the 19th Dynasty.

Observations: (a) Possibly belonging together with Cat. 29 on palaeographic grounds, although the section of the latter is fractionally thicker and narrower. (b) Kertez (1972, 65; Pl. 12:3) restored the name of Twosre, but there is no possibility of restoring the preserved traces as [*stp*]-*n*-*Mwt*.

Eg. Cat. 34*. *Fig.* 31:1. *Pl.* 120:5. Field No. 228/1. Locus 110. Counterpoise of a *menat*-necklace, faience, white frit, blue-green glaze. L. 4cm., W. 2.8cm., Th. 0.8cm. Fragment.
Pierced at the corners for attachment to the necklace, inscribed in black on each face with the beginnings of a royal title. On one side *ntr nfr* --- 'the good god ---' and on the other *s3 R'* ---- 'The son of Re ----'.

Eg. Cat. 35. *Fig.* 33:6. Field No. 206/2. Locus 110. Counterpoise of a *menat*-necklace, faience, white frit, green glaze. H. 2.5cm., W. 3.5cm., Th. 0.8cm. Fragment.
From the upper part of a counterpoise, pierced horizontally along the width for attachment to the necklace. This fragment is badly weathered on its decorated face. What is left shows there was a border frame on three sides consisting of triple wide lines running parallel to the sides, with the central line of the three being segmented by a series of shorter lines set at right angles to the border. Inside the border are several illegible traces of what was possibly a hieroglyphic inscription.

Eg. Cat. 36*. *Fig.* 32:3. *Pl.* 121:7. Field No. 65/1, 253/1, 261/6, 370/1. Loci 102, 103, 110. Counterpoise of a *menat*-necklace, faience, white frit, medium blue glaze. L. 6.7cm., W. 4.8cm., Th. 7cm. Lower portion only.
Restored from four fragments, two of which have lost their glaze. Small, narrow, triangular wedge missing from right side of disk. Design and text on one side only. Design of an open lotus flower with at least thirty petals emanating from the stamen. The individual petals each have a central vein running along their length, this is framed on each side by another vein running parallel to its respective side of the petal. Above, on the neck of the counterpoise, between two vertical border lines, is the hieroglyph *nb* 'gold', above which was a cartouche, now lost.

Eg. Cat. 37*. *Fig.* 33:7. *Pl.* 118:8. Field No. 269/2. Locus 107. Counterpoise of a *menat*-necklace, faience, white frit, traces of blue glaze, decorated with black paint. H. 4cm., W. 4.4cm., Th. 0.8cm.
This fragment of the bottom part of the counterpoise is decorated on both faces with a virtually identical floral pattern, consisting of a quatrefoil of *nymphea caerulia* flowers emanating from a central calyx, alternating with a quatrefoil of petals growing out of the same calyx. On each face only three of the flowers and two of the petals are preserved. The flowers are rendered in outline with several horizontal bands across their length, their tops are indicated by a series of short, vertical strokes. The petals are done in outline with a thick oval inner blob to represent the central vein. On one face (*Fig.* 33:7, right) one of the vertical flowers is hardly recognisable because of the fading and flaking of the paint.

Eg. Cat. 38. *Fig.* 33:8. Field No. T200/2. Surface find. Counterpoise of a *menat*-necklace, faience, white frit, green glaze, decorated with black paint. H. 2.6cm., W. 3.3cm., Th. 1.5cm. Fragment.
This wedge-shaped sherd, comprising approximately two-fifths of the whole, comes from the disk-like finial of the lower part of a *menat*-counterpoise. The decoration, found only on one side, consists of four petals of a rosette, their interstices outlined in black paint.

Eg. Cat. 39. Not illustrated. Field No. 354/16. Locus 111. Counterpoise of a *menat*-necklace, faience, very pale brown frit, bluish-green glaze, decorated with black paint. H. 1.7cm., W. 1.4cm., Th. 0.7cm. Fragment.
The decoration consists of a wide vertical band in the centre, flanked by two bands which are subdivided horizontally into small rectangles, suggesting that this fragment comes from the lower portion of the neck of a counterpoise, just above where the disk-like finial would begin.

Eg. Cat. 40. Not illustrated. Field No. 270/102. Locus 102. Counterpoise of a *menat*-necklace (?), faience, very pale brown frit, green glaze, no decoration. H. 4cm., W. 3cm., Th. 1.7cm. Fragment.
Oval section. Possibly from the middle of a counterpoise, just above the disk-like knob (now lost) at the bottom.

5 Bracelets (Eg. Cat. 41–83)
Whether all of the forty three items listed below were actually bracelets, is not certain. All are incomplete and more than half of them (Cat. 59–83) show no traces of hieroglyphic inscriptions. The remainder (Cat. 41–58) all have inscriptions. It is possible that a single bracelet is represented below by several fragments, but from the standpoint of their dimensions, inscriptions, palaeography and material, this seems unlikely. Since all the fragments are made of faience it is not very probable that they would have stood up to

the stresses of ordinary use as everyday jewellery. Like the *menats*, discussed in the preceding section, they were undoubtedly either of a votive or funerary nature (Hayes, 1953–59, I, 306–9). However, since their inscriptions, as indicated by Cat. 51, took the form of a pious wish beginning 'May the king of Upper and Lower Egypt [and/or another royal title] NN, the Son of Re [and/or another title], the beloved of the Deity X [plus an epithet of the deity] is the one who is granted life forever [and ever]', or similar, it is more probable that they were votive. On the fragment Cat. 55, which mentions a deity other than the sun god Re in the stereotype 'granted life like Re forever and ever' (Cat. 51, cf. Cat. 53 and 54), the deity – as at Serabit el-Khadim – is Hathor (cf. Petrie, 1906, 143–4; Pl. 148). The inscriptions are written either horizontally (Cat. 44–6, 49, 50, 55, 57) or vertically (Cat. 41–3, 47, 48, 51–4, 56). In two cases (Cat. 44, 50) the direction of the writing is reversed, with the signs facing to the left, rather than to the right which is the normal direction.

Eg. Cat. 41. *Fig.* 34:3. *Pl.* 121:4. Field No. 207/1. Locus 102. Bracelet, faience, white frit, green glaze, decorated with black paint. H. 1.8cm., W. 4cm., Th. 1.4cm. Fragment, reassembled from three pieces.
Though damaged, the inscription is quite legible. Vertically and facing to the right, the hieroglyphs preserved, either partially or intact show *T3-wsr-n-mt* clearly to be restored – *T3-wsr(.t)-[stp-]n-m(w)t*, the *nomen* of the later 19th Dynasty ruler, Queen Twosre-setepenmut. Noteworthy, although not rare at this period, are the spellings for *wsr* and for *Mwt*, conforming to normal Late Egyptian usage, as is the omission of the feminine ending .t in *wsr(.t)* which, as it was no longer pronounced, was frequently omitted in writing. Also interesting is that here the name is written without a cartouche. This may not be significant since royal names were occasionally so written, or it may indicate that at the time when the bracelet was inscribed Twosre did not yet exercise the royal power in her own right, but was merely the 'great wife of the king' (*hm.t nswt wr.t*).

Eg. Cat. 42. *Fig.* 34:1. *Pl.* 121:3. Field No. 206/3. Loci 110, 111. Bracelet, faience, white frit, green glaze, inscribed with black paint. H. 2.2cm., W. 2.7cm., Th. 1.1cm. Fragment.
The upper half remains a vertically written cartouche containing the hieroglyphs *Hk3-m3't-R'* 'Hikamare', the first half of the *prenomen* of Ramesses IV, Hika-mare-setepenamun. (For an identical fragment, cf. Cat. 43.)

Eg. Cat. 43. *Fig.* 35:6. *Pl.* 122:5. Field No. 374/1. Locus 106. Bracelet, faience, white frit, blue glaze, inscribed with black paint. H. 3cm., W. 2.9cm., Th. 1.3cm. Fragment.
Inscribed vertically is the upper half of a cartouche containing the hieroglyphs *Hk3-m3't-R'* 'Hikamare', the first half of the *prenomen* of Ramesses IV 'Hika-mare-setepenamun'. (For an identical fragment, see Cat. 42.)

Eg. Cat. 44. *Fig.* 36:3. *Pl.* 122:8. Field No. 245/2. Locus 107. Bracelet, faience, white frit, greenish traces of glaze, inscribed with black paint. L. 1.8cm., W. 1.3cm., Th. 0.6cm. Fragment.
The rear half is preserved of a horizontal cartouche containing the hieroglyphs *hpr-n-R'*, clearly the latter portion of the *prenomen* of Ramesses V of the 20th Dynasty [*Wsr-m3't-R'-*]*hpr-n-R'* '[Usermare-se]kheperenre', since the only other king whose *prenomen* contains the elements *hpr-n-R'*, Thutmose II of the 18th Dynasty, uses a rather different orthography where the *R'* is written in honorific transposition at the beginning of the cartouche.

Eg. Cat. 45. *Fig.* 35:2. Field No. 130/1. Locus 101. Bracelet, faience, brownish frit, traces of blue-green glaze, inscribed with black paint. L. 2.3cm., H. 2.1cm., Th. 1.4cm. Fragment.
The inscription consists of the rear end of a horizontal cartouche facing to the right and containing almost illegible traces of the signs *-hpr-n*. The only possible restoration is [(*Wsr-m3't-R'-s*]*hpr-n[R']* '[Usermare-s]ekheperen[re]', the *prenomen* of Ramesses V (see also Cat. 44 and 46).

Eg. Cat. 46. *Fig.* 35:1. *Pl.* 122:7. Field No. 275/19. Locus 104. Bracelet, faience, brown frit, greenish glaze, inscribed with black paint. L. 2.2cm., W. 1.8cm., Th. 1.1cm. Fragment.
Clearly legible is the rear end of a cartouche, inside which are several minute traces. If the cartouche is vertical, the traces do not fit the hieroglyphic signs of any of the known New Kingdom rulers. However, if read horizontally the traces suit a reading of *hpr*, allowing a restoration of [*Wsr-m3't-R'-s*]*hpr-[n-R']* '[Usermare-se]kheper[enre]', the *prenomen* of Ramesses V. The horizontal *n* beneath *hpr* may be present in the dot-like trace immediately below, or may have been on the part of the bracelet now lost (cf. Cats, 44 and 45).

Eg. Cat. 47. *Fig.* 35:7. *Pl.*.122:6. Field No. 253/2. Locus 102. Bracelet, faience, light brown frit, green glaze, inscribed with black paint. H. 3.2cm., W. 2.8cm., Th. 0.8cm. Fragment.
Preserved vertically is the lower half of a cartouche containing the hieroglyphs *-ms-sw* '-messes', the last element of the name 'Ramesses'. This is followed by the hieroglyph , the initial sign in the word *d.t* 'forever'. The juxtaposition of *d.t* after a royal name is unusual and is probably a carelessly written form of the common stereotypes *'nh d.t* or *di.w 'nh d.t* 'living forever', 'granted life forever' (see now Gruen, 1975, 21–2). The *nomen* Ramesses, without any additional qualifying epithet inside the cartouche and in this spelling, is most commonly attested for Ramesses II (Gauthier 1914, III; 31, 48, 58, 73). In a slightly different spelling, it is attested for Ramesses I, Ramesses IV and Ramesses V (Gauthier 1914, III; 2–8, 180–90, 218). It is most probable that the king in question was Ramesses II, who very commonly omits the epithet within his cartouche, while the others do so only occasionally.

However, because of the indiscriminate interchange at this period of different hieroglyphs for the same sound (see Cat. 48), this attribution cannot be stated unequivocally. It is safer to date the inscription as simply Ramesside.

Eg. Cat. 48*. *Fig.* 36:6. *Pl.* 122:10. Field No. 279/2. Loci 106, 107. Bracelet, faience, white frit, purple glaze, inscribed with black paint. H. 3.5cm., W. 2.7cm., Th. 1cm. Fragment.
Framed vertically between two border lines is the upper portion of a cartouche bearing the royal *nomen*

(⊙⋔⚍ *R'-ms-s(w)* 'Ramesses'.

Observation: In all preliminary publications, this cartouche has been attributed to Ramesses IV, probably on the basis of Gauthier (1914, III; 175, 176:2, 180–90) where it is written with this orthography using the horizontal *s* ⚍. However, this seems an insufficient basis for this attribution, first because the cartouche is incomplete and there is no way of determining whether the second ⚍ was followed by additional elements of a name, e.g. Ramesses-minamum (Ramesses II), Ramesses-Siptah (the earlier *nomen* of Siptah), or Rammesses-hikaon (Ramesses III). Secondly, at this period of Egyptian history there is no special significance in writing the name with a horizontal *s* rather than the vertical *s* ❘, since these forms were interchangeable, often for the same king, depending on whether the name was written horizontally (using the vertical *s*) or vertically (where the horizontal *s* might be preferred). Therefore the name in the cartouche can only be dated as Ramesside.

Eg. Cat. 49. *Fig.* 35:8. *Pl.* 122:4. Field No. 295/1. Locus 101. Bracelet, faience, brownish frit, no traces of glaze, inscribed with black paint. L. 7.2cm., W. 1.6cm., Th. 0.7cm. Reassembled from three pieces.
Running along the length of this small fragment is a horizontal inscription consisting of the title ⚰⫶⫶⫶ *nb h'.w* 'Lord of Diadems', followed by the beginning of a cartouche in which the name begins with a tall, narrow sign, possibly to be restored ❘ *wsr*. The title 'Lord of Diadems' usually, though not invariably, precedes the cartouche containing the *nomen*; but since none of the Ramesside rulers begin their *nomen* with this element, the name in the cartouche can only be a *prenomen*. Thus, the name could be equally restored as *Wsr[-m3't-R']*, *Wsr[-hprw-R']* or *Wsr[-h'.w-R']* and it is impossible to know what was originally written. It should also be noted that there is no trace of, nor room for the sun-disk 'Re ' normally written in the royal names of which it is an element, in graphic honorific transposition at the beginning of the cartouche.

Eg. Cat. 50. *Fig.* 35:3. *Pl.* 122:1. Field No. 248/2. Locus 104. Bracelet, faience, brownish frit, traces of blue glaze, inscribed with black paint. L. 5.8cm., W. 1.4cm., Th. 1cm. Reassembled from two pieces.
The inscription on this small fragment, running horizontally along its length, consists of the rear half of a cartouche containing several virtually illegible traces of hieroglyphic signs. The last three of these appear to be capable of restoration as *hr m3't*. If correct, this would suggest that the full name be restored as [*Mr-n-Pth-htp-*] *hr-m3't*, the *nomen* of the successor of Ramesses II, Merneptah. Although this is not the usual orthography of the name, no other suggests itself.

Eg. Cat. 51*. *Fig.* 36:1. *Pl.* 122:11. Field No. 277/3, 323/1, 371/1. Loci 106, 107, 110. Bracelet, faience, white frit, blue glaze, inscribed with black paint. L. 7.3cm., W. 3.1cm., Th. 1.2cm. Fragment reassembled from three pieces.
The hieroglyphic text which runs vertically between a pair of border lines, contains the end and beginning of the inscription ---*mi [R]'d.t nhh 'nh nswt bity* ---- 'May the king of Upper and Lower Egypt ---- live like Re forever and ever'.

Eg. Cat. 52. *Fig.* 36:2. Field No. 325/4. Locus 110. Bracelet, faience, very pale brown frit, greenish glaze, inscribed with black paint. H. 1.6cm., W. 1.3cm., Th. 0.8cm.
The inscription on this fragment reads *nb h'.w* 'Lord of Diadems' which generally, but not invariably is part of the royal titulary preceding the fifth name – the *nomen* – of the king.

Eg. Cat. 53. *Fig.* 36:4. Field No. 325/3. Locus 110. Bracelet, faience, white frit, green glaze, inscribed with black paint. H. 3cm., W. 2.8cm., Th. 9cm.
The vertical inscription on this fragment, framed between two border lines, consists of the final signs of the word *d.t* followed by *nhh* 'forever and ever', the concluding phrase of the pious stereotype following a king's name 'granted life like Re forever and ever'. (See much better preserved example Cat. 51.)

Eg. Cat. 54. *Fig.* 36:5. Field No. 253/3. Locus 102. Bracelet, faience, white frit, green glaze, inscribed with black paint. H. 3.3cm., W. 1.8cm., Th. 7cm. Fragment.
There are traces of the end of the inscription preserved in the two words *di.w 'nh* '---granted life', which clearly indicates that the original text must have contained a royal name, now lost.

Eg. Cat. 55*. *Fig.* 35:4. *Pl.* 122:2. Field No. 242/2. Locus 110. Bracelet, faience, white frit, green glaze, inscribed with black paint. L. 3.3cm., W. 2.3cm., Th. 0.5cm. Fragment.
Inscribed vertically with a line of hieroglyphs facing to the right. The vertical stroke at the right, half destroyed by the break, is to be restored as the straight end section of a cartouche. Immediately following is the name of the goddess *Ht-hr* followed by the hieroglyph *h*, ⬚❘ clearly the initial in the word ⫶⚬⫶ *hnt* 'Lady', 'mistress' which on the original would have been 'mistress of heaven', 'mistress of the gods', 'mistress of turquoise' or similar. The name of the goddess stood in honorific transposition before the adjective *mry* 'beloved', so that the full original text would have included '---the name of [King X, beloved] of Hathor, mistress of [Y]'.

Eg. Cat. 56*. *Fig.* 35:5. *Pl.* 122:5. Field No. 176/1. Locus 102. Bracelet, faience, white frit, blue-green glaze, inscribed in black. L. 2.9cm., W. 2.7cm., Th. 1cm. Fragment.

Inscribed in black with the hieroglyphs ⸗ --- *nswt bity* --- '--- king of Upper and Lower Egypt ---'.

Eg. Cat. 57*. *Fig.* 34:2. Field No. 327/2. Locus 106. Bracelet, faience, white frit, bluish-green glaze, inscribed with black paint. H. 2.5cm., W. 1.8cm., Th. 0.8cm.

Inscribed horizontally on this fragment are traces of the hieroglyphs ⸗ *nb t3.wy* 'Lord of the Two Lands' ususally (though not invariably) the second element in a title preceding a royal *prenomen*.

Eg. Cat 58. *Fig.* 37:1. Field No. 301/7. Locus 109. Bracelet, faience, white frit, green glaze, no decoration. L. 5cm., W. 1.2cm., Th. 0.9cm. Fragment with square section, reassembled from two pieces.

Eg. Cat 59. *Fig.* 37:2. Field No. 63/5. Locus 110. Bracelet, faience, very pale brown frit, green glaze, no decoration. L. 6.5cm., W. 2.0cm., Th. 1cm. Fragment with rectangular section with rounded corners, reassembled from two pieces.

Eg. Cat 60. *Fig.* 37:3. Field No. 370/18. Locus 110. Bracelet, faience, very pale brown frit, green glaze, no decoration. L. 6.5cm., W. 1.3cm., Th. 0.7cm., Diam. 8.2cm. Fragment with hemispherical section, reassembled from two pieces.

Eg. Cat 61. *Fig.* 37:4. Field No. 245/67, 339. Loci 107, 109. Bracelet, faience, white frit, green glaze, no decoration. L. 7cm., W. 1.5cm., Th. 0.7cm., Diam. 8.0cm. Fragment with oval section, reassembled from two pieces.

Eg. Cat 62. *Fig.* 37:5. Field No. 201/29, 277. Loci 107, 110. Bracelet, faience, brown frit, green glaze, no decoration. L. 7.5cm., W. 0.7cm., Th. 1cm., Diam. 8.0cm. Fragment with triangular section, reassembled from three pieces.

Eg. Cat. 63. *Fig.* 37:6. Field No. 327/33. Locus 106. Bracelet, faience, very pale brown frit, green glaze, no decoration. L. 6.5cm., W. 1.5cm., Th. 1cm., Diam. 8.5cm. Fragment with rectangular section, reassembled from three pieces.

Eg. Cat. 64. *Fig.* 37:7. Field No. 270/103. Locus 102. Bracelet, faience, very pale brown frit, green glaze, no decoration. L. 3.3cm., W. 1.3cm., Th. 1.5cm. Fragment with triangular section.

Eg. Cat. 65. *Fig.* 37:8. Field No. 343/101. Locus 109. Bracelet, faience, very pale brown frit, green glaze, no decoration. L. 4.1cm., W. 1.4cm., Th. 0.7cm. Fragment with oval section, reassembled from two pieces.

Eg. Cat. 66. *Fig.* 37:9. Field No. 353/10. Locus 110. Bracelet, faience, white frit, green glaze, no decoration. L. 3cm., W. 1.2cm., Th. 1cm. Fragment with squarish section.

Eg. Cat. 67. *Fig.* 37:10. Field No. 134/13. Loci 110, 111. Bracelet, faience, white frit, green glaze, no decoration. L. 2.2cm., W. 1.3cm., Th. 1cm. Fragment with rectangular section, with rounded corners.

Eg. Cat. 68. *Fig.* 37:11. Field No. 279/293. Loci 106, 107. Bracelet, faience, very pale brown frit, green glaze, no decoration. L. 1.3cm., W. 1.7cm., Th. 1cm. Fragment with rectangular section, with rounded corners.

Eg. Cat. 69. *Fig.* 37:12. Field No. 237/37. Locus 110. Bracelet, faience, light brown frit, blue glaze, no decoration. L. 2.0cm., W. 1.8cm., Th. 1.1cm. Fragment with trapezoidal section.

Eg. Cat. 70. *Fig.* 37:13. Field No. 133/6. Locus 101. Bracelet, faience, white frit, green glaze, no decoration. L. 3.0cm., W. 2.2cm., Th. 0.8cm. Fragment with rectangular section and rounded edges.

Eg. Cat. 71. *Fig.* 37:14. Field No. 208/45. Locus 102. Bracelet, faience, light brown frit, green glaze, no decoration. L. 3.0cm., W. 1.6cm., Th. 1.2cm. Fragment with almost triangular section.

Eg. Cat. 72. *Fig.* 37:15. Field No. 343/105. Locus 109. Bracelet, faience, white frit, blue glaze, no decoration. L. 2.5cm., W. 1.2cm., Th. 0.4cm. Fragment with rectangular section, rounded on one side.

Eg. Cat. 73. *Fig.* 37:16. Field No. 343/104. Locus 109. Bracelet, faience, very pale brown frit, green glaze, no decoration. L. 2.1cm., W. 1.3cm., Th. 0.5cm. Fragment with oval section.

Eg. Cat. 74. *Fig.* 37:17. Field No. 216/88. Locus 101. Bracelet, faience, grey frit, bluish-green glaze, no decoration. L. 1.5cm., W. 1.4cm., Th. 1.6cm. Fragment with hemispherical section.

Eg. Cat. 75. *Fig.* 37:18. Field No. IJ/16–17/54. Locus 101. Bracelet, faience, brown frit, green glaze, no decoration. L. 1.8cm., W. 0.7cm., Th. 0.7cm. Fragment with elliptical section.

Eg. Cat. 76. *Fig.* 37:19. Field No. 350/3. Locus 110. Bracelet, faience, white frit, green glaze, no decoration. L. 3.1cm., W. 1.5cm., Th. 0.6cm. Fragment with rectangular section and rounded edges.

Eg. Cat. 77. *Fig.* 37:20. Field No. 316/17. Locus 111. Bracelet, faience, very pale brown frit, greenish glaze, no decoration. L. 2.5cm., W. 2.3cm., Th. 0.6cm. Fragment with rectangular section and rounded edges.

Eg. Cat. 78. *Fig.* 37:21. Field No. T-200/22. Surface find. Bracelet, faience, very pale brown frit, bluish-green glaze, no decoration. L. 2.0cm., W. 2.5cm., Th. 0.6cm. Fragment with rectangular section.

Eg. Cat. 79. *Fig.* 37:22. Field No. 231/151. Locus 101. Bracelet, faience, very pale brown frit, blue glaze, no decoration. L. 2.1cm., W. 2.6cm., Th. 0.6cm. Fragment with rectangular section and rounded edges.

Eg. Cat. 80. *Fig.* 37:23. Field No. 316/18. Locus 111. Bracelet, faience, white frit, green glaze, no decoration. L. 2.7cm., W. 2.2cm., Th. 1.0cm. Fragment with rectangular section, rounded edges, concave inner surface and convex outer surface.

Eg. Cat. 81. *Fig*. 37:24. Field No. IJ/16–17/53. Locus 101. Bracelet, faience, brown frit, green glaze, no decoration. L. 2.0cm., W. 3.2cm., Th. 0.7cm. Fragment with rectangular section, rounded edges, convex inner and outer surfaces.

Eg. Cat. 82. *Fig*. 37:25. Field No. 302/11. Locus 101. Bracelet, faience, grey frit, blue glaze, no decoration. L. 2.0cm., W. 3.1cm., Th. 0.8cm. Fragment with rectangular section and rounded edges.

Eg. Cat. 83. Not illustrated. Field No. 337/30. Locus 102. Bracelet, faience, very pale brown frit, blue-green glaze, no decoration. L. 4.2cm., W. 3cm., Th. 0.5cm. Fragment with rectangular section.

Addendum – 1984
In the summer of 1984, during consolidation work at Site 200, several faience fragments were found under the White Floor and the 'red pavement', in Loci 106 and 107, south of the *naos* (cf. II, n.3).

Eg. Cat. 83a. *Fig*.31:7. *Pl*. 125:3. Locus 107. Bracelet, faience, white frit, light blue glaze, decorated with black paint. H. 5.5cm., W. 4.8cm., Th. 0.8cm. Fragment.

The beginning of the inscription is lost. Preserved within a rectangular frame are a pair of vertical cartouches containing the *prenomen* and *nomen* of Seti I. *Mn-m3't-[R']* *Sty-[mr-]n-[Pth]* and beneath these is a vertical line reading *di 'nh w3s mi R'* 'granted life and good fortune like Re'.

Observation: See below, par. 19, for chronological implications of this object.

Eg. Cat. 83b. *Fig*.31:8. *Pl*. 125:2. Locus 106. Bracelet, faience, white frit, blue glaze, decorated with black paint. H. 2.2cm., W. 2.7cm., Th. ca. 0.6cm. Preserved on this small fragment is *Ht-hr* 'Hathor'.

6 Ushabti (Eg. Cat. 84)
The presence of an ushabti, a magical figurine of the deceased which was intended to work in place of the deceased in the Afterworld, is striking. By definition and function, ushabtis were of a mortuary nature and had no cult or other role in a non-mortuary temple. To consider the Timna ushabti purely as a votive offering would be unprecedented. It is possible that the tomb or grave of the owner of the ushabti was somewhere in the vicinity of the shrine, possibly on top of the cliff behind the shrine, where traces of burials have been noted (Rothenberg, 1967; 134, Site 199). Somehow the ushabti was deposited at the shrine, perhaps to ensure for its owner the protection and mediation of Hathor in the Afterworld. The ushabti is certainly not a later, intrusive deposit at the shrine since it is clearly Ramesside in date, and it was during the Ramesside period that the shrine was used by the Egyptians.[5]

Eg. Cat. 84*. *Fig*. 28:2. *Pl*. 119:2. Field No. 257/2.

Locus 102. Ushabti, faience, very pale brown frit, green glaze, inscribed with black paint. H. 4cm. Fragment.
The upper half of the body, just above the knees, is lost and the front part of the feet, as far as the insteps, is missing from the preserved lower half. Three lines of text, starting at the left side of the back and reading from right to left, contain a somewhat garbled and abbreviated, but still recognisable version of most of the standard ushabti formulae, an extract from Chapter 6 of the Book of the Dead. Unfortunately the name and titles of the owner are not preserved, probably having been inscribed on the missing upper half, together with the beginning of the formula. As it stands, the text, which begins in the middle of a word, appears to read:[6]

(X + 1) *-t3 í[r] íp.(t)w.k [r] nw nb ír.t ír<r.w>t m hryt-ntr <r> mh*
(X + 2) *wdb(.w) <r s>rwd sht <r> hnn s't <n> 'imntt <r> í3btt*
(X + 3) *mk.wí k3.í ír hsb.t(w).k m hryt-ntr*

'(X + 1) --- I(f) you are reckoned [at] any time to do what is to be done in the Afterworld <to> inundate (X + 2) the canal-bank(s), <to cause to> grow the fields, <to> transport sand <from> the west <to> the east, (X + 3) behold, I answer. If you are counted in the Afterworld.

The shape and rendering of the individual hieroglyphs suggest a date in the first half of the 19th Dynasty and this seems to be confirmed by the order and arrangement of the individual phrases. On stylistic grounds, the dress and shape of the figure also suggest this dating. The full roundness of the body, with the absence of a base or back pillar, plus the use of the individual strips of the bandages of the mummy as the horizontal registers on which the inscription was written, are all characteristic of the early Ramesside period.

7 Cat Figurines (Eg. Cat. 85–95)
At Timna, as at Serabit el-Khadim, a number of faience figurines of felines were recovered, mostly in a very fragmentary condition. Those from Serabit el-Khadim (see Petrie, 1906; Pl. 153:6–14) have been identified as serval cats, cheetahs and an unknown species. Other representations of felines at the same site, on faience plaques (ibid. Pl. 154) were all identified as serval cats. Since the body markings of the Timna felines are similar to those of both the cats and cheetahs from Serabit el-Khadim, it is difficult to determine what species was intended. For the purposes of this catalogue, the Timna pieces have been labelled, perhaps arbitrarily, as serval cats. Petrie pointed out that the cat had a special relationship with the Hathor cult at Serabit el-Khadim, but that this relationship was not attested elsewhere.[7] Clearly this relationship existed at Timna also and it may be that the connection of the cat with the Hathor cult was peculiar to Asia, where Hathor possessed a special attribute as the patroness of mining camps in her guise of 'Lady of Turquoise', an attribute not attested elsewhere. On the other hand, it should be pointed out that the connection between Hathor and

the cat may have arrived via the cat-headed goddess Bastet (Scott, 1958; 1–7) who, like Hathor, was a goddess of music, dance and love, or via the lioness-headed goddess Sekhmet whose name 'The Powerful One' was also an epithet for Hathor that she assumed in the story of the destruction of mankind. The exact function of the cats is unclear, but they obviously had some cult or votive significance.

Eg. Cat. 85*. *Fig.* 38:1. *Pl.*118:6. Field No. 156/1, 327/1. Loci 101, 106. Figurine, faience, brown frit, blue glaze, brown decoration. L. 7cm., H. 5.4cm., W. 2.3cm.
Feline, with forelegs and forepaws missing, head facing forward, crouching down on all fours on a pedestal. The animal's tail curves around the right rear leg. Its haunches, spine, facial details and ears are finely and carefully modelled. The markings on the pelt are indicated by a series of regularly spaced, parallel dots. It is difficult to determine the exact type of feline intended. While the markings on the body could be those of a cheetah or of a serval cat, the head does not have their prominent ears, eyes or snout and is rather reminiscent of a lioness.

Eg. Cat. 86. *Fig.* 38:2. Field No. 126/1. Locus 101. Figurine, faience, grey frit, blue glaze, brown decoration. L. 3.3cm., H. 1.2cm., W. 2.1cm. Fragmentary.
Only the hindquarters and rear haunches of this feline are preserved. The upper back quarters show that it is crouching low on all fours. However, it is impossible to determine whether the head faced forward or was turned to one side. The end of the spine and the beginning of the tail are modelled. The markings of the pelt, indicated by parallel rows of irregularly shaped dots, are very similar to those of Cat. 85, 88, 90, and may be those of a cheetah or of a serval cat.

Eg. Cat. 87. *Fig.* 38:3. Field No. 241/1. Locus 101. Figurine, faience, very pale brown frit, blue glaze, incised decoration. L. 3.8cm., H. 1.6cm., W. 1.9cm. Fragment.
The head, neck, legs and base of this feline are missing. The head apparently faced forward and it crouched down on all fours. The tail, which hangs straight down at the back, apparently did not curl around either of the rear legs. The spine is modelled with vertebrae and pelt markings indicated by incisions. It is difficult to determine the exact branch of the cat family to which it belonged.

Eg. Cat. 88. *Fig.* 38:4. Field No. 236/1. Locus 107. Figurine, faience, very pale brown frit, blue glaze, brown decoration. H. 1.2cm., W. 2.5cm. Fragment of body.
Of this feline, only a portion of the upper back and right rear haunch are preserved and its original pose cannot be determined. A thin, modelled ridge on the right side may represent the tail as it curves over a haunch, since the spine appears to be indicated by a thick, central line which runs along the length of the fragment. This, and the markings on the pelt are

indicated by brown paint. The specific type of feline cannot be determined.

Eg. Cat. 89. *Fig.* 38:5. Field No. 349/1. Locus 109. Figurine, faience, white frit, blue glaze. L. 2.5cm., H. 2cm., W. 2.2cm. Fragment.
The right front corner of a pedestal is preserved on which the right paw of an animal, probably a feline, is crudely modelled.

Eg. Cat. 90. *Fig.* 33:1. *Pl.*118:7. Field No. 95/1. Loci 110, 112. Figurine, faience, very pale brown frit, blue glaze, black decoration. H. 3.4cm., W. 2cm. Head only.
Feline with long neck and out-thrust head, broken away from the body at the point where the neck joins the shoulders and throat. The angle of the neck and lower jaw suggest that originally the animal crouched down on its rear legs, while resting with its front legs straight so that the back and spine were at a forty-five degree angle to the ground. The details of the face and mane of the animal are carefully indicated with black paint, and the markings on the pelt are rendered by a stippling of black dots arranged in an irregular pattern. The animal appears to be a cheetah or leopard.

Eg. Cat. 91. *Fig.* 33:2. Field No. 278/2. Loci 107, 109. Figurine, faience, white frit, blue glaze, black decoration. H. 3.6cm., W. 1.8cm. Fragment.
All that remains of this feline is the back of the neck and head, the ears and a tiny portion of the animal's forehead. The long neck is curved and slightly bent towards the right, indicating that the feline crouched on all fours, the tail probably curled over one of the rear legs, while the head was turned to the right. No trace of a mane appears on the nape or sides of the neck, so it was neither a cheetah nor a leopard. The prominent, upraised ears and the markings on the pelt, rendered as a series of short, horizontal lines, suggest that it may have been a serval cat.

Eg. Cat. 92. *Fig.* 33:3. Field No. 282/1. Locus 110. Figurine, faience, white frit, blue glaze, brown decoration. L. 2cm., H. 1.7cm., W. 1.7cm.
The markings suggest that this fragment comes from a figurine of a feline. The fragment has the form of a cylinder, diagonally bisected from top to bottom, and suggests that it may have been a rather long neck, the head of which is now lost. The feline would have been crouched on its hind legs, with its head raised up towards the front.

Eg. Cat. 93. *Fig.* 33:4. Field No. 263/2. Locus 103. Figurine, faience, grey frit, blue glaze, brown decoration. L. 3.1cm., H. 2.0cm., W. 2.3cm.
The hindquarters only of a feline, originally crouching on its rear haunches. The angle of the curve of the spine suggests that the now-missing front showed the head thrust out, the animal resting on straight forelegs, half crouching, half standing. The musculature of the rear legs is carefully modelled and the pelt is indicated by a regular stippling of heavy brown dots. The exact type of feline represented cannot be determined.

Eg. Cat. 94. *Fig.* 33:5. Field No. 257/1. Locus 102. Figurine, faience, light brown frit, blue glaze, brown decoration. H. 2.2cm., W. 2.2cm.
Feline with only the head, minus the ears, preserved. Broken off at the neck just below the lower jaw, the line of neck and lower jaw suggests a feline crouched on all fours, with the head facing forward. Facial features modelled and indicated in brown. Pelt on right side of neck indicated by stippling of dots in an irregular pattern. Although the upraised ears are missing, the absence of a mane and the markings on the pelt suggest that the animal is a serval cat.

Eg. Cat. 95. Not illustrated. Field No. 241/2. Locus 101. Figurine, faience, brown frit, blue glaze. L. 3.0cm., H. 2.9cm., W. 1.0cm. Irregular shaped fragment.
It has a rectangular cross-section with a rounded upper portion. The rectangular section undoubtedly had the function of supporting the upper portion of the body. A U-shaped section in the front (?) where the arms of the U are modelled, suggests the forelegs, thrust straight down, and the angle of the back in relation to these then suggests that the figure represented an animal, probably a feline, in a half standing, half crouching attitude with the head raised and facing forwards.

8 Jar Stands (Eg. Cat. 96–101)

Fragments of three inscribed jar stands were found. On two (Cat. 96 and 97) the inscription consisted of a pair of vertical cartouches containing the name of a king. The cartouches on Cat. 96 are virtually intact; on Cat. 97 only the barest traces of one cartouche and the name in it are preserved. The inscription on the third stand (Cat. 98) is written horizontally and consists of a single cartouche containing a royal *prenomen*. At Serabit el-Khadim Petrie (1906, 145–6; Pl. 150:14–16; 151:1–3) likewise found a number of jar stands of which at least three (Pl. 151) were inscribed, two vertically like the Timna pieces and one horizontally. The latter inscription contained only a *prenomen*, but since it was preceded by the title 'Lord of the Two Lands' and was followed immediately by the epithet 'Beloved of Hathor, Lady of Turquoise', it would appear that only the *prenomen* was inscribed. This may not have been the case with Cat. 98.

Eg. Cat. 96*. *Fig.* 31:5. *Pl.*119:3. Field No. 370/2, 353/1. Loci 110, 111. Jar stand, faience, white frit, green glaze, inscribed with black paint. H. 4.8cm., W. 0.9cm., Diam. 7.5cm. Fragment reassembled from two pieces.
This small jar stand is squat, with straight sides and flaring rim, and is inscribed with a pair of vertical cartouches containing the *prenomen Wsr-m3't-R'-mry-'I[mn]* 'Usermare-miamun' and *nomen R'-ms-s(w)-hk3-Iwnw* 'Ramesses-hikaon' of Ramesses III. The *prenomen* is preceded by the title *nb t3.wy* 'Lord of the Two Lands' and the *nomen* by *nb h'.w* 'Lord of Diadems'. Noteworthy, but not uncommon, is the fact that the two cartouches with the *prenomen* at the right, face each other, rather than facing in the same direction.

Eg. Cat. 97. *Fig.* 31:4. Field No. 279/1. Locus 107. Jar stand, faience, white frit, green glaze, inscribed with black ink. H. 5.6cm., W. 7.0cm., Diam. 10cm.
Only a portion of the rim of this stand is preserved, showing a slightly flaring lip and a straight side. High up, starting at the underside of the lip, are traces of a vertical inscription reading *nb t3 [wy]* 'Lord of the Two Lands' , followed by a cartouche of which the outline is quite discernible and in which the hieroglyph *ms* can be read. It is clear that the cartouche originally contained a royal name ending in '-messes'. Since the cartouche is preceded by the title 'Lord of the Two Lands', the cartouche would normally be expected to contain a *prenomen*. Here, the *prenomen* of Amenmesse of the 19th Dynasty immediately suggests itself because the group *Imn* 'Amun' would exactly fill the space above the *ms*, and he was the only king of the Ramesside period whose *prenomen* ends in *ms*. Occasionally however, but not in monumental inscriptions, the title 'Lord of the Two Lands' precedes the cartouche containing the *nomen* and if this were the case here the *nomina* of several Ramesside kings could be restored. Therefore it appears safest to date the cartouche to the Ramesside period, without specifying any particular king.

Eg. Cat. 98. *Fig.* 31:6. *Pl.*121:2. Field No. 245/1. Locus 107. Jar stand, faience, light brown frit, blue glaze, black paint. H. 3cm., W. 0.9cm., Diam. 6.8cm.
This fragment from the lower part of a ring stand, is inscribed with a horizontal cartouche now badly damaged, with about three quarters missing. The preserved traces, depending on the length of the original cartouche, can be restored either as [Wsr]-m3'tR'-stp-n-[R'] 'Usermare-setepenre' the *prenomen* of Ramesses II, or as [Hk3]-m3't-R'-stp-n-[Imn] 'Hika-mare-setepenamun', the full form of the *prenomen* of Ramesses IV. Neither *prenomen* can be indicated with certainty and the safest date for the cartouche is Ramesside (19th–20th Dynasties).

Eg. Cat. 99. *Fig.* 39:4. Field No. –. Locus 101. Jar stand, faience, light brown frit, traces of green glaze, no decoration. H. 3.5cm., Th. 1.2cm., Diam. 4cm. Rim fragment.

Eg. Cat. 100. *Fig.* 39:6. Field No. 257/300. Locus 102. Jar stand, faience, light brown frit, no traces of glaze, no decoration. H. 3.2cm., Th. 1.1cm., Diam. 5.8cm. Rim fragment.

Eg. Cat. 101. *Fig.* 39:7. Field No. 253/11, 3110. Loci 101, 102. Jar stand, faience, light brown frit, blue glaze with green patches, no decoration. H. 3.9cm., Th. 0.9cm., Diam. 7.0cm. Reassembled from two rim fragments.

9 Inscribed Faience Vessels (Eg. Cat. 102–110)

No attempt has been made to arrange typologically the various inscribed vessels described below since, in most cases, they are simply unplaceable sherds. Two inscriptions yield readings of a specific king's name (Cat. 102, 103). In one case (Cat. 104), although traces of both the *prenomen* and *nomen* are preserved, the readings are not unequivocal, and the names can be assigned equally to two different kings of the 19th Dynasty. Of the kings whose names are inscribed on Cat. 105 and, possibly, on Cat. 106, it can only be said that they ended in '-messes'. Cat. 107 and 108 originally bore *prenomens* but only the title 'Lord of the Two Lands' is preserved. The traces on the remaining two fragments (Cat. 109 and 110) likewise indicate that a royal name was inscribed, but they are illegible. The retrograde reading of Cat. 102, inscribed with the name of Merneptah, is paralleled by the arrangement of the same cartouches on a jar stand of Merneptah from Serabit el-Khadim (Petrie, 1906; Pl. 151:2), save that the titles 'Lord of the Two Lands' and 'Lord of Diadems' were not inscribed over their respective cartouches.

Eg. Cat. 102*. *Fig.* 28:3. *Pl.*120:1. Field No. 337/1. Locus 102. Bowl, faience, brown frit, greenish glaze, decoration inscribed in black. L. 7.0cm., W. 5.5cm., Th. 1.4cm. Fragment.
A pair of damaged cartouches, each preceded by a title, and containing respectively, from left to right, a *nomen* and *prenomen*:

which are to be restored *nb-t3.wy B3-n-R'-mry-'Imn nb h['.w Mr-n-]Pth- [htp-hr-m3't]* 'the Lord of the Two Lands, Binere-miamun, the Lord of Dia[dems, Merne] ptah-[hotephimae]', i.e. Merneptah, the successor of Ramesses II. The two names are written retrograde, with the *prenomen* following the *nomen*.

Eg. Cat. 103*. *Fig.* 40:7. Field No. 269/3, 260/1,231, 242. Loci 101, 102, 107, 110. Cup, faience, white frit, blue-green glaze, black decoration. H. 4.8cm., Th. 5cm., Diam. 8.2cm.
Fragment of shoulder and rim of a lotiform cup, decorated in black on both surfaces. On the outside, at the right, surmounted by a sun-disk(?) is a vertical cartouche facing right and containing several hieratic traces which suggest a restoration *Hk3-m3't-R'* 'Hikamare', the short form of the *prenomen* of Ramesses IV. To the left of this are four vertical lotus petals, the outermost of which is rendered by a series of concentric elipses representing the internal veins of the petal. The innermost petal is painted a solid black. The petals are connected to each other by a series of horizontal lines. The top edge of the rim is decorated around its circumference by a series of black rectangles. The inner surface, representing the inside of a lotus flower, shows a series of alternating detailed outer and solid inner petals.

Eg. Cat. 104*. *Fig.* 40:6. *Pl.*122:12. Field No. 339/2, 323. Loci 106, 109. Sherd, faience, very pale brown frit, green glaze, inscribed with black ink. H. 4.8cm., W. 3.5cm., Th. 0.4cm. Reassembled from two fragments.
The exact nature of the vessel cannot be determined. It is inscribed with the right and left halves of a pair of vertical cartouches, each of which surmounts the hieroglyph sign *nb* 'gold', which here has only ornamental significance. The right hand cartouche, with signs facing to the left, contains the traces of inscription which cannot be restored; the uppermost sign is not the round sun-disk, but rather resembles the foot and knee of a seated figure. When taken in conjunction with the second cartouche (which reading from left to right contains traces of four signs, the last three of which are clearly to be read *stp-n-R'*) the probable identity of the king is *Mn-mi-R'-stp-nR' Imn-ms-s(w)* 'Menmire-setepenre Amenesse' of the 19th Dynasty, because he is the only king with one name which ends with *-ms-(sw)* and the other with *-stp-n-R'*. The apparent 'sun-disk' sign in the first cartouche could be restored as the seated figure of the god Amun, and this cartouche would then contain the *nomen*. However, an orthographic objection may be raised in the second cartouche since, while the traces may be restored as *Mn*, there is no room for *mí*, unless it is written above where it does not belong. A second possibility is that there is a sun-disk and the latter element is to be restored *R'-ms-s(w)* 'Ramesses'. In that case, the *-stp-n-R'* in the left hand cartouche suggests the restoration *Wsr-m3't-R'-stp-n-R'* 'Usermare-Setepenre', the *prenomen* of Ramesses II; however, while the vertical stroke at the right could very well be the bottom part of *wsr*, the horizontal stroke at its left does not easily lend itself to any writing of *m3't*. An additional problem is raised by the relative position of the two cartouches since clearly the right hand cartouche contains a *nomen*, while the left hand cartouche contains a *prenomen*. In normal Egyptian orthography, inscriptions are read from right to left and the *prenomen* should therefore precede the *nomen*. All of this suggests careless workmanship by the artist who inscribed the pot. It is safest to date the cartouches simply to the Ramesside period, perhaps a little more explicitly to the 19th Dynasty (see also Cat. 19, 96, 179).

Eg. Cat. 105. *Fig.* 34:4. Field No. 253/103. Locus 102. Inscribed bowl, faience, light brown frit, blue glaze, black paint. H. 5cm., W. 8cm., Diam. 8cm. (at base). On this fragment the lower left hand portion of a cartouche is still preserved. It is vertical and stands over the hieroglyph *nb* 'gold', of which a small trace is visible. Within the cartouche a portion of a *ms* hieroglyph is preserved, indicating that the royal name was most probably Ramesses; but no closer attribution can be made.

Eg. Cat. 106. *Fig.* 41:5. Field No. 70/2. Locus 109. Sherd, faience, light brown frit, greenish glaze, decoration in black paint. H. 8cm., L. 7.5cm., Th. 1.7m.
Fragment from the side or centre of a vessel, perhaps a globular bowl or cup. The decoration consists of two parallel vertical lines, terminating in minute knobs from the beginning or end of the brush strokes and on the left, on a plane with and equidistant from them, traces of a third knob. If they are part of a name, the traces suggest a restoration of *ms*, an element of the name '[Ra]mes[ses]'. However, no name may have been written since there is no trace of

a cartouche, and the strokes could represent the knotted, dangling ends of the strings that were tied at the back of the neck to fasten a collar or necklace (cf. Cat. 146–8).

Eg. Cat. 107. *Fig.* 39:2. Field No. 225/3. Locus 107. Sherd, faience, brown frit, green glaze, decoration in black paint. H. 3.3cm., W. 2.8cm., Th. 1.2m. This fragment from a much larger inscribed object preserves the damaged title *nb t3.wy* 'Lord of the Two Lands', which normally precedes cartouches.

Eg. Cat. 108. *Fig.* 40:2. Field No. 278/143. Locus 107. Sherd, faience, white frit, green glaze, inscribed with black paint. H. 1.7cm., L. 3.5cm., Th. 0.6m. Fragment from some type of vessel; the section suggests a lipless cup, bowl or high-sided dish. Just beneath the rim, on the outside, are traces of inscription which are probably to be restored as *nb h'.w* 'Lord of Diadems'.

Eg. Cat. 109. Not illustrated. Field No. 134/14. Loci 110, 111. Fragment, faience, yellow frit, green glaze, black paint. L. 2cm., W. 1.6cm., Th. 0.7cm. Fragment of a bowl or jar with traces of a sign that may be restored as either ⌡ , or as the back end of a cartouche 〕.

Eg. Cat. 110. *Fig.* 40:3. Field No. 232/25. Locus 101. Sherd, faience, brown frit, green glaze, inscription in black ink. L. 4cm., W. 3.5cm., Th. 0.5cm. Fragment of a small, thin-walled vessel. The inscription consists of part of a vertical cartouche with completely illegible traces of two hieroglyphic signs. No suggestions can be made as to their reading.

10 Decorated Faience Vessels (Eg. Cat. 111–174)
The decorated faience vessels, none of which is complete, have been arranged according to the nature of their decoration. Typologically they fall into two broad classes: (a) bowls, cups and dishes; (b) jars, jugs and juglets. Vessels for drinking wine or beer are usually decorated on both the inner and outer sides, but sometimes on the inner side only. At least four different types of scenes are attested. Cat. 111–112 imitate a flower whose petals unfold, with a motto or wish inscribed in the central calyx. Cat. 113–121 are fragments of one of the two most common vessel motifs, the stylised fish and floral design showing fish swimming in a pond out of which various aquatic plants grow. Cat. 122 belongs to a less common group of motifs which began to come into use in the 19th Dynasty and have as one of their essential features a mildly satirical element of humour. Cat. 123–143 show floral decorations alone, but only those with decoration on the inner surface come from drinking vessels (Cat. 111–118) while others, with floral decoration on the outer surface (Cat. 140–142) come from the jar/jug-type vessel represented by Cat. 144–158. The decoration of the latter took the form of broad collars and necklaces which, together with floral collars, were draped round the necks and shoulders of wine and beer jars on festive occasions.[8]

Cat. 159–163 bear a geometric decoration which has a non-Egyptian look and may be of either Midianite or Aegean inspiration; of these, Cat. 158–159 are drinking vessels. Cat. 163–166 are very tiny sherds whose decoration cannot be satisfactorily classified. Cat. 167–172 are glazed but show no additional decoration, while Cat. 173–174 are jar handles, one decorated with a geometric pattern, the other plain. All the vessels are made of faience and were ceremonial in nature.[9]

Eg. Cat. 111*. *Fig.* 40:8. Field No. 257/282B, 325, 332, 339. Loci 102, 109, 110. Dish, faience, white frit, blue-green glaze, decoration in black paint. L. 6cm., W. 4.5cm., Th. 0.4cm. Fragments.
There are decorations in black paint on the inside. Opening out from the centre are the petals of a lotus flower. The traces of inscription preserved on the flat surface of the centre show the hieroglyph ⌡ *w3s* 'stability' to the right. There was obviously a similar, balancing, sign at the left which is not preserved. Both rested on the hieroglyph ▽ *nb* 'all', so that the full text would be a motto 'all stability and all ----'.[10]

Eg. Cat. 112. *Fig.* 41:8. Field No. 213/8, 328. Loci 105, 106. Wine bowl, faience, white frit, green glaze, decoration in black paint. H. 5.9cm., W. 2.8cm., Th. 0.5cm. Fragment reassembled from two sherds.
The decoration of the interior of this small but deep wine bowl takes the form of a rosette, of which the tips of two and bases of four more petals, all emanating from a central circle, are preserved. Within the central circle is an illegible trace of a hieroglyph. The design is a very simple, stylised floral rosette with a motto or wish inscribed in the centre (cf. Cat. 111).

Eg. Cat. 113. *Fig.* 42 – Wine bowl, faience, very pale brown frit, green glaze, decoration in black paint. H. 7cm., Diam. 36cm. Reconstructed from a number of fragments.[11]
The scene on the inside of the bowl showed a pair of Nile fishes swimming among lotus plants in a pond. Although only one of the pair of fishes is actually preserved, it is clear, both from the position of the fish and the amount of space left, that there were originally two. The pair must have been in a tête-bêche position. On the outside of the bowl, one fragment has two hieroglyphs on it in black ink, reading *ms*; probably the second element in a theophorous name.

Eg. Cat. 114. *Fig.* 43:13. Field No. 278/142. Loci 107, 109, 110. Sherd, faience, brown frit, blue glaze, decoration in black paint. H. 3.5cm., W. 2.0cm., Th. 0.8cm. The decoration consists of two horizontal, slightly curved parallel lines. Under the lower line, to the left, is a much thicker blob slanting obliquely upwards towards the right, which may represent the body and dorsal fin of a fish, together with a body marking (perhaps a gill).

Eg. Cat. 115. Not illustrated. Field No. 293/7. Surface find. Sherd, faience, light brown frit, green glaze, decoration in black paint. H. 2.0cm., L. 2.7cm., Th. 0.6cm.

Since the decoration is on the inner surface, it is probable that the original vessel was some sort of wine bowl or cup. The decoration suggests a fish. The high, ribbed dorsal fin is very prominent, as is the vertical stroke which separates the head from the body. The horizontal stroke at the bottom could represent the lower part of the body or an internal detail.

Eg. Cat. 116. *Fig.* 43:5. Field No. 256/15, 237, 332. Loci 109, 110. Wine bowl, faience, light brown frit, green glaze, decoration in black paint. Three sherds.

The sherds come from the rim of the bowl and are decorated on both the inner and outer sides. On the outside of two of the sherds (256/15 and 237) a single petal of a waterlily (*nymphea caerulia*) is preserved. The decoration on the inner face is more complex: running around the circumference of the rim is a border consisting, from the rim downwards, of two wide and one narrow horizontal bands. In the upper two of these is a checkerboard pattern of dots, one in the upper and two bigger ones in the middle band. The dots in the middle band are connected by a horizontal line. Beneath this decorative border is the main scene which appears to be the stylised floral and fish pattern since at least one pointed leaf of a water-flower is visible and at a right angle to this is a fish's tail (332). On the second fragment (237) the back and dorsal fin(?) of a fish are preserved. On the first fragment are, to the right, two traces which might be part of another plant. If this interpretation of the scene on these fragments is correct, the ornamental border running around the inside of the rim may represent a fence around the pond in which fish and flowers are seen. Similar depictions of enclosure fences are known from some of the tomb reliefs from Amarna dating from the end of the 18th Dynasty.

Eg. Cat. 117. *Fig.* 44:7. Field No. 374/13, 362. Loci 103, 106, 108. Bowl, faience, brown frit, blue glaze, decoration in black paint. Two sherds from the same vessel.

As the two sherds are decorated on the inside, the vessel was probably a wine bowl. On one sherd (362) the left half of a water lily is seen in profile. It consists of three petals, folded one over the other as they grow out of the calyx. Internal detail is sketchily rendered by a single row of dots running the length of each petal. The second sherd (374/13) has five horizontal lines across its middle. These represent the banded central portion of a lotus bud seen in profile. The fact that two different plants are shown on the inside of this wine bowl suggests that the original scene had an aquatic setting such as a fish-pond, and that the scene was one of the common stylised fish and flower designs.

Eg. Cat. 118. *Fig.* 44:3. Field No. 234/150. Locus 101. Wine bowl, faience, white frit, green glaze, decoration in black paint. H. 5.0cm., W. 4.0cm., Th. 0.8cm. Sherd, possibly from the bottom of the vessel.

At the left of the inside can be seen the stem of a plant with three horizontal bands running across it (a papyrus plant?). To its right are two concentric but widely-spaced lines for which no suggestions are made. To the left of the larger of these, following its outline, is a line of dots incised into the sherd. It is possible that the decoration on the inner side of the bowl was a stylised fish and floral pattern. The traces on the outer side are faint but easily recognisable as two flower petals enveloping the vessel, with a V-shaped interstice between them. The petal at the left shows some internal detail, rendered by two vertical lines of blobs running parallel to the sides of the petal.

Eg. Cat. 119. *Fig.* 44:6. Field No. 241/64, 280, 237. Loci 101, 106, 110. Wine bowl, faience, light brown frit, blue glaze, decoration in black paint. L. 7.5cm., W. 4.0cm., Th. 1cm. Two sherds joined together, one (237) not illustrated.

This wine bowl is decorated on the inside and out-side. On the outside two water lily petals can be recognised, one rendered in outline only; the other which overlaps it has internal detail indicated by a series of short dashes. The decoration of the inside shows a stylised fish and flower design. The head of the fish is visible at the lower right, with its round eye and open mouth. In the centre is a lotus flower, its curved stem in front of the fish terminating in a stylised bud near the top of the sherd. At the left of the stem is a curved loop which may represent another flower or an additional shoot from the lotus. To the right of the stem is another plant shoot, directly between it and the mouth of the fish.

Eg. Cat. 120. *Fig.* 43:10.[12] Wine bowl, faience, white frit, green glaze, decoration in black ink. Fragment consisting of 10 sherds, decorated on both the inner and outer faces.

The decoration on the outside of the bowl is simple, consisting of a water lily which is visible on three sherds (262, 362, 366). The central calyx would have been the base of the bowl, with the petals rising outwards on the side. On one sherd in particular the 'V' between two overlapping petals is quite evident although their tips are missing. The decoration on the inner face seems rather more complex, but when viewing the sherds together it is clear that this is a conventional, stylised fish and floral design. Two sherds show the edge of a fishpond (executed by a wide double line) and the tails and rear portions of two fish (one of which has a fin visible). On the other sherds are various plants, water lilies and lotus buds. The detail on the water lilies is rendered by a single vertical row of dots running the length of each petal. The lotus buds are shown with two different types of detail. One is a dot-type decoration very similar to that of the lilies. The other depicts the petals folded over one another obliquely lengthwise since the bud is closed.

Eg. Cat. 121. *Fig.* 27:7. Field No. 257/281B. Locus 102. Sherd, faience, white frit, green glaze, decoration in black paint. L. 3.8cm., W. 2.2cm., Th. 1cm.

The decoration is somewhat enigmatic. It consists of a circle with a dot in the centre, the circle itself being surrounded on one side by a semi-circle of six small dots. At a distance from the semi-circle, and almost obliquely parallel to it, runs a straight line, with another straight line at a right angle to it. These traces suggest an eye and its markings (the circle and semi-circle of dots) and the snout or bill of a creature (bird, fish), marked off by the vertical line parallel to the eye markings, with the mouth then represented by the horizontal line. If this interpretation is correct, this fragment may have come from a wine bowl with a stylised fish and floral design.

Eg. Cat. 122*. *Fig.* 38:6. *Pl.*122:13. Field No. 173/1, 245. Loci 101, 107. Wine bowl, faience, brown frit, green glaze, decoration in black paint. L. 6.9cm., H. 6.9cm., Th. 0.8cm. Fragment reassembled from three sherds.
The fragments come from the inner face of a bowl. The decoration shows, on the left, the head, ears, back of the neck, back, rear and tail of a feline. The body is facing to the left but the head is turned backwards. According to the rather prominent ears, large round eye and the body markings, it is probably a serval cat (cf. Cat. 85–95). At the right, behind its back and over its head, are several chevron-shaped traces which represent a papyrus flower and possibly the bud of a lotus. Clearly the cat is in a thicket of plants. The drawing of papyrus, with its tail dangling down, is a motif represented in Egyptian art as far back as the Old Kingdom. For example, on a 5th Dynasty relief now in the Vatican (Scott 1958; 4) a cat is shown climbing a papyrus stalk, at the end of which rests a nest with three young birds in it. The cat in Cat. 122 has been rendered smoothly and clearly with steady, unwavering strokes which convey an animated sense of activity. This is a departure from the conventional fish and lotus decorations of wine bowls, and suggests a date in the mid-19th Dynasty.[13]

Eg. Cat. 123*. *Fig.* 41:9. Field No. 241/63, 331, 232, 82. Loci 101, 102, 109. Wine bowl, faience, brown frit, green glaze, decoration in black paint. Diam. 24cm. Fragment reassembled from several sherds.
The decoration consists of a band of water lilies running around the inside. The points of the individual petals are flush with the inner edge of the rim. Only two lilies are preserved, each with four petals seen from the side. The detailing of each individual petal consists of a series of vertical dashes running its length.

Eg. Cat. 124. *Fig.* 44:1.[14] Bowl, faience, grey frit, blue-green glaze, decoration in black paint. Fragment reassembled from several sherds.
The decoration on both the inner and outer faces suggests that the vessel was a wine bowl. The designs themselves are difficult to interpret. On the outside of the bowl there may have been a floral pattern. One sherd shows a thick inverted V-shaped line which seems to represent the outline of the tip of a petal. The other sherds show a solid line and on one side a series of shorter dashes which may represent the internal detail of a lily petal, with the solid line representing the outline. The decoration of the inside is more enigmatic. On two sherds (368/5, 319) there seems to be a central circle with blades emanating from it. Both circle and blades are filled with dots, perhaps representing some type of stylised floral design. Since the two sherds cannot be joined, it would suggest that there were several flowers in the design. Another sherd shows a curved line with a zigzag pattern, together with a series of lines and dots (368/5), for which no interpretation is suggested.

Eg. Cat. 125. *Fig.* 44:4. Field No. 279/290B. Locus 107. Wine bowl, faience, white frit, green glaze, decoration in black paint. W. 7.8cm., Th. 0.6cm., Diam. 5.5cm. Sherd from base of vessel.
The decoration represents the central calyx of a water lily and five petal emanating from it. The calyx is rendered by two concentric circles. Inner detail is indicated on several of the leaves.

Eg. Cat. 126. *Fig.* 43:2. Field No. 166/3. Locus 109. Sherd, faience, white frit, green glaze, decoration in black paint. H. 3.0cm., W. 1.8cm., Th. 0.5cm.
The decoration shows the upper parts of two petals of a water lily, each rendered in outline with a vertical central rib.

Eg. Cat. 127. *Fig.* 43:4. Field No. 289/20, 124. Loci 101, 102. Bowl, faience, very pale brown frit, green glaze exterior with patches of green and light blue on interior, decoration in black paint. L. 4.5cm., H. 5.0cm., Th. 0.8cm. Fragment reassembled from two sherds.
The decoration on the outside represents several petals of a water lily. That on the left has its internal detail executed by successively repeating the outline of the petal, from the outside to the centre. The one on the right is rendered with a series of parallel lines running horizontally across its base. The lower end of the V-shaped interstice between the two petals also has a series of lines running horizontally across it, indicating that it is the bottom of a third petal.

Eg. Cat. 128. *Fig.* 44:9. Field No. 297/10. Locus 101. Vessel, faience, brown frit, green glaze, decoration in black paint. H. 2.0cm., W. 1.7cm., Th. 0.9cm. Sherd.
Since the decoration appears on the inner surface of the sherd, it is probable that this came from a vessel for drinking wine. The decoration represents part of a petal of which one side and the internal detail are discernible, rendered by a series of dashes parallel to the side.

Eg. Cat. 129. *Fig.* 44:5. Field No. 302/12, 206. Loci 101, 110, 111. Vessel, faience, light brown frit, green glaze, decoration in black paint. Two sherds.
The decoration on both the inner and outer faces suggests that the vessel was a wine bowl or dish. The traces on the outside of the sherds may be part of the outline (206) and the outline plus internal detail of the petal of a water lily. The decoration on the inner side

of the same sherd appears to be a portion of the calyx (rendered by a double curved line) and a petal (rendered in outline with internal detail shown by a series of dots). The decoration of the second sherd likewise appears to show a petal of the same plant (rendered in the same fashion) alternating with a second petal whose detail is rendered differently.

Eg. Cat. 130. *Fig.* 43:15. Field No. 345/19. Locus 107. Vessel, faience, light brown frit, blue glaze, decoration in black paint. H. 3.2cm., W. 3.5cm., Th. 1cm. The decoration of this sherd is on the inner face, suggesting that it came from a wine bowl. The surface is badly chipped and the design obscure, but it seems to be a petal of a plant drawn in outline, with the internal detail rendered by a series of oblique lines, slanting upwards from left to right. To the right of this petal appears the outline of part of the left side of a second petal, but without any inner detail.

Eg. Cat. 131. Not illustrated. Field No. 373/14. Locus 111. Vessel, faience, brown frit, green glaze, decoration in black paint. H. 2cm., W. 1.5cm., Th. 0.6cm. Since this sherd is decorated on both the inner and outer faces, it probably came from a wine vessel. The design on the inside is possibly the outer edge of a petal, while that on the outside may be the ribbed petal of another flower.

Eg. Cat. 132. Not illustrated. Field No. 216/89. Locus 101. Vessel, faience, brown frit, green glaze, decoration in black paint. H. 6.3cm., W. 4.2cm., Th.0.9cm. The decoration, which is on the inner face of this sherd, may be part of a floral design and suggests that it came from a wine-drinking vessel.

Eg. Cat. 133. *Fig.* 44:10. Field No. 347/11. Locus 108. Sherd, faience, very pale brown frit, green glaze, decoration in black paint. H. 2.9cm., W. 2.0cm., Th. 0.8cm. The decoration, found on one side only, appears to be a portion of the petal of a plant emanating from the calyx. It consists of the outline of the lower half of the left side, to the right of which internal detail has been indicated by a file of vertical dashes running parallel to the outline. The calyx is indicated by a slightly curved line running across the base of the petal.

Eg. Cat. 134. *Fig.* 43:11. Field No. 339/121A. Locus 109. Vessel, faience, very pale brown frit, green glaze, decoration in black paint. H. 4.2cm., W. 3.5cm., Th. 0.5cm. Fragment assembled from two sherds. The decoration on the outer face, suggesting that the sherd came from a wine or beer vessel, apparently represents the petals of a lotus flower radiating out of the calyx. The V-shaped interstice between two of them is clearly marked. Just below it, a hole was pierced through the thickness of each petal, apparently for suspension.

Eg. Cat. 135. *Fig.* 43:12. Field No. 339/121. Locus 109. Vessel, faience, light brown frit, green glaze, decoration in black paint. L. 3cm., W. 3.0cm., Th. 0.5cm. Although the decoration of this sherd is difficult to interpret, the two slightly curved parallel vertical lines on the left suggest the petal of a water lily, the right hand line representing the outer edge, the left hand line indicating the internal detail. To the right, three small dots are arranged in a semi-circle around a fourth, but no suggestion is offered for their interpretation.

Eg. Cat. 136. Not illustrated. Field No. 361/10. Locus 110. Vessel, faience, brown frit, green glaze, decoration in black paint. The vessel from which this sherd comes may have been a cup or jar. The decoration on the outside shows several recognisable petals of a lotus flower.

Eg. Cat. 137. Not illustrated. Field No. 311/188. Locus 101. Vessel, faience, brown frit, green glaze, decoration in brown paint. This fragment may be from a bowl. The design may represent the internal detail of the petal of a flower.

Eg. Cat. 138. *Fig.* 44:2. Field No. 137/21. Locus 110. Wine vessel, faience, brown frit, blue-green glaze, decoration in black paint. H. 4.5cm., W. 2.5cm., Th. 0.9cm. Fragment of a bowl or cup, reassembled from two sherds. The decoration on the outside represents a flower (lotus or lily), the calyx of which would have been the base of the vessel with its petals unfolding and enveloping the side. Parts of two petals can be distinguished, with the V-shaped interstice between them. Internal detail has been indicated on one petal by a series of vertically set blobs running parallel to the side. The decoration on the inside appears to show the interior of the same flower, with at least one petal being discernible.

Eg. Cat. 139. *Fig.* 44:11. Field No. 337/36B. Locus 102. Wine bowl, faience, light brown frit, green glaze, decoration in black paint. H. 4.5cm., W. 5.1cm., Th. 1.0cm. The decoration on the inner surface of this sherd shows two petals of a flower, each with a central vein emanating obliquely from this on each side. The decoration on the outside also represents a flower, its calyx on the bottom and the petals unfolding and enveloping the sides. The V-shaped line represents the right and left sides of two of the petals, with a third petal rising between them in the interstice.

Eg. Cat. 140. Not illustrated. Field No. 297/9. Locus 101. Vessel, faience, brown frit, green glaze, decoration in black paint. H. 1.4cm., W. 1.8cm., Th. 0.7cm. The decoration on the outside of this sherd, consisting of a Y-shaped vertical line at the right and a parallel straight line at the left, suggests three petals of a flower, with the interstice between the first two.

Eg. Cat. 141. Not illustrated. Field No. CD/15–16/3. Locus 101. Vessel, faience, brown frit, green glaze, decoration in black paint. H. 2.0cm., W. 1.5cm., Th. 0.4cm. The thinness of the section suggests that this sherd came from a small vessel, perhaps a cup. This suggestion is reinforced by the decoration on the outside of the sherd which resembles the pointed tip of a flower (lotus ?) petal.

Eg. Cat. 142. *Fig.* 43:7. Field No. 257/282A. Locus 102. Bowl, faience, light brown frit, green glaze, decoration in black paint. H. 3.0cm., W. 2.5cm., Th. 0.5cm. This tiny rim fragment is decorated on the outside with the petal of a water lily whose internal veins are detailed parallel to the outline of the petal.

Eg. Cat. 143. *Fig.* 43:9. Field No. 239/290. Loci 106, 107. Wine bowl, faience, light brown frit, green glaze, decoration in black paint. L. 7.0cm., H. 2.1cm., Th. 0.4cm., Diam. 11.0cm. Rim fragment.
The inside is decorated with an opened water lily of which three petals are preserved. The central petal shows some internal detail rendered by repeating the outline of the petal. The top edge of the rim is also decorated by a series of evenly spaced dots running round its circumference.

Eg. Cat. 144. *Fig.* 37:34. *Col. Pl.* 21. Field No. 242/41, 325. Loci 107, 110. Jar, faience, brown frit, green glaze, decoration in black paint.
The two fragments of this globular jar both come from high on the shoulder. The decoration consists of a stylised floral necklace of the type draped over wine and beer jars at banquets. There are at least two strands of decoration: the upper consists of triangles alternating point up and point down. In the centre of each is a solid dot. The bases of the triangles with the point upwards have a line of several dots overlapping the horizontal border line. A second border line immediately below this marks off the top of the second preserved strand of the necklace. This consists of a frieze of pendants, the upper ends and sides of which are straight, with the lower ends rounded. An oval dot is centred in each pendant just below the line of the upper end and, parallel to each of these dots but below the lower border line of the strand, is a second dot. This second band of dots is then closed off by another horizontal line running round the circumference of the jar. In each register, the round and oval dots may be interpreted as the holes attaching the individual triangles and pendants to the strands of the necklace.

Eg. Cat. 145. *Fig.* 27:9. Field No. 78/21, 257/301. Loci 101, 102. Jar, faience, brown frit, green glaze, decoration in black paint. L. 7.0cm., H. 4.0cm., Th. 0.9cm. Fragment reassembled from two sherds.
The decoration on the outside of the shoulder of the jar represents the outer strand and pendants of a necklace or collar of the type draped over wine and beer jars. Above them is a row of vertical rectangular pendants with rounded ends. Beneath is the lower cord to which they were attached. On this, centred directly under each pendant, is the hole (represented by a dot) by which the pendant was attached. Finally, suspended freely from this cord are three more triangular pendants pointing upwards, acting as counterpoises to the collar so that it would lie flat on the body.

Eg. Cat. 146. *Fig.* 27:10. Field No. 337/2, 237, 276. Loci 102, 110, 111. Jar, faience, very pale brown frit, green glaze, decoration in black paint. Three sherds from the same jar.
Although there are no joins, it is clear that the three sherds come from just above the shoulder of the same jar. The decoration represents a collar. One sherd shares elements with the other two: at the top is a horizontal line representing the lower edge of a cord to which the tops of a row of pendants were attached. These are rectangular with rounded bottoms and set vertically in a row (at the upper right). To their left is a vertical line running down the height of the sherd, representing the cord or binding at the left end of the necklace (if opened out straight). Vertically to the left of this are two wavy lines which thicken towards the bottom. These are the ends of the cords knotted together at the back when the necklace is worn. To the right of the vertical binding of the necklace, and immediately beneath the bottom line of the cord holding the lower ends of the rectangular pendants, are two dots. These are best explained when examining the second sherd (237) on which above, at the right, are the rounded bottoms of three of the row of rectangular pendants. Then comes the cord to which they are attached, then a wide band with a row of dots set just beneath its upper edge, and a second row of dots parallel to these, set just above its lower edge. Since there is no other decoration in this band, it may be that it represents a wide strip of cloth forming the lower row of the necklace. The dots would represent the holes for attaching it to the cords. The presence of the lower row of dots suggests that an additional strand was fixed to the collar. The third sherd (276) shows four complete and traces of two more rectangular pendants from the upper row.

Eg. Cat. 147. *Fig.* 37:31. Field No. 43/1. Locus 109. Jug, faience, very pale brown frit, green glaze, decoration in black paint. H. 5.0cm., W. 5.7cm., Th. 1.0cm. Sherd from shoulder of vessel.
The decoration consists of a horizontal line above, with a circle in the middle, from which three wavy lines hang down, representing the tie and knot of the cords at the back of a broad collar.

Eg. Cat. 148. *Fig.* 37:26. Field No. 141/2. Locus 107. Jug, faience, brown frit, green glaze, decoration in black paint. H. 1.8cm., L. 2.2cm., Th. 0.7cm. Sherd from shoulder of vessel.
Three wavy, almost zigzag lines attached at right angles to a straight line, with a thick projection where two of the wavy lines meet the straight one, indicating that the decoration represents the tie and knot of the cords of a broad collar draped over the shoulder of the vessel.

Eg. Cat. 149. *Fig.* 27:8. Field No. 263/25, 258. Loci 103, 106. Jug, faience, grey frit, blue-green glaze, decoration in black paint. H. 5.0cm., W. 3.0cm., Th. 0.9cm. Fragment reassembled from three sherds.
The decoration on the fragment from the shoulder of a jug takes the form of an ornamental collar, of which

the upper strand can be made out. It consists of two narrow bands separated by a wider band. The two outer bands represent the cords of the necklace to which the ornaments were fastened, consisting of triangles, alternating point up and down and each decorated in the centre by a circle with a dot in the middle. On the outer cords(?) at the base of each triangle, is a line of dots representing, together with the dot in the centre of the triangle, the holes for fixing the ornaments to the cords of the collar.

Eg. Cat. 150. *Fig.* 37:27. Field No. 8/4. Locus 112. Jug, faience, light brown frit, green glaze, decoration in black paint. H. 3.5cm., L. 4.0cm., Th. 1.3cm.
The section of this sherd, from the shoulder and base of the neck, shows that the jug probably had a carinated shoulder and a straight-sided neck. The decoration is on the neck and above the shoulder. A horizontal line encircles the base of the neck. Above it are three irregular blobs. Down the height of the shoulder, at an oblique angle, are a pair of straight lines between which is a file of four dots. At the left and parallel to these, is a third oblique line. This design represents a collar draped over the jug. The thin horizontal and vertical bands with the dots represent some of the cords to which the different components of the necklace were attached, and the dots represent the attachment holes.

Eg. Cat. 151. *Fig.* 37:30. Field No. 289/24. Loci 101, 102. Jug, faience, brown frit, blue-green glaze, decoration in black paint. H. 4.2cm., W. 3.1cm., Th. 0.7cm.
This fragment, from the shoulder of the vessel, has decoration on the outside, running along the upper surface of the shoulder and consisting of a row of pendants which formed a strand of an ornamental collar.

Eg. Cat. 152. *Fig.* 43:14. Field No. 234/151. Locus 101. Vessel, faience, brown frit, green glaze, decoration in black paint. H. 3.3cm., W. 2.4cm., Th. 0.7cm. Sherd from the base of a vessel; the decoration is difficult to interpret.

Eg. Cat. 153. *Fig.* 43:3. Field No. 310/15. Locus 105. Sherd, faience, white frit, blue glaze, decoration in black paint. L. 2.6cm., H. 2.6cm., Th. 1cm. Rim of a bowl.
The decoration may represent a collar necklace. Above are two closely set parallel lines forming an upper border, then a wider band of vertical lines running across the circumference, closed off at the bottom by another horizontal line. Beneath this are two undecipherable traces.

Eg. Cat. 154. *Fig.* 43:1. Field No. 277/109. Loci 107, 110. Sherd, faience, white frit, green glaze, decoration in brown paint. H. 1.6cm., W. 1.9cm., Th. 0.3cm. From the rim of a bowl.
The decoration is part of a collar, showing the upper retaining cord, then the upper parts of two of a row of vertically set rectangular pendants with rounded ends. Centred at the top of each pendant, just below the lower edge of the retaining cord, is a dot representing the hole to attach the pendants to the cord.

Eg. Cat. 155. Not illustrated. Field No. 277/108. Loci 107, 110. Sherd, faience, white frit, green glaze, decoration in black paint. H. 2.0cm., W. 1.9cm., Th. 1cm.
The decoration, on the outside of the vessel, consists of one strand of a collar with three pendants suspended from it.

Eg. Cat. 156. Not illustrated. Field No. 253/112. Locus 102. Vessel, faience, brown frit, green glaze, decoration in black paint. H. 2.3cm., W. 1.8cm., Th. 0.8cm.
The decoration on the sherd consists of two T-shaped lines facing each other and may represent two strands of pendants on a collar, separated by a single cord. This suggests that the vessel was a wine or beer jar.

Eg. Cat. 157. Not illustrated. Field No. 253/111. Locus 102. Vessel, faience, brown frit, traces of green glaze, decoration in black paint. H. 4.5cm., W. 5.3cm., Th. 1cm.
The decoration on this sherd consists of two vertical lines, slightly converging at the top, with the one on the bottom right splitting in a squat, inverted 'V', and suggests a row of pendants from a strand of a collar or necklace. As this decoration appears on the outer side of the sherd, the vessel was probably a wine or beer jar.

Eg. Cat. 158. *Fig.* 44:8. Field No. 245/70. Locus 107. Jug, faience, brownish-white frit, blue-green glaze, decoration in black paint. H. 1.9cm., W. 2.9cm., Th.0.4cm.
The decoration of this sherd, an upper band with three horizontal dots on it, suggests some type of collar or necklace. Three parallel zigzag lines beneath the band could represent the border-thread of the strand or decoration of the strand.

Eg. Cat. 159. *Fig.* 41:6.[15] Wine bowl, faience, very pale brown frit, blue-green glaze, decoration in black paint. Diam. 95cm. Fragment reassembled from numerous sherds.
There is decoration both inside and outside. The latter runs horizontally around and just below the edge of the rim and consisted, to judge from the single trace preserved, of a frieze of water lilies seen from the side. The interior decoration is much more elaborate. Above, its upper edge flush with the inner edge of the rim, is a wide band of interlocking single horizontal S-spirals. Below, separated from this by a much narrower band, is a frieze of flowers, possible encircling the complete inner diameter. Two different types of flowers may be indicated, alternating with one another, since the tips of three petals of a water lily are clearly visible, but set at a distance from them on the same plane, on another part of the band, are the rounded tips of another plant.

Observation: The interlocking S-spiral, while not unknown in Egyptian art in the Old Kingdom, was nevertheless more characteristically an Aegean motif (Kantor, 1947:21). The vessels which Petrie found at Serabit el-Khadim (Petrie, 1906; Pl. 147:9, 10, 12–15) that bore a similar spiral decoration, were described by him (*loc. cit.* 141) as 'foreign ware'. Without going so far, the decoration does have a sufficiently Aegean flavour to be termed 'foreign-inspired'. The same motif also appears on a number of Midianite bowls found at Site 200 (see Chapter III, 3, and. *Fig.* 4:12, 13; 5:1).

Eg. Cat. 160. *Fig.* 41:7. Field No. 260/190, 93, 245, 257, 316. Loci 102, 107, 110, 111. Bowl or cup, faience, very pale brown frit, green glaze, decoration in black paint. Fragment reassembled from several sherds.
A series of small, apparently equi-distant holes was drilled through the thickness, just below the top edge of the rim. These may have run all round the circumference and been intended for hanging the vessel. The decoration running around the outside consists of a double horizontal band, divided at intervals by alternating thick and thin vertical lines.

Eg. Cat. 161. *Fig.* 43:8. Field No. 344/12, 225, 245. Loci 107, 110, 111. Bowl, faience, white frit, blue-green glaze, decoration in black paint. Fragment of rim reassembled from several sherds.
The decoration on the outer side consists of a geometric pattern: a zigzag line fills the centre of a wide horizontal band, framed on top and bottom by a medium-thick, solid line. Under the lower border line runs a second line of equal thickness.

Eg. Cat. 162. *Fig.* 43:6. Field No. 263/1, 319, 339. Loci 103, 106, 108, 109. Bowl, faience, very pale brown frit, green glaze, decoration in black paint. Fragment reassembled from four sherds (two not illustrated).
The decoration which is just below the rim, on the outside, consists of a band of diamond-shaped lozenges framed by a thick border line.

Eg. Cat. 163. *Fig.* 27:5. Field No. 332/7. Locus 109. Vessel, faience, white frit, green glaze, decoration in black paint. H. 3.1cm., W. 2.6cm., Th. 1.1cm.
Although the wall of this sherd is rather thick, there is no way of identifying the type or part of the vessel it came from. The outside decoration consists of three parallel horizontal lines of relatively equal thickness, equi-distant from each other and running round the circumference of the vessel.

Eg. Cat. 164. *Fig.* 27:6. Field No. 257/281A. Locus 102. Sherd, faience, very pale brown frit, green glaze, decoration in black paint. L. 4.3cm., W. 2.5cm., Th. 1.0cm.
The decoration consists of a circle with a dot in the centre, enclosed on one side by a semi-circle of dots, and on the other by two parallel slightly curved lines.

Eg. Cat. 165. Not illustrated. Field No. 276/18A. Locus 111. Vessel, faience, brown frit, green glaze, decoration in black paint. L. 3.1cm., H. 2.1cm., Th. 0.8cm. It is impossible to determine the type of vessel. The decoration consists of four dots along one of its edges.

Eg. Cat. 166. Not illustrated. Field No. 245/72. Locus 107. Vessel, faience, brown frit, no traces of glaze,

decoration in brown paint. L. 4.2cm., W. 2.3cm., Th. 0.9cm. It is impossible to determine the type of vessel. The decoration, a straight line with a darkish triangle at a right angle to it, cannot be interpreted.

Eg. Cat. 167. Not illustrated. Field No. 310/18. Locus 105. Vessel, faience, brown frit, traces of green glaze, decoration in black paint. L. 3.8cm., H. 2.9cm., Th. 0.7cm. It is impossible to determine the type of vessel. The decoration, a line running along the edge, vaguely suggests one side of a cartouche.

Eg. Cat. 168. *Fig.* 41:1. Field No. 239/18. Locus 107. Jar, faience, white frit, green glaze, no decoration. H. 3.5cm., W. 3.8cm., Th. 0.9cm., Diam. 2.6cm. Sherd from a small, globular jar with a straight moulded lip around the rim. This is a characteristic Egyptian shape frequently depicted as being offered by the king.

Eg. Cat. 169. *Fig.* 41:2. Field No. 337/32, 111, 212, 206. Loci 101, 102, 110, 111. Jug, faience, brown frit, blue-green glaze, no decoration. H. 3.7cm., W. 4cm., Th. 0.8cm., Diam. 5cm. Fragment of an open neck of a jug, reassembled from four sherds.

Eg. Cat. 170. *Fig.* 41:3. Field No. 337/31. Locus 102. Bowl, faience, light brown frit, blue-green glaze, no decoration. H. 5.8cm., W. 5.8cm., Th. 0.8cm., Diam. 11.5cm. Fragment reassembled from three sherds. Thick-walled, with a slightly beveled rim, the sides of this vessel flare outwards at a gentle angle from its base.

Eg. Cat. 171. *Fig.* 41:4. Field No. 327/2. Locus 106. Cup, faience, brown frit, green glaze, no decoration. H. 4.4cm., base Diam. 3.2cm., rim Diam. 5.0cm., Th. 0.4cm. Extremely thick base and walls.

Eg. Cat. 172. *Fig.* 45:3. Field No. 303/2. Locus 101. Fragment, faience, white frit, blue-green glaze, no decoration. H. 4.0cm., top Diam. 5.0cm. Fragment. Conical base of a small vessel, probably for ointment or perfume, as suggested by the narrow, circular interior.

Eg. Cat. 173. *Fig.* 37:29. Field No. 319/304. Locus 106. Juglet handle, faience, very pale brown frit, green glaze, decorated with brown paint. L. 2.5cm., W. 1.4cm., Th. 0.4cm. Decorated on the upper side with a ladder-like design.

Eg. Cat. 174. *Fig.* 37:28. Field No. 219/2. Locus 101. Juglet handle, faience, white frit, blue-green glaze, no decoration. L. 2.2cm., W. 1.0cm., Th. 0.8cm.

11 Amuletic Wands ('Magical Knives')
(Eg. Cat. 175–179)

The amuletic wand or 'magical knife' was an apotropaic charm intended to protect the living person by warding off demons and other real and imaginary beasts, particularly serpents. The majority of the known examples, commonly from the Middle Kingdom although they continue throughout Egyptian history, were made of hippopotamus ivory and were elaborately decorated. The Timna pieces, consisting of three head-ends (Cat. 175–177) and two handles (Cat. 178–179), are not so elaborate and would hardly be recognised as amuletic wands if Petrie had not found a number of them at Serabit el-Khadim (Petrie, 1906; 144–5; Pl. 150:1–13), of which two, nos. 2 and 5, had handles similar in shape and decoration to Cat. 178–179. With the handles establishing the presence of the wands at Timna, the three curved heads were recognised as the blades of such wands although, unlike the majority of apotropaic wands, their decoration is very simple, being limited to a slash for a mouth and an eye and eyebrow[16] (see Altenmuller, 1965; Steindorff, 1946, 41–51, 107; and Legge, 1905, 130–52, 297–303; 1906, 159–70).

Eg. Cat. 175*. *Fig.* 45:7. *Col. Pl.* 6. Field No. 283/3. Loci 106, 107. Wand, faience, white frit, traces of green glaze, decorated with black paint. L. 9.6cm., H. 3.3cm., Th. 1.2cm.
This fragment is in the shape of an animal head with a long snout. The mouth is indicated by a thick, horizontal line set approximately halfway between the upper and lower edges of the piece and extending about one third of the distance to the occipital ridge and the eye. The eye is set slightly above the level of the mouth and is rather small, with an eyebrow fully preserved on one side and partially on the other. There are no traces of any other facial details. Superficially the long snout, small eyes set far back and the almost imperceptible ridge over the eyes, give the impression that the creature represented is a crocodile.

Eg. Cat. 176*. *Fig.* 45:5. Field No. 225/1, IJ 16–17 B/1. Loci 101, 107. Wand, faience, very pale brown frit, traces of green glaze, decorated with black paint. L. 9.6cm., H. 3.3cm., Th. 1.1cm.
Of the two pieces which form about two thirds of the object (the rearmost part behind the head is lost) the part representing the snout of the animal retains all its glaze. On the other piece, which has lost virtually all its glaze, the paint indicating the eyes and eyebrows is perfectly preserved. The eyes are somewhat larger than those of Cat. 175 and 177 and the occipital ridge is more pronounced.

Eg. Cat. 177. *Fig.* 45:6. Field No. 278/4, 337/2. Loci 102, 107, 109, 110. Wand, faience, very pale brown frit, traces of green glaze, decorated with black paint. L. 8.2cm., H. 3.6cm., Th. 1.2cm. Fragment in two pieces.
Although both ends are missing, a somewhat rhomboidal shape is preserved with a human eye and eyebrow painted high up on each side, just beneath the occipital ridge. The break between the two fragments occurred vertically at the ridge, with the subsequent loss of detail of both eyes and eyebrows.

Eg. Cat. 178. *Fig.* 45:2. Field No. 362/17, 204, 233. Loci 101, 106, 108, 110. Wand(?), faience, white frit, blue and green glaze, decorated with black paint. L. 4.3cm., W. 2.2cm., Th. 1.0cm.
This fragment was originally the lower part of the handle of an object, but the nature of the object is unclear. It was probably not the counterpoise of a *menat*-necklace since there are no holes at the corners for suspension. Its relatively straight length indicates that it could have been part of a magical wand or knife. At least three such complete wands were found by Petrie at Serabit el-Khadim (Petrie, 1906; Pl. 150:1, 2, 5) and the decoration at one end of Cat. 178, just above the break, is a simple geometric design similar to the decoration on the three Serabit el-Khadim wand handles.

Eg. Cat. 179. *Fig.* 45:4. Field No. 45/1. Loci 110, 111. Wand, faience, light brown frit, green glaze, decorated with black paint. L. 5.0cm., W. 3.0cm., Th. 1.1cm.
This fragment is from the bottom of the handle of an object whose nature is problematical. To judge from the rounded end with relatively straight sides, it was not the counterpoise of a *menat*-necklace. It is also the wrong shape for a pendant end and has no holes for suspension at the attachment end. The decoration consisted of a series of triangles formed on one side by a single line and on the other by a double line. It could be the bottom of the handle of a model sistrum, but it is more likely to be the handle of a wand or magical knife, like Cat. 180. The shape and decoration accord perfectly with a wand from Serabit el-Khadim (Petrie, 1906; Pl. 150:5).

12 Inscribed Glass Sherd (Eg. Cat. 180)
Cat. 180 was the only inscribed glass object found at Site 200 and is dealt with here strictly from the standpoint of the inscription. It is treated from the technical, artistic and stylistic viewpoint by Lehrer-Jacobson in Chapter III, 15.

Eg. Cat. 180. *Fig.* 39:3. Field No. 313. Locus 103. Fragment, green glass. H. 0.9cm., W. 1.8cm., Th. 0.3cm.
The gentle curve of this tiny fragment suggests that it was originally part of a small, thin-walled vessel. The lower portions of a pair of vertical cartouches are legible, framed between a pair of vertical border lines of which only the right hand one is preserved. The signs in the right hand cartouche, which face to the left, appear to be *stp-n-R'*, while in the other cartouche, which faces to the right, is a very tiny *ms*. Since there are two cartouches, it is clear that one contains the *nomen* and the other the *prenomen*, and clearly the latter should be the left hand cartouche with *ms*. Since the cartouche containing the *prenomen* ends with *stp-n-R'* the restoration which suggests itself is [*Wsr-m3 t-R'-*]*stp-n-R'[R'*]*-ms-[sw]*, the *prenomen* and *nomen* of Ramesses II. However this is not the only possibility since the two cartouches could also be restored [*Mn-mi-R'-*]*stp-n-R'[Imn-*]*ms-[sw]* 'Menmire-setepenre Amenmesse', the *prenomen* and *nomen* of a later king of the same dynasty. It is therefore safest to simply assign a date in the 19th Dynasty. It may also be noted that, while the two cartouches are placed in their normal order with the *prenomen* on the right, the hieroglyphic signs in this cartouche are reversed and are read from left to right (cf. also Cat. 19, 97, 104).

13 Scarabs, Scaraboids, Plaques and Seals
(Eg. Cat. 181–194)

Although basically used as a seal, any or all of this class of object could also serve as prophylactic amulets or charms. In many instances this may have been their primary function, and was probably so in the case of scarabs, scaraboids and plaques whose inscriptions invoked, acknowledged or paid allegiance to a deity. In some cases the inscriptions might simply record the name of a god. Equally numerous are inscriptions containing a statement about the relationship of the owner of the object to the god, e.g. 'The god NN is my lord' (Drioton, 1957, 11–33); 'The god NN loves whoever loves him' (Drioton, 1959, 57–68), 'Refuge of heart is not found, save with the god NN' (Drioton, 1960, 3–12), 'May my name endure in the temple of the god NN' (Drioton, 1956, 31–41). Inscriptions of this nature were frequently written as cryptograms, perhaps to increase their potency or prevent their misuse by a non-initiate. In the case of the god Amun, whose name means 'The Hidden One', the power of the charm was certainly increased by punningly or cryptographically hiding his name in the writing, thus materially affecting the identification of the god with the purely physical writing of his name. Such cryptograms of Amun are often easy to discern since they take the form of trigrams, i.e. three hieroglyphs, each representing one of the written consonants of the god's name, i, m, n. In each case the sound value is derived either by acrophony, rebus or by class assimilation of the object represented by the hieroglyph. Where four or five hieroglyphs are written, the cryptogram may be read either as *Imn nb(.i)* or *Imn nb.i*, 'Amun is my lord'. However, there are many possibilities of other readings. The principles of cryptographic readings of Egyptian religious texts have been soundly established for over forty years; however, for some reason they have rarely been applied to the description or study of scarabs found in excavations. At least nine of the scarabs, scaraboids and plaques found at Timna are inscribed with cryptograms (Cat. 182–188, 191, 193). In each case, the god invoked in the inscription is Amun (var. Amun-Re). This is not surprising since it was during the period of the Empire, and particularly during its latter part, the Ramesside period, that Amun was unquestionably the most important and paramount Egyptian deity, and it is precisely at this period that the Egyptian shrine at Timna existed.[17]

Eg. Cat. 181*. *Fig.* 46:1. Field No. 319/2. Loci 106, 108. Scarab, faience, white frit, light blue glaze. L. 1.6cm., W. 1.0cm., H. 0.6cm.
Design incised on base; anatomical details on top and sides neatly and carefully indicated; pierced longitudinally for suspension. Broken, with approximately half the surface of the base missing.
Top: Oval shape. Clypeus, head and prothorax occupy one half of body. Details of clypeus and head summarily outlined by a wide double line. Head separated from prothorax, and prothorax from elythra by double lines. Wing-cases of elythra formed

by a double dorsal line. No V-notches in wing cases. Legs summarily outlined on sides.
Base: Four stylised scrolls and petals attached to a central, circular base with the entire ensemble enclosed in an oval. The type of stylised scroll and petal design is well-attested from the middle of the second millennium BC (the Hyksos period) onwards. Independently, it is not particularly useful for dating purposes.

Eg. Cat. 182*. *Fig.* 46:2. Field No. 278/3. Loci 107, 109, 110. Scarab, steatite, no trace of glaze. L. 1.4cm., W. 1.0cm., H. 0.7cm.
Design incised on base in neatly executed hieroglyphs. Anatomical details on top and sides summarily indicated in outline. Pierced lengthwise for suspension. Broken, about two-fifths of top missing.
Top: Oval shape. Clypeus, head and prothorax only most summarily indicated in outline. Prothorax separated from elythra by wide line. No anatomical details visible on elythra.
Base: At the top is the hieroglyph ☺ representing the sun-disk rising over the horizon and, by class equivalence, it stands for the simple sun-disk. Here it has its normal value of R'. Beneath it is a group of three signs. On the right, one over the other, are the hieroglyphs *mn* representing a gaming board and, by acrophony, to be read *m*, and a long narrow horizontal sign having the value *n*. To the left of these two signs is the hieroglyph representing a hoe ⎂ and normally having the value *mr*, but to be read here *i* by acrophony from *'iknw* 'hoe'. Thus the inscription is a cryptogram reading *'Imn-R'* 'Amun-Re'.

Eg. Cat. 183*. *Fig.* 46:3. *Pl.* 123:3. Field No. 366/1. Locus 109. Scarab, steatite, no trace of glaze. L. 1.9cm., W. 1.3cm., H. 0.9cm.
Inscribed on the base in high relief. Anatomical details on top and sides for the most part carefully indicated. Pierced longitudinally for suspension.
Top: Oval shape. Clypeus and head occupy one quarter of base. Head separated from prothorax by single thick line. No division of prothorax and elythra indicated. No division of elythra into wing-cases, or other markings visible. Legs outlined by incision on sides.
Base: On the left, facing right, is a standing or striding lion whose tail loops overhand towards the centre of the scarab. On its head is a sun-disk. In front of it are the hieroglyphs picturing respectively a feather or plume ⎮ and a uraeus-serpent. The group is clearly a trigram of Amun or Amun-Re. The uraeus serpent = *i* by acrophony from *i'r.t* 'uraeus'; the feather = *m* by acrophony from *m3't* 'truth'; and *n* by acrophony from ⏑ *nb* is 'lord', or by rebus from *ntr*, 'god'. If the sun-disk is to be read separately, it should be read clearly as 'Re'. The complete inscription then reads: 'Amun' or 'Amun-Re'.

Eg. Cat. 184*. *Fig.* 46:10. *Pl.* 123:5. Field No. E14/8. Locus 110. Plaque, faience, white frit, no trace of glaze. L. 1.8cm., W. 1.5cm., Th. 0.5cm.
Inscribed on both faces in sunken relief. Grooved

longitudinally on the sides and pierced longitudinally for suspension.

Obverse: The design of the top represents the hieroglyph *p*, a cushion of reed matting bound together at top and bottom.

Reverse: A falcon-headed human figure, ostensibly the god Re, stands on a groundline and faces to the left. Before him, its base resting on the groundline, is the hieroglyph ∤ , the feather of *m3't* 'truth', 'right'. Over this, and next to the god's head, is the hieroglyph ▽ which pictures the wicker basket *nb*. At first this group appears to be read simply *Nb m3't R'* 'The Lord of Truth is Re', which is the *prenomen* of the 18th Dynasty king Amunhetep III, but may also be understood as a pious declaration about the nature of the god Re. However, the iconography of the falcon-headed human figure is Ramesside rather than 18th Dynasty (Schulman, 1967, 148). It is therefore more probable that this is a cryptogram to be read ∤ = *i*, by group equivalence with 𓀭 = the god Montu = *m* by acrophony; ▽ = *n* by acrophony from *nb*. Thus the entire group is to be read 'Amun'.

Eg. Cat. 185*. *Fig.* 46:11. *Pl.* 123:4. Field No. 269/1. Locus 107. Plaque, steatite, no trace of glaze. L. 1.2cm., W. 1.0cm., Th. 0.6cm.
Inscribed on both faces with crudely incised hieroglyphs. Square, pierced longitudinally for suspension.

Obverse: Two scorpions arranged side by side and *tête-bêche*. The scorpion has two cryptographic values: *m* by acrophony from *mr* 'the grieving one', and *n* by identification with the goddess Neith. It is therefore probable that the *tête-bêche* positioning of the pair of scorpions is a rebus for a word such as *ifn* 'to turn around', 'turn after', *inn* 'turn around', *inh* 'surround', *ink* 'encircle', *ikh* 'set foot on', 'go after', or similar, thus giving by acrophony the value *i*. The inscription is then a cryptogram reading *'Imn* the name of the god Amun.

Reverse: The name of the god Amun spelled with the signs *'I + mn + n*, followed by a sun-disk and an inverted *nb* sign. This is the cryptogram of the formula of allegiance to either Amun or Amun-Re. The name 'Amun' in either case is written in the clear *'Imn*. In the former, the sun-disk has the value *i* by acrophony from *im* 'pupil of the eye', while *nb* has its normal value and the inscription would read *'Imn nb.i* 'Amun is my lord'. In the latter case there is no real cryptogram save for the inversion of the *nb* sign; all signs would have their normal values and the inscription would read *'Imn-R nb(.i)* 'Amun-Re is (my) lord'.

Eg. Cat. 186*. *Fig.* 46:12. *Pl.* 123:2. Field No. 258/1. Locus 106. Plaque, steatite, no trace of glaze. L. 1.6cm., W. 1.3cm., Th. 0.6cm.
Inscribed on both faces with crudely incised hieroglyphs. Rectangular, pierced along the length for suspension.

Obverse: Facing to the right, the god Horus in falcon-shape with the flail projecting above his back, stands on a *nb*-basket. Over his head is a squat, ovalish, sign which could be a sun-disk. To the left, in the field over the falcon's back and at an oblique angle, is an elongated oval whose lower end terminates in a point, probably a crude rendering of the sign ◝ picturing a blob of flesh. The third sign described above, if correct, has the cryptographic value if *i* by acrophony from *iwf* 'flesh'. The ovalish sign over the falcon's head has the value *m* by acrophony from *m33* 'what sees', and the falcon is to be read *n* by acrophony from *ntr* 'god'. Thus the entire group is a trigram of the name of Amun and is to be read *Imn*, 'Amun'.

Reverse: Facing to the right is a standing horse and over its back is a crudely incised, squat *nb* sign which has its normal value. The horse is to be read as *i* by acrophony from *ibr* 'stallion'. Thus the two signs read *nb.i* 'my lord' and together with the name inscribed on the opposite face of the plaque form the sentence *'Imn nb.i* 'Amun is my lord'.

Eg. Cat. 187*. *Fig.* 46:13. *Pl.*123.1. Field No. 135/1. Locus 110. Plaque, steatite, no trace of glaze. L. 1.6cm., W. 1.2cm., Th. 0.6cm.
Inscribed on both faces with crudely incised hieroglyphs. Rectangular, pierced longitudinally for suspension.

Obverse: A quadruped, perhaps a lion, stands or strides, facing to the left. In front of it, facing in the same direction, is a uraeus serpent. Over the animal's back is a long, narrow horizontal line. The whole is enclosed by a rectangular border. The group is clearly a trigram of Amun. The uraeus serpent has the value *i* by acrophony from *i'r.t* 'uraeus' . The lion is to be read *m* by acrophony from *m3i* 'lion' and the long, narrow horizontal sign can only have the value of *n*. The entire group then reads *Imn*, 'Amun'.

Reverse: Identical to the inscription on the reverse of plaque Cat. 185.

Eg. Cat. 188*. *Fig.* 46:14. *Pl.*123:7. Field No. 277/1. Locus 107. Plaque, steatite, no trace of glaze. L.2.0cm., H. 1.6cm., Th. 0.7cm.
Inscribed on both faces with crudely incised hieroglyphs. Rectangualr shape, pierced lengthwise for suspension.

Obverse: Except that the hieroglyph ∤ is on the left rather than the right, the inscription and its interpretation are identical with that of plaque Cat. 185 Reverse.

Reverse: Identical with the obverse.

Eg. Cat. 189*. *Fig.* 46:9. *Pl.*123:6. Field No. 340/1. Locus 107. Seal, faience, very pale brown frit, green glaze. L. 1.9cm.
Loop-type handle, pierced for suspension through the width. Anatomical details on top are in relief while decoration on the base is incised. The decoration on the base of the seal consists of three *mn* signs 𓏠 alternating with two broad, flat signs. This may be a cryptograph but no reading is proposed.

Eg. Cat. 190*. *Fig.* 46:5. Field No. 319/1. Loci 106, 108. Scaraboid, faience, white frit, green glaze. L. 1.1cm. Pierced longitudinally for suspension. No anatomical details are executed on the back or sides. Instead there is a series of irregular planes. The underside of the base has no recognisable decoration, only a lozenge pattern of criss-crossed lines.

Eg. Cat. 191*. *Fig.* 46:6. *Pl.*123:8. Field No. 97/1. Loci 101, 112. Scaraboid, faience, white frit, white glaze. L. 1.4cm.
Incised decoration on underside of base. Pierced longitudinally for suspension. No anatomical details are indicated on either the back or sides. The main element of the design is a higly stylised floral motif consisting of two papyrus stalks set in a quartrefoil pattern with two other unidentified plants whose heads are half-moon shaped. Flanking the latter plants are two disks.

Close study indicates that this motif is in fact a cryptographic inscription where the two disks = *i* and *n* respectively by acrophony from *im* 'eyeball', and by a material variation ⊛ *niwt* 'city'. The floral design is a material variation of ꝏ *mhy.t* 'papyrus' and by acrophony = *m*. The cryptogram, a trigram of Amun, yields the name 'Imn, 'Amun'.

Eg. Cat. 192. *Fig.* 46:4. Field No. 283/1. Locus 107. Seal (?), faience, brown frit, glaze. L. 1.0cm., H. 0.9cm.
Conical, pierced through the back for suspension. The design on the base depicts a single ibex. It may be incorrect to call this object a seal because, while it could be used to make an impression, the design should have some meaning and the ibex alone, unless it is to be read as a word-sign (possibly a name), has no meaning. It may be a conical amulet with the entity the amulet represents being the hieroglyph incised on the base.

Eg. Cat. 193. *Fig.* 46:7. Field No. 598/13. Locus 108. Scarab, faience, white frit, green glaze. L. 1.5cm., W. 1.2cm., Th. 0.8cm. Pierced longitudinally for suspension.
Back: Small clypeus and prothorax. Wing-cases not defined, but with V-shaped notches. Sides and legs summarily rendered.

Base: Facing left vertically 𓂋𓌳𓋴𓇓 *R'-m'sw ḥk3 Iwnw* 'Ramesses-hikaon', the *nomen* of Ramesses III of the 20th Dynasty.

Eg. Cat. 194. *Fig.* 46:8. Field No. 503/6. Locus 109. Scarab, faience, brown frit, traces of greenish glaze. L. 1.2cm., W. 1.0cm., Th. 0.5cm.
Pierced longitudinally for suspension. Back and approximately half of the base missing. What is preserved on the base suggests the prothorax and elytra of a scarab beetle, the wing-cases being separated by a double line and hatched with oblique lines. While the legs are not clearly defined, they may be indicated by what otherwise seem to be traces of a borderline.

Observation: It is probable that the hieroglyph of the scarab beetle is a cryptograph. It could represent any of the deities indicated by the beetle, e.g. Re or Khepri. However, since the hieroglyph has the three values *i, m, n,* used to spell the name of Amun, it may

be read here simply as 'Imn, 'Amun' (cf. also Schulman 1975; 70).

14 Amulets (Eg. Cat. 195–215)
Surprisingly, no amulets of Hathor were found at the site. Among the deities and concepts represented were the cartouches symbolising the power of the kingship (Cat. 195); Pataikos (Ptah the dwarf, Cat. 196–202); Harpocrates (Horus the child, Cat. 203–204); Sekhmet (?) (Cat. 205, 208–209), a *menat* worn by a king (Cat. 206); Khons (Cat. 207) and a mummiform ushbati (Cat. 214–215).

Eg. Cat. 195*. *Fig.* 47:8. Field No. 208/1. Loci 101, 102. Amulet, faience, frit, blue-green glaze. L. 1.7cm., W. 1.0cm., Th. 0.2cm. Lower portion only.
In the shape of a cartouche in which is written vertically with hieroglyphs in sunk relief the royal name *Sty-mr-n-Ptḥ* 'Seti-merneptah'.

Observations: In previous publications (e.g. Rothenberg, 1972, 164; Pl. 48:1) this royal name is identified as Seti I (Menmare Seti-merneptah), the second king of the 19th Dynasty and, on this basis, the object was related to his reign. This is possible but not certain since all that is preserved is the *nomen* Seti-merneptah which by itself is not conclusive since the same *nomen*, with the same spelling, was also used by Seti II (Userkheprure-setepenre Seti-merneptah), the fifth king of the dynasty who reigned almost a century later. Amulets in the form of a cartouche containing either a royal *nomen* or *prenomen* are not uncommon and are attested for both these monarchs (cf. Petrie, 1917; Pl. 26, 27, 33; Pl. xliv:20, 22, 23; Ward, 1902; Pl. vi:436, 31 and 260; Hall, 1913, No. 2010). It should be noted that the amulets with the *nomen* of Seti I which conform to the spelling of Cat. 195 are inscribed on both faces, while those of Seti II are usually inscribed on one face only and with the epithet 'Ptah' following, rather than preceding in honorific transposition the hieroglyph *mr* 'beloved', and occasionally followed by the additional divine name 'Amun' in the secondary epithet 'beloved of Amun'.

Eg. Cat. 196. *Fig.* 48:3. Field No. 135/2. Locus 110. Amulet, faience, brown frit, blue glaze. H. 1.3cm.
Pierced horizontally at the back of the neck for suspension. Pataikos (Ptah the dwarf).

Eg. Cat. 197. *Fig.* 48:2. (Col. Pl. 75). Field No. 127/2. Locus l01. Amulet, faience, light brown frit, green glaze. H. 1.3cm.
Pierced horizontally at the back of the neck for suspension. Pataikos.

Eg. Cat. 198. *Fig.* 48:1. (Col. Pl. 27). Field No. 127/1. Locus 101. Amulet, faience, light brown frit, green glaze. H. 1.3cm.
Pierced horizontally through the ears for suspension. Crude work. Pataikos.

Eg. Cat. 199. *Fig.* 48:10. Field No. 338/1. Locus 109. Amulet, faience, white frit, green glaze. H. 2.5cm.
The head and feet of the amulet are broken away. But since the position of the arms rules out Harpocrates,

and the absence of any characteristic attributes rules out Bes, the deity can only be Pataikos.

Eg. Cat. 200. *Fig.* 48:4. Field No. 339/3. Locus 109. Amulet, faience, white frit, traces of green glaze. H. 1.5cm.
Although the head and legs are missing, this fragment certainly represents Pataikos, since it lacks the characteristic paraphenalia of Bes, while the position of the two arms, pendant at the sides rather than one hand raised to the mouth, rules out Harpocrates (Horus the child).

Eg. Cat. 201. *Fig.* 47:6. Field No. 285/1. Loci 101, 112. Amulet, faience, very pale brown frit, green glaze. H. 2.0cm.
With only the right half of the body preserved, it is difficult to identify the deity represented. It is clearly one of the grotesque types, but the attributes of Bes do not appear to be present, and the right arm at the side rather than at the mouth suggests Pataikos.

Eg. Cat. 202. *Fig.* 48:9. *Col. Pl.* 28. Field No. 234/1. Locus 101. Amulet, faience, brown frit, blue-green glaze. H. 2.8cm.
Pierced horizontally at the back of the head of suspension. Although nude and wearing no head-dress, the deity depicted by the amulet is probably Bes, since he is a squat, grotesque dwarf with a beard and with his tongue hanging out of his mouth. See however Anthes (1965; 122) where a similar amulet (No. 205) is identified as a 'bearded Pataikos'.

Eg. Cat. 203. *Fig.* 48:5. *Col. Pl.* 24. Field No. 127/3. Locus 101. Amulet, faience, light brown frit, green glaze. H. 2.6cm.
Pierced horizontally at the shoulders. Harpocrates.

Eg. Cat. 204. *Fig.* 48:6. *Col. Pl.* 26. Field No. 127/4. Locus 101. Amulet, faience, very pale brown frit, green glaze. H. 1.0cm.
Fragment pierced horizontally at the shoulders for suspension. Although only the upper half is preserved, the gesture of sucking the finger of one hand, plus the sidelock of youth, indicates that the deity represented is Harpocrates.

Eg. Cat. 205*. *Fig.* 48:8. *Pl.* 118:1. Field No. 344/1. Loci 110, 111. Amulet, gypsum. H. 3.8cm.
Looped behind for suspension. The identity of the female deity represented is not certain. She is shown standing in profile, facing to the right. In her left hand she holds a sceptre but the precise details cannot be determined. In her right hand, hanging down at her side, she holds an object which may be an *ankh*. She wears a calf-length skirt (with decoration indicated by cross-hatching) with a long belt pendant in front. Her midriff is bare. Around her neck and shoulders is a broad, triple-stranded collar, and there is a bracelet on her right upper arm. Her head is crowned by a sun-disk surmounting her wig. Unfortunately the details of the face are lost but the outline clearly suggests a feline visage. She could represent any one of the feline-headed goddesses, the two most

common of which were Sekhmet (with the head of a lioness) and Bastet (with the head of a cat). Since the former was almost always shown with a sun-disk, and the latter without one, Sekhmet may be the more probable identification.

Eg. Cat. 206*. *Fig.* 48:11. *Col. Pl.* 22. Field No. 310/1. Locus 105. Amulet, faience, white frit, deep blue glaze. H. 4.8cm.
Looped at the back for suspension. While the disk-like bottom part is missing, the amulet is clearly recognisable as having the shape of a counterpoise on a *menat*-necklace, surmounted by the head of a king wearing the Double Crown of Upper and Lower Egypt on an elaborate wig, and the necklace (aegis).

Eg. Cat. 207*. *Fig.* 48:12. Field No. 129/1. Locus 101. Amulet, faience, brown frit, green glaze. H. 5.0cm.
Pierced horizontally behind the horns for suspension; rib running vertically down the height. In spite of the crude workmanship which almost suggests a female figure, the attributes of the deity, who stands holding a papyrus-headed sceptre, the human body, falcon head, moon-disk (horned disk) and uraeus are those of the moon god Khons.

Eg. Cat. 208. *Fig.* 48:7. Field No. 221/2. Loci 106, 107. Amulet, faience, white frit, blue glaze. H. 2.6cm.
Pierced horizontally behind the ears for suspension. Broken off below the hips, the amulet represents a female, feline-headed deity wearing a sundisk in front of which, between her ears, a uraeus-serpent rears up. The goddess could be any of the feline-headed deities but her face appears to be of a lioness rather than a cat, so she is most probably Sekhmet.

Eg. Cat. 209. *Fig.* 47:4. Field No. 347/1. Locus 108. Amulet, faience, white frit, traces of greenish glaze. H. 1.1cm.
Only the top of the amulet is preserved, consisting of the head of a deity surmounted by a sun-disk and uraeus. Since the face appears to be leonine rather than feline, it is probably Sekhmet.

Eg. Cat. 210. *Fig.* 47:1. Field No. 145/1. Locus 101. Amulet, faience, white frit, green glaze. H. 1.8cm.
The upper half of the amulet is lost, so it is impossible to identify the deity represented. The preserved portion shows only the body from just below the chest. The right hand hangs down at the side and the left hand holds some type of sceptre in front of the body.

Eg. Cat. 211. *Fig.* 47:2. Field No. 137/1. Locus 110. Amulet, faience, light brown frit, blue-green glaze. H. 1.6cm.
Broken off just above the knees, it represents the lower part of a mummiform figure (?Ptah).

Eg. Cat. 212. *Fig.* 47:3a. Field No. 135/4. Locus 110. Amulet, faience, white frit, green glaze. H. 1.7cm.
Broken away above the knees, it represents the feet of a mummiform (?) figure holding a sceptre in front (?Ptah).

Eg. Cat. 213. *Fig.* 47:3b. Field No. 135/4. Locus 110. Amulet, faience, light brown frit, green glaze. H. 1.5cm.
Broken away above the knees, it represents the feet of a mummiform (?) figure holding a sceptre in front (?Ptah).

Eg. Cat. 214. *Fig.* 47:7. Field No. 319/4. Loci 106, 108. Amulet, faience, white frit, green glaze, detail moulded in high relief. H. 2.0cm., W. 1.3cm., Th. 0.9cm.
Possibly part of Cat. 216, this fragment seems to represent part of the right shoulder of a mummiform amulet and part of the hoe that would have been held to the chest with the right hand.

Eg. Cat. 215. *Fig.* 47:5. Field No. 319/5. Loci 106, 108. Amulet, faience, brownish-white frit, blue glaze, detail moulded in high relief. H. 2.4cm., W. 1.3cm., Th. 0.3cm. Fragment reassembled from two pieces.
Pierced horizontally through the width for suspension, it seems to represent a mummiform figure holding a hoe or mattock to its chest. Only the left shoulder, left arm, the implement and the head are preserved. The head is indicated only by a roundish lump with no indications of any facial features or headdress.

15 Miscellaneous Faience Objects (Eg. Cat. 216–228)
Under this heading are included faience sherds of items whose original nature is difficult to determine. Most bear some traces of decoration or inscription. In one case (Cat. 216) the inscription can be dated unequivocally to a specific king. In several other cases (Cat. 217, 220–222) the preserved words and epithets: 'granted life', 'turquoise', 'Hathor, lady of turquoise', '[king]...s [beloved of] Hathor', show that the inscriptions included royal names which are now lost.

Eg. Cat. 216. *Fig.* 40:5. *Pl.* 122:9. Field No. 337/3. Locus 102. Sherd, faience, white frit, faint traces of greenish glaze, inscribed with black paint. H. 4.0cm., W. 1.8cm., Th. 1.1cm.
The flattened, ovalish section of this sherd suggests either a bracelet or a counterpoise. It is inscribed on one face, vertically, with part of the lower half of a cartouche in which one hieroglyphic sign *ḥḳ3*, 'ruler', is legible and which, in view of its position within a cartouche, suggests the name [*R'-ms-s(w)]-ḥḳ3-[Iwnw]* 'Ramesses-hikaon', the *nomen* of Ramesses III.

Eg. Cat. 217. *Fig.* 40:1. Field No. 348/16. Locus 107. Sherd, faience, brown frit, green glaze, inscribed with black paint. H. 2.6cm., W. 1.4cm.
The inscription *diw 'nh* 'granted life' is the stereotyped pious wish usually following a royal name.

Eg. Cat. 218. *Fig.* 37:33. Field No. 310/19. Locus 105. Lid(?), faience, light brown frit, green glaze, no decoration. Th. 1.0cm., Diam. 10.0cm. Fragment.

Eg. Cat. 219. *Fig.* 45:1. *Pl.* 118:5. Field No. 223/1. Locus 101. Gaming piece(?), faience, white frit, green glaze, decoration in black paint. Th. 0.9cm., Diam. 5.2cm.

Pierced through the centre, broken into two pieces. Draughtsmen-like gaming pieces are known from the Ramesside period although the nature of the game is unknown, except that it was a board game. The piece is decorated on both faces with a quartrefoil of papyrus plants.

Eg. Cat. 220. *Fig.* 29:1. Field No. 151/1. Locus 102. Handle(?), faience, white frit, blue glaze, decoration in black paint. L. 4cm., W. 2cm., Th. 1.2cm.
Inscribed with traces of two vertical lines of hieroglyphs. To the right can just be made out the left half of a tall, rectangular sign above which is a narrow horizontal sign. On the left is the word [*m*]*fk3t* 'turquoise'. Clearly this word was part of the expression 'Hathor, the Lady of Turquoise' and this in turn suggests that the rectangular hieroglyph at the right is to be restored 🔲, the normal writing of the deity's name. The narrow horizontal sign above the name of the goddess is probably the flat end of a cartouche which originally contained a royal name.

Eg. Cat. 221. *Fig.* 29:2. Field No. 135/3. Locus 110. Handle(?), faience, white frit, traces of greenish glaze, decoration in black paint. L. 3.3cm., W. 2.5cm., Th. 1.2cm.
Single vertical line of well-shaped hieroglyphs facing to the right and reading ...[*Ht-hr*]*nb.t m*[*fk3t*], '... of Ha[thor], Lady of Tu[rquoise]'.

Eg. Cat. 222. *Fig.* 29:3. *Pl.* 120:4. Field No. 280/2. Locus 106. Handle(?), faience, very pale brown frit, green glaze, decoration in black paint. L. 3.2cm., W. 2.8cm., Th. 1.7cm.
Single vertical line of hieroglyphs, facing to the right, framed on each side by a vertical border line. Above, the lower portion of a cartouche in which appears at the left the end of a tall, narrow sign, probably the vertical *s* 𝍩. Directly beneath this is the damaged but legible name *Ht-hr*. The royal name in the cartouche, ending in *-s* and without any further qualifying epithet, is probably that of Ramesses II, which is usually written in this fashion, and it stood in relationship to the name of the goddess in the phrases '[...Ramesse]s b[eloved of] Hathor [the Lady of Turquoise]'.

Eg. Cat. 223. *Fig.* 37:32. Field No. 154/1, 236/3. Loci 107, 109, 110. Tile (?), faience, brown frit, blue glaze, no decoration. H. 6.5cm., W. 4.6cm., Th. 2.2cm.
Cat. 223 has seven vertical grooves running from top to bottom, but the exact nature of this object is unclear.

Eg. Cat. 224. *Fig.* 40:4 Field No. 311/188. Locus 101. Sherd, faience, white grit, bluish-green glaze, inscription in black paint. H. 2.7cm., W. 1.6cm., Th. 0.7cm.
This tiny sherd could have come from the counterpoise of a *menat*-necklace, a bracelet or a small vessel. Its decoration consists of part of a pair of border lines (only one is clearly preserved) between which stood a cartouche (of which both sides are preserved), inside which are several illegible traces.

Eg. Cat. 225. *Fig.* 39:1. Field No. 277/11. Loci 107, 110. Sherd, faience, brownish-white frit, no traces of glaze, inscription in black ink. L. 2.2cm., W. 1.8cm., Th. 0.6cm.

The clearly legible horizontal and vertical lines which form a right angle suggest a framing border such as occurs on the counterpoises of *menat* necklaces (cf. Cat. 26–40). If this is correct, then the solid black trace beneath the horizontal line could be a damaged ▽ *nb* sign, here introducing a title such as *nb t3.wy*, 'Lord of the Two Lands', or *nb h'.w*, 'Lord of Diadems'.

Eg. Cat. 226. *Fig.* 39:5. Field No. 294/52. Locus 101. Sherd, brown pottery, blue-green glaze, inscription in black paint. L. 2.6cm., H. 2.0cm., Th. 0.8cm.

Inscribed with what appears to be the rear end of a cartouche, inside which can only be seen one completely illegible sign.

Eg. Cat. 227. Not illustrated. Field No. 372/7. Locus 111. Sherd, faience, brown frit, green glaze, decoration in black paint. H. 1.8cm., W. 0.9cm., Th. 1.0cm.

The decoration consists of a curved line and suggests part of the rounded end of a cartouche, but there is no trace of the name it may have enclosed.

Eg. Cat. 228. Not illustrated. Field No. 263/3. Locus 103. Sherd brown pottery, blue glaze. L. 3.5cm., W. 2.0cm., H. 1.8cm.

Irregular chip of glazed pottery, which may have come from a small animal figurine.

16 Various Stone Objects (Eg. Cat. 229–264)

Very few of the stone objects are clearly recognisable as Egyptian, these being Cat. 233–235, and 236–242, various stone vessels, jar lids one of which, Cat. 237, is beautifully decorated with an open waterlily design. The basins (Cat. 229–232) and the offering tables (Cat. 243–245) and stands (Cat. 246–258) are all of relatively crude workmanship and perhaps not all of them made by Egyptians (see also Chapter III, 29). Several of them were used or re-used during the later Midianite phase of occupation of the site.

Eg. Cat. 229. *Fig.* 23:5. Field No. 466/1. Locus 109. Basin. H. 45 cm., Diam. 60 cm., Th. 10 cm. Incomplete.

Eg. Cat. 230. *Fig.* 23:6. Field No. 469/1. Locus 106. Stone fragment. H. 64 cm., W. 51 cm. Found inside basin Cat. 232, perhaps used as a pestle.

Eg. Cat. 231. *Fig.* 23:7. Field No. 467/1. Locus 107. Basin. L. 115cm., H. 45cm., W. 101cm. Incomplete.

Eg. Cat. 232. *Fig.* 23:8. Field No. 468/1. Locus 106. Basin. L. 101cm., H. 50.8cm., W. 71cm. Incomplete. Cat. 230 was found inside.

Eg. Cat. 233*. *Fig.* 22:3. *Pl.* 113:3. Field No. 328/1. Locus 104. Vase, alabaster. H. 5.5cm., Diam. 4.5cm. Tall, gently flaring collar with a pronounced carination where it joins the shoulder. Two loop handles on the sides, pierced for suspension. Base flat.

Eg. Cat. 234. *Fig.* 22:1. Field No. 376/10. Locus 111. Vessel, alabaster. H. 2.3cm., Th. 0.6cm.

Fragment of the rim of a small alabaster vessel, possibly a cup. On the exterior face is an incised decoration consisting of two widely spaced lines running parallel to the edge of the rim, with another vertical incised line at a right angle to the lower one.

Eg. Cat. 235. *Fig.* 22:4. Field No. 258/2. Locus 106. Bowl, alabaster. H. 5.7cm., W. 6.5cm., Th. 1.3cm., Diam. 2.5cm. Fragment. Thick-walled, globular shape.

Eg. Cat. 236. *Fig.* 49:1. Field No. 212/1, 238/1. Loci 101, 111. Jar lid, limestone. Th. 1.2cm., Diam. 7.8cm. Fragment reassembled from two pieces.

The top of the lid is flat with the edge gently curving downwards, and then curving inwards to form a deep groove, with the inner surface of the lid projecting beyond and below this. The groove was intended to receive and hold the rim of the jar.

Eg. Cat. 237. *Fig.* 49:2. Field No. 269/4. Locus 107. Jar lid, limestone, decoration in black paint. Th. 1.2cm., Diam. 9.5cm.

Seen in section, the top surface curves gently down at the edge; the underside of the edge recurves upwards and then flares markedly downwards to the edge of the projecting bottom surface. The groove formed thereby held the rim of the jar. The decoration of the top surface is an open water lily, of which the petals, unfolding from a now-lost central calyx, extend to the edge of the lid.

Eg. Cat. 238. *Fig.* 50:6. Field No. 370/3, 316/1. Locus 110. Jar lid, limestone. Th. 1.3cm., Diam. 12cm.

The upper surface is slightly concave. Its edge curves inwards at the underside, to form a ledge to rest on the rim of the jar.

Eg. Cat. 239. *Fig.* 49:4. Field No. 276/2. Locus 111. Jar lid, limestone. Th. 1.1cm., Diam. 12cm.

Seen in section, the edge forms a lip to rest on the rim of the jar.

Eg. Cat. 240. *Fig.* 49:5. Field No. 309/2. Locus 101. Jar lid, white sandstone. Th. 3cm., Diam. 16cm.

Seen in section, the lid is flat on tip and bottom. The edge is curved with no trace of a lip or groove for attaching it to the jar (cf. Cat. 241).

Eg. Cat. 241. *Fig.* 49:3. Field No. 276/19. Locus 111. Jar lid, red sandstone. Th. 2.3cm., Diam. 10.5cm.

Seen in section, the centre of the lid is slightly thicker than the edge. As there is no lip or groove on the underside to hold the rim of the jar, it is possible that this type of lid was used to seal the vessel either by being fitted to a lip projecting from the inner side of the jar rim, or simply placed over the mouth of the jar and held in place by a mud sealing.

Eg. Cat. 242. *Fig.* 50:7. Field No. 304/17. Loci 106, 108, 109. Jar lid, red sandstone. Th. 2.5cm., Diam. 12cm. Fragment.

Flat underside, convex upper side. No groove or lip for placement on jar mouth.

17 Offering Tables (Eg. Cat. 243–264)

Egyptian offering tables frequently consisted of two elements: the upright base (Cat. 246–258) and the rectangular plates (Cat. 243–245). None were found at Timna as complete units. While a number of bases (stands) were found, mostly in secondary use as part of the Standing Stones in the Midianite phase of the Temple, it is impossible to say which of the plates found belonged to a specific base.[18]

Eg. Cat. 243. *Fig.* 24:3. Field No. 463/1. Locus 107. Offering table, white sandstone. L. 26cm., W. 23cm., Th. 7cm.
Rectangular with thick raised edge on all sides, suggesting that it may originally have been used for libations.

Eg. Cat. 244. *Fig.* 24:2. Field No. 435/1. Locus 107. Offering table, white sandstone. L. 72cm., W. 55cm., Th. 23cm.
Rectangular with raised edges on all sides. This, plus a groove running parallel to the raised edge along one of the shorter sides, suggests that it was originally used for libations.

Eg. Cat. 245. *Fig.* 24:1. Field No. 429/1. Locus 107. Offering table, white sandstone. L. 30cm., W. 23cm., Th. 11cm. Fragment.
Rectangular with thick raised edge on all sides, suggesting that it may have been used for libations.

Eg. Cat. 246. *Fig.* 51:1. *Pl.* 124:7. Field No. 427/1. Locus 106. Offering table stand, white sandstone. H. 12.5cm., Diam. at top 10.5cm., Diam. at base 23cm. Fragment.

Eg. Cat. 247. *Fig.* 51:2. Field No. 446/1. Locus 107. Offering table stand, white sandstone. H. 18cm., Diam. 15cm. Fragment from the neck of the stand.

Eg. Cat. 248. *Fig.* 51:3. *Pl.* 124:2. Field No. 452/1. Locus 106. Offering table stand, white sandstone. H. 27cm., Diam. of top 14cm., Diam. of base 17cm. Fragment including the base and part of the neck of the stand.

Eg. Cat. 249. *Fig.* 51:4. *Pl.* 124:1. Field No. 421/1. Locus 106. Offering table stand, white sandstone. H. 28cm., Diam. of top 16cm., Diam. of base 20.5cm. Fragment. Partly fluted.

Eg. Cat. 250. *Fig.* 51:5. Field No. 423/1. Locus 101 (Wall 1). Offering table stand, white sandstone. H. 26cm., Diam. of top 19cm., Diam. of base 27.5cm. Fragment.

Eg. Cat. 251. *Fig.* 51:6. *Pl.* 124:6. Field No. 422/1. Locus 107. Offering table stand, white sandstone. H. 29.5cm., Diam. of top 14cm., Diam. of base 17cm. Fragment.

Eg. Cat. 252. *Fig.* 51:7. *Pl.* 124:8. Field No. 401/1. Locus 106. Offering table stand, white sandstone. H. 27.3cm., Diam. of top 19cm., Diam. of base 24.8cm.

Eg. Cat. 253. *Fig.* 51:8. *Pl.* 124:3. Field No. 447/1. Locus 106 (Wall 2). Offering table stand, white sandstone. H. 21.0cm., Diam. of top 16.5cm., Diam. of base 28.5cm. Fragment. Fluted on all sides.

Eg. Cat. 254. *Fig.* 51:9. Field No. 417/1. Locus 110. Offering table stand, white sandstone. H. 31.0cm., Diam. 21.0cm. Fragment.

Eg. Cat. 255. *Fig.* 51:10. Field No. 455/1. Locus 101. Offering table stand, white sandstone. H. 31.0cm., Diam. 22.0cm. Fragment.

Eg. Cat. 256. *Fig.* 51:11. *Pl.* 124:4. Field No. 416/1. Locus 107. Offering table stand, white sandstone. H. 57.0cm., Diam. of top 12.0cm., Diam. of base 19cm. Reassembled from two fragments.

Eg. Cat. 257. *Fig.* 51:12. Field No. 460/1. Locus 106. Offering table stand, white sandstone. H. 40.0cm., Diam. of top 21.0cm., Diam. of base 20.5cm. Fragment consisting of narrow central portion which flares outwards at one end.

Eg. Cat. 258. *Fig.* 51:13. *Pl.* 124:5. Field No. 432/1. Locus 107. Offering table stand, white sandstone. H. 37.5cm., Diam. of top 21.5cm., Diam. of base 31.0cm. Fragment. As part of an offering table stand this has an unusual form.

Eg. Cat. 259. *Fig.* 22:7. Field No. 237/8. Locus 110. Offering table (?), white sandstone. H. 6.8cm., W. 6.3cm., Th. 2.7cm. Fragment.
The rectangular (?) shape with a thick raised lip on the upper surface suggests that it may have been a plate of an offering table, possibly used for libations.

18 The Rock Stela[19]

Eg. Cat. 260. *Fig.* 52. *Pl.* 105. A stela carved into the face of the cliff, slightly to the north and about 20 metres above the Temple, measuring 49 x 78 cm.
Within a round-topped frame, it pictures a king wearing the *nemes*-headcloth and uraeus, making offerings to a goddess who faces him. Between the king and the goddess is a single vertical line of hieroglyphs consisting of the titulary of the king, while beneath the scene is the main text consisting of a single line of horizontal hieroglyphs. The titulary comprises a pair of vertical cartouches, separated by the title *nb h',w*, 'Lord of Diadems', containing respectively the *prenomen* and *nomen* of Ramesses III, *Wsr-m3't-R'-mry-Imn R'-ms-s(w)-hk3-Iwnw*, 'Usermaremiamun Ramesses-hikaon'. The goddess, while lacking any of her specific attributes, must be Hathor, the only Egyptian deity associated with the Egyptian shrine of Timna.

The single line of horizontal, somewhat crudely incised hieroglyphs at the bottom of the stela has been seriously damaged by weathering, particularly at the beginning and along the length of the tops of virtually all of the signs. Nevertheless, enough is preserved to restore the text with a reasonable amount of confidence as reading:
h(3)y in wb3 nswt R'-ms-sw-[m]-pr-[R'] <m3'->hrw,

'Coming by the royal butler, the justified Ramesses-[em]per[re]'.

Since this stela has been published in *extenso* (Schulman, 1976; 117–30), only a summary is given here of some of the more salient conclusions. The royal butler Ramessesemperre[20] is a well-known personage, attested by a number of monuments, mainly stelae and fragments from his tomb in the Memphite area. His career began under the latter years of Merneptah and, as the Timna stela indicates, continued well into the earlier years of the 20th Dynasty. That he was of Asiatic origin (as were many of the officials who bore the title 'royal butler')[21] is clearly confirmed by his stela from Abydos where his Semitic name is given as 'Ben-azen of Ziri-Bashan' (*JE*, 3/7/24/7). The Timna inscription suggests that his 'coming' to Timna was an event of some importance. There is a gap in the Egyptian presence at the site from the reign of Twosre' in the 19th Dynasty until that of Ramesses III in the 20th Dynasty – at least none of the intervening rulers are attested there – and the shrine may have been destroyed during this gap. Ramesses III explicitly numbers among his major achievements the sending of his 'messengers to the country of 'Atika' (Harris, I, 78, 1–5; Breasted, 1906; 408) – which has been convincingly identified as Timna (Rothenberg, 1972; 201–3) – 'to the great copper mines which are in this place', and it is probable that the 'coming' by Ramessesemperre commemorated the arrival of that expedition to re-establish the Egyptian presence. This seems more certain in view of the fact that elsewhere in Papyrus Harris I, Ramesses III records that he sent 'butlers and officials to the turquoise country, to my mother Hathor, the Lady of Turquoise' (Harris I, 78; Breasted, 1906, 409), clearly a reference to the region around Serabit el-Khadim in south-western Sinai. Finally, from other inscriptions at Serabit el-Khadim and from the Wadi Hammamat, royal butlers are attested as either themselves commanding, or as being very high among those leading such mining expeditions (Christophe, 1949, 20; Gardiner and Peel, 1952, Nos. 250, 252, 260, 302, 304). Consequently it is not unreasonable to see Ramessesemperre leading the mining expedition to Timna, which Ramesses III boasts of in Papyrus Harris.

19 Concluding Remarks

Before attempting to draw any chronological or historical conclusions based on the material presented in the catalogue of Egyptian finds, several points concerning the nature and reliability of the evidence must be made.

(1) Although the full titulary of an Egyptian king contained five names: the Horus, Nebty, Horus-of-Gold, *prenomen* and *nomen*, it was mainly the last two which were the most important, the former being the king's throne name and the latter his personal name. These two were set off from the others in writing by being enclosed in a cartouche (*snw* 'that which encircles'), the idea being that the king whose name it enclosed ruled all that the sun encircled (see Gardiner, 1956, par. 4–5). Usually one or the other, or both cartouches containing the *prenomen* and *nomen*, were inscribed on objects. Indeed, none of the Timna finds has any of the other names. It frequently happens that more than one king uses the same *prenomen* or the same *nomen*, and the only way to distinguish between them is if the second cartouche-enclosed name is also written. The following list compares the *prenomina* and *nomina* of the 19th and 20th Dynasty rulers, some of which are attested on finds from Timna:

Prenomen	Nomen	Ruler
19th Dynasty Pharaohs		
Menpehtyre	Ramesses	Ramesses I
Menmare	Seti-merneptah	Seti I
Usermare-setepenre	Ramesses-miamun	Ramesses II
Banire-setepenre	Merneptah-hetephimae	Merneptah
Userkheprure-setepenre	Seti-merneptah	Seti II
Menmire-setepenre	Amenmesse-hikawase	Amenmesse
Sekhaenre-setepenre	Ramesses-siptah	Siptah
Akhenre-setepenre	Merneptah-siptah	Siptah (later form)
Sitre-miamun	Twosre-setepenmut	Twosre
20th Dynasty Pharoahs		
Userkhaure-miamun-setepenre	Setnakht-mererre-miamun	Setnakht
Usermare-miamun	Ramesses-hikaon	Ramesses III
Hekmare-setepenamun	Ramesses-hikamae-miamun	Ramesses IV
Usermare-sekheperenre	Ramesses-Amenhikhepeshef-miamun	Ramesses V
Nebmare-miamun	Ramesses-Amenhikhepeshef-neterhikaon	Ramesses VI
Usermare-miamun-setepenre	Ramesses-itamun-neterhikaon	Ramesses VII
Usermare-akhenamun	Ramesses-sethikhepeshef-miamun	Ramesses VIII
Neferkare-setepenre	Ramesses-khaemwase-mereramun	Ramesses IX
Kherermare-setepenre	Amenhikhepeshef-miamun	Ramesses X
Menmare-setepenptah	Ramesses-khaemwase-mereramun-neterhikaon	Ramesses XI

From the foregoing it is clear that not only do some kings share the same *nomen* or *prenomen*, but that in even more cases, they share the same elements of a *prenomen*, e.g. Usermare, Menmare, setepenre, Ramesses, Sety, hikaon, miamun, etc. Consequently, when we have a cartouche with only one of the two names and which is shared by one or more kings or, more frequently, only a portion of the name preserved which is an element common in the names of several kings, we cannot identify which king was meant with any certainty. For purposes of dating, we can only use those cartouches which are unequivocal.

(2) A royal name on an object, particularly a small faience object such as a sistrum, bracelet or amulet, can only be taken as evidence that the object was made during, or even subsequently to the reign of the king in question.

(3) For dating purposes, the weight to be given to the presence of a name should also depend on the nature of the object on which it is found. Thus, if at Timna we have one king's name inscribed on the walls of the shrine as well as on several small votive offerings, and another king, either earlier or later, is attested only by the latter type of evidence, it appears to be more reasonable to connect the first king with the establishment of the shrine.

(4) The earliest royal name attested with certainty at Timna is Seti I (Cat. 83a), next comes Ramesses II (Cat. 1 is certain; Cat. 9, 30 and 47 are probable) and then Merneptah (Cat. 102 is certain; Cat. 50 is possible). Cat. 26 is probably the *prenomen* of Seti II and Twosre is to be read on Cat. 41. As for the 20th Dynasty rulers, the names of Ramesses III (Cat. 96, 216); Ramesses IV (Cat. 27, 42, 43, 103 and probably 28) and Ramesses V (Cat. 44–46) are certain.

(5) Cat. 83a. unequivocally assures that Seti I is the earliest Egyptian king whose name is attested at Site 200, since both his *prenomen* and *nomen* appear on it. However, this does not automatically ensure that Cat. 19, which bears the *prenomen* 'Men---re'' and Cat. 195 which is inscribed with the *nomen* 'Seti-merneptah' also name this king. Furthermore, as noted above, there is no certainty that the object on which Seti I's names are inscribed actually date to his reign and not to the period when his son and successor, Ramesses II, shared the throne with him as co-regent (see Murnane, 1977, 57–87). There is some justification for preferring this later date since the only tangible evidence from the actual shrine to indicate the identity of the king under whose reign the *naos* was built, indicates Ramesses II (Cat. 1) and, as noted above, more weight must be given to this than to the bracelet fragment Cat. 83a.

It should, however, be emphasised that this conclusion is based entirely on the inscriptions found at Site 200 and on their evaluation from the Egyptological point of view. From the stratigraphic viewpoint, the basic fact must be taken into consideration that the vast majority of the inscribed Egyptian objects, including Cat. 1, were found on top of the White Floor, which was apparently laid when the shrine was rebuilt after its first phase had been destroyed.

The Seti I inscription (Cat. 83a), however, was found well below this floor, which may lead to a different conclusion on archaeological grounds (see below, Chapter IV), but it would be difficult to harmonise such a stratigraphic conclusion with the purely Egyptological, historical and epigraphic interpretation and analysis.

(6) As shown by the inscriptions found in the shrine, the Egyptian presence at Timna continued from Ramesses II in an unbroken line of his successors until Twosre', who is attested not only at Serabit el-Khadem, but also at Deir 'Alla in Transjordan (Yoyotte, 1962, 464, 469). There is then a break, during which the earlier 19th Dynasty shrine was probably destroyed, possibly in connection with the chaotic period that ended this Dynasty and is described by Ramesses III at the beginning of the Historical Section of Papyrus Harris I (Breasted, 1906, 398). It now also seems probable that Ramesses'claim further on in the same text, that he despatched an expedition by land and sea to the great copper mines of the land of 'Atika' (= Timna; Rothenberg, 1972, 201–3) refers to the actual re-establishment of the Egyptian presence at Timna, including the rebuilding of the shrine. Though circumstantial, the carving of the rock stela on the face of the cliff above Site 200 by a high-ranking official of Ramesses' court, the royal butler Ramessesemperre, who undoubtedly led this expedition, stands to support this. This phase of the Egyptian presence at Timna therefore lasted until the reign of Ramesses V, after which the Egyptians abandoned or were driven away from the site.

Alan R. Schulman

Notes

1. Many of the Egyptian objects were previously published by the excavator, especially in Rothenberg, 1972, and also in numerous preliminary reports and exhibition catalogues. To save space, we indicate such a previous publication by an asterisk next to the Catalogue number.

2. Kitchen, 1976a, based his interpretation on a previously published poor photograph.

3. Cf. Hayes, 1953–59, II, 45, 232, where he clearly underlines the difference between the votive *menats*, like those found at Timna, and the actual ceremonial *menat*.

4. Cf. Kertesz, 1972, 66; Pl. 13:8. I do not accept her readings.

5. The most recent studies of ushabtis known to me are those of Aubert (1974) and Schneider (1977). The earlier studies of Loret (1883, 89–117; 1884, 70–76), L. Speelers (1923), Petrie (1935), T.G. Allen (1963, 62–73) and Valbelle (1972) are still useful.

6. For technical reasons, the grouping of the hieroglyphic signs in each line of the text has been transcribed arbitrarily in a more normal linear fashion. For the exact relationship of the individual words to each other, see. *Fig. 28:2.*

7. Cf. however, Hayes, 1953–59, II, 316, where he discusses a Late 18th Dynasty Hathor head whose tower head-dress is flanked by a pair of cats. Unfortunately the provenance seems to be unknown.

8. For several very well preserved vessels of this type, cf. Hayes, 1953–59, II, Figs. 150, 187, 205 (18th Dynasty), 257 (Ramesside).

9. Similar faiences were found at Serabit el-Khadim (Petrie, 1906; Pl. 147:8–20; Pl. 155:26–26, 28; Pl. 156). For the types of ceramic decoration of the New Kingdom, see Nagel, 1938; Riefstahl, 1968; Wallis, 1908 and idem, 1900. For foreign motifs in Egyptian ceramics, the most recent study known to me is that of

Smith, 1965, particularly chapter 3, 'The Part Played by the Manufacture of Glazed Faience'.

10. Giveon's translation 'Lord of Life, Stability, Well-being' (in Rothenberg, 1972) is not acceptable.

11. The numerous fragments of this vessel were found dispersed inside as well as outside the temple structure. Several fragments came from Locus 102 and Locus 107.

12. This fragment consisted of 10 sherds found in different loci, inside and outside the Temple courtyard: Field Nos. 368/6, 245, 253, 262, 275, 279, 362, 366, 367, 378; Loci 109, 107, 102, 103, 104, 106, 108.

13. For an excellent, concise study of wine bowls with motifs other than the fish and lotus decoration, see Rogers 1947-8, 154–60.

14. These fragments, which by ware, dimensions and decoration apparently belong to one and the same vessel, were found at different loci inside and outside the Temple courtyard: Field No. 368/5, 257, 265, 319, 341, 345, 363; Loci 109, 102, 103, 106–108, 107, 110–111.

15. As with some of the other faience vessels, the many fragments of this bowl were found dispersed in different loci and layers, inside and outside the Temple courtyard: Field No. 257/283, 177, 206, 237, 276, 277, 289, 319, 332, 343; Loci 102, 101, 110–111, 107–110, 108–106, 109. See below, the discussion of this stratigraphic situation.

16. Also quite similar at first glance to the Amuletic Wands are the apotropaic wands found in the tomb of Tuthmosis IV (Carter-Newberry, 1904, 110–13; Pl. XXV) where they are identified as 'model wands' . A closer scrutiny reveals, however, that while these do have the eyes, there are no 'mouths' as on the Timna wands. Instead, they terminate in a floral design. I am grateful to John Romer for this reference.

17. The study of cryptography on Egyptian scarabs was established by E. Drioton in a number of studies published over two decades, the most important of which are Drioton, 1940, 305–427; idem, 1957, 11–33.

18. For structural – shapes and dimensions – as well as numerical reasons – 3–4 Offering Tables as against at least 25 Stands (or fragments thereof) – the excavator finds it difficult to accept Schulman's reconstruction of a two-part Offering Table (rectangular, large table on top of a small, round, pillar-shaped foot or stand), and tends to accept Petrie's interpretation of these stands as 'incense altars' (see above, II, n. 23). None of the small pillars found in the Temple could possibly have served as a solid base or foot for the heavy and relatively large, rectangular Offering Tables and the former seem to have served another function in the Temple ritual (B.R.).

19. Independently and unknown to me, the stela was also studied by Kitchen with results that are virtually identical to mine, save that he restored the beginning of the line slightly differently: p3 ii 'the coming' (Kitchen, 1976b, 311–12;. Fig. 2).

20. Cf. Ventura 1974, 60–63; Pl. 9. Ventura's translation reads the title incorrectly and makes the names Ramessesemperre into two names, the first 'Ramesses' and the second illegible.

21. Cf. Helck, 1958, 272.

Bibliography

Aldred, C. 1971. *Jewels of the Pharaohs: Egyptian Jewellery of the Dynastic Period*, London.

Allen, T.G. 1963. *A Handbook of the Egyptian Collection*, The Art Institute of Chicago.

Altenmüller, H. 1956. *Die Apotropaia und die Götter*, 24: 19.

Anthes, R. 1965. Mit Rahineh, 1956, *Museum Monographs*, Pennsylvania.

Badawy, A. 1954–68. *A History of Egyptian Architecture*, 3 vols. Berkeley.

Barguet, P. 1953. L'origine et la signification de collier-menat, *BIFAO* 52, 103–11.

Breasted, J.H. 1906. *Ancient Records of Egypt*, Chicago.

Carter, H. and Newberry, P. 1904. *The Tomb of Thoutmosis IV*, London.

Cerny, J. 1955. *The Inscriptions of Sinai from Manuscripts of Alan H. Gardiner and T. Eric Peet, edited and completed. Part II: Translation and Commentary*, Egyptian Exploration Society, 45th Memoir. London.

Christophe, L.A. 1949. La Stele de l'An III de Ramses IV au Ouadi Hammamat (No. 12), *BIFAO* 48, 1–38.

Drioton, E. 1940. Receuil de cryptographie monumentale, *ASAE* 40, 305–427.

—— 1956. Vaeux inscrits sur des scarabees, *MDAIK* 14, 34–41.

—— 1957. Trigrammes d'Amon, *WZKM* 54, 11–33.

—— 1959. Maximes relatifs a l'amour des dieux, *Analecta Biblica* 12, 57–68.

—— 1960. Amon, Refuge du Coeur, *ZAS* 79, 3–12.

Erichesen, W. 1933. *Papyrus Harris I, Bibliotheca Aegyptiaca*, vol. 5.

Erman, A. and Grapow, H. 1955. *Wörterbuch der Aegyptischen Sprache im Auftrage der Deutschen Akademien*, Unveränderter Neudruck, 5 vols..

Gardiner, A.H. 1957. *Egyptian Grammar*, 3rd ed. Oxford.

Gardiner, A.H. and Peel, T.E. 1952. *The Inscriptions of Sinai, 2nd ed. revised and augmented. Part 1: Introduction and Plates*, Egypt Exploration Society, 45th Memoir. London.

Gauthier, H. 1914. *Le Livre des rois d'Egypte*, vol. 3, *Memoires publiés par les membres de l'Institute français d'Archéologie orientale*, vol. 19. Cairo.

Gruen, S.W. 1975. A Note on the Separation of di 'nh from its Complement mi R' on Inscribed Objects, *Göttinger Miszellen*, 18, 21–2.

Hall, H.R. 1913. *Catalogue of Egyptian Scarabs, etc. in the British Museum. Part I. Royal Scarabs*, London.

Hayes, W.C. 1953–9. *The Scepter of Egypt*, 2 vols., New York.

Helck, W. 1958. *Zur Verwaltung des mittleren und neuen Reiches, Probleme der Aegyptologie*, Vol. 3. Berlin.

Hickmann, H. 1949. *Instruments de Musique, Catalogue général des Antiquites égyptiennes du Musée du Caire*. Cairo.

—— 1954. Dieux et Déesses de la Musique, *Cahiers d'Histoire égyptienne* 6, 31–59.

Kantor, H.J. 1947. *The Aegean and the Orient in the Second Millennium B.C.*

Kertez, T. 1972. *A Study of Beads Based on the Timna Sanctuary Beads and the Cult of Hathor in Sinai and the Arabah*, M.A. Thesis, Tel Aviv University (unpublished).

—— 1973. The Beads from the Timna Sanctuary, *YMH* 15/16, 48–51 (Hebrew).

Kiefer, C. and Allibert, A. 1968. Les Ceramiques Blèves Pharaoniques et leur Procedes revolutionnaise d'Emaillage, *Industrie Ceramique* 607, 395–402.

—— 1971. Pharaonic Blue Ceramics – The process of Self-Glazing, *Archaeology* 24, 107–17.

Kitchen, K.A. 1976a. A Thutmoside King at Timna, *Orientalia* 46, 262–4.

—— 1976b. Two Notes on Ramesside History, *Oriens Antiquus*, 15, 311.312.

Legge, G.F., 1905–6. The Magic Ivories of the Middle Empire, *PSBA* 28, 130–52; 297–302; *idem* 29, 159–70.

Lipschitz, O. 1972. Timna, *IEJ* 22, 158.

Loret, V. 1883–4. Les Statuettes funeraires du Musée de Boulaq, *Receuil des travaux relatifs a la Philologie et a l'archeologie egyptiennes et assyriennes*, 4, 89–117; *idem* 5, 70–76.

Lucas, A.E. 1962. *Ancient Egyptian Materials and Industries*, rev. by J.R. Harris, London.

Murnane, W. 1977. Ancient Egyptian Coregencies, *SAOC* 40.

Nagel, G. 1938. *La Ceramique du Nouvel Empire a Deir el Medineh* Vol. I, *Documents de fouilles*, Vol. 10.

Noble, J.V. 1969. The Technique of Egyptian Faience, *AJA* 73, 435–9.

Petrie, W.M.F. 1906. *Researches in Sinai*, London.

—— 1914. *Amulets*, London.

—— 1917. *Scarabs and Cylinders with Names*, London.

—— 1935. *Shabtis*, London.

Randall-MacIver and Woolley, C.L. 1911. *Buhen*, Philadelphia.

Reisner, G.A. 1907–58. *Amulets, Catalogue général des Antiquités egyptiennes du Musée du Caire*, Cairo.

Riefstahl, E. 1968. *Ancient Egyptian Glass and Glazes in the Brooklyn Museum*, Wilbour Monographs, Vol. I, New York.

Rogers, E.A. 1947–8. An Egyptian Wine Bowl of the XIIth Dynasty, *BMMA* 6, 154–60.

Rothenberg, B. 1969a. King Solomon's Mines No More, *ILN* 15, 32–3.

—— 1969b. The Egyptian Temple at Timna, *ILN* 29, 28–9.
—— 1970a. An Egyptian Temple of Hathor Discovered in the Southern Arabah (Israel), *BMH* 12, 28–35.
—— 1970b. Un temple egyptien découvert dans la Arabah, *BTS* 123, 6–14.
—— 1971. *Midianite Timna: Valley of the Biblical Copper Mines*, Exhibition Catalogue, British Museum, London.
—— 1972. *Timna*, London.
—— 1973. *Timna, Tal des biblischen Kupfers*, Katalog aus dem Bergbau Museum Bochum, Nr. 5.
Sachs, C. 1920. Altägyptische Musikinstrumente, *Der Alte Orient*, V. 21.
Schneider, H.D. 1977. *Shabtis*, 3 vols., Leiden.
Schulman, A.R. 1967. The Scarabs, in G. Bass, Cape Gelidonya – A Bronze Age Shipwreck, *Transactions of the American Philosophical Society* 57, 143–7.
—— 1975. Egyptian Scarabs, 17th–16th Century B.C., in M. Novak, *The Mark of Ancient Man; Ancient Near Eastern Stamp Seals and Cylinder Seals*, 68–73, New York.
—— 1976. The Royal Butler Ramessesemperre, *JARCE* 13, 117–30.
Scott, N. 1958. The Cat of Bastet, *BMMA* 17, 1–7.
Smith, W.S. 1965. *Interconnections in the Ancient Near East: A Study of Relationships between the Arts of Egypt, the Aegean and Western Asia*, New Haven.
Soderberg, B. 1968. The Sistrum: A Musicological Study, *Ethnos* 1, 91–133.
Speelers, L. 1923. *Les Figurines funeraires egyptiennes*, Brussels.
Steindorff, G. 1946. The Magical Knives of Ancient Egypt, *Journal of the Walters Art Gallery* 9, 41–57; 106–7.
Valbelle, D. 1972. *Oushebtis de Deir el-Medineh, Documents de fouilles*, Cairo.
Ventura, R. 1974. An Egyptian Rock Stela in Timna, *Tel Aviv* 1, 60–63.
Wallis, H. 1898. *Egyptian Ceramic Art*, London.
—— 1900. *Egyptian Ceramic Art*, London.
Ward, J. 1902. *The Sacred Beetle*, London.
Wilkinson, A. 1971. *Ancient Egyptian Jewellery*, London.
Yoyotte, J. 1962. Un Souvenir du 'Pharaon' Taousert en Jordanie, *Vetus Testamentum* 12, 464–9

7. CATALOGUE OF THE METAL FINDS

Introduction

This catalogue of 514 objects lists most of the non-ferrous metal finds and several objects made of iron, found in the excavation of Site 200. An additional number of copper-based objects were too fragmentary to be identified, drawn and included in the Catalogue, but many of these were also analysed (see below, Chapter III, 8, Table 2). Numerous fragments of iron jewellery were found in the excavation. Several disintegrated immediately on exposure and most were completely corroded. A number of gold objects and a few metal amulets are dealt with in Chapter III, 14.

In this catalogue an attempt is made to identify the metal objects and to describe their shape and appearance, although the actual function of many of the fragments could not be established. Metallographical studies of a number of selected objects are reported in Chapter III, 9–12. Most of the objects listed in this catalogue were systematically analysed at the British Museum Research Laboratory (see Chapter III, 8, Tables 1–5).

1 Figurines (*Fig.* 53)

Met. Cat. No. 1. *Fig.* 53:1, *Col. Pl.* 29, Field No. 339/1, Loc 109, Square H-I 11. Bronze.

Male figurine (L. 42.7, W. 13.4, Th. 15.0mm.). The figurine is of a seated man with elongated body and short stubby legs. The arms are outstretched, two eyes are punched into the head and a non-erect penis is evident.

Met. Cat. No. 2. *Fig.* 53:2, *Pl.* 126: 5, 6, Field No. 48/1, Loc 101, Square D-E 15–16. Bronze.
Male figurine (L. 65.9, W. 18.6, Th. 15.7mm.). The standing figure is of a bearded male with erect penis. The figure is wearing a crown and possibly a pectoral. It was found in an 'as cast' state, as evidenced by part of the mould still *in situ*[1] (*Pl.* 126:6) and by small parts of the flash edging around the figure (*Fig.* 53:2). Among the anatomical parts represented are eyebrows, penis, testicles, buttock and shoulder blades.

Met. Cat. No. 3. *Fig.* 53:3, *Col. Pl.* 11, 12, Field No. 98/1, Loc 110, Square B-C 14–15. Bronze and gold.
Serpent with gilded heads[2] (L. 118.5, W. of wire 3.9, Th. of wire 3.1mm), representing a colubrid snake of the racer type (identified by H. Mendelsohn). The wire used to make up the snake has an oval-shaped cross-section. The eyes are formed by circles engraved into the gold foil of the head. The body of the snake is covered by striations running from head to tail.

Met. Cat. No. 4. *Fig.* 53:4, *Pl.* 126:1, Field No. 51/98, Loc 101, Square E-F 15–16. Copper.
Four-legged 'animal figurine' (L. 91.0, W. 36.1, Th. 7.0mm.). This figure is probably an unintentional product of smelting or casting, solidified on a stone or clay surface. Because of its resemblance to a four-legged animal it was probably brought to the sanctuary as an offering.

Met. Cat. No. 5. *Fig.* 53:5, *Pl.* 126:2, Field No. 51/99, Loc 101, Square E-F 15–16. Copper.
'Rider on animal', probably an unintentional product of smelting or casting, brought to the sanctuary as an offering.

Met. Cat. No. 6. *Fig.* 53:6, *Col. Pl.* 14, Field No. T200/1, Loc 101, Surface O–10. Bronze.
Ram figurine (Ovis aries) (L. 44.4, H. 37.7, W. 12.1mm.). The figurine has a 4 mm hole drilled through the neck, which may indicate that it was used as an amulet.

2 Iron Objects (*Fig.* 54)[3]

Met. Cat. No. 7. *Fig.* 54:1, *Pl.* 128:22, Field No. 295/196, Loc 101, Square F 17–18. Iron.
Ring (D. 10.4, W. 4.6, Th. of wire 2.0mm.). The ring is made of iron wire and is wound two complete turns. The mid-point of the wire is bent so that the thickness of the ring is twice that of the thickness of the wire at any given point along the ring. The terminals are cut straight.

Met. Cat. No. 8. *Fig.* 54:2, *Pl.* 128:24, Field No. 273/3, Loc 110, Square B-C-D 13–14. Iron.
Ring (D. 12.5, W. 7.7, Th. 3.0mm.). The cross-section of the ring is oval, but may have been originally

rectangular and acquired its oval shape due to corrosion. Both terminals are rounded off.

Met. Cat. No. 9. *Fig.* 54:3, Field No. 357/7, Loc 105, 106, Square B-C 6. Iron.
Ring (D. 12.5, W. 5.4, Th. 3.8mm.). Rectangular cross-section. Badly corroded. Both terminals are broken.

Met. Cat. No. 10. *Fig.* 54:4, *Pl.* 128:25, Field No. 322/89, Loc 101, Square F 15–16. Iron.
Ring (frag.) (W. 6.6, Th. 2.0mm.). Rectangular cross-section. Both terminals are broken.

Met. Cat. No. 11. *Fig.* 54:5, *Pl.* 128:23, Field No. 204/8, Loc 109, Square E 13–14. Iron.
Ring (frag.) (D. 13.5 (?), Th. 3.5mm.). Circular cross-section. Both terminals are broken.

Met. Cat. No. 12. *Fig.* 54:6, Field No. 178/2, Loci 110,112 Square A-B 14–15. Iron.
Ring (frag.) (D. 13.0 (?), Th. of wire 3.0mm.). Circular cross- section. Both terminals are broken.

Met. Cat. No. 13. *Fig.* 54:7, Field No. 343/95, Loc 109, Square F.12. Iron.
Ring (frag.) (D. 11.5 (?), Th. of wire 3.2mm.). Circular cross-section. Both terminals are broken.

Met. Cat. No. 14. *Fig.* 54:8, Field No. 138/9, Loc 110, Square C-D 13–14. Iron.
Ring (frag.) (Th. of wire 2.8mm.). Circular cross-section. Both terminals are broken.

Met. Cat. No. 15. *Fig.* 54:9, Field No. 282/29, Loc 110, Square A-B 13–15. Iron.
Ring (frag.) (Th. of wire 2.8mm.). Circular cross-section. Badly corroded. Both terminals are broken.

Met. Cat. No. 16. *Fig.* 54:10, Field No. 319/286, Loc 108, Square G 7–10. Iron.
Ring (D. 6.0, W. 11.2, Th. 5.5mm.). Oval cross-section. Badly corroded. The ring may originally have had a rectangular cross-section which became oval due to the corrosion. The ring is too small for a finger ring.

Met. Cat. No. 17. *Fig.* 54:11, Field No. 256/1, Square G 12. Iron.
Tube (D. 5.4, W. 9.8, Th. 3.2mm.). Rectangular cross-section. The tube is too small for a finger ring and there is no visible evidence of the method of closing.

Met. Cat. No. 18. *Fig.* 54:12, Field No. 221/39, Loc 107, Square E-F 7–8. Iron.
Ring (D. 22.5, Th. of wire 6.0mm.). Circular cross-section. The ring is badly corroded. Both terminals are so badly corroded that it is impossible to tell how they were finished. The ring is too big for a finger ring.

Met. Cat. No. 19. *Fig.* 54:13, Field No. 228/28, Loc 110, Square B-C-D 13–14. Iron.
Rod (frag.)(Th. 5.4, 3.7mm.). Oval cross-section. Both terminals are broken.

Met. Cat. No. 20. *Fig.* 54:14, Field No. 8/48, Loc 112, Square B-C 16–17. Iron.
Rod (frag.)(Th. 3.9mm.). Circular cross-section. Both terminals are broken.

Met. Cat. No. 21. *Fig.* 54:15, Field No. 72/10, Loc 110, Square C-D 11–12. Iron.
Rod (frag.)(Th. 4.8mm.). Circular cross-section. Both terminals are broken. (See Chapter III, 11).

Met. Cat. No. 22. *Fig.* 54:16, Field No. 279/26, Loc 107, 106 Square E-F 7–9. Iron and gold.
Gilded Ear-ring (?) (Th. of iron wire 3.7, Th. of gold foil 0.4mm.). Circular cross-section. (See Chapter III, 11).

Met. Cat. No. 23. *Fig.* 54:17, Field No. 154/3, Loci 101, 109, 110, Square D-E 14–15. Iron and sea shell.
Small iron tube embedded in sea shell. The iron ring has D. 2.5 and Th. 1.4 mm. The sea shell is identified as *vermetus* sp. (see Chapter III, 11).

3 Ear-rings (*Fig.* 55)

Met. Cat. No. 24. *Fig.* 55:1, Field No. 313/4, Loc 103, Square I-J-K 10. Bronze.
Boat-shaped ear-ring (frag.) (L. 6.4, W. 7.4, Th. 1.0–2.0mm.). Circular cross-section except for the bottom, where it is slightly plano-convex.

Met. Cat. No. 25. *Fig.* 55:2, Field No. 201/26, Loc 101, Square D-E 15–16. Bronze.
Boat-shaped ear-ring (frag.) (L. 6.6, W. 7.7, Th. 0.4–1.7mm.). Circular cross-section except for the bottom, where it is slightly plano-convex.

Met. Cat. No. 26. *Fig.* 55:3, Field No. 250/2, Loc 101, Square K 16–18. Bronze.
Boat-shaped ear-ring (frag.) (L. 10.3, W. 9.2, Th. 1.3–3.0mm.). Circular cross-section except for the bottom, where it is slightly plano-convex.

Met. Cat. No. 27. *Fig.* 55:4, *Pl.* 128:14, Field No. 339/10, Loc 109, Square H-I 11. Bronze.
Boat-shaped ear-ring (frag.) (L. 12.3, W. 6.8, Th. 0.6–2.0mm.). Circular cross-section except for the bottom, where it is slightly plano-convex.

Met. Cat. No. 28. *Fig.* 55:5, Field No. 205/15, Loc 110, 111, Square B-C 13–14. Bronze.
Boat-shaped ear-ring (frag.) (L. 13.7, W. 7.2, Th. 1.4–2.0mm.). Circular cross-section except for the bottom, where it is slightly plano-convex.

Met. Cat. No. 29. *Fig.* 55:6, *Pl.* 128:16, Field No. 116/1, Loc 110, 112, Square A-B-C 14–15. Bronze.
Boat-shaped ear-ring (L. 13.7, W. 9.2, Th. 0.5–2.5mm.). Circular cross-section except for the bottom, where it is slightly plano-convex.

Met. Cat. No. 30. *Fig.* 55:7, *Pl.* 128:15, Field No. 319/20 Loc 106, 108, Square G 7–10. Copper.
Boat-shaped ear-ring (L. 15.7, W. 9.0, Th. 1.3mm.). Circular cross-section. Unlike the other boat-shaped ear-rings, which are thick at the bottom and taper at the terminals, this ear-ring is made of wire of uniform thickness.

Met. Cat. No. 31. *Fig.* 55:8, Field No. 263/6 Loc 103 Square J-K 8. Bronze.
Boat-shaped ear-ring (frag.) (L. 10.5, W. 8.5, Th. 0.5–2.4mm.). Circular cross-section except for the bottom, where it is slightly plano-convex.

Met. Cat. No. 32. *Fig.* 55:9, Field No. 245/58 Loc 107, Square A-B 8–10. Bronze.
Boat-shaped ear-ring (frag.) (L. 15.0, W. 7.2, Th. 0.6–2.3mm.). Circular cross-section except for the bottom, where it is slightly plano-convex.

Met. Cat. No. 33. *Fig.* 55:10, Field No. 279/37 Loci 106, 107, Square E-F 7–9. Bronze.
Boat-shaped ear-ring (frag.) (L. 16.9, W. 9.2, Th. 0.5–2.9mm.). Circular cross-section except for the bottom, where it is slightly plano-convex (see Chapter III, 11).

Met. Cat. No. 34. *Fig.* 55:11, Field No. 337/29, Loc 102, Square I-J-K-L 11. Copper.
Boat-shaped ear-ring (frag.) (L. 9.0, W. 4.6, Th. 0.8–1.5mm.). Circular cross-section except for the bottom, where it is slightly plano-convex.

Met. Cat. No. 35. *Fig.* 55:12, Field No. 71/8, Loc 101, Square I-J 17–18. Bronze.
Boat-shaped ear-ring (frag.) (L. 11.3, W. 8.0, Th. 0.4–2.0mm.). Circular cross-section except for the bottom where it is slightly plano-convex.

Met. Cat. No. 36. *Fig.* 55:13, *Pl.* 128:13, Field No. 269/5, Loc 107, Square A-B 8–10. Bronze.
Boat-shaped ear-ring (frag.) (L. 13.2, W. 7.6, Th. 0.7–1.9mm.). Circular cross-section except for the bottom, where it is slightly plano-convex.

Met. Cat. No. 37. *Fig.* 55:14, Field No. 337/21, Loc 102, Square I-J-K-L 11. Copper.
Boat-shaped ear-ring (L. 14.6, W. 11.3, Th. 1.1–3.5mm.). Circular cross-section except for the bottom, where it is slightly plano-convex.

Met. Cat. No. 38. *Fig.* 55:15, Field No. 250/24, Loc 101, Square K 16–18. Bronze.
Composite boat-shaped ear-ring with a decorated drop-shaped pendant attached to the bottom of the ear-ring (ear-ring part L. 23.1, W. 8.0, Th. 1.5 mm; drop pendant W. 0.5, Th. 4.5mm.). The wire used for the ear-ring is rectangular in cross-section and is not thickened at its bottom like the majority of the other boat-shaped ear-rings. The wire tapers to a blunt point with a rectangular cross-section at the top terminal, which is broken off. The drop-shaped pendant has a circular cross-section. The decoration on the drop consists of incised vertical lines. The pendant and the ear-ring were soldered together with gold, as evidenced by a spot of gold at the junction of the pendant and the ear-ring (see Chapter III, 9).

Met. Cat. No. 39. *Fig.* 55:16, *Pl.* 126:4, Field No. 280/1, Loc 106, Square C-D-E 6–7. Bronze.
Boat-shaped ear-ring (L. 62.4, W. 35.1, Th. 1.7–9.3mm.). Circular cross-section at the top and plano-convex at the bottom. The ear-ring weighs 42 grams and seems too big and heavy to be worn. Hammer marks are evident along the surface of the ear-ring.

Met. Cat. No. 40. *Fig.* 55:17, Field No. 239/2, Loc 107, Square C-D 10–11. Bronze.
Chain link (L. 9.3, W. 5.8, Th. of wire 1.3, W. of wire 2.5mm.). Oval cross-section. The terminals are cut straight and flattened.

Met. Cat. No. 41. *Fig.* 55:18, Field No. 319/18, Loc 106, 108, Square G 7–10. Bronze.
Chain link (L. 8.2, W. 6.1, Th. of wire 1.6, W. of wire 3.1mm.). Oval cross-section. The terminals are cut straight and flattened.

Met. Cat. No. 42. *Fig.* 55:19, Field No. 260/184, Loc 102, Square I-J-K 12. Bronze.
Chain link (L. 8.5, W. 6.6, Th. of wire 1.4, W. of wire 2.6mm.). Oval cross-section. One terminal is cut straight and flattened and the other is broken.

Met. Cat. No. 43. *Fig.* 55:20, Field No. 280/5, Loc 106, Square C-D-E 6–7. Bronze.
Chain link (L. 9.9, W. 6.5, Th. of wire 1.2, W. of wire 2.2mm.). Oval cross-section. One terminal is cut straight and flattened and the other is broken.

Met. Cat. No. 44. *Fig.* 55:21, Field No. 13/2, Loc 101, 112, Square C-D 16–17. Copper.
Chain link (L. 9.0, W. 7.6, Th. of wire 1.7, W. of wire 3.1mm.). Oval cross-section. The terminals are cut straight and flattened.

Met. Cat. No. 45. *Fig.* 55:22, Field No. 319/17, Loc 106, 108, Square G 7–10. Bronze.
Chain link (L. 13.3, W. 7.1, Th. of wire 1.9, W. of wire 3.0mm.). Oval cross-section. The terminals are cut straight and flattened.

Met. Cat. No. 46. *Fig.* 55:23, Field No. 233/3, Loc 101, Square I-J 18. Copper.
Chain link (frag.) (L. 9.3, W. 5.4, Th. of wire 1.3, W. of wire 2.5mm.). Oval cross-section. The terminals are cut straight and flattened.

Met. Cat. No. 47. *Fig.* 55:24, Field No. 204/5, Loc 109, Square E 13–14. Bronze.
Chain link (frag.) (W. 9.1, Th. of wire 1.8, W. of wire 3.5mm.). Oval cross-section.

Met. Cat. No. 48. *Fig.* 55:25, Field No. 277/10, Loc 107, 110, Square C-D 10–11. Bronze.
Chain link (frag.) (W. 12.2, Th. of wire 1.7, W. of wire 2.3mm.). Oval cross-section.

Met. Cat. No. 49. *Fig.* 55:26, Field No. 374/12, Loc 106, Square H-I 16. Copper.
Chain link (frag.) (L. 15.5, W. 8.1, Th. of wire 1.2, W. of wire 2.0mm.). Oval cross-section.

Met. Cat. No. 50. *Fig.* 55:27, Field No. 319/10, Loc 106, 108, Square G 7–10. Bronze.
Chain link (L. 34.0, W. 13.9, Th. of wire 2.0, W. of wire 3.8mm.). Oval cross-section. Terminals are cut straight and flattened.

Met. Cat. No. 51. *Fig.* 55:28, Field No. 279/35, Loc 106, 107, Square E-F 7–9. Bronze.
Chain link (L. 23.5, W. 11.0, Th. of wire 1.2, W. of wire 2.5mm.). Oval cross-section. Terminals are cut straight and flattened.

Met. Cat. No. 52. *Fig.* 55:29, Field No. 261/1, Loc 103, Square J-K 10. Bronze.
Chain link (L. 30.4, W. 13.1, Th. of wire 2.0, W. of wire 2.9mm.). Oval cross-section. Terminals are cut straight and flattened.

Met. Cat. No. 53. *Fig.* 55:30, *Pl.* 128:18, Field No. 327/5, Loc 106, Square C 6. Bronze.
Chain (L. of chain 18.3, L. of individual link 7.0, W. of individual link 6.5, Th. of wire 1.0, W. of wire 2.1mm.). Oval cross-section. Terminals are chisel-sheared.

Met. Cat. No. 54. *Fig.* 55:31, Field No. 327/4, Loc 106, Square C 6. Copper.
Copper. Chain (L. of chain 17.7, L. of individual link 8.7, W. of individual link 5.2, Th. of wire 2.6mm.). Oval cross-section. Terminals are cut straight and flattened.

Met. Cat. No. 55. *Fig.* 55:32, Field No. 245/56, Loc 107, Square A-B 8–10. Bronze.
Chain (L. of chain 36.2; L. of individual link 8.2; W. of individual link 5.0; Th. of wire 1.0; W. of wire 1.9mm.). Oval cross-section. Terminals are chisel-sheared (see Chapter III, 11).

Met. Cat. No. 56. *Fig.* 55:33, *Pl.* 128:17, Field No. 319/12, Loc 106, 108, Square G 7–10. Bronze.
Chain (L. of chain 59.0; L. of individual link 10.8; W. of individual link 5.3; Th. of wire 1.6; W. of wire 2.5mm.). Oval cross-section. Terminals are cut straight and flattened (see Chapter III, 11).

Met. Cat. No. 57. *Fig.* 55:34, *Pl.* 128:19, Field No. 208/4, Loc 106, Square C-D-E 6–7. Bronze.
Chain (L. of chain 38.0; L. of individual link 10.1; W. of individual link 6.9; Th. of wire 1.4; W. of wire 3.4mm.). Oval cross-section. Terminals are cut straight and flattened (see Chapter III, 11).

4 Wires, rods, pins, kohl-sticks, sistrum-parts
(*Figs.* 56–58)

Met. Cat. No. 58. *Fig.* 56:1, Field No. 51/25, Loc 101, Square E-F 15–16. Copper.
Wire-made implement (L. 150.1, Th. 2.4mm.). Circular cross-section. Both terminals taper to a point.

Met. Cat. No. 59. *Fig.* 56:2, Field No. 283/7, Loc 106, 107, Square D 7–9. Copper.
Wire (frag.) (Th. 1.6mm.). Circular cross-section. One end tapers to a point and the other is broken (see Chapter III, 11).

Met. Cat. No. 60. *Fig.* 56:3, Field No. 370/15, Loc 110, Square B-C-D 11–14. Bronze.
Pin (?) (frag.) (Th. 2.7mm.). Circular cross-section. One end tapers to a point and the other is broken.

Met. Cat. No. 61. *Fig.* 56:4, *Pl.* 127:7, Field No. 51/17, Loc 101, Square E-F 15–16. Bronze.
Kohl-stick (?) (frag.) (Th. 4.2mm.). Circular to oval cross-section. One end tapers to a point and the other is broken (see Chapter III, 11).

Met. Cat. No. 62. *Fig.* 56:5, *Pl.* 127:8, Field No. 328/2, Loc 105, 106. Bronze.
Kohl-stick (L. 126.7, 3.8 mm per side). Square cross-section. One end tapers to a point with a circular cross-section and the other tapers to a chisel-like edge.

Met. Cat. No. 63. *Fig.* 56:6, Field No. 51/26, Loc 101, Square E-F 15–16. Bronze.
Pin (?) (frag.) (Th. 3.6mm.). Circular cross-section. One end tapers to a point and the other is broken.

Met. Cat. No. 64. *Fig.* 56:7, Field No. 314/48b, Loc 109, Square F 13–14. Copper.
Pin (L. 63.2, Th. 2.0mm.). Circular cross-section. Both ends taper to a point.

Met. Cat. No. 65. *Fig.* 56:8, Field No. 278/123, Loc 107, 109, 110, Square E-F 10–11. Copper.
Wire (frag.) (Th. 2.0mm.). Circular cross-section. One end tapers to a point and the other is broken.

Met. Cat. No. 66. *Fig.* 56:9, Field No. 283/10, Loc 106, 107, Square D 7–9. Copper.
Wire (frag.) (Th. 1.9mm.). Circular cross-section. One end tapers to a point and the other is broken.

Met. Cat. No. 67. *Fig.* 56:10, Field No. 339/11, Loc 109, Square H-I 11. Copper.
Wire (frag.) (Th. 1.7mm.). Circular cross-section. One end tapers to a point and the other is broken.

Met. Cat. No. 68. *Fig.* 56:11, Field No. 245/59, Loc 107, Square A-B 8–10. Copper.
Wire (frag.) (Th. 1.9mm.). Circular cross-section. One end tapers to a point and the other is broken.

Met. Cat. No. 69. *Fig.* 56:12, Field No. 276/17, Loc 111, Square A-B 11. Copper.
Pin (?) (frag.) (Th. 1.9mm.). Circular cross-section. One end tapers to a point and the other is broken.

Met. Cat. No. 70. *Fig.* 56:13, Field No. 295/199, Loc 101, Square F 17–18. Bronze.
Pin (?) (frag.) (Th. 3.1mm.). Circular cross-section. One end tapers to a point and the other is broken.

Met. Cat. No. 71. *Fig.* 56:14, Field No. 51/18, Loc 101, Square E-F 15–16. Bronze.
Pin (frag.) (Th. 3.9mm.). Circular cross-section. One end tapers to a point and the other is broken.

Met. Cat. No. 72. *Fig.* 56:15, Field No. 279/31, Loc 106, 107, Square E-F 7–9. Copper.
Wire (frag.) (Th. 1.7mm.). Circular cross-section. One end tapers to a point and the other is broken.

Met. Cat. No. 73. *Fig.* 56:16, Field No. 339/7, Loc 109, Square H-I 11. Copper.
Wire (frag.) (Th. 1.7mm.). Circular cross-section. One end tapers to a point and the other is broken.

Met. Cat. No. 74. *Fig.* 56:17, Field No. 295/198, Loc 101, Square F 17–18. Bronze.
Pin (?) (frag.) (Th. 2.5mm.). Circular cross-section. One end tapers to a point and the other is broken.

Met. Cat. No. 75. *Fig.* 56:18, Field No. 279/19, Loc 106, 107, Square E-F 7–9. Bronze.
Wire (Th. 1.4mm.). Circular cross-section. One end tapers to a point and the other is broken.

Met. Cat. No. 76. *Fig.* 56:19, Field No. 51/27, Loc 101, Square E-F 15–16. Bronze.
Part of a sistrum (?) (L. 259.2, Th. 4.0mm.). Circular cross-section. Both ends taper to a point.

Met. Cat. No. 77. *Fig.* 57:1, Field No. 279/30, Loc 106, 107, Square E-F 7–9. Copper.
Wire (Th. 3.7mm.). Circular cross-section. Both ends are broken.

Met. Cat. No. 78. *Fig.* 57:2, Field No. 313/3, Loc 103, Square I-J-K 10. Copper.
Wire (Th. 1.7mm.). Circular cross-section. Both ends are broken.

Met. Cat. No. 79. *Fig.* 57:3, Field No. 277/11, Loc 107, 110, Square C-D 10–11. Copper.
Wire (Th. 2.1mm.). Circular cross-section. Both ends are broken.

Met. Cat. No. 80. *Fig.* 57:4, Field No. 331/2, Loc 109, Square G 11. Bronze.
Wire (Th. 2.1mm.). Circular cross-section. Both ends are broken.

Met. Cat. No. 81. *Fig.* 57:5, Field No. 348/2, Loc 107, Square D-E 10. Bronze.
Wire (Th. 2.4mm.). Circular cross-section. Both ends are broken.

Met. Cat. No. 82. *Fig.* 57:6, Field No. 278/124, Loc 107, 109, 110, Square E-F 10–11. Bronze.
Wire (Th. 2.4mm.). Circular cross-section. Both ends are broken.

Met. Cat. No. 83. *Fig.* 57:7, Field No. 283/9, Loc 106, 107, Square D 7–9. Bronze.
Wire (Th. 2.0mm.). Circular cross-section. Both ends are broken.

Met. Cat. No. 84. *Fig.* 57:8, Field No. 245/12, Loc 107, Square D-E-F-G 6–9. Copper.
Wire (Th. 2.0mm.). Circular cross-section. Both ends are broken.

Met. Cat. No. 85. *Fig.* 57:9, Field No. 283/12, Loc 106, 107, Square D 7–9. Copper.
Pin (?) (Th. 3.6mm.). Circular cross-section. Both ends are broken.

Met. Cat. No. 86. *Fig.* 57:10, Field No. 283/214, Loc 106, 107, Square D 7–9. Copper.
Wire (Th. 2.4mm.). Circular cross-section. Both ends are broken. One end slightly tapered.

Met. Cat. No. 87. *Fig.* 57:11, Field No. 269/21, Loc 107, Square A-B 8–10. Copper.
Wire (Th. 4.2mm.). Circular cross-section. Both ends are broken (see Chapter III,11).

Met. Cat. No. 88. *Fig.* 57:12, Field No. 283/222, Loc 106, 107, Square D 7–9. Copper.
Wire (Th. 1.9mm.). Circular cross-section. Both ends are broken.

Met. Cat. No. 89. *Fig.* 57:13, Field No. 283/15, Loc 106, 107, Square D 7–9. Bronze.
Wire (Th. 2.1mm.). Circular cross-section. Both ends are broken.

Met. Cat. No. 90. *Fig.* 57:14, Field No. 279/6, Loc 106, 107, Square E-F 7–9. Copper.
Wire (Th. 1.4mm.). Circular cross-section. Both ends are broken.

Met. Cat. No. 91. *Fig.* 57:15, Field No. 278/6, Loc 107, 109, Square E-F 10–11. Bronze.
Wire (Th. 2.4mm.). Circular cross-section. Both ends are broken.

Met. Cat. No. 92. *Fig.* 57:16, Field No. 319/29, Loc 106, 108, Square G 7–10. Copper.
Wire (Th. 2.9mm.). Circular cross-section. Both ends are broken.

Met. Cat. No. 93. *Fig.* 57:17, Field No. 245/62, Loc 107, Square A-B 8–10. Bronze.
Wire (Th. 2.2mm.). Circular cross-section. Both ends are broken.

Met. Cat. No. 94. *Fig.* 57:18, Field No. 277/13, Loc 107, 110, Square C-D 10–11. Bronze.
Wire (Th. 2.5mm.). Circular cross-section. Both ends are broken.

Met. Cat. No. 95. *Fig.* 57:19, Field No. 116/2, Loc 110, 112, Square A-B-C 14–15. Bronze.
Wire (Th. 2.9mm.). Circular cross-section. Both ends are broken.

Met. Cat. No. 96. *Fig.* 57:20, Field No. 221/6, Loc 106, 107, Square E-F 7–8. Bronze.
Wire (Th. 2.1mm.). Circular cross-section. Both ends are broken.

Met. Cat. No. 97. *Fig.* 57:21, Field No. 277/12, Loc 107, 110, Square C-D 10–11. Bronze.
Wire (Th. 2.0mm.). Circular cross-section. Both ends are broken.

Met. Cat. No. 98. *Fig.* 57:22, Field No. 283/221, Loc 106, 107, Square D 7–9. Bronze.
Wire (Th. 2.3mm.). Circular cross-section. Both ends are broken (see Chapter III, 11).

Met. Cat. No. 99. *Fig.* 57:23, Field No. 262/2, Loc 103, Square J-K 9. Bronze.
Sistrum fragment (?) (Th. 3.6mm.). Circular cross-section. Both ends are broken.

Met. Cat. No. 100. *Fig.* 57:24, Field No. 283/27, Loc 107, Square D 7–9. Copper.
Sistrum fragment (?) (Th. 5.7mm.). Circular cross-section. Both ends are broken.

Met. Cat. No. 101. *Fig.* 58:1, Field No. 283/216, Loc 107, Square D 7–9. Copper.
Wire (Th. 1.4mm.). Circular cross-section. Both ends are broken.

Met. Cat. No. 102. *Fig.* 58:2, Field No. 277/8, Loc 107, 110, Square C-D 10–11. Copper.
Wire (Th. 1.6mm.). Circular cross-section. Both ends are broken.

Met. Cat. No. 103. *Fig.* 58:3, Field No. 221/11, Loc 106, 107, Square E-F 7–8. Bronze.
Wire (Th. 2.3mm.). Circular cross-section. Both ends are broken.

Met. Cat. No. 104. *Fig.* 58:4, Field No. 283/26, Loc 106, 107, Square D 7–9. Copper.
Wire (Th. 1.3mm.). Circular cross-section. Both ends are broken.

Met. Cat. No. 105. *Fig.* 58:5, Field No. 283/24, Loc 106, 107, Square D 7–9. Bronze.
Wire (Th. 1.6mm.). Circular cross-section. Both ends are broken.

Met. Cat. No. 106. *Fig.* 58:6, Field No. 269/17A, Loc 107, Square A-B 8–10. Copper.
Wire (Th. 2.6mm.). Circular cross-section. Both ends are broken.

Met. Cat. No. 107. *Fig.* 58:7, Field No. 345/15, Loc 107, Square D-E-F-G 9. Bronze.
Wire (Th. 2.9mm.). Circular cross-section. Both ends are broken.

Met. Cat. No. 108. *Fig.* 58:8, Field No. 278/7, Loc 107, 109, 110, Square E-F 10–11. Bronze.
Wire (Th. 2.2mm.). Circular cross-section. Both ends are broken.

Met. Cat. No. 109. *Fig.* 58:9, Field No. 279/49, Loc 106, 107, Square E-F 7–9. Copper.
Wire (Th. 2.0mm.). Circular cross-section. Both ends are broken.

Met. Cat. No. 110. *Fig.* 58:10, Field No. 349/27, Loc 109, Square G 11. Copper.
Wire (Th. 1.6mm.). Circular cross-section. Both ends are broken.

Met. Cat. No. 111. *Fig.* 58:11, Field No. 264/6, Loc 103, Square J-K 7. Copper.
Wire (Th. 2.0mm.). Circular cross-section. Both ends are broken.

Met. Cat. No. 112. *Fig.* 58:12, Field No. 260/186, Loc 102, Square I-J-K 12. Bronze.
Wire (Th. 1.7mm.). Circular cross-section. Both ends are broken.

Met. Cat. No. 113. *Fig.* 58:13, Field No. 339/13, Loc 102, 109, Square H-I 11. Copper.
Wire (Th. 1.2mm.). Circular cross-section. Both ends are broken.

Met. Cat. No. 114. *Fig.* 58:14, Field No. 51/31, Loc 101, Square E-F 15–16. Bronze.
Pin (?) (Th. 2.0mm.). Circular cross-section. Both ends are broken.

Met. Cat. No. 115. *Fig.* 58:15, Field No. 343/90, Loc 109, Square F 12. Copper.
Kohl-stick (Th. 2.5mm.). Circular cross-section.

Met. Cat. No. 116. *Fig.* 58:16, Field No. 278/121, Loc 107, 109, Square E-F 10–11. Copper.
Wire (Th. 1.8mm.). Circular cross-section. Both ends are broken.

Met. Cat. No. 117. *Fig.* 58:17, Field No. 345/13, Loc 107, Square D-E-F-G 6–9. Bronze.
Wire (Th. 2.0mm.). Circular cross-section. Both ends are broken.

Met. Cat. No. 118. *Fig.* 58:18, Field No. 283/8, Loc 106, 107, Square D 7–9. Copper.
Wire (Th. 1.9mm.). Circular cross-section. Both ends are broken.

Met. Cat. No. 119. *Fig.* 58:19, Field No. 255/1, Loc 106, 107, Square C-D 7–8. Bronze.
Wire (Th. 2.7mm.). Circular cross-section. Both ends are broken.

Met. Cat. No. 120. *Fig.* 58:20, Field No. 51/28, Loc 101, Square E-F 15–16. Copper.
Rod (Th. 4.6mm.). Circular cross-section. Both ends are broken.

Met. Cat. No. 121. *Fig.* 58:21, Field No. 343/91, Loc 109, Square F 12. Copper.
Rod (Th. 2.3mm.). Circular cross-section. Both ends are broken (see Chapter III, 11).

Met. Cat. No. 122. *Fig.* 58:22, Field No. 314/53, Loc 109, Square F 13–14. Copper.
Wire (Th. 3.0mm.). Circular cross-section. Both ends are broken.

Met. Cat. No. 123. *Fig.* 58:23, Field No. 221/12, Loc 106, 107, Square E-F 7–8. Copper.
Wire (Th. 1.8mm.). Circular cross-section. Both ends are broken.

Met. Cat. No. 124. *Fig.* 58:24, Field No. 344/10, Loc 110, Square B-C 11–12. Iron.
Wire (Th. 1.4mm.). Circular cross-section. Both ends are broken.

Met. Cat. No. 125. *Fig.* 58:25, Field No. 260/187, Loc 102, Square I-J-K 12. Bronze.
Wire (Th. 2.1mm.). Circular cross-section. Both ends are broken.

Met. Cat. No. 126. *Fig.* 58:26, Field No. 319/22, Loc 106, 108, Square G 7–10. Copper.
Pin (?) (frag.) (Th. 1.6mm.). Circular cross-section. Both ends are broken.

Met. Cat. No. 127. *Fig.* 58:27, Field No. 283/29, Loc 106, 107, Square D 7–9. Bronze.
Wire (Th. 3.0mm.). Circular cross-section. Both ends are broken.

Met. Cat. No. 128. *Fig.* 58:28, Field No. 264/1, Loc 103, Square J-K 7. Copper.
Wire (Th. 2.9mm.). Circular cross-section. Both ends are broken.

Met. Cat. No. 129. *Fig.* 58:29, Field No. 279/22, Loc 107, 106, Square E-F 7–9. Copper.
Wire (Th. 2.5mm.). Circular cross-section. Both ends are broken.

Met. Cat. No. 130. *Fig.* 58:30, Field No. 349/26, Loc 109, Square G 11. Copper.
Wire (Th. 2.4mm.). Circular cross-section. Both ends are broken.

Met. Cat. No. 131. *Fig.* 58:31, Field No. 279/7, Loc 106, 107, Square E-F 7–9. Copper.
Wire (Th. 1.4mm.). Circular cross-section. Both ends are broken.

Met. Cat. No. 132. *Fig.* 58:32, Field No. 262/1, Loc 103, Square J-K 9. Copper.
Wire (Th. 1.9mm.). Circular cross-section. Both ends are broken.

Met. Cat. No. 133. *Fig.* 58:33, Field No. 116/3, Loc 110, 112, Square A-B-C 14–15. Copper.
Wire (Th. 2.0mm.). Circular cross-section. Both ends are broken.

Met. Cat. No. 134. *Fig.* 58:34, Field No. 283/22, Loc 106, 107, Square D 7–9. Copper.
Wire (Th. 1.3mm.). Circular cross-section. Both ends are broken (see Chapter III, 11).

Met. Cat. No. 135. *Fig.* 58:35, Field No. 283/215, Loc 106, 107, Square D 7–9. Copper.
Wire (Th. 3.1mm.). Circular cross-section. Both ends are broken.

Met. Cat. No. 136. *Fig.* 58:36, Field No. 279/43, Loc 106, 107, Square E-F 7–9. Bronze.
Wire (Th. 2.3mm.). Circular cross-section. Both ends are broken.

Met. Cat. No. 137. *Fig.* 58:37, Field No. 319/21, Loc 106, 108, Square G 7–10. Copper.
Wire (Th. 2.0mm.). Circular cross-section. Both ends are broken (see Chapter III, 11).

5 Wires, rods, tin and lead objects, needles and rivets (*Fig.* 59)

Met. Cat. No. 138. *Fig.* 59:1, Field No. 339/8, Loc 109, Square H-I 11. Copper.
Wire (Th. 1.7mm.). Circular cross-section. Both ends are broken.

Met. Cat. No. 139. *Fig.* 59:2, Field No. 228/4, Loc 110, Square B-C-D 13–14. Copper.
Wire (Th. 1.7mm.). Circular cross-section. Both ends are broken.

Met. Cat. No. 140. *Fig.* 59:3, Field No. 377/2, Loc 106, Square E-F-G 7. Copper.
Wire (Th. 2.1mm.). Circular cross-section. Both ends are broken.

Met. Cat. No. 141. *Fig.* 59:4, Field No. 279/48, Loc 106, 107, Square E-F 7–9. Copper.
Wire (Th. 1.5mm.). Circular cross-section. Both ends are broken.

Met. Cat. No. 142. *Fig.* 59:5, Field No. 348/7, Loc 107, Square D-E 10. Copper.
Wire (Th. 1.2mm.). Circular cross-section. Both ends. are broken.

Met. Cat. No. 143. *Fig.* 59:6, Field No. 270/95, Loc 102, Square I-J-K 12. Copper.
Wire (Th. 2.3mm.). Circular cross-section. Both ends are broken.

Met. Cat. No. 144. *Fig.* 59:7, Field No. 106/2, Loc 101, Square J-K 15–16. Copper.
Rod (Th. 4.2mm.). Circular cross-section. Both ends are broken.

Met. Cat. No. 145. *Fig.* 59:8, Field No. 221/10, Loc 106, 107, Square J 16–17. Copper.
Rod (Th. 2.9mm.). Circular cross-section. Both ends are broken.

Met. Cat. No. 146. *Fig.* 59:9, Field No. 279/9, Loc 106, 107, Square E-F 7–9. Copper.
Wire (Th. 1.7mm.). Circular cross-section. Both ends are broken.

Met. Cat. No. 147. *Fig.* 59:10, Field No. 343/93, Loc 109, Square F 12. Copper.
Wire (Th. 1.8mm.). Circular cross-section. Both ends are broken.

Met. Cat. No. 148. *Fig.* 59:11, Field No. 263/7, Loc 103, Square J-K 8. Bronze.
Rod (Th. 2.8mm.). Circular cross-section. Both ends are broken.

Met. Cat. No. 149. *Fig.* 59:12, Field No. 275/17, Loc 104, Square D-E 2–5. Bronze.
Wire (Th. 2.3mm.). Circular cross-section. Both ends are broken. Perhaps a bracelet (see Chapter III, 11).

Met. Cat. No. 150. *Fig.* 59:13, Field No. 339/6, Loc 109, Square H-I 11. Bronze.
Wire (Th. 1.4mm.). Circular cross-section. Both ends are broken (see Chapter III, 11).

Met. Cat. No. 151. *Fig.* 59:14, Field No. 277/6, Loc 107, 110, Square C-D 10–11. Copper.
Wire (Th. 1.9mm.). Circular cross-section. Both ends are broken.

Met. Cat. No. 152. *Fig.* 59:15, Field No. 283/218, Loc 106, 107, Square D 7–9. Copper.
Wire (Th. 2.4mm.). Circular cross-section. Both ends are broken.

Met. Cat. No. 153. *Fig.* 59:16, Field No. 204/4, Loc 109, Square E 13–14. Copper.
Wire (Th. 2.2mm.). Circular cross-section. Both ends are broken.

Met. Cat. No. 154. *Fig.* 59:17, Field No. 221/9, Loc 106, 107, Square E-F 7–8. Copper.
Wire (Th. 2.0mm.). Circular cross-section. Both ends are broken.

Met. Cat. No. 155. *Fig.* 59:18, Field No. 279/44, Loc 106, 107, Square E-F 7–9. Copper.
Wire (Th. 2.0mm.). Circular cross-section. Both ends are broken.

Met. Cat. No. 156. *Fig.* 59:19, Field No. 314/51, Loc 109, Square F 13–14. Copper.
Wire (Th. 2.4mm.). Circular cross-section. Both ends are broken.

Met. Cat. No. 157. *Fig.* 59:20, Field No. 279/3, Loc 106, 107, Square E-F 7–9. Copper.
Pin(?) (Th. 2.9mm.). Circular cross-section. Both ends are broken.

Met. Cat. No. 158. *Fig.* 59:21, Field No. 269/9, Loc 107, Square A-B 8–10. Bronze.
Wire (Th. 1.7mm.). Circular cross-section. Both ends are broken.

Met. Cat. No. 159. *Fig.* 59:22, Field No. 283/5, Loc 106, 107, Square D 7–9. Copper.
Wire (Th. 1.7mm.). Circular cross-section. Both ends are broken.

Met. Cat. No. 160. *Fig.* 59:23, Field No. 343/92, Loc 109, Square F 12. Bronze.
Wire (Th. 2.8mm.). Circular cross-section. Both ends are broken.

Met. Cat. No. 161. *Fig.* 59:24, Field No. 279/4, Loc 106, 107, Square E-F 7–9. Copper.
Wire (Th. 1.6mm.). Circular cross-section. Both ends are broken.

Met. Cat. No. 162. *Fig.* 59:25, Field No. 319/288, Loc 108, Square G 7–10. Lead.[4]
Crook shaped pin (?) fragments (two pieces) (Th. 1.5, W. at crook 5.5mm.). Circular cross-section. The ends opposite the crooks are broken (see Chapter III, 11).

Met. Cat. No. 163. *Fig.* 59:26, Field No. 236/2, Loc 107, 109, Square E-F 9–11. Tin.
Droplet (Diam. 13.9, Th. 3.5mm.).

Met. Cat. No. 164. *Fig.* 59:27, Field No. 283/227, Loc 106, 107, Square D 7–9. Lead.
Rod (W. 6.8, Th. 3.0mm.). Oval cross-section. Both ends are broken.

Met. Cat. No. 165. *Fig.* 59:28, Field No. 309/2, Loc 101, Square G-H 19. Lead.
Wire (Th. 1.8mm.). Circular cross-section. Both ends are broken.

Met. Cat. No. 166. *Fig.* 59:29, Field No. 279/21, Loc 106, 107, Square E-F 7–9. Copper.
Crook shaped pin (L. 34.6, W of wire 2.7 – 0.5, W. at crook 6.8mm.). Circular cross-section. The pin tapers to a point.

Met. Cat. No. 167. *Fig.* 59:30, Field No. 319/11, Loc 106, 108, Square G 7–10. Bronze.
Crook shaped pin (L. 53.8, W. at crook 4.2, W. of bar 3.2–2.6, Th. of bar 1.7–0.5mm.). Rectangular cross-section. The bar tapers to a point. The terminal of the crook is rounded.

Met. Cat. No. 168. *Fig.* 59:31, Field No. 319/30, Loc 106, 108, Square G 7–10. Copper.
Needle (Th. of wire 2.1, W. at eyelet 4.0mm.). Circular cross-section. The end of the needle is broken off.

Met. Cat. No. 169. *Fig.* 59:32, Field No. 206/6, Loc 110, Square B-C-D 13–14. Bronze.
Needle (Th. of wire 2.6, W. at eyelet 3.8mm.). Circular cross-section. The end of the needle is broken off.

Met. Cat. No. 170. *Fig.* 59:33, Field No. 319/25, Loc 106, Square G 7–10. Copper.
Needle (Th. of wire 2.1, W. at eyelet 3.1mm.). Circular cross-section. The end of the needle is broken off.

Met. Cat. No. 171. *Fig.* 59:34, Field No. 283/14, Loc 106, 107, Square D 7–9. Bronze.
Needle (Th. of wire 2.9, W. at eyelet 4.9mm.). Oval cross-section. The end of the needle is broken off (see Chapter III, 11).

Met. Cat. No. 172. *Fig.* 59:35, Field No. 269/18, Loc 107, Square A-B 8–10. Bronze.
Head-band (W. 6.8, Th. 0.8mm.). Rectangular cross-section.

Met. Cat. No. 173. *Fig.* 59:36, Field No. 280/91, Loc 106, Square C-D-E 6–7. Copper.
Head-band (W. 6.4, Th. 1.2mm.). Rectangular cross-section. Parallel lines run along the edges of the band. These lines are chiseled into the metal.

Met. Cat. No. 174. *Fig.* 59:37, Field No. 270/93, Loc 102, Square I-J-K 12. Bronze.
Pin (L. 26.1, Max. W. 3.7, Max. Th. 1.3mm.). Rectangular cross-section.

Met. Cat. No. 175. *Fig.* 59:38, Field No. 343/99, Loc 109, Square F 12. Bronze.
Pin (L. 16.0, Max. W. 2.9, Max. Th. 2.6mm.). Rectangular cross-section.

Met. Cat. No. 176. *Fig.* 59:39, Field No. 289/19, Loc 101, 102, Square C 14–15. Bronze.
Pin (L. 16.6, Max. W. 3.6, Max. Th. 1.8mm.). Rectangular cross-section.

Met. Cat. No. 177. *Fig.* 59:40, Field No. 260/188, Loc 102, Square I-J-K 12. Bronze.
Pin (L. 15.0, Max. W. 2.7, Max. Th. 1.5mm.). Rectangular cross-section.

Met. Cat. No. 178. *Fig.* 59:41, Field No. 319/28, Loc 106, 108, Square G 7–10. Copper.
Rivet (L. 15.6, W. of bar 3.0, Th. of bar 2.2, W. at head 5.0, Th. of head 3.0mm.). Rectangular cross-section.

Met. Cat. No. 179. *Fig.* 59:42, Field No. 225/18, Loc 107, Square C-D 9–11. Bronze.
Rivet (L. 13.2, Th. of rod 2.9, Th. at head 4.3mm.). Circular cross-section (see Chapter III, 11).

Met. Cat. No. 180. *Fig.* 59:43, Field No. 337/28, Loc 102, Square I-J-K-L 11. Copper.
Rivet (L. 13.0, W. of bar 4.4, Th. of bar 3.3, W. at head 6.3, Th. at head 6.0mm.). Rectangular cross-section.

Met. Cat. No. 181. *Fig.* 59:44, Field No. 204/7, Loc 109, Square E 13–14. Bronze.
Rivet (L. 11.3, Th. of rod 4.0mm.). Circular cross-section. A rectangular piece of metal is still adhering to the rivet (see Chapter III, 11).

Met. Cat. No. 182. *Fig.* 59:45, Field No. 237/7, Loc 110, Square B-C-D 13–14. Copper.
Rivet (L. 13.4, Th. of rod 3.6mm.). Circular cross-section.

6 Rods and wires (*Figs.* 60–62)

Met. Cat. No. 183. *Fig.* 60:1, Field No. 337/27, Loc 102, Square I-J-K-L 11. Copper.
Rod (W. 6.1, Th. 1.1mm.). Rectangular cross-section. Both ends are broken.

Met. Cat. No. 184. *Fig.* 60:2, Field No. 283/220, Loc 106, 107, Square D 7–9. Bronze.
Wire (W. 2.8, Th. 1.1mm.). Rectangular cross-section. Both ends are broken.

Met. Cat. No. 185. *Fig.* 60:3, Field No. 279/288A, Loc 106, 107, Square E-F 7–9. Bronze.
Rod (W. 5.4, Th. 1.1mm.). Rectangular cross-section. Both ends are broken.

Met. Cat. No. 186. *Fig.* 60:4, Field No. 279/12B, Loc 106, 107, Square E-F 7–9. Copper.
Rod (W. 3.5, Th. 1.6mm.). Rectangular cross-section. Both ends are broken (see Chapter III, 11).

Met. Cat. No. 187. *Fig.* 60:5, Field No. 278/132, Loc 107, 109, 110, Square E-F 10–11. Bronze.
Rod (W. 3.4, Th. 1.2mm.). Rectangular cross-section. Both ends are broken.

Met. Cat. No. 188. *Fig.* 60:6, Field No. 279/11, Loc 106, 107, Square E-F 7–9. Bronze.
Rod (W. 3.7, Th. 1.4mm.). Rectangular cross-section. Both ends are broken.

Met. Cat. No. 189. *Fig.* 60:7, Field No. 370/16, Loc 110, Square B-C-D 11–14. Bronze.
Rod (W. 3.3, Th. 1.4mm.). Rectangular cross-section. Both ends are broken.

Met. Cat. No. 190. *Fig.* 60:8, Field No. 245/57, Loc 107, Square A-B 8–10. Bronze.
Rod (W. 3.8, Th. 1.4mm.). Rectangular cross-section. Both ends are broken (see Chapter III, 11).

Met. Cat. No. 191. *Fig.* 60:9, Field No. 314/52, Loc 109, Square F 13–14. Copper.
Rod (W. 4.7, Th. 2.0mm.). Rectangular cross-section. Both ends are broken.

Met. Cat. No. 192. *Fig.* 60:10, Field No. 279/13, Loc 106, 107, Square E-F 7–9. Bronze.
Rod (W. 3.8, Th. 1.1mm.). Rectangular cross-section. Both ends are broken.

Met. Cat. No. 193. *Fig.* 60:11, Field No. 310/3, Loc 105, Square A-B 4–6. Bronze.
Pin(?)(W. 4.0, Th. 1.3mm.). Rectangular cross-section. Both ends are broken.

Met. Cat. No. 194. *Fig.* 60:12, Field No. 319/13, Loc 106, 108, Square G 7–10. Bronze.
Wire (1.2 mm per side). Square cross-section. Both ends are broken.

Met. Cat. No. 195. *Fig.* 60:13, Field No. 241/10, Loc 101, Square J-K 15. Copper.
Wire (W. 3.1, Th. 1.2mm.). Rectangular cross-section. Both ends are broken.

Met. Cat. No. 196. *Fig.* 60:14, Field No. 265/4, Loc 103, Square J-K 6. Copper.
Pin (?) (W. 4.2, Th. 2.0mm.). Rectangular cross-section. Both ends are broken.

Met. Cat. No. 197. *Fig.* 60:15, Field No. 339/23, Loc 109, Square H-I 11. Copper.
Wire (W. 3.9, Th. 1.4mm.). Rectangular cross-section. Both ends are broken.

Met. Cat. No. 198. *Fig.* 60:16, Field No. 269/20, Loc 107, Square A-B 8–10. Bronze.
Rod, perhaps an awl (W. 4.2, Th. 1.4mm.). Rectangular cross-section. Both ends are broken (see Chapter III, 11).

Met. Cat. No. 199. *Fig.* 60:17, Field No. 339/5, Loc 109, Square H-I 11. Bronze.
Wire (W. 2.3, Th. 1.1mm.). Rectangular cross-section. Both ends are broken.

Met. Cat. No. 200. *Fig.* 60:18, Field No. 332/6, Loc 109, Square E 12. Copper.
Wire (W. 2.1, Th. 1.1mm.). Rectangular cross-section. Both ends are broken.

Met. Cat. No. 201. *Fig.* 60:19, Field No. 269/12, Loc 107, Square A-B 8–10. Bronze.
Wire (W. 2.1, Th. 1.4mm.). Rectangular cross-section. Both ends are broken.

Met. Cat. No. 202. *Fig.* 60:20, Field No. 279/12A, Loc 106, 107, Square E-F 7–9. Bronze.
Wire (W. 3.5, Th. 1.3mm.). Rectangular cross-section. Both ends are broken.

Met. Cat. No. 203. *Fig.* 60:21, Field No. 283/20, Loc 106, 107, Square D 7–9. Bronze.
Wire (W. 4.7, Th. 1.8mm.). Rectangular cross-section. Both ends are broken.

Met. Cat. No. 204. *Fig.* 60:22, Field No. 278/133, Loc 107, 109, 110, Square E-F 10–11. Bronze.
Strip of copper (W. 3.9, Th. 1.0mm.). Rectangular cross-section. One end is broken and the other is rounded off.

Met. Cat. No. 205. *Fig.* 60:23, Field No. 277/7, Loc 107, 110, Square C-D 10–11. Bronze.
Strip of bronze (W. 5.3, Th. 1.0mm.). Rectangular cross-section. One end is broken and the other is rounded off.

Met. Cat. No. 206. *Fig.* 60:24, Field No. 71/7, Loc 101, Square I-J 17–18. Bronze.
Strip of bronze (bent) (W. 9.4, Th. 1.0mm.). Rectangular cross-section. One end is broken and the other is squared off (see Chapter III, 11).

Met. Cat. No. 207. *Fig.* 60:25, Field No. 279/42, Loc 106, 107, Square E-F 7–9. Bronze.
Strip of bronze (bent) (W. 5.5, Th. 0.9mm.). Rectangular cross-section. One end is broken and the other is squared off.

Met. Cat. No. 208. *Fig.* 60:26, Field No. 319/32, Loc 106, 108, Square G 7–10. Bronze.
Strip of bronze (bent) (W. 5.1, Th. 0.7mm.). Rectangular cross-section. One end is broken and the other is squared off.

Met. Cat. No. 209. *Fig.* 60:27, Field No. 319/33, Loc 106, 108, Square G 7–10. Copper.
Strip of bronze (bent) (W. 4.1, Th. 1.0mm.). Rectangular cross-section. One end is broken and the other is rounded off.

Met. Cat. No. 210. *Fig.* 60:28, Field No. 331/3, Loc 109, Square G 11. Copper.
Rod (2.7 mm per side). Square cross-section. Both ends are broken off (see Chapter III, 11).

Met. Cat. No. 211. *Fig.* 60:29, Field No. 279/46, Loc 106, 107, Square E-F 7–9. Bronze.
Rod (2.7 mm per side). Square cross-section. Both ends are broken.

Met. Cat. No. 212. *Fig.* 60:30, Field No. 345/14, Loc 107, Square D-E-F-G 6–9. Copper.
Wire (1.9 mm per side). Square cross-section. Both ends are broken.

Met. Cat. No. 213. *Fig.* 60:31, Field No. 237/4, Loc 110, Square B-C-D 13–14. Bronze.
Rod (W. 5.3, Th. 3.7mm.). Rectangular cross-section. Both ends are broken.

Met. Cat. No. 214. *Fig.* 60:32, Field No. 233/4, Loc 101, Square I-J 18. Copper.
Rod (W. 4.7, Th. 2.7mm.). Rectangular cross-section. Both ends are broken.

Met. Cat. No. 215. *Fig.* 60:33, Field No. 221/4, Loc 106, 107, Square E-F 7–8. Copper.
Bar (7.1 mm per side). Square cross-section. Both ends are broken (see Chapter III, 11).

Met. Cat. No. 216. *Fig.* 60:34, Field No. 338/33, Loc 109, Square G 11. Bronze.
Pin or needle (W. 1.8, Th. 1.0mm.). Plano-convex cross-section. One end is broken, the other tapers to a point.

Met. Cat. No. 217. *Fig.* 60:35, Field No. 265/2, Loc 103, Square J-K 6. Bronze.
Pin (?) (W. 2.1, Th. 1.7mm.). Plano-convex cross-section. One end is broken, the other tapers to a point.

Met. Cat. No. 218. *Fig.* 60:36, Field No. 51/29, Loc 101, Square E-F 15–16. Bronze.
Wire (W. 2.6, Th. 1.5mm.). Oval cross-section. Both ends are broken.

Met. Cat. No. 219. *Fig.* 61:1, Field No. 221/8, Loc 106, 107, Square E-F 7–8. Copper.
Crook-shaped wire (Th. of wire 2.1, W. at head 5.8mm.). Circular cross-section. The terminal opposite the head is broken and the terminal of the head is cut straight.

Met. Cat. No. 220. *Fig.* 61:2, Field No. 221/7, Loc 106, 107, Square E-F 7–8. Bronze.
Crook-shaped wire (Th. of wire 2.2, W. at head 7.4mm.). Circular cross-section. The terminal opposite the head is broken and the terminal of the head is cut straight.

Met. Cat. No. 221. *Fig.* 61:3, Field No. 278/5, Loc 107, 109, 110, Square E-F 10–11. Copper.
Crook-shaped wire (Th. of wire 2.1, W. at head 7.2mm.). Circular cross-section. The terminals are broken.

Met. Cat. No. 222. *Fig.* 61:4, Field No. 279/45, Loc 106, 107, Square E-F 7–9. Copper.
Crook-shaped wire (Th. of wire 1.6, W. at head 5.0mm.). Circular cross-section. The terminal opposite the head is broken and the terminal of the head tapers to a point (see Chapter III, 11).

Met. Cat. No. 223. *Fig.* 61:5, Field No. 283/11, Loc 106, 107, Square D 7–9. Copper.
Crook-shaped wire (Th. of wire 2.5, W. at head 7.2mm.). Circular cross-section. The terminal opposite the head is broken and the terminal of the head is cut straight.

Met. Cat. No. 224. *Fig.* 61:6, Field No. 279/47, Loc 106, 107, Square E-F 7–9. Copper.
Crook-shaped rod (Th. of rod 3.3, W. at head 8.5mm.). Circular cross-section. The terminal opposite the head is broken and the terminal of the head is cut straight.

Met. Cat. No. 225. *Fig.* 61:7, Field No. 279/33, Loc 106, 107, Square E-F 7–9. Bronze.
Crook-shaped wire (Th. of wire 1.7, W. at head 6.0mm.). Circular cross-section. The terminal opposite the head is broken and the terminal of the head is cut straight (see Chapter III, 11).

Met. Cat. No. 226. *Fig.* 61:8, Field No. 279/40, Loc 106, 107, Square E-F 7–9. Copper.
Crook-shaped wire (Th. of wire 2.1, W. at head 5.7mm.). Circular cross-section. The terminal opposite the head is broken and the terminal of the head is cut straight.

Met. Cat. No. 227. *Fig.* 61:9, Field No. 283/213, Loc 106, 107, Square D 7–9. Copper.
Crook-shaped wire (Th. of wire 1.7, W. at head 1.7mm.). Circular cross-section. The terminal opposite the head is broken and the terminal of the head is cut straight.

Met. Cat. No. 228. *Fig.* 61:10, Field No. 277/14, Loc 107, 110, Square C-D 10–11. Copper.
Crook-shaped wire (Th. of wire 1.6, W. at head 6.0mm.). Circular cross-section. The terminal opposite the head is broken and the terminal of the head is cut straight.

Met. Cat. No. 229. *Fig.* 61:11, Field No. 279/8, Loc 107, 106, Square E-F 7–9. Copper.
Crook-shaped wire (Th. of wire 1.5, W. at head 5.3mm.). Circular cross-section. The terminal opposite the head is broken and the terminal of the head tapers almost to a point.

Met. Cat. No. 230. *Fig.* 61:12, Field No. 278/122, Loc 107, 109, 110, Square E-F 10–11. Bronze.
Crook-shaped wire (Th. of wire 1.3, W. at head 3.0mm.). Circular cross-section. The terminal opposite the head is broken and the terminal of the head tapers to a point.

Met. Cat. No. 231. *Fig.* 61:13, Field No. 239/1, Loc 107, Square C-D 10–11. Copper.
Crook-shaped wire (Th. of wire 1.9, W. at head 4.2mm.). Circular cross-section. The terminal opposite the head is broken and the terminal of the head tapers slightly and then is cut straight.

Met. Cat. No. 232. *Fig.* 61:14, Field No. 279/41, Loc 106, 107, Square E-F 7–9. Bronze.
Crook-shaped wire (Th. of wire 1.7, W. at head 5.3mm.). Circular cross-section. The terminal opposite the head is broken and the terminal of the head tapers but is broken.

Met. Cat. No. 233. *Fig.* 61:15, Field No. 339/22, Loc 109, Square H-I 11. Copper.
Crook-shaped wire (Th. of wire 1.7, W. at head 6.1mm.). Circular cross-section. The terminal opposite the head is broken and the terminal of the head tapers to a point.

Met. Cat. No. 234. *Fig.* 61:16, Field No. 82/1, Loc 101, 102, Square J-K 14–15. Bronze.
Crook-shaped wire (Th. of wire 2.0, W. at head 5.5mm.). Circular cross-section. The terminal opposite the head is broken and the terminal of the head tapers slightly and then is cut straight.

Met. Cat. No. 235. *Fig.* 61:17, Field No. 278/115, Loc 107, 109, Square E-F 10–11. Copper.
Crook-shaped wire (Th. of wire 2.2, W. at head 2.2mm.). Circular cross-section. The terminal opposite the head is broken and the terminal of the head tapers slightly and then is cut straight.

Met. Cat. No. 236. *Fig.* 61:18, Field No. 51/32, Loc 101, Square E-F 15–16. Copper.
Crook-shaped wire (Th. of wire 2.4, W. at head 6.7mm.). Circular cross-section. The terminal opposite the head is broken and the terminal of the head tapers to a point.

Met. Cat. No. 237. *Fig.* 61:19, Field No. 205/14, Loc 110, 111, Square B-C 13–14. Copper.
Crook-shaped wire (Th. of wire 2.2, W. at head 4.0mm.). Circular cross-section. The terminal opposite the head is broken and the terminal of the head tapers to a point.

Met. Cat. No. 238. *Fig.* 61:20, Field No. 337/22, Loc 102, Square I-J-K-L 11. Copper.
Crook-shaped wire (Th. of wire 2.1, W. at head 5.7mm.). Circular cross-section. The terminal opposite the head is broken and the terminal of the head tapers to a point.

Met. Cat. No. 239. *Fig.* 61:21, Field No. 371/10, Loc 110, Square B-C 11–14. Copper.
Crook-shaped wire (Th. of wire 2.2, W. at head 4.8mm.). Circular cross-section. The terminal opposite the head is broken and the terminal of the head tapers to a point.

Met. Cat. No. 240. *Fig.* 61:22, Field No. 85/2, Loc 110, 111, Square A-B-C 13–14. Copper.
Sistrum rod (Th. of wire 2.0, W. at heads 4.2 and 4.8mm.). Circular cross-section. Both terminals taper to a point.

Met. Cat. No. 241. *Fig.* 61:23, Field No. 283/6, Loc 106, 107, Square D 7–9. Copper.
Crook-shaped rod (Th. of rod 3.4, W. at head 7.9mm.). Circular cross-section. The terminal opposite the head is broken and the terminal of the head is cut straight.

Met. Cat. No. 242. *Fig.* 61:24, Field No. 51/30, Loc 101, Square E-F 15–16. Copper.
Crook-shaped wire (Th. of wire 2.3, W. at head 4.8mm.). Circular cross-section. The terminal opposite the head is broken and the terminal of the head tapers to a point.

Met. Cat. No. 243. *Fig.* 62:1, Field No. 269/9, Loc 107, Square A-B 8–10. Copper.
Crook-shaped wire (Th. of wire 2.2, W. of the metal at head 3.5, Th. of metal at head 1.0mm.). The wire has a circular cross-section and the head a rectangular cross-section. The terminal opposite the head is broken and the terminal of the head is rounded off.

Met. Cat. No. 244. *Fig.* 62:2, Field No. 51/33, Loc 101, Square E-F 15–16. Bronze.
Crook-shaped wire (Th. of wire 2.2, W. of head 7.2, W. of the metal at head 3.1, Th. of metal at head 1.4mm.). The wire has a circular cross-section and the head a rectangular cross-section. The terminal opposite the head is broken and the terminal of the head is rounded off.

Met. Cat. No. 245. *Fig.* 62:3, Field No. 269/11, Loc 107, Square A-B 8–10. Bronze.
Crook-shaped wire (Th. of wire 2.5, W. of head 6.0, W. of the metal at head 4.7, Th. of metal at head 1.5mm.). The wire has a circular cross-section and the head a rectangular cross-section. The terminal opposite the head is broken and the terminal of the head is rounded off.

Met. Cat. No. 246. *Fig.* 62:4, Field No. 283/4, Loc 106, 107, Square D 7–9. Bronze.
Crook-shaped wire (Th. of wire 2.2, W. of head 4.2, W. of the metal at head 3.2, Th. of metal at head 1.4mm.). The wire has a circular cross-section and the head a rectangular cross-section. The terminal opposite the head is broken and the terminal of the head is rounded off (see Chapter III, 11).

Met. Cat. No. 247. *Fig.* 62:5, Field No. 283/217, Loc 106, 107, Square D 7–9. Copper.
Crook-shaped rod (Th. of rod 3.3, W. of head 3.8, Th. of metal at head 1.9mm.). The rod has a circular cross-section and the head a rectangular cross-section. Both terminals are broken.

Met. Cat. No. 248. *Fig.* 62:6, Field No. 279/10, Loc 106, 107, Square E-F 7–9. Bronze.
Crook-shaped wire (Th. of wire 1.7, W. of head 5.4, W. of metal at head 3.8, Th. of metal at head 0.9mm.). The wire has a circular cross-section and the head a rectangular cross-section. The terminal opposite the head is broken and the head is rounded off (see Chapter III, 11).

Met. Cat. No. 249. *Fig.* 62:7, Field No. 319/284, Loc 108, Square G 7–10. Bronze.
Crook-shaped wire (Th. of wire 2.1, W. of head 6.5, W. of metal at head 2.3, Th. of metal at head 1.1mm.). The wire has a circular cross-section and the head a rectangular cross-section. The terminal opposite the head is broken and the head is rounded off.

Met. Cat. No. 250. *Fig.* 62:8, Field No. 135/5, Loc 110, Square D-E 13–14. Bronze.
Crook-shaped wire (Th. of wire 2.1, W. of head 3.0, W. of metal at head 3.1, Th. of metal at head 1.3mm.). The wire has a circular cross-section and the head a rectangular cross-section. Both terminals are broken.

Met. Cat. No. 251. *Fig.* 62:9, Field No. 239/3, Loc 107, Square C-D 10–11. Copper.
Crook-shaped wire (Th. of wire 2.1, W. of head 2.8, Th. of metal at head 0.8mm.). The wire has a circular cross-section and the head a rectangular cross-section. The terminal opposite the head is broken and the head is rounded off.

Met. Cat. No. 252. *Fig.* 62:10, Field No. 269/17, Loc 107, Square A-B 8–10. Bronze.
Crook-shaped wire (Th. of wire 1.7, W. of head 2.3, Th. of metal at head 1.0mm.). The wire has a circular cross-section and the head a rectangular cross-section. Both terminals are broken.

Met. Cat. No. 253. *Fig.* 62:11, Field No. 367/5, Loc 109, Square F 12–14. Copper.
Crook-shaped wire (W. 4.2, Th. 1.9, W. of head 5.0mm.). Rectangular cross-section. Both terminals are broken.

Met. Cat. No. 254. *Fig.* 62:12, Field No. 332/5, Loc 109, Square E 12. Copper.
Crook-shaped wire (W. 1.9, Th. 1.2, W. of head 4.0mm.). Rectangular cross-section. Both terminals are broken.

Met. Cat. No. 255. *Fig.* 62:13, Field No. 85/1, Loc 110, 111, Square A-B-C 13–14. Copper.
Crook-shaped rod (Sides of rod 3.8, W. of head 9.0, W. of metal at head 4.5, Th. of metal at head 1.1mm.). The rod has a square cross-section and the head a rectangular cross-section. The terminal opposite the head is broken and the terminal of the head is rounded off.

Met. Cat. No. 256. *Fig.* 62:14, Field No. 310/2, Loc 105, Square A-B 4–6. Copper.
Crook-shaped rod (Sides of rod 3.0, W. of head 7.4, W. of metal at head 3.6, Th. of metal at head 1.2mm.). The rod has a square cross-section and the head a rectangular cross-section. The terminal opposite the head is broken and the terminal of the head is rounded off.

Met. Cat. No. 257. *Fig.* 62:15, Field No. 269/19, Loc 107, Square A-B 8–10. Bronze.
Crook-shaped wire (Sides of wire 2.7, W. of head 9.0, W. of metal at head 3.8, Th. of metal at head 1.2mm.). The wire has a square cross-section and the head a rectangular cross-section. The terminal opposite the head is broken and the terminal of the head is rounded off.

Met. Cat. No. 258. *Fig.* 62:16, Field No. 343/86, Loc 109, Square F 12. Bronze.
Sistrum rod (?) (W. of rod 2.7, Th. of rod 1.6, W. of heads 3.5, L. 109mm.). Rectangular cross-section. Both terminals are rounded off.

Met. Cat. No. 259. *Fig.* 62:17, Field No. 225/15, Loc 107, Square C-D 9–11. Bronze.
Crook-shaped wire (W. of wire 2.8, Th. of wire 1.2, W. of head 4.0mm.). Rectangular cross-section. Both terminals are rounded off.

Met. Cat. No. 260. *Fig.* 62:18, Field No. 269/10, Loc 107, Square A-B 8–10. Bronze.
Crook-shaped wire (W. of wire 3.7, Th. of wire 1.3, W. of head 26.2mm.). Rectangular cross-section. The terminal opposite the head is broken and the terminal of the head is rounded off (see Chapter III, 11).

Met. Cat. No. 261. *Fig.* 62:19, Field No. 221/5, Loc 106, 107, Square E-F 7–8. Copper.
Crook-shaped wire (W. of wire 5.6, Th. of wire 1.1, W. of head 5.6, Th. of head 5.2mm.). Rectangular cross-section. Both terminals are broken.

Met. Cat. No. 262. *Fig.* 62:20, Field No. 236, Loc 107, 109, Square E-F 9–11. Bronze.
Crook-shaped wire (W. of wire 2.6, Th. of wire 1.0, W. of head 4.2mm.). Rectangular cross-section. Both terminals are broken.

7 Decorated rings (*Fig.* 63)

Met. Cat. No. 263. *Fig.* 63:1, Field No. 225/16, Loc 107, Square C-D 9–11. Bronze.
Decorated ring (Diam. 7.5, W. 4.0, Th. 1.6mm.). Rectangular cross-section. The rectangular wire tapers from a squared-off end to a small rounded terminal. The ring is decorated with a single incised longitudinal line.

Met. Cat. No. 264. *Fig.* 63:2, Field No. 303/637, Loc 101, Square F 15–16. Bronze.
Decorated ring (Diam. 7.8, W. 5.5, Th. 1.0mm.). Rectangular cross-section. The band tapers from a squared-off end to a small rounded terminal. The ring is decorated with a single incised longitudinal line.

Met. Cat. No. 265. *Fig.* 63:3, Field No. 51/7, Loc 101, Square E-F 15–16. Bronze.
Decorated ring (Diam. 8.4, W. 5.9, Th. 1.4mm.). Rectangular cross-section. One terminal is rounded off and the other consists of a small rectangular tab. The ring is decorated with three parallel incised longitudinal lines. The middle line is deeper and wider than the others.

Met. Cat. No. 266. *Fig.* 63:4, Field No. 250/1, Loc 101, Square K 16–18. Bronze.
Decorated ring (Diam. 6.9, W. 11.1, Th. 0.9mm.). Rectangular cross-section. Both terminals are broken. The ring is decorated with incised cross-hatched lines (see Chapter III, 11).

Met. Cat. No. 267. *Fig.* 63:5, Field No. 279/16, Loc 107, 106, Square E-F 7–9. Bronze.

Decorated ring (frag.) (Diam. 10.7, W. 5.9, Th. 1.0mm.). Rectangular cross-section. Both terminals are broken. The ring is decorated with incised chevrons and flanked by two parallel lines (see Chapter III, 11).

Met. Cat. No. 268. *Fig.* 63:6, *Pl.* 126:30, Field No. 151/2, Loc 102, Square J-K 13–14. Bronze.
Decorated ring (Diam. 17.0, W. at face of ring 1.5, Th. of face of ring 1.5, sides of ring 2.0mm.). The face of the ring has a rectangular cross-section and the sides a square cross-section. The face of the ring is decorated with a rosette pattern flanked on both sides by incised chevrons enclosed by parallel lines. Both terminals are cut straight.

Met. Cat. No. 269. *Fig.* 63:7, Field No. 205/16, Loc 110, 111, Square B-C 13–14. Bronze.
Decorated ring (frag.) (W. 2.1, Th. 2.1mm.). Roughly oval cross-section. The ring is decorated with a rope design. Both terminals are broken.

Met. Cat. No. 270. *Fig.* 63:8, Field No. 283/223, Loc 106, 107, Square D 7–9. Bronze.
Decorated ring (Diam. 4.0, W. 3.0, Th. 1.6mm.). Rectangular cross-section. Both terminals are squared off. The ring is decorated with incised chevrons.

Met. Cat. No. 271. *Fig.* 63:9, Field No. 337/23, Loc 102, Square I-J-K-L 11. Bronze.
Decorated ring (Diam. 12.6, W. 5.6, Th. 1.0mm.). Rectangular cross-section. The ring is decorated with a series of incised chevrons, crosses and small parallel lines, all of which are flanked by two incised longitudinal parallel lines. Both terminals are rounded off.

Met. Cat. No. 272. *Fig.* 63:10, Field No. 275/18, Loc 104, Square D-E 2–5. Copper.
Decorated ring (Diam. 16.4, W. 5.6, Th. 1.4mm.). Rectangular cross-section. The ring is decorated with a number of incised parallel lines, one of which runs the whole length of the ring. Both terminals are rounded off.

Met. Cat. No. 273. *Fig.* 63:11, Field No. 283/208, Loc 106, 107, Square D 7–9. Bronze.
Decorated ring (Diam. 14.8, W. 3.7, Th. 1.0mm.). Rectangular cross-section. The ring is decorated with a single incised parallel line. Both terminals are rounded off (see Chapter III, 11).

Met. Cat. No. 274. Not illustrated, Field No. 319/15, Loc 106, 108, Square G 6–7. Bronze.
Decorated ring (Diam. 18.5, W. 4.2, Th. 1.45mm.). Rectangular cross-section. The ring is decorated with a single incised longitudinal line. Both terminals are rounded off (see Chapter III, 11).

8 Rings (*Figs.* 64–73)

Met. Cat. No. 275. *Fig.* 64:1, Field No. 294/46, Loc 101, Square F 15–16. Bronze.
Ring (Diam. 14.6, W. 2.4, Th. 1.5mm.). Rectangular cross-section. Both terminals are rounded off.

Met. Cat. No. 276. *Fig.* 64:2, Field No. 339/16, Loc 109, Square H-I 11. Bronze.
Ring (Diam. 18.0, W. 4.0, Th. 1.0mm.). Rectangular cross-section. One terminal is broken and the other is rounded off.

Met. Cat. No. 277. *Fig.* 64:3, Field No. 235/1, Loc 109, Square G 12–14. Bronze.
Ring (?) (Diam. 9.1(?), W. 3.4, Th. 1.0mm.). Rectangular cross-section. Both terminals are rounded off.

Met. Cat. No. 278. *Fig.* 64:4, Field No. 278/126, Loc 107, 109, 110, Square E-F 10–11. Bronze.
Ring (Diam. 14.1, W. 2.4, Th. 1.3mm.). Rectangular cross-section. Both terminals are rounded off.

Met. Cat. No. 279. *Fig.* 64:5, Field No. 295/201, Loc 101, Square F 17–18. Bronze.
Ring (frag.) (W. 3.0, Th. 1.2mm.). Rectangular cross-section. One terminal is broken and the other is rounded off.

Met. Cat. No. 280. *Fig.* 64:6, Field No. 265/3,, Loc 103, Square J-K 6. Bronze.
Ring (frag.) (Diam. 13.7, W. 4.0, Th. 1.65mm.). Rectangular cross-section. One terminal is broken and the other is rounded off.

Met. Cat. No. 281. *Fig.* 64:7, Field No. 204/1, Loc 109, Square E 13–14. Bronze.
Ring (Diam. 15.0, W. 3.6, Th. 1.2mm.). Rectangular cross-section. One terminal is broken and the other is rounded off.

Met. Cat. No. 282. *Fig.* 64:8, Field No. 279/17, Loc 106, 107, Square E-F 7–9. Copper.
Ring (?) (frag.) (W. 4.4, Th. 1.1mm.). Rectangular cross-section. One terminal is broken and the other is rounded off.

Met. Cat. No. 283. *Fig.* 64:9, Field No. 257/277, Loc 102, Square I-J-K 13. Bronze.
Ring (frag.) (W. 6.1, Th. 1.0mm.). Rectangular cross-section. One terminal is broken and the other is rounded off.

Met. Cat. No. 284. *Fig.* 64:10, Field No. 269/14, Loc 107, Square A-B 8–10. Copper.
Ring (frag.) (W. 4.2, Th. 1.2mm.). Rectangular cross-section. One terminal is broken and the other is rounded off.

Met. Cat. No. 285. *Fig.* 64:11, Field No. 277/5, Loc 107, 110, Square C-D 10–11. Bronze.
Ring (Diam. 13.1, W. 5.2, Th. 1.4mm.). Rectangular cross-section. One terminal is broken and the other is rounded off.

Met. Cat. No. 286. *Fig.* 64:12, Field No. 319/35, Loc 106, 108, Square G 7–10. Copper.
Ring (frag.) (W. 5.2, Th. 1.5mm.). Rectangular cross-section. One terminal is broken and the other is rounded off.

Met. Cat. No. 287. *Fig.* 64:13, Field No. 233/1, Loc 101, Square I-J 18. Bronze.
Ring (frag.) (Diam. 15.0, W. 8.8, Th. 1.1mm.). Rectangular cross-section. One terminal is broken and the other is rounded off.

Met. Cat. No. 288. *Fig.* 64:14, Field No. 269/15, Loc 107, Square A-B 8–10. Copper.
Ring (frag.) (W. 5.1, Th. 1.5mm.). Rectangular cross-section. One terminal is broken and the other is rounded off.

Met. Cat. No. 289. *Fig.* 64:15, Field No. 319/34, Loc 106, 108, Square G 7–10. Copper.
Ring (frag.) (W. 4.3, Th. 1.2mm.). Rectangular cross-section. One terminal is broken and the other is rounded off.

Met. Cat. No. 290. *Fig.* 64:16, Field No. 319/31, Loc 106, 108, Square G 7–10. Copper.
Ring (frag.) (W. 10.0, Th. 1.15mm.). Rectangular cross-section. Both terminals are broken.

Met. Cat. No. 291. *Fig.* 64:17, Field No. 241/9, Loc 101, Square J-K 15. Bronze.
Ring (frag.) (W. 6.3, Th. 1.4mm.). Rectangular cross-section. Both terminals are broken.

Met. Cat. No. 292. *Fig.* 64:18, Field No. 51/13, Loc 101, Square E-F 15–16. Bronze.
Ring (Diam. 14.2, W. 4.5, Th. 1.6mm.). Rectangular cross-section. Both terminals are broken (see Chapter III, 11).

Met. Cat. No. 293. *Fig.* 64:19, Field No. 245/55, Loc 107, Square A-B 8–10. Bronze.
Ring (Diam. 20.7, W. 8.0, Th. 1.35mm.). Rectangular cross-section. Both terminals are broken.

Met. Cat. No. 294. *Fig.* 64:20, Field No. 51/12, Loc 101, Square E-F 15–16. Copper.
Ring (Diam. 21.0, W. 12.4, Th. 1.0mm.). Rectangular cross-section. Both terminals are broken.

Met. Cat. No. 295. *Fig.* 64:21, Field No. 348/4, Loc 107, Square D-E 10. Copper.
Unbent ring (?) (W. 5.1, Th. 1.25mm.). Crescent cross-section. One terminal is broken and the other rounded off. This band may be an unshaped ring. All the other metal objects from the Temple with crescent shaped cross-section are rings.

Met. Cat. No. 296. *Fig.* 64:22, Field No. 138/1, Loc 110, Square C-D 13–14. Bronze.
Ring (Diam. 7.0, W. 6.5, Th. 1.8mm.). Crescent cross-section. Both terminals are rounded off.

Met. Cat. No. 297. *Fig.* 64:23, Field No. 332/3, Loc 109, Square E 12. Bronze.
Ring (Diam. 4.5, W. 5.0, Th. 1.0mm.). Crescent cross-section. Both terminals are rounded off.

Met. Cat. No. 298. *Fig.* 64:24, Field No. 348/3, Loc 107, Square D-E 10. Copper.
Ring (Diam. 13.1, W. 4.7, Th. 1.3mm.). Crescent cross-section. One terminal is broken and the other rounded off.

Met. Cat. No. 299. *Fig.* 64:25, Field No. 370/17, Loc 110, Square B-C-D 11–14. Bronze.
Ring (?) (frag.) (Diam. 15.7, W. 3.4, Th. 0.9mm.). Crescent cross-section. One terminal is broken and the other rounded off.

Met. Cat. No. 300. *Fig.* 65:1, Field No. 237/5, Loc 110, Square B-C-D 13–14. Copper.
Ring (?) (frag.) (W. 4.4, Th. 0.8mm.). Crescent cross-section. One terminal is broken and the other rounded off.

Met. Cat. No. 301. *Fig.* 65:2, Field No. 319/26, Loc 106, 108, Square G 7–10. Bronze.
Ring (?) (frag.) (Diam. 9.5, W. 5.1, Th. 1.15mm.). Both terminals are broken.

Met. Cat. No. 302. *Fig.* 65:3, Field No. 339/20, Loc 109, Square H-I 11. Bronze.
Ring (?) (frag.) (W. 4.3, Th. 0.9mm.). Crescent cross-section. One terminal is broken and the other rounded off.

Met. Cat. No. 303. *Fig.* 65:4, Field No. 278/128, Loc 107, 109, 110, Square E-F 10–11. Bronze.
Ring (Diam. 7.5, W. 4.0, Th. 1.0mm.). Crescent cross-section. One terminal is broken and the other rounded off.

Met. Cat. No. 304. *Fig.* 65:5, *Pl.* 128:6, Field No. 278/125, Loc 107, 109, 110, Square E-F 10–11. Bronze.
Ring (Diam. 11.3, W. 3.0, Th. 1.0mm.). Crescent cross-section. One terminal is broken and the other rounded off.

Met. Cat. No. 305. *Fig.* 65:6, *Pl.* 128:5, Field No. 343/87, Loc 109, Square F 12. Copper.
Ring with attached cowrie shell (Diam. 16.7, W. 6.0, Th. 1.7mm.). Crescent cross-section. Both terminals are rounded off (see Chapter III, 11).

Met. Cat. No. 306. *Fig.* 65:7, *Pl.* 128:10, Field No. 349/24, Loc 109, Square G 11. Copper.
Ring (Diam. 15.6, W. 7.0, Th. 2.0mm.). Rectangular cross-section. Both terminals are rounded off.

Met. Cat. No. 307. *Fig.* 65:8, Field No. 366/43, Loc 109, Square F 12–14. Copper.
Ring (Diam. 16.6, W. 5.9, Th. 1.5mm.). Rectangular cross-section. Both terminals are rounded off.

Met. Cat. No. 308. *Fig.* 65:9, Field No. 171/1, Loc 101, Square G-H 15–16. Copper.
Ring (Diam. 16.7, W. 5.0, Th. 1.2mm.). Rectangular cross-section. Both terminals are rounded off.

Met. Cat. No. 309. *Fig.* 65:10, *Pl.* 128:7, Field No. 51/11, Loc 101, Square E-F 15–16. Bronze.
Ring (Diam. 22.9, W. 10.6, Th. 1.4mm.). Rectangular cross-section. Both terminals are squared off.

Met. Cat. No. 310. *Fig.* 65:11, Field No. 51/8, Loc 101, Square E-F 15–16. Bronze.
Ring (Diam. 13.0, W. 6.1, Th. 1.0mm.). Rectangular cross-section. One terminals is broken and the other rounded off.

Met. Cat. No. 311. *Fig.* 65:12, Field No. 51/10, Loc 101, Square E-F 15–16. Bronze.
Ring (Diam. 18.8, W. 6.1, Th. 1.0mm.). Rectangular cross-section. Both terminals are rounded off.

Met. Cat. No. 312. *Fig.* 65:13, Field No. 51/9, Loc 101, Square E-F 15–16. Copper.
Ring (Diam. 17.9, W. 5.4, Th. 1.2mm.). Rectangular cross-section. Both terminals are rounded off.

Met. Cat. No. 313. *Fig.* 65:14, Field No. 269/7, Loc 107, Square A-B 8–10. Bronze.
Ring (Diam. 12.5, W. 7.5, Th. 1.0mm.). Rectangular cross-section. Both terminals are rounded off (see Chapter III, 11).

Met. Cat. No. 314. *Fig.* 66:1, Field No. 319/292, Loc 106, 108, Square G 7–10. Copper
Oblong ring (W of wire 2.4, Th. of wire 1.8mm.). Oval cross-section. One terminal is cut straight and the other tapers to a point.

Met. Cat. No. 315. *Fig.* 66:2, Field No. 72/2, Loc 110, Square C-D 11–12. Bronze.
Oblong ring (W of wire 2.7, Th. of wire 1.4mm.). Oval cross-section. One terminal is cut straight and the other tapers to a point.

Met. Cat. No. 316. *Fig.* 66:3, Field No. 227/24, Loc 102, Square I-J-K-L 11. Bronze.
Oblong ring (W of wire 4.8, Th. of wire 2.1mm.). Oval cross-section. Both terminals are rounded off.

Met. Cat. No. 317. *Fig.* 66:4, Field No. J–12/7, Loc 102, Square J 12. Bronze.
Oblong ring (W of wire 2.6, Th. of wire 1.7mm.). Oval cross-section. One terminal is cut straight and the other tapers to a point.

Met. Cat. No. 318. *Fig.* 66:5, Field No. 264/2, Loc 103, Square J-K 7. Copper.
Oblong ring (Th. of wire 2.3mm.). Circular cross-section. Both terminals are cut straight.

Met. Cat. No. 319. *Fig.* 66:6, Field No. 246/1, Loc 101, Square H 18. Bronze.
Oblong ring (Th. of wire 2.1mm.). Circular cross-section. One terminal is cut straight and the other tapers to a point.

Met. Cat. No. 320. *Fig.* 66:7, Field No. 145/3, Loc 101, Square G-H 17–18. Bronze.
Oblong ring (W. of wire 3.0, Th. of wire 2.5mm.). Oval cross-section. Both terminals are cut straight.

Met. Cat. No. 321. *Fig.* 66:8, Field No. J–12/8, Loc 102, Square J 12. Bronze.
Oblong ring (W. of wire 2.6, Th. of wire 1.95mm.). Oval cross-section. Both terminals are cut straight.

Met. Cat. No. 322. *Fig.* 66:9, Field No. 294/47, Loc 101, Square F 15–16. Copper.
Oblong ring (Th. of wire 2.7mm.). Circular cross-section. One terminal is broken and the other tapers to a point (see Chapter III, 11).

Met. Cat. No. 323. *Fig.* 66:10, Field No. 314/49, Loc 109, Square F 13–14. Copper.
Triangular ring (W. of metal 2.7, Th. of metal 1.7mm.). Rectangular cross-section. One terminal is broken and the other tapers slightly and is then cut straight.

Met. Cat. No. 324. *Fig.* 66:11, Field No. 103/1, Loc 109, Square G-H 13–14. Bronze.
Triangular ring (W. of metal 3.4, Th. of metal 0.9mm.). Rectangular cross-section. One terminal is broken and the other is rounded off.

Met. Cat. No. 325. *Fig.* 66:12, Field No. 123/1, Loc 101, Square H-I 16–17. Copper.
Ring (Diam. 5.0, W. 7.8, Th. 1.0mm.). Rectangular cross-section. Both terminals are squared off.

Met. Cat. No. 326. *Fig.* 66:13, Field No. 231/2, Loc 101, Square H 15–17. Copper.
Ring (Diam. 5.6 (?), W. 7.5, Th. 1.8mm.). Rectangular cross-section. Both terminals are squared off.

Met. Cat. No. 327. *Fig.* 66:14, Field No. 279/5, Loc 106, 107, Square E-F 7–9. Copper.
Ring (Diam. 4.6, W. 5.0, Th. 1.0mm.). Rectangular cross-section. Both terminals are squared off.

Met. Cat. No. 328. *Fig.* 66:15, Field No. 257/272, Loc 102, Square I-J-K 13. Bronze.
Ring (Diam. 5.5, W. 4.5, Th. 1.2mm.). Rectangular cross-section. Both terminals are squared off.

Met. Cat. No. 329. *Fig.* 66:16, Field No. 260/181, Loc 103, Square I-J-K 12. Bronze.
Ring (Diam. 4.1, W. 4.0, Th. 1.3mm.). Rectangular cross-section. Both terminals are squared off.

Met. Cat. No. 330. *Fig.* 66:17, Field No. 220/1, Loc 109, Square G-H 12–14. Copper.
Ring (Diam. 4.1, W. 4.7, Th. 1.3mm.). Rectangular cross-section. Both terminals are squared off.

Met. Cat. No. 331. *Fig.* 66:18, Field No. 309/3, Loc 101, Square G-H 19. Bronze.
Ring (Diam. 7.2 (?), W. 6.0, Th. 1.3mm.). Rectangular cross-section. Both terminals are squared off.

Met. Cat. No. 332. *Fig.* 67:1, Field No. 279/38, Loc 106, 107, Square E-F 7–9. Bronze.
Ring (Diam. 4.2, W. 5.7, Th. 1.1mm.). Rectangular cross-section. Both terminals are squared off.

Met. Cat. No. 333. *Fig.* 67:2, Field No. 277/17, Loc 107, 110, Square C-D 10–11. Bronze.
Ring (Diam. 5.4, W. 6.0, Th. 1.3mm.). Rectangular cross-section. Both terminals are squared off.

Met. Cat. No. 334. *Fig.* 67:3, Field No. 211/6, Loc 101, Square G 15–16. Bronze.
Ring (Diam. 3.3, W. 6.7, Th. 1.3mm.). Rectangular cross-section. Both terminals are squared off.

Met. Cat. No. 335. *Fig.* 67:4, Field No. 211/5, Loc 101, Square G 15–16. Bronze.
Ring (Diam. 3.2, W. 7.8, Th. 1.75mm.). Rectangular cross-section. Both terminals are squared off.

Met. Cat. No. 336. *Fig.* 67:5, Field No. 319/24, Loc 106, 108, Square G 7–10. Bronze.
Ring (Diam. 6.9 (?), W. 7.0, Th. 1.3mm.). Rectangular cross-section. Both terminals are squared off.

Met. Cat. No. 337. *Fig.* 67:6, Field No. 233/2, Loc 101, Square I-J 18. Copper.
Ring (Diam. 6.6, W. 3.3, Th. 0.7mm.). Rectangular cross-section. Both terminals are squared off.

Met. Cat. No. 338. *Fig.* 67:7, Field No. 119/1, Loc 101, 110, 112, Square C-D 14–15. Copper.
Ring (Diam. 4.3, W. 2.7, Th. 1.6mm.). Rectangular cross-section. Both terminals are squared off.

Met. Cat. No. 339. *Fig.* 67:8, Field No. 241/8, Loc 101, Square J-K 15. Copper.
Ring (Diam. 5.0, W. 4.0, Th. 1.3mm.). Rectangular cross-section. Both terminals are squared off.

Met. Cat. No. 340. *Fig.* 67:9, Field No. 303/639, Loc 101, Square F 15–16. Bronze.
Ring (Diam. 6.5, W. 6.8, Th. 1.3mm.). Rectangular cross-section. Both terminals are squared off.

Met. Cat. No. 341. *Fig.* 67:10, Field No. 144/1, Loc 101, Square G-H 16–17. Copper.
Ring (Diam. 8.9, W. 8.2, Th. 1.3mm.). Rectangular cross-section. Both terminals are squared off.

Met. Cat. No. 342. *Fig.* 67:11, Field No. 260/182, Loc 102, Square I-J-K 12. Bronze.
Ring (Diam. 4.2, W. 2.3, Th. 1.1mm.). Rectangular cross-section. One terminal is broken and the other is squared off.

Met. Cat. No. 343. *Fig.* 67:12, Field No. 303/638, Loc 101, Square F 15–16. Bronze.
Ring (Diam. 3.8, W. 7.7, Th. 1.4mm.). Rectangular cross-section. One terminal is broken and the other is squared off.

Met. Cat. No. 344. *Fig.* 67:13, Field No. 289/18, Loc 101, 102, Square L 14–15. Copper.
Ring (Diam. 6.0, W. 4.0, Th. 1.5mm.). Rectangular cross-section. One terminal is broken and the other is squared off.

Met. Cat. No. 345. *Fig.* 67:14, Field No. 344/9, Loc 110, Square B-C 11–12. Bronze.
Ring (Diam. 6.4, W. 5.1, Th. 1.0mm.). Rectangular cross-section. Both terminals are squared off.

Met. Cat. No. 346. *Fig.* 67:15, Field No. 303/640, Loc 101, Square F 15–16. Copper.
Ring (Diam. 7.1, W. 8.2, Th. 1.5mm.). Rectangular cross-section. Both terminals are squared off.

Met. Cat. No. 347. *Fig.* 67:16, Field No. 279/34, Loc 106, 107, Square E-F 7–9. Bronze.
Ring (Diam. 3.5, W. 1.9, Th. 0.9mm.). Rectangular cross-section. One terminal is broken and the other tapers to a blunt point.

Met. Cat. No. 348. *Fig.* 67:17, Field No. 279/39, Loc 106, 107, Square E-F 7–9. Bronze.
Ring (Diam. 4.9, W. 1.9, Th. 1.5mm.). Rectangular cross-section. One terminal is broken and the other is rounded off.

Met. Cat. No. 349. *Fig.* 67:18, Field No. 88/1, Loc 102, Square I-J 13–14. Bronze.
Ring (Diam. 4.8, W. 2.9, Th. 2.0mm.). Rectangular cross-section. One terminal is broken and the other is rounded off.

Met. Cat. No. 350. *Fig.* 67:19, Field No. 208/3, Loc 106, Square C-D-E 6–7. Bronze.
Ring (W. 3.9, Th. 1.6mm.). Rectangular cross-section. One terminal is broken and the other is rounded off.

Met. Cat. No. 351. *Fig.* 67:20, Field No. 137/4, Loc 110, Square C-D 14–15. Bronze.
Ring (Diam. 9.9, W. 7.9, Th. 0.9mm.). Rectangular cross-section. One terminal is broken and the other tapers slightly and is rounded off.

Met. Cat. No. 352. *Fig.* 67:21, Field No. 283/219, Loc 106, 107, Square D 7–9. Bronze.
Ring (Frag.) (W. 2.5, Th. 1.1mm.). Rectangular cross-section. One terminal is broken and the other is rounded off.

Met. Cat. No. 353. *Fig.* 67:22, Field No. 257/275, Loc 102, Square I-J-K 13. Bronze.
Ring (?) (Frag.) (W. 6.7, Th. 1.2mm.). Rectangular cross-section. Both terminals are broken

Met. Cat. No. 354. *Fig.* 67:23, Field No. 121/1, Loc 101, Square J 16–17. Bronze.
Ring (?) (Frag.) (W. 5.2, Th. 1.0mm.). Rectangular cross-section. Both terminals are broken.

Met. Cat. No. 355. *Fig.* 67:24, Field No. 303/642, Loc 101, Square F 15–16. Bronze.
Ring (?) (Frag.) (W. 5.0, Th. 1.3mm.). Rectangular cross-section. Both terminals are broken.

Met. Cat. No. 356. *Fig.* 67:25, Field No. 311/186, Loc 101, Square F 15–16. Bronze.
Ring (?) (W. 6.1, Th. 1.2mm.). Rectangular cross-section. Both terminals are broken.

Met. Cat. No. 357. *Fig.* 67:26, Field No. 204/2, Loc 109, Square E 13–14. Copper.
Ring (?) (Diam. 9.0, W. 5.5, Th. 1.7mm.). Rectangular cross-section. Both terminals are broken.

Met. Cat. No. 358. *Fig.* 67:27, Field No. 235/2, Loc 109, Square G 12–14. Tin.
Ring (?) (Frag.) (W. 2.4, Th. 1.6mm.). Rectangular cross-section. Both terminals are broken.

Met. Cat. No. 359. *Fig.* 67:28, Field No. 115/3, Loc 112, Square A-B-C 15–16. Bronze.
Ring (Diam. 8.4, W. 8.3, Th. 1.3mm.). Rectangular cross-section. Both terminals are broken.

Met. Cat. No. 360. *Fig.* 68:1, Field No. 205/17, Loc 110, 111, Square B-C 13–14. Bronze.
Ring (Diam. 2.65, W. 4.0, Th. 1.5mm.). Rectangular cross-section. Both terminals are squared off.

Met. Cat. No. 361. *Fig.* 68:2, Field No. 339/19, Loc 109, Square H-I 11. Bronze.
Ring (Diam. 6.15, W. 6.8, Th. 1.35mm.). Rectangular cross-section. Both terminals are squared off (see Chapter III, 11).

Met. Cat. No. 362. *Fig.* 68:3, Field No. 257/274, Loc 102, Square I-J-K 13. Bronze.
Ring (Diam. 2.5, 1.3 mm per side). Square cross-section. Both terminals are cut straight.

Met. Cat. No. 363. *Fig.* 68:4, Field No. 136/1, Loc 101, Square G-H 18–19. Bronze.
Ring (Diam. 5.6, 2.5 mm per side). Square cross-section. Both terminals are cut straight (see Chapter III, 11).

Met. Cat. No. 364. *Fig.* 68:5, Field No. 286/5, Loc 104, 105, Square C-D 2–4. Copper.
Ring (Diam. 7.9, 2.0 mm per side). Square cross-section. Both terminals taper to a point (see Chapter III, 11).

Met. Cat. No. 365. *Fig.* 68:6, *Pl.* 128:11, Field No. 317/2, Loc 110, Square C-D 12. Bronze.
Ring (Diam. 12.15, 1.9 mm per side). Square cross-section. Both terminals taper to a point (see Chapter III, 11).

Met. Cat. No. 366. *Fig.* 68:7, Field No. 231/147, Loc 101, Square H 15–17. Copper.
Ring (Frag.) (Th. 0.9mm.). Circular cross-section. Both terminals are broken.

Met. Cat. No. 367. *Fig.* 68:8, Field No. 247/1, Loc 109, Square G-H 12–14. Copper.
Ring (Diam. 5.3, Th. 1.4mm.). Circular cross-section. Both terminals are broken.

Met. Cat. No. 368. *Fig.* 68:9, Field No. 228/3, Loc 110, Square B-C-D 13–14. Bronze.
Ring (Diam. 8.1, Th. 1.8mm.). Circular cross-section. Both terminals are cut straight (see Chapter III, 11).

Met. Cat. No. 369. *Fig.* 68:10, Field No. 339/12, Loc 109, Square H-I 11. Bronze.
Ring (Frag.) (Diam. 8.9, Th. 2.2mm.). Circular cross-section. Both terminals are broken (see Chapter III, 11).

Met. Cat. No. 370. *Fig.* 68:11, Field No. 201/24, Loc 101, Square D-E 15–16. Bronze.
Ring (Diam. 8.0, Th. 3.0mm.). Circular cross-section. Both terminals are cut straight.

Met. Cat. No. 371. *Fig.* 68:12, Field No. 302/9, Loc 101, Square F 17–18. Bronze.
Ring (Diam. 5.3, Th. 2.6mm.). Circular cross-section. One terminal is broken and the other tapers slightly and then is cut straight. W

Met. Cat. No. 372. *Fig.* 68:13, Field No. 311/184, Loc 101, Square F 15–16. Copper.
Ring (Diam. 6.6, W. 6.6 – 2.8, Th. 2.0mm.). Rectangular cross-section. One terminal is squared off and the other tapers and then is rounded off.

Met. Cat. No. 373. *Fig.* 68:14, Field No. 296/6, Loc 101, Square L 16–18. Copper.
Ring (Diam. 9.4, W. 8.0 – 2.7, Th. 1.25mm.). Rectangular cross-section. One terminal is squared off and the other tapers and then is rounded off.

Met. Cat. No. 374. *Fig.* 68:15, Field No. 252/1, Loc 104, 105, 106, Square C-D 5–7. Bronze.
Ring (Diam. 8.4, W. 8.0 – 4.2, Th. 1.4mm.). Rectangular cross-section. One terminal is squared off and the other tapers and then is rounded off (see Chapter III, 11).

Met. Cat. No. 375. *Fig.* 68:16, Field No. 303/641, Loc 101, Square F 15–16. Bronze.
Ring (Diam. 7.7, W. 7.2 – 2.5, Th. 1.0mm.). Rectangular cross-section. One terminal is squared off and the other tapers and then is rounded off.

Met. Cat. No. 376. *Fig.* 68:17, Field No. 137/3, Loc 110, Square C-D 14–15. Bronze.
Ring (Diam. 17.75, Th. 2.2mm.). Circular cross-section. Both terminals taper to a point.

Met. Cat. No. 377. *Fig.* 68:18, Field No. 115/2, Loc 112, Square A-B-C 15–16. Bronze.
Ring (Diam. 16.5, Th. 1.7mm.). Circular cross-section. Both terminals taper to a point.

Met. Cat. No. 378. *Fig.* 68:19, Field No. 51/1, Loc 101, Square E-F 15–16. Bronze.
Ring (Diam. 14.5, Th. 1.85mm.). Circular cross-section. Both terminals taper to a point.

Met. Cat. No. 379. *Fig.* 68:20, Field No. 51/4, Loc 101, Square E-F 15–16. Bronze.
Ring (Diam. 16.35, Th. 2.3mm.). Circular cross-section. One terminal tapers to a point and the other is broken.

Met. Cat. No. 380. *Fig.* 69:1, Field No. 51/2, Loc 101, Square E-F 15–16. Bronze.
Ring (Diam. 16.7, Th. 2.2mm.). Circular cross-section. One terminal tapers to a point and the other is broken.

Met. Cat. No. 381. *Fig.* 69:2, Field No. 109/1, Loc 109, 110, Square E-F 14–15. Copper.
Ring (Diam. 15.2, Th. 1.7mm.). Circular cross-section. Both terminals taper to a point.

Met. Cat. No. 382. *Fig.* 69:3, Field No. 72/1, Loc 110, Square C-D 11–12. Bronze.
Ring (Diam. 17.2, Th. 2.0mm.). Circular cross-section. Both terminals taper to a point.

Met. Cat. No. 383. *Fig.* 69:4, Field No. 13/1, Loc 101, 112, Square C-D 16–17. Bronze.
Ring (Diam. 18.2, Th. 1.7mm.). Circular cross-section. Both terminals taper to a point.

Met. Cat. No. 384. *Fig.* 69:5, Field No. 278/9, Loc 107, 109, 110, Square E-F 10–11. Bronze.
Ring (Diam. 12.4, Th. 1.8mm.). Circular cross-section. One terminal is cut at an angle and the other tapers to a point.

Met. Cat. No. 385. *Fig.* 69:6, Field No. 269/16, Loc 107, Square A-B 8–11. Copper.
Ring (Diam. 12.7, Th. 1.6mm.). Circular cross-section. One terminal is broken and the other tapers to a point.

Met. Cat. No. 386. *Fig.* 69:7, *Pl.* 128:9, Field No. 134/1, Loc 110, 111, Square A-B-C 13–14. Bronze.
Ring (Diam. 12.95, Th. 1.5mm.). Circular cross-section. One terminal is cut at an angle and the other tapers to a point.

Met. Cat. No. 387. *Fig.* 69:8, Field No. 277/15, Loc 107, 110, Square C-D 10–11. Bronze.
Ring (Diam. 14.2, Th. 1.95mm.). Circular cross-section. Both terminals taper to a point.

Met. Cat. No. 388. *Fig.* 69:9, Field No. 332/4, Loc 109, Square E 12. Copper.
Ring (Frag.) (Th. 1.9mm.). Circular cross-section. One terminal is broken and the other tapers to a point.

Met. Cat. No. 389. *Fig.* 69:10, Field No. 283/211, Loc 106, 107, Square D 7–9. Copper.
Ring (Frag.) (Diam. 12.2, Th. 2.2mm.). Circular cross-section. One terminal is broken and the other tapers to a point.

Met. Cat. No. 390. *Fig.* 69:11, Field No. 349/25, Loc 109, Square G 11. Copper.
Ring (Frag.) (Diam. 12.5, Th. 1.8mm.). Circular cross-section. One terminal is broken and the other tapers to a point.

Met. Cat. No. 391. *Fig.* 69:12, Field No. 283/224, Loc 106, 107, Square D 7–9. Bronze.
Ring (Frag.) (Diam. 17.5, Th. 2.2mm.). Circular cross-section. One terminal is broken and the other tapers to a point.

Met. Cat. No. 392. *Fig.* 69:13, Field No. 279/288, Loc 106, 107, Square E-F 7–9. Copper.
Ring (Frag.) (Diam. 17.5, Th. 1.9mm.). Circular cross-section. One terminal is broken and the other tapers to a point.

Met. Cat. No. 393. *Fig.* 70:1, Field No. 5/1, Loc 112, Square A-B 15–16. Bronze.
Ring (Diam. 12.7, Th. 1.25mm.). Circular cross-section. Both terminals taper to a point (see Chapter III, 11).

Met. Cat. No. 394. *Fig.* 70:2, Field No. 228/2, Loc 110, Square B-C-D 13–14. Bronze.
Ring (Diam. 16.5, Th. 2.8mm.). Circular cross-section. Both terminals taper to a point.

Met. Cat. No. 395. *Fig.* 70:3, Field No. 51/5, Loc 101, Square E-F 15–16. Bronze.
Ring (Diam. 17.7, Th. 2.2mm.). Circular cross-section. Both terminals taper to a point (see Chapter III, 11).

Met. Cat. No. 396. *Fig.* 70:4, Field No. 283/3, Loc 106, 107, Square D 7–9. Bronze.
Two rings (Ring 'A': Diam. 16.8, Th. 1.3 mm; Ring 'B': Diam. 15.0, Th. 2.0mm.). Ring A has an oval cross-section and ring B has a circular cross-section. Both terminals of ring A taper to a point; one terminal of ring B tapers to a point and the other is hammered flat and rounded off. The two rings were found as shown.

Met. Cat. No. 397. *Fig.* 70:5, Field No. 278/119, Loc 107, Square E-F 10–11. Bronze.
Ring (Th. 1.9mm.). Circular cross-section. Both terminals taper to a point.

Met. Cat. No. 398. *Fig.* 70:6, Field No. 372/6, Loc 111, Square B 11. Bronze.
Ring (Diam. 15.3, Th. 3.0mm.). Circular cross-section. One terminal is broken and the other tapers to a point.

Met. Cat. No. 399. *Fig.* 70:7, Field No. 338/32, Loc 109, Square G 11. Bronze.
Ring (Diam. 17.3, Th. 1.9mm.). Circular cross-section. Both terminals taper to a point.

Met. Cat. No. 400. *Fig.* 70:8, *Pl.* 128:12, Field No. 51/3, Loc 101, Square E-F 15–16. Bronze.
Ring (Diam. 15.0, Th. 2.0mm.). Circular cross-section. Both terminals taper to a point.

Met. Cat. No. 401. *Fig.* 70:9, Field No. 339/4, Loc 109, Square H-I 11. Bronze.
Ring (Diam. 20.0, Th. 2.0mm.). Circular cross-section. One terminal is broken and the other is hammered flat from two sides.

Met. Cat. No. 402. *Fig.* 70:10, Field No. 283/16, Loc 106, 107, Square D 7–9. Copper.
Ring (Diam. 18.9, Th. 1.6mm.). Circular cross-section. Both terminals are broken.

Met. Cat. No. 403. *Fig.* 70:11, Field No. 46/1, Loc 101, Square E-F 15–16. Copper.
Ring (Diam. 17.65, Th. 2.0mm.). Circular cross-section. Both terminals are broken.

Met. Cat. No. 404. *Fig.* 70:12, Field No. 51/6, Loc 101, Square E-F 15–16. Bronze.
Ring (Diam. 21.7, Th. 2.2mm.). Circular cross-section. Both terminals are broken (see Chapter III, 11).

Met. Cat. No. 405. *Fig.* 70:13, Field No. 211/3, Loc 101, Square G 15–17. Copper.
Ring (Diam. 14.1, Th. 2.2mm.). Circular cross-section. Both terminals are broken (see Chapter III, 11).

Met. Cat. No. 406. *Fig.* 70:14, Field No. 278/116, Loc 107, 109, 110, Square E-F 10–11. Bronze.
Ring (Diam. 13.6, Th. 2.3mm.). Circular cross-section. Both terminals are broken (see Chapter III, 11).

Met. Cat. No. 407. *Fig.* 71:1, Field No. 201/25, Loc 101, Square D-E 15–16. Bronze.
Ring (?) (Frag.) (Th. 2.3mm.). Circular cross-section. Both terminals are broken.

Met. Cat. No. 408. *Fig.* 71:2, Field No. 314/48A, Loc 109, Square F 13–14. Copper.
Ring (Frag.) (Th. 1.9mm.). Circular cross-section. Both terminals are broken.

Met. Cat. No. 409. *Fig.* 71:3, Field No. 264/5, Loc 103, Square J-K 7. Copper.
Ring (?) (Frag.) (Th. 3.0mm.). Circular cross-section. Both terminals are broken.

Met. Cat. No. 410. *Fig.* 71:4, Field No. 46/2, Loc 101, Square E-F 15–16. Bronze.
Ring (Frag.) (Th. 1.7mm.). Circular cross-section. Both terminals are broken.

Met. Cat. No. 411. *Fig.* 71:5, Field No. 264/4, Loc 103, Square J-K 7. Copper.
Ring (?) (Frag.) (Th. 3.0mm.). Circular cross-section. Both terminals are broken.

Met. Cat. No. 412. *Fig.* 71:6, Field No. 245/63, Loc 107, Square A-B 8–10. Copper.
Ring (Frag.) (Th. 1.9mm.). Circular cross-section. Both terminals are broken.

Met. Cat. No. 413. *Fig.* 71:7, Field No. 283/19, Loc 106, 107, Square D 7–9. Copper.
Ring (Frag.) (Th. 1.8mm.). Circular cross-section. Both terminals are broken.

Met. Cat. No. 414. *Fig.* 71:8, Field No. 204/3, Loc 109, Square E 13–14. Copper.
Ring (Frag.) (Th. 3.3mm.). Circular cross-section. Both terminals are broken.

Met. Cat. No. 415. *Fig.* 71:9, Field No. 283/18, Loc 106, 107, Square D 7–9. Copper.
Ring (Diam. 13.5, Th. 1.85mm.). Circular cross-section. Both terminals are broken.

Met. Cat. No. 416. *Fig.* 71:10, Field No. 234/137, Loc 101, Square H 15–18. Copper.
Ring (Diam. 16.5, Th. 2.9mm.). Circular cross-section. Both terminals are broken.

Met. Cat. No. 417. *Fig.* 71:11, Field No. 319/16, Loc 106, 108, Square G 7–10. Bronze.
Ring (Diam. 10.1, Th. 1.1mm.). Circular cross-section. Both terminals are broken.

Met. Cat. No. 418. *Fig.* 71:12, Field No. 319/19, Loc 106, 108, Square G 7–10. Bronze.
Ring (Diam. 4.9, W. 2.6, Th. 1.65mm.). Oval cross-section. One terminal is cut straight and the other tapers to a point.

Met. Cat. No. 419. *Fig.* 71:13, Field No. 211/4, Loc 101, Square G 15–17. Copper.
Ring (Diam. 5.6, W. 2.8, Th. 1.5mm.). Oval cross-section. One terminal is cut straight and the other is rounded off.

Met. Cat. No. 420. *Fig.* 71:14, Field No. 207/2, Loc 102, Square I-J 12–13. Bronze.
Ring (Diam. 5.0, W. 1.9, Th. 1.1mm.). Oval cross-section. One terminal is cut straight and the other tapers to a point.

Met. Cat. No. 421. *Fig.* 71:15, Field No. 347/10, Loc 108, Square G-H 9. Copper.
Ring (Diam. 5.6, W. 1.9, Th. 1.3mm.). Oval cross-section. One terminal is cut straight and the other tapers to a point.

Met. Cat. No. 422. *Fig.* 71:16, Field No. 264/3, Loc 103, Square J-K 7. Copper.
Ring (Diam. 8.8, W. 2.9, Th. 1.3mm.). Oval cross-section. One terminal tapers to a point and the other tapers slightly and is broken off.

Met. Cat. No. 423. *Fig.* 71:17, Field No. 283/17, Loc 106, 107, Square D 7–9. Bronze.
Ring (Diam. 3.8, W. 1.5, Th. 0.9mm.). Oval cross-section. One terminal is cut straight and the other tapers to a point.

Met. Cat. No. 424. *Fig.* 71:18, Field No. 260/183, Loc 102, Square I-J-K 12. Copper.
Ring (Diam. 5.7, W. 2.2, Th. 1.1mm.). Oval cross-section. One terminal is cut straight and the other tapers to a point.

Met. Cat. No. 425. *Fig.* 71:19, Field No. 278/117, Loc 107, 109, Square E-F 10–11. Bronze.
Ring (Diam. 8.7, W. 3.4, Th. 1.45mm.). Oval cross-section. One terminal is cut straight and the other tapers to a point.

Met. Cat. No. 426. *Fig.* 71:20, Field No. 14/1, Loc 107, Square E-F 8–9. Copper.
Ring (Diam. 5.5, W. 2.4, Th. 1.1mm.). Oval cross-section. One terminal is cut straight and the other tapers to a point.

Met. Cat. No. 427. *Fig.* 71:21, Field No. 279/32, Loc 106, 107, Square E-F 7–9. Copper.
Ring (Diam. 9.1, W. 2.6, Th. 1.6mm.). Oval cross-section. Both terminals taper to a point.

Met. Cat. No. 428. *Fig.* 71:22, Field No. CD/16–17/1, Loc 101, Square C-D 16–17. Bronze.
Ring (Diam. 8.5, W. 2.1, Th. 1.0mm.). Oval cross-section. Both terminals taper to a point.

Met. Cat. No. 429. *Fig.* 71:23, Field No. 278/127, Loc 107, 109, Square E-F 10–11. Bronze.
Ring (Diam. 3.5, W. 1.5, Th. 1.0mm.). Oval cross-section. Both terminals are broken.

Met. Cat. No. 430. *Fig.* 71:24, Field No. 258/3, Loc 101, Square H 6–7. Bronze.
Ring (Diam. 2.8, W. 2.4, Th. 1.4mm.). Oval cross-section. Both terminals are broken.

Met. Cat. No. 431. *Fig.* 71:25, Field No. 278/129, Loc 107, 109, Square E-F 10–11. Bronze.
Ring (Diam. 4.3, W. 1.8, Th. 1.2mm.). Oval cross-section. Both terminals are broken (see Chapter III, 11).

Met. Cat. No. 432. *Fig.* 71:26, Field No. 237/6, Loc 110, Square B-C-D 13–14. Bronze.
Ring (Diam. 6.4, W. 2.0, Th. 1.6mm.). Oval cross-section. Both terminals are broken.

Met. Cat. No. 433. *Fig.* 71:27, Field No. 154/2, Loc 110, Square D-E 14–15. Bronze.
Ring (Diam. 9.1, W. 2.0, Th. 2.0mm.). Oval cross-section. Both terminals are broken.

Met. Cat. No. 434. *Fig.* 72:1, Field No. 343/89, Loc 109, Square F 12. Bronze.
Ring (Diam. 11.5, W. 1.9, Th. 1.4mm.). Oval cross-section. Both terminals taper to a point.

Met. Cat. No. 435. *Fig.* 72:2, Field No. 348/5, Loc 107, Square D-E 10. Copper.
Ring (Diam. 21.1, W. 2.1, Th. 2.0mm.). Oval cross-section. Both terminals taper to a point.

Met. Cat. No. 436. *Fig.* 72:3, Field No. 245/61, Loc 107, Square A-B 8–10. Bronze.
Ring (Diam. 14.0, W. 2.0, Th. 1.4mm.). Oval cross-section. Both terminals taper to a point.

Met. Cat. No. 437. *Fig.* 72:4, Field No. 277/16, Loc 107, 110, Square A-B 10–11. Copper.
Ring (Diam. 18.3, W. 2.1, Th. 1.1mm.). Oval cross-section. Both terminals taper to a point.

Met. Cat. No. 438. *Fig.* 72:5, Field No. 269/13, Loc 107, Square A-B 8–10. Bronze.
Ring (Diam. 15.2, W. 2.1, Th. 1.5mm.). Oval cross-section. Both terminals taper to a point.

Met. Cat. No. 439. *Fig.* 72:6, Field No. 174/1, Loc 101, 112, Square C-D 17–18. Bronze.
Ring (W. 1.8, Th. 1.0mm.). Oval cross-section. Both terminals taper to a blunt point.

Met. Cat. No. 440. *Fig.* 72:7, Field No. 278/120, Loc 107, 109, 110, Square E-F 10–11. Bronze.
Ring (Diam. 14.0, W. 2.2, Th. 1.7mm.). Oval cross-section. Both terminals taper to a point.

Met. Cat. No. 441. *Fig.* 72:8, Field No. 319/23, Loc 106, 108, Square G 7–10. Bronze.
Ring (Diam. 13.9, W. 1.9, Th. 1.7mm.). Oval cross-section. Both terminals taper to a point.

Met. Cat. No. 442. *Fig.* 72:9, Field No. 278/130, Loc 107, 109, 110, Square E-F 10–11. Bronze.
Ring (Diam. 16.6, W. 2.1, Th. 1.4mm.). Oval cross-section. One terminal is broken and the other tapers to a point.

Met. Cat. No. 443. *Fig.* 72:10, Field No. 343/88, Loc 109, Square F 12. Bronze.
Ring (Diam. 15.4, W. 2.1, Th. 1.0mm.). Oval cross-section. One terminal is broken and the other tapers to a point.

Met. Cat. No. 444. *Fig.* 72:11, Field No. 283/210, Loc 106, 107, Square D 7–9. Bronze.
Ring (Diam. 12.3, W. 2.8, Th. 1.9mm.). Oval cross-section. One terminal is broken and the other tapers to a point.

Met. Cat. No. 445. *Fig.* 72:12, Field No. 277/18, Loc 107, 110, Square C-D 10–11. Bronze.
Ring (W. 1.5, Th. 1.2mm.). Oval cross-section. One terminal is broken and the other tapers to a point.

Met. Cat. No. 446. *Fig.* 73:1, Field No. 348/1, Loc 107, Square D-E 10. Copper.
Ring (Diam. 16.7, W. 2.9, Th. 1.3mm.). Oval cross-section. Both terminals taper to a point.

Met. Cat. No. 447. *Fig.* 73:2, Field No. 278/118, Loc 107, Square E-F 10–11. Bronze.
Ring (Diam. 11.5, W. 2.1, Th. 1.7mm.). Oval cross-section. Both terminals taper to a point (see Chapter III, 11).

Met. Cat. No. 448. *Fig.* 73:3, Field No. 278/8, Loc 107, 109, 110, Square E-F 10–11. Copper.
Ring (Diam. 18.4, W. 2.9, Th. 1.35mm.). Oval cross-section. Both terminals taper to a point.

Met. Cat. No. 449. *Fig.* 73:4, Field No. 279/36, Loc 106, 107, Square E-F 7–9. Bronze.
Ring (Diam. 14.0, W. 3.0, Th. 1.6mm.). Oval cross-section. Both terminals taper to a point (see Chapter III, 11).

Met. Cat. No. 450. *Fig.* 73:5, Field No. 280/6, Loc 106, Square C-D-E 6–7. Bronze.
Ring (Diam. 19.0, W. 3.0, Th. 1.0mm.). Oval cross-section. One terminal tapers to a point and the other is broken.

Met. Cat. No. 451. *Fig.* 73:6, Field No. 279/288B, Loc 106, 107, Square E-F 7–9. Copper.
Ring (Diam. 13.0, W. 2.0, Th. 1.4mm.). Oval cross-section. One terminal tapers to a point and the other is broken.

Met. Cat. No. 452. *Fig.* 73:7, Field No. 234/139, Loc 101, Square H 15–18. Copper.
Ring (W. 2.5, Th. 1.3mm.). Oval cross-section. One terminal tapers to a point and the other is broken.

Met. Cat. No. 453. *Fig.* 73:8, Field No. 277/9, Loc 107, 110, Square C-D 10–11. Copper.
Ring (Diam. 12.7, W. 3.0, Th. 1.8mm.). Oval cross-section. One terminal tapers to a point and the other is broken.

Met. Cat. No. 454. *Fig.* 73:9, Field No. 345/11, Loc 107, Square D-E-F-G 6–9. Copper.
Ring (Diam. 20.6, W. 3.3, Th. 1.55mm.). Oval cross-section. Both terminals taper to a point.

Met. Cat. No. 455. *Fig.* 73:10, Field No. 280/92, Loc 106, Square C-D-E 6–7. Bronze.
Ring (?) (Frag.) (W. 2.2, Th. 1.1mm.). Oval cross-section. One terminal tapers to a point and the other is broken.

Met. Cat. No. 456. *Fig.* 73:11, Field No. 332/2, Loc 109, Square E 12. Copper.
Ring (Dia 15.8, W. 3.8, Th. 1.6mm.). Oval cross-section. One terminal tapers to a point and the other is broken.

Met. Cat. No. 457. *Fig.* 73:12, Field No. 288/5, Loc 101, Square C-D-E 18. Bronze.
Ring (Dia 17.0, W. 2.5, Th. 1.5mm.). Oval cross-section. Both terminals are broken.

Met. Cat. No. 458. *Fig.* 73:13, Field No. 97/2, Loc 101, 112, Square C-D 15–16. Copper.
Ring (Dia 19.6, W. 1.7, Th. 1.6mm.). Oval cross-section. Both terminals are broken.

9 Spatulas and bracelets (*Figs.* 74–75)

Met. Cat. No. 459. *Fig.* 74:1, *Pl.* 127:4, Field No. 176/2, Loc 102, Square J-K 13–14. Bronze.
Spatula (L. 39.5, W. 10.6, Th. 1.2mm.). Rectangular cross-section. Folded socket. Chevrons are incised into the spatula.

Met. Cat. No. 460. *Fig.* 74:2, *Pl.* 127:5, Field No. 319/9, Loc 106, 108, Square G 7–10. Bronze.
Spatula (L. 56.2, W. 6.3, Th. 2.0 mm Th. of tang

2.0mm.). The tang has a circular cross-section and the spatula has a roughly oval cross-section with one side ridged. Zig-zag design incised into the face of the spatula.

Met. Cat. No. 461. *Fig.* 74:3, Field No. 283/28, Loc 106, 107, Square D 7–9. Bronze.
Pin or awl (L. 23.3, W. 2.6, Th. 1.5mm.). Rectangular cross-section. Both terminals taper to a point

Met. Cat. No. 462. *Fig.* 74:4, Field No. 269/8, Loc 107, Square A-B 8–10. Bronze.
Pin or awl (L. 39.0, W. 3.4, Th. 2.0mm.). Rectangular cross-section. Both terminals taper to a point

Met. Cat. No. 463. *Fig.* 74:5, Field No. 128/1, Loc 101, Square I-J 17–18. Copper.
Spatula (L. 80.4, W. 5.2, Th. 1.4 mm Th. of tang 2.6mm.). Spatula has a rectangular cross-section and the tang has a circular cross-section. The terminal of the tang tapers slightly to a blunt point.

Met. Cat. No. 464. *Fig.* 74:6, *Pl.* 127:6, Field No. 221/3, Loc 107, Square E-F 7–8. Copper.
Spatula (L. 81.9, W. 8.0, Th. 0.7 mm Th. of tang 3.0mm.). Spatula has a circular cross-section and the tang has a circular cross-section. The tang tapers to a blunt point (see Chapter III, 11).

Met. Cat. No. 465. *Fig.* 74:7, Field No. 134/2, Loc 110, 111, Square A-B-C 13–14. Bronze.
Point (L. 28.4, W. 4.0, Th. 2.4mm.). Rectangular cross-section. One terminal tapers to a point and the other tapers and then is broken off (see Chapter III, 11).

Met. Cat. No. 466. *Fig.* 74:8, Field No. 234/138, Loc 101, Square H 15–18. Copper.
Point (L. 38.9, W. 3.5, Th. 3.0mm.). Rectangular cross-section. Folded socket.

Met. Cat. No. 467. *Fig.* 74:9, Field No. 128/2, Loc 101, Square I-J 17–18. Bronze.
Point (L. 38.8, W. 2.9, Th. 2.4mm.). Rectangular cross-section. Folded socket.

Met. Cat. No. 468. *Fig.* 74:10, Field No. 51/24, Loc 101, Square E-F 15–16. Bronze.
Bracelet (Dia 39.3, Th. 2.6mm.). Circular cross-section. Both terminals taper slightly to a blunt point.

Met. Cat. No. 469. *Fig.* 74:11, *Pl.* 127:9, Field No. 297/1, Loc 101, Square E 19. Copper.
Spatula (L. 140.1, W. 9.5, Th. 1.2 mm. W. of tang 3.0, Th. of tang 2.2mm.). The spatula has a rectangular cross-section, and the tang an oval cross-section.

Met. Cat. No. 470. *Fig.* 74:12, Field No. 51/23, Loc 101, Square E-F 15–16. Bronze.
Catch type bracelet (Th. 2.1mm.). Circular cross-section. Both terminals are broken.

Met. Cat. No. 471. *Fig.* 74:13, Field No. 51/22, Loc 101, Square E-F 15–16. Copper.
Bracelet (?) Oval cross-section. Both terminals are broken.

Met. Cat. No. 472. *Fig.* 75:1, Field No. D–14/1, Loc 110, Square D 14. Copper.
Spatulated wire (L. 73.5, W. 6.2, Th. 1.5 mm. Th. of rod 3.5mm.). The 'spatula' has a rectangular cross-section and the wire a circular cross-section. The wire tapers to a blunt point and the 'spatula' is cut straight.

Met. Cat. No. 473. *Fig.* 75:2, Field No. 278/137, Loc 107, 109, 110, Square E-F 10–11. Copper.
Spatulated wire (W. 2.0, Th. 0.7 mm. Th. of rod 1.2mm.). The 'spatula' has a rectangular cross-section and the wire a circular cross-section. The terminal of the 'spatula' is rounded off and the terminal of the wire is broken.

Met. Cat. No. 474. *Fig.* 75:3, Field No. 260/185, Loc 102, Square I-J-K 12. Copper.
Spatulated wire (W. 4.4, Th. 1.1 mm. Th. of rod 2.2mm.). The 'spatula' has a rectangular cross-section and the wire a circular cross-section. The terminal of the 'spatula' is rounded off and the terminal of the rod is broken.

Met. Cat. No. 475. *Fig.* 75:4, Field No. 225/17, Loc 107, Square C-D 9–11. Copper.
Spatula (W. 11.7, Th. 1.7 mm. W. of rod 6.3 , Th. of bar 4.6mm.). The spatula and rod both have rectangular cross-sections. The terminal of the spatula is rounded off and the terminal of the rod is broken.

Met. Cat. No. 476. *Fig.* 75:5, Field No. 349/28, Loc 109, Square G 11. Bronze.
Spatulated wire ('Spatula' W. 2.4, Th. 0.9 mm. W. of wire 3.5 , Th. of bar 2.5mm.). Both 'spatula' and wire have rectangular cross-sections. The terminal of the spatula is rounded off and the terminal of the wire is broken.

Met. Cat. No. 477. *Fig.* 75:6, Field No. 245/60, Loc 107, Square A-B 8–10. Bronze.
Spatula (Spatula L. 40.6, W. 7.8, Th. 1.5 mm. Th. of rod 3.7mm.). The spatula has a rectangular cross-section and the rod has a circular cross-section. The terminal of the spatula is cut straight and the terminal of the rod tapers to a blunt point.

Met. Cat. No. 478. *Fig.* 75:7, Field No. 78/1, Loc 101, Square J-K 16–17. Bronze.
Spatula (Spatula L. 38.2, W. 5.2, Th. 1.3 mm. Rod W. 2.6, Th. 3.3mm.). The spatula has a rectangular cross-section and the rod has a diamond shaped cross-section. The terminal of the spatula is rounded off and the terminal of the rod tapers to a blunt point (see Chapter III, 11).

Met. Cat. No. 479. *Fig.* 75:8, Field No. 83/1, Loc 101, Square G-H 18–19. Bronze.
Spatula (Spatula L. 39.2, W. 4.4, Th. 1.4 mm. Rod W. 3.4, Th. 3.7mm.). The spatula has a rectangular cross-section and the rod has a diamond shaped cross-section. The terminal of the spatula is rounded off (?) and the terminal of the rod tapers to a blunt point.

Met. Cat. No. 480. *Fig.* 75:9, *Pl.*126:5, Field No. 343/1, Loc 109, Square F 12. Bronze.
Spatula (Spatula L. 93.4, W. 19.6, Th. 2.0 mm. Th. of tang 4.7mm.). The spatula has a rectangular cross-section and the tang has a circular cross-section. The terminal of the spatula is rounded off and the terminal of the tang tapers to a blunt point. Parallel lines are incised into the face of the spatula.

Met. Cat. No. 481. *Fig.* 75:10, Field No. 260/180, Loc 102, Square I-J-K 12. Bronze.
Spatula (Point W. 9.5, Th. 1.8 mm. Tang W. 2.8, Th. 1.7mm.). Both point and tang have oval cross-sections. The terminal of the tang is broken (see Chapter III, 11).

Met. Cat. No. 482. *Fig.* 75:11, Field No. 339/17, Loc 109, Square H-I 11. Bronze.
Spatula (Point W. 11.7, Th. 2.6mm.). The point has a triangular cross-section and a midrib on one side. The tang end of the point is broken off.

Met. Cat. No. 483. *Fig.* 75:12, Field No. 292/65, Wall 1. Copper.
Spatula (Point W. 9.3, Th. 1.3 mm. Tang W. 5.5, Th. 4.2mm.). The point has an oval cross-section and the tang has a rectangular cross-section. The terminal of the tang is broken.

Met. Cat. No. 484. *Fig.* 75:13, Field No. 51/19, Loc 101, Square E-F 15–16. Copper.
Spatula (Point W. 15.0, Th. 2.7 mm. Tang W. 5.4, Th. 4.3mm.). The point has a diamond shaped cross-section with midribs on both sides and the tang has a rectangular cross-section. The terminal of the tang is broken.

Met. Cat. No. 485. *Fig.* 75:14, Field No. 205/18, Loc 101, 111, Square B-C 13–14. Bronze.
Spatula (Point W. 9.0, Th. 2.7mm.). The point has a diamond shaped cross-section with a midrib on both sides. The tang and bottom part of the point are broken off.

Met. Cat. No. 486. *Fig.* 75:15, *Pl.* 127:11, Field No. T–200/a, Surface (1967 Survey). Copper.
Projectile point (Point W. 12.8, Th. 1.4mm.). The point has an oval cross-section. The projectile has a folded socket.

10 Rods, ferrules (*Fig.* 76)

Met. Cat. No. 487. *Fig.* 76:1, Field No. 319/285, Loc 108, Square G 7–10. Bronze.
Fragment (L. 10.0, W. 6.0 mm. Wire W. 2.6, Th. 1.2mm.). The fragment is a twisted wire which has a wrap-around haft, wrapped around the rod part of the fragment. The rod has an oval cross-section.

Met. Cat. No. 488. *Fig.* 76:2, Field No. 318/1, Loc 109, Square E 12. Copper.
Fragment (L. 19.2, W. 9.0 mm. Wire W. 3.9, Th. 0.9mm.). The fragment consists of two copper wires, one of which is bent. The two wires are attached.

Met. Cat. No. 489. *Fig.* 76:3, Field No. 339/21, Loc 109, Square H-I 11. Bronze.
Barbed rod (W. 3.4, Th. 2.3mm.). Both terminals of the rod are broken.

Met. Cat. No. 490. *Fig.* 76:4, Field No. 334/7, Loc 109, Square F 12. Copper.
Ring with attached crook-shaped rod (Ring Diam. 11.0, W. 2.1, Th. 1.5 mm. Rod Th. 1.5mm.). The ring has an oval cross-section and the rod a circular cross-section. Both terminals of the ring taper to a blunt point. The terminal of the crook is cut straight and the terminal of the rod is broken.

Met. Cat. No. 491. *Fig.* 76:5, Field No. 295/200, Loc 101, Square F 17–18. Copper.
Spatulated fragment with a fragment of wire wrapped round it (Spatula L. 26.0, W. 3.0, Th. 0.7 mm. Tang Th. 1.5 mm. Wire W. 3.2, Th. 1.7mm.). The terminal of the 'spatula' is rounded off and the terminal of the tang tapers to a blunt point. Both terminals of the wire are broken. The 'spatula' has a rectangular cross-section, the tang has a circular cross-section and the wire an oval cross-section.

Met. Cat. No. 492. *Fig.* 76:6, Field No. 226/1, Loc 104, Square C-D-E-F 2–5. Bronze.
Fragment of a spatula (W. 15.2, Th. 2.2mm.). The spatula has a triangular cross-section with a midrib running down one side.

Met. Cat. No. 493. *Fig.* 76:7, *Pl.* 128:2, Field No. 51/15, Loc 101, Square E-F 15–16. Copper.
Ferrule (Diam. 33.8, W. 29.0, Th. 1.4mm.). Rectangular cross-section. Both terminals are squared off.

Met. Cat. No. 494. *Fig.* 76:8, Field No. 337/25, Loc 102, Square I-J-K-L 11. Copper.
Fragment of ferrule with rivet hole (L. 23.8, W. 14.0, Th. 1.6, Diam. of hole 5.2mm.). Rectangular cross-section.

Met. Cat. No. 495. *Fig.* 76:9, *Pl.* 128:1, Field No. 51/14, Loc 101, Square E-F 15–16. Bronze.
Ferrule with rivet hole (Diam. 28.0, W. 12.2, Th. 1.0, Diam. of hole 2.4mm.). Rectangular cross-section. Both terminals are squared off (see Chapter III, 11).

Met. Cat. No. 496. *Fig.* 76:10, Field No. 51/100, Loc 101, Square E-F 15–16. Copper.
Ferrule with fragment of implement still *in situ* (Diam. 17.4, W. 20.9, Th. 5.6mm.). The ferrule has a rectangular cross-section. The remaining fragment of the tool is badly corroded (see Chapter III, 11).

Met. Cat. No. 497. *Fig.* 76:11, *Pl.* 128:3, Field No. 51/36, Loc 101, Square E-F 15–16. Bronze.
Ferrule with rivet *in situ* (Diam. 22.6, W. 33.0, Th. 1.3 mm. Rivet L. 24.3, Th. 3.2mm.). The ferrule has a rectangular cross-section and the rivet has a circular cross-section (see Chapter III, 11).

11 Spirals, hooks, punches, a balance beam (*Fig.* 77)

Met. Cat. No. 498. *Fig.* 77:1, Field No. 257/273, Loc 102, Square I-J-K 13. Copper.
Spiral (Diam. 6.6, Th. 1.4mm.). The metal has a circular cross-section. Both terminals are cut straight.

Met. Cat. No. 499. *Fig.* 77:2, Field No. 208/2, Loc 102, Square I-J 14–15. Copper.
Spiral (Diam. 8.15, Th. 2.2mm.). The metal has a circular cross-section. Both terminals are cut straight.

Met. Cat. No. 500. *Fig.* 77:3, Field No. 348/6, Loc 107, Square D-E 10. Copper.
Spiral (Diam. 7.0, Th. 1.9mm.). The metal has a circular cross-section. Both terminals are cut straight.

Met. Cat. No. 501. *Fig.* 77:4, Field No. 204/6, Loc 109, Square E 13–14. Copper.
Spiral (Diam. 10.3, Th. 2.0mm.). The metal has a circular cross-section. The inner terminal is cut straight and the outer one is broken.

Met. Cat. No. 502. *Fig.* 77:5, Field No. 319/14, Loc 106, 108, Square G 7–10. Copper.
Figure Eight Spiral (L. 12.7, W. 7.7, Th. 1.1mm.). The metal has a circular cross-section. Both terminals are cut straight (see Chapter III, 11).

Met. Cat. No. 503. *Fig.* 77:6, Field No. 283/209, Loc 106, 107, Square D 7–9. Bronze.
Spiral Ring (Diam. 4.0, W. 6.6, Th. 2.0mm.). The metal has an oval cross-section. Both terminals are cut straight.

Met. Cat. No. 504. *Fig.* 77:7, *Pl.* 128:20, Field No. 49/1, Loc 101, Square D-E 15–16. Bronze.
Spiral (Diam. 2.3, W. 3.7, Th. 1.7mm.). The metal has an oval cross-section. The outer terminal is broken and the inner terminal is cut straight.

Met. Cat. No. 505. *Fig.* 77:8, Field No. 257/270, Loc 102, Square I-J-K 13. Copper.
Decorated fragment (L. 33.9, W. 9.3, Th. 4.8mm.). Rectangular cross-section. Cross-hatched lines are incised into the metal (see Chapter III, 11).

Met. Cat. No. 506. *Fig.* 77:9, Field No. 257/271, Loc 102, Square I-J-K 13. Copper.
Bar (W. 6.1, Th. 3.0mm.). Plano-convex cross-section. Both terminals are broken.

Met. Cat. No. 507. *Fig.* 77:10, Field No. 311/185, Loc 101, Square F 15–16. Bronze.
Bar (W. 5.0, Th. 3.4mm.). Diamond shaped cross-section. Both terminals are broken (see Chapter III, 11).

Met. Cat. No. 508. *Fig.* 77:11, Field No. 344/8, Loc 110, Square B-C 11–12. Bronze.
S-shaped hook (L. 43.0, W. 4.1, Th. 1.9mm.). Rectangular cross-section. Both terminals are cut straight.

Met. Cat. No. 509. *Fig.* 77:12, Field No. 145/2, Loc 101, Square G-H 17–18. Bronze.
Fragment (L. 22.4, W. 9.2, Th. 3.0mm.). Rectangular cross-section (see Chapter III, 11).

Met. Cat. No. 510. *Fig.* 77:13, Field No. 339/18, Loc 109, Square H-I 11. Copper.
Chisel (?) (Each side of bar 3.1mm.). Square cross-section (see Chapter III, 11).

Met. Cat. No. 511. *Fig.* 77:14, Field No. 51/16, Loc 101, Square E-F 15–16. Bronze.
Punch (Th. 6.8mm.). Roughly circular cross-section. The working point tapers to 3.1 mm (see Chapter III, 11).

Met. Cat. No. 512. *Fig.* 77:15, Field No. 283/212, Loc 106, 107, Square D 7–9. Bronze.
Punch (sides 7.7mm.). Square cross-section. The working point is 5.8 mm.

Met. Cat. No. 513. *Fig.* 77:16, Field No. 51/20, Loc 101, Square E-F 15–16. Copper.
Sheet fragment (L. 63.6, W. 20.5, Th. 1.0mm.). Rectangular cross-section (see Chapter III, 11).

Met. Cat. No. 514. *Fig.* 77:17, *Pl.* 127:1, Field No. 51/21, Loc 101, Square E-F 15–16. Bronze.
Balance beam (?) with crooked terminals on which a ring is attached to each crook (Bar L. 178.8, W. 4.0, Th. 2.5 mm. Crook W. 4.4, Th. 1.4 mm. Ring A Th. 2.1 mm. Ring B W. 5.2, Th. 0.8mm.). The bar and crooks have rectangular cross-sections. Ring A has a circular cross-section and Ring B a rectangular cross-section. The terminals of the crooks are rounded off. One terminal of Ring A is broken and the other tapers slightly to a blunt point. Both terminals of Ring B are broken.

Notes

1. Part of the mould was found preserved between the legs of the figurine and photographed at the site (*Pl.* 126:6). It was, however, not recognised as such by the conservator at the laboratory of the Institute of Archaeology, Tel Aviv University, and was removed before it could be investigated.
2. Many representations of serpents have been found in Egypt, dating from prehistoric times to the Late New Kingdom (L.E. V, 1984, 643–52), but only the cobra was made of bronze. The other types of serpents were made of flint, carnelian, haematite, red limestone, ivory, glass, red or yellow jasper, faience, agate and (rarely) gold.
3. Most of the iron objects found in Timna contain copper. This fact led to a series of investigations into the origin and method of production of this iron. There can be little doubt that the iron was produced in Timna as a by-product of copper smelting. We assume that iron found in many Bronze Age sites in the Old World originated in a copper smelting furnace and this was most probably the way iron smelting was first discovered (see the detailed investigation in *Researches in the Arabah*, Vol. II).
4. Some of the lead objects encountered at the Temple were found to contain copper (see Chapter III, 8 and 11). Subsequent research showed that these lead objects originated from the Timna copper smelter (see Gale, 1984, and also Vol. II of this publication).

8. THE COMPOSITION OF THE METAL FINDS

Introduction

This is a report on the quantitative analysis of 346 samples of metalwork and the qualitative analysis of a further 145. All but 31 of the quantitative analyses are on metal from the Timna Temple. The remainder are from the large contemporary adjacent smelting camps, Sites 2, 3, 30 and 34 and from the smaller Sites 186 and F2. Also included are the analyses of an ingot from the Roman-Early Islamic smelting camp at Beer Ora (Site 28) south of Timna. The analysis of the ingot found in the Arabah (Rothenberg, 1972, 69), is included as comparative material.

It has been decided to publish together all the available analytical data from the Arabah sites, and thus the discussion will also include the non-Temple material. The bulk of the material analysed is of copper, sometimes alloyed with tin and often containing varying amounts of iron up to several percent. The artifacts of other metals found at the Temple include the following: 24 artifacts of iron, eight of lead, one of silver and two of tin. The presence of copper containing iron, and of iron itself, in areas of copper smelting at this period in the Eastern Mediterranean, has important implications for the origins of iron technology, which will be discussed below. Similarly, the tin objects are of great importance as they are the only metallic tin so far known from a Near Eastern smelting site and are among the few pieces of tin to be scientifically excavated anywhere.

The corpus of analyses represents one of the largest bodies of analytical data ever assembled from one site. It seems therefore appropriate to explain at the outset why a nondescript collection of fragments and, at best, very minor artifacts should merit such attention. Excavated metal production sites from antiquity are rare, but Timna is uniquely important for the study of ancient smelting because of the hundreds of votive offerings brought to the Temple by the miners and metalworkers. These offerings ranged from lumps of ore through to finished cast artifacts, with many hundreds of spills and prills of unrefined metal, manufacturing fragments and trimmings from the various stages of metal production being included. This close association of production centre and product greatly aids the archaeological interpretation of the site and is a vital complement to the experimental reconstruction of the smelting processes. Extensive work on building and operating primitive furnaces to reconstruct the ancient processes, has been conducted by Tylecote and Tylecote *et al.* (1977 and 1978), and more recently by Bamberger (1984; 1985; 1986), Bamberger *et al.* (1986) and Merkel (1983) (see Vol. II of this publication). It is instructive to compare the composition of the copper produced under varying conditions in the experimental furnaces with copper produced in antiquity, enabling the operating parameters of the ancient processes to be more closely defined.

Such a large body of material from one site also enables the consistency of the copper produced to be studied directly, which has hitherto not been possible. This is of interest not just for the study of quality control, but also in the much broader field of provenancing copper based artifacts from their trace element composition. The consistency of copper from one production centre is clearly central to the validity of the whole subject, but in the numerous provenance studies undertaken this has had to be assumed. The general paucity of metal finds from other production sites has so far precluded any meaningful statistical work on the range of composition to

be expected from any one site. This chapter has been restricted to the analysis of the metal finds, but further work has continued on the variation in ore composition from mines at Timna and elsewhere in the Arabah, relating this to the metal composition (Leese, Craddock, Freestone and Rothenberg, 1986).

1 Sampling and analysis

As can be seen on *Figs.* 53–77, most of the artifacts are very small and are often corroded. Thus their sampling and analysis presented considerable difficulties. Where an artifact was very badly or totally corroded no sample was taken, as extensive research has shown that no meaningful analysis can be obtained from such material, the present composition of which bears little relation to the original composition (Craddock 1976, 93–95; Caley, 1964, 1–16; Jedrzejewska 1962). The badly corroded metalwork was, however, surface analysed qualitatively by non-dispensive X-ray Fluorescence, in order to determine the main components and the results were added to the description of the objects in the catalogue. The majority of the uncorroded metalwork was sampled using a portable jeweller's drill with a size 60 (1mm. diameter) hardened steel bit. Previous experimentation has shown no measurable contamination from the bit, an important consideration in the samples which have a high iron content. The first surface drillings containing surface corrosion were discarded and between two and thirty milligrammes of bright metal turnings were collected. Even with this precaution, the collected drillings often contained both metal turnings and corrosion; in order to obtain a satisfactory sample, individual clean metal turnings had sometimes to be picked out of the mixture with a needle while being viewed under a low power microscope. This was a tedious process and it took about half an hour to obtain two milligrammes of drillings. Where an object was too small or corroded to be satisfactorily drilled, samples were taken by scraping off the corroded surface to expose clean metal and then peeling off turnings of metal with a steel scalpel. This method of sampling does rather more visual damage than a single drill hole, but it does ensure a clean sample, even from very small objects. In spite of these precautions, some of the analyses do not total 100% and this is almost certainly due to internal corrosion in the metal. The sample weights varied from about 0.5–30 milligrammes. This poses problems of comparison among the samples, clearly the 30 milligramme sample has a thirty times better detection limit than a 1 milligramme sample. These are, of course, extreme cases and the normal detection limits quoted are based on samples of 5 milligrammes or more, dissolved in acid and made up to 20 millilitres. Sample weights below 5 milligrammes were dissolved with less acid and made up to only 10 instead of the normal 20 millilitres; thus all samples of 2.5 milligrammes or over have a uniform detection limit. Fifteen of the samples weighed less than this limit and their analyses have been marked with an asterisk. The detection limit for these samples is approximately twice as high as that stated for the rest. The general detection limit for each element is better than 0.01 per cent in the metal. The standard deviation is ±2 per cent for the major elements and ±30 per cent for the trace elements (i.e. those below 0.2 per cent in the metal).

The samples were analysed by Atomic Absorption Spectrometry using flame methods for all elements except arsenic, antimony and bismuth which were determined using a graphite furnace attachment for increased sensitivity. The precise details of the methodology have been given in Hughes, Cowell and Craddock (1976), and departures from these methods have been given above.

On the basis of the qualitative and quantitative analyses, each item in the catalogue has been identified metallurgically. The first and main table of quantitative analyses, Table l, is arranged by catalogue entry. Table 2 is of metalwork from Site 200 which is not included in the catalogue. Table 3 covers amulets from the Temple. Table 4 brings together the eight lead, two tin and one silver samples, while Table 5 is of metalwork from other sites in the Wadi Timna and nearby.

2 Analytical Tables

Table 1. Timna Site 200 copper base objects

Met. Cat.	Fig. No.	Cu	Pb	Sn	Ag	Fe	Sb	Ni	Au	Co	As	Mn	Bi	Zn (%)
1	53:1	93.5	.10	5.50	.090	.080		.015		.002	.090		.010	.020
2	53:2	92.5	.44	4.00	.030	1.400	.005	.280		.015	.025		.010	.027
3	53:3	98.0	.38		.005	.520		.030		.035	.500		.010	.015
4	53:4	97.0	.18	3.40	.015	.300		.033		.012	.130		.015	.006
5	53:5	94.5	.40		.100	3.800	.005	.035		.007	.080	.020	.002	.070
6	53:6	84.0	1.10	13.60	.050	.023	.005	.130		.030	.450		.012	
24	55:1*	90.0	.25	3.50	.015	.570					.100		.003	
25	55:2	95.0	.16	4.90	.030	.280	.012	.030		.020	.150			
26	55:3	90.0	1.32	7.90	.015	.770	.010	.017		.005	.075	.001	.007	
28	55:5	93.5	.46	4.30	.020	1.000	.010	.035		.011	.075		.010	
29	55:6	90.5	.20	7.90	.020	.150	.002	.030		.030			.040	
31	55:8	90.5	.10	7.20	.020	.240	.006	.035			.050		.009	
32	55:9	96.0	.28	3.65	.018	.080		.030			.065		.008	
33	55:10	88.0	.41	10.50	.002	.175		.030		.085	.015		.008	.030
35	55:12	91.0	.35	7.00	.015	.850	.011	.035		.015	.130		.009	.015
36	55:13	90.0	.15	9.20	.007	.390	.004	.020		.007	.250		.002	
38	55:15	98.0	.13	1.50	.050	.070		.050	.020		.030		.020	.015
39	55:16	97.0	.36	1.20	.090	.270	.200	.030		.009	.100		.014	.030
40	55:17	94.5	.40	2.30	.062	1.320	.120	.040			.040		.006	
41	55:18	87.5	.80	11.40	.004	.130		.020			.040		.008	
43	55:20*	84.0	.30	10.80	.030	2.000	.040	.040			.250		.050	
45	55:22	96.0	.35	1.90	.055	1.350	.008	.040		.012	.080		.010	
47	55:24	85.5	.32	7.60	.095	2.300	.003	.055		.020	.140		.003	.030
48	55:25	91.0	.28	7.80	.025	.500		.040		.001	.080		.005	
50	55:27	97.0	.37	1.50	.047	1.350		.045			.110		.006	
51	55:28	91.0	.34	7.20	.009	1.300		.040		.015	.050		.002	
52	55:29	95.0	.35	2.10	.065	1.300	.003	.030		.020	.090		.006	
55	55:32*	85.0	.18	4.70	.007	.370					.010		.003	
56	55:33	95.0	.23	3.70	.020	.460	.010	.030		.005	.085		.006	.010
57	55:34	86.0	.35	9.40	.020	2.800	.007	.030		.012	.050		.005	.025
58	56:1	98.5	.44		.008	.090		.025		.009	.045	.005	.005	.005
59	56:2	99.0	.45		.002	.090		.025		.010	.060	.007	.003	.005
60	56:3	92.0	.18	6.70	.065	.300	.003	.250		.010	.050		.005	
61	56:4	98.0	.28	.40	.024	.700		.065		.015	.100		.002	.010
62	56:5	99.0	.13	1.65	.060	.035		.025		.002	.065		.008	
63	56:6	96.5	.50	.95	.035	2.100	.005	.035		.005	.050		.010	.005
65	56:8	99.0	.43		.020	.130		.015		.009	.023	.004	.008	
66	56:9*	96.0	.35			.130	.004	.015			.045		.003	
67	56:10	100.0	.15		.060	.095	.015	.020		.010	.220	.002	.005	
68	56:11	100.0	.45		.002	.100		.020			.030	.002	.006	
71	56:14	97.0	.40	1.90	.015	1.100		.030		.005	.005		.005	.060
72	56:15	99.0	.35			.160		.010		.010	.030	.004	.006	
73	56:16	100.0	.17		.020	.100	.005	.025		.010	.220	.003	.005	
74	56:17	96.5	.07	2.40	.025	.130		.045			.050		.005	
76	56:19	98.0	.17	2.00	.020	.260		.035		.007	.140		.012	.017
77	57:1	98.0	.60		.020	.280		.030		.008	.070		.015	.010
78	57:2	100.0	.50			.100	.002	.030			.035	.004	.012	
79	57:3	98.0	.50		.004	.150		.020			.550	.005	.010	.100
80	57:4	93.0	.18	3.15	.012	2.900	.004	.025			.340		.012	
81	57:5	92.0	.15	3.25	.011	3.700		.025		.012	.160	.005	.010	.017
82	57:6	97.5	.20	1.90	.015	.250	.010	.015			.500		.015	
83	57:7	93.0	.17	3.10	.015	3.600	.002	.030		.012	.400		.002	.006
84	57:8	99.0	.43			.100		.020			.550	.004	.015	

Table 1.   Timna Site 200 copper base objects (continued)

Met. Cat.	Fig. No.	Cu	Pb	Sn	Ag	Fe	Sb	Ni	Au	Co	As	Mn	Bi	Zn	(%)
85	57:9	98.0	.43	.60	.015	.380	.002	.025		.010	1.300		.010		
87	57:11	99.0	.16	.35	.085	.070		.015			.100		.010		
88	57:12	100.0	.40		.005	.100		.040			.100	.003	.013		
89	57:13	97.0	.25	1.90	.015	.100		.025			.500		.010		
90	57:14	99.0	.43	.10	.002	.090		.025		.008	.070	.006	.010	.003	
91	57:15	92.0	.14	2.80	.015	3.700		.030		.011	.250		.015	.010	
93	57:17	96.5	.21	3.10	.013	.090		.020			.100		.008		
94	57:18	95.5	.17	3.25	.015	.100	.001	.020			.070		.015		
95	57:19	93.0	.23	3.40	.012	3.000	.010	.030		.015	.550		.015		
96	57:20	94.5	.22	3.10	.022	.120		.030			.330		.006		
97	57:21	92.0	.18	4.50	.017	3.100	.010	.020			.150		.010		
98	57:22	95.0	.39	6.80	.005	.095	.002	.025		.005	.200	.004	.013		
100	57:24	99.0	.25	.50	.020	.035		.030			.100		.014		
101	58:1	96.5	.50		.002	.090	.010	.030			.040	.003	.005	.090	
102	58:2*	99.0	.45			.200		.025			.050		.002		
103	58:3	91.0	.24	8.50	.040	.380	.010	.040		.020	.100		.006	.035	
107	58:7	98.0	.10	1.25	.025	.460	.003	.020			.050		.010		
108	58:8	92.5	.15	3.20	.013	3.700	.005	.150			.120		.005		
112	58:12	90.0	.50	10.10	.020	.210	.055	.010			.180		.004		
113	58:13	99.0	.40		.002	.800	.003	.022			.030	.005	.008		
114	58:14	97.5	.17	2.10	.015	.210	.005	.020			.090		.020		
116	58:16*	100.0	.15	.80		.260	.008	.030			.080				
117	58:17*	92.0	.20	2.60	.020	3.500	.010	.070			.050		.050		
118	58:18	100.0	.15		.012	.280	.015	.025			.240		.005		
119	58:19	97.0	.30	2.90	.037	.180	.020	.030			.060		.007		
120	58:20	98.0	.15	.60	.035	1.100	.080	.020		.008	.002		.020	.060	
121	58:21	93.0	.15	3.20	.020	2.850	.003	.040			.100		.005		
124	58:24	3.2	.07		.005	97.00	.010	.030		.025	.050		.050		
125	58:25	97.0	.20	2.40	.020	.150	.003	.020			.065		.008		
129	58:29	99.0	.35		.007	.200	.007				.025	.006	.005		
130	58:30	98.0	.45			.360	.007	.015			.030	.003	.005		
131	58:31*	100.0	.50		.005	.050	.025	.070			.020		.002		
132	58:32	100.0	.43			.150		.030		.007	.030	.005	.008		
134	58:34	99.0	.41			.135	.003	.010		.010	.027	.005	.007		
135	58:35	98.5	.41		.005	.130	.003	.025			.033	.005	.007		
136	58:36	91.0	.22	3.85	.020	3.650	.045	.020			.130		.007	.020	
137	58:37	98.0	.40			.210	.003	.025			.055		.005		
139	59:2	100.0	.50			.025	.003	.050			.005				
141	59:4	99.0	.50		.005	.080		.030		.005		.005	.012	.005	
143	59:6	98.5	.50		.003	.070		.025			.020	.005	.004		
144	59:7	97.5	.15		.015	1.100		.040		.005	.040		.010	.080	
148	59:11	96.0	.25	1.80	.007	.080		.035						.040	
149	59:12	92.5	.10	4.70	.035	2.950		.030		.006	.075		.009	.006	
150	59:13	98.0	.18	1.00	.010	.095	.001	.030		.009	.400	.003	.001	.003	
151	59:14	99.0	.50			.055		.045			.030	.010			
152	59:15	98.0	.50			.080	.035			.040	.008	.007			
157	59:20	98.5	.50		.005	.090		.025			.020	.010			
158	59:21	93.0	.15	3.70	.015	2.900	.010	.025		.008	.020		.005	.020	
159	59:22	92.0	.20		.010	.055		.020			.800				
168	59:31*	90.0	.95		.015	3.000		.150			.200			.080	
174	59:37	98.0	.10	1.70	.010	.800		.035			.040		.003		
175	59:38	96.0	.40	1.30	.008	.600		.035			.050			.008	
177	59:40	81.5	.20	8.50	.020	1.800	.005	.025		.030	.050		.003	.070	
178	59:41	97.8	.07		.015	1.300	.015	.025			2.200		.004		
179	59:42	86.0	.09	13.40	.020	.120	.003	.015			.150	.001	.008		
180	59:43	98.0	.30		.002	.015	.005	.010			1.300		.005		
189	60:7	93.5	.25	3.00	.020	4.100		.030			.200			.250	
190	60:8	91.0	.25	2.65	.013	4.850		.008		.006	.100		.008	.020	
192	60:10	93.0	.20	2.10	.007	3.500	.015	.020			.250			.025	
193	60:11	93.0	.25	2.30	.230	4.400	.025	.030			.200			.035	
194	60:12	98.0	.40	1.40	.090	.800	.300	.020			.130				
198	60:16	94.0	.20	2.10	.025	4.300		.015			.095			.015	
202	60:20	89.0	.25	2.90	.010	3.500		.040			.080		.003	.270	
203	60:21	92.0	.30	1.50	.015	5.000	.150	.050			.040			.030	
211	60:29	98.0	.25	2.10	.010	.070	.002	.025			.060		.003		
212	60:30	97.5	.10		.010	.170		.030			.200		.002	.100	
213	60:31	98.0	.10	1.30	.080	.060		.015			.030		.008	.040	
214	60:32	100.0	.25	.40	.005	.010	.001	.045			.040		.003	.005	
215	60:33	99.0	.08	.40	.090	.550		.015		.003	.010		.050	.020	
218	60:36	91.5	.15	7.60	.010	.150	.001	.030			.050	.002	.004		
220	61:2	98.5	.16	1.00	.008	.060		.027		.008	.080	.002	.007	.003	

Table 1. Timna Site 200 copper base objects (continued)

Met. Cat.	Fig. No.	Cu	Pb	Sn	Ag	Fe	Sb	Ni	Au	Co	As	Mn	Bi	Zn	(%)
222	61:4	100.0	.39		.001	.100		.028		.010	.005	.003	.007	.002	
224	61:6	96.0	.50			.080	.200	.030			.030	.004		.010	
225	61:7	94.0	.16	4.15	.020	2.750		.027		.006	.150		.008	.008	
226	61:8	97.0	.45			.028		.020			.030		.003		
229	61:11	100.0	.50			.060		.040		.008	.020	.006	.004		
230	61:12	99.0	.40	1.60		.040	.100	.025			.020				
231	61:13	100.0	.50			.080	.005	.020		.007	.030	.002	.003	.010	
232	61:14	97.5	.15	1.10	.005	.012	.002	.020			.400				
233	61:15	100.5	.50			.055	.005	.015			.020	.007	.003		
234	61:16	88.0	.50	5.40	.015	3.200		.015			.100		.002	.200	
235	61:17	100.0	.50			.070		.020			.030		.005		
236	61:18	99.5	.50		.005	.120	.020	.050		.005	.020	.007	.005		
237	61:19	100.5	.50			.070	.003	.030			.030	.010	.006		
238	61:20	100.0	.50			.080	.005	.015			.030	.010	.004	.010	
239	61:21	100.0	.45		.002	.085		.030		.007	.010	.003	.008	.008	
240A	61:22A	99.0	.50	.20	.001	.080	.015	.025		.007	.060	.006	.007	.005	
240B	61:22B	93.0	.40	6.50	.001	.080	.130	.030		.005	.010	.005	.010	.005	
242	61:24	99.0	.45			.070	.005	.060			.020	.012			
243	62:1	100.0	.45			.050	.030			.030	.008				
244	62:2	90.5	.20	2.10	.007	3.500	2.000	.030			.300		.003	.020	
245	62:3	93.5	.20	3.50	.025	2.400		.025		.010	.090		.002	.060	
246	62:4	93.0	.18	3.30	.015	3.100		.027			.100		.008		
247	62:6	94.0	.10	3.50	.015	3.100		.033		.010	.100		.005		
248	62:7	95.0	.15	3.10	.015	2.350		.020			.200		.003	.020	
249	62:8	94.0	.15	3.60	.050	2.500		.025		.012	.080		.002	.100	
250	62:10	96.0	.20	2.80	.012	.430	.003	.060			.060		.005	.055	
255	62:13	98.0	.05	.09	.012	.480		.005		.003	.005		.007	.020	
256	62:14	98.5	.10	.70	.020	.600	.003	.015		.003	.020		.005	.070	
257	62:15	98.0	.07	1.20	.023	.650		.015		.009	.030		.015	.017	
258	62:16	96.0	.35	2.30	.075	.800		.030		.010	.080		.020		
259	62:17	97.0	.40	1.60	.090	.680	.010	.035		.003	.060		.005	.010	
260	62:18	92.5	.21	3.20	.030	5.100	.002	.015		.010	.120		.008	.020	
262	62:20	98.0	.40	1.50	.050	.800		.040			.150		.003	.300	
264	63:2	92.0	.12	6.90	.015	.140		.025			.100			.100	
266	63:4	95.0	.07	.95	.013	.200	.005	.040		.015	.300		.008		
267	63:5	86.5	.50	10.60	.035	1.100		.012			.020		.006		
268	63:6	89.5	1.90	7.00	.005	.120		.050		.015	.050		.002	.020	
273	63:11	93.5	.23	5.30	.030	.450		.045			.040		.007		
274	Not ill.	89.0	.13	7.20	.008	.600	.005	.035			.100		.006		
292	64:18	92.5	.21	2.65	.014	5.100		.015		.007	.080		.007	.025	
293	64:19	90.0	.15	8.70	.015	.050		.030			.030		.003	.400	
297	64:23	91.5	.40	7.30	.020	1.100		.025			.050		.008		
302	65:3	92.0	.15	7.70	.012	.340		.045			.030		.005		
309	65:10	97.5	.30	1.50	.040	.750	.007	.060		.007	.100		.002	.015	
313	65:14	95.0	.39	3.50	.026	.500		.020		.007	.050		.010		
316	66:3	88.0	.29	8.00	.025	3.050	.004	.030		.020	.150	.001	.010	.020	
317	66:4	88.0	2.50	9.10	.007	1.100		.025		.040	.030		.003	.220	
319	66:6	86.5	.47	7.40	.055	.250	.005	.025			.150		.030		
320	66:7	87.0	.10	10.00	.025	.950		.015			.020		.005	.070	
321	66:8	89.0	.10	4.70	.065	.035		.040			.020		.002	.120	
331	66:18	91.0	.22	5.00	.030	.060	.005	.035			.050		.004	.020	
340	67:9	89.0	.03	9.40	.110	.800		.020		.005	.040		.020		
348	67:17	90.0	.30	6.00	.015	3.900	.002	.045		.005	.100		.005	.050	
358	67:27*	3.9	.07	13.50	.030	.110						.003	.005		
361	68:2	94.5	.75	4.40	.007	.450	.005	.025			.050		.005		
363	68:4	92.0	.18	1.55	.007	6.600	.004	.070		.020	.350		.005	.025	
368	68:9	90.5	.18	3.40	.008	4.000	.012	.060		.008	.100		.007	.015	
369	68:10	90.0	.15	10.00	.015	.190	.004	.035		.003	.130		.010	.014	
370	68:11	92.0	.25	6.20	.005	.500	.005	.040			.050		.005	.040	
371	68:12	92.0	.20	1.30	.002	6.300	.007	.070		.020	.300			.050	
374	68:15	93.0	.23	5.60	.017	.375	.450	.025			.800		.005		
376	68:17	94.0	.43	4.20	.010	2.000	.005	.050		.020	.100			.010	
378	68:19	89.0	2.10	9.10	.006	.120	.002	.040		.020	.020			.012	
379	68:20	90.5	.20	7.60	.015	.300	.003	.035			.100		.005	.010	
380	69:1	93.0	1.30	3.20	.040	.550		.025			.100		.002		
382	69:3	90.0	1.10	7.70		.080	.030			.010	.005				
383	69:4	91.5	.30	5.90	.020	1.150	.030	.030		.010	.060		.003	.020	
384	69:5	93.0	.70	5.80	.130	.200		.070			.080		.003	.160	
385	69:6	98.0	.25	.70	.040	.800	.002	.030		.003	.100		.003	.030	
386	69:7	95.5	.20	3.40	.060	.330	.005	.100		.005	.100			.100	
387	69:8	91.0	.35	6.70	.020	.570		.040			.010			.015	

173

Table 1. Timna Site 200 copper base objects (continued)

Met. Cat.	Fig. No.	Cu	Pb	Sn	Ag	Fe	Sb	Ni	Au	Co	As	Mn	Bi	Zn	(%)
393	70:1	93.0	.22	5.50	.022	1.400	.070	.015			.100		.007		
394	70:2	91.5	.50	5.20	.030	.260	.002	.035			.100				
395	70:3	91.0	.17	7.90	.012	.450	.005	.010			.070		.010		
396	70:4	92.5	.63	6.70	.020	.540		.035		.008	.090		.020		
397	70:5	92.5	.45	6.50	.010	1.100	.010	.070			.050			.030	
398	70:6	87.5	.90	6.10	.020	.720	.020	.045		.010	.120			.020	
399	70:7	92.0	.15	6.80	.015	1.280		.045			.060		.002	.010	
400	70:8	92.0	.60	6.00	.030	.040		.025			.070		.002		
401	70:9	91.5	.25	7.10	.015	.400	.015	.040			.100		.003		
402	70:10	99.5	.50		.003	.070	.002	.040			.020	.010	.005		
404	70:12	92.5	.44	5.10	.120	1.850	.002	.030		.007	.100		.010		
406	70:14	93.0	.70	5.80	.015	1.000	.005	.045		.015	.100			.015	
410	71:4	87.0	.40	2.30	.040	1.000		.045			.100	.004	.003	.080	
413	71:7	97.5	1.50		.003	.190		.050		.030	.050		.003	.020	
415	71:9	98.0	1.30		.002	.200	.005	.060		.045	.050		.005	.010	
418	71:12	87.0	.80	11.20	.005	.020		.040		.010	.040			.130	
421	71:15	94.0	1.00	.90	.005	.080		.045		.015	.030		.003	.100	
425	71:19	93.0	.70	4.30		.100	.002	.020			.030				
431	71:25	88.0	.33	6.40	.009	5.300	.025	.037			.100		.008	.025	
437	72:4	97.0	.30	.50	.025	.400	.300	.040			1.000		.003	.050	
439	72:6	91.5	.52	7.10	.005	.100	.005	.025			.110		.003		
440	72:7	92.5	.10	6.10	.010	.650	.120	.040			.020			.030	
442	72:9	90.0	.60	7.60	.007	.070	.100	.030			.100		.004		
447	73:2	96.5	1.30	1.95	.004	.280	.003	.025		.015	.030			.030	
449	73:4	90.0	.08	8.30	.010	1.060		.025			.025		.007		
453	73:8	98.5	.55	.30	.130	.120	.001	.025		.003	.040		.002	.015	
454	73:9	97.5	1.30	.90	.006	.340	.002	.030		.015	.050			.020	
459	74:1	98.0	.11	1.90	.012	.110		.025		.013	.330		.020		
460	74:2	89.0	.23	9.60	.015	.100		.035		.008	.120		.020		
461	74:3	90.0	.50	8.20	.060	.900	.010	.030			.100		.005	.030	
462	74:4	90.0	.70	7.20	.010	1.100		.050		.015	.110		.025	.010	
465	74:7	97.0	.13	1.70	.035	.250		.030			.025		.010	.015	
466	74:8	99.0	.14	.65	.023	.030		.030			.035		.020		
467	74:9	98.5	.18	.95	.020	.010		.035			.015		.003	.010	
468	74:10	88.0	.30	12.00	.005	.260	.020	.020			.150				
469	74:11	99.0	.33		.020	.570		.030		.003	.080		.014	.020	
470	74:12	98.5	.16	1.30	.010	.130		.050			.360	.002	.025		
471	74:13	97.0	.10			.060	.700	.020			.700				
474	75:3	97.5	.35	.70	.110	.600	.010	.020			.150		.004	.015	
476	75:5	90.0	.50	5.00	.035	.050	.200	.025	.070	.015	.100			.010	
477	75:6	92.5	.15	5.30	.015	.170	.005	.050			.085			.010	
478	75:7	93.0	.12	7.00	.085	.230	.002	.010			.050		.006		
479	75:8	97.0	.25	3.40	.017	.150	.015	.035			.060			.025	
480	75:9	93.0	.14	4.30	.080	1.600		.030			.050		.010	.015	
481	75:10	95.5	.07	3.00	.010	.500		.030		.010	.040		.010	.050	
482	75:11	97.0	.30	1.40	.029	.550		.030		.013	.110		.020	.055	
483	75:12	99.0	.25	.95	.030	.760	.001	.030		.004	.200		.003	.030	
484	75:13	99.0	.30	.85	.040	.830		.033		.010	.045		.010	.045	
485	75:14	95.5	.30	4.10	.060	.400	.002	.025	.010		.060		.003	.005	
486	75:15	97.5	.24	1.20	.015	.045	.010	.030			.050		.015		
487	76:1	96.5	.30	2.00	.005	.030		.020		.020	.050			.030	
489	76:3	97.5	.30	1.30	.017	.180	.600	.050		.008	.100	.002	.004	.040	
492	76:6	98.5	.07	1.30	.013	.040	.002	.025		.004	.030		.020		
493	76:7	93.0	.85		.030	6.600	.005	.070		.030	.210		.010	.030	
494	76:8	98.5	.25		.002	.016		.010			.900		.015		
495	76:9	96.0	1.00	2.10	.005	.280	.002	.040		.025	.040	.001	.030	.020	
496	76:10	96.0	.17	.12	.006	4.150		.070		.030	.060	.005	.010	.025	
497	76:11	92.0	.30	6.30	.010	.500	.001	.035		.010	.020		.002		
504	77:7	94.0	.30	1.30	.015	.650	.050	.040		.006	.130		.010	.030	
505	77:8	98.0	.31	.25	.040	1.260		.030		.010	.060		.007	.045	
506	77:9	96.0	.40	.70	.015	2.000	.001	.050		.010	.050		.002	.045	
507	77:10	97.0	.09	2.50	.025	.250		.010			.024		.006	.010	
508	77:11	93.5	.21	3.30	.015	3.500		.015			.130		.008	.020	
509	77:12	92.0	.40	1.65	.030	5.600		.025		.015	.035		.006	.050	
510	77:13	94.5	.03	1.90	.050	2.250		.035		.020	.030		.006	.033	
512	77:15	97.5	.08	2.50	.005	.210	.002	.050		.012	.050		.002	.010	
513	77:16	94.5	.20	4.80	.020	.130		.025		.006	.600		.010	.010	
514A	77:17 beam	98.0	.06	1.00	.012	.480		.020			.100		.010	.018	
514B	77:17 ring 1	88.0	.28	9.80	.020	1.040		.050		.007	.170		.013		
514C	77:17 ring 2	85.0	.20	8.40	.250	1.420	.005	.025		.007	.120		.010		

Table 2. *Timna site 200 copper base objects not in catalogue and excluding amulets*

Anal. No.	Field No.	Cu	Pb	Sn	Ag	Fe	Sb	Ni	Au	Co	As	Mn	Bi	Zn	(%)
515	80/1	83.0	.08	3.20	.035	3.500	.004	.025	.020		.100		.003	.005	
516	115/13	99.0	.24		.025	.600		.025		.020	.085		.004	.033	
517	124/1	98.0	.12	1.70	.030	.300		.025			.030		.007	.005	
518	154 (Sample 352)	98.0	.15		.025	1.700	.001	.030		.004	.050		.008	.020	
519	240 (Sample 377)	94.0	.90	.30	.150	2.750	.002	.040		.020	.160	.001	.020	.060	
519A	240 (Sample 337)	94.0	.30		.009	5.800		.040		.025	.060	.001	.018	.120	
520	319 (Sample 350)	85.0	1.90		.030	11.80	.001	.030		.013	.100		.002	.110	
521	216/85	95.0	.25		.065	4.100		.035		.007	.170		.010	.090	
522	221/1 (Sample 407)	95.0	.09	.50	.025	.190		.015			.045		.008		
523	228/29 (Sample 615)	95.5	.11	2.85	.015	.570	.003	.015			.020		.006	.100	
524	245/64	88.0	.10	11.30	.020	.370		.030		.035	.040		.015		
525	260/189	95.0	.90	.35	.032	3.300	.005	.025		.027	.060		.015	.040	
526	263/8	99.0	.03		.023	.120		.012			.075	.003	.012		
528	278/134	96.0	.21	3.00	.010	.130	.002	.020		.015	.045		.006		
528A	278/134	97.5	.20	1.00	.010	.120		.030			.050		.007		
528B	278/134	92.0	.20	4.00	.010	3.100	.003	.025		.015	.100		.013		
528C	278/134	98.0	.45	1.10		.520		.030		.010	.025	.006	.008		
528D	278/134	91.5	.15	3.85	.008	4.200		.020			.110		.012		
529	283/21	99.0	.46	.45		.130		.030		.025	.020	.004	.005		
530	283/313	97.0	.50	.90	.008	.055		.035			.030		.002	.040	
531	284/64	83.0	.46		.013	15.20		.085		.055	.120	.002	.010	.070	
532	294/48	96.0	.26	3.40	.065	.600		.040		.015	.040		.012	.120	
533	319/290	91.0	.29	5.40	.180	3.500		.045		.030	.085		.010	.020	
534	319/293	96.0	.35	2.20	.010	.120		.015			.060	.022	.010		
535	339 (Sample 617/1)	97.0	.21		.011	2.700		.027		.020	.017		.006	.070	
536	339 (Sample 617/2)	96.0	.04		.015	3.560		.020		.007	.012		.008	.140	
537	343/94	88.5	.35	8.40	.075	1.500	.003	.035		.045	.070		.020	.020	
538	343/98*	94.5	.31	1.40	.040	3.400	.003	.080		.055	.100		.005		
539	Sample 260	84.0	.62		.090	14.00	.005	.052		.045	.180	.001	.013	.090	

Table 3. *Timna Site 200 – amulets*

Anal. No.	Fig. No.	Field No.	Cu	Pb	Sn	Ag	Fe	Sb	Ni	Au	Co	As	Mn	Bi	Zn	(%)
540	47:9	160/7 (Sample 1014)	96.5		2.50	.007	.040		.060			.040		.002	.020	
541	47:10	295/197 (Sample 1038)	96.0	.43	.35	.035	1.900	.005	.025		.007	.140		.020	.030	
542	47:11	309/1 (Sample 1037)	95.0	.60	3.80	.030	.040		.045			.020		.003	.010	
543	47:12	319/27 (Sample 1036)	96.0	.16	2.80	.025	.230	.001	.040		.003	.070		.002	.010	
544	47:13	339/9 (Sample 1010)	100.0	.50		.005	.120	.020	.020			.030		.003		

Table 4. *Timna lead, tin and silver artifacts*

		Cu	Pb	Sn	Ag	Fe	Sb	Ni	Au	Co	As	Mn	Bi	Zn	(%)
Lead																
Met. Cat.	Fig. No.														
162	59:25 (Sample 265)	.3	99.0		.003	.025							.005		
162A	59:25 (Sample 266) rod	.5	99.0		.005	.020					.900		.001	.020	
162B	59:25 (Sample 266) rod	.4	99.0		.005	.005	.001				1.500		.002		
164	59:27 (Sample 1035)	.2	96.0		.012	.050	.004	.005			.002		.003		
165	59:28 (Sample 267)	.3	99.0		.005	.010	.002				.600		.001		
Anal. No.															
545/1	Sample 623/1 (sheet) Site 2(1)	.0	99.0	.90	.008	.006	.100	.005	.002				.040		
545/2	Sample 623/2 (disc) Site 2(1)	.2	98.5		.011	.010	.080	.008			.020		.045		
545/3	Sample 623/3 (roll) Site 2(1)	.2	99.0		.030	.010	.070	.008	.010	.003			.060		
Tin																
Met. Cat.															
163	59:26 (Sample 300)	.7	.67	99.00				.080	.008					.095	
358	67:27 (Sample 884)	3.9	.07	13.50	.030	.110						.003			
Silver																
Anal. No.															
546	270/96	8.4	.15		87.00	.900	.020		3.200		.005		.005		

(1) Unstratified samples

Table 5. Timna copper base objects, excluding Site 200

Anal. No.	Spl./Field No.	Cu	Pb	Sn	Ag	Fe	Sb	Ni	Au	Co	As	Mn	Bi	Zn	(%)
Arabah (Surface find)															
547A	Sample 1026 (ingot top)	97.0	.21		.002	2.400		.010			.050	.002	.010	.050	
547B	Sample 1026 (ingot bottom)	96.0	.10			2.950		.005			.180	.002	.013	.050	
Beer Ora															
548	Sample 1045	91.0	1.35		.017	7.200	.005	.150		.220	.140	.005	.010	.170	
549	Sample 1045	94.0	1.20		.018	3.100		.120		.100	.025	.003	.008	.200	
Site F2															
550		98.5	.14		.010	1.950	.007	.035		.020	.060		.010	.060	
551	F4/1 (Sample 813) L. 2	96.0	.20		.080	1.300	.700				1.100		.800		
552	F/2 (Sample 817) L. 3	99.0	.10	.15	.045	.160	.004	.017			.075	.003	.005	.015	
Site 186 (Surface)															
553	Sample 1047	95.5	.45	.60	.015	3.800		.055		.020	.135	.001	.012	.032	
Site 2															
554		91.0	.17	6.10	.020	1.700	.015	.020		.010	.015		.005	.250	
555	71/1 (Sample 835)	91.0	.09	9.30	.005	.050		.030		.020	.020		.020	.008	
556	398/3 (Sample 1048)	93.0	.18	5.90	.028	.700	.005	.040			.080		.010	.005	
557	SF1 (Sample 1046)	98.0	.29	.80	.020	1.350		.020		.010	.030		.010	.030	
558	Sample 575	93.0	1.60	.75	.035	4.900	.005	.095		.080	.070		.003	.270	
559	SF67 (Sample 89)	95.0	.10		.015	4.050	.020	.040		.025	.085		.003	.070	
Site 3															
560		98.0	.60		.010	1,850	.005	.150		.075	.045		.004	.200	
561		96.0	.17		.025	2.700	.005	.040		.010	.040	.005	.005	.060	
562		86.0	.07	5.10	.013	.300	.005	.050		.030	.020			.085	
563		90.0	.42		.030	4.100	.010	.040		.020	.100		002	.100	
564	345	85.0	.59		.025	7.153	.047	.049	.005	.038	.141	.005		.088	
565	347	87.3	.07		.013	5.554		.011		.009	.036	.007		.047	
566	348	90.2	.12		.019	.325		.014		.008		.008		.024	
567	349	94.2	.15		.019	.513		.025		.004	.105	.001		.043	
568	350	88.8	.32		.024	2.853	.032	.029		.021	.127	.002		.080	
569	351	83.6	.54		.062	9.720	.037	.049		.024	.112	.010		.064	
570	352	92.1	.40		.012	2.520	.035	.091		.064	.177	.001		.158	
571	353	90.1	.31		.034	4.838	.029	.053		.024	.191	.002		.046	
572		88.9	.01		.026	.715	.036	.130		.011	.201	.002		.046	
573	355	87.6	.22		.032	2.301	.031	.026		.017	.153	.065		.049	
Site 30															
574	804/1 (Sample 796)	95.0	.30	5.50	.015	.540	.002	.040			.050		.010		
575	841/9 (Sample 804)	96.0	.15		.010	2.900	.005	.035		.005	.085	.004	.008		
576	879/1 (Sample 795)	90.5	1.35	8.20	.027	.210	.006	.030			.100		.010		
577	89/2 (Sample 797)	94.0	.75	5.20	.020	.200	.005	.020			.060	.003	.010		
578	935/1 (Sample 803)*	94.5	.65	2.90	.030	1.650	.040	.040		.020	.060	.013		.040	
579	1503A (Sample 1394/A)	91.7	.46	7.33	.022	.131	.023	.040	.009	.007	.217	.005	.014	.006	
580	1503B (Sample 1394/B)	90.8	1.16	1.56	.011	3.385	.021	.083		.045	.079	.019	.042	.060	
Site 34 (Surface)															
581	Sample 629	94.5	.30	2.90	.015	.920	.050	.045		.015	.045	.001	.006	.015	

3 Discussion

(a) The analyses show that the copper is often alloyed with relatively small amounts of tin, and frequently contains large amounts of iron, but otherwise it is reasonably pure. This purity is to be expected as the copper ores from the ancient Timna mines are all secondary weathered deposits from which most of the commonly associated metals: nickel, silver, arsenic, antimony, bismuth and zinc, will have been leached. The secondary ores are mainly malachite nodules with some chalcocite, cuprite, azurite and chrysocolla (Slatkine, 1961; Tylecote *et al.*, 1967, 235–6). The various ore analyses reported (Tylecote and Boydell, 1978, 30–31; Lupu, 1970; 22; and Field in Rothenberg, 1972, 231), concur in that the trace element content is low. By comparison, e.g. the analyses of sulphidic ores from Ireland used in the Bronze Age, typically show trace element contents between ten and one hundred-fold higher than those in the Timna ores, and this is of course reflected in the composition of the Bronze Age metalwork from Ireland (Coghlan, Butler and Parker, 1963, 34–52).

Only one artifact from Timna does not appear to have been made from an oxide ore. This is the needle (sample 813) from site F2 which has 1.1% of arsenic,

0.7% antimony and 0.8% bismuth, and is totally unlike the rest of the metalwork in composition. No figure is quoted for tin or nickel because this was one of the smallest samples analysed; only 0.7 milligramme of sample was available, so the detection limit for nickel was 0.1% and for tin 0.2%. The composition of the needle is typical of copper made from an enriched *fahl* type ore. The copper mace-head from Wadi Zeelim analysed by Key (1971, 243; 1980, 239) was also rich in arsenic and antimony, but low in silver and nickel, suggesting that a similar *fahl* ore was used, which originated from a primary ore deposit. No ancient *fahl* ores are reported at Timna, the ores of which are all secondary. Thus the needle was almost certainly made of imported metal. The Zeelim mace-head is of the Chalcolithic period when the use of *fahl* ore was widespread. The small metalworking site of F2, from which the needle was excavated, was working with much purer metal (see analysis of the prill, sample 817). There was a great deal of Chalcolithic pottery and flintwork at F2, and the needle should be assigned to this period.

(b) Tylecote, Ghaznavi and Boydell (1977 and 1978) performed experimental smelts in a small furnace based on an example excavated at Site 39 in Timna, dated to the Chalcolithic period (Rothenberg, Tylecote and Boydell, 1978), and it is instructive to compare the composition of the copper from these experiments with that extracted by the metallurgists of antiquity.

Two sets of experiments were carried out. In the first only one tuyère was used[1] and in the second two tuyères.[2] With one tuyère, which seems to be typical of Chalcolithic practise, the temperature in the furnace did not rise high enough for a discreet ingot of copper to form at the bottom of the furnace. By using two tuyères, a much higher temperature was achieved throughout the furnace and a copper ingot formed beneath the slag (Tylecote, Ghaznavi and Boydell, 1977, 307, Fig. 2). The ore used by Tylecote was Timna malachite, but the flux was a synthetic iron oxide, which was responsible for at least one important difference in composition (the lead content discussed below). Table 6 shows the composition of

the copper ore used by Tylecote and of the copper produced by furnaces (1) and (2) (Tylecote, *et al.*, 1977, 313, Table 7). Beneath these, for comparison, is an average composition of 52 malachite samples, collected at the Timna mines and camps (Leese *et al.*, 1986), and the average composition of the 326 ancient artifacts analysed. Tylecote (1977) has shown that in the strongly reducing conditions of a furnace for smelting oxide ores, the retention of volatile elements such as zinc, bismuth, antimony and lead is strongly dependent on the temperature. This is clearly shown by comparing smelt (1) with smelt (2) in Table 6; both sets of analyses were performed on the unpurified metal straight from the furnace. The ancient metal would seem to be most similar to that produced in Tylecote's furnace (1) which used only one tuyère, and in which only a relatively low temperature was attained, especially in the slag. As a result of this, much of the copper tended to stay in the slag as discreet globules and prills which had to be mechanically removed by breaking up the cooled slag. Only with a higher slag temperature, as in (2), may a separate ingot form beneath the slag. Whilst the molten copper is moving downwards through the slag at the more elevated temperatures, it can pick up more iron, as has been demonstrated by Tylecote. The manganese content is clearly associated with the iron-rich phase of the metal, as would be expected on metallurgical grounds. However, the relatively high level of manganese in Tylecote's (2) metal is rather surprising as apparently neither the copper ore nor the flux contained detectable manganese (Tylecote *et al.*, 1977, Tab. 1 and 10).

(c) Most copper provenancing studies concentrate on the elements: arsenic, antimony, bismuth, silver and nickel, in order to define groups. Tylecote (1977, 7) has shown that of these elements only the concentrations of silver and nickel in the copper remain relatively unaffected by the smelting process, and this is confirmed by the analytical data for the Timna artifacts.

The standard deviations of the percentages of silver and nickel are approximately equal to the mean concentrations, but for arsenic the standard deviation is

Table 6. Analyses of Timna ores, metal and the products of Tylecote's experimental smelts (figures in parenthesis are in ppm, all others are expressed in %)

		Cu	Pb	As	Sb	Ni	Ag	Zn	Fe	Mn	Bi
Timna ore used in 1 and 2		23.7	0.045	0.05	nd	0.013	tr.	tr.	7.3	–	nd
Experimental smelt 1		–	0.3	0.02	0.024	0.03	0.025	1.2	(15)	–	
Experimental smelt 2		nd	0.2	nd	0.04	–	(50)	2.7	0.07	nd	
Timna ore (52 recent analyses	Average % 20	0.76	0.045	0.04	0.033	0.025	1.0	8.7	0.06	0.01	
Leese *et al.*, 1986)	Median 20	0.14	(70)	(50)	(70)	0.05	0.18	1.8	0.07	0.01	
Ancient Timna copper (average of 326 analyses)			0.37	0.125	0.025	0.035	0.024	0.025	1.2	(12)	0.01

Table 7. Mean and standard deviation of Timna copper

%	As	Sb	Bi	Ag	Ni	Pb	Zn below
Mean	0.125	0.025	0.009	0.025	0.035	0.37	0.01
Standard deviation	0.21	0.135	0.045	0.031	0.035	0.33	0.05

approximately twice the mean, and five times the mean for antimony and bismuth. Only lead had a similar narrow standard deviation, and this is somewhat surprising since earlier work by Tylecote (1977, 7) had suggested that it is also very susceptible to smelting conditions. In fact, the arsenic levels in the ancient metal are enhanced over what would be predicted from the ore analyses, and Merkel (1983) also found this in his metal. The Timna analyses give a good indication of the spread of results to be expected from one orebody worked by a one-step smelting process, and the result is not encouraging for provenancing studies.

(d) Sulphur content (N.D. Meeks and P.T. Craddock): Samples were carefully shaved with a scalpel from thirteen representative items, creating as little disruption as possible to the metallic structure. The shavings were resin mounted for examination and analysis on a Cambridge Stereoscan 600 scanning electron microscope with a wavelength dispersive X-ray spectrometer. *Pl.* 129:3 shows a typical section of the metal (from Met. Cat. No. 235) at ×500 magnification. The white inclusions are lead, slightly distorted by the scalpel, and the black inclusions are copper sulphide. Table 8 shows the overall sulphur and lead contents in the samples, as determined by X-ray spectrometry, together with the lead contents obtained by atomic absorption spectrometry on drillings taken from the same samples. There is good general agreement between the lead contents from the two methods, which suggests that the sulphur

contents for the very restricted volume of the shavings are also representative of the whole. The only exception is the shaving from Met. Cat. No. 415, where an area of relatively low lead content was sampled.

The sulphur content of the metal need cause no surprise, even though the ore is mainly malachite (copper carbonate), since this is formed by the weathering of sulphidic ores and inevitably some unchanged sulphides persist (see above, and Tylecote, Lupu and Rothenberg, 1967). No evidence has been found at Timna to suggest that the ores were subject to a preliminary roasting to oxidise the sulphides. Some of the sulphides would form a matte in the furnace (J. Merkel pers. comm.), some would pass into the slag, and the remainder would pass unchanged into the metal as inclusions. Smiths working with sulphidic ores would have carried out treatments before and during the smelting, to reduce the sulphur content in the resulting metal to approximately the level found in these samples. It follows therefore that the types of ore used cannot be identified solely from the sulphur content of the metal.

(e) The iron contents of the copper is especially interesting since for the first time one can appreciate the amount of iron incorporated into the copper during smelting. The mechanism by which the iron enters the copper during smelting is now well understood (Tylecote and Boydell, 1978). The copper from Tylecote's experimental smelts has a similar iron content to that found at Timna. Under very reducing conditions in the furnace, some of the iron oxide

Table 8. Sulphur and lead contents (in %) as determined by wavelength dispersive X-ray spectrometry (SEM) on individual shavings of metal

Fig. No.	Lead (AAS)	Lead (SEM)	Sulphur (SEM)
7.4	0.5	0.49	0.08
7.14	0.5	0.51	0.18
7.20	0.5	0.31	0.1
9.17	0.5	0.46	0.22
9.22	0.5	0.44	0.1
10.1	0.45	0.44	0.17
10.2	0.4	0.36	0.23
16.19	2.1	1.74	0.28
18.10	0.5	0.54	0.2
19.9	1.3	0.39	0.08
21.9	1.3	1.8	0.38
51/11	0.3	0.2	0.48
Sample 352	0.15	0.3	0.8

Lead content as determined by atomic absorption spectrometry (AAS) on drillings (given for comparison).

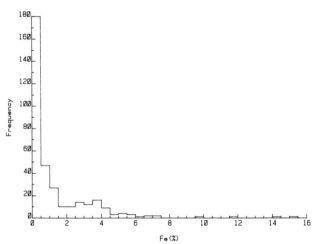

Diagram 1. Iron content of copper-base artifacts from Timna.

added as flux can itself be reduced to iron metal and dissolved into the copper. As the copper solidifies, the iron precipitates out of solution as an iron-rich phase. The iron content of the Timna metalwork must have been a regular feature of ancient copper before purification, for example the Roman-Early Islamic ingot from Beer Ora has 18% of iron, and further afield a lump of Late Bronze Age copper from Ras Al Khaimah has 8% iron (Donaldson, 1985, 99; see also, Cooke and Aschenbrenner, 1975). Iron is not only found in copper castings, as both the cast and hammered components of a Geometric Greek cauldron (Ashmolean Cat. 377) of copper and bronze contained between 4 and 8 per cent of iron in the metal (Boardman, 1961, 78, 160; Craddock, 1977, 115). This cauldron, and much of the Timna metal, indicates that copper can still be satisfactorily used and worked, even when it contains a high percentage of iron.

It is now clear from the study of the Timna metal finds that some metallic iron was usually incorporated into the copper and that it required some deliberate effort to remove it. Tylecote and Boydell (1978) removed it from the experimental copper by a simple remelt in a fireclay crucible in which the iron oxidised and combined with the silica of the crucible to form a slag that could be removed. Merkel was able to achieve almost total removal by directing an air blast across the surface of the molten metal (Merkel, 1982, 334), but analyses of ancient metal show this was rarely performed. However, if the conditions do not permit the iron to oxidise it will be retained. Thus many of the Timna iron-rich coppers are castings which must have been molten in crucibles and moulds, both made from siliceous materials. Furthermore, the tin bronzes at Timna have the same average iron content as the coppers, yet presumably the bronzes must have had an additional crucible melting to make the alloy, during which more of the iron could have been lost if the conditions had been oxidising. Clearly, without deliberate refining the iron remains in the copper. Several fragments of clay moulds for copper ingots have been recovered at Timna (see Rothenberg in Vol. II of this publication), indicating that raw copper rather than finished artifacts was exported from the site. The rough, unfinished nature of the Temple offerings also suggests that the smelters were not skilled at metalworking. Furthermore, the high iron content of the offerings and of the locally made tools, suggest that the copper was not purified on site but was exported impure. This is reinforced by the bun ingot found in the Arabah, Met. Cat. No. 546, which also has a high iron content. Similarly, the Middle Bronze Age ingots from the Hebron Hills and Har Yeruham examined by Maddin and Wheeler (1976) were also extremely impure and contained random but appreciable iron contents between 0.3 and 1.7%. In contrast, however, the finished artifacts of this period in the Near East do not normally have such high iron contents (see, for example Branigan, McKerrell and Tylecote, 1976), suggesting that the unrefined metal which left

the primary production centre, was subsequently purified and alloyed by the metalsmith who made it into artifacts.

(f) Besides iron, manganese oxides such as pyrolusite (MnO_2) were used as fluxes, especially at Timna Site 30 in the Layer I process, post-dating the Temple. Tylecote *et al.* (1977) successfully used a mixture of iron and manganese oxides as flux and produced copper containing 26.6% iron and 1.5% manganese in the iron-rich phase of the metal, and 0.3% iron and 0.088% manganese in the metal overall. As manganese only occurs very sparingly in the ancient Timna copper objects analysed here, and is in proportion with the iron content of the ore, one must assume that manganese oxides were not used as a flux for any of this metalwork. The site 30 Layer I tap slags are virtually iron free, confirming that no Layer I metalwork was deposited at the Temple, and that the tools analysed from Site 30 (Met. Cat. Nos. 572–578) are all of the Layer III-II period.

(g) The tin content of the artifacts has two important and unusual features, the frequency with which tin is present and its low, erratic concentration in the metal (see Diagr. 2).

In the Metal Catalogue (Chapter III, 7), artifacts with more than one percent of tin are described as 'bronze' rather than 'copper'. No less than 240 of the 336 analysed copper-base artifacts have detectable tin. Many of these items are mere spills or fragments of metal (e.g. the large zoomorphic spill, Met. Cat. No. 4) which were previously thought to be debris of primary smelting, but clearly cannot be if they also contain tin. This is also surprising since many of the fragments have a high iron content showing they are of unrefined metal, although they are alloyed with tin.

The tin content of most groups of early bronzes tend to form a normal distribution around a mean of approximately 7% (e.g. Craddock, 1976). The tin content of the Timna bronzes is both low and erratic by contrast. The highest tin content (13.6%) occurs in the figurine of a mountain sheep (Met. Cat. No. 6). This is one of the few competent castings found at

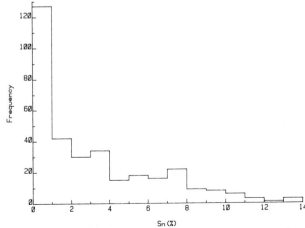

Diagram 2. Tin content of copper-base artifacts from Timna.

Timna; additionally, it contains only traces of iron and on stylistic grounds is believed to be a Midianite import. This contrast also suggests that the majority of the metalwork found at Timna was made locally, using unrefined and poorly alloyed metal, to supply the local needs for artifacts such as mining tools (Met. Cat. No. 552 and 556), personal ornaments and votive offerings.

Many of the artifacts have far too low a tin content to be deliberate alloys (under 1%) and here it is necessary to touch on a broader problem that has concerned archaeologists for some time. Ancient copper artifacts containing a little tin, although rare, are quite widely distributed. A problem is the level at which the tin content is deemed to be a deliberate addition; opinions vary from 0.5 to 3% (Coghlan, 1975, 36). Very low tin contents (under 1%) are usually ascribed to the copper ore (Moorey, 1971, 284-5), but copper ores containing tin (Stannites) are very rare. Tylecote has suggested that the tin may come from the local iron oxide flux, which at Rio Tinto in Spain, for example, contains 0.2% tin. At Timna there are no such problems; neither the copper ores nor the fluxes contain any measurable tin, and tin is absent from the copper prills and primary ingots from the smelter. Directly or indirectly, all the tin has been added after smelting. The solution of one problem poses another: if the very low tin contents are deliberate, why were they added? A tin content of 1% and below has a minimal effect on the physical properties of the copper, so alloying as such can be discounted. However, small amounts of tin would help to melt and purify the copper. Copper melts at 1083°C, tin at 254°C and a 10% tin bronze at about 950°C. It is still common nowadays in small art metal founderies to add a little tin or a high tin bronze to promote the melting of a charge of copper. Furthermore, the addition of small quantities of tin will take oxygen out as stannic oxide which can be removed from the surface. Arsenic will also perform this function, but it is of course absent from the Timna ores. The small quantities of tin found in the Timna metal could have been deliberately added for one or both purposes.

It is not known in what form tin was added to copper to form bronze in antiquity. Either metallic tin or stannic oxide (cassiterite) could have been used since, under reducing conditions, the heat of the molten copper is sufficient to promote the reduction of the tin mineral *in situ* (Marechal, 1963, 41). The almost total absence of tin from ancient sites is puzzling, in view of the large amount incorporated in bronze, and also the references to huge quantities of '*Annaku*', which must surely be tin, made in Mesopotamian trade documents. Charles (1975) has suggested that tin might normally have been traded as cassiterite and thus escape notice by archaeologists. Recently however, the evidence has been transformed by the discovery off the north coast of Israel of many types of ingots of nearly pure tin, weighing between 5 and 25 kilograms each (Galili, Shmueli and Artzy, 1985; Galili, 1986) and dating from the late

second millennium BC to Roman times. The discovery in 1984 of a huge shipload of tin ingots in the Ulu Burun wreck near Kas on the coast of Turkey (Bass 1986) now provides definite evidence for metallic tin as a major trading – and alloying – component in the Bronze Age seaborne metal trade.

Small droplets of tin have been reported in Iberia and Crete from 2000 BC (Forbes, 1972, 161). From Timna itself come two pieces of metallic tin, the droplet (Met. Cat. No. 163) and a small rod, perhaps a ring (Met. Cat. No. 558) (Table 4). The droplet is of tin with traces of lead, copper, antimony and bismuth. These are all metals which are often associated with tin deposits. There are few contemporary tin analyses for comparative purposes but this droplet of tin seems much less pure than the ingots found near Haifa (Maddin *et al.*, 1977). The rod weighs only 0.3 grams and is very corroded, such that only 15% of metallic tin was still present, the remainder being the oxides cassiterite (SnO_2) and romarchite (SnO) with 3% tin. The latter mineral has only been identified recently as a naturally occuring corrosion product. This was on tin pannikins of the 19th century, recently excavated in Canada (Organ and Mandarion, 1971) but its further occurrence at Timna, under very different climatic conditions, suggests that it may be one of the usual forms of tin corrosion on buried objects.

The prills of copper (Met. Cat. No. 551) collected from the excavation of a New Kingdom hearth near the small metalworking site F2 in Timna, contained 0.15% of tin and only 0.16% of iron. Since tin is foreign to the Timna ore deposits, this would suggest that at this period copper was also re-melted for casting at this site.

(h) Almost all the metal contains some lead (Diagr. 3) although not in sufficient quantities to be considered a deliberate addition. The Timna copper ores do contain a small amount of lead, but this is not usually in sufficient quantities to produce the concentrations found in the metal (of the order of 0.25%). However, both the manganese and iron fluxes regularly contain lead. Analyses of the local manganese ores showed

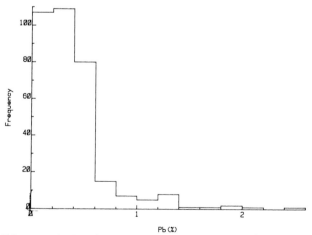

Diagram 3. Lead content of copper-base artifacts from Timna.

them to contain between 0.2–1% Pb (Bodenheimer in Bentor, 1956, 161). Samples of iron ore (from Layer II) and manganese ore (from Layer I, post-Temple, *c.* 10th century BC), from storage installations for flux, found in the excavation of Site 30 and analysed in the British Museum Research Laboratory, were all found to contain lead. Although all of the Timna metal analysed by us appears to have been fluxed with iron, one should expect copper fluxed with manganese from Site 30, Layer I to also contain lead. Furthermore, in Tylecote's experiments where iron and manganese fluxes free of lead were used, the resulting copper was almost free of lead even though the Timna ore used contained traces of lead. This again provides a strong indication that the lead emanates from the flux, not from the copper ore. McLeod, in Rothenberg, 1962, noted the discrepancy between the lead content of the ores and the metal, and Rothenberg (1972, 237) suggested the fluxes as a likely source of the additional lead in the metal, and pointed out the significance of this, if attempting to identify ore sources from artifact analysis.

There are also seven artifacts of lead (Table 4) which are all quite pure. Slatkine (1961, 296) reported wulferite, anglesite, cerusite and plattnerite lead minerals present at Timna and the last three minerals would have been easily reducible. Although no evidence of lead smelting has yet been found, lead isotope analysis has recently shown that the lead is local (Gale and Stos-Gale 1984).

Paul T. Craddock

Bibliography

Bamberger, M. 1985. The Working Conditions of the Ancient Copper Smelting Process, in *Craddock and Hughes* (eds.) 1985, 151–7.

Bentor, Y.K. 1956. The Manganese occurrences at Timna (S. Israel), A Lagoonal Deposit, *Symposium del Manganeso. XX Congreso Geológico Internacional México*, 159–71.

Boardman, J. 1961. *The Cretan Collection in Oxford*, Oxford.

Branigan, K., McKerrell, H. and Tylecote, R.F. 1976. An examination of some Palestinian Bronzes, *JHMS* 10, 15–28.

Caley, E.R. 1964. *Analysis of Ancient Metals*, London.

Charles, J. 1975. Where is the tin?, *Antiquity* 49, 19–24.

Chase, W.T. 1974. Comparative Analyses of Archaeological Bronzes in Beck, C.W. (ed.), *Archaeological Chemistry*, Washington, 148–85.

Coghlan, H.H. 1975. *Notes on the Prehistoric Metallurgy in the Old World*, 2nd Edition, Oxford.

Coghlan, H.H., Butler, J.R. and Parker, G. 1963. *Ores and Metals*, London.

Cooke, S.R.B. and Aschenbrenner, S. 1975. The occurrence of metallic iron in ancient copper, *JFA* 2, 261–72.

Craddock P.T. 1976. The composition of the copper alloys used by the Greek, Etruscan and Roman Civilisations: 1. The Greeks before the Archaic Period, *JAS* 3, 93–113.

—— 1980. Composition of copper produced at Timna, in *Scientific Studies in Early Extractive Metallurgy and Mining*, P.T. Craddock (ed.), London, 165–74.

Donaldson, P. 1985. Prehistoric Tombs of Ras Al-Khaimah, *Oriens Atiquus* 24, 97–101.

Forbes, R.J. 1972. *Studies in Ancient Technology* 9, Leiden.

Gale, N. and Stos-Gale, Z. 1984. Mystery of Timna's iron solved by lead isotope finger-printing, *IAMS Newsletter* 6, 6–7.

Hughes, M.J., Cowell, M.R. and Craddock, P.T. 1976. Atomic Absorption Techniques in Archaeology, *Archaeometry* 18, 19–38.

Jedrzejewska, H. 1962. Sampling precautions in the analysis of antiquities, *Studies in Conservation* 7, 22–32.

Key, C.A. 1980. The Trace-Element Composition of the Copper and Copper Alloy Artifacts of the Nahal Mishmar Hoard, in Bar-Adon, P., *The Cave of the Treasure*, (English Ed.), Jerusalem.

Leese, M.N., Craddock, P.T. Freestone, I.C. and Rothenberg, B. 1986. The Composition of Ores and Metal Objects from Timna, Israel, *Wiener Berichte über Naturwissenschaft in der Kunst* 2.

Lupu, A. 1970. Metallurgical aspects of Chalcolithic copper working at Timna (Israel), *Bull. Historical Metallurgy Group*, 21–3.

Maddin, R. and Wheeler, T.S. 1976. Metallurgical study of seven bar ingots, *IEJ* 26, 170–73.

Maddin, R., Wheeler, T.S. and Muhly, J.D. 1977. Tin in the Ancient Near East, Old questions and New finds, *Expedition* 19, 35–47.

Marechal, J.R. 1963. *Reflections upon Prehistoric Metallurgy*, Lammersdorf.

Merkel, J.F. 1982. *A Reconstruction of Bronze Age Copper Smelting* (unpublished), Ph.D. Thesis, University of London.

—— 1983. Summary of Experimental Results for Late Bronze Age Copper Smelting and Refining, *Masca Journal* Vol. 2, 173–8.

Moorey, P.R.S. 1971. *Catalogue of the Ancient Persian Bronzes in the Ashmolean Museum*, Oxford.

Muhly, J.D., Wheeler, T.S. and Maddin, P. 1977. The Cape Gelidonya shipwreck and the Bronze Age metals trade in the Eastern Mediterranean, *JFA* 4, 353–62.

Organ, R.M. and Mandarino, J.A. 1971. Romarchite and Hydroromarchite, Two new stannous minerals, *The Canadian Mineralogist* 10, 916.

Rothenberg, B. 1962. Ancient Copper Industries in the Western Arabah, *PEQ*, 5–71.

—— 1972. *Timna*, London.

Rothenberg, B., Tylecote, R.F. and Boydell, P.S. 1978. *Chalcolithic Copper Smelting*, Archaeometallurgy, Monograph No. 1, London.

Slatkine, A. 1961. Nodules cupriferes du Negev Meridional (Israel), *Bulletin of the Research Council of Israel* 10, 292–7.

Todd, J.A. and Charles, J. 1977. The Analysis of Non-Metallic Inclusions in Ancient Iron, *PACT* 1, 204–20.

Tylecote, R.F. 1977. Summary of Results of Experimental work on Early Copper Smelting, in Oddy, W.A. (ed.), *Aspects of Early Metallurgy*, London, 5–13.

Tylecote, R.F., Lupu, A. and Rothenberg, B. 1967. A study of Early Copper Smelting and Working Sites in Israel *JIM* 95, 235–43.

Tylecote, R.F., Ghaznavi, H.A. and Boydell, P.J. 1977. Partitioning of Trace Elements Between the Ores, Fluxes, Slags and Metal During Smelting of Copper, *JAS* 4, 305–53.

Tylecote, R.F. and Boydell, P.J. 1978. Experimental copper smelting, in *Chalcolithic Copper Smelting*, Archaeometallurgy Monograph 1, London.

9. EXAMINATION OF SELECTED METAL OBJECTS

Introduction

A cursory examination of the metalwork from Timna indicated that some information about the method of manufacture of various items could be obtained from a more detailed investigation of a few of the pieces.

1 Evidence for Soldering

From among the very large number of metal objects from Timna, only two composite objects have been recorded. The first is the snake (Met. Cat. No. 3) on which a layer of base gold has been applied over the head. The second is an ear-ring (Met. Cat. No.38) which was originally thought to have been gilded. Investigation under the microscope has shown that the assumed 'gilding' is most probably solder which has been used to construct the ear-ring from two separate pieces of metal. A qualitative analysis of the solder by X-ray fluorescence spectrometry indicated the presence of significant amounts of copper, silver and gold, together with a very small amount of tin. The analysis of the lower part of this ear-ring (see Chapter III, 8, Tab. 1) showed that it was made from a low tin (1.5%) bronze, containing the usual trace elements with, in addition, a trace of gold. Although the ear-ring is one of only three of the Timna bronzes to contain a trace of gold, the concentrations of the other minor components of the alloy are comparable with those in the rest of the bronzes, and therefore no inferences should be drawn from the presence of this extra element, which might only represent contamination of the sample by the solder.

If it can be assumed that the tin, which was detected in the solder, represents contamination from the surrounding bronze, the solder may be regarded as a ternary alloy of copper, silver and gold. This is a hard solder which would have a melting point above red heat. Although very little is known about solder alloys before the Roman period, the presence of gold must be regarded as unusual in an alloy to be used for soldering bronze. One Roman method of joining bronze was to use soft solder (i.e. a mixture of tin and lead), but this practice seems to have originated only in the Iron Age, perhaps early in the first millennium BC. Hence, although lead and tin were available at Timna, it would be surprising if they had been used to make a solder. In contrast, the history of hard soldering is much older, going back to at least 3000 BC in Mesopotamia.

2 Evidence for casting

It is clear from crucibles which have been found at Timna that metals were cast on site and it may be inferred that the crucibles were used either in the final stage of copper production, i.e. for purifying the metal, or for alloying the copper with tin. The absence of any recognisable mould fragments from the site, together with the general appearance of the metal finds which suggests that they were almost all made by working, might be taken to indicate that these crucibles were not used in the manufacture of cast objects but only for the production of ingots. However, microscopic examination of the surface of a large number of the Timna finds has revealed two objects on which dendritic patterns are very clear, three others on which they are just recognizable, and a number of similar items on which dendrites may be present although the appearance of the surface is such that it is impossible to be certain.

The spatula (Met. Cat. No. 482) and the pointed, square sectioned, rod (Met. Cat. No. 62) both have very clear dendritic patterns on the surface (*Pl.* 130:1, 2). This indicates that these were cast approximately to shape but, since the dendrites are slightly distorted, the objects must have been finished by working and annealing to obtain the final dimensions.

A taper section was polished on the broken tang of the spatula and this was examined under the metallurgical microscope. The metal was found to contain a large number of pores, which is a further indication that the object was cast. However, etching with alcoholic ferric chloride revealed a pattern of twinned equiaxial crystals superimposed on a dendritic structure and this again indicates that the metal has been worked and annealed. Although some deformed annealing twins are visible, showing that the final operation was working, the essentially undistorted dendritic structure which is visible is evidence that the post-casting working was not very extensive.

Apart from the two objects on which the dendrites are quite clear, there are some indications for a cast structure on the surfaces of a piece of thick rod with spatulate end (Met. Cat. No. 475), and ear-ring (Met. Cat. No. 29) and a spatula (Met. Cat. No. 484), and possibly slight evidence for casting on the surfaces of a number of other pieces of rod of various thicknesses. It has also been postulated (Chapter III, 10) that some (or most) of the bronze wire artifacts may have been made from cast wire. Indeed, the ear-ring mentioned above could have been fashioned from a piece of thick cast wire.

The evidence for casting adds a new dimension to the range of metalworking operations at Timna. Hitherto it had been assumed that the artifacts were mostly wrought from ingots; but it now seems clear that some mould-making process was regularly carried out. None of the shapes are very intricate, and the spatulas and thick square rods could have been cast in open moulds.

However, if the round sectioned rods and, especially, the bronze wire are in fact cast, then the use of closed moulds is indicated. The absence of any flash-lines on the surviving objects would suggest that a lost-wax process was used, rather than piece-moulds.

W.A. Oddy

Bibliography

Tylecote, R.F. 1968. *The Solid Phase Welding of Metals*, London.
Tylecote, R.F. and Boydell, P.J. 1978. Experimental copper smelting, in *Chalcolithic Copper Smelting*, Archaeometallurgy Monograph No. 1, 27–51.

10. WIRE ARTIFACTS

Introduction

Among the metal finds from the Temple are a large number of pieces which are best described as wire. In many cases the artifacts are short lengths of wire which do not seem to have been produced for any particular purpose, but which may represent offcuts from longer pieces. In other cases the wire has been made into a recognisable object such as a finger ring, chain-link or an ear-ring, or has had one end bent over into a loop to form an implement of uncertain use (e.g. see Met. Cat. Nos. 219–262).

Almost all the scrap wire and objects of uncertain use are approximately round in cross-section, but many of the fabricated objects such as finger rings and pieces of chain have been made from wire with an oval cross section. A large number of the metal finds from the Temple site were examined briefly and from the artifacts which could be described as wire, 66 items were selected for a more thorough microscopic examination with a view to determining how the wire was made. The items selected covered both the round and oval wire, and included what appeared to be offcuts as well as implements. Pieces of wire in metals other than copper were also included in the survey.

1 Wire in antiquity

Among recent studies of wire-making in antiquity is Oddy, 1977, which includes a bibliography of earlier work. The questions posed in that paper were (a) when was wire-drawing invented, and (b) how was wire made before the invention of drawing?

With regard to the question of the drawing of wire, all the available evidence points to the use of the draw-plate becoming widespread in the early medieval period, although earlier claims have been made for its invention (Rump, 1968; Thomsen and Thomsen, 1976; Epprecht and Mutz, 1974–5; Furger-Gunti, 1978). In a recent survey of ancient gold jewellery from all over the Old World, which was carried out at the British Museum, the earliest pieces of undoubtedly drawn wire which have been identified are dated to the 6th or 7th centuries AD and they came from sites as far apart as Sweden and Egypt. The survey of British Museum material was restricted to gold wire because its lack of susceptibility to corrosion preserves the marks on the surface which can be associated with the processes of manufacture; it may be significant that in cases where drawn wire has been claimed for earlier periods the wire in question has been of copper or one of its alloys (Epprecht and Mutz, 1974–5; Anastasiadis, 1950). The well-known piece of wire cable from Pompeii, which is often quoted as an example of Roman drawn wire (Aitchison, 1960; Feldhaus, 1931) is now known to be relatively modern (Dickmann, 1962: Jüngst, 1977) and, similarly, the gold wire on the Persian rhyton in the Metropolitan Museum, which was previously thought to have been drawn (Maryon, 1956; Thomsen and Thomsen, 1976) is now thought to be hand-made (Meyers, 1978).

A number of claims that wire was drawn in the Roman period and earlier, rest on the findings of what appear to be wire drawing plates in archaeological contexts (Rump, 1968; Thomsen and Thomsen, 1974). The problem with this type of evidence is that nail-makers use a tool which is very similar in appearance to a wire-drawing plate and thus, in the absence of corroboration in the form of archaeologically stratified wire which is undoubtedly drawn, the finding of an iron plate pierced with small holes cannot be taken as unequivocal evidence of wire-drawing.

As far as Timna Site 200 is concerned, it is the techniques of making wire by hand which are important. Oddy (1977) identified four main techniques called respectively *hammering*, *block-twisting*, *strip-twisting*, and *strip-drawing*. These can normally be identified and differentiated by characteristic marks on the surface which are the result of the different manufacturing processes, although these may become indistinct as a result of wear, abrasion or the products of corrosion. Also, if the wire has been finished to a high degree of perfection the characteristic marks may disappear. However, since these marks are not normally visible to the naked eye, the wiremaker did not usually undertake surface finishing and most hand-made wire does, in fact, preserve some evidence of the technique used to make it.

Wire made by *hammering* is usually characterised by having a faceted surface rather than a smooth one, and short irregular 'creases' are often visible on the surface resulting from the way the metal has stretched under the effect of the hammer. In some cases hammered wires are found which are very smooth and regular in cross-section, and in these cases it is possible that a swage block was used to make the wire (Tickle, 1978).

Block-twisting is the result of cutting a strip of metal from the edge of a thick sheet, so that the strip has a square cross-section, twisting it to give a 'barley-sugar' effect, and then rolling between two hard flat surfaces. This gives a solid wire which has two independent helical creases along its length. In describing this technique previously (Oddy, 1977) it was stated that block-twisted wire has four independent helical creases on the surface, but this is only true if the original square wire has an absolutely regular cross-section. With handmade wire this is not normally the case, during twisting and rolling two of the faces of the square-wire become dominant, resulting in only two spiral creases on the finished wire.

Strip-twisted wire is made by spirally winding a long strip of metal foil round another piece of wire (as if making a drinking straw), withdrawing the wire from the centre and tightening the spiral. Characteristic features are a *single* spiral crease and the fact that the wire is hollow. *Strip-drawn* wire can look very similar to this in that it has a cavity down the centre and a single longitudinal crease, but the crease may either run spirally round the wire or more or less longitudinally along it. The wire is made by pulling a strip of metal foil through a series of holes of decreasing diameter so that it curls up longitudinally into a hollow tube which gradually tightens up as the wire is pulled through smaller and smaller holes. If the wire is rotated as it is pulled, the crease will form a spiral round the wire. This technique must not be confused with wire-drawing, since the process of pulling the strip through the holes does not lengthen the wire, but only decreases its diameter. Also the resulting wire will have a cavity down the centre and this is obviously not present in drawn wire.

2 The wire artifacts from Timna

The 66 wire artifacts selected for detailed investigation were examined under a binocular microscope at magnifications ranging from ×6 to ×25. It quickly became apparent that none of the wire showed any evidence whatsoever of having been drawn and, although one fragment has a striated surface (Met. Cat. No. 371), it has a non-uniform cross-section, which obviously precludes drawing. The striations on this fragment appear to be the result of unusual corrosion effects on an alloy which probably has a lamellar structure as a result of working.

Virtually all the wire was of copper or a copper-based alloy, with the exception of five of the fragments. Met. Cat. No. 124 is a piece of thick iron wire which may be modern although the surface of the wire suggests that it was made by hammering, and not by drawing, as would be expected if it was a modern intrusion. However, a pseudo-hammered surface could well result from the flaking away of corrosion products and therefore the question of its origin must rest upon a metallographic examination rather than on either its composition or the appearance of the surface. The other four non-copper wires are all made of lead. Met. Cat. No. 165 is a piece of thick lead wire which has a longitudinal seam. This would normally be taken to indicate that it was made by *strip-drawing* but in this case there is reason to believe that a handmade technique not previously identified was used. It is proposed to describe this as *folding* (see below). The other three fragments of lead wire are thinner and are marked collectively as Met. Cat. No. 162. One piece is tightly rolled into a ball and details of the surface are obscured by the corrosion products; nevertheless there is slight evidence that it was also made by *folding*. The other two pieces clearly exhibit a longitudinal crease, again indicating *folding*.

When the copper alloy wires were examined it became apparent that they could be divided into two main groups. One group consisted mainly of manufactured items, mostly finger rings and fragments of chain, which have a smooth surface and a round or oval cross-section. Care has been taken in 'finishing' the wire in this group and no striations or creases, which might indicate how it was made, are visible on the surface. The only clues are from Met. Cat. No. 167 and 119, which appear to have a dendritic structure on the surface revealed by the etching effects of the corrosion process, and from Met. Cat. Nos. 57 and 395, which have been metallographically examined by Tylecote (Chapter III, 11) and shown to have been 'superficially cold-worked' and to have 'a twinned equiaxed structure' indicating that the wire was 'worked and annealed'. It can be postulated that the wire in this group has been made by casting roughly to shape, followed by annealing and working to the final dimensions. The alternative explanation is that the wire was forged from small ingots; but if the patterns in the surface of Nos. 167 and 119 really are dendrites, then these pieces at least must have been cast roughly to shape. It is interesting to note that dendrites definitely do occur on the surface of some of the larger items from Timna (e.g. the spatula, Met. Cat. No. 482, and the square-sectioned pointed rod, Met. Cat. No. 62) and there is little doubt that in these cases the objects were cast roughly to shape and then finished by working and annealing.

The wire of this first group must be regarded as a variation of the hammered type (Oddy, 1977) in which casting roughly to shape may have been used to decrease the amount of hammering which would otherwise have been required in making the wire from an ingot.

The second group of copper-based wires had two characteristic surface features. First, the surface was usually rough and uneven, although this might have been partly due to the corrosion processes and, second, a single longitudinal crease was present running the length of the wire. In some cases the crease was only visible along short stretches of the wire, having become obscured by corrosion products along much of its length. There is, however, no doubt that all the wires in this group were made by the same technique.

A single longitudinal crease is usually regarded (Oddy, 1977) as characteristic of *strip-drawn* wire, but in the case of the wire from Timna careful examination showed that this method of manufacture could not have been used. The seam or crease on *strip-drawn* wire is the result of the rolling up of the original strip of metal, which was always very thin in comparison with its width, whereas in the wire from Timna the longitudinal crease clearly delineates a butt-join between the edges of a thick sheet of metal. In many cases a split is visible running along the crease and this split penetrates to the centre of the wire. On some of the wires, particularly those which have accidentally been broken in modern times, the end section resembles a letter 'C' with very thick walls, and it is clear that the wire has been made by *folding* longitudinally a strip of metal whose width must have been approximately twice its thickness. The resulting 'wire' was then rounded by hammering. A typical transverse section of one of the pieces of wire made by *folding* is illustrated in *Pl*. 130:4 and the postulated scheme for the manufacture of the wire is shown in *Ill*. 28. A blunt chisel has been introduced as a method of scoring the metal strip so that it is easier to fold, but there is no evidence for this in the cross-section.

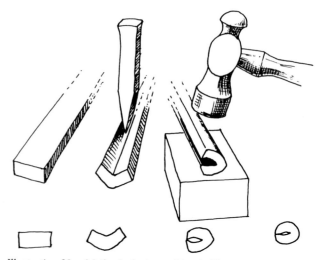

Illustration 28.   Method of wire making in Timna.

One distinctive fact about the two groups of wire which it is assumed were made by *casting and working* and by *folding* respectively, is that they differ not only in the way they were made, but also in the composition of the alloy used. Of the 18 wires with smooth surfaces and round or oval cross-sections which were made by casting and working, all except two were made of bronze, or at least of copper containing a deliberate addition of tin, i.e. between 0.5 and 10% tin. The two exceptions are Met. Cat. Nos. 438 and 444, in which tin is completely absent. Of the 29 pieces of wire which were made by *folding*, 23 contain no tin but the other six pieces contain between 0.8 and 6.8% tin.

Examination of the analytical results shows that the iron content, as well as the tin content, differs according to whether the wire is made of copper or bronze. The bronze wire artifacts all have iron contents $\geq 0.1\%$ (13 analyses altogether) which is unusually high for Bronze Age metalwork, while the *folded* copper wires, with three exceptions, have iron contents $\leq 0.1\%$ (18 analyses altogether).

The differences between the two groups of copper-alloy wire can be summarised as follows, although it must be remembered that there are some exceptions to the classification:

Wire made by *folding*	Wire made by (*casting ?*) and *hammering*
Short lengths without recognisable use	Finished objects such as finger rings and chains.
Made of copper	Made of bronze (0.5% < [Sn] < 10%)
Iron content ≤ 0.1%	Iron content ≥ 0.1%

Craddock has commented (see above, Chapter III, 8, 3 (a)) on a possible reason for the presence of small amounts of tin (<1%), which he considers too small to result from deliberate alloying. He regards the tin as originating in the 'fluxing' in the crucible of copper from which the objects were made, in order to initiate the melting. This somewhat arbitrary classification is not, however, altogether supported by the results of the examination of the wire. Although the tin contents of the (cast and) worked group of wire objects lie either within the ranges 0.5–1.5% (3 analyses) or 5–10% (10 analyses), too much emphasis must not be placed on this apparent separation into groups which may be the result of too few quantitative analyses to be statistically significant. In view of the total spread of the results (from 0.5 to 10%) it seems inappropriate to differentiate in nomenclature between those alloys containing less than 1% of tin and those containing more than 1%, and it is better to regard those wires which contain any tin at all as bronzes.

The presence, or absence, of tin in the wire artifacts also correlates with whether the wire appears to be a deliberately manufactured object (i.e. the cast and hammered bronze artifacts) or merely scrap material (i.e. the folded wire). It could be postulated that the folded copper wire represents ingot material which was prepared in a form suitable for weighing out into a crucible for making an alloy; but in that case it is

difficult to understand why such trouble was taken to make it into wire first.

It should, however, be noted that the correlation of the tin content with the type of artifact does make metallurgical sense. Objects such as finger and ear-rings are much more serviceable when made in bronze while, on the other hand, one of the essential properties of a wire is that it should be flexible, and for this unalloyed copper is better than bronze. Hence it is not really surprising to find that the *folded* wire usually contains no tin while the *cast and hammered* wire artifacts are made of bronze.

The obvious corollary to the fact that, as a general rule, the objects which have been deliberately manufactured into finger rings, ear-rings and chains are made of bronze and contain an unusual amount of iron, is to ask whether the explanation lies in the fact that these bronze objects were imported into Timna. This is a possibility because of the complete absence of tin at Timna, both in the copper ore and the fluxes. The question is also especially pertinent when it is remembered that, although some at least of the bronzes found at Timna were cast roughly to shape, no fragments of moulds have been recognised among the other excavated finds. (However, see Met. Cat. No. 2 and Chapter III, 7, note 1 – the Editor).

If the bronzes were not imported, then the analytical data indicates that the process of alloying the copper with imported tin has usually also introduced more iron into the alloy, and this effect is contrary to normal metallurgical expectation.

The clue to whether the bronzes were imported or made on site comes from an examination of the six exceptional *folded* wires which do contain tin. Five of these also have iron contents which are $\geq 0.1\%$ and so correlate with the iron contents of the *hammered* bronze wire artifacts. On the face of it, it seems more sensible to suggest that all the *folded* wire had a common origin, presumably at or near Site 200, and thus there is no need to postulate that the bronze wire artifacts were imported.

In addition to the copper wires which were almost certainly made by *folding*, there are eleven other copper wires for which it was impossible to establish a method of manufacture because the surface was either devoid of characteristic marks or obscured by corrosion products. Also there are four pieces of wire which differ in some respect from either of the two main groups. Met. Cat. No. 364 is made of copper and has a square cross-section with no evidence of seams or creases. It was probably made by *hammering* and this agrees with observations on the metallographic structure reported by Tylecote (Chapter III, 11). Met. Cat. No. 194 has an approximately square or trapezium-shaped cross-section which is very irregular. It appears to be a strip cut from the edge of a piece of sheet bronze. Met. Cat. No. 491 is made of copper foil which has been rolled longitudinally to form a hollow wire. This resembles *strip-drawn* wire, but in this case the wire must have been made entirely by hand as it varies in diameter along its length. One end is flattened, presumably as a result of cutting the wire. Finally, Met. Cat. No. 82 is made of bronze and has *two* longitudinal seams running along opposite sides of the wire. It appears to have been made by hammering two strips of bronze together, but this would be without a known parallel and seems to be a very strange way of making wire, unless it represents a piece of folded wire which split longitudinally during manufacture.

3 Summary Table

The following table summarises the conclusions arising from the microscopic examination of 66 of the objects made of wire which were selected for detailed examination.

(1) Objects made by *folding*
   Made of copper:
   Met. Cat. Nos. 30, 58, 67, 78, 84, 93, 113, 118, 133, 137, 143, 147, 151, 154, 157, 222, 226, 231, 233, 235, 237, 239, 242.
   Made of bronze:
   Met. Cat. Nos. 91, 98, 116, 225, 232, 246.
   Made of lead:
   Met. Cat. Nos. 162, 165.

(2) Objects made by (*casting ?*) and *hammering*
   Made of bronze:
   Met. Cat. Nos. 47, 53, 57, 167, 218, 319, 320, 369, 371, 378, 379, 394, 395, 437, 457, 504.
   Made of copper:
   Met. Cat. Nos. 438, 444.

(3) Objects whose method of manufacture is uncertain
   Met. Cat. Nos. 88, 102, 104, 122, 130, 141, 168, 241, 381, 413, 415.

(4) Miscellaneous
   Made of iron:
   Met. Cat. No. 124.
   Made of copper or bronze:
   Met. Cat. Nos. 82, 194, 364, 491.

W.A. Oddy

Bibliography

Aitchison, L. 1960. *A History of Metals, Vol. 1*, London, 214.
Anastasiadis, E. 1950. Bronze Welding, Riveting and Wiremaking by the Ancient Greeks, *Metal Progress* 58, 322–4.
Dickmann, H. 1962. Neue Beiträge zur Geschichte des Drahtseiles, *Stahl und Eisen* 82, 166–9.
Epprecht, W. and Mutz, A. 1974–5. Gezogener römischer Draht, *Jahrbuch der Schweizerischen Gesellschaft für Ur-und Früh-geschichte (Basel)* 58, 157–61.
Feldhaus, F.M. 1931. *Die Technik der Antike und des Mittelalters*, Potsdam, 215.
Furger-Gunti, A. 1978. Gezogener Draht an keltischen Fibeln des 1 Jahrh. v. Chr., *Draht* 29, 727–30.
Jüngst, H. 1977. Personal communication by letter, August 1977.
Maryon, H. 1956. Fine Metal-work, in *A History of Technology* Vol. II (ed. Singer, C. *et al.*), Oxford, 449–92.
Meyers, P. 1978. Personal communication, March 1978.
Negbi, O. 1976. *Canaanite Gods in Metal*, Tel Aviv.
Oddy, W.A. 1977. The Production of Gold Wire in Antiquity, *Gold Bulletin* 10, 79–87.
Rump, P. 1968. Beitrag zur Geschichte des Drahtzieheisens, *Stahl und Eisen* 88, 53–7.
Thomsen, E.G. and Thomsen, H.H. 1974. Early Wire Drawing through Dies, *Journal of Engineering for Industry* (Transactions of the American Soc. Mech. Engineers. Series B) 96, 1216–21.
—— 1976. An analysis of Wire Making in Antiquity, in *Proceedings of the Fourth North American Metalworking Research Conference* (ed. Altan, T.), Columbus, Ohio, 140–46.
Tickle, M. 1978. Personal communication

11. METALLURGICAL NOTES ON SELECTED METAL OBJECTS

The objects dealt with below were received from H.G. Bachmann, who had carried out a qualitative X-ray fluorescence examination on them. All but three were found to be copper-base materials; half of these had substantial amounts of tin and were therefore classified as bronzes, while the rest were pure or impure coppers. On the whole there was very little lead present and it seems that there was no attempt to dilute the bronzes with this element; all traces of lead must have been introduced from the ores or fluxes.

Met. Cat. No.	Arabah Expedition Anal. Sample No.	
21	404	Nearly all rusted iron (?) with some spots of Cu.
22	638	Gold plated mineral probably originally iron. As usual the gold tells us very little. The hardness is 79 HV1 which suggests that it is not pure gold but 18 carat.
23	642	Iron compound within shell. Traces of magnetite in an iron corrosion product. No metallic iron remains.
33	296	A dilute corroded bronze with elongated grains but without signs of final cold working. Hardness 106 HV.
50	334	A fine-grained copper with a little slag. The grain size is variable in places, the grains are twinned and there are signs of intense deformation near the surface. Hardness 125 HV1.
55	291	Chain. A copper-base solid solution sectioned through the direction of slag/oxide inclusions. The grain size is coarse and the shape equiaxed; it is twinned with slip bands. The hardness is 97 HV1. A dilute tin bronze which has been worked and annealed and finally cold worked.
56	639	Chain links. A copper-base solid solution which is badly corroded leaving pores which may have contained lead. The structure is equiaxed and there are traces of coring which suggests that it has not had much heat and work since casting. The hardness varies from 133 to 99 HV1

186

Met. Cat. No.	Anal. Sample No.	
		which indicates an appreciable tin content (6 to 10%).
57	338	Corroded copper-base wire showing deformation markings near surface. Superficially cold worked.
59	635	Pointed wire. Another piece of copper-base solid solution with twinned equaxed grains, but little slag and no slip bands. Hardness 103. Probably a worked and annealed tin bronze.
61	298	Very hard bronze wire with both oxide and slag inclusions.
87	353	Copper with a few oxide inclusions; has a worked equiaxed grain structure with twins. Hardness 140 HV1.
98	337	Copper with a rather coarse structure which is twinned and has had a considerable amount of cold work near the edges. The centre has a hardness of 93 HV1.
121	399	Almost fully corroded; some equiaxed residual copper.
134	354	Impure copper rod with cuprous oxide inclusions. Equiaxed structure. Hardness 100 HV1.
137	401	Copper with a little grey slag but no Pb. Cold worked.
149	343	Badly corroded worked bronze wire. Equiaxed. Hardness 138 HV.
150	355	Wrought and annealed very dilute bronze wire. Hardness 80 HV.
162	266	Lead. When etched there were large areas of primary lead with small amounts of copper. The hardness was less than 6 HV.
171	368	An equiaxed bronze, twinned with deformation markings. Some slag but no delta nor lead. Hardness 113 HV1. Heavily worked dilute bronze.
179	370	A bronze with fine grain, much corroded along the slip bands. Hot worked and finally cold worked to give a hardness of 150 HV.
181	372	Mostly oxide, stemming from an equiaxed copper.
186	396	A dilute bronze; corroded with equiaxed worked structure.
190	400	Copper showing intergranular corrosion near the surface. Fine grained with twins. Some large cuprous oxide dendrites and fine precipitate of same in grains. Well worked. Hardness 168 HV1.
198	341	An awl. Fairly pure copper with slag and cuprous oxide. It has a fine elongated grain structure suggestive of wire drawing. The hardness was 156 HV which indicates a fully hardened copper or dilute alloy.
206	365	A heavily corroded, heavily worked bronze showing slip markings and precipitated copper in the corroded grain boundaries. Hardness exceeds 186 HV.
210	406	Very heavily worked copper with elongated complex oxide inclusions. Fine grain twins and slip markings. Hardness 168 HV1. Hot and cold worked.
215	373	Heavy copper bar with cuprous oxide in grains and in the boundaries. Equiaxed and twinned structure with heavy distortion at the edges. Hardness in centre 57. Superficially cold worked annealed copper.
220	637	Wire loop. Very similar to Met. Cat. No. 222, but the hardness is only 79, suggesting less tin and definitely no final working.
222	636	A piece of wire with some slag and equiaxed and twinned grains. Hardness is 103 HV.
225	402	Well oxidized bronze with a fine equiaxed structure and some elongated grains showing twins. Wrought bronze.
246	342	Very similar to Met. Cat. No. 198. It has a fine equiaxed structure at the edges and is generally more equiaxed and less worked. This is confirmed by the lower hardness of 125 HV1.
248	397	Bronze wire containing a lot of oxide stringers. Well worked, elongated grain structure. Hardness 234 HV1.
255	633	Square rod. A copper-base solid solution with lots of blue stringers of slag, oxide or sulphide, more probably the latter. The grains are small, equiaxed and twinned and the hardness is 125 HV, all of which suggest some final cold work imposed on an otherwise hot worked structure.
260	405	A dilute bronze with oxide inclusions; heavily worked with twins and slip markings. Hardness 153 HV1.
266	333	This is a wrought bronze with twins and slip markings. There is no delta nor lead but a beautiful destannified copper grain boundary nearly going right through the section. This would suggest a fair amount of water in the deposit. The hardness is 186 HV1.
267	395	Almost no metal left, but is clearly a worked or drawn structure with slag stringers showing direction of working.
273	335	Corroded bronze strip. Equiaxed worked structure with deformation

Met. Cat. No.	Anal. Sample No.	
		markings. Hardness 168 HV1. Finally cold worked.
274	411	Heavily corroded fine equiaxed bronze with some twinning. Heavy cold working.
292	363	Well-worked dilute bronze containing oxide. Fine grained with twins and slip markings.
305	643	Ring with cowrie shell. A well-corroded copper-base solid solution with equiaxed grains. A wrought copper-base alloy.
313	336	Bronze in good condition but some intergranular corrosion near the surface. Fine equiaxed grains with twins and some slip bands near the surface. Hardness 143 HV1. Worked and annealed bronze with superficial cold working.
322	634	Wire. A copper-base solid solution with equiaxed grains with twins and slip bands. The slag particles are small and it would seem that this piece has been mostly cold worked and annealed followed by final cold working to give a hardness of 100 HV1.
361	348	A very dilute bronze in annealed state with a fine grained equiaxed structure with twins. Hardness 86 HV; seems to confirm absence of final cold work.
363	293	Very elongated slag stringers as in Met. Cat. No 368. The metal has fine grain but no slip bands. Hardness is 123 HV. Worked and annealed.
364	344	Copper with intergranular corrosion. Heavily worked with twins and slip bands. Hardness 106 HV.
365	646	A double wire ring. A copper-base solid solution much corroded. The grains are equiaxed and twinned. Hardness 117 HV.
368	360	A bronze with a considerable amount of slag in the form of elongated stringers which are evidence of considerable hot working. The grain size is fine and the twins are bent. This, and the hardness of 153 HV show that the piece has been finally cold worked.
369	645	A rod. Copper-base solid solution with equiaxed grains, twins and slip bands. The hardness is 145 HV.
374	644	A thin sheet ring. A copper-base solid solution with traces of the original delta constituent which shows that it has not been completely homogenised or that the tin content

Met. Cat. No.	Anal. Sample No.	
		exceeds 13%. The grains are equiaxed and twinned and final cold work is shown by the presence of slip bands. The hardness is 117 HV which is more consistent with a dilute tin bronze.
393	362	Bronze wire ring. Equiaxed worked structure with no visible twins but heavily elongated slag stringers. Hardness 133 HV1. Hot worked bronze.
395	346	A bronze with some grey slag and no Pb. Shows a twinned equiaxed structure but no delta. Hardness 133 HV1. Worked and annealed.
404	374	Corroded dilute bronze with equiaxed structure.
431	409	Small, very hard bronze ring with complex oxide stringers. The structure is twinned and has a hardness of 239 HV1 showing clearly that it is still in the cold worked state. Severely hot and cold worked bronze.
447	351	A dilute bronze with slag. Has an equiaxed structure with corroded grain boundaries near the edges; the grains here show signs of heavy deformation. Inside, the grains are twinned with some deformation markings. There is some slag but no delta nor lead. Hardness 132 HV1. Very dilute α-bronze with considerable cold work.
449	347	A cold-worked bronze with corrosion along the slip bands. No Pb. The grains are equiaxed and twinned; no delta. Hardness 150 HV1.
464	356	A heavily worked and corroded dilute bronze.
465	359	Awl? of hard dilute copper alloy. Equiaxed worked structure with twins. Very fine grain. Hardness 178 HV1. Heavily worked.
478	366	Flattened dilute bronze rod with a twinned equiaxed structure. Marked evidence of destannification in the grain boundaries. No slip markings and the hardness of 97 HV1 suggests that it is still in the annealed condition or has been only lightly worked.
481	358	A bronze spatula. Equiaxed grains with intercrystalline corrosion. Twins and slip markings show it to have been heavily worked at the edge. Hardness, however, is only 75 HV1, no doubt due to the corrosion.
495	640	Copper-base alloy strip with some slag. The grains are twinned and

Met. Cat. No.	Anal. Sample No.	

equiaxed and the slag stringers are long, suggesting hot working. Hardness is 127 HV1.

| 496 | 361 | Distorted grains with complex oxide inclusions and twins. Hardness 133 HV1. |

| 497 | 641 | Tube with iron rivet. A well-corroded copper-base solid solution with twinned equiaxed grains, some slag but no slip bands. The hardness is only 76 HV1 which suggests a fully annealed dilute alloy. Rivet fully oxidized to magnetite, etc. |

| 502 | 340 | Very corroded. Equiaxed structure with evidence of slip markings. Worked, annealed and finally cold worked copper. |

| 505 | 371 | Wrought annealed copper with cuprous oxide. Very fine equiaxed grains with twins. Hardness 70 HV1. |

| 507 | 369 | Fine grain bronze with intercrystalline corrosion near surface. No Pb. The grains are twinned but no slip bands are visible. |

| 508 | 339 | A very dilute bronze consisting of fine, elongated grains and slag. The tin content must be very low. Hardness 90 HV1. Worked annealed α-bronze. |

| 509 | 410 | A dilute bronze with complex oxide inclusions and destannification. Fine equiaxed grains with twins and slip markings. No lead. Hardness 121 HV1. Cold working. |

| 510 | 367 | Copper with a very fine precipitate which might be oxide. A twinned and distorted structure with slip bands near the surface. Hardness 148 HV1. Heavily worked impure copper. |

| 511 | 457 | Corroded bronze with twinned equaxed structure and strain markings. Cold working. |

| 513 | 454 | Piece of magnetic copper (Pl. 125:3). Copper with 'globules' of iron and some cuprous oxide. On etching the iron particles were found to be complex, probably due to the precipitation of the 8% copper that is soluble at 1100°C, on cooling. There may also be some carbide present as sorbite, but this is unlikely in the presence of the cuprous oxide phase in the copper. The iron had behaved in a ductile manner during working as might be expected. The particles had distorted during the working of the strip. The hardness was 95 HV1 showing that iron gives appreciable |

Met. Cat. No.	Anal. Sample No.	

second phase and solution hardening to copper. At room temperature the equilibrium amount of iron in solution is less than 0.1%, but it is unlikely that equilibrium is obtained under normal conditions, and the effective solubility may be as high as 1 or 2%.

Items not in Met. Cat.

| 350 | (Field No. 319, Loci 106, 108). Solidified copper with at least two other phases within it, with dendritic and grain boundary distribution. One is light grey, probably iron or sulphide and the other dark brown which darkens further on etching to give a 2–phase structure. There is no fayalite slag so the phases are probably matte and iron. A small amount of lead may be present. |

| 352 | (Field No. 154). A cast copper showing some coring with a good deal of pro-eutectic cuprous oxide and some oxide precipitate. Lead may also be present; some holes. Hardness 61 HV1. (See Chapter III, 8, Table 3). |

| 357 | (Field No. 253/4, Locus 102) (Ill. 29; Fig. 84:128). Gold. This piece is made from three pieces of gold foil. The major part consists of two pieces welded or brazed intermittently along the length and one end has been bent into a loop. About halfway along the artifact a third piece has been joined on – this is now very short, like a tag. |

It is difficult to decide whether the joins have been made by hammer welding which is comparatively easy on gold (e.g. the gold boxes in the Museum at Dublin) or whether they have been made by soldering with copper, a technique known since the

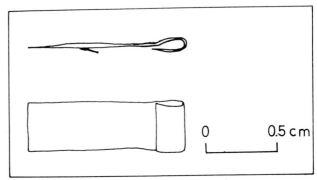

Illustration 29.   Gold foil; hammer welded or brazed.

Met.	Anal.
Cat.	Sample
No.	No.

EBA. If the latter process has been used it has been done very well, so that there is no residual copper and most of the latter has been diffused into the gold. But we should have expected the join to be more complete and less intermittent and, for this reason, think it is a good example of pressure welding.

The metal has an equiaxed structure. The microhardness was 92 HV (50g) and it is most likely natural gold with less than 1% copper and 10–20% silver. (See Chapter III, 14, part 2, 2).

377   (Field No. 240, Locus 109). Runner. Cast with cored structure with cuprous oxide and more complex precipitate. No delta eutectoid and the hardness of 84 HV suggests an impure copper (see Chapter III, 8, Table 3).

407   (Field No. 221/1, Locus 101). Worked copper sheet with equiaxed grains and complex oxide inclusions (Cu_2O + SnO_2?). The grain boundaries are often corroded but there is a subgrain structure within which probably stems from recrystallisation after working (see Chapter III, 8, Table 3).

Conclusions

Seventy-five objects were examined and seventy of these were copper-base. The remainder were of gold, lead and iron respectively. Most of the copper-base objects were in a sufficiently good state of preservation for the determination of structural details. With the exception of the two obviously cast copper lumps, 350 and 352 (see also Chapter III, 8, Table 2), all had been wrought and most of these were in a work-hardened condition. The hardness limits of the copper objects were between 70 and 156 HV, while the bronzes were between 86 and 239 HV.

None of the copper-base objects nor the lead or iron showed any special peculiarity, all being worked objects of utility or adornment. No welds were detected in these and the tin content of the bronzes did not exceed 13%. The lead content of the copper-base materials did not exceed about 1%.

The small piece of gold (Anal. Sample No. 357) showed evidence of joining along its length as though it had been made of two foils with a hollow at one end (now closed) and a welded-on 'tag' in the middle (see *Ill.* 29). The hardness would suggest a natural gold without admixture of copper. The 'weld' was made either by hammer welding or by copper brazing (Tylecote, 1968, 189 et seq.). As no sign of copper concentration could be seen in an electron probe examination carried out by Dr H. McKerrell,

we tend to favour the solid phase welding hypothesis, although it is just possible that a post-braze diffusion was sufficient to disseminate the copper and remove any detectable concentration left over after brazing.

As the period in question was a transitional period between the Late Bronze Age and the Early Iron Age, it is not surprising to find some small pieces of iron such as the rivet in Met. Cat. No. 497 and the (?) iron core of Met. Cat. No. 22 with gold wrapped round it. Met. Cat. No. 23 also seems to contain iron, although this may merely be a deposit. It is unfortunate that no metallic iron remains apart from the alpha iron 'globules' in Met. Cat. No. 513.

The production of copper by direct reduction of oxidised ores with the aid of a ferruginous flux, can yield iron as well as copper, and it is possible (Tylecote and Boydell, 1978, 27–51) to recover this iron in a usable form. There is no proof, however, that this technique has been used in this case.[1]

Craddock has calculated the mean composition of the copper found on the site, which shows 1.5% Fe. Experimental smelting work on Timna ores (Tylecote and Boydell, ibid.) showed that such an iron content is easy to obtain. In some cases, however, this iron is in the form of an iron silicate slag and in others it is present as the iron oxide, magnetite or hematite due to its complete corrosion during burial.

It is clear that impure copper was made and used on the site and the overall composition reflects that of the orebody, especially the silver and nickel contents. In order to make bronze, tin would be imported, in the form of metallic tin or bronze scrap. Work-hardening has been extensively used where necessary.

R.F. Tylecote

Note

1. Since this chapter was written (1979), much further work has been done on this subject by an IAMS research group, and there is now evidence for the origin of the iron objects found in the Timna Temple, from a copper smelting furnace at Timna. See IAMS *Newsletter*, No. 6, 1984, 6–7, and N.H. Gale, H.C. Bachmann, B. Rothenberg, A. Stos-Gale and R.F. Tylecote, 'The Adventitious Production of Iron in the Smelting of Copper in Timna', in Vol. II of this publication. (B.R.)

12. ON THE "LOST-WAX" CASTING TECHNIQUE OF A PHALLIC FIGURINE

Introduction

The figurine (Met. Cat. No. 2; *Fig.* 53:2; *Pl.* 126:5–6), probably a fertility idol, was made by the classic 'lost wax' process. When it was discovered, it still had small parts of the mould adhering to it, but these were unfortunately removed and lost (see Chapter III, 7, no. (1)).

The figurine displays a noticeable lack of anatomical proportion but this is common to many votive figurines and idols discovered in the Ancient Near East (Negbi, 1976). Perhaps its most unique and revealing feature is that the right leg is distinctly shorter than the left and its foot is not only twisted but is also reversed, with its heel towards the front.

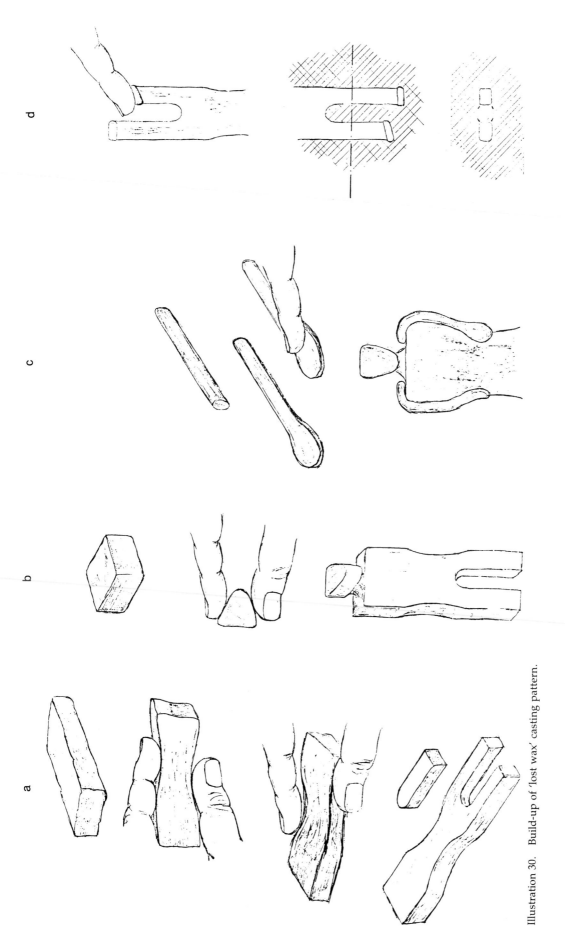

191

Illustration 30.   Build-up of 'lost wax' casting pattern.

The fact that the ancient founder did not completely clean the casting may be taken as proof that upon discovering the idol had a serious defect it was discarded, perhaps put aside for future remelting. It is hardly likely that anyone would have associated himself with a 'crippled' fertility idol.

1 The making of this figurine can be broken down into three basic operations: 1, Making the wax pattern; 2, Preparation of the mould; 3, Casting. The wax pattern was not made by sculpting but was 'built up' in approximately the following way:
(a) The torso was formed from a piece of wax which had been worked into a rectangular pad. The waist was formed by squeezing between the forefinger and thumb. Next, the buttocks were formed by pinching the wax towards the center of the rectangular form. The legs were then formed by removing wax from the lower portion of the form (*Ill.* 30a).
(b) The head was made by pinching a small cube of wax between the thumb and forefinger of the right hand, resulting in the triangular shape of the head (when viewed from above). This rough shape was then pressed onto the torso, the neck being formed by blending the wax with a small modelling stick. The join of the head to the torso is visible on the back of the casting. Finally, a small pad of wax was applied to the top of the head. This is described as a 'crown' and was probably also the downgate through which the metal was poured (*Ill.* 30b).
(c) The arms were made by rolling a small piece of wax into a small diameter 'rod'. Its end was then flattened to form a hand. This is especially evident on the left hand. The arms were then pressed onto the shoulders of the torso. The joins can be clearly seen, especially on the left shoulder (*Ill.* 30c).
(d) The feet were formed by joining two small pads of wax to the ends of the legs. At this stage both legs were of equal length.
(e) Finally the phallus was added, testicles formed and the rest of the anatomical details were added, using a small modelling stick.

2 The wax pattern was then ready for investing with the mould material, doubtless a clay slurry, when a 'mishap' occurred: the right foot pad fell off. The worker must have applied the first coat of slurry by suspending the pattern upside down, gripping it by the foot. He hurriedly pressed the foot back into position with such force that he compressed the leg. He also failed to notice that he had replaced the foot 'back-to-front'. (It is possible that pattern making and subsequent 'repair' were not carried out by the same person.)

When viewed from the front, the right foot appears at a very sharp angle to the horizontal. This would have occurred if the foot was replaced by pressing it into position with the forefinger using a rocking motion. This must have occurred just prior to or at the first stage of investing and for some reason the distortion went unobserved (*Ill.* 30d).

There is a 'flash' build-up around the inside of the legs which seems to be due to fissures in the clay mould, caused by shrinkage cracks formed during the firing to remove the wax and to pre-heat the mould prior to casting.

Eli Minoff

13 DEBRIS FROM METALLURGICAL ACTIVITIES AT SITE 200

Introduction

In all strata of the site, and in most loci inside and outside the Temple, some debris of metallurgical activities – furnace parts, crucible and tuyère fragments and slag of different types – were found, some obviously stray finds out of context, others close to the location of metalworkings. Altogether there were about 3 kg of debris, i.e. evidence for rather small scale metallurgical activities.

The excavations established the existence of two centres of metal working: 1. Stratum V, dated to the Sinai-Arabah Copper Age – Early Phase. At the very bottom of the site's stratification, close to bedrock, and probably related to several rock-cut pits, there were localized concentrations of charcoal ashes, heat-crazed rocks and many small, broken pieces of slag, indicating prehistoric, pre-Temple metallurgical activities at Site 200. Most of these debris were found in the area of Loci 101–102, extending also southwards, underneath the *naos* and *pro-naos* structures. Some of the slags found here, although by their physical appearance similar to casting slag, turned out to be proper copper smelting slag related to the prehistoric, pre-Temple occupation of the site. It should be mentioned here that similar prehistoric occupation debris, also including habitational structures and cult installations, i.e. primitive walling, flint objects, pottery and slag, were found during the Timna Survey (Rothenberg, 1967; idem, 1983) further along the east and south sides of the Timna massif, at the foot of 'The Pillars'.[1]

Locus 109, stratum III (*Illus.* 31), dated to the Ramesside Period (19th–20th Dynasties), was the location of intensive metalworking, apparently a Temple workshop for the casting of votive objects, one of which – a phallic idol (*Pl.* 126:6) – was actually found with part of the casting mould of clay still adhering to it.

Built onto the White Floor were two casting furnaces, Fu I and Fu II, which were functionally and stratigraphically connected with a thick deposit of woodash and other features. Here was found a concentration of metallurgical debris – besides some hard burned and slagged furnace parts, tuyère and crucible fragments (*Illus.* 32–4) with dross incrustations and adhering slag. A brief description of these melting-casting furnaces and their stratigraphic location (see *Illus.* 6, 17 and 31) is included in the excavation report of Locus 109 (above, p. 64). The present chapter gives further details, mainly of Fu I, which was sufficiently preserved to allow at least a tentative reconstruction, and deals with the related metallurgical finds from Site 200.

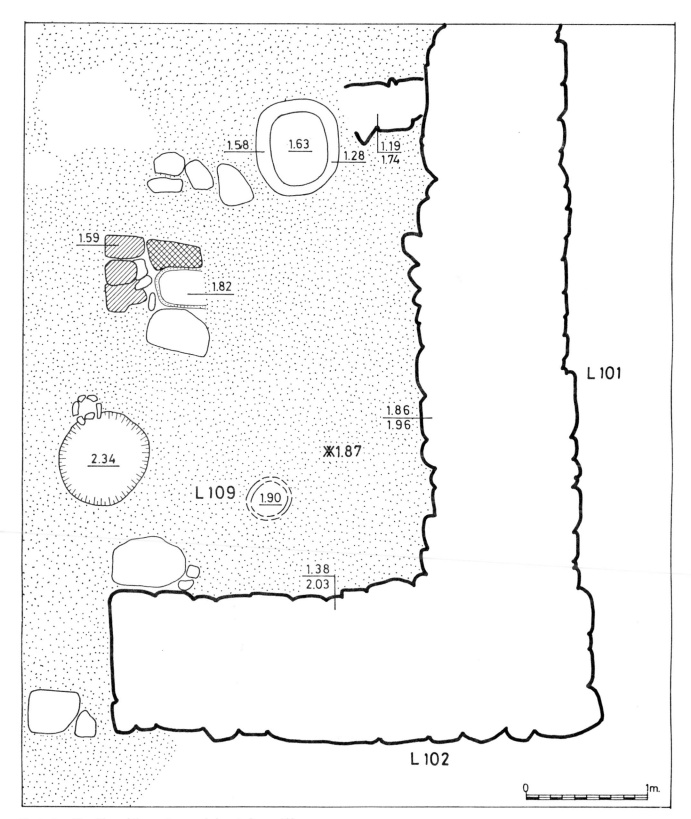

Illustration 31. Plan of the casting workshop in Locus 109.

The physical appearance of the finds in Stratum V is very similar to that of the debris of the cottage-industry type copper smelting activities found during our surveys at many of the habitation sites of the Sinai-Arabah Copper Age – Early Phase in the vicinity of the Timna Valley (and southern Sinai; see Rothenberg, 1972; idem, 1967; idem, 1967a; idem, 1979) whilst the installations and debris at Locus 109 suggested a casting workshop. The characteristic metallurgical installations, debris and products found at all New Kingdom copper smelting camps of Timna were totally absent at Site 200.

A relatively large quantity of metallurgical process debris was found at Loci 101–102 Stratum II, which suggested a third, and somewhat later centre of metallurgical activities. However, as most of these finds – like all the numerous artifacts found together with the metallurgical debris – appeared to have come from the inside of the Temple and were obviously 'dumped' here, it seems more plausible to assume that, in spite of the high frequency of metallurgical debris in these loci, no separate metallurgical activities took place there.

As a first step in the investigation of the metallurgical installations and debris, a representative sample collection was studied by D.T.A. and wet chemistry at the Haifa Technion (1970/71).[2] Some years later (1979), after metallic tin had been identified among hundreds of metal artifacts and fragments and it became apparent that alloying with metallic tin for tin-bronze took place at the Temple (see Chapter III, 7–8), a random sample of objects – crucible fragments, slag and metal droplets – was checked for tin. The complete slag samples were ground and examined as loose powder by X-ray diffraction (from Fe upwards); the crucible fragments were examined whole, as much as possible without the ceramic body, to determine the composition of the inner surface deposits. These XRF scans were made at the Geomet Laboratories, Chessington, England.

Our excavations at smelting camp Site 3 in Timna uncovered a store containing a considerable quantity of irregular shaped, small copper ingots and prills which had been mechanically recovered from primary smelting slag and collected for further processing. Since many such prills and spills were found in the layers of Site 200, often together with other metallurgical debris, typical samples were included in this chapter as samples of the raw copper used in the casting operations in the Temple courtyard (Table 6).[3]

1 The Casting installation Fu I
The installation Fu I (*Illus.* 6, 17, 31, *Pl.* 68–9) was set up at the time the White Floor of Stratum III was laid. A thick, whitish mortar setting for the furnace stones (*Illus.* 17) was actually part of this floor.

Fu I was almost square (90 × 80cm.) and was constructed of red and white sandstone and granite blocks. It had an inner lining of clayey material, burnt to a grey hard crust, found partly *in situ*. This lining contained much sand and even small stone frag-

Table 1.   Slagged furnace fragments (from Fu I)

Sample No.	159	302/2	302/4
Field No.	220	247	247
Locus	109	109	109
Stratum	II-III[5]	III	III
%	D.T.A.		XRF
SiO$_2$	81.80	83.04	
FeO	2.70	2.14	Ma
Cu	0.83	Tr	Mi
Al$_2$O$_3$	8.96	9.45	
MnO	1.90	0.77	
P$_2$O$_5$	1.37	0.34	
CaO	0.92	1.67	
MgO	1.19	0.49	Tr
Zn	0.03	–	
K$_2$O	–	0.44	
Na$_2$O	–	0.42	
Ba			Tr
Zr			Tr
Sr			Tr
Pb			Tr

Ma = Major component
Mi = Minor component
Tr = Trace

ments. The actual hearth (33 x 28cm.) was located in the centre, apparently open at its front (north). The bottom of the hearth was found covered by red burned, fused, clayey material. The immediate surrounding of Fu I showed yellowish heat discolouration, gradually fading out with increasing distance from the installation. There was a lot of woodash everywhere inside and around Fu I, a fact which by itself indicated crucible melting rather than smelting.[4]

According to its shape and contents, Fu I was an installation set up to produce the conditions necessary for crucible melting of metal and it could not possibly have been a smelting furnace. Fu I did not show the kind of heavy slagging commonly found on the walls of smelting furnaces, but a very thin and porous, dark incrustation of the clayey lining best described as sintering or vitrification due to the fluxing effect of woodash. Some of the 'slagging' could well be dross formed in the melting crucibles which was spilled onto the furnace wall.

2 Tuyères (*Illus.* 32)
Only a few fragments of red clay-tuyères were found (Boxes 259, 275, 310, 321). These tuyères are smaller (diam. 6.3–7.0cm., air hole 1.6–1.8cm., length 2.0–3.5cm.) than the tuyères of smelting furnaces (diam. 6.0–7.5cm., air hole 2.0cm., length 6.0–8.0cm.) (Rothenberg, 1972; idem, in Conrad and Rothenberg, 1981), found in large numbers in all smelting camps of the New Kingdom in the Arabah (and Sinai). Instead of the complex construction in several (at least three) layers of clay, refractories and reed netting of the hemispherical New Kingdom smelting tuyères (Rothenberg in Vol. II of this publication), there is a shorter, simple round clay nozzle which tapers towards the air hole. The slight, glassy slagging is preserved on the tuyère top, which must have protruded inside the hearth, whilst the rest of the

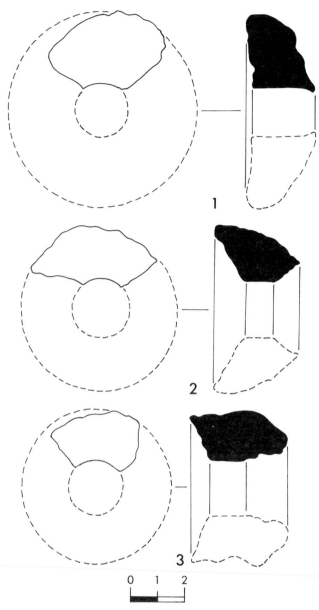

Illustration 32.   Tuyère fragments found in the Temple courtyard.

Table 2.   *Tuyères (from Fu I)*

Sample No.	321/1		321/2	
Field No.	257		257	
Locus	102		102	
Stratum	III-V[5]		III-V	
%	Ceramic D.T.A.	Slag D.T.A.	Ceramic D.T.A.	Slag D.T.A.
SiO$_2$	62.51	53.51	70.53	61.05
FeO	6.10	5.28	6.10	5.28
Cu	0.31	2.30	0.48	2.32
Al$_2$O$_3$	16.25	14.52	18.00	17.00
MnO	0.13	0.13	0.15	0.31
P$_2$O$_5$	0.36	0.31	0.20	0.32
CaO	2.52	9.00	2.12	4.06
MgO	1.06	2.53	0.49	0.62
K$_2$O	2.57	2.17	1.35	1.42
Na$_2$O	0.54	0.97	0.27	0.27

3 Crucibles (*Illus.* 33 and 34)

Many crucible fragments were found at Site 200, most of them rather small pieces which made a reconstruction of the original vessels quite difficult.[6] However, a number of rims and bases allowed the certain reconstruction of three crucible types:

(1) Small hemispherical, bowl-shaped vessel with round bottom and a protruding spout for pouring the molten metal (diam. 10–20cm., thickness of wall 1–2cm.);

(2) Small circular, angular vessel with thick flat base and protruding spout (diam. *c.* 16cm., thickness of wall 1.2cm., thickness of flat base 1.2–2.0cm.).

These two crucible types, found also in the New Kingdom smelting camps of Timna (Rothenberg, 1972, 80, Pl. 41; Tylecote, Lupu and Rothenberg, 1967, 241; see also Tite *et al.* in Vol II of this publication) showed almost no signs of vitrification on the outside and most have been heated from the top (see above, the related location of the tuyères). Under the binocular, the ceramic body of the crucibles of both types shows a high proportion of coarse- and fine-grained quartz and tiny bits of grog or shale. The quartz may have been added as temper to improve the refractory properties of the crucibles (temperatures of minimum 1100°C), although it could also be intrinsic to the local clay of Timna.

The inside of the crucibles was almost completely covered by a thin film of glossy vitrification and near the rim, especially next to the spout, adhered a thicker, spotty layer of vitreous, slaggy material similar to crucible slag, many lumps of which were also found at the site (see below).

A number of crucible fragments of both types, and the adhering dross, were chemically analysed. Some XRF analyses were made in order to identify possible tin-bronze casting crucibles (Sample No. 518, 520). Three crucible fragments (Field No. 257, 280, C-D 16–17) were analysed by atomic absorption spectrography (AAS) (British Museum Research Lab., court. M. Tite) (Table 3).

tuyère's side is just hard-burnt clay clear of any slagging. As the slagging ends towards the back of the tuyère's side in a straight line around the circumference of the tuyère, indicating where part of the tubular tuyère was protected from the heat by the furnace wall, the angle of this line to the horizontal indicates the angle of penetration of the tuyère through the furnace wall and into the hearth: 0°–20° (the smelting tuyères 35°–45°), i.e. the air was blown almost horizontally across the top of the tuyère standing inside the hearth and was not directed towards the centre and bottom of the hearth (as would be the case in a smelting furnace).

Two tuyères were chemically investigated, whereby the (slagged) ceramic body and the outside slagging were separated (Table 2).

195

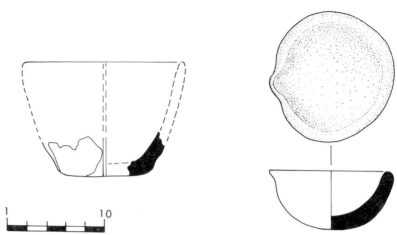

Illustration 33.   The two crucible types common at the Arabah New Kingdom smelting sites (drawn from samples from Site 2 Layer II).

(3) The third crucible type is a large, angular vessel with a thick, flat base (*Illus.* 34) (diam. 12.5cm., height of vessel *c.* 6.0cm., thickness of wall 1.8cm., thickness of base 2.0cm.), two fragments of which were found right next to Fu I (Box 278, Locus 109, Stratum III).[7] The surviving parts of this crucible, although representing a full section of the vessel, from the well-preserved rim to the base, do not show any dross incrustation on the inside, nor is a pouring lip nor spout preserved on the existing fragments.

The outer surface is very uneven, with bits of the non-plastic temper protruding, the inner surface is quite smooth, fired to a light grey. The crucible fabric shows a massive tempering with angular crushed slag fragments and small pieces of stones, some up to a diameter of several millimetres.

Surprisingly, this crucible, the only one of its kind found in the Arabah, is heavily slagged at its base, i.e. it must have been heated from outside and from below. As a matter of fact, the heat was so intense that the base of the crucible started to slag and form

Table 3. Crucible fabric and dross incrustation (from Fu I)

Sample No.	150/1		150/2		151		153			158/4			331[8]	
Field No.	263		263		211		366			270			–	
Locus	103		103		101		109			102			–	
Stratum							III			III-V[6]			–	
%	Ceramic	Dross	Ceramic	Dross	Ceramic D.T.A.	Dross	Ceramic	Dross	XRF	Ceramic	Dross	XRF	Ceramic + Dross	XRF
SiO₂	n.a.	61.56	n.a.	55.66	58.70	n.a.	n.a.	51.80		n.a.	58.06		64.31	
FeO		7.00		6.14	9.13			12.90	Ma		8.51	Mi	2.86	Mi
Cu		2.57		3.27	0.25			2.04	MI		3.26	Mi	5.10	Mi
		(2.2)						(0.88)						
Al₂O3		13.32		15.15	20.51			11.30			11.04		3.82	
MnO		0.10		0.12	0.15			0.27			0.11		5.28	Mi
P₂O₅		0.73		0.28	0.13			0.07			0.97		0.30	
CaO		0.53		6.09	3.02			8.07			6.92		5.15	
MgO		0.18		0.70	0.19			1.84			1.18		0.80	
Zn		0.02		0.03	0.06			0.09	Tr			Tr		Tr
K₂O											1.32			
Na₂O											0.35			
Ba									Tr			Tr		Mi
Zr												Tr		Tr
Sr									Tr			Tr		Tr
Rb									Tr			Tr		
Pb									Tr					
Cr									Tr					Tr
Ti														
Sn														
Bi														
Co														
Ni														

( ) metallic copper mechanically recovered
n.a. = not analysed

Table 3. *Crucible fabric and dross incrustation (from Fu I) (Continued)*

Sample No.	516	517	518	519	520	–	–	–
Field No.	–	280	253	293	257	257	280	C-D 16–17
Locus	–	106	102	Surface	102	102	106	101
Stratum	–	II(V)	III	Wall 3	III	III	II-V	–
%	Dross XRF	Dross XRF	Dross XRF	Dross XRF	Dross XRF	Ceramic (AAS)	Ceramic (AAS)	Ceramic (AAS)
SiO_2						59.6	56.4	55.9
FeO	Ma	Mi	Mi	Mi	Mi	6.35	7.52	4.83
Cu	Mi	Mi	Mi	Mi	Mi	0.180	0.600	0.070
Al_2O_3						16.3	13.5	14.9
MnO	Tr	Tr	Tr	Tr	Tr	0.071	0.069	0.067
P_2O_5								
CaO						8.01	7.33	6.10
MgO						2.64	3.23	5.19
Zn		Tr	Tr	Tr	Tr	<0.01	<0.01	<0.01
K_2O						3.48	3.30	3.57
Na_2O						0.825	1.74	0.694
Ba		Tr			Tr			
Zr	Tr	Tr	Tr	Tr	Tr			
Sr	Tr	Tr	Tr	Tr	Tr			
Rb	Tr	Tr	Tr	Tr	Tr			
Pb		Tr		Tr	Tr	<0.01	<0.01	<0.01
Cr	Tr	Tr	Tr		Tr	0.014	0.011	0.009
Ti	Tr	Tr	Tr	Tr	Tr	0.780	0.850	1.08
Sn	n.a.	Tr	Mi		Mi			
Bi				Tr				
Co					Tr			
Ni						0.003	0.009	0.022

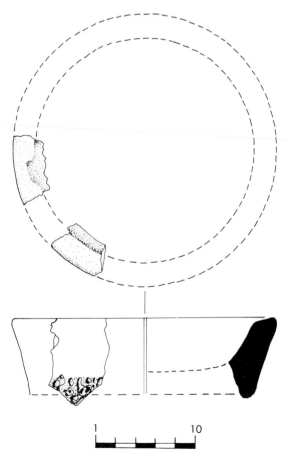

Illustration 34. Slag-tempered shallow crucible, heated from below.

lengthy slag drops. This slagging covered about 0.5cm. of the bottom end of the crucible wall, where it formed an almost straight line, indicating that from here upwards the crucible was protected from the heat, i.e. it sat in some sort of protecting container which was open at the top and bottom, and held the crucible like a wide ring-shank (well-known from modern crucible technology (Tylecote, 1986, 100), or, alternatively, a two-tiered installation with a circular aperture in the horizontal separation between the two tiers which served as a holding device for the crucible. (See Appendix to this chapter.)

4 Slag

Most of the 2–3kg. of metallurgical debris found at the site consists of small to tiny bits of slag which are either unrelated, often solitary stray finds from different layers of the site, or found in relatively large concentrations in two centres of metal working: in Stratum V of the northern half of the site, dated to the Copper Age – Early Phase (Chalcolithic), and in Stratum III of Locus 109, dated to the Ramesside Period. Contrary to Stratum V, where patchy charcoal concentrations, heat-crazed stones and, foremost, the physical appearance of the slag indicated primitive copper smelting, the installations Fu I and Fu II, and the related tuyères, crucibles and slag in Locus 109 Stratum III and some probably related slag finds in adjacent loci, present the clear picture of a melting-refining-casting and alloying workshop in the Temple courtyard.

The following catalogue gives a short description of the physical appearance, under the binocular, of the slag finds.[9] Also indicated is the slag type to which, according to its chemistry, the respective sample relates: Smelting Slag (SS), Crucible-casting slag (CS) (the problem of the definition of these slag types is discussed below in paragraph 5).

Catalogue of the slag finds

Field No. 80, Locus 102, Stratum II – a small lump of platy slag (*c*. 10mm.).

Field No. 82, Sample No. 312, Locus 102, Stratum II – a small lump of slag, many tiny gas holes. (SS)

Field No. 115, Sample No. 313, Locus 112, Stratum – small slag piece, porous but quite dense. (SS)

Field No. 172, Sample No. 315, Locus 101, Stratum – bits of thin (0.5–1.0mm.) platy slag, tiny gas holes. (SS)

Field No. 204, Sample No. 161/5–7, Locus 110, Stratum II – several bits of slag, many gas holes, visible copper prills (161/7). (SS)

Field No. 207, Sample No. 317/1, Locus 102, Stratum III – dross-like slag. (CS)

Field No. 207, Sample No. 317/2, Locus 102, Stratum III – a small slag piece, quite dense. (SS)

Field No. 219, Sample No. 155/1–2, Locus 101, – small pieces of rather viscous slag. (SS)

Field No. 228, Sample No. 318/1–2, Locus 110, Stratum II – small bits of platy slag. (SS)

Field No. 325, Locus –, Stratum II – small pieces of platy slag (5mm.), quite solid; several small drop-like slags.

Field No. 233, Sample No. 162/11, Locus 101, – a solid lump of slag, very black. (SS)

Field No. 234, Sample No. 163/1–2, 4–5, Locus 101, Stratum III – viscous, nodular slag pieces, some with many gas holes. (CS)

Field No. 234, Sample No. 163/3, Locus 101, Stratum III – platy slag lump, very low density, lots of gas holes. (SS)

Field No. 235, Sample No. 303/1, 3–5, Locus 109, Stratum II-III – small lumps of slag, many gas holes. (SS)

Field No. 246, Sample No. 156/1–2, Locus 101 – small platy lumps of slag. (SS)

Field No. 247, Sample No. 302/1, Locus 109, Stratum III – drossy, porous slag bits. (CS)

Field No. 253, Sample No. 319/1–2, Locus 102, Stratum III – small lumps of slag, quite solid. (SS)

Field No. 257, Sample No. 321/3–4, Locus 102, Stratum III(-V) – small platy lumps of slag, gas holes. (SS)

Field No. 260, Sample No. 322/1–5, Locus 102, Stratum II – small lumps of somewhat dense slag. (SS)

Field No. 261, Sample No. 323/1–2, Locus 102, Stratum II – small lumps of slag, many gas holes. (SS)

Field No. —, Sample No. 331/2, Locus — – small slag piece. (SS)

Field No. 270, Sample No. 158/2–3, Locus 102, Stratus III(V) – platy and nodular slag pieces, gas holes but quite solid. (SS)

Field No. 278, Locus 109, Stratus III (next to Fu I) – quite a quantity of small, platy and nodular slag pieces, also drop-like lumps; some look like con-cretions of small drops of slag fused together, which gives the slag piece a granular impression; many copper prills visible in broken bits of viscous looking slag.

Field No. 287, Locus 102 – small nodular and drop-like slag pieces, also tiny fragments of nodular slag, some still adhering to crucible bottom fragment (= incrustation).

Field No. 289, Loci 101–102 – several drop-shaped small bits of slag.

Field No. 295, Locus 101 – some slag nodules, some with very many copper prills. Two platy slag pieces (*c*. 10mm. thick), quite solid with few gas holes.

Field No. 297, Locus 101, Stratum III-V – more solid slag pieces (up to 3cm. long), one platy (*c*. 10mm.), apparently fused together from slag prills, the others are viscous, porous slag.

Field No. 301, Locus 109, Stratum II – extremely porous, dross-like slag bits, found together with small crucible fragments, one apparently a slagged furnace fragment.

Field No. 303, Locus 101, Stratum II – drop-like, nodular slag pieces with very many tiny metallic copper prills, found together with small, very dark, charred crucible fragments.

Field No. 304, Sample No. 154/1–2, Loci 106–108–109, Stratum I – small slag pieces, platy with many gas holes. (SS)

Field No. 305, Locus 101, Stratum V – a round lump of granular slag (diam. 2.5cm.) with numerous tiny (2–3mm.) metallic copper prills.

Field No. 319, Loci 106–108, Stratum II – two small, perfectly ball-shaped slag nodules (diam. 10 and 18mm. respectively); many dross prills and tiny, platy (3mm.) bits of slag, all showing numerous gas holes.

Field No. 322, Locus 101, Stratum V – small drop-shaped slag lumps with numerous metallic copper prills.

Field No. 334, Locus 109, Stratum III – furnace fragments heavily fused to look like very porous slag; also drop-shaped slag nodules showing numerous metallic copper prills, still sticking to clay fragments. Many furnace fragments of hard-baked sandy clay show patches (up to 4 × 5 cm.) of slagging, but these look more like non-metallurgical fused ceramic.

Field No. 337, Locus 102, Stratum III – somewhat bigger and more solid, platy slag pieces, also small nodular pieces.

Field No. 343, Locus 109, Stratum III – tiny slag prills and spills and very thin platy slag, also some nodular (diam. 1–1.5cm.) slag. Found next to a few small charred and slagged crucible fragments.

Field No. 339, Locus 109, Stratum II-III – relatively large number of slag pieces: small nodular and broken slag, also tiny platy slag bits, very similar to prehistoric smelting slag. Also a few somewhat bigger, nodular slags (2 × 4cm.). One round, granular slag piece (diam. 3 × 1.5cm.) looks like fused-together single slag nodules, giving the impression of a more substantial sort of skimmed-off dross.

Field No. 349, Locus 109, Stratum II – few small to tiny slag nodules and bits of broken, porous slag.

Field No. 352, Locus 103, Stratum – thin (3–5mm.) plate-like pieces of slag with many embedded copper prills, look like fused-together tiny bits.

Field No. 364, Locus 110 – one small (1.5 × 3cm.) piece of more solid concretion of fused slag drops, many gas holes, quite dense.

Field No. 366, Sample No. 152, Locus 109, Stratum III – small piece of slag, quite dense, green oxidation. (SS)

Field No. 367, Locus 109, Stratum III – several tiny drop-shaped pieces of slag but also some bigger slag lumps.

Field No. 369, Locus 106, Stratum II – thin (3–5mm.) platy slag with many green spots and prills of metallic copper. The slag has a particularly rough, broken character, perhaps a concretion of fused small slag bits.

Field No. C-D 16–17, Sample No. 311, Locus 101 – rough, porous dross-like thin pieces of slag, found together with crucible fragment. (CS)

Tables 4-5 record the results of the chemical analyses of a representative group of slag samples.

Table 4. Smelting and re-melting slag from Site 200

Element or Compound	152	155/1	155/2	156/1	156/2	158/2	158/3	158/4	161/5	161/6	161/7	162/1	162/2	163/3
Sample No.	152	155/1	155/2	156/1	156/2	158/2	158/3	158/4	161/5	161/6	161/7	162/1	162/2	163/3
Field No.	366	219	219	246	246	270	270	270	240	240	240	233	233	234
Locus	109	101	101	101	101	102	102	102	110	110	110	101	101	101
Stratum	III	–	–	–	–	III-V	III-V	III-V	II	II	II	–	–	III
%				D.T.A.				XRF				D.T.A	D.T.A	D.T.A
SiO_2	36.00	33.23	22.63	40.23	30.30	37.61	34.76		33.69	35.87	45.54	23.63	51.40	53.46
FeO	42.05	47.34	1.29	42.62	54.80	43.79	40.13	Ma	44.10	35.00	35.02	0.71	35.55	33.39
Cu	3.52	1.85	1.95	0.81	0.77	1.57	1.90	Mi	1.58	0.37	4.01	2.27	0.91	0.47
Al_2O_3	3.02	1.78	1.22	2.65	1.25	1.87	1.31		1.81	3.24	0.33	2.73	0.62	0.28
MnO	0.30	0.38	40.30	0.14	0.16	0.41	0.11		0.29	0.32	0.56	47.94	0.075	0.06
P_2O_5	1.17	0.59	n.d.	0.39	0.52	2.68	2.13		0.15	2.04	0.13	0.58	n.d.	0.28
CaO	5.80	7.20	8.40	8.31	4.40	6.47	9.67		11.64	11.82	6.83	0.14	1.38	n.d.
MgO	0.80	0.90	0.80	0.74	0.40	1.27	0.39		1.98	1.19	0.47	0.05	n.d.	n.d.
Zn	0.11	0.60	0.12	0.33	0.38			Tr				0.14		
K_2O						0.52	0.42							
Na_2O						0.71	0.82							
Ba								Tr						
Zr								Tr						
Sr								Tr						
Rb														
Pb								Tr						

Table 4. Smelting and re-melting slag from Site 200 (Continued)

Element or Compound	303/1	303/3	303/4	303/5	312	313	315	317/2		318/1	318/2	319/1
Sample No.	303/1	303/3	303/4	303/5	312	313	315	317/2		318/1	318/2	319/1
Field No.	225	235	235	235	82	115	172	207		228	228	253
Locus	109	109	109	109	102	112	101	102		110	110	102
Stratum	II-III	II-III	II-III	II-III	II	–	–	II		II	II	III
%									XRF			
SiO_2	44.16	65.25	35.53	33.42	36.68	43.76	45.91	37.97		50.28	47.24	37.78
FeO	36.60	27.65	42.13	51.20	37.37	33.60	32.00	35.74	Ma	32.62	37.02	44.50
Cu	0.59	Tr	1.68	2.17	1.30	2.67 (0.26)	2.70	2.33	Mi	2.14 (0.24)	1.61 (0.24)	1.31
Al_2O_3	1.63	0.70	1.49	1.53	1.82	1.31	1.06	1.41		2.28	1.56	1.29
MnO	2.45	0.25	0.37	0.33	0.36	5.10	4.82	5.41		0.12	0.22	
P_2O_5	1.91	0.45	1.76	1.36	1.97	1.93	0.85	1.82		1.59	1.93	2.15
CaO	2.38	1.94	5.52	5.70	12.47	6.82	7.34	6.96		6.08	5.41	6.15
MgO	0.10	0.10	0.18	0.18	2.21	0.98	0.68	1.49		0.39	0.20	0.86
Zn									Tr			
K_2O					0.49	0.40		0.50		0.61	0.62	0.38
Na_2O					0.67	0.26		0.28		0.27	0.27	0.46
Ba									Tr			
Zr									Tr			
Sr									Tr			
Rb												
Pb									Tr			

( ) metallic copper mechanically recovered

Table 4. Smelting and re-melting slag from Site 200 (Continued)

Element or Compound	Sample No. Field No. Locus Stratum	319/1 253 102 III	319/2 253 102 III	321/3 257 102 III-V	321/4 257 102 III-V	322/1 260 102 II	322/2 260 102 II	322/3 260 102 II	322/4 260 102 II	323/1 261 103 III	323/2 261 103 III	XRF	331/2 – – –
%													
SiO$_2$		37.78	37.92	36.24	42.07	40.20	41.43	43.06	30.36	40.66	40.38	Ma	39.21
FeO		44.50	44.30	37.98	28.62	42.09	53.95	34.54	51.00	37.32	45.61	Mi	44.89
Cu		1.31	0.65	1.68	4.13	3.08 (1.05)	7.51	2.73	1.35	0.44	1.57		0.77
Al$_2$O$_3$		1.29	1.91	1.68	1.98	2.27	3.58	3.20	1.88	2.76	2.40		1.58
MnO			0.24		2.44	0.13	0.23	0.16	0.16	1.20	0.17		0.24
P$_2$O$_5$		2.15	2.23	1.59	1.81	1.92	1.89	1.86	2.15	1.27	1.29		1.59
CaO		6.15	8.10	13.74	9.67	6.67	9.28	8.21	7.73	5.87	4.86		6.52
MgO		0.86	0.40	2.18	2.67	1.58	1.20	1.21	0.50	0.58	0.49		1.38
Zn												Tr	
K$_2$O		0.38	0.26	0.39	0.53	0.26	0.42	0.26	0.40				
Na$_2$O		0.46	0.27	0.42	0.30	0.27	0.32	0.23	0.44				
Ba												Tr	
Zr												Tr	
Sr												Tr	
Rb													
Pb												Tr	

( ) metallic copper mechanically recovered

Table 5. Casting slag

Element or Compound	Sample No. Field No. Locus Stratum	234/1 234 101 III	234/2 234 101 III	234/4 234 101 III	234/5 234 101 III	302 247 109 III	311 C-D 16–17 101 –	XRF	317 207 102 II D.T.A
%									
SiO$_2$			44.72	55.07	49.80	57.85	76.58	41.54		75.22
FeO			7.98	10.64	7.16	9.56	3.49	8.74	Mi	1.43
Cu			0.75	1.74	0.79	0.17	0.31	1.37	Mi	3.86
Al$_2$O$_3$			11.03	14.88	18.34	19.62	8.28	14.40		7.80
MnO			0.13	0.12	0.08	0.08	2.66			0.15
P$_2$O$_5$			0.54	0.92	0.60	0.64	1.02	1.00		n.d.
CaO			6.22	5.29	6.42	1.53	3.67	15.20		0.28
MgO			3.23	1.88	0.79	1.68	0.68	5.49		0.20
Zn									Tr	
K$_2$O							1.1	1.42		2.76
Na$_2$O							0.36	0.73		0.23
Ba										
Zr									Tr	
Sr									Tr	
Rb									Tr	
Pb									Tr	
Sn									Mi	

5 Metal spills

A number of copper droplets and spills were found in the Temple courtyard and their analyses are given in Chapter III, 8. Two metal fragments, the shape of which is typical for metal spills of a casting operation, are added here (Table 6, analyses by Lupu).

6 Discussion

(1) Most of the metallurgical debris, comprising mainly small-to-tiny pieces of slag, even when found in definable layers, must be considered stray finds out of stratigraphic as well as functional (metallurgical) context. This conclusion can easily be understood considering the total height of the five-strata debris at the site, which is altogether only 0.30–1.5m., and consists of sandy, often quite loose fill layers. In addition, we have to consider the obviously seasonal

Table 6. Metal spills

Element or Compound	Sample No. Field No. Locus Stratum	160 206 110–111 II	303 235 109 III
%			
SiO$_2$		0.23	–
Fe		12.66	1.93
Cu		79.17	86.37
Al$_2$O$_3$		0.03	–
Pb		0.40	–
Mn		n.d.	–
Ca		n.d.	–
Mg		Tr	–
Zn		0.03	–
Sn		–	7.10

n.d. = not detected

occupation of the site and the many clearing as well as levelling operations which took place in the 150 years during which the Temple was functioning.

For this reason, the evaluation of the metallurgical finds is mainly based on their physical appearance and chemistry and, also, wherever possible, on their functional connection with excavated installations (Fu I and II) and working places (Stratum V of the north-east half of the site). The metallurgical find context, i.e. other metallurgical debris such as ashes, crucibles, furnace lining, ore bits, metal drops, etc., was, of course, often decisive for the interpretation of the analytical results.

(2) The slagged furnace parts do not have the appearance of slagged walls of a smelting hearth, but are just slagged or fused clay and/or sandstone. This is also confirmed by Table 1, showing very high silica, fairly high alumina, low iron and very low copper. All furnace fragments originated without doubt from the New Kingdom Fu I or Fu II, which have to be considered casting and not smelting furnaces, i.e. a hearth in which a metal-containing crucible is being heated until its content is liquidized and can be cast. The walls of such a casting furnace would usually have no direct contact with the metal or slag content of the crucible, except when some of it is spilled accidentally or by 'skimming' off the impurities from the top of the molten metal (see below).

(3) The tuyères used for the operation of Fu I and Fu II are smaller and much less sophisticated in their construction than the tuyères used in the contemporary copper smelting camps of Timna (and the Arabah and Sinai). The analyses of the slagged tuyères (Table 2) show the consistency of aluminium silicate or alkali aluminium silicates, with some iron and copper, which is typical for copper casting debris and the same as the slag incrustations in the crucibles (see below), i.e. the tuyères must have been in contact with a crucible in action. Reading the slagging condition and borderline of the tuyères, and their chemistry, we may assume that these were placed near or on the rim of the crucible standing in the furnace, in order to blow air across the surface of the molten metal.

(4) The hemispherical, round-bottomed as well as the flat-bottomed, angular crucibles (*Illus.* 31), were common in the New Kingdom mining and metal working camps in the Arabah and in excavations were found together in the same strata (e.g. Site 2, Area F). It was somewhat surprising to find two different types of vessels with exactly the same function in the same strata. The only explanation which occurred to us is their introduction and use by metallurgists originating from different technical traditions, i.e. in our case probably Egyptians and Midianites respectively.

Both types of crucible show much fine and coarse grain quartz in the clay matrix. They were heated from the top and we may assume that the crucibles stood in the furnace either in a bed of charcoal-ash or sand. It is this position in the furnace and the low heat conductivity of the clay which protected the vessels from the high temperature needed to liquidize the metal and avoided the possible collapse of its clay fabric.

The chemistry of the slagged crucibles and dross incrustation (Table 3) is, as expected, very similar to the mainly aluminium silicate consistency of proper casting slag (see below), though the copper contents (mostly not in solution and often visible as tiny metal prills) is usually somewhat higher than in the slag.

The third type of crucible (*Illus.* 34) – angular with thick walls and flat base – is the only crucible-type vessel known to us from the Near East which was evidently heated from underneath. In fact, the clay matrix of the lowest part of the crucible evidently began to collapse and liquidize. This vessel, although obviously used in a pyrotechnical process (see note 7) may not be a copper melting-casting crucible. Could it be a cupel-type of vessel used in connection with the production of lead – at least as a byproduct of copper smelting – in Timna,[10] perhaps even in the Temple courtyard?

The crushed slag temper of this vessel is of considerable interest as, to our knowledge, this is the earliest of its kind found so far and testifies to a very advanced pyrotechnology. The use of slag temper in metal technology is otherwise known in Timna from the 10th century BC (Rothenberg 1980a, 198; Bachmann and Rothenberg 1980, 220), when the 22nd Dynasty Egyptians returned to Timna for a short time (Site 30, layer I) and smelted copper in slag-tempered clay furnaces. However, this technology has not been met with in any of the many Egyptian smelting camps of the New Kingdom investigated by us in the Arabah and in Sinai and, especially in the light of recent findings in the area of Fenan, south-east of the Dead Sea, in an Edomite context (Hauptmann *et al.*, 1985), we tend to see here a technological 'import' from the East (Arabia) rather than from the South (Egypt).

(5) There are basically two slag types: (a) smelting slag, (b) casting or crucible slag.
(a) Smelting slag, mainly ferrous slag of fayalitic composition. The slag, which is the collector of the impurities from the ores, flux, furnace wall material and woodash, shows the typical high Si_2O_2 and Fe, and fairly high CaO content, is low in alkali and phosphorus and contains many prills of metallic copper. The high Fe content indicates an iron oxide flux, the common flux of the Arabah smelters, but some slag samples (S.N. 155/2 and 162/1) show that manganese oxide was occasionally used as flux.[11] It seems most likely that this manganese oxide derived from crushed iron-manganese concretions commonly used as flux in the Arabah (see below Chapter III, 29 and *Pl.* 155) and was not intentionally introduced as in later periods. It is mainly the high iron (manganese) content that characterises this type of slag as smelting slag, the like of which is found in all smelting plants of the Arabah (and Sinai).

This immediately raises the question whether, and in which strata, copper smelting actually did take place at Site 200. Summing up the evidence from the excavations at the site, there is some indication of primitive copper smelting in Stratum V, of the Sinai-Arabah Copper Age – Early Phase (Chalcolithic era), but no smelting installations nor signs thereof were found in any of the phases of the Egyptian Temple. Furthermore, the physical appearance of the smelting slag found at Site 200 is quite different from that of the common smelting slag found in all the New Kingdom smelting camps. Besides large slag cakes of 30kg. and more, or parts thereof, we found in the New Kingdom smelting camps large heaps of broken up platy slag, all of which is characteristic tapped slag, whilst the slag at Site 200 is definitely not tapped.

The slag finds at Site 200 can be classified according to some common characteristics into two main, though rather inhomogeneous, groups: one group of slag, derived from a primary smelting process is related to Stratum V and is comprised of small pieces of slag which were evidently broken off from a larger piece of slag and crushed in order to recover mechanically the copper prills entrapped in the slag. Many of these slag pieces are quite solid and dense, but show many gas holes and some gangue and charcoal inclusions, and do not appear to have been tapped. However, most of this quite viscous looking slag must have been liquidized during the smelting process before it was left to cool down, prior to its racking out of the smelting hearth. This type of smelting slag is typical for the earliest copper smelting sites in the Arabah, especially Site 39 (Rothenberg, Tylecote and Boydell, 1978) and Site F2 (see Vol. III of this publication) and is also found at most prehistoric copper smelting sites in the Near East and Europe published to date.

The second group of slag of fayalitic composition is more nodular, often with a granular surface indicating that the slag pieces are in fact conglomerates of not quite liquidized slag prills and drops, i.e. not quite molten slag prills fused together. There are also drop-like lumps of various sizes, as well as very thin, crust-like slag which appeared to have been 'skimmed off' some molten mass of metal. Obviously these slag lumps, although of a composition typical for smelting slag, could not have been tapped out of a smelter and take such shapes and structure. We therefore assume that these lumps of proper copper smelting slag, which contained a large quantity of visible copper prills of various sizes, were remelted to recover the metallic copper, i.e. they were brought to the Temple, perhaps even as votive gifts, to provide metallic copper for the casting workshop.[12] This would explain why so many of the small slag fragments described above in the Catalogue of the slag finds, contain such an extraordinarily large quantity of visible copper prills (see Sample No. 161/7; Field Nos. 278, 295, 303 and, especially, Field No. 315), and fit well the chemical composition of many of the slag fragments found in the Temple (Table 4).

(b) Casting or crucible slag (see Table 5) is basically an alkali aluminium silicate that contains some iron, metallic copper and copper oxides (not in solution, often giving a red tint to the slag). It is formed by the reaction of the crucible fabric with the slag inclusions and metallic iron (a second phase in the original metal) in the copper as well as fuel ash (Tylecote, 1982, 239) and the silica flux (sand from the granites). The latter (analyses by Lupu) has the following composition: $68–71\%$ SiO_2; $14–17\%$ Al_2O_3; $8–11\%$ $Na_2O + K_2O$ and next to the fuel ash, which also adds lime and alkalies, is a major contributor to the casting slag.

A part of the casting slag is in fact dross which remains as an incrustation on the inside of the crucible, especially around the spout, but some of the slag is skimmed off the surface of the molten metal and discarded. It is found as small lumps, and in quite small quantities, at the casting site.

(6) The presence of tin in many of the copper-base objects (see Chapter III, 8, Table 1) and, above all, the finding of metallic tin, including a droplet of metallic tin (Met. Cat. No. 163) right next to the casting installation in Locus 109, raises the question of possible tin-bronze alloying and casting in the Temple workshop. For this reason a number of crucible fragments, slag lumps and some metal droplets, found in the vicinity of the casting installation of Locus 109, were checked for tin (the crucibles and slag by XRF, the metal fragments by DTA). We are listing here the items which contained tin:

Crucible fragments, Sample No. 518 and 520 (XRF: Minor)
Casting slag, Sample No. 311 (XRF: Traces)
Metal droplet, Sample No. 303 (DTA: 7.10% Sn, 86.3% Cu, 1.93% Fe).

The search for tin in the range of metallurgical debris, from crucibles to slag and metal product, showed the presence of significant quantities of tin in a ratio to be expected in a workshop where tin was intentionally added to cast tin-bronze. If in addition to these positive results we consider the finds of many tin-containing metal spills (see Met. Cat. No. 4) we may conclude that many of the tin-bronze artifacts made of locally produced copper and offered as votive gifts to Hathor, were produced in the Temple workshop. It now seems more than likely that even the low and erratic tin content found in numerous artifacts in the Temple, is an intentional addition to the copper in the casting crucible, probably for metal-technological reasons (see above Chapter III, 8).

Notes

1. Details of those sites will be published in Vol. III of this publication.

2. The chemical analyses of many metallurgical samples were performed by (late) Professor Alexandru Lupu at the Department of Mineral Engineering, Haifa Technion, Israel Institute of Technology.

Although the present chapter is mainly based on Professor Lupu's analyses (Tables 1–6), its author (B.R.) is solely responsible for the archaeological and archaeo-metallurgical interpretations.

3 See also further analyses of such raw copper fragments in Chapter III, 8 above.

4. Very little loose woodash is found at smelting sites and no ash whatsoever inside a smelting hearth, as all woodash from the burnt charcoal would be absorbed into the smelting slag.

5. Because of the locally, sometimes unclear, detailed stratification, some of the find boxes contain material from more than one stratum. However, in most cases it was possible to reconstruct the original context – in the case of Sample 159 it might have been Stratum III.

6. Many of the crucible fragments found at Site 200 were too fragmentary to be drawn, e.g. the five fragments sent to Geomet for analysis were just bodysherds (from boxes 253, 257, 280 and 293). *Illus.* 33 represents an endeavour to extract the maximum information from these fragments, albeit sometimes rather tentative. This reconstruction was also based on crucible fragments from Site 2.

7. These crucible fragments were unfortunately mislaid after the excavation and only located at the time when this volume was already near its final stages. We are therefore unable to include here further analytical or petrographic information, which will be published elsewhere.

8. Sample 331 is a crucible fragment, part of it was analysed (together with the adhering dross) at Haifa, the other part was investigated by XRF by Geomet which did separate the incrustation from the ceramic body.

9. Not many details of the physical appearance of the slag samples ground to power and analysed by the late Alexandru Lupu can be given, because the related notes could not be found among Lupu's papers. However, since the slag samples which were not chemically investigated and not ground up (i.e. without Sample No.) are of the same types, their fuller description gives an adequate picture of the different slag types present at Site 200 and of the problems of their interpretation.

10. See Gale, Bachmann, Rothenberg, Stos-Gale and Tylecote in Vol. II of this publication.

11. For the use of manganese ore as flux in 10th cent. BC Timna, Site 30, Layer I, see Bachmann and Rothenberg 1980.

12. In the light of these conclusions it will be necessary to reconsider other similar findings in temples and shrines, as at Kition (Cyprus) (Karageorghis 1976, 72–6), characteristically considered to be evidence for smelting of copper ores in crucibles.

Appendix
Notes on the examination of two crucible fragments from Site 200

The two fragments (Box 278) are extensively vitrified on their under sides showing they were heated from beneath. This practice is unusual at Timna but crucibles heated from beneath are known from other early metalworking sites in Europe (Tylecote, 1986, 97; Lamm, 1977, 107) and 1st millennium BC India and Africa (personal observation by P. Craddock).

Both the inner and outer surfaces of the crucibles were analysed by non dispersive X-Ray fluorescence and had the same qualitative composition: iron, manganese and silicon with some calcium, potassium, zinconium, strontium, titanium and copper. As the copper content on the inner surface of the fragments is no greater than on the outer surface, and the whole environment is permeated with copper salts, this is no indication by itself that the crucible were used for melting copper. However the only other possible candidate is lead which tends to penetrate the clay body much more readily than copper. If lead had

been melted, then one should have expected to pick up lead in the ceramic and, anyway, there would have been no need to employ such high temperatures. Thus it seems most likely that the crucibles were used for melting copper. The ceramic body is olive-grey and well vitrified. It is a vessiculated fabric containing many large inclusion, up to several millimeters across, of crushed iron-manganese rich slag (some of which have melted, giving a slagged appearance to parts of the crucible) and of shale.

There are also frequent small fibre impressions in the clay. Organic temper is very common in early refractories, grass-fibres are very common. However, these are much finer and anyway grass is a rather rare commodity at Timna. It is possible that hair was being used as a substitute.

The general fabric is similar to that of other crucibles from Timna.

Paul T. Craddock and Ian Freestone

Bibliography

Tylecote, R.F. 1986. *The Prehistory of Metallurgy in the British Isles*, London.
Lamm, K. 1977. 'Early medieval metalworking on Helgo, Central Sweden', *Aspects of Early Metallurgy* (ed. Oddy, W.A.), London.

14 BEADS AND PENDANTS

1 The Beads

Introduction

The beads found in the Temple area amounted to 5457 and were parts of collars, necklaces, belts, pectorals, bracelets and ear-rings, which were apparently brought as votive offerings to Hathor. Of the total, 65% were found at Loci 101 and 102 outside the Temple courtyard; the rest were found scattered all over the Temple area. Only rarely were groups of beads found strung together in necklaces. Since they were found together with inscribed objects and cartouches, the beads can be dated with certainty to the 19th–20th Dynasties; this makes it possible to define the most common types of beads of this period for further comparative studies.

We base our classification of the Timna beads on H.C. Beck (1928), but further subdivided each type according to its material. The main bead types are:

a. Disc Beads
b. Short Beads
c. Standard Beads
d. Long Beads

In addition to these general types of beads, there are special types which will be described later.

The materials used for the Timna beads are:
(1) Natural materials (unprocessed) including: stones, semi-precious stones, crystals, shells, corals and bones.
(2) Artificial materials: faience, clay, glass, copper, bronze and gold.

Beads and pendants found in the Temple:

1) Beads (total): 5457

   a. Disc Beads 3324
   b. Short Beads 1106
   c. Standard Beads 265
   d. Long Beads 102
   e. Special Shell Beads 647
   f. Gold Beads 7
   g. Spacer Beads 6

2) Pendants (total): 48

a Disc Beads

1 *Disc Beads made of natural materials*[1]

Disc Beads made of conus shells (*Fig.* 78: 1–4 and *Col. Pl.* 19:3):

Many Disc Beads were made from the conus shell (*conus quercinus*, Solander, 1786) which is found on the Red Sea coast. The ends of the conus shell were filed until only a flat disc remained, which was perforated and strung for necklaces or bracelets.

406 Disc Beads made of shell were found, 38% of them at Loci 101 and 102, and the remainder mainly in Loci 106, 107, 109–111. The diameter varies between 8 and 20mm. and thicknesses between 1 and 6mm. 80% of the Disc Beads are made from the smaller type of conus shells and the rest from larger shells. The perforation is plain, Type IV. This type of bead was continuously in use since prehistoric times. It was found in the Megiddo tombs (Guy, 1938, T.1100, T.37) and also in Strata V and III (Lamon, Schipton, 1939, Pl. 91:76, 77). One Disc Bead was made of mother of pearl (diameter 15mm., thickness 1mm.) and has a plain perforation of Type IV.

According to Petrie (1914, 22, 27) the use of shells had a protective function against the evil eye. Beads of this type were found at Serabit el-Khadem (Petrie, 1906, Pl. 179).

Disc Beads made of ostrich egg shells (*Fig.* 78:5):

Six small, circular Disc Beads, of the size of the conus shell discs, were found on the site. They were made from ostrich egg shells and used together with the conus shell beads.

Disc Beads made of stone (*Fig.* 78:6–21, *Fig.* 79:22):

Fourteen beads are circular discs, six of which are irregular shaped flat pebbles with naturally formed holes (*Fig.* 78:12–15). At Megiddo, similar stones were found in Stratum VIII (Loud, 1948, Pl. 213:78). According to Petrie, at Tell el-Ajjul such pebbles were specially brought to the site from the seashore a couple of miles away, and we may assume that the pebble beads of the Temple were brought there from a nearby wadi bed.

Fourteen disc-shaped beads were made of carnelian, a red chalcedony found in the surrounding sedimentary rocks (*Fig.* 78:16–21, *Col. Pl.* 19).[2] Diameter varies between 3–10mm. and thicknesses between 0.1–0.3mm. They are of the barrel disc type (A 1b) and the perforation is the plain Type IV.

Three Disc Beads are of flint and one of onyx (*Fig.* 78:6–8). Their diameter ranges from 5 to 15mm.

150 beads were made of mica schist in the shape of thin, circular discs (*Fig.* 79:22). 60% were found in Locus 101. The mica beads have a diameter of 8–10mm. and a thickness of 0.5mm. Their perforations are the plain Type IV, 2mm. wide. Mica schist occurs in the Timna area. It is so thin that it could be peeled from the rock and cut with a knife to circular shapes, and then perforated.[3]

Circular Disc Beads made of copper (*Fig.* 79:23, 23A-B):

Three copper beads of the A1b type were found, one is of very thin (0.01mm.) copper leaf, with a diameter of 11mm.; the other two are 1.2mm. thick with a diameter of 6mm.

2 *Disc Beads made of artificial materials*

Disc Beads made of faience (*Fig.* 79:24–31):

The most common beads are the Disc Beads of types A1a and A1b, made of faience.[4] 2700 beads of this type were found scattered all over the Temple area, 45% of them in Locus 101. Some of the beads were found still strung together. Their average diameter is 4–5mm. and their thickness 1.5–2mm. They are green, light brown, white, brown, turquoise blue, all of a smooth, velvety appearance but not glazed. Some 3% of them have a diameter of 8–20mm. and a thickness of 1mm., and are coloured dark green, blue and black.

Beads of this type were found in many excavations in Palestine: in Lachish (Tufnell and Harding, 1940, Pl. 34:3–7; Pl. 36:104, 105), in all of the Late Bronze Age levels of the Fosse Temple (types A1a, A1b). According to the excavators, there was a large scale influx of faience beads during the 18th Dynasty, to the exclusion of beads of stone and metal. Faience beads were also used at Iron Age Lachish (Tufnell and Harding, 1958, Pl. 29:1, 2), in the Megiddo tombs (Guy, 1938, Tb. 550, Pl. 175:16), in Megiddo Stratum V and Stratum IV (Lamon *et al.*, 1939, Pl. 91:34–36, 38), and in 18th–19th Dynasty Tell el-Ajjul, together with a gold ring dating from the period of Ramesses II (Petrie, 1933, Pl. X). In Sinai Petrie found the same type of Disc Beads (Petrie, 1906, Pl. 159).

In Egypt, similar circular disc-shaped faience beads were very widely used from early periods onwards. The Three Princesses of the royal harem of Tuthmosis III wear three strings of blue faience Disc Beads *c.* 160cm. long and pale to rich blue in colour (Winlock, 1948, Pl. XV/A). In Tutankhamun's tomb (Carter and Mace, 1963, Pl. 20:8), they are also seen in three decorative rows, or strung in long necklaces. Royal high officials customarily wore Disc Beads made of precious metals (Lange and Hirmer, 1968, Pl. 209).

Circular gadrooned Disc Beads made of faience (*Fig.* 79:32–35; *Col. Pl.* 19:10–11, 16):

Twenty-six faience beads in the form of daisy flowers and gadrooned (edged with decorative fluting or reeding) of type A3a, are among the Timna beads. Their diameter is 5–7mm. and they are 1mm. thick. One bead of black glazed faience has a 16mm. diameter and is 2.5mm. thick. The colours of the 25 faience beads vary from red-brown to yellow and green. All of these beads are made in a one-piece mould, with the exception of the large black faience bead which was made in a two-piece mould.[5]

Petrie found necklaces made from combinations of this type of bead at Serabit el Khadem (Petrie, 1906, Pl. 159) and they were also found at Megiddo (Guy, 1938, Tb. 912B, Pl. 132:26) in a LB II context. In the Lachish Fosse Temple area 50 beads of this type were found, in the same colours as those of Timna (Tufnell *et al.*, 1940, Pl. XIV). A bead of this type was also found in Tell el-Ajjul (Petrie, 1932, Pl. XXV:71).

The gadrooned Disc Beads use the daisy flower motif found in ornaments of the New Kingdom: on the Saqqara tomb ceiling in Tomb No. 24 of Beckeranef of the 22nd Dynasty, at Abd El Gurna near Thebes, on the ceiling of Tomb No. 68 of Nespeneferhor of the 21st Dynasty; on the ceiling decoration of an unknown tomb in Tell el Amarna; in the decoration of Tomb No. 48 of Neferhotep of the 19th Dynasty at Thebes; on the ceiling of the tomb of an unknown at Abd El Gurna – Thebes of the 17th Dynasty, on the ceiling of Tomb No. 21 of Urire from the same period and place, and on a Deir el-Medineh tomb ceiling of an unknown person of the 18th Dynasty. Combined with lotus flowers, this motif appears on tomb ceilings at Abd el Gurna of the 18th and 19th Dynasties, and in many sarcophagus decorations of the New Kingdom Period (Pavlova-Fortova Samalova, 1963, Pls. 46–48).

b Short Beads

1 *Short Beads made of natural materials*

Short Beads made of stones (*Fig. 79:36–41*):

According to Beck's nomenclature, a Short Bead is a regular bead in which the length is more than one third the diameter of the bead, and less than ninetenths its diameter. Short Beads of type B1b and B1d were found, carved from limestone and chalcedony. Their diameter is between 3 and 7mm., thickness 2–4mm. The perforation is not always plain (Type IV) but sometimes Type III, i.e. bored from both ends. The sides of the holes are parallel or only slightly conical. The Short Beads were made by the same methods as the Disc Beads (see above).

The Short Beads are barrel shaped (B1b) and short truncated convex cones (B1d). Seven of the beads found in the Temple are of onyx (black chalcedony), five are brown chalcedony, 60 are carnelian (*Fig. 79:38–41*) and the rest are of brown, yellow and white limestone. The carnelian beads go from light red (usually the smaller beads) to deeper red, their diameter 6–8mm., their thickness 3–5mm.

Beads of this type, made of carnelian, were found in the Megiddo tombs in a LB II context (Guy, 1938, Pl. 95:22–26; Tb. 62, Pl. 168:11; Tb. 37; Pl. 137:13c; Tb. 20, Pl. 135:15) in Stratum VIII at Megiddo (Loud,

1948, Pl. 213:61) and in the Lachish Fosse Temple Area (Tufnell *et al.*, 1940, Pl. XXXIV: 15).

2 *Short Beads made of artificial materials*

Short Beads made of faience (*Fig. 79:42*):

A total of 975 faience Short Beads were found at Site 200, 487 of them (50%) were found in Locus 101, the others were scattered all over the Temple area. They are of Type B2b (short cylinder), the same as the faience Disc Beads in colour and material. Their diameter is 2–3mm. and thickness 1.1–1.5mm., with perforation of Type IV-III.

Beads of this type were found in the Lachish Fosse Temple (Tufnell *et al.*, 1940, Pl. XXXIV:1, 2, 9–15, 20–22) and in the Megiddo tombs (Guy, 1938, Pb.37, Pl. 138:5; Tb. 20, Pl. 135:15 E). Petrie also found beads of this type at Tell el-Fara (Petrie, 1930). The technology of these beads is the same as that of the Disc Beads.

Short Beads made of glass (*Fig. 79:43–45*):

Thirty-four glass Short Beads were found in the area, ten of them green, six blue and the rest brown and yellow-brown. The diameters are 4–7mm. and the lengths 1–3.5mm.

On 16 of the Timna glass Short Beads the 'wire-wound' technique is discernible (Sleen, 1967, 22–5), the beginning and end points of the glass rod are clearly visible as short nibs left over after the separation of the bead from the rod. The other glass Short Beads are made by the 'cone' technique (Beck 1928, 60). These beads are blue, green, brown, turquoise, yellow and black, and their diameter is 1–5mm.

Glass Short Beads were found at Megiddo (Loud, 1948, Pl. 215:108, 109; *Pl.* 209:32) in Strata VIIIA and IX; at Tell el-Fara dated to the 20th Dynasty, and among the Tel Jemma beads (Petrie, 1928, Pl. XXII:194). Petrie found many glass beads in the glass factory at Tell el-Amarna, together with glass cones ready for bead making (Petrie, 1894, 26–27).

c Standard Beads

1 *Standard Beads made of natural materials*

Standard Beads made of stone (*Fig. 79:46–50*):

The Standard Bead is a regular bead with a diameter approximately the same as its length.

Eighty-three beads made of different stones were found on the site, 52% in Locus 101. 22 of these beads were of limestone and are yellow, green, brown, white and black. Diameter and length are 3–8mm. 45 beads are of chalcedony of different shades of brown, light brown and black. There are two garnets, deep brown in colour, and fifteen carnelians (*Fig. 79:48–50*).

Fifty of the beads are of type C1b (barrel), 28 of type C1f (truncated convex bicone), 5 of type C1a (circular). The technique for making this type of bead is the same as described above for the Short and Disc Beads made of stone. The perforations are of Types IV and III.

Beads of this type were found at almost every excavation in this region, e.g. at Megiddo (Guy, 1938 Tb. 1100C, Pl. 147:12a), and at Tell el-Ajjul Petrie found a garnet bead of the same type as the Timna examples (Petrie, 1934, Pl. XXI:208).

Semi-precious stones were used in Egypt from archaic times. The tombs of the late 11th Dynasty contained a great many of the most varied stones. The necklace belonging to Queen Nefru, the wife of King Mentuhotep, was composed of 27 large, semi-precious stones of various shapes and colours (Vilmikova, 1969, 20–22).

The Timna region is rich in minerals and it may be assumed that the stone beads were locally made, with the possible exception of the carnelian beads whose more elaborate grinding and polishing technique indicates their origin in Egypt.

2 *Standard Beads made of artificial materials*

Standard Beads made of faience (*Fig.* 80:51):

The most common shape of the faience Standard Beads is the standard barrel type C1b. 61 faience beads of this type were found on the site. In many cases the glaze has altered its lustre and colour and sometimes the patina has produced a new surface colour. Most of the beads were turquoise coloured and there are also black, yellow and light brown faience beads of this type.[6]

Faience beads were in use in Egypt from predynastic times (Brunton, 1937, 85, Pl. XXXIX) and have been found in almost every excavated site, also in Palestine.

Undecorated glass Standard Beads (*Fig.* 80:52, 53):

Fifty-seven undecorated circular Standard Beads made of glass (Type C1a, circular) were found on the site, with diameter of 5–10mm. and length approximately the same as the diameter. Two of these beads are of a beautifully transparent blue glass, somewhat elongated in shape. Nine are an imitation of turquoise and the others are green, brown and black.

Fifteen beads of this type were made by the wire-wound glass technique (Sleen, 1967, 22–5), and two of them were still stuck together. The others were made by the folded cone technique (Beck, 1928, 60).

Decorated circular glass Standard Beads
(*Fig.* 80:54–61; *Col. Pl.* 19:13–15, 21):

Sixty-four decorated, circular Standard Beads and nine broken pieces of this type were found on the site. 13 are decorated by the simple dot or simple eye spot method (*Fig.* 80:55).[7] The diameter is 5–10mm., with length approximately the same as the diameter. The matrix of eight of these beads is white, with brown, red or black and blue dots. One of the beads, which is especially well preserved, has a white matrix and three blue dots.

Standard Beads with twisted thread decoration (*Fig.* 80:54):

Three beads – with brown, turquoise and yellow matrices – are decorated with threads twisted round the body. The brown matrix bead has a yellow band (diameter 11mm.), the turquoise matrix has a red band (diameter 10mm.) and the yellow matrix a black twisted band (diameter 10mm.).

According to Eisen (1916, 3) the straight bands appear on glass beads during the 19th to 20th Dynasties and were found at Tell el-Fara (Starkey, 1930, G30).

Standard Beads with composite coil decoration:

Nine beads of this type were found in the Temple. Seven have a yellow matrix (diameter 12–17mm.) and the composite coil is white and black. Another bead has a brown matrix and the composite coil is white and black; and the last is black with a white and red composite coil and a black dot (*Fig.* 80:56).

According to Eisen (1916 10; Pl. 1:29, 35, 48) the earliest examples of this type date from the 19th Dynasty. Beads of this type in the Lisht collection (excavated by Arthur Mace) are dated not later than the 19th–20th Dynasties.

Standard Beads with stratified eye decoration (*Fig.* 80:57–59):

Ten beads found in the Temple are decorated by the stratified eye technique.[8] The matrixes are green (2), brown (5), blue (3), the eyes are black with white rings. The diameter of the beads is 12–15mm. On one of the brown beads the black dot has fallen out and one can see the white zone underlying it.

The earliest eye beads of this type are reported from Tell el-Jehudiyeh (Griffith, 1890, 47; Pl. XV, XVI) and are described as variegated yellow, white and blue 'glass beads' with 'red eyes'. Some of these are from Tumulus IV. Eight were found with a glazed steatite scarab of Thuthmosis III (18th Dynasty). Others from Tumulus IV, 2, were described as 'beads of glass, opaque blue and greenish-white with red eyes' found with a scarab of Ramesses III (20th Dynasty).[9] Petrie (1892, Fig. 10, Pl. VIII) described a bead of mixed eye-spots and twisted threads which was found together with a scarab of Ramesses II. At Ehnasya Petrie (1905, 34, Fig. 16, Pl. XL) found two beads with a black base and white spots. According to Petrie, these beads exhibit a style in vogue during the reign of Thuthmosis III. However, had these beads been so early, similar beads should have been found in the Palace of Amenhetep III at Thebes or in Akhenaton's city at Tell el-Amarna. Since none were found in either place, it seems probable that the beads mentioned by Griffith and Petrie in connection with Thuthmosis III were not really made during his reign and the scarab with his cartouche was a commemorative scarab made after his death. We assume therefore that the eye-spot type has to be dated later than the Tell el-Amarna period, i.e. the earliest date for these beads is the 19th Dynasty, the time of the Timna Temple. Eye beads were found at Tell el-Fara in the context of Ramesses II (Starkey, 1930, 15, 17, 20, 24, 26) and at Megiddo in Strata VA, V and IV (Loud, 1948, Pl. 218; Lamon *et al.*, 1939, Pl. 92:34, 50).

Crumbed glass Standard Beads (*Fig.* 80:61):

Eleven crumbed glass beads[10] were found on the site. The matrix is brown or black and the glass crumbs melted into the matrix are white, yellow, red, blue and green. Diameter is 10–17mm. Crumbed beads were found by Petrie at Tell el-Fara (Starkey 1930, K 8) and dated by Starkey to the 22nd Dynasty. They also continue into later periods (Sleen, 1967, 46, 48).

Horned Standard Beads (*Fig.* 80:60):

Three of the beads found in the Temple have strongly projecting 'horned' eyes. One has a yellow matrix, a diameter of 5mm. and a black projecting horn. The second has a green matrix, a diameter of 8mm. and two projecting black horns placed opposite each other. The third has a brown matrix and a diameter of 14mm. with two black horns.

The Tell el-Fara collection contains a horned bead dated to the 20th Dynasty; it has a brown matrix and three projecting black eyes surrounded by a white zone (Starkey 1930, K30, 17). A horned bead was found at Tell Abu Huwam in Stratum III, which is dated to about 1100 BC (Hamilton, 1935, Pl. XXXIV: 135).

d Long Beads

According to Beck, Long Beads are regular beads in which the length is more than one and one-tenth times the diameter of the bead. They may be ellipsoidal long beads, long barrel, long convex, convex bicone, cylindrical or other shapes. We have also included the segmented beads in this group.

1 *Long Beads made of natural materials*

Long Beads made of stone (*Fig.* 80:62–64; *Col. Pl.* 19:12, 31):

The Long Beads found at the site are of two types: cylindrical (D2b) and barrel (D1b). Nine are of Type D2B and twelve of Type D1b. Most of the beads are made of yellow or light brown limestone, five are of carnelian (*Fig.* 80:64), barrel shaped (D1b), 7–12mm. long and 3–5mm. wide; one is of flint (*Fig.* 80:62) and one is of grey chalcedony (*Fig.* 80:63).

Carnelian long barrel beads were called 'seweret' beads in Egypt, where they were often over 15.5mm. long and 9.5mm. in diameter, and strung on a thick twisted cord, either singly or flanked at either end by a green cylindrical or spherical bead of faience or felspar. Such strings of Long Beads were worn tied closely to the throat. Representations of similar beads are common on coffins, where they are always painted red in the coffin vignettes and on the throats of masks (of anthropoid coffins). Their names are written ⟨hieroglyphs⟩ and ⟨hieroglyphs⟩. Their characteristics stone must have been carnelian, as this is the material usually named in the rubrics. So typical had carnelian beads of this type become, that it is not uncommon to find them labelled simple '*hrst*-carnelian', ⟨hieroglyphs⟩; ⟨hieroglyphs⟩; ⟨hieroglyphs⟩. The rubrics direct that they shall be attached 'to the neck' ⟨hieroglyphs⟩ or to the breast ⟨hieroglyphs⟩; ⟨hieroglyphs⟩ (Winlock and Mace, 1916, 62–63, Pl. XXVI).

The craftsmanship exhibited by the carnelian long barrel 'seweret' beads found in the Temple, is super-ior to Long Beads made of other materials, and it is therefore reasonable to assume that they were brought from Egypt to Timna.

At Megiddo (Guy, 1938, Tb. 877:B1; Pl. 95:14, 15; Tb. 111Oc, Pl. 147:12c) long barrel 'seweret' beads were found in a LB II context, and beads of this type were found in a similar context in most excavations in the region.

Long barrel beads made of marble were found in Meggido together with an amethyst long barrel bead which is dated to LB I, i.e. earlier than the Timna beads (ibid, Tb. 37, Pl. 136:199; Tb. 20, Pl. 135:150). A tomb contained a chalcedony long barrel bead from the same period, and in Stratum VIII (1479–1350 BC) (Loud, 1948, Pl. 213:58) a long barrel stone bead was found. Many cylinder-shaped Long Beads of white limestone were found in the Meggido tombs (Guy, 1938, Tb. 73, Pl. 161:15; Tb. 37, Pl. 136:19D).

Three beads found in the Timna Temple are made of greyish-green calcite. They are barrel-shaped, 10mm. long with a diameter of 5mm. and have a white strip around the middle. The material of the strip looks like white marble (*Fig.* 80:63).

2 *Long Beads made of artificial materials*

Long Beads made of faience (*Fig.* 80:65–67; *Col. Pl.* 19:23, 30):

Twenty-nine Long Beads of this type – twenty of type D2b (long cylindrical) and five of Type D1b (long barrel) were found together with other beads. Their diameter is 2–5mm. and length 6–20mm.; the colours are green, turquoise, white, black and brown. The perforation, according to Beck (1928, 51), is of Type VII (tubular). One cylinder shaped, green faience Long Bead has two red lines at each end. This is also the largest of this type, its diameter is 14mm. and its length 20mm.

According to Hayes (II, 1959, 394–5) the cylindrical and long barrel faience beads are typical of the Ramesside Period. Cylindrical faience beads of this type were found in the Lachish Fosse Temple (Tufnell *et al.*, 1940, Pl. XXXIV:31, 39), one black and one blue. The Tell el-Fara collection (Starkey, 1930, A:25, 45, 50, 60) also has some long cylindrical faience beads, dated to the 19th and 20th Dynasties.

Faience barrel beads of type D1b were also found at Lachish (Tufnell, *et al.*, 1940, Pl. XXXVI:103), in the Megiddo tombs (Guy, 1938, Tb. 877, B1, Pl. 95:7; Tb. 1100, Pl. 147:12) dated LB II and LB I, at Tell el-Fara (Starkey, 1930, D:50, 64, 105, 134) dated 20th–22nd Dynasties.

Segmented or multiple Long Beads made of faience (*Fig.* 80:68; *Col. Pl.* 19:24):

Twenty-three small, segmented, green and light brown Long Beads made of joined beads of Type A1a were found at the site. They are short faience beads of 2–5mm. diameter, joined together purposely or accidentally. The longest single bead consists of six short beads.

Beads of this type were found in the Lachish Fosse Temple (Tufnell *et al.*, 1940, Pl. XXXV:70–78), at Tell el-Ajjul (Petrie 1932, Pl XXV:126), at Tell el-Fara (Starkey 1930, B:10, 25, 46, 55, 76), dated 19th–22nd Dynasties, and at Beth Shean (Rowe, 1940, Pl. XXXVI:31).

Decorated Long Beads made of glass (*Fig.* 81:69–71B)

Twenty-five Long Beads of glass, four of them cylindrical (Type D2b) and 21 barrel-shaped (Type D1b) were found on the site. They are decorated with coloured glass spirals around the body of the bead by the 'wire wound method' (see Beck 1928). Sometimes two different coloured glass spirals are applied. The matrix colours of the Timna Temple glass Long Beads are blue, black and brown and the coloured glass wires are yellow, white and red. Two big cylindrical glass beads of Type D2b are decorated white on black by the 'irregular wire drawn' scallop beads method (Beck 1928, 7, 8; Fig. 35:Ate): Longitudinally striped canes are applied to the matrix in a zigzag pattern to imitate wire-wound or combed chevron beads (*Fig.* 81:69–70). They are 50mm. long and 20mm. in diameter.

Two beads of this type, also black on white, were found at Megiddo in Strata VI and VII (Loud, 1948, Pl. 216:118; Pl. 214:89), Beth Shean produced another bead of this type in the Ramesses III Temple (Rowe, 1940, Pl. LXVI:A:4).

Also found at Timna was a single, large, blue Long Bead or pendant (broken into three parts), made of glass and incompletely perforated, with a decoration of white lines and a composite coiled thread of white and brown. It is 50mm. long and approximately 20mm. in diameter. A broken piece of an irregular wire-drawn scalloped black bead with yellow decoration, the shape of which cannot be identified, also seems to belong to this group of beads.

In the Late Bronze Age, decorated Long Beads of this type, made in Egypt, were imported into Canaan wherever Egyptian strongholds existed. In Megiddo (Guy 1938, Pb. 37, Pl. 161:6, 7) in a tomb of this period, two long barrel-shaped glass beads (diameter 8mm., length 13, 14mm.) were found, one with a blue matrix and light ash blue spirals, the other of Indian ink colour with white spirals. Similar beads were also found at Tell Abu Huwam (Hamilton, 1935, Pl. XXXIV:42, 44, 135) in Late Bronze Age Stratum II and in Stratum III dated *c.* 1100 BC (Wenanum). In the Lachich Fosse Temple, beads of this type were also found in LB II context (Tufnell, *et al.*, 1940, Pl. XXXIV:25–30, 40, 41).

e Special Types of Beads

1 *Melon Beads* (*Fig.* 81:72, 73), Lenticular Beads (*Fig.* 81:74), and Collared Beads (*Fig.* 81:75)

Regularly shaped beads can be modified by gadrooning them or by adding collars. Two gadrooned Melon Beads are in the collection, made of green coloured faience with diameter of 5mm. and 10mm. respectively and the same length. The basic shape is C1a.

Melon-shaped Beads are typical of the Later Bronze Age and are found all over Canaan and Sinai. In the Lachish Fosse Temple (Tufnell *et al.*, 1940, Pl. XXXV:50, 54, 55; Pl. XXXVI:104) they are made of blue and green faience. At Megiddo (Loud, 1948, Pl. 212:53; Pl. 208:21) they were found in Strata IX and XI, from the end of LB II; in Hazor (Yadin *et al.*, 1961, Pl. CCXCIV:2; Pl. CCCXXXVII:4–6) at level 1B in an Egyptian context, and in Stratum 1A. In Sinai (Petrie, 1906, Pl. 159) melon-shaped 'glazed beads from 18th and 19th Dynasties' were found at Serabit el-Khadem.

We have only one Collared Bead made of glass from Timna. It is green, 4mm. long and circular with a short (1mm.) collar. Three Lenticular Beads of shape 1B1, with a vertical instead of a longitudinal perforation, are made of faience. Their length is from 10–15mm., with diameter 3–5mm. They are white and green, and the largest has a black spot on one side.

2 *Shell Beads* (*Fig.* 82:79-90; *Col. Pl.* 19:32–39)

Beads were made from a variety of shells, some used whole and others only in part. For example, the conus shell was cut into discs and used as Disc Beads (see above a.(1)).

The *Conus Quercinus Solander* 1786 (*Fig.* 82:83):

Of the Conidae family and found on the shores of the Red Sea, it was used in its original shape as a bead, after perforation for stringing. The largest of this type of perforated shell is 35mm. long. 46 perforated shells of this type were found at the site, 75% of them in Locus 101.

Petrie (1914, 27, Pl. XIV:110a) found this type of shell used for necklaces at Zowaydeh Hawara, from prehistoric times up to the 23rd Dynasty.

Forty-nine *Engina Mendicaria Linnaeus* 1788 shells of the Buccinidae family (*Fig.* 82:89) were found, with brown and white stripes, 5–8mm. in diameter and 10–12mm. in length. They are perforated for stringing and come from the Red Sea (*Fig.* 82:89; *Col. Pl.* 19:36).

Three-hundred-and-twelve Shell Beads are made from *Cypraea Carneola Linnaeus* 1758 of the Cypraeidae family (*Fig.* 82:87; *Col. Pl.* 19:38). Their diameter is 12–14mm. and length 18–20mm., and each is perforated for stringing. There are some large shells of this type (30–50mm. long) used as pendants.

Petrie (1914, 27, Pl. XIV:a-e) mentions the use of this shell as an amulet for protection against the evil eye and witchcraft, and it is found in Egypt from prehistoric to Roman times. The Egyptians imitated this shell in gold and wore necklaces of gold cypreae shells such as the one found at Dashur from the 12th Dynasty (Morgan, 1895, Pl. XVII). At Megiddo (Loud, 1948, Pl. 217:129; Pl. 227:4) perforated shells of this species were found in Strata VA and IX.

Clanculus Pharaonis Linne 1758 Shell (*Fig.* 82:86; *Col. Pl.* 19:34):

These shells are of the Trochidae family. The average length of 39 beads found in the Temple is about 20mm. and they are perforated for stringing. Petrie dated the use of this kind of Shell Bead to the 25th

Dynasty, but it was probably also used earlier, together with other Shell Beads. Its provenance is also the Red Sea.

Neria polita Linnaeus 1758 shells (*Fig. 82:90; Col. Pl. 19:32*):
Of the Neritidae family, 186 perforated beads made of these shells were found in the Temple. Their diameter is 15–25mm., with a length of 15–30mm.

Petrie found necklaces of these shells, combined with carnelian beads, in Serabit el-Khadem and dated them to the 12th Dynasty. These shells come from the Red Sea.

Among the Shell Beads, the following species were also found:
One *Drupa Tuberculata Blainville* 1832 (*Fig. 82:84, Col. Pl. 19:33*): It has beautiful brown dots and is perforated.
Five *Terebra cerithina Lamarck* 1822 shells of the Terebridae family. One is a *Terebra subulata Linnaeus* 1767 (*Fig. 82:85*), and three are *terebra* sp. of the same family.
All these shells originate from the Red Sea and were perforated for use as beads.

The Strombidae family is represented by one *Strombus gibberulus Linnaeus* 1758 shell (*Fig. 82:80*), and the Cimatidae family by two *Cymatium pileare Linnaeus* 1758 (*Fig. 82:81*), all perforated and originating in the Red Sea.

There are also four shells belonging to the Cerithidae family, *Ceritium erithraeonense Lamarck* 1822 (*Fig. 82:79*), 45–50mm. long and perforated for stringing. They are also Red Sea shells.

According to Petrie (1914, 9), shells were used as amulets against external hostile agencies. Giving these amulets to the goddess Hathor was a sacred ritual. Apparently the Timna Temple was also used by non-Egyptian workers from Midian and local inhabitants from the Arabah and the Negeb, who picked up sea shells and strung them to make necklaces as votive gifts for Hathor.

Red Sea shells were used from the Badarian period onwards (Brunton, 1937, 52). During the 18th Dynasty, Disc Beads of conus shells went out of use in Egypt but not in Canaan and Sinai. In Egypt they were replaced by similar beads of faience, as may be seen in the tomb of Tutankhamun, where many thousands of faience beads, but none of shells were found. The Shell Beads began to come into use again during the 19th Dynasty and were still being made in the 22nd Dynasty (Hayes, 1959, 179).

f Gold Beads (*Fig. 82:91-97*)

Seven small gold beads were found in the Temple: four hemispherical Standard Circular Beads (3–5mm.); one small Disc Bead made of an 0.01mm. thick gold circlet (diameter 4 mm); one Disc Bead of a thin gold crenelated circlet (0.02mm. thick, 4 mm diameter); one Standard Bead of granulated gold (3mm. long, 3mm. diameter). The gold beads found in the Timna Temple were of hammered gold leaf, like similar beads found in Egypt.

g Spacer Beads (*Fig. 83:98-103*)

Beads of faience or stone, multi-tubular with the axes in one plane, of Type A2a, and Group XVII of Beck's classification, were used as spacers. Also Type A3 of the same group are rectangular spacing beads in which the axes of the perforations are parallel. From Timna we have one bead of this type, in which the transverse section is an ellipse with flattened sides (Type A3b1) (*Fig. 80:103*). Of the 61 spacers found in the Temple, two (Type A2a) are made of carnelian (*Fig. 83:100, 101*). Both are 20mm. long and are shaped like two long cylinders fused together on their longitudinal perimeters. One is 3mm. thick and the other 5mm. Both have multiple perforations Type X. According to Beck, these spacers are typical of the 19th Dynasty.

Four spacers made of bone were found. One is a light brown rectangular example, 20mm. long, 12mm. wide and 4mm. thick, with three rows of Type X multiple perforations parallel to its width. The second spacer is of the same type and colour, but its perforations are parallel to its length. It is 18mm. long, 11mm. wide and 3mm. thick. Another spacer is made of a thin rectangular plate 1.5mm. thick, 21mm. long and 13mm. wide. The fourth spacer is also rectangular in shape, 18mm. long, 12mm. wide and 3.5mm. thick. It has two Type X perforations parallel to its width, one on the upper part and the second on the lower. This spacer has a decoration of 8 small incised circles, each 2mm. in diameter with a central dot (*Fig. 83:99*). Bone spacers of this type were found in Megiddo (Loud, 1939, Pl. 218:135; Pls. 193–4), in the Megiddo Tombs (Guy, 1938, Pl. 153:1) in LB II context, and in Tell el-Ajjul (Petrie, 1933, Pl. XXIX).

2 The Pendants

1 Pendants made of natural materials
(*Fig. 83:104–117*)
Pendants are included in the study of beads because in many cases it is difficult to distinguish between the two, the only difference being in the perforation. Most of the Timna pendants have simple forms, derived from regular beads, but pierced so as to make them into pendants. The pendants were strung together with the beads to make a necklace, or were simply attached to a cord passed through the perforation.[11]

Simple Pendants:
The simplest pendants are those made of stone, rectangular, prismatic or cylindrical, with the perforation at one end. A 36mm. long, 7mm. wide, prismatic hematite pendant (*Fig. 83:104; Col. Pl. 20:9*) of Type B4a, is similar to the pendant published by Petrie (1914, Pl. XV:123(b.k)). Another pendant (*Fig. 83:105*), cylindrical in form and made of grey hematite, 20mm. long, 7mm. diameter, with a horizontal perforation on one end, was found strung together with beads. There are also oval-shaped flat pendants of limestone, imitating the so-called ' drop-pendant' of Type BaC (*Fig. 83:106; Fig. 83:107; Col. Pl. 20:7*)

with a scraped texture on the outer side. Seven flat mica schist pendants (*Fig. 83:108, 109; Col. Pl. 20:15–17*) with rectangular rounded corners and perforated upper ends, are 15mm. long and 8mm. wide. Since the mica schist has a thin laminal cleavage, these pendants are extremely thin and breakable.

Ten carnelian pendants have a flat, prismoidal form (*Fig. 83:110*), are 6–20mm. long, with a maximum base width of 3–10mm. They are perforated at their narrow ends.

A pendant made of rose quartz (*Fig. 83:111*) is 30mm. long and 12mm. in diameter. It has a long barrel-shape (D1b) and a perforation on its upper part.

A Lotus-Bud Pendant (*Fig. 83:112; Col. Pl. 19:29*):

A lotus bud or 'lotus seed vessel' pendant, made of carnelian, represents a special type of New Kingdom pendants. According to Hayes (1959, 395) 'in the Ramesside Period.. the forms of pendants are as varied as those of the beads with which they were combined. Besides as simple shapes as drop, bulla, lozenge, pear.. petal, leaf and inverted knob, we find at this period necklace and bracelet pendants in the form of 'inverted lotus flowers', lilies and palmettes, small vases . . .'. These 'small vases' are the 'lotus bud pendants' of Type B–3d, Group XXVII in Beck's classification. They were found in most excavations of this period in Egypt and Canaan, in the B.A. Lachish Fosse temple (Tufnell *et al.*, 1940, Pl. XXXV:6; ibid, 1958, Pl. 29:43), in L.B. Megiddo (Guy, 1938, Tbs. 989, 912, 877 and others).

Similar pendants were found at Tell el-Fara in Stratum VII A (19th–20th Dynasties) (Macdonald, 1932, Pl. XLIX:922; Pl. LVI:984; Pl. LI; Petrie, 1930, Pl. XXVII:552), in Tell el-Ajjul (Petrie, 1932, Pl. VIII: 185, 186) in the temple of Amunhotep III, at Beth Shean (Rowe, 1940, Pl. XXXIII:78; Pl. XXXIV:68), in Egyptian excavations (Vernier 1907–27, No. 53011), at Buhen in the 18th Dynasty temple (Randall *et al.*, MCM: Pl. 54).

Bone Pendants (*Fig. 83:113–117*):

Six bone pendants were found on the site, five of them in Loci 101 and 102, and one in Locus 109. The largest of them is a simple arrowhead shaped pendant (*Fig. 83:113*) of Type B3, Group XXXVIII, with two perforations on its upper part (broken) It is 53mm. long, 0.5mm. thick and 20mm. wide at the top, and tapers to a point.

The second pendant is of carved bone (*Fig. 83:114*), 38mm. long, 15mm. wide and approximately 4mm. thick. It has a natural bone colour and the shape of an arrowhead, with a decoration of parallel lines incised 5mm. apart.

A fragment of bone and many small bone pieces are the broken parts of an amulet 46mm. long and 2mm. thick. It has a perforation on its upper part and the form of an oval-shaped arrowhead.

The fourth pendant is a carved piece of bone (*Fig. 83:116*), 15mm. wide. The rounded upper part shows three holes (two in a line, the third symetrically between the upper two but lower).

There are also two fragments of annular bone pendants (or rings) of Group XX, Type A1a with a diameter of 22mm. and 26mm., and 6mm. and 10mm. thick respectively.

Several bone pendants were found at Megiddo. One pendant similar to the second one described above was found in Strata V and IX (Loud, 1948, Pl. 218:128; Pl. 211:49) in the tombs in an L.B. II context (Guy, 1938, Tb. 45, Pl. 111; Tb. 42, Pl. 109).

A bone pendant similar to the fourth example described was found in Megiddo at Stratum VA (Loud, 1948, Pl. 218:135). Annular bone pendants, perhaps also used as ear-rings, and others made of ivory, faience and other materials, were found at Beth Shean in the Ramesside temple (Rowe, 1940, Pl. XXIX:1–11). A ring made of mother of pearl, of the same type, was found at Tell el-Ajjul (Petrie, 1934, Pl. XXXVII). Young servants are seen wearing similar earrings on Egyptian wall paintings (Desroches Noblecourt, 1962, Pl. 10) from the 18th Dynasty, and on Egyptian tomb scenes from the same Dynasty (Lange and Hirmer, 1968, Pls. 221, 229).

2 *Pendants made of artificial materials (Fig. 84:118-137)*

Pendants made of copper-base metal (*Fig. 84:133–137*):

Five pendants made of copper or bronze were found. Four were only fragmentary but the fifth (No. 136) is 24mm. long, 12mm. wide and 1.3mm. thick. One of the pendants (No. 137) has an incised decoration on both sides, showing a simple plant (*Pl. 127:3*). No. 134 had three perforations on its upper part, one broken away; No. 135 has three perforations, one on top and two at its lower end.

Pendants made of pottery and faience:

A simple cylindrical pendant made of pottery (*Fig. 84:118*) 30mm. long, was found on the site. It is brown and has a plain perforation on its upper part.

Lotus Flower Pendants (*Fig. 84:119–121; Col. Pl. 19:26–28*):

Three faience pendants 15mm. long, two green glazed and one red, were found in Locus 101. They are of Type B1c, Group XXVI and represent lotus flowers.

In the Lachish Fosse Temple similar pendants were found in an 18th Dynasty context (Tufnell *et al.*, 1940, Pl. XXXVI:103; Pl. XXXV:88). They are blue glazed with red. At Megiddo (Loud, 1948, Pl. 216:115, 121) similar pendants were found in Stratum VIB (Early Iron I). In Egypt there were lotus flower pendants among the jewellery of the Three Princesses of the 18th Dynasty (Winlock, 1948, Pl. X) and in the Deir el Medineh excavations (Bruyère, 1937, 65).

Another Lotus Flower Pendant, classified by Beck as 'conventional lotus pendants of Type B6b' was found in the Temple. Made of green glazed faience, this pendant is 25mm. long, with a maximum width of 12mm.

A Hawk Pendant (*Fig. 84:122*):

A small fragment of a Horus pendant, 5mm. in

length, in green faience and perforated longitudinally, was found on the site.

A Faience Pendant in the form of a beetle:

A green glazed pendant in the form of a beetle, 10mm. long, is perforated longitudinally. Its back has a protruding line showing the prothorax, the head and the clypeus; the elytra and tybias are not shown.

An Egyptian 'White Crown' pendant (*Fig.* 84:123):

A faience pendant in the form of an Egyptian White Crown of Upper Egypt was found at the site. The perforation is on the upper part but only traces of it are preserved. It is 20mm. long and 15mm. wide, made of a frit paste with a white glaze, but the latter lost its glassy aspect and the colour is now undefinable. The White Crown was commonly used as a pendant or amulet in Egypt (Petrie, 1914, Pl. IV:48a–d; Reisner, 1907, Pl. VI:5858).

Gold Leaf Pendants (*Fig.* 84:124–132):

It is not always possible to define with certainty the function of the small gold leaf objects found at the Temple. They are small flower petals worked in repouss technique, which seem to have adorned head bands or belts. Eight small gold leaf objects and part of a headband were found.

No. 124 is a leaf of flower petal shape used as a pendant, 15mm. long, 6mm. wide and 0.03mm. thick. It is decorated with incised lines and the perforation is on the upper end.

No. 125 is a small leaf, perhaps part of a pendant, of rectangular elongated form, with the upper part folded over for stringing. It is 11mm. long, 4mm. wide and 0.05mm. thick. It has a border decoration of dots. A similar leaf object was found at Beth Shean in the temple of Amenhetep III (Rowe, 1940, Pl. XXXIV:5–8). (See Chapter III, 8, Analytical Sample No. 357.)

No. 126 is a gold leaf in the form of two flower petals bound together (which could also be two wings of a golden fly with the body missing). This leaf is 10mm. long, 0.05mm. thick and has a maximum width of 5mm. for each petal. Several fine lines are incised longitudinally into both petals. Gold flies were very common in Egypt from the New Kingdom period onwards (Wilkinson, 1971, 98–9). Six gold flies were given by Thutmosis I to the soldier Pen-Nekhebet, and three enormous flies attached to a gold chain were found on the mummy of Queen Ahhotpe of the 17th Dynasty.

Nos. 127–130 are small parts of gold leaves, broken off from larger objects and were probably fragments of pendants. No. 130 is part of a larger pendant with repouss decoration of circles and dots, and measures 2.2 × 1.6mm.

No. 131 is oval-shaped, 18mm. long, 13mm. wide and 0.05mm. thick. It has repoussé decoration in the form of patterns of fine, perforated small dots in concentric circles. Since it has two perforations in the upper part, it was probably worn as a pendant.

A Gold Head Band (*Fig.* 84:132)

A fragment of a gold head band or diadem was found at Locus 102. It is very thin (0.05mm.), 26mm.

long and 10mm. wide, and has an incised decoration of zig-zag lines.

A similar head band, but larger and wider, was found at Megiddo in stratum IX (Loud, 1948, Pl. 227:5) and at Tell el Ajjul (Petrie, 1933, Pls. XIV, XV). In Egypt, princesses wore similar diadems and head bands as the Three Egyptian Princesses from the harim of Tuthmosis III (Winlock, 1948, Pl. VII).

Metal Amulets or Pendants of an Open Human Hand (*Fig.* 47:9–13)

Five metal amulets or pendants of this type (Beck 1928, Group XXXI, Fig. 28:B5) were found at the site. They represent an open human hand, the fingers indicated by incised lines. The upper end of the pendants is folded over to facilitate its attachment to a chain or string.

No. 9: bronze, 4.2cm. long, maximum width 1.7cm., 2mm. thick. Two fingers are preserved, the rest is broken (Anal. Table No. 540).

No. 10: copper, 2cm. long, maximum width 1.2cm., 1mm. thick. Five fingers are visible (*Pl.* 127:2; Anal. Table No. 541).

No. 11: bronze, 4.3cm. long, maximum width 2cm., 15mm. thick. Five fingers are visible (partly broken) (Anal. Table No. 542).

No. 12: bronze, 2cm. long, maximum width 7cm., 1mm. thick. Three fingers are preserved (Anal. Table No. 543).

No. 13: bronze, 2.3cm. long, maximum width 1.8cm., 1mm. thick. Five fingers are visible (Anal. Table No. 544).

The human hand, mostly made of faience or bone, was often used in Egypt as a magic charm (Petrie, 1914, Pl. 1:a–g; Reisner, 1907, Pl. IX:1211–1213; see now also Helck and Otto, 1977, Vol. II, 938–43).

Trude Kertesz

Notes

1. We are most grateful to Professor Barash and Dr Danin for the identification of the various shell beads, and to Dr G. Lehrer-Jacobson for the identification of the glass beads.

2. The ancient method of making stone beads in Egypt is described by Reisner, 1923, 93–94.

3. See Quibell and Green, 1902, 12, for possible flint drills for beads in Egypt.

4. See Thomas, 1956, on methods for provenance studies of faience beads. See also Newton and Renfrew, 1970, 199–206.

5. See Petrie, 1894, 28, Pl. XVIII:414, 415, 425, 426, 428, for bead moulds from Tell el-Amarna.

6. On the technique for making beads of this type, as well as similar types described above, see Lucas, 1962, 44, 45. See also Wulff, 1968, 98–107, and for Egyptian blue glazes, Kiefer and Allibert, 1971, 107–17.

7. On the technique for making decorated glass beads, see Eisen 1916.

8. On the stratified eye technique see Eisen, 1916, 17.

9. See Eisen (1916, 7) who dates the eye beads from the 19th Dynasty to the Ptolemaic Period.

10. On the technology of the Crumbed Glass Beads, see Sleen, 1967.

11. The natural flat stones, described above, can also be classified as pendants.

Bibliography

Beck, H.C. 1928. Classification and Nomenclature of Beads and Pendants, *Archaeologia* 77, 1–76.

—— 1934. Glass before 1500 B.C., *Ancient Egypt and the East*, Part I, 7–21, London.

Bruyère, B. 1937. *Rapport des Fouilles de Deir el Medineh*, Cairo.

Brunton, G. 1937. *Mostagedda*, London.

Brunton, G. and Thompson, G. 1948. *The Badarian Civilisation*, London.

Carter, H. and Mace, A.C. 1963. *The Tomb of Tut-ankh-Amen*, II–III, New York.

Desroches Noblecourt, C. 1962. *Egyptian Wall Paintings*, New York.

Duncan, J.G. 1930. *Corpus of Dated Palestinian Pottery*, London.

Eisen, G. 1916. Eye-beads from the Earliest Time to the Present, *AJA* 20, 1–27.

Griffith, F.L. 1890. Tell el Jehudieh, *EEFM* VII, London.

Guy, P.L.O. 1938. *Megiddo Tombs*, Chicago.

Hamilton, R.W. 1935. Excavations at Tell-Abu-Hwam. *QDAP* 4, 1–70.

Hayes, W.C. 1953–9. *The Scepter of Egypt*, 2 vols., New York.

Kiefer, C. and Allibert, H.A. 1971. Pharaonic blue Ceramics, The process of self glazing, *Archaeology* 24, 107–17.

Lamon, R.S. and Shipton, G.M. 1939. *Megiddo I*, Chicago.

Lange, K. and Hirmer, M. 1968. *Egypt. Architecture, Sculpture and Painting in Three Thousand Years*, London and New York.

Loud, G. 1939. *The Megiddo Ivories*, Chicago.

—— 1948. *Megiddo II*, Chicago.

Lucas, A. 1962. *Ancient Egyptian Materials and Industries*, rev. by J.R. Harris, London.

Macdonald, E. 1932. *Beth Pelet II*. Prehistoric Fara (with Starkey, J.L. and Harding L.), London.

Morgan, de J. 1895, 1903. *Fouilles à Dahchour I*, 2 vols., Vienna.

Newton, R.G. and Renfrew, C. 1970. British Faience Beads Reconsidered, *Antiquity* 44, 199–206.

Pavlova-Fortova-Samalova. 1963. *Das Aegyptische Ornament*, Prague.

Petrie, W.M.F. 1892. *Illahun*, London.

—— 1894. *Tell el Amarna*, London.

—— 1902. *Abydos*, London.

—— 1905. *Ehnasya*, London.

—— 1906. *Researches in Sinai*, London.

—— 1914. *Amulets*, London.

—— 1920. *Prehistoric Egypt*, London.

—— 1923. *Arts and Crafts*, Edinburgh.

—— 1928. *Gerar*, London.

—— 1930. *Beth Pelet I*, London.

—— 1931. *Ancient Gaza I*, London.

—— 1932. *Ancient Gaza II*, London.

—— 1933. *Ancient Gaza III*, London.

—— 1934. *Ancient Gaza IV*, London.

Petrie, F., Mackay, E.J. and Murray, M.A. 1952. *City of the Shepherd Kings* and *Ancient Gaza V*, London.

Quibell, J.E. and Green, F.W. 1900. *Hierankopolis I*, Chicago.

—— 1902. *Hierankopolis II*, Chicago.

Randall-MacIver, R.D. and Woolley, C.L. 1911. *Buhen*, Philadelphia.

Reisner, G.A. 1907. *Amulets*, *Catalogue Générale des Antiquités Egyptiennes du Musée du Caire*, Cairo.

—— 1923. *Excavations at Kezma* IV-V, Cambridge, Mass.

Rowe, A. 1940. *The Four Canaanite Temples of Beith Shan II*, Philadelphia.

Sleen, van der W.G.N. 1967. *A Handbook of Beads*, Liège.

Starkey, J.L. 1930. The Beth Pelet beads, in Duncan, *Corpus of Palestinian Pottery*, London.

Thomas, L.C. 1956. Notes on the Spectrochemical Analysis of Faience, in Stone, J.F.S., The Use and Distribution of Faience, *PPS* 22, 63–84.

Tufnell, O.C. and Harding, G.L. 1940. *Lachish II (Tell ed-Duweir)*, Oxford.

—— 1953. *Lachish III (Tell ed-Duweir)*, Oxford.

—— 1958. *Lachish IV (Tell ed-Duweir)*, Oxford.

Vernier, E. 1907–27. *Bijou et Orfévres*, *Catalogue Générale des Antiquités Egyptiennes du Musée de Caire*, Cairo.

Vilmikova, M. 1969. *Egyptian Jewellery*, Prague.

Wilkinson, J.G. 1878. *Manners and Customs of the Ancient Egyptians*, Vol. II, London.

Winlock, H.E. 1948. *The Treasure of Three Egyptian Princesses*, New York.

Winlock, H.E. and Mace, A.C. 1916. *The Tomb of Senebtisi at Lisht*, New York.

Wulff, H.E., Wulff, H. and Koch, L. 1968. Egyptian Faience, *Archaeology* Vol. 21, 98–107.

Yadin, Y. *et al.* 1961. *Hazor III-IV*, Jerusalem.

15 EGYPTIAN GLASS

Introduction

Site 200 yielded a great number of glass fragments, most of which are related to its Egyptian phase and only a very few belong to the Roman period. There are over 150 fragments of 50–60 cored vessels, and also fragments of a flat inlay piece and of a bracelet or ring stand. Glass beads were also found, but these are discussed in Chapter III, 14. These glass finds probably represent the largest group of Egyptian glass found outside of Egypt and are of particular importance because the context is closely dated by inscriptions to the 19th and 20th Dynasties.

The following vessel types have been identified: 1) bowl; 2) krateriskos; 3) lentoid flask; 4) amphoriskos. There were also fragments of globular vessels, like the pomegranate bottles,[1] and of pear-shaped vessels.

Except for one green bottle, all the vessels were made of light blue glass, predominately opaque turquoise, or of dark blue glass (cobalt and dark turquoise). On the light blue vessels the decorative garland, festoon, zig-zag and feather patterns are white, yellow and dark blue. White, yellow and light blue patterns decorate the dark blue vessels. Decorations may appear in a single colour, as a combination of two, or all three hues together. A feature common to many of the vessels is a strip made of twisted white and apparently black (probably in reality dark blue or dark brown) glass threads applied on the rim and sometimes also on the shoulder and lower part.

Fragments from two different bowls are blue, with a yellow strip applied on the rim. Of particular importance is a fragment bearing the impressed double cartouche of Ramesses II (Egypt. Cat. No. 180, Fig. 39:3).

The extremely fragmentary condition of the glass must be pointed out. Despite the large number of fragments, it was difficult to find even two or three pieces from the same vessel which could be fitted together, and not a single complete profile was restorable. Similarly, most of the other finds at the site (faience, pottery, beads) were also fragmented. This fragmentary condition of the finds was also noticed by Petrie (1906, 138–9) at Serabit el-Khadem in Sinai and he suggested that the offerings were intentionally broken and scattered by Bedouins. Another explanation is offered by T. Kertesz, who suggests that the objects were ritually broken as part of the funereal aspect of the Hathor cult (Kertesz, 1976, 134–6). However, it seems more likely that in remote and isolated mining sites such as Timna and Serabit el-Khadem, even fragments made of valuable materials were brought as offerings. During the New Kingdom, glass was considered almost as precious as gold, silver or gemstones, with which it was often combined (see Nolte, 1968, 10–12). The close similarity between blue glass and turquoise made even such glass fragments worthy as an offering to Hathor. This would explain the unusually large number of glass fragments in the Timna Temple, matched only by finds in Egyptian royal tombs.

Recently, special attention was paid to the nature of the core on which cored vessels were built. Chemical and petrographic analyses, carried out by Bimson and Werner (1968, 121–2) and Brill (1968, 123–4) have established that the Egyptian glass vessels of the New Kingdom were built around cores made of a mixture of clay and plant material, and not of sand as was previously assumed. According to Bimson and Werner, the clay and plant core was covered by a layer of limewash, still detectable on many fragments. Brill (ibid), who did not find this limewash on all the fragments, concluded however that this layer is 'a secondary deposit formed in the core residue after long burial in soil'. Our examination of the fragments found at Timna established that all have a smooth white-grey inner surface, indicating the presence of a coating.

Another interesting feature which can throw light on the manufacturing method used for these vessels, appears at the junction of neck and shoulder: on the narrow-necked vessels (bottles, amphoriskoi), a rough, irregular, wavy strip protrudes on the inside, as though the soft glass had been pushed in when the neck was added to the body (*Col. Pl.* 10). On vessels with wider necks (krateriskoi) the strip was smoothed over after removal of the core.[2]. These neck-marks seem to confirm Labino's reconstruction of the core technique (Labino, 1966, 125). Experimenting with the 'trailing-on' method, he succeeded in producing 'Egyptian-type' vessels. By this method the glass is trailed onto the core with a dipstick, starting from the bottom and working upwards. The entire body is built up by repeated trailing, tooling and heating. Afterwards, according to Labino, the core was returned to the heat and more glass added to the end near the rod, to form the neck. Thus the vessel was built by a repeated sequence of operations, one of which added the neck to the body. This may explain the strip at the junction of the two parts.

F. Schuler (1962, 35)[3] rejected the 'trailing-on' method, claiming that '. . . the process should be simple enough to execute, but it would tend to trap air and give a spiral structure to the glass, if the glass applied had bubbles or striae'. However, Labino (ibid) solved this problem of trapped air by tooling. Regarding the spiral structure, this is indeed visible on our bowl fragment No. 1.

For stratigraphic reasons (see below, Chapter IV) it is impossible to make a division between the glass fragments belonging to the 19th and 20th Dynasty. The twisted black and white strip (sometimes accompanied by parallel monochrome strips) bordering the body pattern, the pattern itself, and the occasional marvered blobs which are characteristic features of most of the Timna fragments, are also common amongst the vessels from Gurob[4] dated mainly to the reign of Ramesses II. This dating is now also confirmed by the fragment bearing the cartouche of Ramesses II (No. 22b).

Also comparable are the glass vessels from the latest phase of the Fosse Temple (Structure III) at Lachish (Tufnell and Harding 1940, 24, 64, Pl. XXIV) dated to approximately the end of the 18th–19th Dynasty (1325–1223 BC). A few fragments from blue monochrome bowls and a krateriskos (Nos. 1, 2 and 13) have parallels at el-Menshiyeh and Lisht, dated to the later part of the 19th–20th Dynasty (Hayes, 1959, 403, Pl. 255; Nolte, 1968, 24, 25, No. 28, 75).

1 Bowls

Fragments of two hemispherical bowls were found. Both bowls were made of light blue glass, probably monochrome, with a yellow strip applied on the outer edge of the rim.

1. *Fig.* 85:1; *Col. Pl.* 7:4, Field No. 289/17, H. 2.6cm., W. 3cm.
Fragment of rim and wall. The yellow strip, placed on the edge, forms a narrow ledge with the rim. There is no trace of the core but the fragment clearly shows a spiral structure which would indicate that the glass was wound or trailed onto the core (see Labino, 1966, 125). The glass is porous and has a matte surface. It is virtually free of decay, showing only slight brown encrustation in the pores.

2. *Col. Pl.* 7:3, Field No. 206/55–57, H. 2.3cm., W. 2.4cm.
Fragment of rim and wall from a similar bowl, but with a thicker wall. Badly weathered and quite brittle. The bowl is not a very common form among core-made vessels and is unknown in the later Persian and Hellenistic cored groups. From the eighth century BC onwards bowls were made by casting in moulds.

There are only a few known examples of cored bowls. Three are of the variegated type (dark glass with thread decoration): one in the Sangiorgi Collection (Collezioni di Vetri Antichi, Milano-Roma, 1914, Pl. 1), a fragmentary bowl in the British Museum (Nolte, 1968, Taf. XX:4), and a shallow bowl in the Manchester Museum, from a house-deposit in Gurob dated to the time of Ramesses II (Petrie, 1839–40, Pl. XVIII:18; and Harden, 1969, Pl. III:D).

A fourth bowl (in the Metropolitan Museum) is the closest parallel to our bowl fragments. Of blue monochrome glass with a yellow strip on the rim, it was found in a 20th Dynasty factory near el-Menshiyeh (Hayes, 1959, 403, Fig. 225; 403, Fig. 225; Nolte, 1968, Taf XXVIII:50). Hayes mentions a number of monochrome bowls from the later New Kingdom, but gives no further details. Another blue bowl with yellow rim from el-Menshiyeh is mentioned by Nolte (1968, 75). Our fragments thus belong to the el-Menshiyeh monochrome type.

2 Krateriskoi

The decorations on the necks of the krateriskoi consist mostly of zig-zag, ogee or festoon patterns. A characteristic feature of these vessels is a twisted strip of 'black' and white glass, not only around the rim but also on the shoulder or the lower part of the body. In two cases this twisted strip is placed between two plain coloured white and/or yellow strips. Only one fragment seems to belong to a monochrome turquoise krateriskos.

3. *Fig. 85:2; Col. Pl.* 7:2, Field No. 203/34, H. 2.6cm., W. 2.2cm., Diam. of mouth ca 4.8 cm.
Fragment of rim and neck. Turquoise with twisted strip on rim. Zig-zag pattern of yellow and white.
Col. Pl. 7:1, Field No. 203/35
Fragment of shoulder and neck, possibly from the same vessel as 203/34. Turquoise, with twisted strip marvered flush on shoulder. Beginning of loose feather pattern in white and yellow on body. Brownish encrustation.

4. *Fig.* 85:3; Field No. 269/73–74, H. 2.8cm., W. 2.3cm.. and H. 1.8cm., W. 1.5cm.
Two fragments of rim and neck. Opaque light blue. Remains of twisted strip on rim. Widely spaced ogee pattern of white threads. Whitish weathering.

5. Field No. 353/8, H. 1.5cm., W. 3.5cm.
Fragment of neck. Opaque light blue with zig-zag pattern of white and yellow. Whitish weathering.

6. Field No. 353/10.
Small fragment of neck. Turquoise with zig-zag pattern of white and yellow.

7. Field No. 353/9, H. 2.0cm., W. 1.2cm.
Fragment of neck. Deep blue (turquoise). Remains of twisted strip on rim. Festoon pattern of white, yellow and dark blue.

8. *Fig.* 85:4; *Col. Pl.* 7:7, Field No. 245/39, H. 3.5cm., W. 4.0cm.
Fragment of neck and shoulder. Translucent turquoise with festoon of dark blue on neck. Straight white line at base of neck. Twisted strip on shoulder showing the point where the two ends meet. Beginning of yellow and white feather pattern on body. Whitish weathering.

9. *Fig.* 85:6; *Col. Pl.* 7:6, Field No. 227/51, H. 3.0cm., W. 3.5cm.
Fragment of shoulder. Opaque light blue with parallel strips: yellow, twisted, white and (probably) another yellow.

10. *Fig.* 85:5; *Col. Pl.* 7:5, Field No. 227/49–50, H. 2.5cm., W. 4.5cm.. and H. 1.5cm., W. 1.5cm.
Two fragments of shoulder and beginning of neck. Opaque light blue with traces of zig-zag pattern on neck. White and twisted strips on shoulder.

11. *Col. Pl.* 7:8, Field No. 151/3, H. 2.8cm., W. 3.0cm.
Fragment of a large-mouthed vessel, possibly a krateriskos. Deep blue glass. At base of neck a yellow strip; vestige of yellow pattern on neck. Remains of a yellow handle. Whitish weathering.

12. *Fig.* 85:7; *Col. Pl.* 7:10, Field No. 241/59, H. 2.5cm., W. 2.3cm.
Fragment of body, possibly from a krateriskos. Turquoise with garland pattern of dark blue and white. Twisted strip marvered flush below the pattern. Brownish weathering.

13. *Col. Pl.* 7:9, Field No. 209/39, H. 2.6cm., W. 3.4cm.
Fragment of body. Monochrome turquoise glass. The surface is cracked. Traces of tooling.

A close analogy to the decoration of Nos. 9 and 10 is provided by the fragmentary krateriskos from a house-deposit at Gurob from the time of Tutankhamun (Petrie, 1889–90, Pl. XVIII:37, and Nolte, 1968, Taf. XVI:1 known only from a drawing). The same decoration of twisted black and white and two other monochrome strips appears on a bottle of unknown provenance in the Victoria and Albert Museum. The bottle is dated by Nolte to the Tutankhamun-Ramesses II period.

A proximal parallel to Nos. 3, 8 and 12, which only have the twisted strip bordering the body pattern, is a krateriskos in the Freer Gallery (Eisen, 1927, *Pl.* 2, last on right; Ettinghausen 1926, *Fig.* 16; Nolte, 1968, Taf. XVI:4. Nolte dated the vessel to the period between Tutankhamun and Ramesses II).

No. 13 probably belongs to a monochrome krateriskos with rather thick walls. There are two known variant types: one, with or without handles, has a high foot; the other is footless (with a flat base) and usually has no handles. It is impossible to tell to which variant our fragment belongs. Parallels to both variants are found in Nolte, 1968, Taf. XXIII:7 and 8; Taf. XXIV:11 and 13. The exact provenance of these vessels is unknown. However, fragments of similar monochrome vessels (blue and other colours) found at the site of a glass factory at el-Lisht were dated to the Late Ramesside period (Hayes, 1959, 403).

3 Lentoid Flasks

Fragments from three lentoid flasks were identified. No. 14 (three fragments) is sufficiently well preserved to enable positive identification. The other two fragments, Nos. 15 and 16, are ascribed to lentoid flasks because of their narrow curvature; they are probably parts of the narrow side.

14. *Fig.* 86:1; *Col. Pl.* 8, Field No. 278/110, H. 5.0cm., W. 4.8cm.
Fragment of body. Opaque light blue. Dark blue and white parallel strips on shoulder and loose feather pattern of dark blue on body. Dark blue glass blobs, marvered flush, appear between the shoulder strips and the body pattern. Remains of lower part of a small handle on shoulder.

Field No. 278/111, H. 1.6cm., W. 0.7cm.
Fragment of a small handle, of same opaque light blue glass.
Field No. 278/112, H. 2.0 cm., W. 2.2 cm..
Fragment of shoulder and neck. Although there is no common join, it clearly belongs to the same vessel as 178/110. Same opaque blue glass with a horizontal strip on the shoulder.

15. Field No. 239/9, H. 2.0cm., W. 2.0cm.
Fragment of body. Dark blue glass with traces of a feather pattern, only the grooves remain and the threads have completely disappeared. Brownish weathering.

16. Field No. 157/9, H. 2.3cm., W. 2.3cm.
Fragment of body. Opaque light blue with a feather pattern as on No. 15. Traces of an applied twisted strip. Brownish weathering.

A close parallel to No. 14 in shape, colour combination and shoulder decoration is a lentoid flask of unknown provenance in the Myers Museum at Eton College, included by Nolte in her chronological group Tutankhamun-Ramesses II (Nolte, 1968, 116; Taf. XVIII:19). The way in which the feather pattern is applied on our fragment (a few widely spaced lines meeting in a point) is almost identical to the pattern on a krateriskos of the same period in the Ashmolean Museum (Nolte 1968, 115; Taf. XVII:17). Similar blobs marvered flush are seen on a miniature lentoid flask (in the British Museum) from a house-deposit at Gurob dating to the reign of Ramesses II (Petrie, 1889–90, Pl. XVIII:17; and Nolte, 1968, Taf. XVI:5).
The tightly drawn feather pattern on Nos. 15 and 16 is comparable to a flask from the Walters Art Gallery (Nolte, 1968, Taf. XVIII:26) and to another in the Louvre (Nolte, 1968, Taf. XX:6), both of unknown provenance.
A twisted strip on the lower part of the body, as on No. 16, appears on two miniature lentoid flasks from a house-deposit at Gurob (Petrie, 1889–90, Pl. XVII:l7 and 20:11; Nolte, 1968, Taf. XVI:2 and 5), dated to the period from the end of the 18th to the beginning of the 19th Dynasties.

4 Amphoriskoi

Fragments of four amphoriskoi have been identified. All have elongated bodies tapering to narrow rounded bases. Handles are missing on the preserved fragments but No. 17 has traces of a handle on the shoulder. A broken handle, No. 39, found at the site, could have belonged to Nos. 20 or 21.

17. *Fig.* 86:5. Two fragments with no common join may belong to the same vessel.
Field No. 245/41, H. 2.3cm.
Fragment of shoulder. Dark blue (cobalt) with white threads, traces of handle.
Field No. 245/43, H. 2.0cm.
Fragment of narrow, rounded base. Same dark blue as above and white thread.

18. *Fig.* 86:4. Three non-joining fragments from the same vessel.
Field No. 316/13, H. 1.8cm.
Fragment of mouth and neck (distorted). Dark blue glass weathered to dull grey, with twisted strip on rim. Festoon pattern of yellow and light blue on neck.
Field No. 316/14, H. 2.0cm., W. 3.0cm..
Fragment of body (restored from two smaller fragments). Dark blue glass, as above, with garland pattern in what was probably white and yellow. Dull grey weathering.
Field No. 316/15, H. 3.2cm.
Fragment of rounded base (restored from three smaller fragments). Same dark blue as above. Two or three horizontal strips, probably originally white, on upper part of fragment. Dull grey weathering.

19. Field No. 337/19, H. 3.0cm., W. 3.0cm. and H. 1.5cm., W. 1.5cm.
Two body fragments from the same vessel. Dark blue glass with a feather pattern of white and another light colour which has disappeared completely in places, leaving only the grooves.

20. *Fig.* 86:2, Field No. 245/40, H. 4.0cm., W. 2.2cm. (the larger fragment).
Two fragments of body and shoulder from the same vessel. Opaque light blue, with three white lines on shoulder and elaborate feather pattern of white and lighter blue. Brownish weathering.

21. Field No. 157/7, H. 4.8cm., W. 1.0cm.
Fragment of body, same colours and decoration as No. 20. Brownish weathering.

A possible parallel for No. 18 is an amphoriskos from the British Museum (Nolte 1968, Taf. XVII:11). The neck decoration is different, in this case zig-zag, but the body pattern of festoons or garlands and the two strips which border it are similar to those on our fragment. It is included by Nolte in the group of the Tutankhamun-Ramesses II period. The decoration on No. 19 is similar to that on an amphoriskos from Abydos (Amlineau 1898, 59, not illustrated; Nolte, 1968, Taf. XIX:37, 119), and possibly also to that on the amphoriskos from Gurob (Brunton and Engelbach 1927, Pl. LIII, Group 705 F, Petrie, ascribed to the Ramesside period; Nolte, 1968, Taf. XIX:39).
Nos. 20 and 21 can be compared to the same amphoriskoi mentioned above (Nolte, 1968, Taf. XIX:37, 39). A similar combination of patterns also appears on the amphoriskos from the cache in the Fosse Temple at Lachish (Tufnell and Harding 1940, 64, Pl. XXIV:77). This vessel was compared by the excavators to the amphoriskos from Gurob and dated to the early Ramesside period (19th Dynasty, Structure III).

5 Pomegranate Bottle

Three fragments of opaque green glass: a rim fragment, a body fragment with a cartouche, and a fragment of a rounded base, may belong to one and the same pomegranate-like bottle.

22. (a) *Fig.* 86:7, Field No. 80/11, H. 2.0cm., W. 1.5cm.
A 'petal-shaped' rim fragment. Brownish weathering.
(b) *Fig.* 39:3, Field No. F.13, H. 1.0cm., W. 2.0cm.
Small body fragment with the lower part of an impressed double cartouche bearing the name of Ramesses II (see Egypt. Cat. No. 180). *Fig.* 39:3.
(c) Field No. 325/13, W. 4.0cm.
Fragment of rounded base. Brownish weathering.

A few monochrome pomegranate bottles are known. These include a yellow bottle from el-Menshiyeh (Hayes, 1953, Fig. 255 top right; Nolte, 1968, Taf. XXVII:42) and an opaque green bottle with only a yellow strip bordering the rim. The latter, of unknown provenance, is in the Louvre (Nolte, 1968, Taf. XXVII:40I) and is the closest parallel to our vessel. There is also a double vessel made of greenish-white glass, consisting of a lentoid flask and a pomegranate vessel joined together. This vessel was found at Darb-Esbeida, together with ushabtis of Ramesses VI (Nolte, 1968, 76, Taf. XXVI:39). Also mentioned by Nolte is another group of smaller monochrome green pomegranate vessels of unknown origin (Nolte, 1968, 128, not illustrated). The occurrence of a royal cartouche on glass vessels is not unique. In Egypt several examples are from the 18th Dynasty, and the cartouche was cut, scratched, painted or inlaid (Nolte, 1968, 13). The body fragment with the impressed double cartouche of Ramesses II dates the Timna vessel to the 19th Dynasty.

6 Various Fragments

Fragments which could not be attributed to a specific type of vessel have been categorized into four groups:
(a) Necks and rims, all from narrow-necked vessels;
(b) Body fragments; (c) Bases; (d) Handles.

(a) Necks and Rims:

23. Field No. 334/5. Two fragments from the same neck. H. 2.2 and 2.0cm.
Light blue with white and yellow ogeed zig-zag pattern. Twisted strip of dark brown and white glass on rim. Whitish weathering.

24. Field No. 343/83. Two fragments from the same neck and rim. H. 3.0 and 2.0cm..
Dark blue glass with twisted strip on rim and festoon pattern of white and yellow. Brownish weathering.

25. Field No. 245/41, H. 1.5cm.
Fragment of neck. Dark blue with festoon pattern of blue and white.

26. *Fig.* 86:6, Field No. 51/97, H. 2.3cm.
Fragment of neck and shoulder. Opaque light blue with zig-zag pattern of white and yellow.

27. Field No. 245, H. 1.0cm.
Small fragment of rim and beginning of neck. Twisted strip on rim. Opaque light blue with white zig-zag pattern. Brownish weathering.

28. Field No. 253/50, H. 0.8cm.
Rim of dark blue glass with twisted strip. Brownish weathering.

29. Field No. 211/267, H. 1.0cm.
Fragment of rim, similar to No. 28.

(b) Body fragments:

30. *Fig.* 86:3, *Col. Pl.* 7:6, Field No. 273/1. Three fragments, possibly from the same vessel. H. 4.0, 2.8 and 2.5cm.
Dark translucent turquoise with garland pattern of white and yellow. Yellow horizontal strip above pattern. Thickness of wall increases downwards.

31. Field No. 304/15, H. 1.4cm., W. 3.0cm.
Opaque light blue with large feather pattern of dark blue and white. Whitish-brownish weathering.
For similar decoration: Nolte, 1968, Taf. XIX:36.

32. Field No. 376/7. Three small fragments of the same vessel. H. 3.0, 1.5 and 2.0cm.
Dark blue glass with two horizontal strips marvered flush and feather pattern of white glass. Greyish weathering.

33. Five small fragments from different vessels.

(c) Bases:

34. Field No. 208/43, W. 4.0cm.
Fragment of flat base with part of wall tapering upwards. Dark blue glass.

35. Field No. 370/10, H. 2.5cm., W. 2.7cm.
Fragment of a rounded base. Dark blue glass. Brownish weathering.

36. Field No. 273/4.
Fragment of a rounded base. Dark blue glass with yellow horizontal strip marvered flush.

37. Field No. 371/9, H. 1.0cm., W. 2.0cm.
Fragment of base, probably knobbed. Light blue opaque glass. Brownish weathering.

(d) Handles:

38. Field No. 125/5, H. 3.5cm., W. 1.3cm.
Fragment of broad handle of dark turquoise glass. Similar handle: Tufnell and Harding 1940, Pl. XXIV:88.

39. Field No. 157/10, H. 1.5cm., W. 0.8cm.
Small loop handle attached to body fragment, possibly from an amphoriskos. Light brown or purple glass with white thread applied lengthwise. The body fragment is opaque light blue with a feather pattern. Brownish weathering.

7 Miscellaneous

40. Field No. 373/12, 2.3 × 3.0 × Th. 1.3cm.
Fragment of thick inlay plaque of dark blue glass with a curving yellow inlaid strip.

41. Field No. 372/4, H. 2.0cm., W. 1.8cm.
Fragment of bracelet or ring stand of opaque green glass with a stamped rectangular element.

Gusta Lehrer-Jacobson

Notes

1. The identification of this fragment was suggested by Dr. D. Barag.
   The attribution of fragments to vessel types remains somewhat tentative because of the small size of the fragments.
2. This feature has been instrumental in determining whether a fragment from the shoulder-neck area can be ascribed either to a wide- or a narrow-necked vessel.
3. Schuler (1962, 35) experimented with a method involving mouldcasting on a core.
4. Petrie 1889–90, Pl. XVII, XX; Brunton and Engelbach, 1927, Pl. LIII, group 705F; Nolte, 1968, 73–74, III. Nolte includes them in her 'Werkkreis', covering the period between Tutankhamun and Ramesses II.

Bibliography

Amlineau, E. 1899. *Fouilles à Abydos I*. Paris.

Bimson, M. and Werner, A.E. 1968. Problems in Egyptian cored glasses, *Studies in Glass History and Design*, London, 121–2.

Brill, R. and Wosinski, J. 1968. A petrographic study of Egyptian and other cored vessels. *Studies in Glass History and Design*. Sheffield. 123–4.

Brunton, G. and Englebach, R. 1927. *Gurob*, London.

Eisen, G. 1927. *Glass I*, New York.

Ettinghausen, R. 1962. *Ancient Glass in the Freer Gallery*, Washington.

Harden, D.B. 1969. *Ancient Glass I; Pre-Roman*, London.

Hayes, W. 1953–59. *The Scepter of Egypt*, 2 vols., New York.

Kertesz, T. 1976. The Breaking of Offerings in the Cult of Hathor, *Tel Aviv* Vol. 3, 134–6.

Labino, D. 1966. The Egyptian sand-core technique: a new interpretation, *Journal of Glass Studies* 7, 125.

Nolte, B. 1968. *Die Glasgefässe im Alten Agypten*, Berlin.

Petrie, W.M.F. 1889–90. *Illahun, Kahun and Gurob*, London.

—— 1906. *Researches in Sinai*, London.

Rothenberg, B. 1972. *Timna*, London.

Schuler, F. 1962. Ancient Glassmaking Techniques, The Egyptian Core Vessel Process, *Archaeology* 15, 32–7.

Tufnell, O and Harding, J.L. 1940. *Lachish II (Tell ed-Duweir)*, Oxford.

16 THE EXAMINATION OF SOME EGYPTIAN GLASS OBJECTS

1 Chemical Analyses[1]

Eight samples of cored vessels and a piece of waste glass from Timna have been analysed. The cored vessel fragments are various shades of blue, while the waste glass is green opaque. Quantitative determinations were made for nine of the major and minor elements by atomic absorption and phosphorus was analysed colourimetrically. Seventeen additional elements were estimated by semi-quantitative emission spectrography. The silica was estimated by differences from 100%.

The Glass Samples

Anal. C.G.M No.	Field No.	
3370	342/69	Fragment of cored vessel. Dark blue transparent glass with yellow threaded decoration. Some remains of core residue. Moderately weathered.
3371	354/69	Fragment of cored vessel. Light blue transparent glass. Some remains of core residue. Moderately weathered.
3372	227/51	Fragment, probably of cored vessel. Light blue opaque glass with white and black threaded decoration. Heavily weathered. (See Chapter III, 14, Glass Cat. No. 9).
3373	157/7	Fragment of cored vessel. Light blue translucent glass with weathered remains of decoration of unknown colour. Little or no remains of core residue. Heavily weathered.
3374	273/69	Fragment of cored vessel. Light blue transparent glass with yellow threaded decoration. Remains of core residue. Moderately-heavily weathered.
3403	1454	Fragment of cored vessel. Light blue opaque glass with yellow and white threaded decoration. Some remains of core residue. Lightly weathered.
3405	1454	Fragment of cored vessel. Medium blue opaque glass with yellow and white threaded decoration, and speck of white opaque with purple glass. Some remains of core residue. Moderately weathered.
3408	1456	Irregularly shaped but rounded dripping. Possibly waste glass from a manufacturing operation. Light green opaque glass. Accretion adhering. Heavily weathered.
3409	206/51	Fragment of cored vessel. Medium blue transparent glass with yellow threading at rim. Very heavily weathered.

As expected, the glasses are all of the soda-lime type. Cobalt oxide is the colourant of the single dark blue glass, whereas the seven light blue glasses are coloured with copper oxide. The five opaque and translucent cored vessel samples contain antimony, probably in the form of calcium antimonate, as a white opacifier, and the waste glass is coloured with copper and yellow lead antimonate. There is nothing unusual about any of the compositions.

The most interesting observation regarding the cored vessels is that their analyses are very similar to the analyses of twenty-two glass samples from Amarna and eleven glasses from Malkata.[2] In fact, the analyses of the Timna, Amarna and Malkata glasses can be regarded as being viturally indistinguishable. The agreement is so close that one is led to conclude that the cored vessel glasses from all three sites were made from very similar materials,

Table 1. Chemical Analyses of Some Glasses from Timna

		dk. blue transp. 3370	lt. blue transp. 3371	lt. blue opaque 3372	lt. blue transl. 3373	lt. blue transp. 3374	lt. blue opaque 3403	med. blue opaque 3405	lt. green opaque 3408	med. blue transp. 3409
SiO_2	Δ	58.1	62.2	61.0	64.8	63.2	63.4	66.5	62.1	68.4
Na_2O	a	19.3	21.6	18.8	16.2	18.2	18.5	17.0	19.2	15.3
CaO	a	7.82	7.74	7.54	8.00	8.00	7.51	7.02	7.94	8.50
K_2O	a	0.68	1.38	1.78	1.70	2.70	1.67	1.90	2.34	1.74
MgO	a	6.02	2.64	5.14	4.69	4.18	4.95	4.23	4.26	2.83
Al_2O_3	a	6.17	1.67	0.97	1.01	1.05	1.01	0.97	0.92	1.46
Fe_2O_3	a	1.09	1.09	0.93	0.89	0.96	0.47	0.45	0.94	0.61
TiO_2		0.20	0.15	0.10	0.10	0.10	0.08	0.10	0.05	0.10
Sb_2O_5	a	nf	nf	1.78	0.88	nf	0.84	0.50	0.50	nf
MnO	a	0.18	0.24	0.11	0.085	0.15	0.042	0.029	0.023	0.21
CuO	a	0.005	0.85	1.36	1.20	0.95	1.17	0.98	0.63	0.60
CoO		0.08	nf	nf	nf	nf	nf	nf	nf	nf
SnO_2		nf	nf	0.09	0.09	0.01	0.10	0.10	0.03	nf
Ag_2O		0.008	0.008	0.006	0.006	0.001	0.001	0.001	0.001	0.001
PbO		0.001	0.005	0.01	0.03	nf	0.001	0.003	0.77	0.001
BaO		<0.01	<0.01	<0.01	<0.01	<0.01	0.03	0.01	0.05	0.03
SrO		0.15	0.15	0.15	0.15	0.15	0.10	0.10	0.10	0.10
Li_2O		nf	nf	nf	nf	nf	<0.001	<0.001	<0.001	<0.001
Rb_2O		nf	nf	nf	nf	nf	nf	nf	nf	nf
B_2O_3		0.05	0.02	0.01	0.02	0.01	0.02	0.01	0.01	0.02
V_2O_5		<0.005	<0.005	nf	nf	nf	nf	nf	nf	nf
Cr_2O_3		<0.005	<0.005	<0.005	<0.005	<0.005	nf	nf	nf	nf
NiO		0.07	0.04	0.04	0.005	0.10	nf	nf	nf	nf
ZnO		0.031	0.018	0.0051	0.0076	0.0026	0.015	0.008	0.090	0.018
ZrO_2		nf	nf	nf	nf	nf	<0.01	<0.01	<0.01	<0.01
Bi_2O_3		nf	nf	nf	nf	nf	nf	nf	nf	nf
P_2O_5	c	0.085	0.20	0.17	0.16	0.24	–	–	–	–
As_2O_5		–	–	–	–	–	0.05	0.05	nf	0.05

nf Sought but not found
– Not sought
a Analysis by atomic absorption
c Analysis by colourimetry, all other values are by semi-quantitative emission spectrography
Δ SiO_2 estimated by difference from 100.0%
    All analyses by Dr. Brant Rising and co-workers at Lucius Pitkin Laboratories, New York City.

using very similar batch preparations, and following procedures which could not have differed markedly from one another. One can reasonably conclude that there was an association of some sort between the craftsmen who made the three groups of glasses and/or the factories where they were made. Although the glasses are not thought to be contemporaneous, such an association could exist in the form of technological traditions which were handed down from one generation to another.

There are two noteworthy chemical features. The fragment of the dark blue cored vessel, Anal. C.G.M. 3370, which is coloured with cobalt, contains about the same level of nickel oxide as the dark blue glasses from Amarna; but it also contains a considerably higher amount of alumina, somewhat more magnesia, and what we believe to be a significantly higher amount of boron. (We have frequently seen evidence of an association between boron and cobalt). The indication is that, although the glasses from Timna in general bear a remarkably close relationship with those of Amarna in their fundamental compositions,

the actual batch of cobalt colourant used for colouring this particular Timna glass differed somewhat from that used for the analysed dark blue glasses from Amarna. One should not conclude, for example, that these glasses 'came out of the same pot on the same day'.

A similar situation arises with the light blue Timna glasses. These glasses are coloured with copper and contain no cobalt. Apparently the copper came from a source which also introduced some nickel, and possibly some manganese. The blue copper colourant does not appear to have been derived from a bronze-based substance, since two of the light blue glasses do not show the presence of sufficient tin. The two which do contain about the right amount of tin also contain antimony, which undoubtedly came from a calcium antimonate white opacifier. The tin in these two glasses, which are both opaque, could have come in with antimony. Their light blue counterparts from Amarna are coloured with copper, but contain much less manganese and no nickel detectable by our spectrographic technique. Once again, the base glass compositions are in good agreement, but the colourant sources were different. As with the cobalt colourant, these findings suggest some difference in time and/or location of manufacture between the Timna and Amarna glasses.

The analyses of the beads from the smelting camp, Site 2[1] differ somewhat from those of the cored fragments. The differences are 'borderline', but the author is inclined to see them as being significant, indicating that there is something different about the glasses from the two different locations.

2 Lead-Isotope Determinations

A lead-isotope ratio determination was run on a sample of one of the glasses. This sample (Pb–1079) came from the yellow band in the same core-formed fragment from which Anal. C.G.M. No. 3374 was taken. This glass is presumed to have been coloured with the same lead-antimonate colourant opacifier which was commonly, perhaps exclusively used during that period (Brill, 1968). The colour is due to flakes of $Pb_2Sb_2O_7$ suspended throughout the glass. This pigment was prepared separately and added to the softened glass. Therefore it was not necessarily prepared at the glass factory itself. Lead was extracted from a small sample of the glass and its isotope ratios determined.

This determination was made because lead-isotope ratios can often be used to locate the possible mining regions which might have produced the lead found in ancient objects. The method has been published extensively[3] and we need explain here only that the technique is based on attempting to match lead-isotope ratios from objects with the ratios determined for galena (lead sulphide) ores from ancient mining regions. Within the limitations of certain assumptions, it is possible to rule out many lead-mining regions as the possible sources of lead for any given object. Where a close match is established between the lead in an object and that from some mining region, it is then possible to infer that the lead came either from that region or from some other region which is so similar geologically that it yields ores with the same isotope ratios.

In the previous publications, results have been presented for numerous samples of various types of ancient materials containing lead.

Table 2. Lead Isotope Ratios †

Sample/Group	Pb^{207}/Pb^{206}	Pb^{208}/Pb^{206}	Pb^{204}/Pb^{206}
Early Egyptian **	= 0.780–0.820	= 1.98 –2.04	= 0.0500–0.0524
Group L **	= 0.830–0.833	= 2.055–2.067	= 0.0529–0.0532
Limit Group X *	= 0.834	= 2.062	= 0.0530
Late Egyptian Pb–208 *	= 0.834–0.839 0.8392	= 2.064–2.082 2.0745	= 0.0530–0.0535 0.0530
Limit Group X **	= 0.842	= 2.086	= 0.0537
Group E **	= 0.841–0.849	= 2.078–2.096	= 0.0536–0.0544
Group S *(?)	= 0.853–0.865 (?)	= 2.09–2.11 (?)	= 0.0544–0.0553
Fowakhir *(?)	0.8693	2.0947	0.05609
	(Falls off trend)	(Falls off trend)	
Pb–643	0.86862	2.1192	0.055559
Pb–1229	0.87072	2.1199	0.055648
Pb–831	0.87347	2.1254	0.055761
Pb–159	0.87422	2.1238	0.055944
Pb–1079	0.87447	2.1254	0.05583
Pb–82**2	0.87552	2.1273	0.055682
Pb–139	0.8992	2.143	0.05750
		(Falls off trend)	(Falls off trend)

The samples are listed in approximate order of ascending Pb^{207}/Pb^{206} ratios. Therefore proximity in the table indicates similarity in isotopic composition.

** indicates a great difference in ratios.

* indicates a difference but not a great one.

Double space indicates samples are close but not 'identical'.

† All samples except two were analysed at the National Bureau of Standards by Dr I. Lynus Barnes and his colleagues, M. Diaz, E. Deal, J. Gramlich, L. Powell and K. Sapenfeld. The analysis of the ore from Fowakhir was reported by Stacey *et al.* (1980, 175–88). The analysis of Pb–139 was done in 1964 by J.M. Wampler at Brookhaven National Laboratory (see reference in note 9).

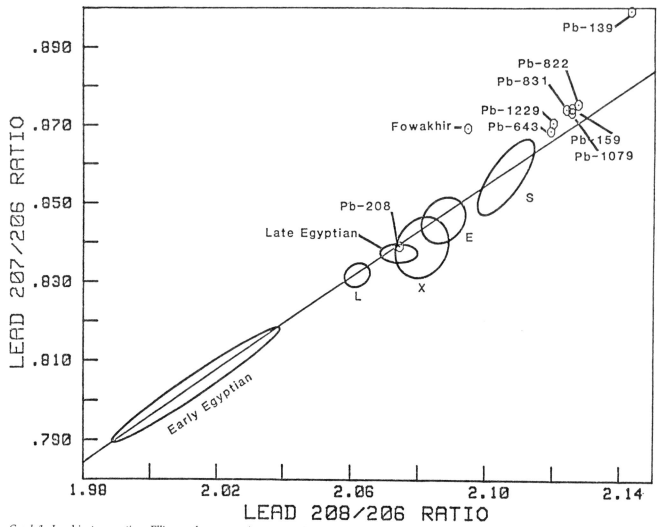

Graph 1. Lead-isotope ratios. Ellipses show groupings summarising results on several hundred leads and ores run previously. Early Egyptian group contains about thirty 18th Dynasty (and earlier) glasses, glazes and kohls from Amarna and elsewhere in Egypt. Late Egyptian group (tentatively drawn) contains twenty-two leads, sixteen of which are from Egypt dating later than the 18th Dynasty. (Remaining samples are of unknown place of manufacture.) Pb–1079 is a yellow opaque glass from Timna. (Note reversal of x and y axes when comparing with previous publications.)

The loops and groupings illustrated in Graph 1 summarise results for about 900 samples, some of which are glasses and glazes coloured with lead antimonate. Data for many other samples still await publication. Altogether, about eighty specimens of early glasses and glazes have been run.[4] These came from Egypt, Mesopotamia and elsewhere throughout the Ancient Near East and the eastern Mediterranean. In addition to early samples, examples of Hellenistic and Roman glasses are included.

The leads in the lower left-hand corner of the graph are of special interest here. This group contains about 30 ancient ores and objects from Egypt. For the most part, these objects are yellow glasses and faience glazes coloured with lead antimonate, and a large number of kohl samples (Brill, Barnes, Adams, 1974). In ancient times kohl from Egypt was often finely ground galena ore. Among the many hundreds of other samples we have run, no other samples have yet fallen into this part of the graph (except for a single red glass from Persepolis). The group contains

only objects and materials of 18th Dynasty or earlier Egyptian provenances, together with a few galena ores from the Eastern Desert. These results provide a useful, but not necessarily infallible, guideline. It appears that early Egyptian lead-antimonate glazes and glasses usually, if not always contain lead of this type. Although there is no perfect match between the kohl samples and the few lead ores we have been able to obtain, they have generally similar isotope ratios. Therefore it seems reasonable to assume for the time being that the lead used to manufacture glasses and glazes, and perhaps other materials, in ancient Egypt, came from mines located somewhere in the Eastern Desert. However, this generalisation only holds through about the 18th Dynasty. Samples we have studied of Egyptian glass[5] dating from the 6th–3rd century BC – small, brightly coloured figurines – contain high concentrations of lead of a different isotopic make-up. These, together with a few examples of Ptolemaic and Roman glasses and some other samples of Egyptian origin, have isotopic compositions falling within a rather narrow range in the Group X region of the graph. This is tentatively drawn in Graph 1 as a Late Egyptian group. At present the evidence is only slender, but it appears that as far as glassmaking was concerned, at some time between the end of the 18th Dynasty and the 6th century BC, the particular mines providing lead of the type found in the earlier glasses were no longer worked and new sources were opened up. This does not mean that the 'new' mines were located at a great distance from those worked earlier. The geology of the Eastern Desert may be such that outcroppings of galena ore of significantly different geological dates may occur relatively close to each other.

One of the most important findings of this earlier work is that the yellow opaque glasses from what we have loosely termed 'Mesopotamia' – an area also including the surrounding regions – contained types of lead entirely different from what was found in the early Egyptian glasses. It has long been recognised that although cored vessels of Egyptian and Mesopotamian manufacture closely resemble one another in the manufacturing method and, superficially, in style, on careful examination the glasses from these two regions are very distinct. Moreover, the core residues remaining inside the glasses are entirely different (Brill and Wosinski, 1968) and the chemical compositions of the glasses are markedly different from each other. Unpublished results[6] on ancient Mesopotamian glasses and glazes show four or five clusters of isotope ratios between the lower limits of Group L on the graph, up close to the beginning of Group E. Some also lie above Group L. (Group L represents Laurion-like leads while Group E contains leads from England and Europe.)

Generally, the Hellenistic and Roman glasses are quite variable in their isotope ratios. They usually fall between Group L and Group E. Towards the lower end of Group E are several glasses (as well as bronzes and metallic leads) which are associated with Constantinople.

In the upper right-hand corner of the graph is a group of leads and ores labelled S which once designated leads from Spain, but is now known to contain leads from other sources, e.g. Sardinia and Wales. As is explained below, we now know that certain ores from Egypt also fall near Group S.

The above is an introduction to explain the reason for running the lead-isotope ratios on the yellow glass in the cored vessel fragment from Timna: it was expected that the association implied by the compositional relationships found between the Timna glass and those from Amarna would be confirmed. It was suspected that the lead in the Timna specimen might resemble the 18th Dynasty examples. However, the results were entirely unexpected.

The lead-isotope ratios of the Timna glass, sample Pb–1079, are recorded in Table 2 and plotted in Graph 1. Surprisingly, the lead in the Timna glass does not resemble any of the Egyptian glasses run earlier – neither the 18th Dynasty nor the later specimens – nor does it resemble any of the Mesopotamian glasses. Instead, the lead in the sample falls towards the upper right-hand corner of the graph, just beyond Group S.

We are not inclined to associate the lead in this glass with the mines in Spain. We also believe it would be premature to associate the glass with ores from Sardinia, or Kouchke, a lead mine some 25 km. north-east of Bafq in Iran, despite the fact that the glass and these two ores are all close matches for each other.[7] Strictly speaking, none of these connections can be ruled out; however, we believe there may be a different lesson to be re-inforced in this case. Clearly, the coincidence of the two ores is an example of the overlapping effect, in which ores from widely-separated regions have similar isotope ratios because they are geologically similar. There is now evidence that there may be ores in the Eastern Desert, in the Negev, and possibly in the Arabian peninsula, which isotopically resemble the lead in the Timna glass.

Stacey *et al.* (1980) reported data on an ore from Fowakhirs. They[8] (and also Hassan and Hassan (1981), who discuss the same data) describe it as coming from a quartz vein in Precambrian granite. The isotopic composition of this lead ore is very different from those of any of the other early or late Egyptian glasses we have studied. While this ore does not provide a match for the Timna glass, it does fall into the same area of the graph, suggesting that other mines in that region might contain an ore which would match the glass. The argument is weakened, however, by the fact that the Fowakhir ore falls off the trend of the data as plotted here and above the general trend on a Pb^{204}/Pb^{206} versus Pb^{207}/Pb^{206} plot.

The only other Egyptian ore we have analysed which is at all similar to this glass is Pb–139, an ore from Um-Samiuki.[9] This sample falls well off to the right of the Timna glass and, among the samples we have run, resembles only artifacts and ores from China, Japan and India. Like the Fowakhir ore, it falls off the general trend of the data. Nevertheless, this still strengthens the possibility that somewhere in Egypt ores could exist which would match the Timna lead. That picture should not, however, be oversimplified. Fowakhir is located along the Wadi Hammamat (about 80 km. by land plus 200 km. by water route from Timna), while Um-Samiuki is much further up the Nile, opposite Aswan. The geology of Egypt is so complex that questions of the sort posed by the Timna glass will not be answered reliably until many more Egyptian ores which would have been accessible to early metallurgists, have been analysed.

List of Lead Samples

Lead Sample No.

Pb–139   Galena ore with sphalerite, pyrite and chalcopyrite. From Um-Samiuki (24°16′N, 34°49′E). Donated by A.M. Abdul-Gawad, 1964.

Pb–159   Galena ore. From Montepuni, Sardinia (level unknown). Donated by Paul E. Desautels, Smithsonian Institute, 1964.

Pb–208   Lead-copper slag from the Wadi Nimrah at the south end of the Timna Valley. From Site 37 on survey map, Arabah Expedition, *Palestine Exploration Quarterly*, 1962, 14.

Pb–643   Pellet of copper with many oxide or sulphide globules. Donated by Lawrence Cope; Cope No. SF 66 ii. See R.F. Tylecote, 'A metallurgical investigation of material from early copper working sites in the Arabah', *Bulletin of the Historical Metallurgy Group*, 2, No. 2, 1968.

Pb–822   Lead-zinc ore. From Kouchke, Iran (25 km. north-east of Bafq). Donated by J. Pereira, 1965.

Pb–831   Galena ore with pyrite and calcite. From Iglesias, Sardinia, Montepuni and Montevecchio Company. Said to be from near Roman workings.

Pb–1079   Yellow glass believed to have been coloured with lead- antimonate. Timna Temple excavation. From same fragment as Anal. CMG No. 3374.

Pb–1229   Copper triangular ingot from Har Yeruham in the north-eastern Negev, south of Beersheba. Probably of Middle Bronze Age I date (c 2200–2000 BC). Israel Department of Antiquities and Museums, No. 64–883. For further information see: Robert Maddin and Tamara Stech Wheeler, 'Metallurgical Study of Seven Bar Ingots', *Israel Exploration Journal*, 26, No. 4, 170–73, 1976. This publication reports that the ingot contains 4.8% lead. The ingots are described as having been made from weathered copper ores containing galena.

There are also interesting and useful comparisons to be made with other samples. These are: Pb–208, a lead-containing copper slag from the Timna Valley; Pb–643, a copper pellet from the Arabah (Tylecote, 1968); and Pb–1229, a copper bar ingot from Har Yeruham which contains 4.8% lead (Maddin and Wheeler 1976). The lead in the slag falls in Group X and is quite unlike any of the other samples discussed individually in this report. However, it does fall close to the group of about 20 later Egyptian glasses mentioned above – those which contained lead from mines other than those providing the lead used for the 18th Dynasty glasses and glazes. The copper pellet and the ingot are very close to one another and are not very different from the Timna glass, although they do not by any means match it. The significance is twofold. First, this is evidence that in this important early centre of metallurgical activity there are leads somewhat similar to that found in the Timna glass. Hence, so far as we know, the yellow pigment could just as well have been made locally (relative to Timna) as in Egypt. The answer may be found by not only looking at more lead-isotope data for ores and finds from these regions, but also by considering which centre is in closer proximity to the antimony source. The second point of significance is that this is again a reminder of the complications introduced by overlapping, and that there can be considerable isotopic variation among leads from within a given mining region.

Unfortunately we have not yet had an opportunity to complete our lead determinations for another set of glass samples from Malkata and Lisht in Egypt, although the chemical analyses have been completed and are referred to above. Chemically, the Malkata glasses are indistinguishable from the Amarna glasses, and it will not be surprising if the lead in the Malkata glasses turns out to resemble that in the Amarna glasses. However, the glasses from Lisht, which are of uncertain but probably later date, differ compositionally from the Amarna glasses. When their lead isotope results are compared with those from Timna and Amarna, it will be interesting to see whether the picture becomes clearer or more complicated.

3 Conclusions

We have considered two types of evidence. The chemical analyses of the eight cored-glass fragments from Timna very closely resemble glasses from Amarna. They appear to have been made from very similar materials, following similar batch recipes and by manufacturing techniques which did not differ markedly from those employed at Amarna. The only noticeable differences are in the colourants, implying that different sources of cobalt and copper were used for making the glasses from the two sites. The lead-isotope data prove that there was something quite different about Timna glass and the Amarna glasses. The lead in the yellow colourant opacifier in the Timna glass came from a different ore deposit from the lead found in the Amarna glasses. (It also came from a different ore deposit from those leads found in the tentatively described group of Late Egyptian leads.) The lead in the Timna glass could have come from a different mine or mining region in Egypt or, more likely at the moment, from a mine closer to Timna somewhere in the Arabah. (Less likely candidates are Kouche and Sardinia.) Continuing analyses of lead ores and artifacts from the region might very well pin down the sources of lead found in all of the glasses discussed here.

Although the glasses are different from one another, they are in a sense 'more alike than they are different'. Overall, the glassmaking activities at Timna – at least in regard to the few specimens analysed – clearly have stronger ties with the technological traditions of Egypt than with the traditions of any other contemporary glassmaking regions.

Robert H. Brill and I. Lynus Barnes

Notes

1. The following is part of the Corning Glass Museum's programme on chemical analyses of early glasses. Besides the eight samples from the Timna Temple, discussed here, two glass beads from smelting camp Site 2 in Timna were analysed, but those results will be published in the excavation report on Site 2.
2. Unpublished results by the Corning Museum of Glass.
3. Some relevant references are:
I.L. Barnes, T.J. Murphy, J.W. Gramlich and W.R. Shields, 'Lead Separation by Anodic Deposition and Isotope Ratio Mass Spectrometry of Microgram and Smaller Samples', *Analytical Chemistry*, 45, No. 11, 1973, 1881–3.
Robert H. Brill and William R. Shields, 'Lead Isotopes in Ancient Coins', *Science in Numismatics*, Special Publication No. 8, Royal Numismatic Society, Oxford University Press 1972, 279–303.
Robert H. Brill, 'Lead and Oxygen Isotopes in Ancient Objects', *The Impact of the Natural Sciences on Archaeology*, London, The British Academy, 1970, 143–64.
Robert H. Brill, 'Lead Isotopes in Ancient Glass', *Proceedings of the IV International Congress on Glass, Ravenne-Venise*, Liege, International Congress on Glass, 1969, 255–61.
4. These results were presented in a paper at the annual meeting of the Archaeological Institute of America in Atlanta in December 1977 (Brill, Barnes and Adams, 1974, 9–27).
5. Unpublished results. For descriptions see Sidney M. Goldstein, *Pre-Roman and Early Roman Glass in the Corning Museum of Glass*, Corning, 1979, Object Nos. 345, 382, 430, 435, pp. 159–66.
6. Unpublished results by the Corning Museum of Glass.
7. The Timna glass, and also samples Pb–643 and Pb–1229, are also all exact matches for some white lead pigments found in Italian Renaissance paintings (unpublished data from the C.G.M laboratory). This is regarded as purely coincidental. The white leads must come from lead in Sardinia or other Italian mines yielding lead with the same ratios.
8. The site has several different spellings. Its location is 2600'N, 33°38'E, along the Wadi Hammamat.
9. Robert H. Brill and J.M. Wampler, 'Isotope Studies of Ancient Lead', *American Journal of Archaeology*, 71, 1969, 63–77. Ores from Um-Samiuki have also been published by Z. Stos-Fertner and N.H. Gale in 'Chemical and lead isotope analysis of Ancient Egyptian gold, silver and lead', *Archaeo Physika*, 10, 1979, 299, 314. Their data are reproduced by Hassan and Hassan (1981) and agree with our data.

Bibliography

Brill, R.H. 1968. The Scientific Investigation of Ancient Glasses, in *Proceedings of the VIII International Congress on Glass*, London, Sheffield, 47–68.

Brill, R.H., Barnes, I.L. and Adams, B. 1974. Lead Isotopes in Some Ancient Egyptian Objects, *Recent Advances in Science and Technology of Materials* 3, New York, 9–27.

Brill, R.H., Wosinski, J.F. 1968. A Petrographic Study of Egyptian and Other Cored Vessels, *Proceedings of the VIII International Congress on Glass, London*, Sheffield, 123–4.

Hassan, A.A. and Hassan, F.A. 1981. Sources of Galena in Predynastic Egypt at Nagada, *Archaeometry* 23, No. 1, 77–82.

Maddin, R. and Stech-Wheeler, T. 1976. Metallurgical Study of Seven Bar Ingots, *IEJ* 26, 170–73.

Stacey, J.S., *et al.* 1980. A Lead Isotope Study of Mineralization in the Saudi Arabian Shield, *Contributions to Mineralogy and Petrology* 74, 175–88.

Tylecote, R.F. 1968. A metallurgical investigation of material from early copper working sites in the Arabah, *BHMG* 2.

17 COLOURANTS IN EGYPTIAN GLASS

The range of colours in the glass from the Temple at Timna is relatively limited. The fragments of core vessels are all of translucent blue, with the exception of two examples, one of which is green and the other brown, appearing black by reflected light. Many of these fragments are decorated with trails of white, yellow and 'black' glass. The beads are of translucent blue, opaque yellow and 'black' glass and decorated with white, yellow, 'black' and red.

Colour in glass and glazes may be due to the presence of crystalline or colloidal material held in suspension in the glass, or to coloured ions 'in solution' in the glass matrix. X-ray diffraction analysis of the samples of glass from Timna, including both beads and fragments of core vessels, indicated that the opacifiers were the same as those found in the Egyptian material previously examined (Mavis Bimson and A.E. Werner, *Annales du 4e Congrès des Journes Internationales du Verre*, 1968, 262–6). Lead antimonate was found to be the opacifying agent in the samples of yellow and of green glass and calcium antimonate was found to be present in the samples of opaque white and opaque blue glass. The colour and opacity of the red areas in the beads were shown to be due to metallic copper and not cuprous oxide.

Qualitative spectrographic analysis of the light blue beads and fragments of core vessels showed the presence of copper, which is thus responsible for the colour; but in several fragments of darker blue translucent glass from core vessels, cobalt was also found to be present and undoubtedly is responsible for this darker colour. Qualitative spectrographic analysis of the brown glass showed that the colour is due to the presence of iron and manganese. In all these cases the colour is due to the elements being present as ions in solution in the glass matrix. In the case of the translucent green fragment of core vessel, it was found that the colour is due to a copper-blue glass modified by the presence of a small amount of lead antimonate which has a yellow colour. Quantitative analysis of the glass with regard to the elements responsible for the green colour gave the following results: 0.55% CuO, 0.8% Sb_2O_5, O.6% PbO.

A.E. Werner

18 IDENTIFICATION OF TEXTILE REMNANTS

Clods of earth contained very small pieces of cloth and cordage. The cloth adhered so firmly to the soil that its removal, cleaning from heavy incrustation of dirt and salts, was almost impossible without causing damage. The majority of the threads and fibres were in a state of deterioration and spores of the celluloiytic fungus *Chaetomium* Sp. were found. Handling difficulties had to be overcome since the material disintegrated to dust when touched by forceps. Single thread-pieces were carefully isolated from the soil, mounted and embedded in lactophenol stained by cotton-blue and analysed microscopically. Most specimens were also examined in polarized light.

Fibres analysis of fabrics from Site 200

(box)	Field No.	Context	Fibre	Description
1.	323	A-B 6–7, Loc. 105,106 on top of Wall 3 in debris	Wool	Scales visible, lumen present
2.	124	J-K 16–17, Loc. 101	Flax (polarized light)	Dislocations and nodes present
3.	173	E-F 15–16, Loc. 101	Flax (polarized light)	Dislocations and nodes present
4.	327	C 6, Loc. 106. Behind standing stones and altars, animal bones in cloth.	Flax (polarized light)	Dislocations and nodes present
5.	319 (3 samples)	G 7–12, Loc. 106,108. Sand layer in red-grey fill down to olive green-grey	Wool ?	The fabric is pinkish-blue-green
6.	204	E 13–14, Locl 110	Flax	Rope and cloth
7.	311	F 15–16, Loc. 101. Reddish and olive green-grey material close to Wall 1	Wool	Blue-coloured
8.	357	B-C 6, Loc.105–106. Above and between collapsed sandstone blocks.	Wool	Yellow and red cloth in excellent condition
9.	Sq. E 14	Loc. 110. together with corroded iron.	Wool	
10.	229	E 16, Loc. 101.	Flax	Fine fabric, dislocation and nodes present
11.	320	A-B 1–5, Loc. 105.	Wool	Yellow and red cloth
12.	320		Wool	Brittle cloth (dense)
13.	280	C-E 6–7, Loc. 106. At Wall 3	Flax	Cloth, dislocations clear
14.	303	F 15–16, Loc. 101.	Wool ?	Very fine net. Weft thinner than warp.

Chaya Frydman

19 TEXTILES AND TEXTILE IMPRESSIONS ON POTTERY

Introduction

The following Catalogue comprises a representative cross-section of the textile finds in the Temple. Technical aspects, such as fibres, spinning direction, weave types, etc., were examined in order to try to identify the use and provenance of the textiles. These were also compared with the textile-impressed sherds found in the Temple.

When found, the textiles adhered to lumps of earth (*Pl.* 134:1, 2) and during the excavation no attempt was made to separate the cloth from the earth. However, to prevent disintegration of the earth-encrusted textiles, they were covered with a solidifying liquid adhesive.

Our investigations began with the contents of four randomly selected boxes.[1] Every item (fibres, threads, textile-imprinted lumps of earth and minute scraps of cloth) was given a serial number in addition to its field (box) number. After being photographed, a number of the items disintegrated into tiny shreds of cloth and pellets of earth. Since it was feared that exposure to air would completely disintegrate the fabrics, as happened with the Chalcolithic and Roman material from Alishar Hyk (Kendall, 1937, 334 note 1), it was decided to study at first only the items from the four boxes. The other textile finds remained in temperature and humidity controlled storage. A year later these remaining boxes were opened. Although some contained larger fragments than those found in the previously opened four boxes, our random samples (Table 1) proved to be representative.

1 Textiles Catalogue[2–3]

Text. Cat. No. 1. *Pl.* 133:2, Field No. 357/1a+b. Loci 105, 106.
Two fragments of the same cloth, Frag. a I 2 × 1.4cm.; Frag. b II 2 × 0.8cm. Fibre: Wool. Count: 10 × 32 (both fragments). Twist: S (both directions).

Free of adhering earth. One of the few fragments that could be examined from both sides. On Frag. I it was possible to count the number of twists on the threads of Direction I (24 twist per cm.). These threads appeared stronger spun than Direction II. Since the fibre is wool, the much higher thread count in one direction makes it reasonable to assume that the direction of the high count (II) was the weft and of the low count the warp. The weave type is therefore weft-faced (*Ill.* 35:2b). the most common in wool.

An earlier impression that the weave might be twill (*Ill.* 35:2f), because the rows looked diagonal and weft seemed to pass over two warps and under one, was not confirmed when a weft thread was deliberately undone to expose the warps.

Text. Cat. No. 2. Field No. 357/2.
Fragment of red coloured fabric, 1 × 0.5cm. Fibre: Wool. Count: undetermined. Twist: undetermined.

Disintegrated completely after being photographed. The rapid disintegration of this fragment on exposure may be due to the red dyestuff. Pfister

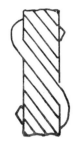
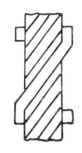

1

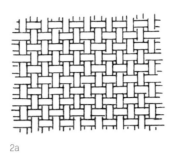

2a

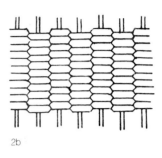

2b

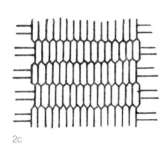

2c

2d

2e

2f

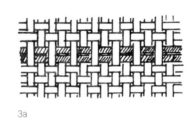

3a

3b

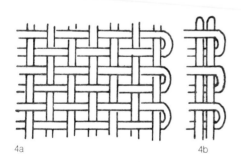

4a   4b

4c

Illustration 35.   Textile weaving techniques.
1. S and Z twist
2. Weaving technique
   a. Tabby
   b. Weft-faced
   c. Warp-faced (Rep)
   d. Extended Tabby (basket weave)
   e. Extended Tabby (half-basket weave)
   f. Twill (2 × 1)

3. a. Self-band
   b. Soumak
4. Selvages
   a. Simple
   b. Strengthened
   c. Fringes

(1951, 49, n.3) noted a similar disintegration related to red coloured textile from Halabiya. It may be assumed that the dye used was either made from *Rubia tinctorum*. Madder extracted from the roots of *Rubia Tinctorium*, common in the Mediterranean region, was found in fabric from the tomb of Tutankhamon (Lucas, 1962, 153).

Text. Cat. No. 3. Field No. 375/3.
Strands of fibres, single threads and two scraps of fabric (I, II). Frag. I 1 × 0.5cm. Frag. II 0.5 × ?cm. Fibre: Wool. Count: 10 × 16–20 (Frag. I only). Twist: S (both single threads and fragments).

Technique: plain weave. Type: weft-faced (*Ill.* 35:2b) with the first weft passing over one warp and under the second warp. The second weft reverses the process. More weft than warps.

Text. Cat. No. 4. Field No. 357/4.
Bundle of yellow threads and a small red scrap of textile Fibre: Wool. Twist: S (all threads).

When taken from the box, the threads adhered to lumps of earth and scraps of textile. It is not clear if the threads were naturally yellow or dyed this colour. Among the scraps of textile was one coloured red. Perhaps these were threads ready for weaving, or warp ends, possibly the fringes of a cloth or garment (*Ill.* 35:4c). Comparisons: Hazor (Yadin, 1961, 339, Pl. CCCXXIX, CCCXXX); Tell el Farah (Petrie and Tufnell, 1930, 19, Pl. 55); Megiddo (Loud, 1938, Pl. 38:173).

Fringes at the hem of garments are usually related to the process of weaving. After a piece is woven on the loom, the bottom warp ends are generally left free to some length. To prevent undoing of the wefts, and at the same time to put the expensive threads to some use, the ends of the cloth were reinforced and often decorated. The wearing of fringed garments was characteristic for royalty and priests, as attested by ancient Near Eastern art and in the Bible (e.g. the *ephod* robe, Exodus 28:33–34). The wool threads of this item could have been warp or weft threads twisted into fringes, perhaps all that remained of a priest's robe.

Text. Cat. No. 5. Field No. 303/5, Locus 101.
Frag. 2 × 2.9cm. Fibre: Wool. Count: 8 × 20. Twist: Weft-faced tabby (*Ill.* 35:2b).
Fragment adhering to lump of earth from which it could not be separated.

Text. Cat. No. 6. Field No. 303/6, *Pl.* 134:1.
Frag. 4 × 3cm. Fibre: Probably wool. Count: 10 × 14 (centre section only). Twist: S both directions.
Cloth folded into several layers, covered with earth and solidified into one lump. What can be seen of the warps and wefts indicates plain weave. However, the weave may have been weft-faced with the wefts later becoming partially undone. A thickening at one end indicates selvage.

Text. Cat. No. 7. Field No. 303/7.
Frag. 2 × 2.5cm. (large lump). Fibre: not identified. Count and twist not determined.

Two lumps of earth with threads and textile impression. Examination of a tiny segment showed weave type might have been extended tabby (*Ill.* 35:2d).

Text. Cat. No. 8. Field No. 303/8.
Frag. 1.5 × 1.5cm. Fibre: Probably wool. Count: 8 × 16. Twist: Direction I – S; Direction II – Z.
Although weave type is hard to identify, it seems to be tabby (*Ill.* 35:2a) or twill (*Ill.* 35:2f). Wefts (D. II) seem to pass over two warps, but this was not confirmed experimentally since the textile could not be undone at any point to check on how the threads went.

Text. Cat. No. 9. Field No. 303/9.
Lump of earth with textile imprint.
No determination of size or weave type was possible since the lump completely disintegrated when examination began. Under the microscope, copper ore grains showed among the sand grains.

Text. Cat. No. 10. Field No. 303/10.
Two fragments, probably from the same sheet. Frag. I 0.5 × 0.5cm.; Frag. II 0.4 × 0.4cm. Fibre: Not identified. Count: 10–14 × 20–24 (double threads). Twist: Both warp and weft are S twist.
The weave type is extended tabby = semi-basket (*Ill.* 35:2e) using single warps and double wefts, i.e. with each throw two wefts were passed under or over single warps.

Text. Cat. No. 11. Field No. 303/11.
Fragments, 2.0 × 2.0cm., adhering to lump of earth. Fibre: Probably wool, not positively identified. Count: Not determined. Twist: A few single threads were found to be Z-twist.

Text. Cat. No. 12. Field No. 303/11a, *Pl.* 134:2.
Frag. 2.0 × 2.5 cm.. Fibre: Probably wool,. Count: 15 × 35. Twist: D I – S; D II – Z.

Text. Cat. No. 13. Field No. 303/12/1, *Pl.* 133:4.
Part of a rope made from three strands of threads. Length 2.4, diameter 0.7cm. Fibre: Unidentified plant fibre, definitely not flax.
Each thread is made up of a large number of fibres, though it is not clear if these were twisted together or placed side by side. Rope was made by braiding together fibres, hair or strips of leather, or by spinning and twining these materials (Gilbert, 1956, 451). In the Predynastic Period there were already many different ropes made from various materials such as flax, grass, palm fibre and camel hair (Lucas and Harris, 1962, 134). Palm fibre and grass are still used in rope making in Egypt (Mackay, 1916, 126; Lucas and Harris, 1962, 135).

Text. Cat. No. 14. Field No. 303/12.2.
Textile wrapping lump of earth. Frag. 1.5 × 1.0cm. Fibre: Not identified. Count: 16 × 20 (single threads). Twist: Not determined.
Both warps and wefts are double. The weave type is extended tabby (*Ill.* 35:2d).

Text. Cat. No. 15. Field No. 303/12.3.
Textile wrapping lump of earth. Frag. 1.0 × 1.0 cm..
Fibre: Not identified. Count: 8–9 × 30–32. Twist: Not
determined.

The number of threads in D II is greater than in D I,
only the threads of D II are visible. Probably the
weave is weft-faced tabby but may also be twill.

Text. Cat. No. 16. Field No. 303/12.4.
Frag. 0.8 × 0.9cm. Fibre: Not identified. Count: Not
determined. Twist: One thread S-twisted.

Text. Cat. No. 17. Field No. 303/12.5.
Frag. 1.0 × 1.7cm. Fibre: Not identified. Count: D I –
8, D II too dense to be counted.

The weave type is probably weft-faced tabby, but
could also be twill.

Text. Cat. No. 18. Field No. 303/12.6.
Frag. 1.7 × 1.4cm. Fibre: Wool. Count: Not determined.

The weave type was not identified.

Text. Cat. No. 19. Field No. 280/13. Locus 106.
Two fragments (probably from the same sheet). Frag.
I 1.0 × 0.8cm.; Frag. II 0.5 × 0.7cm. Fibre: Flax.
Count: 20 × 16.

The weave type is tabby.

Text. Cat. No. 20. Field No. 324/30. Loci 105, 106.
Frag. 2.0 × 2.0cm. Fibre: Flax. Count: 12 × 12. Twist:
S (both directions).

The weave type is tabby (*Ill.* 35:2a) with equal warp
and weft count. A 'true linen weave' (Crowfoot, 1955,

22). The fragments were woven from threads of equal
thickness.

Text. Cat. No. 21. Field No. 324/31. *Pl.* 134:3.
Frag. 3.5 × 2.0cm. Fibre: Flax. Count: 12 × 12. Twist:
S (both directions).

Tabby weave with equal warp and weft count.

Text. Cat. No. 22. Field No. 324/32.
Fragment of creased but clean textile, 2.0 × 2.5cm.
Fibre: Flax. Count: 12 × 12. Twist: S (both directions).

The weave type is tabby with equal warp and weft
count. Fragments 324/31–32 seem to be from the same
piece of textile.

Text. Cat. No. 23. Field No. 324/33. *Pl.* 134:4.
Fragment with adhering earth, only one side visible,
2.0 × 1.5 cm. Fibre: Unidentified. Count: 5 × 12 (D II
probably greater than D I). Twist: S (both directions).

The 'ribs' produced by the weft threads were recognisable.

Text. Cat. No. 24. Field No. 324/34.
Fragment adhering to lump of earth, 1.3 × 2.3cm.
Fibre: Flax. Count: 15 × ?. Twist: Not determined.

A slight thickening at one end is perhaps a hem or
selvage (*Ill.* 35:4a, 4b).

Text. Cat. No. 25. Field No. 324/35. *Pl.* 133:5.
Fibres adhering to a piece of wood.

The fibres are inside the earth stuck to the wood.
The piece is worked on two sides and may have split
off from a square seam.

Table 1. Summary of Textiles Investigated[3]

Text Cat. No.	Field No. Direction:–	Dimensions (in cm.) I	II	Fibre	Thread Count (per cm.) I	II	Twist I	II	Remarks
1	357a/1	2	1.4	W	10	32	S	S	Weft-faced
	357b/1	2	0.8	W	10	32	S	S	Weft-faced
2	357/2	1	0.5	W	–	–	–	–	Red colour
3	357/3	1	0.5	W	10	16–20	S	S	Tabby
4	357/4	Fibres		W	–	–	S		Red and yellow
5	303/5	2	2.9	W	8	20	S	Z	Weft-faced
6	303/6	4	3	W(?)	10	14	S	S	Tabby
7	303/7	2	2.5	–	–	–	–	–	Basket weave
8	303/8	1.5	1.5	W	8	16	S	Z	Weft-faced
9	303/9	Textile imprints							
10	303/10	0.5	0.5	–	10–14	20–24	S	S	Half basket weave
11	303/11	–	–	W(?)	–	–	Z	–	
12	303/11a	2	2.5	W(?)	15	35	S	Z	Weft-faced
13	303/12.1[4]	2.4(L)	0.7	Plant fibres					Rope
14	303/12.2	1.5	1	–	16	20	–	–	Basket weave
15	303/12.3	1	1	–	8–9	30–32	–	–	Weft-faced
16	303/12.4	0.8	0.9	–	–	–	S	–	Basket weave(?)
17	303/12.5	1	1.7	–	8	–	–	–	Weft-faced
18	303/12.6	1.7	1.4	W	–	–	–	–	
19	280/13	1	0.8	F	16	20	S	S	Tabby
20	324/30	2	2	F	12	12	S	S	Tabby
21	324/31	3.5	2	F	12	12	–	–	Tabby
22	324/32	2	2.5	F	12	12	–	–	Tabby
23	324/33	2	1.5	–	5	12	S	S	?
24	324/34	1.3	2.3	F	15	–	–	–	?
25	324/35	Fibres and wood							
26	324/36	3.2	1.5	–	–	–	–	–	

W = Wool
F = Flax

Text. Cat. No. 26. Field No. 324/36.
Fragment adhering to lump of earth, 3.2 × 1.5cm.
Fibre: Unidentified. Count: Not possible. Twist: Not
determined.

2 The Fibre Materials

The textile remains included items made from both
animal and plant fibres: eight items were definitely
wool, three others probably wool, and five items of
flax.

We know from the evidence of ancient documents
(Jacobsen, 1970, 217; Kramer, 1963, 104; Driver and
Miles, 1955, 31 ff.) and ancient art (Parrot, 1948, Pl. 5,
VIa, 16) as well as from excavations (Woolley, 1950,
50; Mellaart, 1963, 99 ff.; Bellinger, 1962a, 13 ff;
Astrom, 1965, 112) that wool was extensively used in
Anatolia, Mesopotamia and Cyprus. The use of linen
is known from Mesopotamia (Kramer, 1962, 28–9;
Jacobsen, 1970, 219), doubtfully from Anatolia (Ken-
dall, 1937, 335; Jacobsen, 1970, 219; Helbaek, 1963, 44;
Ryder, 1965; Burnham, 1965, 170) and from Cyprus
(Aström, 1965, 112, 113; Pieridou, 1967, 26 ff.).

Documentary evidence indicates that the peoples
of Israel used both linen and wool. However, very
few textile remains or impressions of textiles have
been discovered in Israel up to the Iron Age. Chalcolithic textiles were found in the Cave of the Treasure
(Bar-Adon, 1971, 159 ff.) and at Ghassul (Crowfoot,
1956, 432). Bronze Age textiles were found only in
Jericho (Crowfoot, 1960, 519 ff.; idem, 1965, 663 ff.)
and impressions in Nahariya (Dothan, 1956, 22) and
Gezer (Macallister, 1912, 76). From the Iron Age the
material available is also very meagre. Textile impressions were found in Hazor (Yadin, 1960, Pl. CLXI:12,
14), Gezer (Macallister, 1912, 76), Megiddo (Loud,
1948, Pl. 228, 229), Arad (Aharoni, 1967, 272) and on
the bases of Negev-ware pots (Sheffer, 1976). The
only textile remains are from Kuntilat Ajrud (Sheffer,
1978).

In Egypt, on the other hand, the number of woollen finds is very small. From the Predynastic Period
some items were found in Naqada and Kahum (Petrie
and Quibell, 1896, 24; Petrie, 1890, 28) and Tel el
Amarna (Pendelbury, 1951, 246). Riefstahl reports
that all these woollen finds are of doubtful antiquity
(1944, 29, 53, 54 n, 74) because there is no text relating
to the production of wool, nor is there a known word
of the Pharaonic period meaning wool. We must also
note the Egyptian taboo against the use of wool in
temples and religious ceremonies, which apparently
is echoed in the biblical Joseph story (Genesis 46:34).
This taboo also continued into the later periods
(Herodotus II:81).

The Egyptians used linen, from which they produced a great range of items from coarse material for
containers and awnings to delicate, fine fabrics, from
earliest Neolithic times, in the Fayum (Caton-Thompson and Gardiner, 1934, 46, 49, Pl. XXVIII:3), through
Proto- and Pre-dynastic Periods (Midgley, 1928, 64–7)
and the patterned coloured linens of the New Kingdom period, to the magnificent linens in the royal

tombs of Tuthmosis IV (Carter and Newberry, 1904,
143–4) and Tutankhamun (Crowfoot and Davies,
1941; Carter and Mace, 1963, 170–4; Lucas and Harris,
1963, 185–6).

Thus in the Western Asian countries wool was the
fibre most widely used. Egypt was the classical linen
country of antiquity.

3 Spinning Techniques

Animal and plant fibres need to be spun in order to
make threads long enough for weaving. Spinning is
done by drawing and twisting a few fibres either to
the left or to the right, giving the thread a left 'S' or
right 'Z' twist (*Ill.* 35:1). Flax fibres are wetted for
spinning. When drying they rotate naturally to the
left and are therefore usually spun in this direction.
Wool does not rotate naturally and its spinning direction usually conforms to the natural rotation of the
predominant plant fibre of the region. However, in
regions where wool was the main fibre, it was spun
either to the left or to the right, according to custom
or individual preference (Bellinger, 1959, 3).

Of the twenty-seven fragments examined, it was
possible to determine the spinning direction in only
fourteen, and not always of both warp and weft
threads. The threads of the linens were all S-spun.
The wool and unidentified threads showed a greater
variation: on some both warp and weft threads were
S-spun, while on others the warp might be S-spun
and the weft Z-spun or vice versa.

In Egypt, S-spinning was practised from the earliest times and became so firmly established that it
persisted throughout all periods. In Syria and along
the Syrian coast, however, spinning traditions were
mainly Z. For Palmyran woollens, S-spinning is the
norm, except for the purple-dyed wool thread which
was imported from the Syrian coast and was invariably Z-spun (Pfister, 1934, 22; 1937, 40–41; Crowfoot,
1945, 38). Probably the Z-spinning tradition was
brought by nomad migrants to Transjordan and the
adjacent regions, where it still persists (Crowfoot,
1944, 129, n. 3; idem, 1945, 38).

Spinning tradition appears to have remained
unchanged through the ages. Along the Syrian coast,
in Jordan and north-western Arabia, wool is Z-spun
even today, while in Egypt it continues to be S-spun.
As shown above, Z-spinning of wool is predominant
at the Timna Temple and it may therefore be assumed
that its woollen textiles came from a region where
Z-spinning is standard.

4 Weaving Technique

Spinning direction is not the only criterion of provenance and weaving techniques must also be considered.

Both warp and weft could be counted on sixteen of
the twenty-seven Timna Temple textiles and either
warp or weft on only two. In order to establish the
weaving type of any fibre, the direction of warp or
weft needs to be known also. However, the warp or
weft direction of the Timna fibres could not be definitely established, since borders or selvages (*Ill.*

35:4a-b) are lacking, but considering that warp threads are generally stronger and finer spun (Bellinger, 1950a, 2; Crowfoot and Crowfoot, 1961, 59 n. 1) the finer thread was treated as warp.

We also tried to determine warp and weft directions by comparisons with material from countries of the regions adjacent to the Arabah.

(a) Linen

It was possible to count warp and weft threads in four of the five linen items found in the Temple. In three items twelve threads were found in both directions. Egyptian linen usually shows a ratio of three or less wefts to four or more warps, although in some New Kingdom linens the warp count was double (and more) the weft (Braulik,1900, 44 ff.; Wilkinson, 1878, 161).

The question of provenance of Roman linen from Nessana, Qumran and Muraba'at was discussed by Bellinger and Crowfoot (Bellinger, 1962b, 92; 1950b, 108; Crowfoot, 1955, 22) who decided that those linens are not Egyptian because of their low thread count and the attempt at a 'true linen weave' (an equal number of warps and wefts per square centimetre, *Ill.* 35:2a). However, we found the same characteristics already in the Timna linen.

(b) Wool

Since warps and wefts of the woollen fragments could not be definitely identified, we based our identifications on fibre characteristics: since wool fibres are covered with scales which tend to catch, and because of the characteristic 'crimp' of the material, the warps of wool are widely spaced on the loom and held under constant tension. The wefts in wool weaving are beaten tight and therefore exceed the warp threads in number. This weft-faced weave, with its widely spaced and concealed warps and tightly beaten wefts (*Ill.* 35:2b) was mostly used for wool fabrics in our region, including Egypt and Syria, from the Chalcolithic to the Roman period (Bar Adon, 1971, 3; Lapp, 1974, 73–5; Bellinger, 1962a, 14 ff.; Yadin, 1963, 193; Kendrick, 1920, 3; Wilson, 1933, 4; Pfister and Bellinger, 1945, 2; Crowfoot and Crowfoot, 1961, 55 ff.).

Extended tabby (*Ill.* 35:2c) i.e. warp and weft threads used in pairs, appears in three items from Timna, while another shows a semi-basket weave with only one direction paired. Apparently extended tabby is little known from any period (Bar Adon, 1971, 248; Yadin, 1963, 198; Pfister and Bellinger, 1945, 2).

5 Impressions on Pottery – Catalogue

Text. Imp. Cat. No. 1. Field No. P/2, Locus 112, 110, Square A-B 14–15.
Light grey sherd, clear, well defined textile impression. Spinning direction of weft threads is Z (defined from plasticine impressions).

Text. Imp. Cat. No. 2. Field No. 241/65, Locus 109, Square G-H 12–14.
Reddish sherd, poorly baked and crumbling, impression indistinct.

Text. Imp. Cat. No. 3. Field No. 280/23, Locus 106, Square C-D-E 6–7.
Brownish sherd, indistinct impression, only sporadically visible, spinning direction of weft threads is Z (identified from plasticine impressions).

Text. Imp. Cat. No. 4. Field No. 154/19, *Pl.* 129:1, 2, Locus 101, Square C-D 14–15.
This sherd is especially important since it furnishes definite evidence that the impressions (prints) on the Negev-ware bases were caused by textiles and not by mats. It is a small, grey sherd with impressions of threads in one direction only, as if stretched on a loom. From the sherd emerges a single white flax thread, tightly S-spun, 12 twists per cm. (in ordinary wool thread there are only 3–5 twists per cm.). The twists of the thread impressions are reversed in direction. Lines of carbonized threads, running parallel with thread impressions, are visible on the reverse side of the fragment. Since the threads are fairly thick and in one direction only, it is likely that they are warp ends.

Text. Imp. Cat. No. 5. Field No. 157, Loc 102, Square J-K 13–14.
Reddish sherd impression, indistinct, number of threads is defined from plasticine impressions. Base was set on worn textile.

6 Textile Impressed Pottery Bases

Five impressed sherds from the Temple were examined in order to identify the material which made them, and to compare them with the textiles found.

In the excavations at Tell el-Kheleifeh, in the southernmost Arabah, such impressions were found on sherds of pots described as 'crude, hand-made, friable, smoke-blackened, built up on a mat' (Glueck, 1938, 14; idem, 1940, 17). Rothenberg, who found

Table 2. Summary of Textile Impressions on Pottery

Text. Imp. Cat. No.	Field No. of sherd	Impression on Sherd Max. Length (cm.)		Thickness of Sherd	No. of Threads per cm.		Spinning Direction	
		Warp	Weft		Warp	Weft	Warp	Weft
1	P/2	1.5	3	0.5	3–4	16	–	Z
2	241/65	3	2.5	0.5	4–5	16(?)	–	–
3	280/23	3	5	0.8	4	18–20	–	Z
4	154/19	1.5	1.8	0.1–0.2	10–15	–	–	S
5	157	2	1.5	0.6	3–4	16(?)	–	–

similar impressed sherds in his surveys and excavations in the Arabah, called this type of pottery 'Negev-ware' (1967, 20). The same pottery was also found in surveys of the Negev (Aharoni, 1960, 98 ff.; Evenari, 1958, 241; Cohen, 1976, 44; and many others).

Amiran explains the impressions on the pot bases as resulting from the potter's use of mats as a kind of turntable ('bat') when making the pot (Amiran, 1963, 19). The general acceptance of Amiran's explanation probably accounts for the fact that later investigators invariably identified the impressions on the Negev pot bases as mat impressions. However, our investigation showed that the pots were put on material woven from spun threads. This material could thus not have been a mat, since mats are invariably plaited from unspun fibres or unprepared rushes or leaves, generally by hand without any implement, while cloth can be woven only from prepared and spun threads, with one set (warps) fixed and stretched on the loom and the other set (wefts) interlaced at right angles.

We know that the practice of drying wet pots on cloth survived at least to the beginning of the twentieth century, as the hand-formed pottery made by the women of the Arab villages near Jerusalem was usually placed to dry on old fabrics spread on their roofs (Güthe, 1908, 24). As a further test, plasticine imprints were taken of the sherd impressions and it was then possible to make out the warps and wefts, and in a few cases even the spinning direction.[5]

It was found that the textiles differed widely as to weaving density and thickness of threads. In most cases the number of threads in one direction greatly exceeded the number in the other direction, which made it most likely that the fibres used were wool. This is also confirmed by the impressions on the sherds, which show a greater thread count in one direction than in the other. We recall here the fact mentioned above, that most of the linen from Timna shows an equal thread count, while in the wool textiles the wefts exceed the warps in number.

On the assumption that wool was used, we prepared a number of experimental textiles resembling as nearly as possible in thread thickness, weaving density, etc., the textiles impressed on the pot bases. We then made plasticine imprints of our experimental examples and compared these with the impressions on the sherds. As expected, the spinning direction on the copies was the reverse of the original (thus if the thread is S-spun it will show as Z-spun, and vice versa), and for more than four warps and wefts per centimetre, the thread count is simply the number of grooves in a row. Above four warps and wefts per centimetre, the number of grooves counted must be doubled in order to arrive at the true count since each groove indicates a thread passed under another, perpendicular to it, while further on the same thread passed over does not show in the impression, Furthermore, with a loose weave and obviously even number of threads, one needs only to count the grooves in order to arrive at the true number of threads in any direction.

7 Conclusions

(1) On a large number of the Timna textiles the thread is Z-spun. Since Z-spinning is entirely absent in Egyptian fabrics of the early periods, the Z-spun fabrics of the later periods may have been either imports or made by foreign weavers in Egypt.

Almost no woollens from the fourteenth to twelfth centuries BC are known from excavations in Egypt and, moreover, it is extremely unlikely that any woollen fabrics would be found inside an Egyptian temple, since the use of wool in connection with ritual was taboo in ancient Egypt. Yet, in Timna a large number of woollen textiles were found inside the Temple.

The linen fragments demonstrate the use of tabby weave, with approximately the same number of warps and wefts to the square centimetre (equal count). However, in Egyptian linen of all periods the warps greatly exceed the wefts in number, and sometimes the weft threads were nearly obscured by them.

We therefore conclude that most of the Timna textiles did not originate in Egypt, but were possibly brought to the site, or locally produced, by the mine workers recruited, according to the archaeological evidence, from the local Negev tribes or from Northwest Arabia (Midian).

(2) Although the above data can tell us little about function, the coarser and heavier textiles may have been tent cloth (for awnings, partitions, etc.) or perhaps curtains; the fine textiles are probably remnants of clothes or priests' sacred vestments which were kept permanently in the Temple. There are also pieces of cord which could have served many purposes; they might have been tent ropes or even belts.

Avigail Sheffer and Amalia Tidhar

Notes

1. The extremely delicate task of separation of the textile fragments from the lumps of earth, as well as the laboratory test and fibre determinations, were undertaken by Dr. Chaya Frydman (see also Chapter III, 18) of the Israel Fiber Institute, Jerusalem.

2. The terminology used in the Catalogue is in accordance with the terms used by the Textile Museum of Washington. Some of the definitions are given here:

Fibres The fundamental unit in the fabrication of textile yarns and fabrics American Society for Testing Materials (ASTM).

Yarn = thread A generic term for an assemblage of fibres or filaments, either natural or manufactured, twisted or laid together to form a continuous strand suitable for use in weaving, knitting or otherwise intertwining to form textile fabrics (ASTM).

Twist = spinning direction A yarn has an 'S' twist if when held in a vertical position, the spiral conforms in the direction of slope to the central portion of the letter 'S', and a 'Z' twist if the spirals conform in the direction of slope to the central portion of the letter 'Z' (ASTM).

Warp 1. The yarn running lengthwise of a woven fabric. 2. The sheet of yarns laid together on a beam (ASTM).

Weft Yarn running from selvage to selvage at right angles to the warp in woven fabric.

Tabby = cloth A technique in which the first weft passes under one warp or group of warps and over the second warp or group of warps. The second weft reverses the process. All odd numbered wefts follow the course of the first weft, and all even numbered wefts follow the course of the second weft.

Balanced tabby A type of cloth weave in which there are relatively the same number of warps and wefts to the centimetre and in which both warp and weft bend.

Weft-faced A type of cloth weave with less warp than weft to the centimetre. The weft often turns back within the row and is not thrown from selvage to selvage.

Extended tabby = basket weave A cloth weave with both warps and wefts used in pairs.

Twill A type of weave in which the weft does not retrace its course more often than every third time. The underpasses are in echelon right or left or chevron.

Soumak The weft is carried manually over a group of warp threads and then passed under and back around part of the group.

Selvage The longitudinal edge of a textile closed by weft loops.

Self-band Several weft threads passing through one shed.

3. (a) The dimension indicated in the Catalogue is always of the longest thread in a given direction.

(b) The Thread Count shows the number of threads per centimetre. The first number (Direction I) stands for the warp threads; the second number (Direction II) is the count of the weft threads.

The data relating to thread counts were quite often determined on the basis of fragments of less than one square centimetre. They must therefore be treated with some caution, especially since no second count was possible in the great majority of cases.

4. The fragments 303/12 1–6 were found stuck together in a bundle and thought at first to belong to a single sheet. When freed from adhering sand and earth, and separated, it was found that they were six independent items. Therefore each item was given a separate number.

5. This method was already used by Holmes (1884,, 397 ff.) who made the same experiments with mat and textile impressions from prehistoric sites in the United States, and also obtained more information from the positive copies than from the original impressions.

Bibliography

Aharoni, Y. 1960. The Ancient Desert Agriculture of the Negev V. An Israelite Agricultural Settlement Ramat Matred, *IEJ* 10, 97–111.

—— 1967. Arad, *IEJ* 17, 270–72.

Amiran, R. 1963. *The Ancient Pottery of Eretz-Israel*, Jerusalem (Hebrew).

—— 1969. *The Ancient Pottery of the Holy Land*, Jerusalem.

Astrom, P. 1965. Remains of Ancient Cloth from Cyprus, *OA* 5, 111–14.

Bar-Adon, P. 1971. *The Cave of the Treasure*, Jerusalem (Hebrew) (English ed. 1980).

Bellinger, L. 1950a. Textile Analysis: Early Techniques in Egypt and the Near East, *Washington Textile Museum, Workshop Notes*, 2.

—— 1950b. Report upon a Fragment of Cloth from the Dead Sea Scroll Cave, *BASOR* 11, 9–11.

—— 1959. Craft Habits Part II. Spinning and Fibers in Warp Yarns, *Washington Textile Museum, Workshop Notes* 20.

—— 1962a. Textile from Gordion, *The Bulletin of the Needle and Bobbin Club* 46, 5–33.

—— 1962b. Textile, in Colt, D.H. ed. *Excavations at Nessana*, London, 92–105.

Braulik, A. 1900. *Altägyptische Gewebe*, Stuttgart.

Burnham, M. 1965. Catal-Hüyük, The Textile and Twine Fabrics, *AS* 15, 169–74.

Carter, H. and Mace, A.C. 1963. *The Tomb of Tut-Ankh-Amen* I, New York.

Carter, H. and Newberry, P.E. 1904. *The Tomb of Tuthmosis IV*, London.

Caton Thompson, G. and Gardiner, E.W. 1934. *The Desert Fayum*, London.

Cohen, R. 1976. Excavations at Hurvat Halukim, *Atiqot* 11, 34–50.

Crowfoot, E. 1960. Report on Textile, in Kenyon, K.M., *Excavation at Jericho I*, London, 519–626.

—— 1965. Textile Matting and Basketry, in Kenyon, K.M., *Excavation at Jericho II*, London, 662–3.

Crowfoot, M.G. 1944. Handicraft in Palestine: Jerusalem Hammock Cradles and Hebron Rugs, *PEQ* 76, 121–30.

—— 1945. The Tent Beautiful. A Study of Pattern Weaving in Transjordan, *PEQ* 77, 34–46.

—— 1955. The Linen Textiles, in Barthelemy, D. and Milik, J.T., *Qumran Cave I, Discoveries in the Judean Desert I*, Oxford, 18–40.

—— 1956. Textile, Basketry and Mats. in Singer, C., Holmyard, E.J. and Hall, A.R. *History of Technology* I, 413–55.

Crowfoot, M.G. and Crowfoot, E. 1961. The Textiles and Basketry, in Benoit, P., Milik, J.T. and Vaux, R. de, *Les Grottes de Muraba'at, Discoveries in the Judean Desert I*, Oxford, 51–63.

Crowfoot, M.G. and Davies, N. de G. 1941. The Tunic of Tut'ankhamun, *JEA* 27, 113–30.

Dothan, M. 1956. The Excavations at Nahariyah, Preliminary Report, *IEJ* 6, 14–25.

Driver, G.R. and Miles, J.C. 1955. *The Babylonian Laws*, Oxford.

Evenari, M. 1958. The Ancient Desert Agriculture of the Negev III. Early Beginning, *IEJ* 8, 231–68.

Gilbert, K.R. 1956. Rope Making, in Singer, C., Holmyard, E.J. and Hall, H.R. *A History of Technology* I, Oxford, 451–5.

Glueck, N. 1938. The first Campaign at Tell el Kheleifeh (Ezion-Geber), *BASOR* 71, 13–18.

—— 1940. The Third Season of Excavations at Tell el Kheleifeh, *BASOR* 79, 2–18.

Güthe, D.H. 1908. *Palästina (Land und Leute)*, Leipzig.

Helbaek, H. 1963. Textiles from Catal Hüyük, *Archaeology* 16, 39–46.

Holmes, W.H. 1884. Prehistoric Textile Fabrics of the United States, Derived From Impressions on Pottery, *Smithsonian Institute Bureau of Ethnology*, 393–425.

Jacobsen, T. 1970. On the Textile Industry at Ur under Ibbi-Sin, in Moran, W.L. ed., *Toward the Image of Tammuz*, Cambridge, 216–24.

Kendall, A.I. 1937. Chalcolithic Textile Fragments, in von der Osten, H.H., *The Alishar Hüyük. Seasons 1930-32* III, Chicago, 334–5.

Kendrick, A.F. 1920. *Catalogue of the Textiles from Burying Grounds in Egypt* I, London.

Kramer, S.N. 1962. The Biblical Song of Songs and the Sumerian Love Songs. *Expedition*, 5–28.

—— 1963. *The Sumerians*, Chicago.

Lapp, P.W. and Lapp, N.L. 1974. Discoveries in the Wadi el-Daliyeh, *AASOR* vol. XLI.

Lucas, A. 1963. The Chemistry of the Tomb, in Carter, H. *The Tomb of Tut-Ankh-Amun* II, New York, 162–8.

Lucas, A. 1962. *Ancient Egyptian Materials and Industries* rev. by J.R Harris, London.

Loud, G. 1938. *The Megiddo Ivories*, Chicago.

—— 1948. *Megiddo II*, Chicago.

Macalister, R.A.S. 1911. *The Excavations of Gezer I*, London.

—— 1912. *The Excavations of Gezer II*, London.

Mackay, E.J. 1916. Note on a New Tomb (No. 260) at Drah Ab'l Naga, Thebes, *JEA* 3, 125–6.

Mellaart, J. 1963. Excavations at Catal-Hüyük 1962. Second Preliminary Report, *AS* 13, 43–103.

Midgley, T. 1928. The Textile and Matting, in Bruton, G. and Caton-Thompson, G., *The Badarian Civilisation*, London.

Parrot, A. 1948. *Tello*, Paris.

Pendelbury, J.D.S. 1951. *City of Akhnaton* (III), London.

Petrie, W.M.F. 1890. *Kahun, Gurob and Hawara*, London.

Petrie, W.M.F. and Quibell, J.H. 1896. *Nagada and Ballas*, London.

Petrie, W.M.F. and Tufnell, O. 1930. *Beth-Pelet I*, London.

Pfister, R. 1934. *Textiles de Palmyre*, Paris.

—— 1937. *Nouveaux Textiles de Palmyre*, Paris.

—— 1951. *Textiles de Halabiyeh (Zenobia)*, Paris.

Pfister, R. and Bellinger, L. 1945. *The Excavations at Dura Europos II, The Textiles*, New Haven.

Pieridau, A. 1967. *Pieces of Cloth from Early and Middle Cypriote Periods*, 25–9.

Riefstahl, E. 1944. *Patterned Textiles in Pharaonic Egypt*, New York.

Rothenberg, B. 1967. *Archaeology in the Negev and the Arabah*, Tel Aviv, (Hebrew).

—— 1972. *Timna*, London.

Ryder, M.L. 1965. Report of Textiles from Catal Hüyük, *AS* 15, 175–6.

Sheffer, A. 1976. Comparative Analysis of 'Negev Ware' Textile Impression from Tel-Masos, *Tel Aviv* 3, 81–8.

—— 1978. The Textiles, in Meshel, Z., *Kuntillat Ajrud, A Religious Centre from the Time of the Judean Monarchy on the Border of Sinai*, Israel Museum, Jerusalem.

Wilkison, J.G. 1878. *The Manners and Customs of Ancient Egyptian II*, London.

Wilson, L.M. 1933. *Ancient Textiles from Egypt in the University of Michigan Collection*, Ann Arbor.

Yadin, Y. 1963. *The Finds from the Bar Kochba Period in the Cave of Letters*, Jerusalem.

Yadin, Y., *et al.* 1960. *Hazor II*, Jerusalem.

—— 1961. *Hazor III-IV*, Jerusalem.

20 WOOD

Introduction

Except for pollen grains and hard seeds and fruits, wood is the best preserved of all plant tissues in archaeological remains. This is due to its thick-walled, usually lignified cells, which retain their shape even when the wood is burned to charcoal.

The anatomical structure of wood varies, both qualitatively and quantitatively, depending on the genus or even the species. These variations are in the types of cells, their shapes and dimensions, in their relative proportions and in the patterns of their arrangements. On the basis of these characteristics it is possible to identify woody material from archaeological excavations (Fahn, 1975; Liphschitz and Waisel, 1976).

The wood material found in the Timna Temple consists of two types: (a) short segments of twigs of various diameters; (b) pieces of worked wood. Both types were found as wood or charcoal.

Identification of the wood remains can give information on the sources of trees and shrubs used for fuel, building, the preparation of tools, etc., and whether they grew locally or were imported from other countries (Täckholm and Drar, 1941). This in turn may provide information about trading and other relations. The form in which the wood material was found may give some indication of whether it was worked on the spot or elsewhere.

The Timna valley is covered by a shallow layer of coarse sand mixed with gravel covering a configuration of Nubian Sandstone. The vegetation is poor but trees such as species of *Acacia, Haloxylon persicum* and single trees of *Pistacia atlantica* (Danin, 1977) can be found in the region.

1 Materials and Methods

In the excavations of the Temple, only one or two wood segments but many charcoal pieces were found in several loci, most of the pieces not exceeding about 2 cm.^{3} in size. The segments of unburned wood remains were sectioned free-hand with a razor blade, in transverse and in longitudinal (radial and tangential) directions, for examination under the light microscope. Part of the charcoal pieces were examined, in the same orientations, with the scanning electron microscope (Fahn, 1975). Where many charcoal fragments were found in the Find Boxes, one or two samples selected at random were examined. The proportion of the wood to the charcoal is therefore much higher in the following Table than in the excavated material.

Table 1. Identification of the plants.

Box No.	Loc.	Square	Description	Plant Name
252	104, 105, 106	C-D 5–7	Charcoal	*Pinus* sp.
279	106, 107	E-F 7–9	Item I: segment of a young branch Item II: piece of a plank	*Pinus halepensis*
270	102	J 12	Piece of a plank	
265	103	J-K 6	Flattened piece of wood, 4mm. thick, painted greyish-blue on one side	
157	102	J-K 13–14	Piece of a plank	*Pinus nigra* or *Pinus sylvestris*
289	101, 102	L 14–15	Item II: piece of worked wood painted turquoise on one side	
296	101	L 16–18	Piece of a plank 11cm. long	Probably *Abies* sp.
319	106, 108	G 7–12	Piece of wood	*Cupressus sempervirens*
289	101, 102	L 14–15	Item I: flat sheet of wood	*Populus euphratica*
315	101	F 15–16	Segment of a branch	Probably *Haloxylon persicum*
211	101	G 15–17	Item I: segment of a young branch	Probably *Suaeda palaestina*
240	109	G-H 12–14	Charred segment of a branch	Chenopodiaceae
0			Piece of a young branch	
268	101	D-E 15–17	Charcoal	
322	101	F 15–16	Charcoal	
252	104, 105, 106	C-D 5–7	Item II: charcoal	*Acacia* sp.
374	103, 106	H-I 6	Charcoal	
230	101	C-E 18	Piece of a young branch ca. 10mm. in diameter	
306	101	B 20–21	Piece of a branch	
209	101	G 15–18	Piece of a young branch	Probably *Acacia gerrardii* var. *negevensis*
201	101	D-E 15–16	Piece of a branch	
319	106, 108	G 7–12	Items I and II: Charcoal	
283	106, 107	D 7–9	Piece of a charred branch ca. 8mm. in diameter	
303	101	F 15–16	Charcoal	*Pistacia atlantica*
257	102	I-K 13	Piece of a charred branch ca. 5mm. in diameter	
279	106, 107	E-F 7–9	Segment of a young branch	*Tamarix* sp.

2 Results
The following wood characteristics served for identification:

Pinaceae

Pinus L. sp. – Pine
No vessels; axial system consisting of tracheids only. No axial parenchyma except in connection with resin ducts. Resin ducts with thin-walled epithelial cells are normally found: vertical ducts in each growth ring (*Pl.* 135:1) and radial ducts in rays (*Pl.* 135:4, 5). Rays, except for those containing the resin ducts, are uniseriate, heterocellular with ray tracheids marginal or within ray.

Pinus halepensis Mill. – Aleppo pine
Walls of ray tracheids usually thin, sometimes slightly dentate. There are 1–4 pinoid, medium-sized pits in a cross-field (*Pl.* 135:5). *P. halepensis* grows on hills and mountains. It occurs at present in remnants of natural forests, e.g. in Israel (Samaria, the Judaean mountains) and Jordan (Gilead). In the past it occupied much wider areas (Zohary, 1966). It is suitable for use as timber.

Pinus nigra Arnold – Austrian pine or
P. sylvestris L. – Scots pine
Walls of ray tracheids thick and dentate (*Pl.* 135:3). There are 1–2 very large fenestriform pits in a cross-field (*Pl.* 135:2) (Jacquiot 1955). It is impossible to distinguish between *P. nigra* and *P. sylvestris* by the anatomical characteristics of the wood (Hubber and Rouschal, 1954). *P. nigra* grows naturally in southern Europe (Flora Europaea, 1964), Cyprus, Crimea, western Caucasus, Balkans and the Carpathians, and western Syria (Davis, 1965). *P. sylvestris* is found in north and central Europe, extending southwards in the mountains to southern Spain, northern Italy, Macedonia (Fl. Eur., 1964) and Asia Minor (Davis, 1965). It is more likely that the examined items belong to *P. nigra* since they could be found closer in western Syria and Cyprus.

Abies Mill. sp. – Fir
No vessels; axial system mainly consisting of tracheids. Normally there are no resin ducts. Bordered pits of tracheids uniseriate sometimes biseriate. Rays homocellular, uniseriate; pits 1–2 sometimes up to 4 small in a cross-field, taxodioid in early wood; ray cells with horizontal and tangential walls thick, strongly pitted, sometimes containing crystals; height of rays usually 3–15 cells but higher rays also present (Greguss, 1955; Jacquiot, 1955) (*Pl.* 135:6; 136:7, 8).

Species of *Abies* grow naturally in Algeria, Greece and islands of the Ionian Sea, Yugoslavia, Spain, Asia Minor, West Caucasus and the Armenian plateau. *A. cilicica* Carr. is found in Syria and Lebanon (Mouterde, 1966).

Cupressaceae

Cupressus sempervirens L. – Cypress
No vessels; axial system consisting mainly of tracheids. No resin ducts, either normal or traumatic. Rays homocellular, uniseriate, height of rays in older twigs 1–20 cells, but rays of up to 40 cells can also be found, shorter in young twigs; tangential (end) walls of ray cells relatively thin, oblique or perpendicular, sometimes concave, smooth or sometimes faintly beaded; horizontal walls mostly thin; pits aperture in cross-field included and narrow (*Pl.* 136:9, 10).

C. sempervirens grows naturally in Jordan (Edom and Gilead), in South Sinai, as well as in Lebanon and Crete. Yields timber and tannins (Zohary, 1966).

Salicaceae

Populus euphratica Oliv. – Poplar
Diffuse porous. Pores single or in radial multiples of 2–3, occasionally in clusters. Vessels with simple perforation plates; pits rhomboidal, pit aperture horizontally elongated; pitting alternate. Fibres with rather thin walls. Rays homogeneous, uniseriate. Ray-vessel pits large, round, slightly oval or rectangular (*Pl.* 136:11–13).

This is the only wild poplar species growing in the area. It grows on river banks and near springs, in Israel in the central Negev, the upper and lower Jordan Valley and in the Dead Sea area (Zohary, 1966).

Chenopodiaceae – beet family
Wood of unusual structure – with phloem strands included in it. In charcoal many of the phloem cells are destroyed and only holes remain (*Pl.* 137:16). Rays are often missing or scarce. Axial parenchyma storied.

Haloxylon persicum Bge.
Pores in large clusters. Conjunctive parenchyma in narrow concentric short bands. Wood parenchyma scanty, in longitudinal short rows containing crystals. Rays sporadic, very short in cross section. Fibres thick-walled (*Pl.* 137:14, 15). In the specimen examined the cross-sectioned material split into short bands, apparently where the conjunctive parenchyma was (*Pl.* 137: 14). These characteristics cannot exclude a few other species of the Chenopodiaceae family; however, *H. persicum* is the only one of these which grows very near Timna. It grows mainly on sand. It is found along the Arabah Valley, in North and South Sinai and the desert parts of Jordan (Gilead, Ammon, Moav and Edom) (Zohary, 1966; Danin, 1977).

Suaeda palaestina Eig et Zohary
Pores in medium sized clusters of various shapes. Conjunctive parenchyma in concentric bands alternating with wide bands of thick-walled fibres; storied, many of its cells contain rhomboid crystals. No rays. Since the examined item is a thin branch, and sectioning it is difficult, the identification cannot be conclusive. *S. palaestina* grows on desert salines in the lower Jordan Valley and Dead Sea area.

Item 240 differs from the two species above in the following anatomical characteristics: Pores mainly single or in very small clusters. Conjunctive parenchyma in concentric layers occasionally also extending radially. Occasional rays. Fibre walls rather thin. This specimen has not been further identified.

Mimosoideae

Acacia Wild.
One of the species growing in the Negev: *A. gerrardii* Benth. var. *negevensis* Zoh., *A. raddiana* Savi or *A. tortilis* (Forssk.) Hayne.
Diffuse porous. Pores mostly solitary, occasionally in small multiples or clusters. Gum-like substances filling vessels in some specimens. Wood parenchyma paratracheal, aliform, mainly confluent, forming narrow to broad concentric bands alternating with bands of thick-walled gelatinous fibres. Long vertical rows of parenchyma cells frequently containing rhomboidal crystals. Rays mostly 3–6 cells wide, of various length, uniseriate rays also present (cf. Chudnoff, 1956; Fahn, 1959) (*Pl.* 138:21–23, 25). In young branches rays are narrower and longer. In part of the specimens growth rings were distinct. This feature was not found in *A. tortilis* and *A. raddiana* as described by Fahn (1959). Growth rings, however, were observed in a few samples of *A. gerrardii* var. *negevensis* examined by the author (*Pl.* 138:24). It may, therefore, be concluded that some of the above items belong to this species. However, this problem needs further investigation and therefore a final conclusion regarding the species of *Acacia* cannot yet be reached.

All three species are restricted to wadi beds in the Negev and Sinai deserts. *A. tortilis* grows in the Arabah Valley; *A. raddiana* forms a limited diffused belt in the northern Negev; *A. gerrardii* var. *negevensis* forms a limited diffused belt in the northern Negev, in the Arabah Valley and Jordan (Edom), and in isolated stands in northern Sinai (Halevy and Orshan, 1972; Zohary, 1972).

Anacardiaceae

Pistacia atlantica Desf.
Growth rings distinct. Ring porous, one or a few rows of mainly solitary very large pores at the beginning of a growth ring, the rest of the pores mainly in radial multiples. Vessels, except for the larger ones, have spiral thickening. Rays multiseriate up to six cells wide and uniseriate, heterogeneous; many of the cells at the marginal tiers of the rays contain crystals. Radial resin ducts within multiseriate rays are concentrated in certain areas (Grundwag and Werker, 1976) (*Pl.* 139:26–29).

This species grows in the central Negev mountains; single trees grow not very far to the west of Timna.

Tamaricaceae

Tamarix L. sp. – Tamarisk
Diffuse porous. Growth rings hardly distinguishable. Pores mostly solitary, sometimes in multiples or clusters (in first growth ring of twigs). Wood parenchyma storied, paratracheal vasicentric. Rays heterogeneous, with mainly square and procument cells; up to 8 cells wide, some are higher than 80 cells (*Pl.* 136:19, 20). The identification of *Tamarix* species is most difficult even in fresh material. Moreover, the specimen examined was a small twig, and in woody plants very often not all the wood features of the first growth rings coincide with those of later years. Therefore no further attempt to identify the species has been made.

There are two species of *Tamarix* which are fairly common in the vicinity: *T. nilotica* in the Negev, the Arabah and in Jordan (Edom); and *T. aphylla* in the negev, the Arabah, Sinai and all the deserts of Egypt (Zohary, 1972; Täckholm, 1956). Other species, such as *T. negevensis* which grows in the Arabah (Eilat, Ein Yahav) are uncommon or even rare (Zohary, 1972).

3 Discussion

One type of wood remains, i.e. pieces of young and old branches, are almost all from trees and shrubs which are local or grow not very far from Timna. These are: *Haloxylon persicum*, *Acacia* sp. and *Tamarix* sp. The fact that they were unworked and that part of the *Acacia* sample and the unidentified chenopod were found as charcoals, indicate that the above mentioned species were collected mainly for fuel. This is true since there is no reason to assume that the Temple was abandoned due to a fire (Rothenberg, 1972). It is of interest to note, however, that both the wood of *Acacia* and *Tamarix* could have been used for other purposes as well. *Pistacia atlantica* was found in two locations in the vicinity of their fruits, which are edible (cf. Kislev, Chapter III, 18). The tree was also deified and adored by the ancients (Zohary, 1972).

The second type of wood remains is of pieces of worked wood, very little of which was found charred. None of the trees which supplied the raw material for this group is native in this locality: *Pinus halepensis* occupied in the past much larger areas than at present, forming natural forests the remains of which stand today on Israel's hills and mountains (the Coastal Galilee, Mount Carmel, Samaria, Judaean mountains) and in Jordan (Gilead) (Zohary, 1966). It can be used as timber and apparently was used as such in the Late Bronze Age. *Cupressus sempervirens* does not grow wild in the area but some trees remain in Jordan (Gilead and Edom). *Populus euphratica* grows on river banks and springs, the nearest location to Timna being the central Negev and the lower Jordan River Valley. Its timber is soft and resilient due to its many scattered vessels and thin-walled fibres, and it is therefore suitable for some kinds of carpentry. In the Timna Temple it was found in the form of a flat sheet.

Pinus nigra (or *P. sylvestris*) and *Abies* sp. had to be brought from further away, perhaps as finished objects from Egypt or north-western Arabia. *Pinus nigra* grows naturally in Cyprus and western Syria and southern Europe (*P. sylvestris* – in Asia Minor and Mediterranean European countries) and species of *Abies* are found in Syria and Lebanon, Asia Minor, Mediterranean European countries and North Africa. The fact that an unworked young branch of *Pinus halepensis* has been found may hint that at least some of the cut trees were brought to Timna as such and were worked on the spot.

The specimens most abundant in the excavation were those of *Acacia* sp., a tree growing in the near vicinity. Of the worked wood, specimens of the two pine species were the most abundant. Although *Haloxylon persicum* grows abundantly in the area, only one item, a segment of a thin branch, was found.

Some of the species identified in the Temple have also been found in other archaeological excavations in the region at various periods: charred material found in Timna near the copper mines was identified by Fahn as *Acacia* of one of the species growing in the Negev (Rothenberg, 1962); bowls found in the 'Cave of the Letters' near Ein Gedi were identified by Fahn as *Acacia* sp., either *A. raddiana* or *A. tortilis*, and an arrow pole found in the latter location he identified as *Tamarix* sp. (Yadin, 1963). *Pinus halepensis* and *Tamarix* sp. were identified by Fahn and Zamski (below, Chapter III, 19). *Pistacia atlantica* and *Pinus halepensis* were found in Tel Arad and Tel Sheva (Liphshitz and Waisel, 1971a, b, 1972a); these species and *Populus euphratica* in Ein Bokek (Liphshitz and Waisel 1972b); and *Cupressus sempervirens* in addition to Tel Arad, Tel Sheva and Ein Bokek also in St Catherine's Monastery in southern Sinai (Lipshitz and Waisel, 1976).

In ancient Egypt items made of *Pinus halepensis*, *Abies cilicica* and *Cupressus sempervirens* were found (Täckhol and Drar, 1941).

Pinus nigra (or *P. sylvestris*) has apparently been identified in the region for the first time.

Ella Werker

Bibliography

Chudnoff, M. 1956. Minute anatomy and identification of the woods of Israel, Publication of Forest Research Station, *Ilanoth* 3, 37–52.

Danin, A. 1977. *The Vegetation of the Negev (North of Nahal Paran)*, Tel Aviv, (Hebrew).

Davis, P.H. 1965. *Flora of Turkey and the East Aegean Islands*, Vol. I, Edinburgh.

Fahn, A. 1959. Xylem structure and annual rhythm of development in trees and shrubs of the desert, II, *Acacia tortilis* and *A. raddiana*, *BIRC* 7, 23–8.

—— 1975. A burned wood specimen from an archaeological excavation in Jerusalem, *IAWA Bulletin* 1975/2, 23–4.

Greguss, P. 1955. *Xylotomische Bestimmung der heute lebenden Gymnospermen*, Akadémiai Kiadó, Budapest.

Grundwag, M. and Werker, E. 1976. Comparative wood anatomy as an aid to identification of *Pistacia* L. species, *Israel J. Bot.* 25, 152–67.

Halevy, G. and Orshan, G. 1972. Ecological studies on *Acacia* species in the Negev and Sinai, *Israel J. Bot.* 21, 197–208.

Huber, B. and Rouschal, C. 1954. *Mikrophotographischer Atlas Mediterraner Hölzer*, Berlin-Grunewald.

Jacquiot, C. 1955. *Atlas D'Anatomie des Bois des Conifères*, Centre Technique du Bois, Paris.

Liphschitz and Waisel. 1971a. *Dendroarchaeological Investigations I, Tel Sheva*, Tel Aviv, (Hebrew).

—— 1971b. *Dendroarchaeological Investigations II, Tel Arad*, Tel Aviv, (Hebrew).

—— 1972a. *Dendroarchaeological Investigations VI, Tel Sheva*, Tel Aviv, (Hebrew).

—— 1972b. *Dendroarchaeological Investigations IV, Ein Bokek – trees and shrubs*, Tel Aviv, (Hebrew).

—— 1973. Analysis of the botanical material of the 1969–1970 seasons and the climatic history of the Beer-sheba region, in *Beer Sheba I*, ed. Aharoni, Y., Tel Aviv.

—— 1976. Dendroarchaeological Investigations in Israel (St Catherine's Monastery in Southern Sinai), *IES* 26, 39–44.

Mouterde, P. 1966. *Nouvelle Flore du Liban et de la Syrie*, Vol. I, Beirut.

Rothenberg, B. 1962. Ancient copper industries in the Western Arabah, *PEQ* 94, 5–65.

—— 1972 *Timna*, London.

Täckholm, V. 1956. *Students' Flora of Egypt*, Cairo.

Täckholm, V. and Drar, M. 1941. *Flora of Egypt*, Vol. I, Cairo.

Tutin, T.G., *et al.* 1964. *Flora Europea* Vol. I, Cambridge.

Yadin, Y. 1963. *Judean Desert Studies: The Finds from the Bar-Kokhba Period in the "Cave of Letters"*, Jerusalem.

Zohar, M. 1966. *Flora Palaestina* Vol. I, Jerusalem.

—— 1972. *Flora Palaestina* Vol. II, Jerusalem

21 EXAMINATION OF SOME WOODEN OBJECTS

Below is a summary of the identifications of some wooden objects from the Timna Temple.

(1) A small disc of wood (Field No. 265/1) which was stained green on one side, was made of the wood of *Pinus*. *Pinus halepensis* is known to have grown in the hills and mountains of Israel (Galilee, Carmel, Judea, Samaria) and in Jordan (Gilead and Ammon).

(2) The comb (Field No. 211/1; *Pl.* 140:1) was made of the wood of *Buxus sempervirens*. *Pl.* 140:2 and 3 show the characteristic scalariform perforation plates of vessel members of this species. A comb found by Y. Yadin in the 'Cave of Letters' was made from the same wood. (In Yadin 1963, 136, Pl. 51, the wood of the comb was first mistakenly identified by me as *Arbutus*, but this identification should now be rectified). The wood of *Buxus sempervirens* is still used today in many countries, e.g. Turkey, for comb making. *Buxus sempervirens* grows in Europe, North Africa and Asia, but not in Israel and Egypt.

(3) Disintegrating pieces of wood, brown in colour (Field No. 166) are of the *Tamarix* species. The storied wood parenchyma (*Pl.* 141: 6 and 7), the type of rays, the crystals in the ray cells and other anatomical features (*Pl.* 141:4 and 5) make the identification of the genus possible. The identification of the species is almost impossible in *Tamarix*, even in fresh material. Several species of *Tamarix* grow in the Arabah Valley. The walls of the fibres, vessels and parenchyma of the archaeological specimen are very thin, especially when compared with fresh wood material of *Tamarix* (*Pl.* 142:8 and 9). Specific stains revealed that the thin walls retained the cellulose and lost all the lignin. This indicates that the material examined was exposed for a long period to the action of lignin-digesting micro-organisms.

(4) Charcoals, Field No. (Boxes) 34, 56, 101, 259, 367 and 368, are of the wood of Acacia species. We arrived at this conclusion by comparing paraffin embedded cross-and longitudinal-sections of the charcoals and of sections of fresh wood Acacia species growing in the Negev and Arabah Valley (*Pl.* 142:10 and 11). We were unable to identify the exact species of the *Acacia*.

A. Fahn and E. Zamski

22. FRUIT REMAINS

Introduction[1]

In the 1969 excavation season, some fruit stones were uncovered and collected by hand or with the aid of sieves. They included grapes (*Vitis vinifera* L.), pistachio (*Pistacia atlantica* Desf.), olives (*Olea europaea* L.), and dates (*Phoenix dactylifera* L.), which were preserved desiccated due to the dry climate. Surfaces, however, were sometimes worn, probably due to abrasive action by sand.

The botanical remnants were generally found more than 0.5m. below the surface; some olive and date stones were also collected at higher levels. Grape pips were found in Strata IV-III, and remains of pistachio, olive and date in Stratum II. Those from boxes 211, 294 and 204 were found at Locus 101, outside the Temple courtyard.

The fruit remains (Table 1) included one larger sample of grape and very small samples of the other species scattered all over the site, but concentrated near the walls of the sanctuary.

In the following paragraphs a detailed description of grape pips and pistachio nuts will be presented. A characterisation of the plant type may help develop not only a better understanding of the strains of plants used and cultivated in ancient times, but also a more accurate knowledge of processes such as domestication or spread of cultivated plants.

1 Grape

Seven hundred and three pips were found scattered on the floor of the Temple, near its south-eastern corner (Find Box 363) in Squares G-H 6–9, 83–92cm. below the surface. The pips, which were preserved in a dry state without noticeable shrinkage or change in shape, allow us to make a thorough investigation of their morphology without the bias caused by the process of charring. Only the thin skin of the seed coat had disappeared.

Viticulture has been practiced for several millennia and yet the origin of the vine is not completely understood. Most investigators (e.g. De Candolle,

1855, 872; Zohary and Spiegel-Roy, 1975) are of the opinion that the cultivated grape is derived from the related wild grape, *Vitis sylvestris* C.C. Gmelin (= *V. vinifera* L. ssp. *Sylvestris* (C.C. Gmelin) Hegi). However, Sosnovskii (1949, 528) is convinced that both *V. sylvestris* and *V. vinifera* L. (= *V. vinifera* L. ssp. *vinifera*) developed independently from some now extinct ancestors that were later completely dissolved in diverse strains of the cultivated grape.

V. sylvestris is a dioecious climber with late ripening, globose fruit, pleasantly acidic in taste. Seeds, usually three to the berry, are subglobose, with a short truncate beak. This wild species is distributed in the south and temperate regions of Europe and Asia from the Atlantic Coast to the western Himalayas.

The location of the most southern true wild grape stands in the Near East is not yet clear. *V. sylvestris* has been reported in the north-western corner of Syria (Mouterde 1970); the Dan valley and the Golan Heights in north-eastern Israel have escaped wild-like wine grape (Eig, Zohary and Feinbrun, 1948). Moreover, three pips found in Epipalaeolithic Natufian and Kebaran levels in Nahal Oren, Carmen (Noy, Legge and Higgs, 1973), though a unique find, may indicate that *V. sylvestris* grew naturally in the northern parts of Israel.

The cultivated *V. vinifera* has hermaphrodite flowers and sweet globose to ellipsoid fruits. Seeds, usually 0–4 per fruit, are pyriform with a rather long beak. It has a typically eastern Mediterranean distribution reaching to the south, in special habitats, in the Negev mountains, e.g. the vineyard, mentioned in the Nessana Papyri, No. 31, 6th century (Kraemer, 1958; Evanari, Shanan and Tadmore, 1971); in Jordan (Edom) (Oppenheimer, 1931, 85) at Jericho (Antoninus Martyr, 6th century) and in Egypt (Keimer, 1924).

The grape pip with its peculiar morphology (*Pl.* 131:1–4) gives us an excellent opportunity to consider the character of the beak, scutum and fossettes, in addition to its dimensions. The shape of the pips is quite variable. The natural variability caused by environmental and genetical factors is greater in natural populations and primitive cultivated plants than in modern plants. In order to secure reliable data for a definitive characterisation of the findings, it is necessary to study quite a large sample, perhaps 100 specimens. The variations in the shape of grape pips is partially explained by the non-constant number of pips in the berry. The regular dorso-ventral character of the pip in a one-seeded berry becomes distorted in the two-seeded berry. Pips from the latter have a greater width. In multi-seeded berries, pips and beaks tend to be longer (Terpó, 1976). It is supposed that each part develops independently; this means that dimension and shape are not linked in a simple manner (*Ill.* 36, *Pl.* 131:2–4).

The diagnostic value of the dimensions and sculpture of vine pips on the specific and varietal level is primary. In order to understand pip morphology, details of its development from the ovule stage (based principally on Pratt, 1971, and Terpó, 1976, 1977) need to be introduced.

Table 1. List of fruit remains

Field (Box) No.	Locus	Squares	Plant Name	No. of items	Depth (cm)
37	106	E-F 6–7	Olive	1	0–10
204	110	E 13–14	Pistachio		32–53
211	101	G 15–17	Olive	1	20–36
214	106–107	E-F 7–8	Olive	1	8–53
293	106	D-H 6	Date	1	*
294	101	F15–16	Olive	1	–
296	101	L 16–18	Olive		0–17
319	106–108	G 7–12	Olive	2	10–13
319	106–108	G 7–12	Date		10–13
323	105–106	A-B 6–7	Date		54–63
328	106	E-F 6	Pistachio	13	*
363	106-108	G-H 6-9	Grape	703	83-92

*Found under a huge rock which fell on Wall 3.

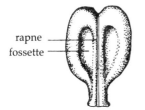 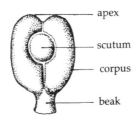

Illustration 36. Development of the vine pip from the ovule stage.

The ovule is anatropous, having two integuments which cover a massive nucellus. The distal part of the funiculus, which connects the placenta to the ovule, lies parallel to and fuses with the ventral side of the outer integument to form the raphe. The region where the funiculus ends and enters the nucellus is called the chalaza. In young stages it is found at the apex of the ovule. Meristematic activity in the chalaza adds to its thickness and forms the scutum. After fertilization, the shape of the ovule changes as a result of local meristematic growth in the chalaza region and various parts of the outer integument. At the micropyle, the small opening of the ovule, the outer integument thickens and elongates to form the beak. On the ventral side of the ovule, the middle layers of the outer integument thicken and form two longitudinal, more or less parallel projections, which enter the nucellus on either side of the raphe taking with them the inner integument. These projections, which are pocket-like, form two shallow, narrow cavities as seen from outside, called fossettes. In fresh seeds they are almost completely filled with parenchymatous tissue which usually disappears in very old samples to form two deep, narrow furrows. With the development of the fossettes, the region between them and the chalaza enlarges considerably, lengthening the pip, forming its apex and shifting the chalaza to the dorsal surface of the pip.

The pips found in Timna are small (Table 2, *Pl.* 131), elongated or sometimes almost round to obovate or pentagonal, with the apex clearly incised. In a side view (*Pl.* 131:4) the pip is obovate. The beak is straight, cylindrical and quite long; its surface is covered with warts and its apex shallowly bilobate. The scutum is oval, slightly concave and the fossettes are almost parallel, L-shaped or hooked, ending in the upper third of the corpus, i.e. the body of the pip excluding its beak (*Pl.* 131:3).

The dimensions and ratios of the whole pip are usually the only data available in the literature. From the work of Stummer (1911) and thereafter, it has generally been accepted that in extant strains it is possible to distinguish cultivated from wild vine by the breadth/length ratio. Though overlapping exists, the breadth in the former is about half its length, whereas that of the latter is about three-quarters. The beak length is also considered a very valuable characteristic in distinguishing cultivated from wild vines. In the course of evolution, the beak lengthens proportionately faster than the corpus length under the effects of cultivation and selection (Negrul, 1960; Terpó, 1976). For the sake of accuracy, the beak length is measured twice, from the dorsal side and the profile, and the average calculated (Table 2). Beak/pip length ratio is presented in Table 3. The data on the chalazal scutum, especially its shape, its width/pip width ratio and its place along the pip, provide good criteria. Data on dimensions and ratios of fossettes, and fossette/pip length ratio, are given in Tables 2 and 3. For the sake of accuracy and better comparison, only the left fossette was measured with the beak orientated downwards.

For several reasons, the grape pips found in Timna should be considered as representing a cultivated form, though none of them is conclusive by itself:

Table 2. Measurements (in mm) of grape pips from Timna 200 (N = 100)*

	Pip			Beak		Scutum		Fossette	
	Length	Breadth	Thickness	Length	Breadth	Length	Breadth	Length	Breadth
Minimum	4.2	2.7	2.2	0.6	0.5	1.0	0.8	1.6	0.4
Average	5.09 ± .09	3.21 ± .05	2.58 ± .03	1.08 ± .04	1.07 ± .03	1.50 ± .04	1.07 ± .03	2.30 ± .05	0.66 + .02
Maximum	6.1	4.2	2.9	1.6	1.5	2.2	1.5	3.4	0.9

Table 3. Ratios (× 100) of measurements given in Table 2

	$\frac{B}{L}$	$\frac{L}{T}$	$\frac{LB}{L}$	$\frac{LS}{L}$	$\frac{BS}{B}$	$\frac{LS}{BF}$	$\frac{LF}{LC}$	$\frac{BF}{B}$	DC	DL
Minimum	52	161	14	20	25	248	44	13	37	26
Average	63 ± 1	198 ± 3	21 ± 1	29 ± 1	33 ± 1	355 ± 12	58 ± 1	21 ± 1	50 ± 1	38 ± 1
Maximum	78	259	27	37	45	558	77	29	69	51

(B = breadth of pip, BF = breadth of fossette, BS = breadth of scutum, DC = distance of centre of scutum from the apex/corpus length, DL = the same distance/pip length, L = length of pip, LB = length of beak, LC = length of corpus, LF = length of fossette, LS = length of scutum, T = thickness of pip)

* The quantity to the right of the ± sign was computed according to the formula (1.96 x the standard deviation/$\sqrt{N}$) where N is the sample size. The sum of the mean and this quantity is the upper limit of an interval whose lower limit is the difference between the mean and this quantity. This interval is called a confidence interval. Statistically, it contains the true value of the population mean with 95% confidence (corresponding to a P-value of 0.05).

(1) The average pip breadth/pip length ratio is between those of recent wild and cultivated forms. Also the beak length is among the highest of the wild vine types (Terpó, 1976), and amongst the lowest of the cultivated vines (Potebnja, 1911). This intermediate situation also seems to hold true for other criteria (Tables 2, 3).

When comparing pips of recently cultivated forms with that of wild ones, it should be kept in mind that about five millennia of viticulture have passed and pip shape did not change abruptly at the moment of domestication. On the contrary, it seems that the changes occurred slowly and possibly at different rates in various strains and regions. Therefore it is not surprising to find pips of cultivated vine which are intermediate between the wild and cultivated types.

(2) The nearest wild vine today is hundreds of kilometres distant from Timna, whereas the cultivated forms might have been grown in the vicinity.

(3) The find of pips inside the Temple suggests that the fruit might have been an offering. It is likely that worshippers would have chosen a superior quality for this purpose.

(4) There were strong links between Timna and North-west Arabia (Midian), and perhaps also Edom, and in none of these countries is wild vine found. Except for Egypt (latitude 29°50'), Timna is the southernmost place with finds of grapevine pips.

Cultivated grape remnants begin to appear from the Early Bronze Age onwards. Finds from this period are known in the Levant from Jericho (Hopf, 1969); Arad (Hopf, 1978); Lachish (= Tell ed-Duweir; Helbaek, 1958); and Hama (north-west Syria; Helbaek, 1948, 1958). A Middle Bronze Age find is known from Ta'anach (Liphschitz and Waisel, 1972); and Late Bronze Age from Kamid el-Loz in Lebanon (Behre, 1970). From Egypt, grape vine pips are known from First Dynasty Abydos and Nagada (the latter being very small, 4–4.5 × 2.8 × 1.8mm.; Keimer 1924), from 3rd Dynasty Saqqara (Lauer, Takholm and Aberg, 1950) and from El-Omary, dating from about the same period (Helbaek, 1963). Finds from later historical periods are of course much more abundant.

2 Pistachio

Thirteen dry nuts were found (Find Box No. 328), under a huge rock fall in Squares E-F 6 (*Pl.* 132:1). Another half nutshell (Box No. 204) came from Squares E 13–14. The nuts are dry, still hard and not brittle, sometimes slightly cracked, and seedless. They are either lentil-like or resemble a flattened bolster and show big holes which sometimes extend over half their surface, probably caused by rodents. Usually only one of the two wide faces of the nutshell was gnawed and teeth marks can still be seen. Sometimes remnants of the parenchymatous tissue which filled the hilum is preserved. Only five nuts could be measured, but because of the holes no accurate thickness measurements could be obtained. The length

Table 4. Dimensions (in mm.) and indices of pistachio
(N = 5)

	L	B	T	100 L/B	100 T/B
Minumum	5.0	4.7	–	90	–
Average	5.2	5.6	4	94	71
Maximum	5.4	6.0	–	109	–

was measured along the axis beginning in the hilum (Table 4).

It is difficult to identify species of wild pistachio from the nut size and shape because of their similarity, except *Pistacia lentiscus* L. whose fruit is significantly smaller. The nut of the wild pistachio species is subglobular or laterally compressed as a lentil or stretched like a bolster; it may be longer than broad, or broader than long. The other species growing in the Near East, namely *P. palaestina* Boiss., *P. atlantica* Desf. and *P. khinjuk* Stocks are found in mountains and have a wide range of nut size. All are aromatic dioecious, deciduous bushes or trees with pinnate leaves. The edible fruits grow abundantly in clusters of loose panicles on the distal part of branches from the previous year. They are consumed roasted as a delicacy and are sold in the markets of the Near East. Immature fruit clusters are sold for pickling (Helbaek, 1966b). The red, unripe fruit turns to a turquoise colour and the thin flesh shrinks, causing the skin to wrinkle in an approximately net-like pattern.

P. palaestina is more mesophilic in requirements, abundant in the Mediterranean maquis of Israel, Jordan, Lebanon and Syria, Cyprus and Anatolia. The most southern districts are the Judaean mountains and Edom (Zohary, 1952). It is also found thinly distributed in a few places along the Red Sea coast of Arabia (Migahid and Hammouda, 1974). Generally it is a bush or small tree, and only rarely do its dimensions exceed 10m.

P. khinjuk is a small tree restricted to high mountains of semi-dry or dry regions, sometimes forming open or steppe-like forests (in Anatolia, Iraq, Iran and Afghanistan). The nearest regions are Jordan, the mountains of Southern Sinai and Gebel Galala in Northeast Egypt (Zohary, 1952).

P. atlantica is a conspicuous tree which may live for hundreds of years and reach a height of 20 m. Its natural habitat is in semi-arid regions, but it occasionally extends to desert wadis. Such wadis are frequently named after the tree because of its conspicuousness, edible fruits and shade. The nearest location of *P. atlantica* is in Nahal Botem, about 5km. west of Timna (Danin and Orshan, 1970). In this area there are three other wadis with similar names: Nahal Botmim (= Wadi Umm Buttuma), Wadi Umm Butma and Wadi Abu Buteima. All are small wadis on the high calcerous plateau (about 600m. above sea level) of the southern Negev and the adjacent part of Sinai.

In order to establish the identification of the pistachio by comparative morphology of the nut, the hilum was chosen as a diagnostic characteristic.[2] The hilum, the region where vascular bundles enter through the base of the stony endocarp and nourish the developing seed, is a more or less shallow crater-shaped depression. In *P. palaestina* it is very shallow or even flat or slightly convex (*Pl.* 132:5–6); in *P. atlantica* the hilum is a more or less deep crater (*Pl.* 132:3) and in *P. khinjuk* the smooth face of the crater is disturbed by a more or less conspicuous radial ridge, stretching from near the hole through which the bundles enter the endocarp to near the hilum rim, and is a continuation of the keel encircling the nut (*Pl.* 132:4). The nut opens along this line during germination.

For geographical and morphological reasons, the nuts from Timna should be identified as belonging to *P. atlantica* (compare *Pl.* 132:1 with *Pl.* 132:3).

Remains of fruits and nuts of pistachio are known in the Mediterranean region from the earliest periods.

Charred nuts were reported from the Franchthi cave in southern Greece from the Palaeolithic to the end of the Neolithic (Hansen, 1978). A substantial find is reported from the late Natufian period of Tell Abu Hureyra, North Syria (Hillman, 1975). Some small charred nut fragments were found in the Pre-Pottery Neolithic A period in Mureybit, North Syria (van Zeist, 1970); hundreds of nutshell fragments came from Tell Aswad, near Damascus (van Zeist and Bakker Heeres, 1979). In PPN B there are reports of similar finds from sites all over the Fertile Crescent: more than 50 imprints and some five gallons of carbonized nuts were found at Beidha, in Jordan (Helbaek, 1966a); charred nut remnants at Ramad, south-west Syria (van Zeist and Bottema, 1966); nut-shells from Pre-Ceramic Cape Andreas-Kastos, north-east Cyprus (van Zeist, 1981); some remains which were identified as probably pistachio from Aceramic Khirokitia, southern Cyprus (Waines and Price, 1977); hundreds of charred nuts from Çayönü, south-east Anatolia (van Zeist, 1972); others from Jarmo, north-east Iraq; Sarab, Iranian Kurdistan (Helbaek, 1964) and Ali Kosh, south-west Iran (Helbaek, 1969); 17 nutshell fragments from late PPNB or the early Ceramic Neolithic were found at Tell Es-Sinn, eastern Syria (van Zeist, 1980a); charred nuts are also reported from Ghediki and Sesklo, Thessaly, Greece (Renfrew, 1966). In the Pottery Neolithic Period one charred fruit was uncovered at Catal Hüyük, Anatolia (Helbaek 1965); two pistachios from Dhali-Agridhi, Central Cyprus (Stewart, 1974); one nut and one fruit from Hacilar, north-west Anatolia (Helbaek, 1970).

The Chalcolithic excavations at Nahal Mishmar in the Judean Desert (Zaitschek, 1980), Girikihaciyan, south-east Anatolia (van Zeist, 1980b) and Tepecik and Korucutepe (van Zeist and Bakker-Heeres, 1975) should also be mentioned. From the Early Bronze Period pistachio has been found in Israel at Lachish (Helbaek, 1958) and Arad (Hopf, 1978); in western Cyprus at Kissonerga Mylouthkia, also two nutshells of *P. lentiscus* (Colledge, 1980), and Korucutepe. One nut imprint of *P. terebinthus* L. came from Middle Bronze Age Knossos, Crete (Aström and Hjelmqvist, 1971) and five fruit imprints from Late Bronze Age Kalopsidha, Cyprus (Helbaek, 1966b).

It is apparent that except for the finds at Knossos,

and perhaps also from the Franchthi cave and a few others, almost all finds belong to *P. atlantica*.

There are several reasons for the relative abundance of *P. atlantica* at archaeological sites:

(1) The vast area of distribution of this plant, extending over a quarter of the circumference of the globe, from the Canary Islands in the Atlantic Ocean to Pamir in Central Asia.
(2) Its significance as a principal constituent of forests.
(3) Its large size, which implies greater quantities of fruit.
(4) Its rather thick and hard nutshell, which means better survival in ancient sites.

Furthermore, the new differential characteristic of the hilum, mentioned above, may make identification easier, especially where the shape and size of the nuts are not characteristic.

3 Olives

Five dry, whole olive kernels in a good state of preservation, and one and a half abraded kernels were found (Boxes 37, 296) in the uppermost layers (*Pl.* 131:5–7). Others were found within the Temple (Boxes 37, 214 and 319). Those from boxes 296, 211 and 294 come from the 'dump' outside the north wall (Locus 101). The kernels are long and straight, three of them ending in a sharp point. All are smooth, with branching grooves which are most prominent at the blunt base, with the exception of No. 319 (1) which is densely covered with small warts and has a rough appearance. Within the grooves, veins of fruit were sometimes visible.

Olive remains are known all over the Near East, from the Prepottery Neolithic Period onwards (Noy, Legge and Higgs, 1973).

4 Dates

One complete and two halves of dry date seeds were found at the site, one (Box No. 293) on top of the southern wall (W. 3) and the others (Boxes Nos. 319 and 323) at Loci 105–106–108. The seed from Box No. 293 measured 14.4 × 8.0 × 6.9mm. Date finds are very abundant in Egypt and the Near East.

5 Discussion

The fruit remains found at the Timna Temple raise two major questions: where did the fruit come from?

Table 5. Dimensions (in mm) and Index of Olive Kernels

Field (Box) No.	Length	Diameter	Length/ Diameter × 100
211	17.4	7.7	226
214	18.6	7.5	248
294	14.0	7.7	182
319(1)	17.4	6.9	252
319(2)	13.9	7.0	199
Average	16.3	7.4	221

And was there any connection between the fruits and the cult offerings?

The four fruit species may be divided into two groups, one consisting of the wild pistachio nuts and the other comprising the cultivated grapes, olives and dates. Wild trees of *P. atlantica* grow today in the vicinity of Timna, so it is quite safe to suppose that the finds originated in this location. It is also possible that the *P. atlantica* trees were more abundant in the past.

The dates might have come from an oasis or plantation in Egypt, Sinai, southern Israel, Jordan or Arabia. It is more difficult to determine the origins of Mediterranean fruits. The grape pips and olive kernels appear to be remains of whole fruits introduced as food, rather than for processing to wine or oil. The grapes may have come as raisins from Egypt or Arabia (Midian), but it is not impossible that they were raised in an arid region with the aid of irrigation and brought to Timna as fresh grapes. Concerning the latter possibility, water is relatively plentiful in the Arabah itself, e.g. at Ein Yotvata, 15km. north of Timna. However, the winter here is rather too warm for the normal development of vines. In places with more suitable temperatures, as in the surrounding mountains, irrigation sources are rare. The considerations regarding the olive tree are similar to those of the vine.

The find of a charred twig of *P. atlantica* (cf. above Chapter III, 20) at a distance of about a metre and at a difference of depth of 5cm. from the pistachio nuts, may be evidence that they were brought to Timna as twigs with clusters of fruits. The different states of preservation of the nuts and the twig, however, raise some reservations regarding this supposition.

The absence of cereals and, especially, legumes might be thought to support the idea that the fruits were part of the Temple ritual. However, since all the food remains found at Timna are inedible because of their hardness or bitterness, it is obvious that small animals and insects could easily have eaten any edible grains and seeds after the sanctuary was abandoned.

Mordechai E. Kislev

Notes

1. The sincere thanks of the author to Mrs Sara Shalom for the patient measurements, Dr M. Snyder for assistance with the statistics, Ofra Feuchtwanger for the photography, U. Yoran and I. Langsam for the SEM micrographs, and E. Huber for the drawings of the grape pips.
2. This method of identification of pistachio nuts from archaeological contexts, used here for the first time, will be described in detail in a forthcoming publication.

Bibliography

Antonius Martyr. 1881. *Antonini Placentini Itinerarium* (trans. J. Gildemeister), Berlin.

Astrom, P. and Hjelmqvist, H. 1971. Grain impressions from Cyprus and Crete, *Opusc. Athen.* 10, 9–14.

Behre, K.E. 1970. Kulturpflanzenreste aus Kamid el-Loz, in Hachmann, R., *Kamid El-Loz 1966/67*, Saarbrücker Beitrage zur Altertumskunde 4, Bonn, 59–69.

Candolle, A. de. 1855. *Géographie botanique raisonée*, Paris and Geneva.

Colledge, S. 1980. Plant species from Kissonerga Mylouthkia – feature 16.3, in Peltenburg, E.J., Lemba archaeological project, Cyprus 1978, in *Levant* 12, London, 18–20.

Danin, A. and Orshan, G. 1970. Distribution of indigenous trees in the northern central Negev highlands, *La-yaaran* 20, 91–119 (Hebrew with English summary).

Eig, A., Zohary, M. and Feinbrun, F. 1948. *Analytical Flora of Palestine*, 2nd ed., Jerusalem (Hebrew).

Evenari, M., Shanan, L. and Tadmore, N. 1971. *The Negev: The Challenge of a Desert*, Cambridge, Mass.

Helbaek, H. 1948. Les empreintes de cereales, in Riis, P.J., *Hama* II, 3, Copenhagen, 205–7.

—— 1958. Plant economy in ancient Lachish, in Tufnell, O., *Lachish* IV, London, 309–17.

—— 1963. Late Cypriote vegetable diet at Apliki, *Opusc. Aten.* 4, 171–86.

—— 1964. First impressions of the Catal Hüyük plant husbandry, *AS* 14, 121–3.

—— 1966a. Pre-pottery Neolithic farming at Beidha, in Kirkbride, D., Five seasons at the Pre-pottery Neolithic village of Beidha in Jordan, *PEQ* 98, 61–6.

—— 1966b. What farming produced at Cypriote Kalopsidha, in Astrom, P. et al., *Excavations at Kalopsidha and Ayios Iakovos in Cyprus*, Stud. Medit. Archaeol. 2, Lund, 115–26.

—— 1969. Plant collecting, dry farming, and irrigation agriculture in prehistoric Deh Luran, in Hole, F., Flannery, K.V. and Neely, J.A., *Prehistory and Human Ecology of the Deh Luran Plain*, Mem. Mus. Anthrop. Univ. Mich. 1, Ann Arbor, 383–426.

—— 1970. The plant husbandry of Hacilar, in Mellaart, J., *Excavations in Hacilar*, Edinburgh, 189–244.

Hillman, G. 1975. The plant remains from Tell Abu Hureyra: a preliminary report, in Moore, A.M.T. et al., The Excavation of Tell Abu Hureyra in Syria, *PPS* 41, 70–73.

Hopf, M. 1969. Plant remains and early farming in Jericho, in Ucko, P.J. and Dimbleby, G.W. (eds.), *The Domestication and Exploitation of Plants and Animals*, London, 355–9.

—— 1978. Plant remains, strata V–1, in Amiran, R., *Early Arad*, Jerusalem, 64–82.

Keimer, L. 1924. *Die Gartenpflanzen im alten Agypten*, Hamburg.

Kraemer, C.J. 1958. Non Literary Papyri, in H.D. Colt, *Excavations at Nessana* Vol. III, Princeton.

Lauer, J.P., Täckholm, V.L. and Aberg, E. 1950. Les Plantes découvertes dans les souterrains de l'enceinté du roi Zoser à Saqqarah, *Bull. Inst. Egypte* 32, 121–7.

Liphschitz, N. and Waisel, Y. 1972. *Dendroarchaeological investigations: 8. Tel Ta'anach, mimeographed report*, Tel Aviv.

Migahid, A.M. and Hammouda, M.A. 1974. *Flora of Saudi Arabia*, Riyadh.

Mouterde, P. 1970. *Nouvelle Flore du Liban et de la Syrie* 2, Beirut.

Negrul, A.M. 1960. Evolucija razmera semjan i jogod u vinograda, *Izv. timiryazev. sel.-khoz. Akad* 2, 167–76.

Noy, T., Legge, A.J. and Higgs, E.S. 1973. Excavations at Nahal Oren, Israel, *PPS* 39, 75–99.

Oppenheimer, H.R. 1931. *Flora Transjordanica* (Reliquiae Aaronsohnianae, I), Bull. Soc. Bot. Genève Ser. 2, 126–409; 23, 510–19.

Potebnja, A. 1911. Die Samen von Vitis vinifera und ihre Bedeutung für die Klassifikation der Sorten, *Trudy prikl, Genet. selek.*, Ser. 1, 3, 147–65, St Petersburg (Russian with German translation).

Pratt, C. 1971. Reproductive anatomy in cultivated grapes – a review, *Amer. J. Enol. Vitic.* 22, 92–109.

Renfrew, J.M. 1966. A report on recent finds of carbonized cereal grains and seeds from prehistoric Thessaly, *Thessalika* 5, 21–36.

Sosnovskii, D.I. 1949. Vitaceae, in Shishkin, B.K., *Flora of the U.S.S.R.* 14, Moscow and Leningrad (Trans. Lavoot, R., 1974, Jerusalem, 516–43).

Stewart, R.T. 1975. The plant remains from Tell Abu Hureyra: a preliminary report, in Moore A.M.T. et al., The Excavation of Tell Abu Hureyra in Syria, in *PPS* 41, 70–73.

Stummer, A. 1911. Zur Urgeschichte der Rebe und des Weinbaues, *Mitt. Anthropol. Ges. Wien* 41, 283–96.

Terpó, A. 1976–77. The carpological examination of wild-growing vine species in Hungary, *Acta Bot. Acad. Sci. Hungar.* 22, 209–47; 23, 247–73.

Waines, J.G. and Price, N.P. 1977. Plant remains from Khirokitia in Cyprus, *Paleorient* 3, 281–4.

Zaitschek, D.V. 1980. Plant remains from the Cave of the Treasure, in Bar-Adon, P., *The Cave of the Treasure*, Jerusalem, 223.

Zeist, W. van. 1970. The Oriental Institute Excavations at Mureybit, Syria: Preliminary report on the 1965 campaign, part III, the palaeobotany *JNES* 29, 167–76.

—— 1972. Palaeobotanical results of the 1970 season at Cayönü, Turkey, *Helinium* 12, 3–19.

—— 1980. Examen des graines de Tell es Sinn, *Anatolica* 7, 55–9.

—— 1980. Plant remains from Girikihaciyan, Turkey, in *Anatolica* 7, 75–89.

—— 1981. Plant remains from Cape Andreas-Kastros (Cyprus), in Le Brun, A., *Un Site néolithique précéramique en Chypre: Cap Andreas – Kastros*, Paris, 95–9.

Zeist, W. van and Bakker-Heeres, J.A.H. 1975. Prehistoric and early historic plant husbandry in the Altinova plain, southern Turkey, in van Loon, M.N., *Korucutepe I* Amsterdam and Oxford, 223–57.

—— 1979. Some economic and ecological aspects of the plant husbandry of Tell Aswad, *Paleorient* 5, 161–9.

Zeist, W. van and Bottema, S. 1968. Palaeobotanical investigations at Ramad, *Ann. Archéol. Arab. Syriennes* 16, 179–80.

Zohary, D. and Spiegel-Roy, P. 1975. Beginnings of fruit growing in the Old World, *Science* 187, 319–27.

Zohary, M. 1952. A monographical study of the genus Pistacia, *Palest. J. Bot.* 5, 187–228.

23. FISH REMAINS

Introduction

Fish remains were found in the Timna Temple (Site 200) and in two smelting camps (Site 2 and Site 30).[1] As Timna is situated in the desert, *c.* 30km north of Eilat, any fish consumed there had to be imported over long distances, presumably in a dried condition.

Materials and Methods

Careful dry sieving of the excavated material was carried out, especially at the Temple, which made it also possible to collect and identify the small fish remains.

The fishbones were defined by comparison with recent bone material. Nomenclature and methods of measurement used by Goodrich (1958) and Lepiksaar and Heinrich (1977) were adopted.[2] Measurements were expressed in millimeters with a tolerance of 0.1 mm.

The vertebral column is divided into an anterior part, the truncal (precaudal or abdominal section) and the caudal section, beginning with the first vertebra with fused haemal apophyses forming a haemal spine. The vertebrae are numbered serially from head to tail. Wherever serial numbers of vertebrae are specified they must be considered as only approximations.

The body length of the subfossil fish was estimated, based on corresponding measurements of recent fish bones. The minimal number of individuals (M.I.N.) was estimated for each family or species of fish according to frequency of the most frequent skeletal element. Differences in size, age and location were taken into consideration.

1 Excavated Fishbones

(1) Meagre, *Argyrosomus regius* (Asso, 1801)
Family: *Sciaenidae*. Croakers, Drums. *Teleostei*.
Identification: (*Pl.* 143 and *Pl.* 144:1–13)

Os operculare – gill-cover bone.
Two postarticular ribs (*cristae*) can be identified on the medial surface of the gill-cover bone, instead of the single one found in many other families. The two ribs end with projecting spikes (*processus spinosi*) at the posterior rim of the operculum (*Pl.* 144:10, two arrows; 12, arrows a).

A characteristic pattern of the fine wavy lines can be seen on the lateral surface of the flat *operculum* (*Pl.* 144:10a, 11a), similar to the wavy lines on the Sciaenid vertebra (see below). The rostral view displays a cavity of rectangular form (*fovea articularis*) which articulates with the hyomandibular bone (*Pl.* 144:11 arrow; 12 arrow b).

Otolith or *statolith* – ear stone (*Ill.* 37).
The *otoliths* of the Sciaenid genera show resemblances in many characteristics. Thus Schmidt (1968) speaks of 'Sciaenid-type' *otoliths*. They are unusually large and bulky; the rims have no serration; the medial (inner) surface is smooth and shows a depression – the *sulcus* which is composed of the anteriorly situated flat *ostium* and the long, deep posterior *cauda*. Viewed from the lateral aspect, a conglomerate of knob-like protrusions projects from the posterior part of the outer surface (*Pl.* 143:7a, 7b arrow; 8a, 8b arrow; 9a, 9b). The different members of the Sciaenid family have developed highly distinct forms of *otoliths*. The two excavated specimens and, for comparison, a recent one from the Mediterranean Sciaenid *Argyrosomus regius*, show the above mentioned characteristics of the Sciaenid *otolith*. In addition, they show the following characteristics: They have an elongated shape, with parallel dorsal and ventral rims (*Pl.* 143:7, 8, 9), and a rounded sharp anterior rim. The depression on the convex medial surface has the form of a 'tadpole'. The *ostium*, the head of the 'tadpole', is not hollowed into the *otolith* but only outlined on the surface. It is wider than usual and occupies most of the anterior part of the medial surface (*Pl.* 143:7 arrow a; 8 arrow a; 9 arrow a). The *cauda*, the tail of the 'tadpole', is a deep strongly bent notch (*Pl.* 143:7, 8, 9 arrow b). Its pointed end – anteriorly directed – reaches but does not meet the ventral rim of the *otolith* (*Pl.* 143:7, 8, 9) (Schmidt, 1968; Frost, 1927). This last feature is thought to be typical for the genus *Argyrosomus* (FAO, 1972).

The vertebrae:
The *Sciaenidae* have distinctive surface patterns of the vertebrae. The surface of the truncal vertebrae is covered with an embroidery-like texture of fine threads and filaments. Also, the broad lateral pillar in the middle part of these vertebrae appears like a cord of filaments running horizontally (*Pl.* 143:1, 2). Another feature of the truncal vertebrae is a ventromedial notch formed by two parallel ledges. These ledges help to differentiate *Sciaenidae* from *Serranidae*.

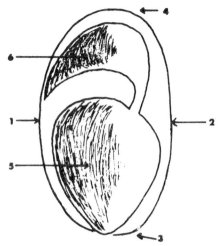

Illustration 37. Left Sciaenid otolith — Medial surface (See *Pl.* 143:7.8.9)

1. Ventral Rim   2. Dorsal Rim
3. Anterior Rim   4. Posterior Rim
5. Ostium   6. Cauda

In the caudal vertebrae the lateral pillar is very small in its cranial part and gradually becomes broader caudally. It is also covered with a pattern of filaments and threads. The lateral pits adjoining the pillar from above and below show a porous texture of small cellular elements (*Pl.* 143:3–6).

Skeletal elements identified as *Sciaenidae*:

Operculum	2
Otolith	2
Vertebrae	13
Total	17 (17.9% of all fish remains)

Topographical distribution of the specimens:
One *otolith* was found in the Temple (M.I.N. = 1).
Sixteen specimens were found in Camp Site 2 (M.I.N. = 7) and Site 30 (M.I.N. = 1): one *otolith*, two *opercula* and 13 *vertebrae*.

Measurements:

Table 1: Otoliths of Sciaenidae – Summarising the measurements of the two excavated otoliths, *and of one recent* otolith *of Argyrosomus regius for comparison.*

	Site 200 Box 319	Site 2, Pit K Box 413	Recent Argyros. regius (Meagre)
Total length (cranio-caudal)	12.2mm	15.7mm	10.2mm
Height (dorso-ventral)	6.7mm	8.2mm	5.8mm
Depth (latero-medial)	5.8mm	7.3mm	4.0mm
Side	Left	Left	Left
Estimated total length of the fish(*)	34.2cm	44.0cm	28.6cm
Measured total length of the recent fish			27.0cm
Pl. 143	No: 8,8a,8b	7,7a,7b	9,9a,9b

* This estimation is made by multiplying the length of the otolith by a factor of 28 (Schmidt, 1968). Note the good correspondence of estimated and measured length of the recent fish.

Estimation of the length of the Timna *Sciaenidae*:
The measurements given for the *otoliths* allow an estimation of the total length of *Sciaenidae* excavated in Timna to be between 34 and 44cm.

Zoogeographical distribution and economical significance:
The Meagre, *Argyrosomus regius*, is found throughout the Mediterranean, the eastern Atlantic and the east coast of South Africa. It is absent from the northern part of the Red Sea and from the Gulf of Eilat. The collections of fish in the Universities of Tel Aviv and Jerusalem have no specimens of *Sciaenidae* from the Red Sea (Dor, 1978). The Meagre inhabits the shallow, warm coastal waters of sandy shores. A pelagic predator, it feeds on smaller fish. It has the ability, like other fish of the same family, to produce loud drumming underwater noises by contracting and expanding its swim bladder. This phenomenon was already known in antiquity (Thompson, 1947) and might have been related to the story of the Sirens. The unusually large otoliths (ear stones) of these fish may have some role in the reception of these noises. The Meagre is caught with nets as well as with lines.

In Timna, 17.9% of the excavated fish remains were identified as *Sciaenidae* (M.I.N. = 8). As previously described, these remains can be identified as *Argyrosomus regius* with a high degree of certainty. It may be assumed that these Meagres were caught near the northern coast of the Sinai peninsula, probably near the Bardawil lagoons.

(2) Gilthead, Sea Bream, *Sparus auratus* (Linnaeus 1758)
Silverbream, *Rhabdosargus haffara* (or some close relative) (Klunzinger, 1870)
Family: *Sparidae*. Sea Breams, Porgies. *Teleostei*.
General Information and Identification:
(*Pl.* 145:36–40)
Both genera belong to the type of the *Sparidae* which feed on hard material, mainly mussels (Gregory, 1959). Their jaws and teeth are adapted accordingly. Three long curved and conical anterior teeth, 'canines', situated on each jawbone, serve to rip loose mussels off rocks. The hard matter is then crushed between rows of 'molars', the so-called 'pavement teeth' (Owen, 1840–45; Adamicka, 1972). Their shape is rounded and obtuse, yet the external-lateral teeth are longer and more oval and conical. Four aboral molars stand out on each side: two in the third (from lateral) row of the upper jaw, and two in the second row of the lower jaw. In the mature dentition these two teeth are replaced by one elliptical, much larger tooth with its long axis parallel to the axis of the body. There are reserve teeth – successor molars – underneath the sockle of these 'large molars'. These produce a swelling at the aboral part of the dorsal rim of

the premaxillary bones (*Pl.* 145:32 arrow; 33 arrow) which are characteristic for these species. This 'large swelling' helps to differentiate them from other similar Sparids such as *Pagrus* and *Acanthopagrus*.

Description of excavated specimens:

Os Praemaxillare – part of the upper jaw.

Os Dentale – part of the lower jaw.
Most of these bones were unbroken and well-preserved with only the teeth missing. In some specimens the 'large molars' were still attached to the dental bone (*Pl.* 145:39 arrow, 39a arrow, 40 arrow). Many loose large molars were found as well.

The jaw bones have rows of 'sockles' connecting the teeth to the bone. The form of the sockles makes it possible to analyse the types of the missing teeth. On the rostral end of the excavated premaxillary and dental bones one finds three deep *alveola*-like sockles for the canine teeth (*Pl.* 145:33a arrow). Parallel to the lateral borders there are deep long oval depressions which constitute the sockles for the lateral row of conical molars (*Pl.* 145:38a arrow). Rows of shallow depressions medial to this lateral row belong to the round flat molars. Here the floor of the sockle is composed of a flat bone lamella which is perforated by numerous *foramina*, whereas the sides are grooved with numerous radiating lines (*Pl.* 145:32a arrow, 35 arrow). Sometimes there is only one extra long oval depression instead of two. This means that the jaw bone belongs to a fully developed adult fish in which one 'large molar' tooth replaced the two smaller molars (*Pl.* 145:34 arrow, 35 arrow). The two recent specimens of *Rhabdosargus haffara* which were at my disposal had two large molars but not an extra large molar – replacing the two smaller ones. All the excavated premaxillary jaw bones show the typical 'large swelling' described above.

Os Maxillare – (not bearing teeth) – part of the upper jaw. The 'large swelling' over the large molars of the premaxillary bone determines the shape of the overlying *os maxillare*. On the medial aspect of the body the maxillary bone forms a deep cavity which accepts the 'large swelling' (*Pl.* 145: 36 arrow, 37 arrow). By fitting into each other, the two bones act as one to exert extra grinding pressure.

Vertebrae – The general form of the vertebral bones of the Sparids conforms to the shape of the fish, namely, they are high and laterally compressed. The truncal vertebrae have a high base of the neural apophysis nearly forming a complete tube around the spinal cord. Most of the excavated vertebrae were badly damaged.

Summary of the Sparid finds:

In Timna, 75 bones and teeth were identified as Sparidae (79% of all fish remains). Twenty-eight specimens were found in the Temple (Site 200) (M.I.N. = 15) and the rest in two smelting camps: 28 in Site 2 (M.I.N. = 11) and 19 in Site 30 (M.I.N. = 17).

Skeletal elements identified as *Sparidae*:

Head	*Premaxillare*	17
	Dentale	10
	Detached teeth (large molars)	27
	Maxillare	1
Vertebral column		
	Vertebrae	20
	Total	75 (79% of all fish remains)

Measurements:

Premaxillary bones (17 were measurable, 7 left and 10 right). Total length variation 14.4–26mm.
Dentary bones (10 were measurable, 4 left and 10 right). Total length variation 20.0–31.8mm.
Detached teeth (large molars) (24 were measurable):

Length in mm	6	7	8	9	10	11
No. of teeth	7	3	7	3	3	1

Maxillary bone (one, left). The premaxillary process is broken (*Pl.* 145:36). Total length: 30.0mm.

Total length of the Timna Sparids:

This was estimated using the ratio of the length of recent Sparids and the length of their jaw bones. Estimated total length: 20–32cm.

Differentiation between *Sparus auratus* and *Rhabdosargus*:
The excavated specimens of premaxillary, maxillary and dental bones have been identified as jaw bones of a Sparid-type fish, as described above. These specimens are very similar to jaw bones of recent *Sparus auratus* as well as recent *Rhabdosargus haffara*. The same is true for the excavated vertebrae. Differentiation between *Sparus auratus* and *Rhabdosargus* was therefore not possible.

Zoogeographical distribution and economical significance:
The Gilthead seabream, *Sparus auratus*, is common throughout the Mediterranean and also present on the eastern Atlantic but never found in the Red Sea. Only in the last several years has the Gilthead invaded the Bitter Lakes through the Suez Canal. It feeds on molluscs and crustaceans along the coastal waters of the continental shelf down to a depth of about 60m., as well as in saline littoral lagoons. It is being caught today mainly in the autumn, when the fish leave the littoral lagoon of Bardawil, at the northern coast of the Sinai Peninsula on their way to the Mediterranean.

Silver Bream, *Rhabdosargus haffara*, is common in the Red Sea and in the Indian Ocean, and lives mostly in shallow waters over sandy ground, using its anterior teeth for digging in the sand, seeking molluscs and crustaceans (Porter, 1973).

(3) Nile catfish *Clarias gariepinus* (Burshell 1882), previously named *Clarias lazera*
Family: *Clariidae*. African catfish. *Teleostei*.

General Information and Identification:
Vertebrae: The vertebrae of *Clariidae* are readily identified by very deep, transverse oblong 'upper lateral pits' and 'lower lateral pits' on each side of the 'lateral pillar' (*Pl.* 146:41, 42). The surface of the lateral pillar is covered with fascicles of fine filaments running horizontally (*Pl.* 146:42).

Description of two excavated Vertebrae:
Two *clariid* vertebrae were excavated (2.1% of all fish remains; M.I.N. = 2), one truncal – 5.2mm. long, and one caudal – 7.4mm. long. Both were found in the smelting camp Site 30 (*Pl.* 146:41, 42).

Zoogeographical distribution and economical significance:
Clariidae are freshwater fishes living in swamps, ponds and slow running streams. A typical feature of the family is an auxilliary air breathing organ behind the gills which enables survival in poorly oxygenated waters. *Clarias gariepinus* is found in the Nile (Boulenger, 1907), in Syria (Teugels, 1982) and in Israel (Goren, 1983). It is common in the coastal area of Israel and also along the Jordan system (Goren, 1974) almost as far down as the Dead Sea. Fish remains, excavated in Tell Hesbon, on the Jordanian mountains east of the mouth of the Jordan, were defined as coming from *Clarias* (Lepiksaar, 1978). In the southern part of Jordan occurrence of *Clarias* is not known in recent times (Krupp, 1986).

Clarias was abundant in ancient times in the Nile and was depicted on numerous reliefs and wall paintings in ancient Egyptian tombs (Wild 1953). Bones of *Clarias* were also excavated in northern Sinai (by the Northern Sinai Expedition, Professor Eliezer Oren, Ben Gurion University, Beer Sheba, Israel), along the south of the 'Pelusiac Branch' of the Nile which comprised the 'Eastern Delta' that has since dried and turned to desert (Sneh, Weissbrod and Perath, 1975).

The dry desert-like biotop of Timna excludes the occurrence of *Clarias* in its vicinity. The import of catfish from the north, hundreds of kilometres away, is not very probable and no other indication of contact with the north has been found in the excavation of the Temple or the camps of Timna (see Chapter IV). It can therefore be assumed that the Timna catfish, *Clarias gariepinus*, was imported from Egypt, probably from the area of the 'Eastern Delta' of the Nile.

(4) The Timna shark-bead
Family: *Carcharhinidae* – Blue sharks. Order: *Selachii* – Sharks. Class: *Chondrychtes* – Cartilaginous fish.
General Information and Identification: (*Ill.* 38 and 39)
The shark family of *Carcharhinidae* belongs to the cartilaginous fishes. Their internal skeleton (skull and vertebrae) is cartilaginous and not bony. The cartilage of the vertebrae is partly calcified during the lifetime of the fish by deposition of lime salts (Apatit). This calcification makes it possible to find vertebrae of cartilaginous fish in archaeological excavations. The *Carcharhinid* vertebra has a biconcave centre built as from a calcified 'double cone'. The external surface of

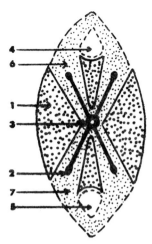

Illustration 38. Transverse section through middle of centre of Carcharhinid vertebra (See *Pl.* 146: 46,47)

1. Intermedial	2. Diagonal lamella
3. Double cone	4. Neural canal
5. Haemal canal	6. Neural arch (Cartilaginous)
7. Haemal arch (Cartilaginous)	

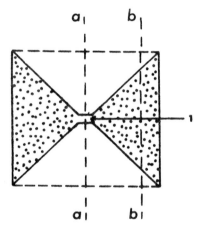

Illustration 39 Longitudinal section through the centre of the vertebra.

a–a: Transverse section cut midway between the anterior and posterior ends of the centre in the direction of the line a–a.
b–b: Transverse section taken near the end of the centre in the direction of the line b–b.

the centre is covered with a thin calcified layer. Longitudinal calcified plates, the 'radiating lamellae', lie in radial planes between the double cone and the surface of the centre (*Ill.* 38).

A transverse section was made in the midst between the anterior and the posterior end of the centre – in the direction depicted in *Pl.* 146:44 arrow; *Ill.* 39 a-a. Grounded 'thin sections' (transparent cuts) were prepared from the archaeological material and from recent shark vertebrae for comparison. A transverse section through the middle of the vertebra displays a 'calcification pattern' which characterises distinct families and in some cases distinct species of sharks. The 'radiating lamellae' are seen in the section in a form of a star with rays (*Ill.* 38). There are two different types of lamellae. The 'intermedialia' (Ridewood, 1921) appear in the sections as two vertical and two horizontal wedges which together form a

'Maltese cross' (*Pl.* 146:46). The 'diagonal lamellae' appear as oblique rays between the vertical and the horizontal wedges of the intermedialia (*Ill.* 38; *Pl.* 146:46) The 'double cone' appears as a small ring in the middle of the cross.

Description of the excavated specimen:
A bead made of a pierced shark vertebra (see also Kertesz, 1972, and above Chapter III, 14)

(a) Transverse section through the middle of the bead (*Pl.* 146:44 arrow and 46). The section was prepared as previously described and a characteristic pattern of rays was found (*Ill.* 39 a-a).

Intermedialia – The four wedges are completely calcified, with marks of circular structure. All the wedges appear massive, especially the lateral ones. The dorsal wedge is broader than the ventral one. Both have an indentation on their bases for the neural and haemal canals respectively (*Pl.* 146:46).

Diagonal lamellae – The four thin diagonal 'rays' are distinctly visible. At the distal end of two of them a typical swelling is recognizable. The length of the diagonal lamellae is more than half the length of the wedges of the intermedialia. The hole that was drilled through the middle of the bead almost completely obliterated the 'ring' of the double cone (*Pl.* 146:46).

(b) Transverse section made near the posterior end of the bead (*Pl.* 146:47; *Ill.* 39, b-b).

The pattern of the intermedialia is similar to the middle section (see above). The diagonal lamellae which are developed only near the middle of the vertebra do not show. The double cone appears as a broad ring (*Pl.* 146:47).

Description of the surface of the shark bead and measurements (*Pl.* 146:43, 44, 45):

The bead consists of a centrum of a shark vertebra. This centrum is typically biconcave. Unlike bony fish, there are no articular joints between the centra of adjoining vertebrae. There are four deep pits for the cartilaginous apophyses, two dorsal pits to fit the neural arch and two ventral ones, closer to each other, for the haemal arch (*Ill.* 38:6, 7; *Pl.* 146:44, 45). At the bottom of each of these pits there is a medial septum – the diagonal lamella (*Ill.* 38:3).

Measurements:
The length of the vertebral centre is 4.8mm. and the horizontal and vertical diametres of the contact face with the adjoining vertebrae are 9.2 and 9.3mm. respectively. A hole was drilled excentrically through the double cone with a diameter of 2.0mm. This bead was found in smelting camp Site 2 in Square D, pit 42.

Definition of the shark vertebra:
The calcification pattern of the sections of the bead as described above (*Pl.* 146:46, 47) allows the definition of the shark vertebra down to the *carcharhinid* family. In this family the wedges of the intermedialia appear compact on thin sections as they are completely calcified (Hasse 1879, 1882, 1885; Ridewood, 1921; Rauther, 1933). Other families of the same shark suborder (*Galeoidaea*) like *Isuridae* (Porbeagles), *Scy-*

liorhinidae (cat sharks) and others have also both types of radiating lamellae, but the wedges of the intermedialia are incompletely calcified, which is typical for these families. Among the numerous members of the *Carcharhinid* family, most have very short diagonals visible on thin sections of their vertebrae. Other species have the same long diagonals with end-swellings as were found in the Timna bead. Definition of the exact species of the Timna shark was not possible because there was not enough material available for comparison.

Conclusions

The 95 bones and teeth of fish excavated in Timna were identified as belonging to four different fish families: *Sciaenidae, Sparidae, Clariidae* and *Carcharhinidae*. These fish were caught in seawater (the Mediterranean and the Red Sea) and in freshwater. Most of the genera and species were seafish, whereas only two vertebrae belonged to freshwater fish (catfish) well known and appreciated in ancient Egypt, as evidenced by numerous wall paintings showing catfish-catching scenes along the Nile.

Sciaenids, Silurids, and possibly also the *Sparids,* were brought to Timna from far away, preserved by drying. A puzzling question is why the inhabitants of Timna imported part of the fish from far away instead of bringing all of them from the Red Sea, only 30km. away. One could explain this fact by referring to the work-pay relationships of the time. In Egypt workers were often partially paid with fish (Helck, 1964, 816) and if so they would prefer fish they were used to from their homeland. Egyptian workers would therefore possibly get fish imported from the Nile or the Mediterranean coast.

Mediterranean fish were certainly not imported to Timna during the 'Midianite' phase of the Temple and miners' camps as the 'Midianites' came from the Arabian peninsula, far from the Mediterranean.

Hanan Lernau

Notes

1. Because of the relatively small quantity of fish remains found in the Timna excavations, it was decided to publish in this study all the finds including those from the contemporary smelting camps, Site 2 and Site 30 in the Timna Valley.

2. As this paper was received on 27th August 1978, the methods of A. Morales and K. Rosenlund (1979) could not be taken into account in this study.

I am deeply grateful to Dr Johannes Lepiksaar, Goteborg, who patiently introduced me years ago into the difficult field of fish-osteology. Sincere thanks to Dr Kurat and Dr Niedermayer of the Mineralogical Department of the Natural History Museum of Vienna for the opportunity to use the methods of the Institute and to Mr Sverak who kindly prepared most of the grounded 'thin sections' of shark centres.

Bibliography

Adamicka, P. 1972. Funktionsanatomische Untersuchungen am Kopf von Akanthopterygiern (Pisces, Teleostei). Teil III Chrysophrys aurata. *Zool. Jb. Anat.,* 89.
Boulenger, G.A. 1907. The fishes of the Nile, in Anderson, *Zoology of Egypt,* London.
Dor, M. 1978. Personal communication.
FAO. 1972. *Species identification sheets for fishery purposes. Eastern Indian Ocean,* Volume III, Sciaenidae, p. 1 and 4, Rome.

Frost, G. Allen. 1927. A comparative study of the Otoliths of the Neopterygian Fishes. *Ann. Mag. Nat. Hist.* XX.

Goodrich, E.S. 1958. *Studies on the structure and development of vertebrates.* Vol. I, New York.

Goren, M. 1974. The freshwater fishes of Israel. *Journal of Zoology* 23.

—— 1983. *Freshwater fishes of Israel.* Biology and Taxonomy, Tel Aviv (Hebrew), 48–50.

Gregory, W.K. 1959. *Fish skulls.* Laurel, Florida.

Hasse, C. 1879, 1882, 1885. *Das natürliche System der Elasmobranchier auf Grundlage des Baues und der Entwicklung ihrer Wirbelsäule,* Jena.

Helck, W. 1964. *Materialien zur Wirtschaftsgeschichte des Neuen Reiches.* (Teil V).

Kertesz, G. 1972. *A study of beads based on the Timna Sanctuary beads and the cult of Hathor in Sinai and the Arabah.* M.A. Thesis, Tel Aviv University.

Klunzinger, C.B. 1870. *Synopsis der Fische des Roten Meeres.*

Krupp, F. 1986. Personal communication.

Lepiksaar, J. and Heinrich, D. 1977. Untersuchungen an Fischresten aus der frühmittelalterlichen Siedlung Haithabu, in *Ausgrabungen in Haitabu,* Bericht 10, Neumünster.

Lepiksaar, J. 1978. Personal communication.

Owen, R. 1840–45. *Odontography: Treatise on the comparative anatomy of the teeth,* Vol. I, II.

Porter, C. and Ben Tuvia, A. 1973. *Commercial fishes of the Gulf of Aqaba and the Gulf of Suez* (Hebrew).

Rauther, M. 1933. *Echte Fische, in Bronns, G.H. Klassen und Ordnungen des Tierreichs,* Bd. 6 Buch 2.

Ridewood, W.D. 1921. On the Calcification of the Vertebral Centra in Sharks and Rays. *Philos. Trans. Royal Society,* London, Series B, Vol. 210.

Schmidt, W. 1968. Vergleichende morphologische Studie über die Otolithen mariner Knochenfische. *Archiv für Fischereiwissenschaft.* Jahrgang 19.1, Beiheft 34.

Sneh, A., Weissbrod, T., and Perath, I. 1975. Evidence for an ancient Egyptian frontier canal. *American Scientist,* 63.

Teugels, G.G. 1982. Preliminary results of a morphological study of five African species of the subgenus Clarias (Pisces, Clariidae), *J. of Nat. History* 16, 439–64.

Wild, H. 1953. Le Tombeau de Ti II, Le Chapelle. *Memoirs de l'Institut Français d'Archéologie Orientale du Caire,* 65.

24 MAMMALIAN REMAINS[1]

Introduction

Research Methods

This report deals with mammalian bones found in the Timna Temple (Site 200). The earth in and around the Temple structures was carefully sieved and yielded a total of several thousand bones, mainly small fragments. This material was measured whenever possible and an attempt was made to define the species, age, sex and withers height of the animals, as well as the minimal number of individuals (Minimum Individual Number: M.I.N.). Measurements were determined in millimetres for mature bones only (with few exceptions – with a tolerance of 0.5mm) (Boessneck *et al.* 1971; von den Driesch, 1976).

Caprovine bones were differentiated into goats (*Capra hircus*) and sheep (*Ovis aries*), according to guidelines laid down by Boessneck, Müller and Teichert (1964). This distinction is difficult and in cases of doubt the term 'goats or sheep' was used. The age of the animals was estimated according to data for recent late maturing breeds (Silver 1969). Withers heights for sheep were calculated according to Haak

(1965, 60) and for goats according to Schramm (1967). The M.I.N. was estimated according to Chaplin (1971).

1 Results

3146 caprovine bones and fragments were identified. A large number of small bone splinters, about 0.5kg., could not be identified, but probably belonged to caprovines too.

Table 1 shows the distribution of the bones according to skeletal elements, species and M.I.N. Table 2 and 3 show the distribution of skull and limb bones respectively according to age groups and M.I.N.

1 Cranial skeleton (Tables 1–3)

35.4% of the excavated caprovine bones were parts of the cranial skeleton. 588 bone fragments were parts of brain skulls (*neuro-cranium*), horncores and mandibles. 525 were tooth fragments.

Supraoccipital bones (Squama occipitalis): 20 fragments of this bone, a part of the occipital bone, were found: five goats (M.I.N. = 5); four sheep (M.I.N. = 4) and 11 'goats or sheep' (M.I.N. = 11). Sixteen (80%) belong to young animals, six to infantiles and ten to juveniles (M.I.N. = 16) (Table 2).

The flat occipital bone of the young animal is much thinner than the adult bone and the sutures around it are still discernible. Supraoccipital bones of goats and sheep of different age groups are shown in *Pl.* 148:10–14.

Petromastoid bones (*perioticum*): 47 specimens of this bone were identified (M.I.N. = 25). The hard petrous bone which surrounds the inner ear is usually preserved intact; the *mastoid* bone which is fused to it is often broken (*Pl.* 149:15–16).

Sixteen bones (34%) probably belong to goats (M.I.N. = 9); 21 (44.7%) to sheep (M.I.N. = 11) and ten to 'goats or sheep' (M.I.N. = 5).

Horncores (*Processus cornualis*): 82 fragments of caprovine horncores were found (none measurable), of which nine fragments belong to juveniles, 26 to infantiles (74.3%). Nineteen horncores belong to goats (eight adults and eleven infantiles). Of the adult horncores, seven are male and one female. Four horncores belong to sheep (one adult male and three juveniles). Fifteen fragments of infantile cores and 44 additional splinters belong to 'goats or sheep'.

The horncore of the goat may be identified by the extension of the frontal sinus almost to the tip of the core, by the screw-like twist of the male core and by the scimital shape of the female core (*Pl.* 147:4–5). The horncore of the sheep (*Pl.* 147:6) is identified by its triangular cross-section. Horncores of different age groups are shown in *Pl.* 148:7–9.

Mandibles: 105 fragments of caprovine mandibular bones were found. None is measurable. Differentiation into goats and sheep is impossible. There are 36 fragments of the articular or the muscular processes of the mandibular ramus (M.I.N. = 28) (*Pl.* 149:17–22). Of these, 19 fragments (52.8%) belong to infantiles (M.I.N. = 16) and 17 (47.2%) to juveniles or

Table 1: Caprovines: Distribution of the bones according to skeleton elements

	Caprovines			Goats		Sheep	
	n	%	MIN	n	MIN	n	MIN
Horncores	82	2.6	–	19	–	4	–
Neuro-cranium	401	12.8	25	16	9	21	11
Mandible	105	3.3	28	–	–	–	–
Teeth	525	16.7	–	–	–	–	–
Atlas	16	0.5	–	–	–	–	–
Axis	14	0.5	–	–	–	–	–
Cervical Vertebrae	40	1.3	–	–	–	–	–
Thoracic Vertebrae	105	3.3	–	–	–	–	–
Lumbal Vertebrae	161	5.1	–	–	–	–	–
Ribs	383	12.2	–	–	–	–	–
Scapula	85	2.7	22	–	–	–	–
Humerus	153	4.9	46	5	5	3	3
Radius	42	1.3	19	8	5	6	4
Ulna	7	0.2	7	–	–	–	–
Carpus	47	1.5	9	–	–	–	–
Os coxae	98	3.1	15	–	–	–	–
Femur	123	3.9	16	1	1	2	2
Patella	14	0.5	8	–	–	–	–
Tibia	110	3.5	16	4	4	1	1
Astragalus	30	1.0	18	9	6	3	2
Calcaneus	12	0.4	8	6	3	4	3
Tarsus, remaining	16	0.5	6	–	–	–	–
Metapodia (Metacarpus and Metatarsus)	409	13.0	–	39	–	43	–
Phalanx 1	88	2.8	–	10	–	8	–
Phalanx 2	44	1.4	–	12	–	7	–
Phalanx 3	36	1.1	–	13	–	11	–
	3146		46	142	9	113	11
	100%			4.5%		3.6%	

Table 2: Goats or Sheep – Age grouping of skull bones and M.I.N

	Age Group	Estimated Age (Years)	L	R	L + R	M.I.N.
1. Horncore	infantile	less than 1/2	–	–	26	74.3%
	juvenile/adult	more than 1/2	–	–	9	–
2. Squama occipitalis Bone wall:						
very thin	infantile	less than 1/2	–	–	6	30% 6
thin	juvenile	more than 1/2	–	–	10	50% 10
thick	sub-adult/adult	more than 2	–	–	4	4
3. Mandible	infantile	less than 1/2	7	12	19 52	8% 16
	juvenile/adult	more than 1/2	5	12	17	12

adults (M.I.N. = 12) (Table 2). Anterior parts of mandibulars of different age groups are shown on Pl. 150:23–30A.

Teeth: 525 teeth of goats or sheep were found. They are mostly broken. No permanent teeth are measurable, but 35 can be identified. The majority of the teeth belong to young animals. There are many deciduals and unerrupted premolars.

2 Post cranial skeleton (Table l)
The post cranial part of the skeleton accounts for 64.6% (2033) of the excavated caprovine bones. A high percentage belongs to young animals. The differentiation between goats and sheep is thus hindered because the characteristic structural differences in these bones mostly develop in adult animals.

Scapular bones (Shoulder blades): 85 fragments of scapular bones were identified (M.I.N. = 22). None is measurable. Twenty fragments can be age-classified, and 12 of them belong to infantile animals less than six months old (Table 3. 1).

Humerus: 153 fragments of the humerus were identified (M.I.N. = 46). Five belong to goats (M.I.N. = 5) and three to sheep (M.I.N. = 3). The widths of the distal end of three goat humeri were measured and found to be 34.5, 34.5 and 29.5 mm. The one humerus of a sheep was 34.0mm.
The age of 66 humeri was determined (M.I.N. = 46) and 29 of them (43.9%) belong to infantiles or juveniles less than nine months old (Table 3. 2).

Radius: 42 fragments of radial bones were found

Table 3: Goats and Sheep – Age grouping of limb bones and Minimum Individual Numbers (M.I.N.)

State of fusing of epiphysis	Size	Age Group	Estimated Age	Caprovines		Goats		Sheep	
				n	M.I.N	n	M.I.N	n	M.I.N
1. Scapula									
Tuber not fused		inf.	less than 1/2	12	9	–	–	–	–
Tuber just fused		juv.	1/2 – 3/4	3	3	–	–	–	–
Tuber fused		juv./ad.	more than 3/4	5	4	–	–	–	–
				20					
2. Humerus									
Dist. not fused, prox. ?		inf./juv.	less than 3/4	19	12	–	–	–	–
Dist. just fused		juv.	3/4	10	8	2	2	1	1
Dist. fused, prox. ?		juv./ad.	more than 3/4	32	21	2	2	1	1
Prox. just fused		subadult	3 1/2	3	3	1	1	–	–
Prox. fused		adult	more than 3 1/2	2	2	–	–	–	–
				66					
3. Radius									
Prox. not fused, dist. ?		inf.	less than 3/4	3	3	1	1	–	–
Prox. fused, dist. ?		juv./ad.	3/4 – 3 1/2	16	16	4	4	4	4
Dist. not fused, prox. ?		juv./ad.	less than 3 1/2	6	–	3	–	1	–
Dist. just fused		subadult	3 1/2	1	–	–	–	–	–
Dist. fused		adult	more than 3 1/2	1	–	–	–	1	–
				27					
4. Femur									
Prox. not fused, dist. ?	small	inf.	less than 1	15	5	–	–	–	–
Prox. not fused, dist. ?	large	juv./subad.	1 – 2 1/2	15	8	–	–	–	–
Prox. fused		adult	more than 3 1/2	3	3	1	1	2	2
Dist. not fused, prox. ?		juv./ad.	less than 3	7	–	–	–	–	–
Dist. fused		adult	more than 3	1	–	–	–	–	–
				41					
5. Tibia									
Dist. not fused, prox. ?		inf./juv.	less than 1 1/2	11	8	–	–	–	–
Dist. fused, prox. ?		subad./ad.	more than 1 1/2	7	7	–	–	–	–
Prox. not fused, dist. ?		inf./subad.	less than 3 1/2	23	–	–	–	–	–
Prox. fused		adult	more than 3 1/2	1	1	–	–	–	–
				42					
6. Calcaneus									
Tuber not fused		inf./juv.	less than 2 1/2	4	3	3	2	1	1
Tuber fused		adult	more than 2 1/2	1	1	–	–	1	1
				5					
7. Metapodia (Metacarpus and Metatarsus)									
Dist. not fused		inf./juv.	less than 2	82	–	31	–	25	–
Dist. fused		subad./ad.	more than 2	26	–	8	–	18	–
				108					
8. Phalanx 1									
Prox. not fused		inf./juv.	less than 1 1/2	62	–	6	–	4	–
Prox. fused		juv./ad.	more than 1 1/2	17	–	4	–	4	–
				79					
9. Phalanx 2									
Prox. not fused		inf./juv.	less than 1 1/2	21	–	4	–	2	–
Prox. fused		juv./ad.	more than 1 1/2	15	–	8	–	5	–
				36					

prox. = proximal end of the bone
dist. = distal end of the bone
inf. = infantile
juv. = juvenile
subad. = subadult
ad. = adult
? = broken up – therefore unknown

Table 4: Goats and Sheep (measurements in mm)

Astragalus	Goats							n	max.	min.	mean	Sheep			n	max.	min.
1. Greatest length of the lateral half	31.5	30.5	30.0	–	28.5	27.5	27.0	6	31.5	7.0	29.2	29.5	31.5	–	2	31.5	29.5
2. Greatest length of the medial half	28.5	29.0	30.0	26.5	26.5	25.5	25.0	7	30.0	25.0	27.3	28.0	28.0	28.0	3	28.0	28.0
3. Depth of the lateral half	–	–	17.0	–	15.0	14.5	14.5	4	17.0	14.5	15.3	17.0	16.5	16.0	3	17.0	16.0
4. Depth of the medial half	16.5	16.5	–	17.5	14.5	16.0	15.5	6	17.5	15.5	16.0	18.0	17.5	17.5	3	18.0	17.5
5. Breadth of the distal end (head)	–	–	21.0	–	17.5	18.5	18.0	4	21.0	17.5	18.8	18.5	–	18.5	2	18.5	–
6. Sex	m	o	o	m	f	m	f					f	m	m			
7. Withers height (cm) (Factor 22.68) (Teichert 1974)												66.9	66.9	(70)			

Table 5: Goats and Sheep (Measurement in mm)

Metacarpus	Capra Goat					Ovis Sheep						
	Site 200	Site 200	Site 200	Spain[3] Bronze Age	Recent Capra ibex nubiana	Site 200	Site 200	Site 200	Jericho[4] Middle Bronze Age	Jericho[5] Middle Bronze Age	Germany[6] Recent Breed Heid-schnucke	Merino
Field No.[2]	310	167	303			118	310	252				
1. Greatest length	116.0	–	107.0	102–117	123.5	145.0	–	–	139–153	132–134	119–139.5	135–164
1a. Greatest length without distal ephphysis	99.0	99.0	89.0	–	–	128	126	–	–	–	–	–
2. Greatest breadth of proximal end	27.0	25.5	25.5	20.5–27.5	27.0	25.5	26.5	22.5	20–22	22.8–23.6	20.8–24.2	27.0–32.9
3. Smallest breadth of diaphysis	17.5	16.0	15.5	13–17	16.5	15.5	14.5	–	12–14	13.6–14.3	11.3–14.1	15.3–20.0
4. Greatest breadth of distal end	31.5	–	29.0	24–32.5	30.0	–	–	–	24.5–25.0	24.8	22.5–27.8	28.0–37.1
5. Dorso-volar diameter of peripheral trochlea section	10.0	–	10.25	–	–	11.3	–	–	–	–	–	–
6. Dorso-volar diameter of verticillus	17.5	–	17.5	–	–	17.0	–	–	–	–	–	–
7. Index $\frac{5 \cdot 100^2}{6}$	57.1	–	58.5	–	–	66.5	–	–	–	–	–	–
8. Sex	male	male	male	–	female	female	female	–	–	–	female	female
9. State of fusing[1] distal diaphysis	+	–	–	–	–	–	–	–	+	+	+	+
10. Age	ad.	juv.	juv.	ad.	–	juv.	juv.	–	adult	adult	adult	adult
11. Withers height (cm.) Factor 4.85 (Haak)						(70.3)	(70.0)	–	–	64.5	62.6	72.8
Factor 5.75 (Schramm)	66.7	66.0	61.5	60–65								
12. Pl.147: No.	1	–	–	–	–	2	–	–	–	–	–	–

[1] + Fused – Not fused
[2] Boessneck 1969
[3] von den Driesch 1972
[4] Nobis, 1968
[5] Clutton-Brook, 1971
[6] Haak, 1965

(M.I.N. = 19), eight belonging to goats (M.I.N. = 5) and six to sheep (M.I.N. = 4).

Carpal bones: 47 carpal bones were found (M.I.N. = 9), and are identifiable as *os carpi radiale*, *os carpi intermedium*, *os carpi ulnare*, *os carpi secundum-tertium* and *os carpi quartum*.

Ox coxae: 98 fragments of caprovine *os coxae* were found (M.I.N. = 15), two belonging to sheep. There are six female and two male bones.

Femur: 123 fragments of the femoral bones (M.I.N. = 16) were found. None was measurable. One fragment belongs to an adult goat and two belong to adult sheep. Nineteen loose epiphyses of *femur* heads (*caput femuris*) were found, together with the fitting proximal ends of femoral shafts of infantile-juvenile animals. Pl. 150:31–35 shows loose epiphyses of different age groups, together with some fitting shafts. Fifteen fragments (36.6%) out of the 41 which could be age-analysed belong to infantiles younger than one year (Table 3. 4).

Tibia and Patella: 110 fragments of caprovine tibial bones (M.I.N. = 16) and 14 *patellae* (M.I.N. = 8) were found. The widths of two adult proximal tibial ends measure 40.1 and 42.0mm., and of one distal end 26.0mm. There were 18 loose epiphyses, nine proxi-

Table 6: Goats and Sheep (Measurement in mm)

Metatarsus	Capra Goat				Ovis Sheep			
	Site 200	Site 200	Recent Capra ibex nubiana	Site 200	Jericho[2] Middle Bronze Age	Jericho[3] Middle Bronze Age	Germany[4] Recent Breed Heidschnucke	Merino
Field No.	211	310		303				
1. Greatest length	–	–	134.0	158.0	139–153	131.3–136	127–150	140.3–180
2. Greatest breadth	23.5	20.0	22.5	21.0	20–22	17.5	18.2–21.3	23.3–29.0
3. Smallest breadth of the diaphysis	–	–	13.5	12.0	12–14	10.5–12.1	10–12.5	13.2–16.2
4. Greatest breadth of the distal end	–	–	26.5	26.0	24.5–25.0	21.9–25.4	25.3–28.8	26.2–33.1
5. Dorso-volar diameter of the peripheral trochlea section	–	–	10.5	10.8	–	–	–	–
6. Dorso-volar diameter of the verticillus	–	–	16.5	16.5	–	–	–	–
7. Index $\frac{5 \cdot 100}{6}$ (Boessneck 1969)	–	–	(63.6)	(65.4)	–	–	–	–
8. Sex	–	–	female	female	–	–	female	female
9. State of fusing of the distal diaphysis[1]	–	–	+	–	+	+	+	+
10. Withers height (cm.) Factor 4.55 (Haak)	–	–	–	71.8	66.4	60.8	63.3 mean adult	3.3 mean adult
11. Age	–	–	adult	juvenile	–	–		
12. Plate 147: No.	–	–	–	3	–	–	–	–

[1] + fused – not fused   [3] Clutton-Brock, 1971
[2] Nobis 1968   [4] Haak, 1965

Table 7: Age-groups of Caprovines according to state of fusing of the distal epiphyses of the metapodia

Metapodia / Age	Infantiles	Juveniles distal epiphyses not fused less than 2 years old	Subadult/Adult distal epiphysis fused more than 2 years old	Total
Goats	–	31	8	39
Sheep	–	25	18	43
Goats or Sheep	10	16	–	26
Caprovines	10	72	26	108

mal and nine distal. The distal widths of five of these range between 24.5 and 28.5mm. Two fitting diaphyses were found as well. Age grouping for 42 *tibiae* shows that 26.2% belong to infantile and juvenile animals (younger than 18 months) (Table 3. 5).

Tarsal bones: 58 tarsal bones were found. They included *astragali, calcanei*, centrotarsal bones (*scapho-cuboid*) and the second-third tarsal bone.

Os Astragalus: 30 *astragali* were found (M.I.N. = 18), nine belonging to goats (M.I.N. = 6), three to sheep (M.I.N. = 2) and 18 to 'goats and sheep' (M.I.N. = 10). Ten *astragali* are measurable. The lateral length variation of seven goat astragali is 31.5 to 27.0mm. (mean = 29.2mm.). Three sheep *astragali* have a lateral length of 29.5mm. Their wither height was estimated at 66.9cm. according to Teichert (1974) (Table 4).

Os Calcaneus: 12 *calcanei* were found, six belonging to goats (M.I.N. = 3), four to sheep (M.I.N. = 3) and

two to 'goats or sheep' (M.I.N. = 2). None is measurable. Four bones belong to young animals (less than 2.5 years old) and one belongs to an adult (Table 3, 6). In addition, 11 centro-tarsal bones (M.I.N. = 6) and five second-third tarsal bones (M.I.N. = 5) could be identified.

Metapodia (metacarpal and metatarsal bones): 409 fragments of *metapodia* were found. 175 fragments include a proximal end, a 169 a distal end, and 65 only the corpus. The bones are usually split along their longitudinal axis, and rarely transversally. This was presumably done in order to get the marrow. Sixty fragments are not split and some of their proximal or distal surfaces are measurable. Only two of the excavated *metapodia* show evidence of fire.

Age differentiation is possible for 108 distal epiphysis. Some are loose and some are fused. 72 (66.7%) belong to juveniles (younger than two years); 26 (24.1%) to sub-adult adults (older than two years) and ten (9.3%) to infantiles. In this group of bones, 39 belong to goats, 43 to sheep and 26 to 'goats or sheep' (Table 3. 7) (Table 7).

Measurements of 48 distal *epiphysis* allow differentiation between goats and sheep. The results confirm the found morphological characteristics of goats and of sheep. Thus twelve *metacarpi* and eight *metatarsi* belong to goats, and twelve *metacarpi* and 16 *metatarsi* to sheep (Table 8).

Measurements of individual goat *metapodia* with estimations of the withers height:

(a) A complete right *metacarpus* of a subadult-adult male goat more than two years old. The measured

Table 8: Differentiation of distal ends of metapodia in goats and sheep bones by use of index* (Measurements in mm.)

Distal end	Goat n	Variation of Index Max.	Min.	Mean	Sheep n	Variation of Index Max.	Min.	Mean	Total
Metacarpus	12	63.6	51.7	58.9	12	74.2	64.7	67.8	24
Metatarsus	8	62.1	58.0	59.4	16	75.0	62.1	67.5	24
									48

* See Tables 5 and 6, Index (No. 7)

length is 116mm. and the estimated withers height is 66.7cm. (*Pl.* 147:1) (Table 5).
(b) A left *metacarpus* of a juvenile male goat (less than two years old) without the distal *epiphysis*. The measured length is 99.0mm. and the estimated withers height is 66cm. (Table 5).
(c) A right *metacarpal* shaft of a juvenile male goat with a fitting, loose distal *epiphysis*. The measured length is 107.0mm. and the estimated withers height is 61.5cm. (Table 5).
(d) Two proximal ends of goat *metatarsals* have a proximal breadth of 22.5mm. and 20.0mm. (Table 6). It may be added that in a recent female *Capra ibex nubiana* a *metacarpus* has a length of 123.5mm. and a *metatarsus* of 134mm.

Measurements of individual sheep *metapodia* with estimates of the withers heights:

(a) A left *metacarpal* fragment of a juvenile female sheep without an *epiphysis*. The measured length including the fitting loose distal *epiphysis* is 145.0mm. The estimated withers height is 70.3cm. (*Pl.* 147:2) (Table 5).
(b) A left *metacarpal* fragment of a juvenile female sheep with no distal *epiphysis*. The measured length is 126.0mm. The estimated withers height is 70.0cm. (Table 5).
(c) A left *metatarsus* of a juvenile female sheep without *epiphysis*. The measured length including the loose fitting distal *epiphysis* is 158.0mm. The estimated withers height is 71.8cm. (*Pl.* 147:3) (Table 6).

Phalanges
168 caprovine *phalangi* were found.
Phalanx 1: 88 first *phalangi* were found, including 27 loose proximal *epiphyses*. Ten belong to goats and eight to sheep. Age grouping was possible for 79 bones and 62 (78.5%) were found to belong to infantiles-juveniles (younger than 18 months), six being goats, four sheep and 52 'goats or sheep' (Table 3. 8). The measured peripheral length of three first *phalangi* of goats is 40.0–37.0mm., and for three sheep bones 36.0 – 34.5mm.
Phalanx 2: 44 second *phalangi* were found (including five loose proximal *epiphyses*), 12 belong to goats, seven to sheep and the rest to 'goats or sheep'. Age grouping was possible for 36 bones (Table 3. 9) and 21 (58.3%) were found to belong to infantiles-juvenile animals (younger than 18 months). Measurements of six second *phalangi* of goats show a peripheral length of 24.5–21.5mm. Four second *phalangi* of sheep have a measured length of 22.5–21.9mm.
Phalanx 3: 36 third *phalangi* were found, 13 belonging

to goats, 11 to sheep and 12 to 'goats or sheep'. The measured diagonal length of the sole of eight goat *phalangi* is 33.0–26.0mm. (Mean = 29.8mm.), and for five sheep *phalangi* 30.5–25.5mm (Mean = 28.0mm.).

2 Discussion

The state of preservation of the excavated bones
The bones excavated in the Timna Temple are mostly broken, yet they have been well preserved by the dry desert climate. This is not only true for hard teeth, but also for thin and fragile skull and limb bones of young animals which were found in large quantities (*Pl.* 148–150). Visitors to the Temple apparently crushed the fine bones on the floor into a multitude of splinters.

Distribution of the bones according to skeletal elements
The excavated material (3146 bone fragments) includes practically all the skeletal elements of goats and sheep (Table 1). When grouped into body sections, their distribution is almost equal: 35.4% of the bones belonging to the head, 34.8% to the forequarters and 29.8% to the hind-quarters. In several instances adjoining bones were found together. Thus the three *phalanges* were found at the same spot, as well as bones with fitting loose *epiphyses*. The same is true for elements of wrists and ankles. It therefore seems plausible that the bones were found at their primary place of deposition and that they were not brought into the Temple secondarily.

Distribution of the bones according to species
Out of over 3000 caprovine bones (M.I.N. = 46) only 142 can be classified as goats and 113 as sheep, with only an insignificant difference in the M.I.N. (9 and 11 respectively). When the distribution of different adult skeletal elements is considered, it is again evident that the numbers of goats and sheep which may be identified are similar.

Horncores of young animals belong to goats or sheep, but otherwise estimation of the numbers of kids and lambkins is not possible.

Distribution of the bones according to sexes
Some adult skeletal elements allow differentiation between males and females. This is true for horncores, *ossa coxae* and *astragali* (see corresponding descriptions). Bones of young goats and sheep have no sex characteristics, but it may be safely assumed that most young bones belong to male kids and lambkins, for the simple reason that young female caprovines were reared for breeding purposes and not sacrificed.

Distribution of bones according to age groups
About 40–50% of the excavated bones belong to very young animals. This can be deduced from the small size and other features of horncores, occipital bones and mandibles, and also from the state of fusion of *epiphyses* of various bones such as the *scapula*, *radius*, etc. (see description of corresponding bones). There are relatively few adult bones in the excavated material. It should also be remembered that bones of younger animals are more likely to be destroyed and lost. Therefore it seems that in the Timna Temple the sacrificed animals were predominantly very young.

3 Profile of the Timna Caprovines
The excavated bones give us a fair clue to the nature of the Timna goats and sheep.

The Timna goat: The Timna male goat has screw-shaped horns and an estimated withers height of 66.7–61.5cm. The female had scimitar-shaped horns and was somewhat lower. These goats have been bred in south-west Asia since the Early Bronze Age.

It may be interesting to compare our data with those of goats from the Bronze Age in Spain given by v.d. Driesch (1972). Based on measurements of meta-carpal bones, she estimated their withers height to be 60–65cm. (Table 5).

The goats of Timna are fairly large considering the fact that their withers height was mainly estimated by measurements of young bones.

The Timna sheep: The Timna male sheep was horned, of the ammonshorn-bearing type. There is no information about the female being horned or not. Their withers height is estimated at 70.3–71.8cm.. This may be compared with sheep described elsewhere: sheep from Jericho from the Middle Bronze Age had an estimated withers height of 64.5cm. (Clutton-Brock, 1971). Female Bronze Age sheep from Spain had an estimated withers height of 60–65cm. (v.d. Driesch, 1972). A recent primitive breed from Germany, the Heidschnucke, has a withers height of 62.6cm. (Haak, 1965)(Table 5 and 6).

It may be concluded that the Timna sheep was a medium-sized breed. It probably had long, slender legs characterising desert sheep that have to walk long distances to find their food (e.d. Driesch, 1972).

4 Conclusions
A large number of bone fragments was found in the small area of the Timna Temple, measuring about 8 x 10m. This concentration of bones must be the result of the activity of man. The excavated bones come from an approximately equal number of goats and sheep, mostly from younger ones.

The bone material includes bones of all edible parts of the body, as well as bones of feet which are not edible. This fact shows us that the animals were brought alive and slaughtered in or near the Temple, otherwise the feet would not have been transported there. The carcasses as a whole were cut up in the Temple or nearby. The equal distribution of bones from different parts of the body also serves to show that no special parts of the animals were brought to the Temple as offerings. The animals seem to have been eaten at the Temple site, as is shown by the remnants of fire places found on the floors, especially in Strata II and V, and by the splitting of the metapodia in order to get out the marrow. The animals were not sacrificed as 'burned offerings' since there is almost no evidence of fire on the bones.

There are several possibilities of how the sacrificial rite took place. In the excavations, a group of Standing Stones was found, belonging to Stratum II, the Midianite phase. Worshippers would sprinkle these stones with the blood of the sacrificed animals which they later consumed on the spot. Another possibility is that parts of the sacrificed animals, probably heads and feet, would be set as offerings in front of the Standing Stones or on the 'offering bench' along Wall 2. The fact that the fine, brittle skull bones of the kids and lambkins have been preserved to date, would suggest that they were set in front of the Standing Stones raw, unboiled. Following the sacrificial rite, the shrine must have been guarded for some time to avoid removal of the remains by jackals, hyenas, scavenging birds and the like. Otherwise, in the desert, the bones would have had to be buried deep in the sand in order not to be scattered.

When the Egyptians left the area, in the middle of the 12th Century BC, the Midianites remained and turned the Egyptian sanctuary into a Semitic desert shrine. They apparently changed the rite, since Hathor is not known to have received animal sacrifices in Egypt, and offered in this tent-covered sanctuary the best of their possessions, namely the young animals of their herds. Similar offerings have been reported from Serabit el Khadem, the Hathor sanctuary in southern Sinai, where heads and other body parts of geese and oxen were sacrificed (Schafik Allam, 1963).

Hanan Lernau

Notes
1. This paper was received on 27 August 1978.
2. The Field Numbers in this chapter are in fact the find box numbers, since the individual bones could not be marked with an object number. The significance of the find circumstances of the bones is, of course, mainly in the loci and strata, which are indicated by the Field (= Box) number. (See list of Find Boxes.)

Bibliography
Boessneck, J. 1969. Osteological Differences between Sheep (Ovis aries Linné) and Goat (Capra hircus Linné), in Brothwell, D. and Higgs, E. (eds.) *Science in Archaeology*, 2nd ed., London.
—— 1971. Die Tierknochenfunde aus dem Oppidum von Manching, in Kramer, W.(ed.) *Die Ausgrabungen in Manching 6*. Wiesbaden.
Boessneck, J., Müller, H.H. and Teichert, M. 1964. Osteologische Unterscheidungsmerkmale zwischen Schaf (Ovis aries L.) und Ziege (Capra hircus L.), *Kühn-Archiv 78*, 1–129.
Chaplin, R.E. 1971. *The Study of Animal Bones from Archaeological Sites*, London.
Clutton-Brock, J. 1971. The Primary Food Animals of the Jericho Tell from the Proto-Neolithic to the Byzantine Period, *Levant 3*, 41–55.
Von den Driesch, A. 1972. Osteo-Archaeologische Untersuchungen auf der Iberischen Halbinsel, *Studien über frühe Tierknochenfunde von der Iberischen Halbinsel 3*, Munich.

—— 1976. A Guide to the Measurement of Animal Bones from Archaeological Sites, *Peabody Museum Bulletin* No. 1.

Haak, D. 1965. *Metrische Untersuchungen an Rohrenknochen bei deutschen Merinoschafen und Heidschnucken*, Dissertation, Munich.

Schafik, A. 1963. Beitrage zum Hathor-Kult, *Münchner, Aegyptologische Studien* 4, Berlin.

Schramm, Z. 1967. Long Bones and Height in Withers of Goat, *Roczniki Wyzszej Szkoly Rolniczej w Poznaniu* 36, 89–105, (Polish with English summary).

Silver, I.A. 1969. The Ageing of Domestic Animals, in Brothwell, D. and Higgs, E. (eds.) *Science in Archaeology*, 2nd ed., London, 283–302.

C. STRATUM I

25 THE POTTERY OF THE ROMAN PERIOD

Introduction

In the following pages an effort will be made to establish the identity of the 'Roman Pottery' found in the Temple of Timna.[1]

Since the overall number of sherds from the late periods was small (about 40 fragments), all those sherds with sufficiently distinct features, were studied in detail. The rest, which were mainly bodysherds, may be assigned, owing to their fabric and processing, to the same period as those treated in the following pages, with possibly five or six Byzantine pieces.

1 The Finds and their Analysis

Rom. Pot. Cat. No. 1. *Fig.* 87:1, Field No. 552/3, Locus 111, Square TT 58.
Rim and body fragment of thick-walled flat bowl. Clay (10YR 7/4): mediocre levigation, pale yellow-brown in section. Firing mediocre. Slip (10YR 7/3): same as clay with pinkish patches. Possible (see comparisons) with flat, strong cut base.

Comparisons: Alayiq (Kelso and Baramki, 1955, 27, Pl. 23: A79). Although the parallel is not exact, our bowl pertains to Type 12 or 13 which are both 'Roman'; Beth Zur: (Sellers *et al.*, Fig. 18: 18), pottery cache in stratum III (640–587 BC); Mampsis Necropolis: (Negev, 1977, Fig. 10: 67), Phase II; Qumran III: (de Vaux, 1956, Fig. 3: 2), Period Ib (50–31 BC); Tel Mevorakh: (Stern, 1978, Fig. 4: 9), Persian.

This bowl is difficult to date exactly without having recourse to its context. The comparisons point to its Persian origin and to its occurrence as late as the second century BC.

Rom. Pot. Cat. No. 2. *Fig.* 87:2, *Pl.* 151:3, Field No. 226/13, Locus 104, Square D-E-F 2-5.
Thin-walled carinated bowl. Clay (10R 5/6): well levigated red ware. Excellent firing. Straight rounded rim, slightly inturned. Strongly pronounced carination immediately below rim. Ring base.
Comparisons: Hammond, 1962: The rim section of Form I2(c): 1, 3, 6 has similarities with that of our bowl. According to Hammond (idem 175), this form has a life span between the end of the Hellenistic period and some time in the Roman. Our specimen has a very pronounced, practically rectangular 'step' that connects between rim and body. This step is typical of many examples of Form I3 which seems to

have a longer life span than I2 in definitely Roman times (idem); Mampsis Necropolis: (Negev, 1977, Fig. 66, 48). Phase II – Last quarter of the first century AD to 150/200 AD; Petra: (Parr, 1970, Fig. 5: 66–69, 71). These bowls pertain to Phases X-XI which belong to the end of the independent Nabataean Kingdom and include many later intrusions (pp. 364–6). All these have certain similarities to our bowl. A close resemblance exists to Fig. 7: 109 that belongs to Phase XIII, which is Trajanic (p. 366). Cf. also Number 99; Petra Theatre: (Hammond, 1965, Pl. LIX). There is no exact parallel to our bowl, but the typical stepped transition from rim to body is a common feature (nos. 11, 16, etc.). These bowls belong to the mid-second century AD (p. 65); Sbaita: (Crowfoot, 1936, Pl. IV. 6), first to second century AD

This is one of the most common Nabataean and Early Roman bowl shapes. Because of its abundance and the manifold variations, only more closely fitting comparisons were cited. The life span of our subtype, as shown so far, is well within the second century AD In this respect the evidence from Mampsis and the Colonade Street in Petra (Parr, ibid.) is the most relevant.

Rom. Pot. Cat. No. 3. *Fig.* 87:3, Field No. 120/7, Loci 101, 102, Square A-B-C 17–18.
Rim fragment of small and deep bowl or cup. Clay (5YR 6/6): dark grey core, small black grits. Firing: mediocre to well done. Slip: brownish-yellow, probably only on rim.

Comparisons: Ain el-Ghuweir: (Bar-Adon, 1971, Fig. 9: 16). 50 BC–68 AD; Dibon: (Tushingham, 1972, Fig. 3: 1, 3, 4, 35, 36), Nabataean–10 AD Wares A(3) and B; Jerusalem (Dominus Flevit): (Bagatti and Milik, 1958, Fig. 27: 15. Grotta 151). Herodian; Hammond (1962, 178–9, Class II, i(c)), first century AD and possibly later (better wares than our example); Nessana: (Colt, 1965, Shape 30), Hellenistic (different ware); Petra: (Hammond, 1965, Nos. 58–78, 80–82, etc.), Nabataean. Form-wise our family of vessels, but ware is better. No. 73 has similar slip; Petra: (Parr, 1970, Fig. 6: 97), Phase XII; (Murray and Ellis, 1940, Pl. VIII: 38–67, etc. Pl. XXXI), South Cave complex. Much better ware than our example; (Hammond, 1965, Pl. LIX: 18–19), Phase Ic. Better quality ware than our example; Jerusalem, Temple Mount: (Mazar, 1971, Fig. 18: 6, 7), Herodian.

Our fragment is not sufficient to reconstruct with certainty the exact shape of this vessel. As to its chronology, Hammond correctly pointed out that this type of cup was '. . . in a limited number of variations common throughout Syro-Palestine from the (Late ?) Hellenistic period . . . even into the Early Islamic Period' (Hammond, 1973, 31). This is, of course, correct only so far as the form is concerned and we must be very careful not to assign this vessel according to this criterium alone. These observations are not helpful in establishing a correct sequence by basing ourselves upon the published material as it stands. Still, we may point to Mampsis and Petra (the Theatre, the Colonade (Parr, 1970) and Street, as evidence of a continued production of the Nabataean cup after the Roman conquest. Even at Dibon, Tushingham concludes that the 'plain non egg shell ware type B also . . . continued in use' (ibid. 54).

Most of the above cited published examples continue the tradition of the thin walled Nabataean cups. Our example is, however, of a somewhat coarser variety, which is picked up on sites in the Negev and represents a somewhat later offshoot of the Nabataean prototypes. A similar development could be established in relation to another family of cups at Migdal Tsafit (Gichon, 1974b, Nos. 25, 32, 84, 85, 87, 91).

Rom. Pot. Cat. No. 4. *Fig.* 87:4, Field No. 325/23, Locus 110, Square C-D 12.
Rim and shoulder of cooking pot. Relatively thin-walled. Clay (2.5YR 5/6): pink with small black grits. Firing: good. Levigation: good. Slip: outside dark grey-pinkish; inside pink. Surface: thinly and shallowly ribbed. Straight and rounded rim. Neck bulges convexly and is correspondingly concave on the inside. A shallow groove terminates the neck above the drop shoulder.

Comparisons: Gerasa: (Kraeling, 1938, Fig. 41: X58. Tomb 8). First century – Arabic; Jaffa: (Kaplan, 1904, Fig. 2:9). Close parallel; Nahal Hever: (Aharoni, 1961, Fig. 10:9), second century AD (lip different); Petra High Place: (Cleveland, 1960) Fig. 5:1–5 – first to second centuries; Fig. 6:2 – second century; Fig. 7:1–2 – third century; Petra: (Parr, 1970, Fig. 8:126, 130. Phases XVII and XVIII); Qumran: (de Vaux, 1954, Fig. 3:22. Niveau I) 'de la fin de l'époque hellenistique'; Samaria: (Hennessy, 1970, Fig. 9:14, phase D), Late Hellenistic; Jerusalem, Temple Mount: (Mazar, 1971, Fig. 17:8). Hasmonean-Herodian (first century BC); Ashdod: (Dotan, 1971, Fig. 24:1). 100–75 BC.

While the pots from Qumran and Samaria are examples for the Hellenistic prototypes of our vessel, the specimen brought by Parr enables us to follow the continuation of its production well into the Byzantine period. His Fig.8:126 represents the typical neck shape and Fig. 8:130 the generally similar form from an even later context.

An exact counterpart to our pot comes only from the Petra High Place, where Cleveland has been able to establish a relatively secure chronology for the typological development of the Nabataean closed cooking pot. Our vessel is dated to the second century and has antecedents in the first century BC. It

continues, with small changes in relative dimensions, into the third century.

Rom. Pot. Cat. No. 5. *Fig.* 87:5, Field No. 233/13, Locus 101, Square I-J 18.
Rim and neck of jar or jug. Well levigated pink clay (2.5 YR 5/8) with small black grits. Firing: good. The rim is somewhat thickened and everted. The neck is double-grooved.

Comparisons: Alayiq: (Kelso and Baramki, 1955, Pl. 24:X61). Form 21 (Hellenistic form), Roman – first century BC (Pl. 25:X85) Form 28, early Roman; Bethany: (Saller, 1957, Fig. 39:4073), Persian-Early Hellenistic; Petra: (Parr, 1970, Fig. 4:51), Phase VIII; Samaria: (Hennessy 1970, Fig. 6:5), Phase A (Iron Age-first century AD).

This rim differs from the vast amount of Hellenistic to Roman rims with thickened moulded lips, insofar that it has a double grooved neck. This type is rather scarce and has exact counterparts only in the above quoted examples that pertain to the first century BC, or an even earlier date.

Rom. Pot. Cat. No. 6. *Fig.* 87:6, Field No. 325/24, Locus 110, Square C-D 12.
Mouth and neck of jug. Reddish-pink well levigated clay (2.5 YR 5/8). Firing: good. Creamy slip on exterior. Everted rim. Round collar on cylindrical neck beneath rim. Handle, with sharp central ridge, elliptical in section, attached to lip.

Comparisons: Bethany: (Saller, 1957, Fig. 58:4134). Rare specimen, from cave 41, undated; Heshbon: (Saller, 1957, Fig. 1:20), Early Roman, 63 BC–AD 135; Judean Desert: (Aharoni, 1962a, Fig. 2:15), Bar Kokhba period; Ramat Rahel: (Aharoni, 1926b, Fig. 6:24), Herodian; Jerusalem, Temple Mount: (Mazar, 1971, Fig. 18:21), Herodian.

Most of the similar jugs, pertaining to the period from the first century BC to the second and third centuries AD, have a handle fixed to the neck well below the rim. Confining ourselves to these vessels, we have two instances with a definite dating (Heshbon and the Judean Desert caves), the latter from the Bar Kokhba war (AD 132–135) and the former dated 63 BC–AD 135. A related type of jug, but with a pinched spout, is dated in the Petra Theatre to the second century AD (Phase Ic), (Hammond, 1965, Pl. LIX, 24).

Rom. Pot. Cat. No. 7. *Fig.* 87:7, *Pl.* No. 151:5, Field No. 337/41, Locus 102, Square I-J-K-L 11.
Jug of well levigated light yellowish-brown clay (2.5 Y 6/2). Pale pinkish-brown exterior slip (10 YR 8/3). The vessel is of the wide-mouthed pitcher type with one flat (?) handle, similar to bottles of the same shape. The rim is thickened and everted. Neck and shoulder merge and drop conically to the straight, strong walled and ribbed body. Ribbing round, pronounced and wide spaced. Flat base with central cavity with a small umbo in its middle. The inside is distinctly ribbed owing to marks left by potter's tool or hand.

Comparisons: Dibon: (Winnet and Reed, 1964, Pl. 69:12), Nabataean-Roman; Jerusalem, Dominus Flevit: (Bagatti and Milik, 1958, Fig. 30:4. Da. 358), Herodian; Nazareth: (Bagatti, 1969, Fig. 220:12), (with two handles), from Silo 46 that yielded first to second century AD; Petra: (Hammond, 1973, No 21), Nabataean; Samaria: (Hennessy, 1970, Fig. 10:8), Phase E Upper (different base).

There is very little published comparative material and no exact replica. The examples quoted above indicate a Herodian-Nabataean date. The typical base continues into the Late Byzantine–Early Islamic period at Bethany: (Saller, 1957, Fig. 59:4825, Fig. 60:3586). An intermediate stage is represented by the not-exactly parallel base: Gerasa (Kraeling, 1938, Fig. 45:X1).

Rom. Pot. Cat. No. 8. *Fig.* 87:8, *Pl.* No. 151:1, Field No. 325/25, Locus 110, Square C-D 12.
Small bottle, rim missing. Red clay (2.5 YR 5/6), well levigated. Firing: good. Long cylindrical neck, pointed convex base.

Comparisons: Jericho: (Pritchard, 1958, Pl. 59:23), Herodian; Petra: (Hammond, 1973, Nos. 45, 46, 49, 52, 53), Nabataean; Petra, Street: (Murray and Ellis, 1940, Pl. IX:47; Pl. XXXII:138). Terminus ad quem: mid-second century AD (cf. ibid. pp. 10, 30).

This bottle is very commonly represented. Our specimen, defined by the pointed base, is only one of many sub-types. Its special features may have chronological as well as regional significance. The examples from Petra occur in first and second century AD contexts.

Rom. Pot. Cat. No. 9. *Fig.* 87:9, Field No. 236/10, Locus 107, 109, Square E-F 9–11.
Rim fragment of small bottle or juglet similar in ware (2.5 YR 5/6) and execution to No. 8.

Rom. Pot. Cat. No. 10. *Fig.* 87:10, Field No. 325/26, Locus 110, Square C-D 12.
Rim and neck fragment of small bottle or juglet. Reddish-yellow, well-levigated clay (5 YR 6/6). Small black grits. Firing: good. Everted rim. Cylindrical narrow neck.

Comparisons: Caesarea: (Riley, 1975, No. 61); Jericho: (Pritchard, 1958, Pl. 59:22, 29), Roman; Jerusalem, French Hill: (Strange, 1975, Fig. 15:38), first half of first century AD; Jerusalem, Tombs: (Avigad, 1967, Fig. 17:6), first century AD; (Rahmani, 1961, Fig. 5:17), Hasmonean-Herodian; Petra: (Hammond, 1973, Nos. 51, 57), Nabataean; Silet-edh-Dhahar: (Sellers and Baramki, 1953, Fig. 30:17), Herodian; Tel en-Nasbeh: (Wampler, 1947, Pl. 75:1735).

The smallness of this fragment makes comparison rather conjectural. The general impression from the published specimen is that this bottle(?) rim is of first century AD date. However, because of its rather common, utilitarian shape, it is also possible that it comes from either before or after that century. The great homogeneity of the late period finds from Timna, however, renders the allocation of our frag-ment to either earlier Hellenistic or later Roman and Byzantine times improbable.

Rom. Pot. Cat. No. 11. *Fig.* 87:11, *Pl.* No. 151:2, Field No. 337/140, Locus 102, Square I-J-K-L 11.
Goblet. Clay (2.5 YR 6/6), light coral red to grey core. Small black grits. Well levigated. Firing: good. Slip: red. Rim stongly everted. Conical body. Ring base with carved out small inside disc. Above base, a shallow scraped out groove is partly filled-in with translucent slip or paste that bears traces of rough rouletting. On the inside concentric bands of dark coral slip cover the lower half of the body.

Comparisons: Gerasa: (Kraeling, 1938, Fig. 37:4. Tomb 5), beginning of third century AD; Mampsis Necropolis: (Negev and Siria, 1977, Fig. 3:19), Phase I – first century AD; Petra: (Hammond, 1973, Nos. 88, 89, 91), Nabataean. Class IIIa; Petra: (Horsfield, 1939, Fig. 52:385).

This type of fine goblet was produced in Oboda during the first century AD. It was discovered in the Mampsis Necropolis in tombs of a similar period. The form is typical Nabataean, compared with the somewhat earlier Judaean counterparts. The form survived, however, up to the third century AD, according to finds from Gerasa. The classification at Petra by Hammond as Class IIIa ware probably also indicates a second century date.

The band of rouletting is reminiscent of a similar decoration upon a large cup from Petra described by C.M. Bennett (1973).

Rom. Pot. Cat. No. 12. *Fig.* 87:12, Field No. 325/21, Locus 110, Square C-D 12.
Lower shoulder and body piece of globular jug. Red clay (2.5 YR 5/8). Firing: good. Distinct exterior gloss. Decorated with rouletting. Nabataean sigilatta ware.

Comparisons: Dibon: (Winnet and Reed, 1964, Pl. 60:10; Pl. 70:1), Nabataean; Petra Street: (Murray and Ellis, 1940, Pl. XXXII:139), Form 68, second century AD; Sbaita: (Crowford, 1936, Pl. III:3. Deposit 1), second century AD.

If this rouletting belongs to the Nabataean oino-choe which has been described in extenso by Murray and Ellis (1940, 20–21) we have a form that continues into the first half of the second century, if not later. Both the ware and the rouletting point in that direction.

Attention must be called to the fragment from Tsafit (Gichon, 1974b, No. 43, Fig. 16:2), since the rouletting on our sherd is somewhat more akin to it than to the Petra jugs. This brings us again well into the second century AD.

Rom. Pot. Cat. No. 13. *Fig.* 87:13, Field No. 337/42, Locus 102, Square I-J-K-L 11.
Body fragment of large jug or jar. Reddish yellow clay (5 YR 6/6). Firing: good. Exterior pale brown slip (10 YR 7/3). Incised, wavy line combing. Appliqué rope design extends vertically downwards from the lower part of the handle.

Comparisons: Tel Goren: (Mazar, 1966, Fig. 28:12. Stratum I), third to fifth centuries AD; Jerusalem Citadel: (John, 1950, Fig. 20:5), late third to early fourth centuries AD.

Band combing does appear on Nabataean vessels. Cf. Petra (Murray and Ellis, 1940, Pl. XXV:15; Parr, 1940, Fig. 6:78, Phase XII). Single or double wavy lines are more common: Petra (Murray *et al.*, 1940, Pl. 29:98, etc.; Horsfield, 1938, Pl. IX:18; Parr, 1940, 80); Mampsis (Negev *et al.*, 1977, Fig. 4:25, 26 (Phase I).

Elaborate combing is typical of the Byzantine period, although the author has picked up sherds with intricate combing from purely pre-Byzantine sites. The combination of wave combing with band combing, sometimes partly overlapping each other, is more typical for Byzantine Israel, including the Negev, than evidenced by published material. This decoration occurs often on deep, flat-bottomed bowls, akin to the Eastern mortaria as well as upon jugs and jars. For example, see Bethany: (Saller, 1957, 267; Fig. 51; Fig. 41:7310, 7038), Jerusalem North Wall: (Hamilton, 1944, Fig. 22:32), Ramat Rahel: (Aharoni, 1962a, Fig. 3:10–11, Fig. 4:4, 17, 23), Tel Mevorakh: (Stern, 1978, Fig. 1:1, 2).

Moulded, thumb impressed, appliqué bands are also more typical, especially for the later (?) Byzantine period, than proven by the publications. See however Ramat Rahel: (Aharoni, 1964, Fig. 7:15, Fig. 13:12–13), Jerusalem, North Wall: (Hamilton, 1944, Fig. 16:6). Together with wavy line decor they appear at Bethany: (Saller, 1957, Fig. 54:3817).

The occurrence of appliqué bands upon wavy band combing is proven for the middle and late Roman period by the above cited comparisons for our specimen. This, and the fact that the appliqué decor can be traced throughout all of the Early Islamic and Ottoman period, permits the assumption that our fragment is a later intrusion. However, appliqué bands occur on Nabataean wares too, such as in Petra (Murray *et al.*, 1940, Pl. IX:17). There (ibid.) Pl. XXXII:152 shows a vertically attached thumb-impressed rope design, not unlike our specimen, while other vessels (Nos. 156, 158, 160) are decorated with broad, narrowly combed wavy bands. Since the context of the late finds from Locus 102, to which our fragment belongs, runs from the first century BC to the second or third centuries AD, we tend to assign our fragment to the same period.

Rom. Pot. Cat. No. 14. *Fig.* 88:1, Field No. 226/3, Locus 104, Square D-E-F 2–5.
Complete lower half and a small fragment of the upper part of a mould-made lamp. Clay: 2.5 YR 6/6 light red. The poor preservation of the upper part makes it impossible to assign this sherd to a definite type. Although it could possibly be later than the more complete lamps (Nos. 15–18), it is definitely pre-Byzantine.

Rom. Pot. Cat. No. 15. *Fig.* 88:2, *Pl.* No. 151:6, Field No. 245/5, Locus 107, Square A-B 8–10.
Disc-shaped oil lamp fragment. Moulded clay, very pale brown (10 YR 7/3). Small grits. Firing: good.

Large fill-hole with lentoid design. Pierced knob-handle. The nozzle might be reconstructed like No. 18 below, or somewhat more elaborately like No. 200 in Sussman (1972).

Comparisons: Beit Nattif: (Baramki, 1936, Pl. VI:9). The same form as ours, decoration varies. Third-fourth centuries AD; Bethany: (Saller, 1957, Fig. 34:2). Diverse common traits. Late Roman lamps; Jerusalem, Dominus Flevit: (Bagatti and Milik, 1958, Fig. 25:16–18). Diverse common traits; Kennedy: (1963, Pl. XXIII:516 (7). Type y, second-third centuries AD; Rosenthal-Sivan: (1978, Type Nos. 336–338. Design: 393), first-second centuries AD; Silet edh-Dhahar: (Sellers and Baramki, 1953, Fig. 34:315. Type 1b), second century AD; Sussman: (1972, Nos. 45, 198, 200, 220). Terminus ad quem of all the material dealt with, is the Bar Kokhba War (first half of second century AD), but no exact dating was attempted. Some of the material is probably later second century AD.

This type of lamp begins not later than the first century BC. Chronologically it may be divided into an earlier variety, and a later one that continues the earlier prototype (Smith, 1961). If we take the double ring, with undecorated zone in between, around the fill-hole as a criterion for dating, our example should be allocated to the earlier variety (Sussman, 1972: No. 45). This is not, however, an absolute rule since examples from Beit Nattif and Jerusalem, Dominus Flevit may belong to the later period. Doubtless this lamp existed in the second century AD.

Rom. Pot. Cat. No. 16. *Fig.* 88:3, Field No. 337/37, Locus 102, Square I-J-K-L11.
Moulded disc-shaped oil lamp fragment. Clay, yellowish red (5 YR 5/6). Small white grits. Firing: good. Decorated with ribs radiating from the large fill-hole, surrounded by one raised concentric circle. Shaped in mould.

Comparisons: Nazareth: (Bagatti, 1969, Fig. 235:33, Pattern); Sileth edh-Dhahar: (Sellers and Baramki, 1953, Fig. 41:63. Type VIII), fourth to early fifth centuries AD; Siyar el Ghanam: (Corbo, 1955, 25, Fot. 72:2), 'Periode erodiano e bizantino'; Sussman: (1972, No. 164).

Radiating ribs are the trademark of the Byzantine 'Jerusalem Slipper' and the related wares (Rosenthal-Sivan, 1978, 112 ff). They are, however, far from uncommon also in earlier periods (cf. our No. 17). It is usually very easy to differentiate between the early and late lamps with this decoration, both by shape and ware of the vessel and the exact pattern of the ribs.

There is a connecting link between the Hellenistic-Early Roman ribbed lamps and the late types cited in Rosenthal–Sivan, 1978 as No. 452 which, to my mind, may be assigned to the fourth century AD. The earlier link, clearly belonging to the third century AD, is their No. 402. To these, the transitional type No. 515 should be added.

The upper part of our lamp is concave in shape and the fill-hole is at its apex. It seems to belong to the earlier intermediate types of the second-third centuries AD.

Rom. Pot. Cat. No. 17. *Fig.* 88:4, *Pl.* No. 151:4, Field No. 325/1, Locus 110, Square C-D 12.
Disc-shaped, moulded oil lamp. Reddish-yellow clay (5 YR 7/6). Small quartz grits. Dark grey exterior slip (5 YR 4/1). Firing: good.
Small fill-hole with two concentric raised circles on disc. Decorated with ribs radiating in two tiers from each of two circles. Small, differently shaped rosettes fixed upon tier of ribs.

Comparisons: Athenian Agora: (Howland, 1958, Pl. 50. Type 50B). Second century BC to first century AD; Corinth: (Broneer, 1930, Pl. VI:303, 304). Type XVIII – second century BC-first century AD Type XXII and Type XXIV – second half of the first century AD (Prototypes); Oboda: (Negev, 1974, Pl. 17:88). Nabataean lamp – first century AD (cf. Nos. 87 and 89); Petra, High Place: (Cleveland, 1960, 71–72, Pl. 18A); Petra: (Hammond, 1973, Nos. 93–94). Nabataean; Petra, Street: (Murray and Ellis, 1940, Pl. XXXVI:18), first century AD; Petra, Sela: (Horsfield, 1939, No. 42. Pl. XI:42), first century AD.

This is a typical first century AD Nabataean lamp. We know for certain that it was produced at Oboda (Negev, 1974, 28). However, production at Oboda may well have continued into the second century AD. One proof is the production of the wheelmade 'Herodian' lamp in its potter's workshop (ibid., No. 86). This being the case the lamps of our type, coming from the same waste heap, may be contemporary with it and may also have been produced at Oboda in the second century.

Rom. Pot. Cat. No. 18. *Fig.* 88:5, *Pl.* No. 151:7, Field No. 238/2, Locus 111, Square B-C 11.
Disc-shaped oil lamp with rounded nozzle. Clay, reddish yellow (7.5 YR 6/6), light grey core. Small white grits. Firing: good. Large fill-hole set into two concentric rings like No. 15 above. Upper disc decorated with moulded wheel design. Knob handle.

Comparisons: Beth Shean Synagogue: (Zori, 1967, Fig. 12:7). Wheel design. Third century AD; Sussman: (1972, Nos. 104–106). Wheel design (the type of lamp is different); Rosenthal and Sivan: (1978, No. 339), as above, similarity of the design.

This disc-shaped lamp of the Roman period is characterised by its large fill-hole and the small rounded nozzle. In the publications dealing with Roman period lamps in Israel, these lamps are dated to the third and fourth centuries AD (Kennedy, 1963, Type 8) and the fourth-fifth centuries AD (Sellers and Baramhi, 1953, Type V). Most of the comparative material has differences in small particulars such as the fill-hole and its surrounding ring. The wheel shaped design has its exact counterparts already on 'Herodian Lamps' (c.f. comparisons above). Taken together with the evidence from the rest of the pottery, including the lamps, our specimen seems to belong to the earliest examples of its kind.

2 Discussion
The pottery examined which, as mentioned above, includes a considerable quantity of small fragments not published here, forms with a few exceptions a single homogeneous group belonging to the period between the end of the first century and the third century AD. Looking for an historical frame, we suggest the period from the conquest of Nabataea by Trajan in AD 105/6 and the anarchy following the end of the Severan dynasty (AD 235), although actually the mines may have ceased production earlier.

As shown above, most of the vessels discussed could have also been assigned to an earlier date. Yet, all the painted egg shell ware and the typical unpainted pottery that has come to light in pre-conquest levels of Nabataean sites, are missing from Timna. Among the vessel types absent are those assigned by A. Negev (1971, Figs. 2–4) to his Phase I of the Nabataean Necropolis of Mampsis, which were also picked up from other sites such as Mojet Awad (Moa ?), Shaar Ramon (Moahila ?), Migdal Ef'eh and Tamara.[2] The absence of these vessels from Timna speaks against a significant Nabataean presence prior to the Roman occupation in the second century AD.

Of special importance are the lamps. Both the Herodian and pre-conquest Nabataean lamps, such as the Sileth edh-Dhahar type Ia which appears in practically every first century excavation, and lamps typical of Petra and other Nabataean sites such as Broneer types XVIII, XXII, XXVIII, etc., are absent from our site. No. 17, the earliest specimen, is a local product inspired by these Corinthian prototypes that could have been produced in Oboda as late as the Bar Kokhba period (first half of the second century AD) and later. The latest lamp is No. 18 which is typical for the third century AD. The occurence of the earlier types at Roman Timna must thus be explained as the continued production of earlier types that continued, often with only small changes, into the Roman and even Byzantine period.[3] A similar phenomenon for different types of pottery has been established by the excavation of the burgus of Tsafit, located between Tamara and Mampsis and inhabited between 105–6 and the Diocletian period (late 3rd century).[4]

We can explain the longevity of certain cooking pot types by the conservatism of both the producer, i.e. the local craftsmen, and the buyer, i.e. the house-wife. We may add the functional perfection, the pleasing form and the good quality of these products. Even if the latter deteriorated during the post-conquest period, the vessels with these forms had already acquired sufficient repute to ensure continuing demand.

At Timna this continuity can be followed by means of the bottles Cat. Nos. 8 and 9, which started to appear in the first century BC and continued in the first century AD (cf. Lapp, 1961, 199, Type 92, especially F-G; Kahana, 1952–3), or the small bowl, Cat. No. 2 which appears in Hellenistic times and can be traced from that period onwards (cf. Hammond, 1962: Type I–2(c); Hammond, 1973,: Nos. 83–87). Another vessel that traces its origin to the Hellenistic period is the small, deep bowl, Cat. No. 3, as shown by the comparisons. So far it seems as if there was a hiatus in its production in the third and fourth centuries but it reappeared in the fourth and fifth centuries AD as the main specimens of the 'Fine Byzantine Wares' (cf. Gichon,, 1974, 119 ff, especially Types γ and δ).

The historical implication of the above provides further evidence for the Roman occupation of the Negev and the Arabah immediately upon the annexation of Arabia.[5] Without going into the history of the area as a whole, we may assume that some of the Timna mines were reactivated by the Romans, possibly under Trajan. There is no conclusive ceramic evidence for a Nabataean phase. If the mines were used at all in the Nabataean period, such use must have been sporadic and mainly by local tribesmen who did not leave any pure pre-conquest material traces behind, such as painted wares. In any case, the existence of the mines must have been known before and this knowledge was apparently passed on to the Romans. The quick reopening of the copper mines of Timna and Nahal Amram, as well as Nahal Tuweiba-Taba in the south and those in the Feinan area in the northern Arabah, and of many other sites of the eastern Arabah, was another important economic aspect of the Roman conquest of these parts.[6]

The very few sherds of a possible, although doubtful Byzantine date (Cat. No. 13 and non-published fragments) do not prove the operation of the Timna mines in Byzantine times, but this conclusion does not exclude their sporadic exploitation, as well as that of the smelting camps, by local nomads during Byzantine and later times.

Besides exploring formerly used mines and smelting camps, the Romans opened new mines immediately upon, or soon after the annexation. In this respect it should be stressed that the Roman pottery from Beer Ora (Rothenberg, 1972, 212–23[7]) has clear affinities with the Timna finds on one hand and the burgus of Tsafit on the other and, as far as other comparisons go, may be assigned to a period analogous to Roman Timna, i.e. the second and possibly the third century AD (Lamp Cat. No. 18 has no earlier parallels than from the third century).

The body responsible for the operation of the western Arabah and the Gulf of Eilat mines may well have been the Legio III Cyrenaica, as proven *inter alia* by the Atilius Turbon inscription, although it may not have been the first on the spot.[8] Autochthon troops such as the *Alae* and *Cohortes Ulpiae Petraeorum* will have participated in guard duties. As their name proves, these had already been raised by Trajan, probably from among former regular Nabataean army units.[9] This and the probable employment of local manpower for actual work in the mines, and most certainly for the furnishing of tools, vessels, etc., goes a long way to explaining the continuity of production of the local pottery and its unbroken

development at sites like Timna.[10] At the same time, the loss of independence, the military operation involved in the Roman takeover and its economic consequences, amply explain the disappearance of the painted and unpainted Nabataean 'egg-shell' wares. Even if there were no sudden cessation in their product, as we know now, rough industrial sites and the military posts protecting them would not be furnished with this fine table ware. This may be taken as a good criterion for dating sites in the Negev. Whatever their unimportance or ruggedness, and whichever their task, during the period of Nabataean independence the various sites always used the typical exquisite painted and unpainted table ware. In contrast, after the Roman takeover, these wares obviously became a luxury product that was neither issued, nor offered for sale to the troops, gangs of workmen or the like. What with the competition of production centres in southern Judaea, such as Gaza and Hebron, or even Egypt, all these must have weakened the Nabataean potters and restricted their capability to serve and supply faraway outposts.[11]

Mordechai Gichon

Notes

1. The author wishes to thank his former assistants, Mrs Hedva Larisch-Zur, Dr Moshe Fischer and Mr Zvi Shaham for their valuable aid during various stages of the work.
[The detailed comparisons with pottery from excavations in Israel often involved the use of its local chronology, used by the authors quoted. For the convenience of the reader, we are quoting here from the relevant section of a more general chronological table (M. Avi-Yonah and E. Stern (ed.), *Encyclopaedia of Archaeological Excavations in the Holy Land*, Jerusalem, 1977):

Hellenistic I	332–152 BC
Hellenistic II (Hasmonean)	152–37 BC
Roman I (Herodian)	37 BC–70 AD
Roman II (incl. Bar Kokhba war: 132–135 AD)	
	70–180 AD
Roman III	180–324
Byzantine I	324–451
Byzantine II	451–640
Early Islamic	640–1099

(*B.R.*)]

2. Gichon, M., 1976. 'Excavations at Mezad Tamar, 1973–1974', *IEJ* 26:188–94; idem, 1976b, 'Excavations at Mezad Tamar – "Tamara" –', 1973–1975', *Saalburg Jahrbuch* XXXIII: 80–94.

3. A good example of this continuity, or of the reappearance of an earlier type of vessel, is the first to third century cooking pot, found inter alia at the Petra High Place (above, No. 4), that was a common form in the Late Byzantine and even more so Early Islamic layers of the excavations conducted by R. Cohen in the Arabah and southern Negev. (I take this opportunity to express my gratitude to Mr Cohen for discussing with me the pottery from Rothenberg's survey, to be published in Vol. III of this publication).

4. Gichon, 1974b, 34 ff.

5. For the Roman occupation of the Negev cf. literature quoted in Gichon, 1972. Further evidence was furnished by A. Negev's explorations in the Negev, such as his excavation of Mampsis which proves its Roman occupation under Hadrian at the latest (Negev, 1971). See also note 8 below.

6. The processing, by the author of this chapter, of the pottery from Rothenberg's Arabah Surveys (to be published in Vol. III of this publication), proved the existence of 'Early Roman' (first to second centuries AD) and also of Early Islamic (seventh century AD) pottery on many mining sites.

7. The Beer Ora pottery was lately re-investigated by the author of this chapter and will be published in Vol. IV of this publication.

8. See Rothenberg, 1972, 223, on mining in the Gulf of Eilat and the Legio III Cyrenaica. Cf. also Alt, 1935, 62–4 and the recent

alternative suggestions: Kollmann, 1972, 145–6. On the problem of the early garrisons of Arabia: Bowersock, 1970, 38 ff; idem, 1971, 219–42 and bibliography of both papers; Speidel, 1974, 934–9; Keppie, 1973, 859–64. The latter is a comprehensive summing up of the legionary movements around the period of the annexation. How much the soldiers did actual mining work there is open to argument. That stone cutting was among their duties we know from a legionary's personal account, cf. Pap. Mich. VIII, 466, 18 ff.

9. Gichon, 1972, 49–50, No. 43. For the incorporation of existing allied troops into the Roman order of battle in the course of incorporating their country into the Empire, cf. Parker, 1928, 89; Cheesman, 1914, 59.

10. The close affinity between the autonomous Nabataean pottery and that of the post-conquest provincia Arabia wares, is well demonstrated by our hesitation to commit ourselves finally in the preliminary report (Rothenberg, 1972, 177–9). There we designated our wares as 'Nabataean Roman'. Only in the final analysis did we arrive at a more definitive dating.

11. Gichon, 1972 and no. 43a. Proof of the continuation of the use of Nabataean pottery after 106 inter alia: Petra, Theatre: (Hammond, 1973, 64–5 and plates); Petra, Street: (Murray and Ellis, 1940, 27–30 and Plates); Mampsis, Necropolis: (Negev, 1971 – Phase II). Tsafit furnishes a good example of the cessation of the painted 'egg shell' ware in Roman military sites. Its Nabataean predecessor, the adjacent tower of Efeh, abounds in painted 'egg shell' ware. When the latter was destroyed, the former was built (Gichon, 1974a, 24). Cf. the Beer Ora Report. The survey of the *limes Palaestinae* proves its inclusion in southern Judaea as far as the pottery is concerned. The pottery from Tsafti (Gichon, 1974), Tamara (Gichon, in preparation) and En Boqeq (Gichon, in the press) i.e. excavated pottery covering the second to the seventh centuries AD proved this fact. The petrographic analyses of Roman pottery from Rothenberg's Arabah survey, which was prepared by J. Glass, provided evidence that some of these Roman wares were made from Nilotic clays i.e. came from Egypt.

Bibliography

Aharoni, Y. 1961. The caves of Nahal Hever, *Atiqot* III, 148–62.
—— 1962a. The Expedition to the Judaean Desert 1961. Expedition B – The Cave of Horror, *IEJ* 12, 186–99.
—— 1962b. *Excavations at Ramat Rahel. Seasons 1959-1960*, Rome.
—— 1964. *Excavations at Ramat Rahel. Seasons 1961-1962*, Rome.
Alt, A. 1935. Aus der Araba III, 2, *ZDPV* 58, 62–4.
Avigad, N. 1967. Jewish Rock-Cut Tombs in Jerusalem and in the Judaean Hill-Country, *EI* 8, 119–42 (Hebrew).
Bagatti, B. 1969. *Excavations in Nazareth, Vol. I, From the Beginning till the XII Century*, Jerusalem.
Bagatti, B. and Milik, J.T. 1958. *Gli Scavi del "Dominus Flevit" (Monte Oliveto - Gerusalemme). Parte I, La Necropoli del Periode Romano*, Jerusalem.
Bar-Adon, P. 1971. Another Settlement of the Judaean Desert Sect at Ain el-Ghuweir on the Dead Sea, *EI* 10, 72–89, (Hebrew).
Baramki, D.C. 1936. Two Roman Cisterns at Beit Nattif, *QDAP* V, 3–10.
Bennett, C-M. 1973. An Unusual Cup from Petra (Southern Jordan), *Levant* V, 131–3.
Bowersock, G.W. 1970. The Annexation and Initial Garrison of Arabia, *PEQ* 102, 38 ff.
—— 1971. A Report on the Provincia Arabia, *JRS* LXI, 219–42.
Broneer, O. 1930. *Corinth IV, 2. Terracotta Lamps*, Cambridge, Mass.
Cheesman, G.L. 1914. *The Auxilia of the Roman Imperial Army*, Oxford.
Cleveland, R.L. 1960. The Excavation of the Conway High Place (Petra) and Soundings at Kh. Ader, *AASOR* XXXIV-XXXV.
Colt, H.D. (ed.). 1962. *Excavations at Nessana* I.
Corbo, V. 1955. *Gli Scavi di Kh. Siyar el-Ghanam (Campo dei Pastori) e in Monasteri dei Dintorni*, Jerusalem.
Crowfoot, G.M. 1936. The Nabataean Ware of Sbaita, *PEQ QST*, 14–27.
Dothan, M. 1971. Ashdod II-III. The Second and Third Seasons of Excavations 1963, 1965, Soundings in 1967, *Atiqot* IX-X.
Gichon, M. 1972. The Plan of a Roman Camp Depicted Upon a Lamp from Samaria, *PEQ* 104.

—— 1974a. Fine Byzantine Ware from the South of Israel, *PEQ* 106, 119–39.
—— 1974b. Migdal Tsafit, A Burgus in the Negev (Israel). *Saalburg-Jahrbuch* XXXI, 16–40.
—— 1976a. Excavations at Mezad Tamar 1973–1974, *IEJ* 26, 188–94.
—— 1976b. Excavations at Mezad Tamar-"Tamara", *Saalburg-Jahrbuch* XXXIII, 80–94.
Hamilton, R.W. 1944. Excavations Against the North Wall of Jerusalem 1937-8, *QDAP* X, 1–54.
Hammond, P.C. 1962. A Classification of Nabatean Fine Ware, *AJA* 66, 169–80.
—— 1965. *The Excavations of the Main Theatre at Petra 1961-1962, Final Report*, London.
—— 1973. Pottery from Petra, *PEQ* 105, 27–49.
Hennessy, J.B. 1970. Excavations at Samaria-Sebaste 1968, *Levant* II, 1–21.
Horsfield, G. and A. 1939. Sela Petra. The Rock of Edom and Nabatene. IV. The Finds, *QDAP* IX, 105–204.
Howland, H.R. 1958. *The Athenian Agora, Vol. IV. Greek Lamps and their Survivals*, Princeton, N.J.
Johns, C.N. 1950. The Jerusalem Citadel, *QDAP* XIV, 121–90.
Kahana, P. 1952–3. Pottery Types from the Jewish Ossuary Tombs around Jerusalem, *IEJ* 2, 125–39, 176–82; *IEJ* 3, 48–54.
Kaplan, J. 1964. Two Groups of pottery of the first century A.D. from Jaffa and surroundings, Tel Aviv-Jaffa (Hebrew).
Kelso, J.L. and Baramki, D.C. 1955. The Excavations at New Testament Jericho and Khirbet en-Nitla, *AASOR* XXIX-XXX.
Kennedy, C.A. 1963. The Development of the Lamp in Palestine, *Berytus* XIV, 67–115.
Keppie, L.J.F. 1973. The Legionary Garrison of Judaea under Hadrian, *Latomus* XXXII, 859–64.
Kollman, E.D. 1972. A Soldier's Joke or an Epitaph, *IEJ* 22, 145–6.
Kraeling, C.H. (ed.). 1938. *Gerasa-City of the Decapolis*, New Haven.
Lapp, P.W. 1961. *Palestinian Ceramic Chronology 200 BC-70AD*, New Haven.
—— 1974. Discoveries in the Wadi ed-Daliyeh, *AASOR* XLI.
Mazar, B. 1966. En Gedi 1961–1962, *Atiqot* V.
—— 1971. *The Excavations in the Old City of Jerusalem near the Temple Mount*, Jerusalem.
Murray, M.A. and Ellis, J.C. 1940. *A Street in Petra*, London.
Negev, A. 1971. The Nabatean Necropolis of Mampsis, *IEJ* 21, 124.
—— 1974. *The Nabatean Potter's Workshop at Oboda*, Bonn.
Negev, A. and Siva, R. 1977. The Pottery of the Nabatean Necropolis at Mampsis, *Rei Cretariae Romanae Fautorum* XVII/XVIII, 109–31.
Parker, H.M.D. 1928. *The Roman Legions*, Cambridge.
Parr, P.J. 1970. A Sequence of Pottery from Petra, in Sanders, J.A. (ed.) *Near Eastern Archaeology in the Twentieth Century. Essays in Honour of N. Glueck*, New York, 348–81.
Pritchard, J.B. 1958. The Excavations at Herodian Jericho, *AASOR* XXXII-XXXIII.
Rahmani, L.Y. 1961. Jewish Rock-Cut Tombs in Jerusalem, *Atiqot* III, 93–120.
Riley, J.A. 1975. The Pottery from the First Season of Excavations in the Caesarea Hippodrome, *BASOR* 218, 25–63.
Rosenthal, R. and Sivan, R. 1978. Ancient Lamps in the Schloessinger Collection, *Qedem* 8.
Rothenberg, B. 1972. *Timna*, London.
Saller, S.J. 1957. *Excavations at Bethany*, Jerusalem.
Sauel, J.A. 1973. *Heshbon Pottery 1971*, Berrien Springs.
Sellers, O.R. *et al.* 1968. The 1957 Excavations at Beth Zur, *AASOR* XXXVIII.
Sellers, O.R. and Baramki, D.C. 1953. A Roman Byzantine Burial Cave in Northern Palestine, *BASOR Supplementary Studies* 15–16.
Smith, R.H. 1961. The "Herodian" Lamp of Palestine: Types and Dates, *Berytus* XIV, 53–65.
Speidel, M.P. 1974. Exercitus Arabicus, *Latomus* XXXIII, 934–9.
Stern, E. 1978. Excavations at Tel Mevorakh (1973–1976). Part One: From the Iron Age to the Roman Period, *Qedem* 9.
Strange, J.F. 1975. Late Hellenistic and Herodian Ossuary Tombs at French Hill, Jerusalem, *BASOR* 219, 39–67.
Sussman, V. 1972. *Ornamented Jewish Oil Lamps. From the Fall of the Second Temple through the Revolt of Bar Kochba*, Jerusalem (Hebrew).
Tushingham, A.D. 1972. The Excavations at Dibon (Dhiban) in Moab, *AASOR* XL.

de Vaux, R. 1954. Fouilles au Khirbet Qumran, *RB* 61, 206–36.

—— 1956. Fouilles au Khirbet Qumran, *RB* 63: 533–77.

Wampler, J.C. 1947. *Tel en-Nasbeh. Vol. II - The Pottery*, Berkeley, Calif.

Winnet, F.W. and Reed, W.I. 1964. The Excavations at Dibon (Dhiban) in Moab, *AASOR* XXXVI-XXXVII.

Zori, N. 1967. The Ancient Synagogue at Beth Shean, *EI* 8, 149–67 (Hebrew).

26 ROMAN GLASS

Roman glass is represented by fragments of three or four small bottles, including two mould-blown vessels.

(a) *Fig.* 86:18. Field No. 317/7. Locus 110. H. 3cm., Diam. of mouth 1.8cm.
Neck and rim (restored from two fragments). Rim outsplayed and folded inside. Cylindrical neck. Brown glass. Partial enamel-like weathering.

(b) *Fig.* 86:9. Field No. 344/7. Locus 110, 111. H. 2.5cm.
Lower half of a handle of a small juglet (?). Drawn handle with deep groove. Brown glass. Enamel-like weathering. Attached to a mould-blown body fragment with vertical ribs of same colour. A few small fragments of mould-blown body with vertical ribs of the same colour and pattern as the fragment attached to the handle, were also found. They belong to the same vessel and suggest that the whole body was decorated with ribs. A close parallel is in the Haaretz Museum, Tel Aviv, collection, No. 90658.

(c) Field No. 344/7. Loci 110, 111 (not illustrated).
Small fragment from another mould-blown ribbed bottle. Brown glass. Milky weathering.

All three fragments (a–c) belong to the so-called Sidonian mould-blown group, dating to the second half of the first century to early second century AD.

Gusta Lehrer-Jacobson

D. MISCELLANEA

27 RECENT INVERTEBRATES AS VOTIVE GIFTS

Introduction

The present study covers just over 100 marine and fresh-water invertebrates found within the Temple,[1] including *c.* 20 corals, 85 marine shells, five freshwater shells and one sea urchin. While most of the invertebrates are probably local in origin, a few are definite imports. Most of the shells could come from the Red Sea. Below is a species by species analysis of the Temple shells. (All measurements are in millimetres.)

1 Marine Invertebrates

(1) *Phylum Coelenterata*[2]
Class: Anthozoa (corals and sea-anemones); Subclass: Hexacorallia; Order: Madreporaria (True corals or stone corals)

   5 *Stylophora pistillata* (Esper, 1795) (*Pl.* 152:2)

4–5 *Pocillopora verrucosa* (Ellis and Solander, 1786) (*Pl.* 152:5). A photograph of another specimen has already been published (Rothenberg, 1972, Pl.108, centre row, left).

   4 *Millepora dichotoma* (Forskal, 1775) (*Pl.* 152:4)

2–4 *Platygyra lamellina* (Ehrenberg, 1834) (*Pl.* 152:3)

   1 *Acropora* sp. (*Pl.* 152:7)

   1 *Coscinarea monile* (Forskål, 1775) (*Pl.* 152:1). This specimen has been re-crystalized and leached, and is probably older than the other remains, although they are all found living today in the Red Sea.

Coral was also found at the Deir el-Bahri Temple (see note 1) and is known from Hyksos to Ptolemaic Tell el-Maskhuta in the Egyptian Delta (Patricia Crawford, personal communication), from the Roman and Islamic port of Quseir al-Qadim, due east of Luxor on the Red Sea in Upper Egypt (personal analysis), and from the Pre-Islamic site of Hajar bin Humeid in South Arabia (Van Beek, 1969, 289, Pl. 55c). A Graeco-Roman coral amulet from Egypt is in the Petrie collection in University College London (Petrie, 1914, Pl. XIV, 106).

(2) *Phylum Mollusca*
Class: Gastropoda (snails); Subclass: Streptoneura (Prosobranchia); Order: Diotocardia (Archaeogastropoda); Superfamily: Trochacea; Family: Turbinidae (Turban shells).
10 *Turbo* spp. (*Pl.*153:10). Only the thick calcareous operculum, the 'cat's eye', are present, making specific attribution impossible. Two species, or two age groups, are suggested based on the following opercula measurements:

Maximum Diameters	Maximum Thickness	Comments
19.75 × 18	9	slightly worn
19 × 17	7.25	fresh
17.5 × 16	7	slightly worn
17	7	worn, half preserved
16	7	worn, half preserved
13 × 12.5	5	badly worn and pitted
10 × 8.75	3.5	worn and pitted
9.5 × 8.5	4	worn and pitted
9 × 8.25	3.5	worn and pitted
8.25 × 7.5	3	slightly worn and pitted

Turban shells are often found living on the outer edges of coral reefs. The opercula has often been used by man: those from the largest species, the Green Turban (*Turbo marmoratus* L.) have been used as paperweights and may weigh ½kg. They are often used also as personal ornaments, Simmonds (1879,

303) notes their use as 'necklets and pins, studs and solitaires'.

One *Turbo* operculum is known from Koptos in Egypt (Petrie, 1914, 28, Pl. XV:120). One burnt *Turbo* operculum, but no shells, was found at 7th century BC Umm el-Biyara near Petra in southern Jordan (personal analysis). This specimen has a diameter of 21 × 19mm. and a maximum thickness of 9mm., slightly larger than the Timna specimens. Excavations by Peter J. Parr in Petra have produced a number of shells (analysed by Dr S. van Benthem Jutting but unpublished), including two *Turbo* shells and three opercula. At least three opercula are known from Tell el-Maskhuta and Quseir has produced large numbers of *Turbo* shells and opercula.

Family: Trochidae (Top shells)

1 *Clanculus (Clanculus) pharonium* (Linnaeus, 1758) (Strawberry top).
Only one small body fragment of this species is present, but its bright red colour and regularly spaced black and white dots make its identification certain. The species attains a maximum height of 25mm. and is a vegetarian found in the shallow tropical waters and intertidal zone of the Indian Ocean, apparently not being present in the Red Sea (Oliver, 1975, 32) although Mienis (1977, 349) reports it from this Sea.

One complete shell of the 25th Dynasty is in the Petrie collection (Petrie, 1914, 28, Pl. XV, 119). Two shells, one holed, are known from the Harifian (9th millennium BC) village site of Abu Salem in Israel (Mienis, 1977, 349) and it is also known from Tell Jemmeh in Israel (Paula Hesse, personal communication) and Tell el-Maskhuta.

Order: Monotocardia; Suborder: Taenioglossa (Mesogastropoda); Superfamily: Cerithiiacea; Family: Cerithiiacea.

6 *Cerithium erythraeonense* (Lamarck 1822) (Pl. 153:9)
Five of these specimens are almost complete, while one is a small fragment. Three of the complete shells are holed on the lowest body whorl, possibly for stringing. This common Red Sea species attains a maximum length of 70mm. The largest in the collection is 60mm. long and 24mm. wide. Four of the shells are naturally worn. Detailed information on the holed shells is given here:

Length	Maximum Width	Hole Diameter	Comments
47.7	21	11.5 × 9	worn (Pl.156:9, right)
33.5+	17	8 × 9.5	2 holes, no apex
		8 × 10	
40+	24.5	2 worn holes	no apex

One shell is known from Umm el-Biyara, 52mm. long and 21.5mm. wide, with a 6 × 5.5 mm, nicely cut, almost rectangular hole opposite the mouth. This shell is also known from Tell el-Maskhuta and Quseir.

1 *Rhinoclavis* spp. (Pl. 153:8)
This almost complete shell is 37mm. long and has a maximum width of 12mm. The lip of the mouth is broken and it is worn at the apex, but is not otherwise modified.

Superfamily: Strombacea; Family: Strombidae (True conchs, Strombs)

7 *Strombus (Gibberulus) gibberulus albus* (Humped conch) (Morch, 1850) (Pl. 153:5)
This subspecies is today common in the Gulf of Eilat-Aqaba and attains a maximum length of 56.1mm. and a width of 28.5mm. The largest Timna specimen is 46mm. long and 26.5mm. wide. Four specimens have definite holes in the body whorl and two other apical fragments were probably holed here as well. The seventh shell lacks the apex and part of the body. The holes range in size from 6 × 8.5mm. to 8.5 × 19mm. One unholed specimen, *c.* 43mm. long and 25mm. wide, comes from Umm el-Biyara and it is also present at Petra and Quseir.

1 *Strombus (Lentigo) fasciatus* (Born, 1778) (Pl. 153:6).
This species, common in the Gulf of Eilat-Aqaba in shallow waters in sandy or muddy areas, attains a length of 50mm. and a width of 26mm. The specimen is 33mm. long and 20.5mm. wide, and has some colour bands still preserved. There is a hole in the centre of the body whorl 9.5 × 18 mm. One shell of this type, holed on the ventral side, is in the Petrie collection (Petrie, 1914, XIV, 110c).[3]

One shell holed on the body whorl and lacking an apex, as well as another *Strombus* species, are known from Abu Salem (Mienis, 1977, 349). One specimen is known from Nimrud and two from Petra. A different *Strombus* species is known from Iron Age IA (1200–1050 BC) Tell Masos in Israel (Volkmar Fritz, personal communication).

1 *Lambis (Lambis) truncata sebae* (Kiener, 1843) (Truncate spider conch) (Pl. 153:4).
There are two fragments, probably of one individual. The larger lip fragment (illustrated) is 111 × 59mm. in size and the other is a much smaller apical fragment. The species reaches a maximum length of 30cm. and width of 20cm.

Two examples are known from Petra and it is common, probably as a food source, at Quseir. The species is distributed from Eilat to Polynesia. It lives in colonies on sandy, algal and coral rubble in the vicinity of coral reefs, usually in shallow waters down to about 40 feet.

Superfamily: Cypraeaceae; Family: Cypraeidae (Cowries)

2 *Cypraea (= Erosaria) turdus wickworthi* (Schilder and Schilder, 1936) (Pl. 153:1 (right) and 3).
One unmodified shell of this subspecies is 39mm. long, 26mm. wide and 19mm. high (Rothenberg, 1972, Pl. 108, bottom row left; and Pl. 153:1 (right)). The other specimen, 30.5mm. long and 21.5mm. wide, has an open dorsum (Pl. 153:3). This Red Sea cowrie attains a maximum length of 45mm. and, unlike most cowries, seems to favour areas in which living coral is abundant (Burgess, 1970, 236). Today the chief population of this form is in the western Red Sea, in waters 3 to 130 feet deep.

One example, with colour preserved, is known from Natufian Rosh Zin (D16) in the Negev, Israel (Tchernov, 1977, 72). They are also known from Pre-Pottery Neolithic (2) and Early Bronze Age (1) Jericho in the Jordan Valley, here also with dorsum removed (Biggs, 1963, 127, *Fig*. 2d). Another specimen with the dorsum removed was found at a Byzantine site (No. 5, Nayel) in the Wadi el-Hasa survey of Jordan, over 1100 miles from the Red Sea (Burton McDonald, personal communication and personal analysis).

1 *Cypraea* (= *Monetaria*) (*Ornamentaria*) *annulus* (Linnaeus, 1758) (Gold-ringer, Money cowrie) (*Pl.* 153:1, left)

This worn specimen is 28mm. long, 21mm. wide and 13.5mm. high. It has a ground-down area on the side of the dorsum measuring 13 × 9mm., and a hole in its centre 8 × 6mm. The species grows to 30mm. and is found in the Red Sea and waters east in shallow tidal pools, reef flats and in shallow waters under and among vegetation.

This 'money cowrie' is known from many sites including Predynastic graves at Abydos in Egypt; one from prehistoric Naqada in Egypt, holed on the side of the body (Petrie, 1914, Pl. XIV, 107b); Jericho; Petra; East Karnak (Temple of Akhenaten), Upper Egypt, where the dorsum is always removed (personal analysis); Umm el-Biyara (most with the dorsum removed); Tell el-Maskhuta where over 200 have been found, some with an open dorsum, and one from Roman Gheyta in Egypt with the dorsum removed (Petrie, ibid, 107c).

C. moneta is also very widely distributed, with one shell reported from Natufian Wad B of the Wady el Mughara caves, Mount Carmel, Israel (Garrod and Bate, 1937, 224), Tell Hesban in northern Jordan (Crawford, 1976, 172) and specimens with an open dorsum from Archaic Greek tombs at Salamis in eastern Cyprus (Demetropoulos, 1970, 301, 302, Fig. 2).

1 *Cypraea tigris* (Linnaeus 1758) (Tiger cowrie)
Cypraea (= *Lycina*) *pantherina* (Panther cowrie) (Lightfoot, 1786) (*Pl.* 153:2)

The three restorable fragments of this cowrie, still not complete, yield a probably length of 78mm. The fragmentary nature of the specimen and vague colour patterns make specific assignation impossible.

The Tiger cowrie reaches a maximum length of 13–14cm. It is probably the best known of all cowries. Its natural distribution is the southern Red Sea and areas to the east of it (Burgess, 1970, 230).

The Panther cowrie reaches a maximum length of 80mm. and seems to favour areas in which living coral is abundant and lives on dead coral where there is abundant plant growth. This cowrie has been reported from Naqada, where a specimen has an open dorsum (Petrie, 1914, Pl. XIV, 107a) and from Tell Jemmeh and Petra.

7 *Cypraea* spp. (13 fragments)
It is not possible to specifically attribute the remaining cowrie fragments. There are about 40 other Red Sea cowrie species besides those already noted. Two of the fragments are of *C. tigris/C. pantherina* size. One other unmodified fragment has a restored length of *c.* 50mm. Five other individuals have half the body preserved and an open dorsum, and four of these have restorable maximum lengths: 21, 23, *c.* 30 and *c.* 40mm.

The removal of the cowrie dorsum is a common feature, as the above examples illustrate. Other instances of such modification for ornamental use include one shell from Late Natufian Rosh Horesha, Central Negev (Mienis, 1977, 385); five from Abu Salem (ibid., 349–50); numerous examples from the Iron Age IA burial Cave A4 in the Umm ad-Dananir region of the Baq'ah Valley, Jordan (McGovern, personal correspondence and personal analysis); from Phoenician to Roman Tell Keisan in the Upper Galilee, Israel (personal analysis), and Ashdod in southern Israel (personal analysis). The removal of the dorsum is still practiced today in East Africa (McMillan, 1968; Cambridge, 1969).

In Ancient Egypt cowries were also modelled from silver, carnelian and porphyry (Petrie, 1914, Pl. XIV, 107f–h, j–k; XLIV, 1071; XLVI, 107m) and gold.

Many cultures see the cowrie as a fertility symbol, possibly since the ventral side of the shell resembles the female genitalia. Cowries have often been used to confer fertility and to protect against sterility. In the Egyptian Middle Kingdom concubine figures in faience, clay and other materials are sometimes shown wearing a cowrie shell girdle. Such girdles have been excavated, and are characteristic of the 12th Dynasty. Girdles with cowries made of gold are known for Princess Sit-Hathor-Yunet from Lahun (Aldred, 1971, 191, Pl. 35); Princess Sit-Hathor from Dahshur (ibid., Pl. 33) and for Queen Mereret (ibid., 196, Pl. 45). Princess Sit-Hathor and Queen Mereret also had pectorals with gold cowries. Even today Nilotic women wear aprons with cowrie shells sewn onto them to protect against sterility (ibid., 16).

Cowries are often seen as symbols of love and to have the power to increase sexual potency. In some areas they are given as bridal gifts. The scientific name *Cypraea*, derives from Cyprus where the worship of Aphrodite is thought to have begun.

Cowrie shells have also been perceived as the half-closed but ever watchful human eye fringed with lashes, and as a prophylactic against the evil eye. Cowries were used for the eyes of the Pre-Pottery Neolithic plastered skulls from Jericho, and are sometimes found in Egyptian mummies to ensure good eyesight in the afterlife.

Aldred (1971, 16) sees the cowrie as the precursor of the udjat eye, with both serving as protection against the evil eye: 'The cowrie shell and cowroid design were used in ancient Egypt for the same purpose though in time they were supplemented and replaced by the *wedjet* [= udjat] eye of the sky-god Horus'. *Udjat* eyes and cowries have sometimes been found together, as in the amulets on a knotted cord from Kafr Ammar in Egypt (Petrie, 1914, 29, Pl. XVII, 131 b, c, f; XIX, 131g).

Superfamily: Tonnacea; Family: Cymatidae, Colubrariidae, Bursidae
1 worn fragment which is not identifiable further.

Family: Cassidae (Helmet shells)
1 small, distal end fragment. Not identifiable further.

Suborder: Stenoglossa (Neogastropoda); Superfamily: Muricacea; Family: Thaididae (Rock shell, Dog whelk).

1 *Thais* (= *Purpura*) *hippocastanus savigny* (Deshayes) (*Pl.* 153:13)
This is a badly eroded specimen, 28mm. long and maximum width 21mm. The species reaches a maximum of 60mm. and is an active predator which creeps about on rocks, feeding on barnacles and bivalves. It is found in shallow waters, often in the intertidal zone. Two specimens of this shell are known from Petra.

Superfamily: Buccinacea; Family: Columbellidae (Dove shells)

2 *Columbella* cf. *fulgurans* (Lamarck)
Both specimens lack the apex and could have been strung. They are 9–10mm. long (preserved length) and 7–7.5mm wide. Dove shells live on coral or sand on the shore and on into deeper waters. Most are carnivorous but some feed on algae and detritus. They have a long history of ornamental use. Holed examples are known from Upper Paleolithic Ksar 'Akil in Lebanon (Altena, 1962, 89; Inizan and Gaillard, 1978, Figs. 2, 5) and Natufian Hayonim Terrace near Nahariya in Israel (personal analysis).

Family: Fasciolariidae (Tulip, Band or Spindle shell)

1 *Latirus* (Latirus) *polygonius* (Gmelin, 1791) (*Pl.* 153:14)
This very worn specimen (Rothenberg, 1972, Pl. 108, bottom row, right) has a preserved length of 48mm. and a restored length of *c.* 55mm. and is 26.5mm. wide. It has a hole on the side of the body whorl, near the mouth, 13mm. long and 6mm. wide. This is a carnivorous gastropod which reaches a maximum length of 60–70mm. and lives in shallow waters, often on coral reefs.

Superfamily: Conacea (Toxoglossa); Family: Conidae (Cone shells)[4].

2 *Conus* (Puncticulis) *arenatus* (Sand-dusted cone) (Hwass in Bruguière, 1792)
Only two fresh apices were found, one with a maximum width of 21mm. The species attains a maximum length of 75mm.

1 *Conus* (Hermes) *terebra* (Born, 1780 = *clavus* Linnaeus, 1758)
This specimen is 42.5mm. long and 23mm. wide (*Pl.* 153:12). It grows to a maximum of 100mm.

1 *Conus* (Chelyconus) *catus* (Hwass in Bruguière, 1792)

This complete and worn shell (Rothenberg, 1972, Pl. 108, bottom row, centre, and *Pl.* 153:11 here) is 35mm. long and 21mm. wide. It attains a length of 40mm.

1 *Conus* (Puncticulis) *arenatus* (Hwass in Bruguière, 1792)

Conus (Virroconus) *ebraeus* (Linnaeus, 1758)
This badly worn and abraded shell might be a gerontic *C. ebraeus*. It is 45mm. long and 29mm. wide. *C. ebraeus* has a maximum length of 50mm.

3 *Conus* spp. (9 fragments)
The additional nine Cone fragments are not specifically identifiable. Three fragments are badly worn. Two shells have apical diameters of 23 and *c.* 25mm. Most Cones live in shallow waters, often associated with coral and Giant clams. Seven shells are known from Abu Salem and they are also known from Tell Jemmeh; Tell Masos; Umm el-Biyara (1); Petra (4) and Quseir. One complete shell is known from Zowaydeh in Egypt (Petrie, 1914, Pl. XIV, 110a).

Family: Terebridae (Auger shells)

5 *Terebra* spp. (*Pl.* 153:7)
All specimens are fragments lacking apices and distal ends. Their shape is very similar to the fossi internal moulds of turriculate shells (Price, below, Chapter III, 28). Their measurements are:

Preserved Maximum Length	Maximum Width	Comments
39	11	fresh (*Pl.*153:7)
35	16	badly worn
29.5	12	fresh
28.5	8	worn
15	7	worn, burnt

There are at least eight *Terebra* species found in the Red Sea. All are carnivorous and live in fine sand and often in the vicinity of coral reefs. One *Terebra* is known from prehistoric Naqadeh, and also from Koptos (Petrie, 1914, Pl. XV, 122). One shell with a broken aperture is known from Abu Salem and it was also found at Early Bronze Age Jericho (Biggs, 1963, 127, Fig. 2p) and Tell el-Maskhuta.

Class: Bivalvia (Pelecypoda); Subclass: Pteriomorphia; Order: Arcoida;[5] Superfamily: Arcacea; Family: Arcidae (Ark shells)

1 *Barbatia* (Gray, 1842) spp. (*Pl.* 154:3)
This is a worn, but complete valve and is 69mm. long and 48mm. high. This Ark shell lives attached to rocks and in cliff crevices.

1 *Anadra* (Gray, 1847) spp. (*Pl.* 154:2)
This item is a worn hinge fragment. *Anadra*, some possibly used in some way, are known from Hajar bin Humeid in South Arabia (Van Beek, 1969, 289, Pl. 56a). They are also known from Quesir.

Order: Pterioida; Suborder: Pteriina; Superfamily: Pteriacea (Pearl, Wing and Hammer oysters); Family: Pteriidae (Pearl oysters).[6]

8 fragments of *Pinctada margaritifera* (Linnaeus, 1758) (Black Lip Pearl oyster).

This Indo-Pacific species grows to 25cm. in length. Three of the eight fragments from the Temple have been used in some manner. Fr. R. Woodward[7] noticed abrasion on two fragments and suggests that they may have been utilised as scrapers (*Pl.* 152:8, 9). Another fragment has been shaped into an elongated 'plate', probably of an ornamental nature (*Pl.* 152:10).

In the Middle Kingdom large trimmed Pearl oysters were worn as pendants, often bearing the protective name of the reigning Pharaoh. They may be of Nubian inspiration (Aldred, 1971, 196) and were worn in Nubia until very recently (ibid., 177). Examples are known in the original shell (Petrie, 1914, 27, XLIV, 112a), electrum (ibid., Pl. XIV, 112c), carnelian (ibid., Pl. XIV, 112e) and gold (ibid., *Pl.* XIV, 112d; Aldred, 1971, 196, 213, Pl. 45, 79).

Shells of this species are known from Tell Jemmeh and Qadim in Israel. These shells do produce pearls, and the nacre layer ('mother-of-pearl') has been used ornamentally.

Subclass: Heterodonta; Order: Veneroida; Superfamily: Tridacnacea; Family: Tridacnidae (Giant clams)

10 *Tridacna* (Chametrachea) *squamosa* (Lamarck, 1819) (Scaly or Fluted clam, Squamose Giant clam) (15 fragments) (*Pl.* 154:1)
There are probably ten valves of ten individuals of this species, four of which are badly worn. The only shell with a complete height is 88mm. with a width of 68+mm. (*Pl.* 154:1, right). The largest fragment has a height of 89+mm. (*Pl.* 154:1, left). The fragment with the greatest length is 78+mm. long.

This species is not found today in the northern Red Sea, but is found in its southern part (Rosewater, 1965) and it is this species which was engraved in the ancient Near East (Stucky, 1974; Amiet, 1976) and traded west as far as the Greek mainland and Italy. This species attains a maximum length of about 40cm.

9 *Tridacna* (Chametrachea) *maxima* (Roding, 1798) (15 fragments) (*Pl.* 153:15)
There are probably nine individuals present, four of which are badly worn. There are two complete valves, one 111mm. long and 68.5mm. high (*Pl.* 153:15 left) and the other 96mm. long and 61.5mm. high (Rothenberg, 1972, Pl. 108, top row, centre; and here *Pl.* 153:15 right).

This species is found throughout the Red Sea and attains a maximum length of 35cm. The largest shell in the Timna collection probably comes from an individual just over 150mm. in length.

5+ *Tridacna* spp.
The remaining 29 fragments are not specifically assignable but probably belong to one of the two species already noted. At least five individuals are present and four fragments are badly worn.
Unworked *Tridacna* shells are known from Tell

Jemmah; Tell Masos; Petra (2); Pella in Jordan (1, personal analysis); Tell Hadidi on the Euphrates in Syria (1, personal analysis); Tell el-Maskhuta, and Quseir, as well as various sites in Jordan (Crystal M. Bennett, personal correspondence). It is possible that they were used as offering plates or bowls in the Timna Temple.

Superfamily: Tellinacea; Family: Tellinidae (Tellin)

1 *Quidnipagus palatum*[8]
One small fragment preserved.

(3) *Phylum Echinodermata* (Sea urchins, Starfish, Brittlestars)
Class: Echinoidea (Sea urchins).[9]

1 *Phyllacanthus imperialis* (Lamarck)
There are numerous plate fragments (*Pl.* 152:6) and three small spine fragments, probably of one individual. As the specific name implies, this is an impressive animal and would make a handsome votive object. Two remains are known from Petra.

2 Fresh-Water Molluscs

5 *Aspatharia* (Spathopsis) *rubens* (Lamarck, 1819) (Fresh-water mussel).
There are probably five individuals present in the Temple collection (based on valve size) and these shells must be imported from Egypt and the Nile River system. As this species is not found in the Jordan fresh-water system, it must also have been imported into Israel where it is known from Arad; Jerusalem; Chalcolithic Ben Shemen; Gezer, and Gudr near Hebron (according to Mienis). It is also known from Tell Jemmeh; Tell el-Fará (North) in Israel (personal analysis), and Tell el Dab'a in the Egyptian Delta (Boessneck, 1976, 18).

3 Conclusions

All of the shells were brought at least 30km. to the Temple and are forms found associated with coral reefs or in shallow inter-tidal waters, easily collected. However, some of the shells must come from the Indian Ocean (possibly *Clanculus*), from the southern Red Sea (*Tridacna squamosa*) and from Egypt (*Aspatharia*), and must have been brought to the Temple.

Cats are very common objects found in the Hathor Temples and it is possible to interpret the *Turbo* opercula as 'cat's eyes' as they are commonly referred to today. The larger Timna cowries, when in a fresh state, exhibit a cat-like decoration and it is possible that they were selected particularly because of this resemblance. Both opercula and cowries are known from many sites as personal ornaments; cowries certainly have a long history of votive usage, but the feline similarities may be of particular significance here.

In many areas associated closely with Hathor, cowries have a special connection with love, women and childbirth. If the cowrie is interpreted as representing the female genitalia it is worth noting that a number of phallic objects were found in the Timna Temple

(*Fig.* 30:6; 92:5) and a number of roughly shaped wooden phalli were found at the Hathor Temple at Deir el-Bahri (Currelly in Naville, 1913, 31). The five Terebra shells (*Pl.* 153:7) and three turriculate gastropod fossils (*Pl.* 154:11) found at Timna also resemble the phallus.

As Hathor is also the goddess of dancing and music, it may be reasonable to think that some of the shells were used as musical instruments. The larger shells, particularly the *Tridacna*, may have been used as clappers, instruments which often have Hathor heads on them.

Many of the shells (*Cerithium*, *Strombus*, *Cypraea*, *Columbella*, *Latirus*) are holed and some have been found *in situ* strung as ornaments. *Tridacna* shells may have been used as offering plates or incense burners.

The marine and fresh-water invertebrates from the Temple, though not numerous, may have an importance well beyond their numbers in our understanding of the Hathor cult.

David S. Reese

Notes

1. Although shells have not been carefully studied from any of the other Hathor Temples, they have in fact been found at Deir el-Bahri: 'Interesting relics of the XVIIIth Dynasty are the shells found in the deposit of votive offerings. The fruits comprise dum, date, fig and nutmeg. The last must have come from the East, and it, with several of the shells, may well be relics of Hatshepsu's expedition to Punt. One or two bits of coral and anti-gum also found, can with little doubt be ascribed to this source' (Hall in Naville *et al.*, 1913, 18).

2. Thanks to John W. Wells, Department of Geological Sciences, Cornell University, Ithaca, New York, for these identifications.

3. In the Petrie Collection (University College London) this shell is incorrectly called 'Conus'.

4. I thank R.P. Scase of Leatherhead, Surrey, England, and A.P.H. Oliver of Crowhurst, Sussex, England, for help with the identification of the Cones.

5. Thanks to Joseph Rosewater, Department of Invertebrate Zoology, Smithsonian Institutions, Washington D.C., U.S.A., for examining the Ark shells and Giant clams.

6. Fred R. Woodward, Department of Natural History, Art Gallery Museum, Kelvingrove, Glasgow, Scotland, kindly studied the Pearl oyster and Fresh-water bivalve remains.

7. F.R. Woodward prepared a report on some Molluscan Remains from the Temple, published as an addendum to the present chapter.

8. Thanks to William E. Old, Jr., Department of Invertebrates, American Museum of Natural History, New York City, U.S.A., for this identification.

9. Thanks to David L. Pawson, Department of Invertebrate Zoology (Echinoderms), N.M.N.H., for this identification.

Bibliography

Aldred, C. 1971. *Jewels of the Pharaohs: Egyptian Jewellery of the Dynastic Period*, London.

Altena, C.O. van Regteren. 1962. Molluscs and Echinoderms from Paleolithic Deposits in the Rock Shelter of Ksar 'Akil, Lebanon, *Zoologische Mededelingen*, Vol. XXXVIII, No. 5, 87–99.

Amiet, P. 1976. Tridacna trouves a Suse, *Revue d'Assyriologie et d'Archeologię Orientale*, 70.

Biggs, H.E.J. 1963. On the Molluscs collected during the Excavations at Jericho, 1952–1958, and their Archaeological Significance, *Man* No. 153, 215–18.

Boessneck, J. 1976. *Tell el-Dab'a III, Die Tierknochenfunde, 1966-1969*, Vienna.

Crawford, P. 1976. The Mullusca of Tell Heshan, *Andrews University Seminary Studies*, Vol. XIV, Michigan.

Demetropoulos, A. 1970. Marine Molluscs, Land Snails, etc., in Karageorghis, *Excavations in the Necropolis of Salamis*, Vol. II (Text), Nicosia.

Garrod, D.A.E. and Bate, D.M.A. 1947. *The Stone Age of Mt. Carmel: Excavations at the Wady el-Mughara*, Oxford.

Inizan, M.L. and Gaillard, J.M. 1978. Coquillages de Ksar-Aqil: Eléments de Parure?, *Paléorient*, Vol. 4, 295–306.

McMillan, N. 1968. The Preparation of Cowry-shells for Use as Counters, *The Conchologists' Newsletter*, Vol. 25.

Mienis, H.K. 1977. Marine Molluscs from the Epipaleolithic Natufian and Harifian of the Har Harif, Central Negev, Israel, in Marks, A.E. (ed.) *Prehistory and Paleoenvironments in the Central Negev, Israel*, Vol. 2, Dallas.

Naville, E., Hall, H.R. and Ayrton, E.R. 1907. *The XIIth Dynasty Temple at Deir el Bahri*, London.

Naville, E., *et al.* 1913. *The XIIth Dynasty Temple at Deir el Bahri*, Part III, London.

Petrie, W.M.F. 1906. *Researches in Sinai*, London.

—— 1914. *Amulets*, London.

Reese, D.S. 1978. Molluscs from Archaeological Sites in Cyprus: 'Kastros', Cape St. Andreas and other Pre-Bronze Age Mediterranean Site, *Fisheries Bulletin*, No. 5, Nicosis, 3–112.

Rosewater, J. 1965. The Family Tridacnidae in the Indo-Pacific, *Indo-Pacific Mollusca*, Vol. 1, No. 16, 347–96.

Rothenberg, B. 1972. *Timna*, London.

Simmonds, P. 1879. *The Commercial Products of the Sea*, London.

Starr, R.F.S. 1939. *Nuzi*, Vol. I, Cambridge.

Stucky, R.A. 1974. The engraved Tridacna shells, *Dedalo*, Sao Paulo, Brazil.

Tchernov, E. 1977. Some Late Quaternary Faunal Remains from the Avdat/Aqev Area, in Marks, A.E. (ed.) *Prehistory and Paleoenvironments in the Central Negev, Israel*, Dallas, 69–73.

Van Beek, G.W. 1969. *Hajar bin Humeid: Investigations at a Pre-Islamic Site in South Arabia*, Baltimore

4 Addendum: Report on some Molluscan Remains

Field No. 206. Small nacreous fragment from disc region of a bivalve mollusk, most probably belonging to the African freshwater mussel, *Aspatharia* (Spathopsis) *rubens* (Lamarck).

Field No. 234. Large umbanal fragment of right valve of the African freshwater bivalve *Aspatharia* (together with a fragmented umbanal fragment of a smaller juvenile right valve of a further individual).

Field No. 339. Large antero disc fragment of left valve of the African freshwater bivalve *Aspatharia* (together with a smaller indeterminate disc fragment).

Field No. 50. Fragment of dorsal valve margin of the Marine Pearl Oyster, *Pinctada margaritifera* (Linnaeus, 1758).

Field No. 85. Antero-dorsal large fragment of valve of the Marine Pearl Oyster, *Pinctada*.

Field No. 93. Two nacreous disc fragments, the larger being that of a left valve, of a freshwater bivalve, most probably *Aspatharia*.

Field No. 99. Nacreous disc fragment most probably belonging to the Marine Pearl Oyster, *Pinctada*.

Field No. 228. Large antero-dorso umbanal fragment of the valve of the Marine Pearl Oyster, *Pinctada*. The ventral fragmental margin shows evidence of abrasion possibly indicating that it was actively utilised as a scraper (*Pl.* 152:8).

Field No. 277, 279. Both are indeterminate disc fragments from the Marine Pearl Oyster, *Pinctada*. Fragment 277 shows abrasion of its longest margin, indicating that it may have been utilised as a scraper (*Pl.* 152:9). Fragment 279 shows definite shaping into an elongated 'plate' most probably of an ornamental nature (*Pl.* 152:10).

Field No. 278. Small indeterminate nacreous disc fragment most probably belonging to the Marine Pearl Oyster, *Pinctada*.

Field No. 316. Indeterminate nacreous disc fragment most probably belonging to the Marine Pearl Oyster, *Pinctada*.

Field No. 348. Three small nacreous disc fragments of an indeterminate bivalve mollusc.

Remarks: The presence of valve fragments of the Marine Pearl Oyster, *Pinctada margaritifera* (Linnaeus) indicates their intentional collection. One individual, Field No. 279, exhibits definite shaping to produce an elongate plate, almost certainly for ornamentation. In addition, some samples, especially Field No. 228, indicate having been actively used as a kind of scraper.

The presence of fragments of the African freshwater bivalve, *Aspatharia* (*Spathopsis*) *rubens* (Lamarck) may have been for votive or ornamental purposes, but also possibly for food, the genus still being widely consumed throughout Africa.

Fred. R. Woodward

28 MINERALS AND FOSSILS

Twenty-seven of the votive objects from the floor of the Temple investigated by us are geological in nature. Four are small samples of mineral material, the rest are fossils. Of the 23 fossils, nine are single valves or valve-fragments of bivalve shells, mainly oysters, seven are echinoid tests and seven are limestone internal moulds of gastropod shells.

1 Minerals

One of the mineral specimens is a small block of whitish selenite crystals, roughly 4 × 2 × 3cm. The block has been subjected at some stage of its history to extensive sand-blasting which has eroded out weaker areas of the crystalline mass, leaving a rough irregular surface.

A second specimen, roughly 4.5 × 1 × 2cm. shows cream-coloured micritic limestone passing into well-spaced thick laminae of calcite stained pink and with the interlaminal spaces now weathered out but probably originally comprising mud-rich layers. The coarsely laminated region passes through more closely spaced laminae into white crystalline calcite. The laminated and crystalline layers appear to be dripstone, probably formed within a joint or fissure of some limestone body. Small areas of green staining within the laminated region are probably traces of the mineral malachite.

A third specimen, 2.5 × 2 × 0.75cm., comprises a thick crust of red-brown haematite overlying, and in the boundary region impregnating, a thin layer of medium to coarse grained, rather poorly sorted pink sandstone. The free surface of the crust takes the intriguing form of a series of ridges roughly semicircular in transverse section, each 2–3mm. wide and around a cm. long, separated by sharp troughs within which the haematite is seen to have a fibrous or acicular structure. The distinctive and unusual form of this crust may result from mineralisation along a bedding plane within the sandstone where sedimentary structures such as load-casts were preserved, but the author knows of no such occurrences elsewhere.

The fourth mineral specimen is a small, irregularly lobate concretion of buff to grey coloured chert whose maximum dimension is about 2cm. The concretion has a lighter surface patina which on some of the lobes has been broken through; there is no adhering matrix.

2 Fossils

Fossil bivalves
Two specimens merit individual description. One is a small fragment, 3 × 1 × 0.5cm. of cream-coloured micritic limestone, superficially stained pink and bearing a fragment of pectinid shell, apparently the distal portion of a right valve. The shell is probably referable to the genus *Neithea* (*Pl.* 154:6). *Neithea dutrugei* (Coquand) is known from the Cenomanian Stage (upper Cretaceous) of the Israel-Jordan area and the fragment could possibly belong to this or a related species. The second distinctive specimen is a fragment of very thick (rudist?) bivalve shell whose outer layers have been extensively bored by clionid sponges (borings of which are placed in the ichnogenus *Entobia*). In this specimen again a small patch of bright green staining may indicate traces of the mineral malachite.

The remaining specimens in this category are all fossil oysters (*Pl.* 154:9). They are small single valves or fragments of larger valves, none of which is strictly determinable in isolation even to generic level. They may, however, be compared with ostreids known rocks of Cenomanian-Turonian age in Israel and Jordan. These include *Exogyra mereti* (Coquand) and *Exogyra flabellata* (Goldfuss). At least two smooth fragments from the Timna collection, one of which has a hollow conical form, would be compatible with the former species, while three or four other fragments, arcuate in form and with traces of plication, could belong to the latter species. One or two of these become more distinctive specimens by virtue of an epifauna of smaller ostreid shells and encrusting serpulid worm tubes.

Fossil echinoids
These are useful in that they are more complete and better preserved than the other fossil material and can therefore be identified more closely. Five can be identified with reasonable confidence as belonging to

Heterodiadema libycum (Desor) (Rothenberg, 1972, *Pl.* 108, top row, left; *Pl.* 154:2 here). Three of these are reasonably complete tests, the other two show signs of having undergone surface weathering and considerable abrasion. The largest specimen is about 3.25cm. in diameter and 1.5cm. in height, the smallest specimen is about 5mm. less in diameter and of similar height. On the remaining two specimens which are smaller (2.4 × 1.3cm. and 1.7 × 1cm.) the test has been considerably abraded; the identification of these must remain more tentative – ? *Coenholecty pus larteti* (Cotteau) (*Pl.* 154:4). Both forms are well known from the Cenomanian-Turonian rocks of the Middle East.

Fossil gastropods

Since these are simply internal moulds and in most cases incomplete, identification can only be attempted in the most general terms. Three internal moulds are from only very gradually expanding turriculate shells such that the incomplete portions represented are almost parallel sided. The overall form is suggestive of genera such as *Nerinea* and *Nerinella* (*Pl.* 154:10). The specimens are 1, 1.4 and 2.3cm. in diameter and 3.5, 2 and 3.5cm. long respectively. Two other specimens, each of about 2cm. maximum diameter and respectively 3.5 and 5cm. long, are fusiform varieties with spiral angles of around 25° but are not suggestive to the author of any particular genus (*Pl.* 154:7). The same is true of a sixth specimen, about 3cm. in maximum diameter and of similar length, which is trochiform in overall morphology with a spiral angle of about 40°. This is a form which would perhaps have appeared as 'Trochus' or 'Trochalia' on older faunal lists but on the basis of the present material no modern generic designation can be given (*Pl.* 154:8). The final specimen is very small, its maximum dimension about 8mm., and may represent only the proximal region of a larger shell. The overall morphology in this case appears to have been turbiniform.

3 Comments

The identities of the echinoid specimens indicate derivation from rocks of Cenomanian-Turonian (upper Cretaceous) age. Their internal matrix, the matrix of one of the bivalve specimens and that of the gastropod internal moulds is a fine-grained, cream to buff micritic limestone. This would suggest derivation from the rocks, mainly limestones, dolomites and marls, of the Judea Group of Cenomanian-Turonian age, the nearest outcrops of which are in the Timna cliffs. None of the fossil material is inconsistent with derivation from this source. Most of the bivalves, though fragmentary and some at least of the gastropods, would be compatible with (though they cannot be positively identified as) taxa known from the Judea Group or from rocks of similar age in the Israel–Jordan area (cf. faunal lists in Kafri, 1972). Chert concretions are present at various horizons in the Judea Group. Specimens like the laminated calcite described above would be very likely to occur in

joints and fissures in limestones like those of the Judea Group and the traces of malachite on this and other specimens would tie in well with derivation from a source near to the Timna copper deposit. The haematite crust may have come from the sandstones of the lower Cretaceous Kurnub Group. All this is to argue that the majority, possibly even all the geological objects from the Timna Temple could have had a quite local derivation. Derivation from further afield cannot be ruled out but neither is it necessary, or in the circumstances very likely.

Most of the echinoid tests are abraded and a few of these and other specimens appear to have undergone weathering at the surface. All the fossils, with the exception of the pectinid shell fragment, are completely free of external matrix. All this suggests that the fossils are likely to be naturally weathered-out specimens, perhaps picked up from scree deposits or from the beds of wadis. There is no indication of any human modification to any of these specimens.

As to the significance of the specimens as votive objects, that is most likely to relate to their unusual and intriguing nature. This probably applies in particular to the four mineral specimens. It may also apply to the internal moulds of the three turriculate gastropods where any animal affinities would probably not have been apparent. With most of the shell fragments, however, the echinoids and the fusiform, trochiform and turbiniform gastropods, similarities to living invertebrate animals would have been far more apparent and in some cases quite obvious.

David Price

4 Addendum: Fossils as Votives from Archaeological Sites

Fossils are frequently found at archaeological sites, but are rarely saved or studied. Kenneth P. Oakley has written extensively on the archaeological occurrence and historical uses of fossils (Oakley, 1965; 1973; 1975) and other authors have discussed various uses of fossils by man (Kennedy, 1976; Reese, 1975).

Fossils from archaeological sites may have been weathered from building stone or collected as curiosities and chance specimens. Sometimes they are found in definite votive contexts, particularly in burials or sanctuaries, and it is these that are particularly relevant here.

Fossil shark and carnivore teeth have been found in an Upper Palaeolithic burial in Moravia (Kennedy, 1976, 43); fossil gastropods from the Neolithic temples of Malta (ibid., 44); fossil deer from a Minoan shrine at Knossos on Crete (Bate, 1918, 221), fossil molluscs from the Minoan peak sanctuary on Mt Juktas inland from Knossos (personal analysis); fossil oysters and echinoids from the Greek sanctuary at Kommos in southern Crete (personal analysis), and a fossil elephant molar from the Asklepeion on the Greek island of Cos (Barnum, 1926).

Of particular relevance to the echinoids from the

Timna Temple is a *?Coenhalectypus* found by Crystal M. Bennett at 7th century BC Umm el-Biyara near Petra in Jordan. Her more recent excavations in Jordan have produced other fossil echinoids, some of which are votives (personal communication). One fossil echinoid, holed through the centre, was found by Patrick McGovern at Cave A4, Umm ad-Dananir region of the Baq'ah Valley, Jordan, associated with burials and dating to the Iron Age IA period (personal communication).

David S. Reese

Bibliography

Bate, D.M.A. 1918. On a new genus of extinct Muscardine Rodent from the Balearic Islands, *Proc. Zool. Soc. London*, 209–22.

Brown, B. 1962. Is this the Earliest Known Fossil Collected by Man?, *Natural History*, Vol. XXVI, No. 5, 535.

Kafri, U. 1972. Lithostratigraphy and environment of deposition Judea Group, Western and Central Israel, *Bull. Geol. Surv. Israel* 54, 1–56.

Kennedy, C.B. 1976. A Fossil for What Ails You, *Fossils Magazine*, Vol. 1, 42–57.

Oakley, K.P. 1965. Folklore of Fossils, *Antiquity*, Vol. XXXIV, 9–16, Figs. 1–7, Pls. I-II; Vol. XXXIX, 117–25, Figs. 8–11, Pls. XXI-XXVI.

—— 1973. Fossil shells observed by Acheulian man, *Antiquity*, Vol. XXXXVII, 59–60, Pl. XIa.

—— 1975. Decorative and Symbolic Uses of Vertebrate Fossils, *Occasional Papers on Technology*, No. 12, Pitt Rivers Museum, Oxford.

Reese, D.S. 1975. Men, Saints or Dragons?, *Expedition* Vol.17, No. 4, 26–30.

29 NOTES ON SOME STONE IMPLEMENTS AND MINERAL VOTIVE GIFTS

1 Most of the architectural elements made of stone were dealt with above, in the Catalogue of the Egyptian Finds (Chapter III, 6) and in the excavation report (Chapter II). However, a few stone objects from the Temple structure and furnishings, which were not identified as such by the area supervisors among the mass of debris, and not processed originally as find objects, are dealt with here.

Pl. 111:2-4 Fragments of Square Pillars, find location not clear. *Pl.* 111:4 shows vague features of Hathor's face.

Pl. 114:1 Fragment of statuette, white sandstone, found at Loc. 107 among the debris from the *naos*. It is the high socle of a crouching animal, sculptured in the round, the hindquarters of which are clearly distinguishable. The proportions of the fragment suggest the statuette of a sphinx.

Pl. 114:2 Field No. 402/1 Fragment of white sandstone, found at Loc. 106. This object is too fragmentary to be identified, but it could be the base of a seated statue.

Pl. 114:3 Field No. 407/1, Square A-B 9–10, Loc. 107. Found in the heap of building debris of Loc. 107, this fragment of white sandstone shows two almost parallel lines cut straight across. We have no suggestions to offer on the possible significance of these lines.

Pl. 114:4 White sandstone fragment found at Loc. 107. The lower part of this fragment, which appears to be the high socle of a statuette, shows fine work, but its upper parts are destroyed beyond recognition.

Pl. 116:2 White sandstone base (25 × 25cm.) of a statuette found in Loc. 107. Two feet are partly preserved, the right one forward, indicating that we are dealing with a fragment of a standing figure. Could this be a fragment of a cult sculpture originally located in the centre niche of the *naos*?

Pl. 116:4-5 Sandstone pebble (6.5 × 7.5cm.) with engraved signs, so far unidentified, on both of its flat sides. The find circumstances of this pebble were, unfortunately, not recorded and it must therefore be considered as unstratified.

Pl. 155:1-3 In layers III-II of the Temple, numerous heavy black nodules of haematite, often containing manganese also, were found. There can be no doubt that these nodules were brought intentionally to the Temple, since many look like voluminous female figurines, often very much like a mother-and-child. It would seem most likely that these 'figurines' were votive offerings.

Pl. 155:4-8 Found together with the haematite nodules were many small, yellowish-red botryoidal quartz concretions, the like of which occur in the sandstone formations of the Timna Valley. Since those often have the shape of small figurines, sometimes of phallic character, they were obviously also brought to the Temple as votive gifts.

Pl. 155:9 Whilst the botryoidal concretions and nodules were offered in the Temple as found in nature, the phallic likeness and significance of this sandstone pebble was emphasised by grinding out a shallow groove around the 'glans'. This 'phallic technique' is known from many ancient sites in the Near East and similar objects were also found at Site 200 (see *Figs.* 30:6 and 92:5).

2 A fair number of working implements, mainly for crushing and grinding, were found in the layers of Site 200. Since the great majority was found in Stratum III, i.e. the White Floor horizon, at least some of these implements and tools may have been used in the metal workshop in the Temple courtyard. It should be mentioned, however, that contrary to the finds of numerous complete grinding bowls, mortars and querns in the smelting camps of Timna (Rothenberg, 1972, Pls. 23–25; idem, 1962, Pl. XIII), all such implements found in the Temple, with the exception of *Fig.* 50:1, which was found in the upper fill layer of Loc. 101 (*Pl.* 8), were found broken and of little practical use. Since not even one mortar or quern could be restored from these fragments, it is possible that the latter were brought to the Temple as votive offerings already 'killed', according to the widespread Hathor ritual of 'breaking the offerings' (Kertesz, 1976).

Saddle-backed querns
(a) Three saddle-backed, oval querns of coarse-grained red sandstone were found in Stratum V, underneath the Ramesside Temple structure (*Fig.* 50:2; 89:2 and 8). These are roughly shaped working implements similar to others found at prehistoric sites in the Arabah, dated to the Sinai-Arabah Copper Age – Early Phase (Chalcolithic, see Rothenberg, 1962, Pl. XIV; *idem*, 1972, Pl. 10). Since in this stratum evidence for copper smelting was found, those implements, together with other pounding and grinding tools (see below) may have been used for the preparation of the smelting charge.

(b) Most of the thirteen fragments of saddle-backed querns (*Figs.* 50:3–5; 89:1, 3–7, 9–12) found in Strata III-II were made of fine-grain red sandstone and are well shaped. The same type of quern was found in all the Ramesside smelting camps in the Arabah (Rothenberg, 1962, Pl. XIII: 7–8, 33–36; *idem*, 1972, Pls. 23–24). Unless these quern fragments were votive offerings, it is difficult to suggest any use for them in the Temple, as there would be no metallurgical function for such tools in a casting workshop (see above, Chapter III, 13).

Mortars, grinding bowls and anvils
(a) Two small mortars (*Fig.* 91:4–5) made of white sandstone, could not have been working implements because of the soft material of which they were made. We must therefore assume that they were votive offerings or of some ritual use.

(b) The three small grinding bowls (*Fig.* 90:1–3) made of white or red sandstone, are of a domestic or votive nature. The same would apply to the rather shallow bowl of red sandstone (*Fig.* 90:5) which shows unusually fine surface treatment and was probably especially made as a votive gift by an Egyptian stone-mason. We have not found a similar bowl in any of our excavations in the Arabah.

(c) The shallow, oval basin (*Fig.* 90:4) is a common feature in the Ramesside sites of Timna, but here in the Temple it should be considered as of ritual use.

(d) The fragment of a rotary grinding stone made of fine-grain granite (*Fig.* 90:6) is a rather unusual find in Timna. Rotary grinding tools were indeed found inseveral of the Ramesside smelting camps (Rothenberg, 1962, Pl. XII: 20; *idem*, 1972, Pl. 25), but these were rather crudely made and of red sandstone. Since the fragment was a surface find, it could well be Roman or even a later intrusion at Site 200.

(e) Small anvil stones which were used on all sides, as evidenced by shallow ground-out 'cup marks', are rather common in the Ramesside camps of Timna (Rothenberg, 1962, Pl. XIII). Some of these implements were, in fact, pestles or hammer stones, but others, like the Temple specimen (*Fig.* 91:1), were used as anvils for the crushing of ores or slag, and invariably show shallow round depressions. *Fig.* 92:2–3 show parts of large anvils.

The Temple specimens (*Figs.* 21:1 and 92:2–3) could well have been used in the metallurgical workshop of the Temple (see above, Chapter III, 13) for the breaking of slags and metal.

Pestles and hammerstones
(a) Two ball-shaped hammerstones of granite and limestone respectively, were found in Stratum V (*Fig.* 91:2–3) and are in fact a common prehistoric tool type. Another, larger, round hammerstone (*Fig.* 92:6), found in Stratum V, was also made of granite and showed clear signs of heavy use. These tools may have been used in Stratum V in the local smelting operations, although they could also have been used as grinding tools, even for simple domestic purposes, in connection with the rock-cut pits of Stratum V.

(b) Two heavy pestle-type tools (*Fig.* 92: 1, 4) were found in the Ramesside layers of the Temple, near the casting-melting installation and may well have served as pounding tools for the crushing of slag and/or metal (see above, Chapter III, 13).

IV. The Archaeological History of Site 200

1 The Strata of Site 200

Based on the sequences of superimposed layers and interfaces, horizontal and vertical standing features in relation to the artifacts found in their culture-historical context, as well as the hieroglyphic inscriptions, the following strata could be identified at Site 200:

STRATUM	CULTURE-HISTORICAL CONTEXT	DATE
I	Roman	2nd–3rd centuries AD
II	Midianite	12th century BC
III	New Kingdom 19th–20th Dynasties	14th century, BC
IV		
V	Sinai-Arabah Copper Age — Early Phase (Chalcolithic–Early Bronze I)	4th millennium BC

2 Stratigraphical and Chronological Problems

(1) The Timna Temple was almost continuously occupied, but this was evidently only a seasonal occupation, as was also found in the smelting camps of Timna. During the periods of seasonal abandonment considerable damage was done to the structures, perhaps by marauding bedouins and, consequently, there were many repairs and changes, disturbed features and interfaces. Good examples of such changes, sometimes apparently including also a change of function, can be seen in the closing of the passage in the east end of Wall 3 and the enlarging of the whole of the courtyard by extending both Wall 1 and Wall 3 eastwards. Of especially far-reaching effect on the stratigraphy of Site 200 was the almost complete demolition of the first Temple structures and the rebuilding of the Temple in its White Floor related occupational phase, accompanied by secondary use of many building elements and standing features of the previous phase.

Especially devastating, and of dire consequences for the early stratigraphy of the Temple, was the fact that not only did the soft white sandstone building elements (which had to be brought some distance to the site) quickly deteriorate when in use, but many of these elements of the pre-White Floor phase (Stratum IV) were crushed into small pieces to serve as material for the White Floor of Stratum III. Furthermore, there were repeated heavy rockfalls onto the southern part of the site, perhaps caused by the frequent light to medium wave earthquakes in the Timna Valley. These rockfalls demolished features of the Temple and often introduced a confusing sequence of red sandstone boulder tumble layers into the stratigraphy of the Temple. It therefore became necessary to try to integrate these occurrences into the main occupational phases and to elicit their chronological significance by minute co-ordination of the fill layers, occupational deposits and interfaces with the standing features, architecture, the numerous small finds and the hieroglyphic evidence. This effort resulted in the establishment of five distinct archaeological strata.

(2) All the finds in the Temple strata, with the exception of a few metal objects and thousands of beads widely dispersed in all layers, were found in a fragmentary condition, often apparently deliberately broken. The repeated disturbance of the very soft sand layers without solid interfaces (except the bedrock and White Floor), as well as major levelling operations in preparation for the laying of the White Floor, was evidently accompanied by considerable movement of sand containing small finds from Strata IV and V. Foremost here was the dumping over the wall and into Locus 101 of masses of discarded, fragmentary objects and this caused extensive stray movements of small, tiny objects throughout the layers of Site 200. It was therefore extremely difficult to date the ever-changing features of the Temple in relation to individual finds, although in one case – the cartouche of Seti I – the circumstances of the find were decisive for the chronological determination of an archaeological stratum.

To demonstrate this major problem, we list here the find records of the fragments of four (partly) reconstructed objects. The find levels of some of the individual pieces were individually measured in the field and do not always conform with the levels listed for the whole findbox. Brackets ( ) indicate possible stratum.

1. *Faience wine bowl*, Eg. Cat. No. 120, *Fig.* 43:10.
15 Fragments (10 fragments actually fitting together).

Field No.	Locus	Level	Stratum
368/6	109	225	IV–V
367	109	199–209	III } on solid
366	109	199	III } White Floor
362	106, 108	209–218	IV
262	103	180–227	III
279	106	180–196	II
245	107	126–153	III (II)
275	104	140–203	III
378	106	180–222	III–V
253	102	193–215	III–V
319	106, 108	184–187	II
257	102	190–225	III–V
363	106, 108	218–227	V (IV)
345	107	206	IV
265	103	200–249	(III)

2. *Midianite goblet, Fig. 6:22.*

Field No.	Locus	Level	Stratum
345/16	107	206	V
42	111	0–10	Surface
245	107	126–153	III (II)

3. *Faience bracelet, Eg. Cat. No. 51, Fig. 36:1.*

Field No.	Locus	Level	Stratum
277/3	107	162	III
234	101	187	(V)
323	109	144–165	II (III)
371	110	178	IV

4. *Faience bowl, Eg. Cat. No. 103, Fig. 40:7.*

Field No.	Locus	Level	Stratum	
260/1	102	180–213	II (III)	
269/3	107	155	III (II)	
279	107	180–196	(IV)	
231	101	179–185	II (III)	outside
243	101	203–210	(III) V	courtyard
232	101	201–211	V	in dump

There was, however, a definitive clustering of masses of small finds in certain levels and areas (see below the highly informative list of find boxes according to levels and loci), which clearly indicated habitation interfaces where other remains provided insufficient evidence. In other words, although artifactual analysis did not allow a detailed chronological determination for the individual fill layers, repairs, changes and reconstructions, the overall stratigraphy of the main strata could be ascertained, based on phase-related interfaces and features and groups of artifacts of chronological significance.

(3) A major problem, which was also of considerable historical significance, was the changeover from Stratum IV to Stratum III since the Temple structures of Stratum IV were almost totally obliterated, many of its architectural elements re-used in another context and also crushed to serve as material for the White Floor. However, the fortunate find of an inscribed fragment of a faience bracelet in an unquestionable stratigraphic location below the White Floor could be seen as evidence for the chronological determination of the pre-White Floor horizon – Stratum IV – to Seti I. Furthermore, a large building stone bearing a Pharaonic cartouche, found on top of the White Floor, was a clear indication for the construction of White Floor-related Stratum III during the reign of Ramesses II.

(4) The dusty layer of fine sand containing relatively numerous small finds which occurred directly below the solid White Floor, was also an intriguing stratigraphical problem. Since it was generally only a thin layer 'attached' to the bottom of the White Floor, or found where the White Floor had been removed (as in the area of the red sandstone rockfall in the south-west corner of the Temple courtyard), we had the problem of how to relate the finds from this 'attached' layer.

Since in most areas the White Floor was a hard, cemented layer of crushed white sandstone, the 'attached' dusty sand layer below it could not be considered as the bottom part of the White Floor itself. It was the result of seepage of fine white dust from the very fine-grained crushed sandstone layer onto the medium-grained red sand layer below, i.e. the dusty layer was in fact an interface below the White Floor, stratigraphically belonging to Stratum IV.

(5) We refrained from a reconstruction of the *naos* structure because there was not enough information left to go on, from both the archaeological and architectural points of view. The large pile of structural elements from the destruction of Stratum III, found on the White Floor, must have been dumped there in total disorder by the occupants of Stratum II, who 'built' their temporary tented shrine from the available building elements of the previous, destroyed, Egyptian Hathor Temple.

3 Summary of the Archaeological Evidence for the Strata of Site 200

In the following paragraphs we sum up the stratigraphic evidence for Site 200 as a whole, which has so far only been discussed in relation to the individual loci. We also discuss here the problems of interpretation of some features which are of significance for the all over stratigraphy of Site 200.

Stratum I Roman sherds, discrete, trampled interfaces and small fireplaces were found in the upper fill layer of wind-carried sand to a considerable depth. Obviously, at the time of the Roman occupation of Site 200 the Temple courtyard was still a shallow depression, bordered by sand-covered walls. A few Roman sherds were also found underneath the huge rocks which lay on top of the south-west part of the site, evidence for heavy rockfall long after the Roman occupation of the site had come to an end. A few stray Roman sherds were also found in the shallow sand layer north of the Temple structure, up to the rockface of the Pillars (D-J 19–25). A Roman treasure-hunting trench deeply penetrated the layers of the centre part of the *naos* and *pro-naos*, dug after much of the heavy stone pavement of the *pro-naos* was lifted and piled up at the side of the trench. This deep trench caused a great deal of disturbance: the hard-trampled interface of Stratum I, high up in the layers of the *naos*, produced a mixed group of finds, consisting of Roman sherds, Egyptian faience and beads, Midianite pottery, etc. Roman sherds were also found deep in the disturbed layers under the (removed) *pro-naos* pavement of Stratum III and deep down in the fill layers and on (disturbed) interfaces inside the *naos*.

The Roman presence at Site 200 was of a short lived, temporary nature and no structural features were connected with it.

Stratum II The dominant horizontal feature of Stratum II was the olive green-grey interface which was found, superimposed over the White Floor of Stratum III, in most of the loci. It was not an intentionally laid 'floor', nor was it really a continuous layer of homogeneous material. This distinct interface seemed to have been formed in or on the upper levels of the habitation layers of Stratum III, by continuous, intensive use and the cluster of numerous metallurgical fragments, ashy deposits and copper based artifacts and debris found on it. This interface was connected with drastic changes in the overall lay-out of the Temple site, its architecture and, most important, its cult character. The latter was signified by the erection of special vertical featurs like the Standing Stones and basins, the Offering Bench and the 'Cell of the Priest', all covered by a tented roof of thick woollen cloth.

The olive green-grey interface – architecture – finds

The specific character of Stratum II was emphasized by a particularly large quantity of Midianite sherds, including some vessels of votive character, as well as exclusively Midianite sherds intentionally inserted into Wall 2, all along and at the level of the Stratum II Offering Bench. There was also a conspicuously large number of metal objects and fragments, including copper-base metal figurines and the gilded copper serpent.

The olive green-grey interface ran against the inside and outside of Wall 1, as well as against the inside of Wall 2 (south) and Wall 3. It was also clearly discernible amongst and above the big pile of architectural elements from the destroyed Stratum III *naos* in Loci 106–107, in clear stratigraphic superimposition over the White Floor in the south-west section of the Temple courtyard. It was particularly substantial in the area north of the *naos* structure (Locus 110).

The olive green-grey interface continued through the doorway of the Temple courtyard and spread outside into Locus 102. Here it ran against the outside of Wall 2, accompanied by numerous small finds, and could still be seen at a distance of 3–4 metres from the wall. The olive green-grey interface was not discernible south of the doorway, outside the Temple courtyard, but here a hard interface (with only very few finds) may have been related to the adjacent interface of Stratum II. The olive green-grey interface was not found in most of the area of the *pro-naos* or inside the *naos*, a fact which may have had a functional reason, but may also be the result of disturbances in the Roman period.

Special features in Stratum II

The naos and pro-naos The upper course of the *naos* structure, as excavated, was built of white sandstone blocks (including a stela with traces of a hieroglyphic inscription), which were in secondary use in Stratum II.

A thick layer of lime plaster had been applied onto this upper course of building stones, and also onto the adjacent pavement stones of the *pro-naos*. Lime plaster was also used on the upper course of the north 'wall' of the *naos*. The fact that lime plaster was not known in Egypt at this time could, by itself, indicate that this part of the structure belonged to a non-Egyptian phase of the Temple, i.e. to Stratum II. It could also be possible, though not very likely, that the lime plaster was made by the local Midianites for use in the Egyptian Temple. However, the upper course of building stones belonged to the post-White Floor phase of the *naos* structure. This was clearly indicated by the position of the olive green-grey interface in relation to this upper course in B 13 and also by the distinctive difference in quality between Stratum II and the earlier masonry and building techniques, and by the obvious functional changes of some of the building elements, such as the 'corner stones' (see above, p. 83).

The *pro-naos* was heavily damaged by a wide and deep Roman trench which had also badly disturbed the layers below the pavement. However, remains of the olive green-grey interface, accompanied by numerous finds, especially of copper-base metal fragments, were found in Squares D-F 9–14, on the western side of the *pro-naos*.

The Offering Bench This standard element in the schema of a Semitic shrine was built against the inside of Wall 2, on both sides of the entrance. A similar bench was also found in Timna, in another non-Egyptian shrine (Smelting Camp Site 2, Area A).[1] It was set into the olive green-grey layer which here was running against the face of Wall 2. As previously mentioned, there was a particularly large quantity of small finds, mainly of metal, and Midianite sherds in the olive green-grey layer, around and adjacent to the Offering Bench. Finely decorated Midianite sherds, together with bones of young sheep and goats, had been deliberately inserted into Wall 2, behind the Offering Bench, perhaps as votive offerings.

The row of Standing Stones and related features Similar to the Offering Bench, Standing Stones (*mazeboth*) are a standard element of Semitic shrines. The row of Standing Stones, including the basin and granite boulder, which was put up along the south wall of the Temple courtyard, in the olive green-grey layer, clearly belongs stratigraphically, and functionally, to Stratum II. The olive green-grey interface was discernible running against the inside of Wall 3, all along Locus 106, with the bottom of the Standing Stones sunk in it. It is significant that some of the Standing Stones which had fallen down from their stone 'pedestal', were lying on the olive green-grey interface. However, the basin (in E 6–7) which belonged to the row of Standing Stones was here in secondary use in Stratum II, although it still stood in its original position on the White Floor of Stratum III. Similar secondary use of a basin can be seen in Loci B-C 9–10, where a large basin, still in its original Stratum III position, has been re-used in Stratum II, having been repaired with lime plaster.

Wall 3, behind the row of Standing Stones and closer to the Pillar, was badly crushed by a pre-Stratum II rockfall (which may have been the cause for the damage or destruction of the *naos* of Stratum III), and was partly repaired for use in Stratum II.

However, as the western part of the wall, immediately below the overhanging rock ledge, had been totally destroyed and was just a pile of broken red stones, a row of medium-sized white rocks was laid along its top, up to the face of the Pillar, to serve as a 'demarcation line' marking the boundaries of the Stratum II shrine.

The 'Cell of the Priest' A rubble wall (Wall 4) was put up after the end of Wall 1, close to the Pillar, had been removed to serve as an entrance to the 'addition' (Locus 112). The stratigraphic position of this wall is quite well established by its relation to the interfaces of Site 200, showing that its construction level was related to the olive green-grey interface, i.e. Stratum II. It seemed that the still standing western end·part of Wall 1 had been damaged and was roughly repaired at the time Wall 4 was constructed. An Egyptian Offering Stand was found inserted into this section of Wall 1 as a simple building stone.

Fitting this 'addition' into the schema of a typical Semitic shrine, it may be interpreted as the 'Cell of the Priest', also found in the non-Egyptian shrine of Site 2, mentioned above.

Wall 5 This was just a short, rough wall, with an Egyptian Offering Table in secondary use as a conveniently shaped building stone. A small, damaged, white sandstone basin, also obviously in secondary use, stood in front of the short wall. Both the wall and the basin sat on top of a thin layer of habitation fill, lying in the White Floor, and were related to activities on the olive green-grey horizon, i.e. Wall 5 belonged to Stratum II.

Piled up in the niche formed by Wall 5, Wall 1 and the basin, was a solid mass of rich copper ore nodules mixed with numerous metal objects and some pottery, obviously votive gifts collected from the Egyptian Temple by the occupants of Stratum II.[2] It would seem plausible to assume that this collection was made by the Midianite metallurgists, pending further working or reworking of the ores and metal fragments. A very similar hoard of small finds mixed with other votive gifts, beads and pottery, was found in Locus 101.

The tented roof of Stratum II Masses of folded-over woollen cloth were found along the outside of Wall 1 (Locus 101) and the inside of Wall 3 (Locus 106) in an obvious olive green-grey interface context. This fits well the prevailing view that wool would not have been used in an Egyptian Sanctuary (see above, Chapter III, 18) and the few pieces of finer cloth found in the earlier Egyptian strata of the Temple were indeed made of flax.

The folded masses of textile in Stratum II were at first rather difficult to explain. However, a reasonable solution presented itself with the subsequent discovery of two postholes in the middle of the Temple courtyard, strongly suggesting a tented roof, at least over the central part of the courtyard not covered by the overhanging rockshelf of the Pillar.

Metallurgical debris on the olive green-grey floor A special feature of Stratum II was the existence of a small metallurgical workshop area on the olive green-grey interface in Locus 101 (D-E 16–17). Next to it lay a particularly rich hoard of mixed small finds, mainly of metal but also Egyptian faience beads, Midianite and other pottery, and strikingly shaped ore nodules, apparently a depository (or *bothros*) for discarded votive gifts. This seemed to have served as a source of material for the adjacent small workshop. Patches of ash, pieces of typical crucible slag and crucible fragments, indicated refining and casting operations on a small scale, a type of workshop often met with in shrines and temples of different periods (see also Stratum III).

Similar remains of metallurgical activities, including crucible fragments and ashes, were found in the olive green-grey interface in B-D 13–14 (Locus 110) (see above Chapter III, 13). It is stratigraphically significant that the olive green-grey interface ran over the top of the small casting installation in Stratum III in the Temple courtyard (F 12).

Stratum III The dominant horizontal feature of Stratum III was the White Floor, made of crushed white sandstone (some pieces still showed traces of masonry), apparently laid after a thorough levelling operation. The White Floor was stratigraphically related to most of the still perserved architectural features (though some of them were altered by the occupants of Stratum II). In fact, it could be established in the excavations that most of the Hathor Temple of Stratum III was actually built simultaneously with the laying of the White Floor.

The vast majority of small finds recorded at Site 200 came from Stratum III, the White Floor horizon. Here numerous Egyptian faience fragments, many hieroglyphic inscriptions and Pharaonic cartouches, and also ordinary, plain, as well as decorated Egyptian Nile-ware pottery, were found together with many Midianite and also some Negev-ware sherds. Most of the votive gifts referred to above came from Stratum III, which represents the major phase of the Hathor Temple.

The White Floor – Architecture – Finds The White Floor in the northern half of the Temple courtyard was well-preserved up to a thickness of 22 cm., but it was badly disturbed or even broken off, in the southern half of the courtyard. Some of this damage was caused by activities related to Stratum II, e.g. by the building of the Offering Bench in the south-eastern corner of the courtyard; but in other areas, such as under the large pile of white building debris in Loci 106–107 (west), the White Floor could not be distinguished as such, and most of it was removed by the excavators together with the debris. Even in this disturbed area, however, traces of the White Floor remained discernible, either as small remaining pockets or as a spread of White Floor material.

It is important to emphasise that on top of the red sandstone boulders in Loci 106–107 (which originated in repeated rockfalls from the upper reaches of the

Pillar – including the boulder with the shallow, large cupmark in B-B 9–10), there was clear evidence that the White Floor had been laid over it. During the laying of the White Floor (in Locus 107), some of the red sandstone rockfall had to be levelled to form a solid, flat base for the new floor. Whilst the White Floor in the courtyard and below some of the walls (see below) was a solid pavement, on the slope outside the Temple (Locus 103 and part of Locus 102) it was only a thin spread of White Floor material which could well have been an unintentional dispersion of this material. As such it indicated the occupational surface of Stratum III for some of the area outside the Temple.

The stratigraphic relation of the White Floor to the architecture of the Temple not only established the context of Stratum III, but also provided some evidence for the changes which occurred during the transition from Stratum IV to Stratum III.

Both the western ends of Walls 1 and 3 were found badly damaged and stratigraphically indeterminable, but the middle section of both walls – D-G (half) 14–15 in Wall 1; E-G 6 in Wall 3 – were found intact and of considerable stratigraphic significance. Both were built on a hard interface of red sand above bedrock, mainly of red sandstone boulders, and the White Floor ran against their inside. At several spots close to Wall 1 the White Floor even went over debris that had fallen from the wall. These sections of Walls 1 and 3, which also showed a much better construction than the rest of these walls, evidently predated the White Floor. Furthermore, the end of the earlier section of Wall 1 had a well-built, head-and-stretcher laid end-face which appeared to have been the actual end of Wall 1 in an earlier wall (of Stratum IV).

Locus 101, outside this section of Wall 1, showed a more complex stratification than the area further east, and there appeared to be an additional interface and another layer of finds.

A similar stratigraphic situation was found on the exactly opposite section of Wall 3, also built on a hard red sand layer above bedrock, but, because of later disturbances, mainly by rockfall and Stratum II activities, the earlier stratigraphy which also existed here was more difficult to define in detail.

The whole of the eastern side of the courtyard, i.e. the eastern end of Walls 1 and 3 and all of Wall 2, were built of white sandstone, and by a different method of masonry; they were built onto the White Floor, which was here up to about 10cm. thick. The fact that this thick floor ended precisely under the outer face of the walls, and also in a straight line across the outer edge of the doorway, must be considered conclusive evidence that they and the White Floor were contemporary. The builders of Stratum III, after the preliminary levelling of the ground, laid the White Floor over the whole of the courtyard area and then built the white sandstone wall on top of it.

There is structural and also some stratigraphical evidence (in G-H 14) that the white sandstone walling of the courtyard was an enlargement of the previous courtyard (of Stratum IV), a conclusion also streng-

thened by the stratigraphy outside these walls, which showed under the olive green-grey interface only one habitation layer (on the White Floor or related interface) with many finds, including a cartouche of Ramesses II.

Special features in Stratum III

The naos and pro-naos   The lower of the two preserved stone courses of the *naos* walls, with the exception of the two square pillar bases, sat on a hard surface with a distinct scatter of White Floor material. A thick White Floor was laid against the outside on both the north and south side of the *naos*, and at exactly the same level of the upper surfaces.

The thick White Floor also continued at the same upper level along the pavement stones of the *pro-naos*, though here a thin layer of White Floor material was also found below this pavement. It seems that after the construction of the *naos* walls, the *pro-naos* pavement was 'inserted' into the White Floor and, like the *naos* foundation, represented the ritual centre of the Hathor Temple of Stratum III.

The *naos* of Stratum III was definitely a reconstruction of an earlier stratum because some of its elements were evidently in secondary use in the Stratum III structure. Furthermore, the two square pillar-bases left *in situ* from the Stratum IV *naos*, aligned with the two top niches in the face of the Pillar behind the *naos*, did not really 'fit' as pillar bases into the Stratum III structure.

Insufficient structural evidence remained *in situ* in Stratum III to allow a detailed reconstruction of the original plan and features. This was particularly difficult because a large number of building elements, including finely shaped and decorated building stones – one bearing a cartouche of Ramesses II – as well as a Hathor sculpture, were found in a heap on top of the White Floor in Loci 106–107, next to the *naos*. It seems that most of the Temple *naos* building of Stratum III was actually demolished and/or deliberately dismantled and its elements dumped in total disarray into a corner of the courtyard and onto the Stratum III floor. For sound stratigraphic reasons we relate this operation to the complete change in plan of the subsequent tented shrine of Stratum II, already discussed above.

The refining-casting installations of Stratum III   The stratigraphic context of the metallurgical workshop in the north-east corner of the Temple, its installations (Fu I, II and large adjacent pits) and debris (slag, crucibles, ashes) could be firmly established (in 1974) when a section dug through the edge of Fu I showed that it was set up together with the laying of the White Floor. The large pit next to the furnace was definitely part of the workshop arrangements, despite its somewhat strange location close to the entrance to the Temple, and because of its relation to the White Floor and the build-up of wood ash around it, as well as the crucible slag found inside it.

The basins in the courtyard   Two of the large basins (in E-F 6–7 and B-C 9–10), in secondary use in Stratum II, were found still *in situ*, standing on the White Floor.

Stratum IV Since there was no dominant feature of Stratum IV left after the total obliteration of its architecture and the levelling operations prior to the laying of the White Floor, the stratigraphic evidence for Stratum IV had to be accumulated by painstaking, systematic investigation of the still preserved standing features and by a search in the sand and rubble layers below the White Floor. Although some of the individual pieces of evidence thus secured may appear to be circumstantial, the main aim of this search, the existence of a Stratum IV phase of the Temple, has undoubtedly been achieved.

The reliance on individual finds below the Stratum III layer as evidence for Stratum IV, was rather tentative because for several reasons, explained above, a number of obviously stray finds – often fragments of vessels the main parts of which were found on or even above the White Floor – infiltrated the lower layers.

With the exception of the stratigraphic significance of the cartouche of Seti I, which will be further discussed below, it was often the existence of a conspicuous clustering of finds at a certain level, which indicated an interface in a loose sand layer below the White Floor.

Remains in situ Only very few, but rather significant structural remains could be stratigraphically determined as belonging to Stratum IV and those contribute some information towards a reconstruction of the original Temple.

In Wall 1, the middle section, D(half)-G(half) 14–15, was built on a hard, stony interface of a still earlier stratum (V). Piles of red stones, lying all along the inside of Wall 1 in Squares F-H 14, which were obviously debris from the wall, had the White Floor of Stratum II running over them and against the wall.

From the finely built end of this section in the middle of G 14–15, a straight strip of major disturbance, about a metre wide, underneath the White Floor, containing building debris and a non-indigenous white sandstone fragment, could be followed a considerable way towards the opposite wall (Wall 3). Here a similar middle section of red sandstone built on a hard sand interface was apparently left *in situ* by the rebuilders of the Stratum III Temple. We therefore concluded that the original front wall of the Temple ran along this narrow strip, i.e. the Temple courtyard of Stratum IV was about 2.5m. shorter than the courtyard of Stratum III. Judging by the texture of the debris and the stones used for the newly built Wall 2 – and the lack of any stray or dumped red sandstone boulders in the vicinity – we may assume that the original Temple also had a front courtyard wall of white sandstone.

The two square pillar bases found in obvious secondary use in the front foundation wall of the Stratum III *naos* stood *in situ* on a hard interface of medium-grained sand, well below the White Floor-related interface of Stratum III. On this interface a number of New Kingdom finds were recorded, including Midianite sherds. This hard interface was also found *in situ* under the Stratum III stone pavement of the *pro-naos*, and inside the *naos*, between the White Floor-related interface and the brownish sand fill related to Stratum V.

Little stratigraphic evidence for Stratum IV was found in the southern half of the Temple courtyard, probably because this area was repeatedly disturbed and the lower layers were not always protected by the White Floor above. The northern half of the courtyard provided some evidence in the form of a locally preserved, hard interface and the dusty fine sandlayer with few finds below the White Floor. However, reliable evidence for a Stratum IV related interface, well below the White Floor and well above the Chalcolithic occupation stratum, came from underneath the *naos* and *pro-naos* structures.

The find circumstances and the location of a bracelet bearing the cartouche of Seti I also provided decisive stratigraphic evidence for Stratum IV. The White Floor of Stratum III was found to have extended also over the top of a layer of red sandstones, in D-E 9–10 (Locus 107), from the frequent rockfalls onto this part of the site. This red sandstone layer was fairly compact and apparently levelled before the White Floor was laid over it. In fact, it first gave the impression of a separate 'red pavement', perhaps used instead and beside the White Floor. Subsequent trial trenches through this 'pavement', however, established its nature as a levelled tumble of rocks. The Seti I bracelet fragment was found below one of these red stones and could not have infiltrated to this location from above after the levelling of the red stone mass and the laying of the White Floor.

Remains of Stratum IV in secondary use Although almost all of the White Sandstone building elements used in the construction of the *naos* and *pro-naos* appeared to be in secondary use, this assumption was difficult to ascertain. However, the centre stone slab in the front wall of the *naos* showed markings cut into its surface which clearly could not have had any connection with the use of the slab in its present position in the Stratum III *naos* foundation. Further evidence for a structural element which originated from the Stratum IV Temple was provided by the large white pillar base in C 13, found standing on edge, inserted into the White Floor. Since the podium-like pillar base protruding from the centre of the surface of this large stone could, of course, not be used in its present position, and as its fine, decorative masonry was mostly invisible as it stood on its edge, this element was quite certainly in secondary use in Stratum III.

Stratum V All over the area of Site 200, and also in its surrounding areas up to the face of the Pillar formation north of the Temple, a habitation layer was clearly discernible at the base of the stratification, close to bedrock, containing a typical prehistoric assemblage of flint tools, debitage of flint workings and stone tools, handmade pottery, bones and bits of ostrich egg shells, pieces of wood and twigs of shrubs and acacia trees. There were also numerous tiny fragments of copper ore and a quantity of small pieces of copper smelting slag.

This habitation layer, Stratum V, was of course totally unconnected with the later, superimposed Temple strata, though a number of New Kingdom small finds did in fact infiltrate down into Stratum V.

The prehistoric habitation layer under the Temple consisted mainly of fine-grained sand mixed with quartz pebbles of orangy or brownish colour, an apparent admixture of the original wind-carried loess with disintegrating red sandstone bedrock.

The interface of Stratum V was often hard trodden or burnt, and showed remains of fireplaces, heat-crazed small rocks and ashy spots, always accompanied by flints and sherds (datable to the Sinai-Arabah Copper Age – Early Phase). The habitation layer of Stratum V was found below all the walls of the Temple courtyard, and inside and underneath the doorway to the Temple in I 11, where a small installation had Sinai-Arabah Copper Age – Early Phase sherds associated with it. Another installation was found below the remains of Wall 3 (in A 6), also accompanied by early sherds, flints and flint pestles and a small fragment of a copper tool. This installation was inserted between huge red sandstone boulders from an early rockfall, which clearly pre-dated even the Stratum V habitation of Site 200.

The concentration of shallow, pit-like cupmarks in the bedrock inside, below and next to the *naos* may be explained by the fact that this particular spot was almost permanently in the shade of the overhanging rock ledge of the Pillar. It is also quite possible that similar cupmarks, typical of early habitation sites in the region, can be found at other spots along the Timna massif which have not been excavated. It appears likely that the shallow pits of Site 200 were connected with small scale copper smelting activities, remains of which have been located in Stratum V and also at other contemporary, Sinai-Arabah Copper Age – Early Phase habitation sites close by and, indeed, all over the region of the southern Arabah, adjacent to Timna.

4 The History of Site 200

The Earliest Occupation of Site 200 – Stratum V

The remains found in the bottom layer of Site 200, which are dated to the Sinai-Arabah Copper Age –Early Phase (Chalcolithic-Early Bronze I – Pre-Dynastic) represent a short-lived, squatter type occupation of the site, evidently with some copper smelting – part of the same type of occupation found all along the southern side of the Timna massif. This is a group of the very extensive indigenous Sinaitic population, the earliest copper miners in the region, which our surveys[3] found to have occupied South and Central Sinai and the Arabah, and which developed their own flint industries ('Elatian' and 'Timnian') and ceramic traditions. Large clusters of small settle-

ments were located around the copper ore bearing areas of Sinai and the Arabah, especially the Timna Valley and it close surroundings. We do not have enough archaeological material to accurately place the finds of Stratum V within the Sinai-Arabah Copper Age – Early Phase sequence,[4] but the existing evidence points to a late 4th millennium BC date.

It is difficult to define accurately the nature of the Sinai-Arabah Copper Age – Early Phase occupation of Site 200 because no architecture was found and the shallow, pit-like cupmarks found there are very common in cult as well as ordinary habitation sites of this population. Site 200, which for most of the day is in the shade of the huge overhanging rock-ledge of the Pillar, would make a fine habitation or camping site. The fact that a short distance away there existed a shrine of this period, with a huge rock altar (Site 112), makes it unlikely that Site 200 was a cult site.

The First Hathor Temple in Timna – Stratum IV

According to the available evidence, Stratum IV saw the erection of the first Hathor Mining Temple, apparently by a mining expedition during the reign of Seti I. Comparing the Egyptian pottery of the Temple with the pottery of Layers III–II of the Egyptian smelting camp Site 30 in Timna, it became apparent that the Temple was not erected at the time of the initial Egyptian copper mining in the Timna Valley. Pottery evidence from Site 30, Layer III,[5] proved that there was a pre-Temple phase of Egyptian activities at Timna. This could indicate that the beginning of Egyptian involvement in the Arabah may have preceded the time of Seti I (1318–1304 BC), but such an assumption is not verifiable on the strength of the available evidence, nor is it possible at this stage to identify such a pre-Seti I Pharaoh.

Midianite pottery found in Stratum IV indicated that already in this first phase of the Temple the cult partnership, which we see throughout the existence of the Hathor Temple, was present from the beginning. We know from Serabit el-Khadim that a similar cult partnership existed between the Egyptians and local Semitic tribes who called Hathor by the Semitic name of Baalat.[6]

It is tempting to relate the Biblical traditions of the Kenites-Midianites as traditional metallurgists, inhabitants of North-West Arabia, to the 'Midianites' from the Hejaz, who worked with the Egyptians in the copper smelting camps of the Arabah and also brought offerings to the Mining Temple, but these aspects will be dealt with in a future publication.

According to the evidence of the local 'Negev-ware' pottery in Stratum IV, members of the local tribes who worked in the Egyptian mines and smelters apparently also took part in the Temple rites. We are led to suggest that these local inhabitants of the Arabah and the Negev are the Amalekites mentioned in the Biblical narrative.[7]

It is difficult to establish the time and mode of the abandonment or destruction of the Stratum IV Hathor Temple because the builders of the new Temple of Stratum III almost totally obliterated the previous Temple – perhaps after an earthquake and/or heavy rockfall from the overhanging rock-shelf of the Pillar had demolished the earlier structures.

The Hathor Temple of Ramesses II – Stratum III

The Hathor Temple of Stratum III, with its newly laid White Floor and newly built central *naos* and *pro-naos* and courtyard walls, was evidently erected during the reign of Ramesses II (1304–1237 BC) and remained in use, with periods of abandonment, re-use and frequent renovation, until the time of Ramesses V (1160–1156 BC).

One of the outstanding features of the Temple of Stratum III was the casting workshop in the Temple courtyard, where raw copper from the smelters was refined and alloyed with imported metallic tin and cast into votive objects. A considerable number of iron finger-rings and bracelets, at least some of which were gilded, could be proved to have been manufactured from iron produced as a by-product in the copper smelters of Timna – the first archaeological evidence for the origin of iron as an adventitious by-product of copper smelting in the Late Bronze Age.[8] There were also fragments of lead of local origin which had apparently been worked in the Temple workshop.

As in the previous Temple of Stratum IV, the Midianites and, to a much lesser degree, the Negevites also, worked in the mines and smelters and continued their participation in the ritual activities of the Stratum III Hathor Temple.

It was evidently during the reign of Ramesses V that the Mining Temple, as well as the huge Egyptian copper industry of the Arabah, was abandoned by the Egyptians. Although during the 22nd Dynasty (10th–9th centuries BC) an Egyptian expedition returned for a short period to the smelters of Timna – Site 30, Layer I – the Egyptians did not return to the Hathor Mining Temple, perhaps because by that time the Temple was already devastated and partly covered by a huge rockfall (on top of Stratum II).

The Midianite tented shrine – Stratum II

The Midianites who worked with the Egyptians in the Arabah copper mines evidently also worshipped Hathor in the Timna Temple and were left in sole control of the copper industries of the Arabah after the final departure of the Egyptian mining operations in the 20th Dynasty. It was not possible to ascertain whether it was an earthquake or a particularly large rockfall from the top of the Pillar, which crushed the south wall (Wall 3) of the Temple courtyard and perhaps also caused the collapse of the *naos* structure. There is clear evidence that the Temple had been badly damaged before the Midianites turned the Egyptian Hathor Temple into a Semitic tented desert shrine – Stratum II.

Stratum II is characterised by a basic change in the very concept of the Temple rites, represented by the architecture and standing features of Stratum II as compared with Stratum III. Whilst the layout of Stratum III was clearly that of an Egyptian shrine, the layout of Stratum II, with its woollen tent cover, Offering Bench, Standing Stones (*mazeboth*) and attached 'Cell of the Priest' was built according to the typical schema of a Semitic Temple. The tent cover and hangings made of wool, the preponderance of Midianite pottery – exclusive as votive offerings inserted into the Temple Wall (Wall 2) behind the Offering Bench – and the numerous bones of young goats and sheep, completed the picture of the Midianite desert shrine.[9]

The reconstruction of the historical significance of Stratum II, implying the presence of people from Midian = Hejaz – the 'Midianites' – in Timna throughout the 14th–12th centuries BC, is not only based on the specific character of the features of Stratum II, but also on additional artifactual evidence:

(1) The proven Midianite (Hejaz) origin of large numbers of ordinary, often finely decorated kitchen-ware vessels (in addition to large storage vessels) including little bowls, jars and juglets for daily use, characteristic of all layers of the New Kingdom smelting camps in the Arabah.

(2) The find of numerous camel bones in the New Kingdom smelting camps of Timna,[10] which must have come with people from Arabia as camels were not used in Egypt until Roman times. A drawing of a camel has been found in Qurayya, on a typical Midianite sherd [11] and many drawings of ostriches were found on the Midianite pottery in the Temple, as well as in the Hejaz. It should be noted that, contrary to the large number of camel bones found in the smelting camps, no camel bones were found in the Temple itself, even in the Midianite Stratum II context. This might indicate that the camel was taboo as a votive offering for the Midianites also.

(3) Pottery found in the Timna Temple provided evidence for the meeting in the Arabah of two different technological attitudes, i.e. different traditions in pottery making which could be identified as Egyptian and Midianite respectively,[12] working next to each other in the same area. The same contemporaneous use of two different technological traditions could be established in the smelting furnace techniques of Timna[13] where Egyptian and Midianite metallurgists worked together. It was highly informative, in relation to the problem discussed here, that there was also pottery as well as a smelting furnace that showed the integration of both technological attitudes in the production of intermediate types of vessels and furnaces, i.e. Egyptians and Midianite simply learned from each other whilst working together in the same workshop.

The stratigraphic location of the olive green-grey interface of Stratum II, right on top of, or even in the habitation and destruction remains of Stratum III, without any intervening fill layers to indicate a period of abandonment, provided evidence for an almost direct continuation of the occupation of Site 200, after the Egyptians left the area. Further evidence for the uninterrupted continuation of the Midianite presence in Timna, i.e. a date for Stratum II in the middle of the 12th century, can be seen in the smelting camps, where no separate Midianite layer of activities could be distinguished.

After the Egyptians abandoned their copper workings in the Arabah, the Midianites, for a short time, carried on with their work in the smelters, until they too left the area. The Midianite presence in Timna is well documented for the period of the 19th–20th Dynasties, i.e. the end of the 14th to the middle of the 12th centuries BC, for Strata IV–III, with Stratum II immediately following as a further, shortlived phase of the Timna Temple.[14]

Although the excavations at Site 200 did not provide a definite date for the end of the Stratum II occupation, the tented shrine of Stratum II was undoubtedly only a shortlived, makeshift establishment. However, it was still standing when a huge rockfall descended from the top of the Pillar, covering most of the south-west section of the site, including the row of Standing Stones and also part of the *naos*. This rockfall signified the final end of the Temple of Timna.

Roman squatters in the Temple courtyard – Stratum I

The abandoned structures of the Temple gradually became covered by wind-borne sand, but the outlines of the outer Temple wall were still recognisable when a new group of miners arrived in the Timna Valley. This is dated by pottery to the Roman period, more accurately to the period from the end of the 1st to the 3rd century AD. Small groups of these miners repeatedly camped for short periods amongst the sand-covered walls of the Temple, in the welcome shade of the Pillar. At some stage, the newcomers must have recognised the nature of the sand-covered remains underneath, and dug a robbers' trench into the *naos* and *pro-naos*. It is probably due to these Roman diggings that some architectural elements and a square Pillar fragment from the Egyptian Temple came to be lying on the surface of the hillock of Site 200.

Historically, Stratum I relates to the period from the Roman conquest of Nabataea in AD 105/6, to the end of the Severan Dynasty in AD 235, during which time the copper mines in the Arabah were again active. There is no evidence for Nabataean activities in the mines of the Timna Valley, nor in the Wadi Amram. Roman mining in the Timna Valley was on a fairly small scale, mainly because the copper deposits had already been intensively exploited by the Egyptian mining expeditions.

After the Roman period no further occupation of Site 200 could be identified. Barely visible in Roman times, the Temple structure slowly disappeared under the cover of wind-borne sand and at some stage another huge rockfall descended onto the southern half of the low, sandy hillock of Site 200.

Notes

1. Rothenberg, 1972.
2. A number of such rich ore nodules, many of peculiar shape, were also found in Stratum II and IV, and had obviously been brought to the Temple as votive gifts (see Chapter III, 27–29).
3. See preliminary reports in Sinai 1981.
4. Radiocarbon dates: 5th–4th millennium BC; see above, p. 18, note 75.
5. Rothenberg in Conrad and Rothenberg, 1981, 210.
6. Petrie, 1906, 129–32; Albright, 1969, Chapt. III.
7. Rothenberg, 1972, 153–4.
8. N.H. Gale, H.G. Bachmann, B. Rothenberg, and R.F. Tylecote in Vol. 2 of this publication.
9. 'Midianite' is here mainly understood as a geographical term, i.e. the Hejaz (North-West Arabia) commonly identified by Biblical geographers as Midian. We refrain in this volume from the Biblical aspects of the problem of people from Midian = Hejaz in Timna of the Late Bronze Age. This is, however, undoubtedly of considerable importance for the understanding of the Biblical scene at the time of the Exodus.
10. According to H. Lernau, who investigated the camel bones of our excavation, the camels were used as transport animals, as well as food.
11. Ingraham, M. *et al.* 1981, pl. 79:14.
12. J. Glass, above, III, 4, pp. 105–8.
13. Rothenberg, *Researches in the Arabah*, Vol. 2, Chapt. 1.
14. It should be noted that in the Late New Kingdom Layer I of Site 30, dated 10th–9th centuries BC, no Midianite pottery was found.

Abbreviations

AAAS	Annales Archéologie Arabienne Syriennes
AASOR	Annual of the American School of Oriental Research
ACTA BOT. ACAD. SCI. HUNGAR	Acta Botanica Academiae Scientiarum Hungaricae
AJA	American Journal of Archaeology
AMER. J. ENOL. VITIC.	American Journal of Enology and Viticulture
ANN. MAG. NAT. HIST.	Annual Magazine of Natural History
AS	Anatolian Studies
ASAE	Annales du Service des Antiquités de l'Egypte
BA	Biblical Archaeology
BASOR	Bulletin of the American School of Oriental Research
BASOR SUPLE. STUD.	Bulletin of the American School of Oriental Research. Supplement Studies
BER. DT. BOT. GES.	Berichte der Deutschen Botanischen Gesellschaft
BHMG	Bulletin, the Historical Metallurgy Group
BIAL	Bulletin of the Institute of Archaeology, London
BIFAO	Bulletin de l'Institut Français d'Archéologie Orientale, Chicago
BIRC	Bulletin, Research Council of Israel
BMH	Bulletin, Museum Haaretz, Tel Aviv
BMMA	Bulletin of the Metropolitan Museum of Art
BMOP	British Museum Occasional Paper
BTS	Bible et Terre Sainte
BULL. INST. EGYPTE	Bulletin de l'Institut d'Egypte
BULL. SOC. BOT. GENEVE	Bulletin de la Societé Botanique de Genève
CG	Catalogue Général del Antiquités egyptiennes du Musée de Caire
EEFM	Egypt Exploration Fund Memoirs
GM	Göttinger Miszellen, Beiträge zur ägyptologischen Diskussion
HIST. MET.	Historical Metallurgy (Journal of the Historical Metallurgy Society)
IEJ	Israel Exploration Journal
ILN	Illustrated London News
ISRAEL J. BOT.	Israel Journal of Botany
ISRAEL JOUR. OF EARTH SCI.	Israel Journal of Earth Science
ISVEST. TSHA.	Izvestia Timiryazevskoi sel'sko-khozyaistvennoi akademii
JARCE	Journal of the American Research Center in Egypt
JAS	Journal of Archaeological Science
JEA	Journal of Egyptian Archaeology
JFA	Journal of Field Archaeology
JHMS	Journal of the Historical Metallurgy Society
JIM	Journal of the Institute of Metals, London
J. MORPH	Journal of Morphology
JNES	Journal of Near Eastern Studies
LE	Lexikon der Agyptologie, I-VI, 1972–1986
MDAIK	Mitteilungen des Deutschen Archäologischen Institutes, Abteilung Kairo
MEM. MUS. ANTHROP. UNIV. MICH.	Memoirs of the Museum of Anthropology, University of Michigan
MITT.ANTHROPOL. GES. WIEN	Mitteilungen der Anthropologischen Gesellschaft in Wien
OAS	Oposcula Athenesis
PALEST. J. BOT.	Palestine Journal of Botany
PEQ	Palestine Exploration Quarterly
PPS	Proceedings of the Prehistoric Society
PSBA	Proceedings of the Society of Biblical Archaeology
QDAP	The Quarterly of the Department of Antiquities in Palestine
RB	Revue Biblique
VT	Vetus Testamentum
Wb	Erman, A. and Grapow, H., 1955, Wörterbuch der Agyptischen Sprache, im Auftrag der Deutschen Akademien, 5 Vol. (Neudruck)
WZKM	Wiener Zeitschrift für die Kunde des Morgenlandes
YMH	Yearbook, Museum Haaretz, Tel Aviv
ZAS	Zeitschrift für Aegyptische Sprache und Alterstumskunde
ZDPV	Zeitschrift, Deutscher Palästina Verein
ZEIT. DEUTSCH. GEOL. GES.	Zeitschrift der Deutschen Geologischen Gesellschaft

General Bibliography

Aharoni, Y. 1960. An Israelite Agricultural Settlement at Ramat Matred (The Ancient Desert Agriculture of the Negev), *IEJ* l0, 97–111.

—— 1966. King Solomon's Mines – to Yadin's question, *Haaretz*, 10.4.66 (Hebrew).

—— 1978. *The Archaeology of Eretz Israel*, Jerusalem (Hebrew).

Albright, W.F. 1941. New Light on the Early History of Phoenician Colonization, *BASOR*, 83, 14–22.

—— 1949. *The Archaeology of Palestine*, London.

—— 1969. *The Proto-Sinaitic Inscriptions and their Decipherment*. Cambridge, Mass.

Alexander, J. 1970. *The Direction of Archaeological Excavations*, London.

Alt, A. 1935. Aus der Araba II, *ZDPV*, 58, 1–78.

Amiran, R. 1969. *Ancient Pottery of the Holy Land*, Jerusalem.

Amiran, R., Beit-Arieh, I. and Glass, J. 1973. The interrelationship between Arad and Sites in Southern Sinai in the Early Bronze Age II, *IEJ* 23, 193–7.

Avigad, N. 1963. The Yotham Seal from Elath (Hebrew), *Elath*, Jerusalem, 21–5.

Avi-Yona, M. 1940. *Map of Roman Palestine*, (1:125000), London.

Bachmann, H.G. and Hauptmann, A. 1984. Zur alten Kupfergewinnung in Fenan und Hirbet en-Nahas im Wadi Arabah in Südjordanien, *Anschnitt*, 35, 110–23.

Bachmann, H.G. and Rothenberg, B. 1980. Die Verhüttungsverfahren von Site 30, in Conrad and Rothenberg 1980, 215–36.

Bamberger, M. 1984. Shape and Microstructure of copper produced in a reconstructed ancient smelting process, *Hist. Met.* Vol. 18, 31–4.

—— 1985. The Working Conditions of the Ancient Copper Smelting Process, in Craddock and Hughes (eds.), 1985, 151–7.

Bamberger, M., *et al.* 1986. Ancient Smelting of Oxide Copper Ore, Archaeological Evidence at Timna and Experimental Approach, *Metall* 40, 1166–76.

—— 1988. Mathematical modeling of Late Bronze Age/Iron Age smelting of oxide copper ore, *Metall*, 42.

Barker, P. 1977. *Techniques of Archaeological Excavation*, London.

Bartura, Y., Hauptmann, A., Schöne-Warnefeld, G. 1980. Zur Mineralogie und Geologie der antik genutzten Kupferlagerstätte im Timna-Tal, in Conrad and Rothenberg, 1980.

Bass, G.F. 1986. A Bronze Age Shipwreck at Uln Burun (Kas): 1984 Campaign, *AJA* 90, 269–96.

Beit-Arieh, I. 1977. *South Sinai in the Early Bronze Age*, (Ph.D. Thesis), Tel Aviv.

—— 1980. Settlements of the Early Bronze Age II in South Sinai, in Meshel and Finkelstein (eds.) *Sinai in Antiquity* (Hebrew), Tel Aviv, 295–312.

—— 1981. A Pattern of Settlements in Southern Sinai and Southern Canaan in the Third Millennium B.C., *BASOR* No. 243, 31–55.

—— 1981a. An Early Bronze Age II Site near Sheikh 'Awad in Southern Sinai, *Tel Aviv*, 95–127.

—— 1982. An Early Bronze Age II Site near the Feisan Oasis in Southern Sinai, *Tel Aviv*, 146–56.

Bercovici, A. 1978. Flint implements from Timna Site 39, in Rothenberg, Tylecote, Boydell (eds.) 1978, 16–20.

Cohen, R. 1980. The Iron Age Fortresses in the Central Negev, in *BASOR* 236, 61–78.

Conrad, H.G. and Rothenberg, B. (eds.). 1980. *Antikes Kupfer im Timna-Tal*, Der Anschnitt, Beiheft 1, Bochum.

Craddock, P.T. 1980 (ed.) *Scientific Studies in Early Mining and Extractive Metallurgy*, British Museum Occasional Paper No. 20, 165–75.

—— 1980a. The Composition of Copper Produced at the Ancient Smelting Camps in the Wadi Timna, Israel, in Craddock (ed.) 1980.

Craddock, P.T. and Hughes, M.J. (eds.). 1985. *Furnaces and Smelting technology in Antiquity*, British Museum Occasional Paper No. 48.

Flinder, A. 1969. Jeziret Fara'un (MSS report, unpublished).

—— 1985. *Secrets of the Bible Seas*, London.

Frank, F. 1934. Aus der Araba I, *ZDPV* 57, 191–280.

Gale, N.H. 1984. Mystery of Timna's iron solved by lead isotope "fingerprinting", *IAMS Newsletter* No. 6, 6–7.

Galili, E. 1986. Tin and Copper ingots from the depths of the sea, *IAMS Newsletter* No. 9.

Galili, E., Shmueli, N. and Artzy, M. 1985. Bronze Age ship's cargo of Copper and Tin, *Nautical Archaeology*, 14.4.

Glueck, N. 1935. Explorations in Eastern Palestine, II, *AASOR*, 15.

—— 1938. The First Campaign at Tell el-Kheleifeh (Ezion Geber), *BASOR*, 71, 3–17.

—— 1939. The Second Campaign at Tell el-Kheleifeh (Ezion Geber: Elath), *BASOR*, 75, 8–22.

—— 1940. The Third Season of Excavation at Tell el-Kheleifeh, *BASOR*, 79, 2–18.

—— 1940. *The Other Side of the Jordan*, New Haven.

—— 1959. *Rivers in the Desert*, New York.

—— 1960. Archaeological Exploration of the Negev in 1959, *BASOR*, 159, 3–14.

—— 1965. Ezion Geber, *B.A.* 28, 2–18.

—— 1967. Some Edomite Pottery from Tell el-Kheleifeh, *BASOR* 188, note 40.

—— 1969. Some Ezion Geber: Elath Iron II Pottery, *Eretz Israel*, 9; 51–9.

—— 1977. Tell el-Kheleifeh, *Encyclopedia of Archaeological Excavations in the Holy Land*, Jerusalem, Vol. 3.

Gourdin, W.H. and Kingery, W.D. 1975. The Beginnings of Pyrotechnology: Neolithic and Egyptian Lime Plaster, *JFA* Vol. 2, 133–50.

Harris, E.C. 1979. *Principles of Archaeological Stratigraphy*, London.

Hauptmann, A., Weisgerber, G. and Knauf, E.A. 1985. Archaometallurgische und bergbauarchäologische Untersuchungen im Gebiete von Fenan, Wadi Arabah (Jordanien), *Anschnitt*, 37, 163–95.

IAMS Newsletter. 1979–. The Institute for Archaeo-Metallurgical Studies, Institute of Archaeology, University College London.

Ingraham, M., Johnson, T., Rihani, B. and Shatla, I. 1981. Preliminary Report on a Reconnaisance Survey of the Northwestern Province, *Atlal, Vol. 5, 59–84*.

Kalsbeek, J. and London, G. 1978. A Late Second Millennium B.C. Potting Puzzle, BASOR 232, 47–56.

Karageorghis, V. 1976. *Kition*, London.

Keeley, A. 1976. *The Pottery of Ancient Egypt*, Toronto.

Kenyon, K. 1961. *Beginning in Archaeology*, London.

—— 1971. An Essay on archaeological techniques: the publication of results from the excavations of a tell, *Harvard Theological Review*, 64, 271–9.

Kertesz, Tr. 1976. The Breaking of Offerings in the Cult of Hathor, *Tel Aviv* 3, 134–6

Kozloff, B. 1974. A Brief Note on the Lithic Industries of Sinai, *Yearbook, Museum Haaretz Tel Aviv* 15/16, 35–49.

Lucas, A.E. 1962. *Ancient Egyptian Materials and Industries*, 4th ed. rev. by J.R. Harris, London.

Lupu, A. and Rothenberg, B. 1970. The Extractive Metallurgy of the Early Iron Age Copper Industry in the Arabah, Israel, *Archaeologia Austriaca*, 47, 91–130.

McLeod, B.H. 1962. The Metallurgy of King Solomon's Copper Smelters, *PEQ*, 94, 68–71.

Merkel, J. 1983. *A Reconstruction of Bronze Age Copper smelting, Experiments Based on Archaeological Evidence from Timna, Israel* (unpublished) Ph.D. Thesis, Inst. of Archaeology, University of London.

—— 1983a. Summary of Experimental Results for Late Bronze Age Copper Smelting and Refining, *MASCA Journal* Vol. 2, 6:176.

Meshel, S. 1975. Notes to the Problem of Tell el-Kheleifeh, Elat and Ezion Geber (Hebrew), *Eretz Israel*, 12, 49–56.

280

Milton, C., *et al.* 1976. Slag from an Ancient Copper Smelter at Timna, Israel, *JHMS* 10/1, 24–33.

Muhli, J.D. 1976. *Copper and Tin*, New Haven.

Munsell Color Company Inc. 1970. *Munsell Books of Color*, Baltimore, Maryland.

Musil, A. 1908. *Arabia Petraea*, Vol. 2/2, Vienna.

Parr, P.J., Harding, G.L. and Dayton, J.E. 1970. Preliminary Survey in N.W. Arabia, 1968, *BIAL*, 23–61.

Petherick, J. 1861. *Egypt, the Soudan and Central Africa*, Edinburgh-London.

Petrie, W.M.F. 1906. *Researches in Sinai*, London.

Pinkerfeld, J. 1942. Excavations at Ezion-Geber, *Kedem* I (Hebrew series).

Pratico, G. 1982. A Reappraisal of Nelson Glueck's Excavations at Tell el-Kheleifeh, *ASOR* Newsletter 6, 6–11.

—— 1983. *Tell el-Kheleifeh 1937-1940*: A reappraisal of Nelson Glueck's Excavations with Special Attention to the Site's Architectual and Ceramic Traditions, Ph.D. Thesis, Cambridge Mass.

—— 1985. Nelson Glueck's 1938–1940 Excavations at Tell el-Kheleifeh: A Reappraisal, *BASOR*, 259.

Preiss, M. 1967. Notes to the Geology of the Copper Ore Deposits in Timna and Nahal Amram, in Rothenberg, 1967a.

Ronen, A. 1970. Flint implements from Sinai, *PEQ* 102, 20–37.

Rothenberg, B. 1959. König Salomons Kupfergruben, *Frankfurter Allgemeine Zeitung*, 5.8.59.

—— 1960. King Solomon's Mines, A New Discovery, *ILN*, 3.9.60.

—— 1962. Ancient Copper Industries in the Western Arabah (with contributions by Aharoni, Y. and McLeod, B.H.), *PEQ*, 94, 5–65.

—— 1965. How did King Solomon's Mines disappear, *Haaretz*, 10.12.65 (Hebrew).

—— 1966. The Chalcolithic Copper Industry at Timna, *BMH*, 8, 86–93.

—— 1967. Archaeological Sites the Southern Arabah and the Eilat Mountains, in Ron, Z. (ed.), *The Eilat Survey*, Vol. II, 283–331, Eilat Regional Council, Yatvata, Vol. III, (Hebrew).

—— 1967a. *Zefunot Negev: Archaeology in the Negev and the Arabah* (Hebrew), Tel Aviv.

—— 1967b. Excavations at Timna, 1964–66, *BHM* 9, 53–70.

—— 1969. King Solomon's Mines no more, *ILN*, (Archaeology 2323) 15.11.69, 32–3.

—— 1969a. The Egyptian Temple of Timna, *ILN*, (Archaeology 2324) 29.11.69, 28–9.

—— 1970. An Egyptian Temple of Hathor discovered in the Southern Arabah (Israel), *BMH* No. 12, 28–35.

—— 1970a. Un temple égyptien découvert dans la Arabah, in *Bible et Terre Sainte*, Paris, 6–14.

—— 1971. *Midianite Timna - Valley of the Biblical Copper Mines*, Exhibition Catalogue, British Museum, London.

—— 1971a. The Arabah in Roman and Byzantine Times in the Light of New Research, *Roman Frontier Studies 1967*, Tel Aviv, 211–23.

—— 1972. *Timna, Valley of the Biblical Copper Mines*, London.

—— 1978. Timna, *Encyclopedia of Archaeological Excavations in the Holy Land*, Masada, Jerusalem, 1184–1203.

—— 1979. *Sinai*, Bern.

—— 1980. Introduction, in Conrad and Rothenberg 1980, 17–32.

—— 1980a. Die Archäologie des Verhüttungslagers Site 30, in Conrad and Rothenberg 1980, 187–214.

—— 1983. Explorations and Archaeo-Metallurgy in the Arabah Valley (Israel) *Bulletin of the Metals Museum*, Japan Institute of Metals, 1–16.

—— 1983a. Metal-working in the 'Dark Age' of the Near East, *IAMS Newsletter* No. 5 (unsigned news report), 3.

—— 1985. Copper Smelting Furnaces in the Arabah, Israel: The Archaeological Evidence, in Craddock, P.T. and Hughes M.J. (eds.) 1985, 123–50.

—— 1986. Radiocarbon (C14) Dating helps to solve riddle of Timna's Late Bronze Age Smelting Furnaces, *IAMS Newsletter* 10,7, 1986.

Rothenberg, B. and Bachmann, H.G. 1973. *Timna, Tal des Biblischen Kupfers*, Exhibition Catalogue, Bergbau Museum, Bochum.

Rothenberg, B. and Blanco, A. 1981. *Studies in Mining and Metallurgy in South-West Spain*, (Metal in History, Vol. I), London.

Rothenberg, B. and Glass, J. 1983. The Midianite Pottery, in Sawyer, J.F.A. and Clines, J.A. (eds.) *Midian, Moab and Edom*, Sheffield, 65–124.

Rothenberg, B. and Ordentlich, I. 1979. A comparative chronology of Sinai, Egypt and Palestine, *BIAL* No. 16, London, 233–7.

Rothenberg, B., Tylecote, R.F. and Boydell, P.J. 1978. *Chalcolithic Copper Smelting*, Archaeo-Metallurgy One, IAMS Monograph (N, 1), London.

Slatkin, A. 1961. Nodules Cuprifères du Negev Meridional (Israel), *Bull. Counc. Israel*, Sect. G 10, 292–301.

Tylecote, R.F. 1982. Metallurgical Crucibles and Crucible Slags, *Archaeological Ceramics*, 231–42.

Tylecote, R.F. and Boydell, P.J. 1978. Experiments on Copper Smelting based on Early Furnaces found at Timna, in Rothenberg, B. *et al. Chalcolithic Copper Smelting*, Archaeo-Metallurgy, IAMS Monograph (No. 1), London, 27–49.

Tylecote, R.F., Lupu, A. and Rothenberg, B. 1967. A Study of Early Copper Smelting and Working Sites in Israel, *Journal of the Institute of Metals*, London, 235–43.

Tylecote, R.F. and Merkel, J.F. 1985. Experimental smelting techniques: achievements and future, in Craddock and Hughes, 1985, 3–20.

Weisgerber, G. 1975. Ägyptische Kupfergewinnung in Timna (Südisrael), *Archäologisches Korrespondenzblatt*, 35–40.

Wright, G.E. 1961. More on King Solomon's Mines, *BA* 24, 59–62.

Yadin, Y. 1961. Hazor, Gezer and Megiddo in the days of Solomon, in Malamad (ed.) *The Kingdoms of Israel and Judah*, (Hebrew), Jerusalem.

—— 1965. King Solomon's Mines – how did they disappear? *Haaretz*, 3.12.65 (Hebrew).

—— 1966. More on Solomon's Mines, *Haaretz*, 26.4.66 (Hebrew).

List of Loci, Levels and Find Box Numbers

LOCUS 101

LEVEL	BOX NUMBERS

LEVEL	BOX NUMBERS
100	
	4 9 10 12 13 16 17 19 20 22 31 39 40
110	
	306
120	
	58 60 97 100 119 294 306 307
130	
	46 48 49 57 100 114 118 119 122 155 174 294 306 307 308
140	
	47 50 51 92 100 107 111 112 113 118 119 139 146 159 163 167 171 294 295 307 308
150	
	54 84 94 100 101 108 109 110 118 119 132 137 144 146 172 203 288 294 295 303 308 312 230
160	
	81 83 90 105 117 118 120 123 131 146 147 156 160 173 201 209 211 212 217 218 219 229 230 288 295 297 303 308 309 311 312
170	
	66 67 68 74 75 81 83 128 129 136 148 170 175 211 212 216 217 218 219 223 227 230 231 246 268 285 303 308 309 311 312 322
180	
	68 69 71 75 76 78 80 81 82 87 106 121 127 130 143 209 211 212 216 224 227 231 233 234 241 246 250 254 268 302 309 311
190	
	71 104 124 125 126 208 224 227 233 234 241 250 254 268 302 315
200	
	202 224 232 233 241 243 250 254 289 290 296 315 234
210	
	202 241 243 250 289 290 296 254
220	
	289 369
230	

LOCUS 102

LEVEL	BOX NUMBERS
100	
	339
110	
	339
120	
	339
130	
	70 102 339
140	
	70 102 339
150	
	70 102 339
160	
	70 81 102 284 339
170	
	81 284 339
180	
	80 81 82 88 151 260 284 337 339
190	
	207 208 253 257 260 337 339
200	
	253 260 270 287 337 339 289
210	
	253 260 270 287 337 289
220	
	270 287 337 289
230	
	270 337
240	

LOCUS 103

LEVEL	BOX NUMBERS
170	
	261 248
180	
	374 261 262 248
190	
	263 264 374 261 262 263 248 272
200	
	263 264 265 374 261 262 263 248 272
210	
	263 264 265 374 261 262 263 272
220	
	263 264 265 374 261 262 263 313 321 272
230	
	264 265 352 321 352 272
240	
	265 321 272
250	

LOCUS 104

LEVEL	BOX NUMBERS
140	
	275
150	
	249 275
160	
	266 275
170	
	248 266 275
180	
	226 266 275 248
190	
	226 266 272 275 333 248
200	
	226 248 266 267 272 275 333 286
210	
	226 266 272 333 267 286
220	
	226 266 272 333 267 286
230	
	226 266 272 267
240	
	226 267 272

LOCUS 105

LEVEL	BOX NUMBERS
80	
	310
90	
	252 310
100	
	252 310
110	
	252 310
120	
	252 310
130	
	252 310 324
140	
	252 274 310 323 324 329 330 356
150	
	274 323 330 356 357 252
160	
	274 320 323 330 355 356 357
170	
	274 320 330 357
180	
	274 320 357
190	
	357 274
200	
	286 357
210	
	286 357
220	
	286
230	

LOCUS 106

LEVEL	BOX NUMBERS
0/10	
	21 27 28 30 35 37 44
80	
	305
90	
	252 305
100	
	252 280 305
110	
	252 305 280
120	
	252 305 280
130	
	213 214 252 305 324 327 304 280
140	
	214 244 252 255 273 305 323 324 327 329 330 304 213 280
150	
	214 244 252 255 258 271 273 305 323 327 330 357 304 213 280
160	
	214 244 255 258 271 273 283 323 327 330 342 355 357 304
170	
	214 244 255 258 271 273 330 357 304 283
180	
	214 221 244 258 259 271 273 319 357 374 377 304 378 283 279
190	
	357 374 377 360 279 378
200	
	345 357 362 374 377 378
210	
	357 362 363 374 378
220	
	363 374 378
230	

LOCUS 107

LEVEL	BOX NUMBERS
0/10	
	11 14 26 29
100	
	53 168
110	
	59 62 158 239
120	
	62 86 93 158 222 239 245
130	
	62 140 141 142 158 214 222 225 239 245
140	
	214 245 236 239 242 245 251 255 269 277 278
150	
	214 244 236 239 242 245 251 255 269 277 278
160	
	214 236 251 255 269 277 278 283 348
170	
	214 236 255 269 277 278 348 283
180	
	214 221 278 279 348 283
190	
	279
200	
	345
210	

LOCUS 108

LEVEL	BOX NUMBERS
130	
	304
140	
	304 244
150	
	304 244
160	
	304 244
170	
	304 244
180	
	304 259 319 244
190	
200	
	362 345 347
210	
	362 363
220	
	363
230	

LOCUS 109

LEVEL	BOX NUMBERS
0/10	
	18 24 25
10/20	
	43
100	
	301 339
110	
	301 339
120	
	60 61 100 301 318 334 339
130	
	55 70 73 77 100 102 103 140 301 304 318 334 339
140	
	70 73· 77 79 100 102 103 304 318 326 334 339
150	
	70 73 77 79 94 99 100 102 103 108 109 135 278 301 304 318 326 332 334 339 204 236
160	
	70 79 102 204 220 278 301 304 326 331 332 334 339 350 236
170	
	204 220 240 278 301 304 314 331 332 334 339 343 235 236
180	
	204 240 247 259 278 298 299 304 314 331 338 339 343 349 235 220
190	
	247 256 299 338 339 367 366 235
200	
	256 299 300 338 339 367
210	
	256 300
220	
	256 300 368
230	

LOCUS 110

LEVEL   BOX NUMBERS

0/10
    1   2   3   4 11 18 23 26 36 38 85

10/20
  85 91

110
  45 62 116 158 239 317

120
  56 58 60 62 63 72 93 100 116 119 158 205 239 317

130
  55 56 62 63 72 100 116 119 158 205 225 238 239 317 325

140
  56 63 72 95 100 116 119 205 225 238 239 325

150
  98 99 100 116 119 135 137 138 205 206 210 225 228 238 239 242 251 277 281 325

160
210 236 237 251 277 325 344 350 361 370 228

170
210 215 236 277 282 353 359 361 365 370 375

180
134 210 215 282 359 365 370 371 375

190
282 358 359 365

200
358 359

210

LOCUS 111

LEVEL   BOX NUMBERS

0/10
  38 42 52 85

10/20
  52 85

100
316

110
  45

120
  93 205

130
205 238

140
205 238 276

150
205 206 210 238 276 281 228

160
210 237 276 344 361 370 228

170
210 215 276 282 353 354 361 370 372

180
134 210 215 276 282 354 370 371 373 376

190
282 358 376

200
358 376

210
376

220

LOCUS 112

LEVEL	BOX NUMBERS
0/10	
	2 3 4 5 6 7 8 9 10 12 13 15 16 17 19 20 31 32 33 34 36 39 40 41
10/20	
	91
110	
	116
120	
	58 97 116 119
130	
	57 115 116 119 122 174
140	
	95 115 116 119
150	
	98 115 116 119 137
160	
	120 218 335
170	
	218 282 285 335
180	
	282 335 336
190	
	282 336
200	

List of Find Boxes[1]

No.	Square	Level	Locus	Stratum
1	CD 13–14	110–120	110	
2	AB 14–15	108–118	112, 110	
3	BC 14–15	135–145	112, 110	
4	CD 14–15	115–125	110	
5	AB 15–16	118–128	112	
6	CD 15–16	126–136	101, 112	
7	AB 16–17	0–10	112	
8	BC 16–17	0–10	112	
9	CD 16–17	125–135	101, 112	
10	CD 17–18	124–134	101, 112	
11	CD 10–11	107–110	107, 110	
12	CD 16–17	125–135	101, 112	
13	CD 16–17	125–135	101, 112	
14	EF 8–9	128–138	107	
15	BC 15–16	160–170	112	
16	CD 15–16	126–136	101, 112	
17	CD 15–16	126–136	101, 112	
18	DE 11–12	154–164	109, 110	
19	BC 17–18	160–170	101, 112	
20	AB 17–18	150–160	101, 112	
21	DE 6–7	0–10	106	
22	DE 18–19	151–161	101	
23	CD 11–12	110–120	110	
24	FG 11–12	0–10	109	
25	EF 11–12	113–123	109	
26	CD 10–11	110–120	107, 110	
27	EF 6–7	136–146	106	
28	CD 6–7	84–94	106	
29	EF 8–9	170–180	107	
30	DE 6–7	0–10	106	
31	CD 17–18	124–134	101, 112	
32	AB 15–16	0–10	112	
33	ABC 15–16	128–138	112	
34	BC 16–17	0–10	112	
35	CD 5–6	84–94	106	
36	BC 14–15	135–145	110, 112	
37	EF 6–7	128–138	106	
38	BC 11–12	173–183	110, 111	
39	BC 17–18	160–170	101, 112	
40	AB 17–18	150–160	101, 112	
41	AB 16–17	0–10	112	
42	AB 11–12	110–120	111	+
43	EF 13–14	125–145	109	
44	FG 6–7	174–184	106	
45	BC 11–12	114	110, 111	
46	EF 15–16	139	101	
47	EF 15–16	140	101	
48	DE 15–16	165	101	
49	DE 15–16	134	101	
50	DE 15–16	143	101	
51	EF 15–16	147	101	II
52	AB 11–12	173–175	111	
53	AB 8–9	108	107	
54	EF 15–16	157	101	
55	DE 13–14	138	110	II(I)
56	CA 13–14	120–140	110	
57	CD 15–16	136	101, 112	
58	CD 14–15	125	110	
59	BC 8–9	118	107	
60	DE 14–15	127	110	
61	EF 11–12	123	109, 110	
62	CD 10–11	117–134	107, 110	
63	CD 11–12	120–142	110	II
64	IJ 15–16	175	101	
65	BC 13–14	125–154	110	
66	IJ 15–16	175	101	
67	IJ 16–17	172–182	101	
68	IJ 15–16	172–181	101	
69	IJ 18–19	188–189	101	
70	HI 13–14	139–164	109	
71	IJ 17–18	188–192	101	
72	CD 11–12	120–145	110	II
73	GH 12–13	136–150	109	I
74	IJ 16–17	172	101	
75	HI 18–19	180–176	101	
76	HI 17–18	181		
77	GH 13–14	131–153	109	
78	JK 16–17	189		
79	EF 13–14	145–160	109, 110	
80	JK 14–15	185	102	II
81	IJ 14–15	168–187	102	II
82	JK 14–15	184–185	102	II
83	GH 18–19	168–171	101	
84	HI 15–16	158	101	
85	ABC 13–14	158	101	
86	CD 9–10	123	107	
87	IJ 16–17	123	107	V—III
88	IJ 13–14	182	102	II
89	IJ 13–14	184	102	II
90	GH 17–18	163–166	101	
91	ABC 14–15–16	10–20	110, 112	
92	DE 15–16	143	101	
93	ABC 8–12	120	107, 110, 111	
94	EF 14–15	157	109, 110	
95	BC 14–15	145	110, 112	(II)
96	I 15–16	181	101	
97	CD 15–16	126	101, 112	
98	BC 14–15	155	110, 112	II
99	DE 13–14	155	110, 112	
100	DE 14–15	127–152	110	
101	HI 15–16	158	101	
102	HI 13–14	139–164	109	
103	GH 13–14	131–154	109	
104	JK 17–18	192	101	
105	GH 18–19	168	101	
106	JK 15–16	185	101	
107	GH 15–16	145	101	
108	EF 14–15	157	109, 110	
109	EF 14–15	157	109, 110	
110	EF 15–16	152	101	
111	EF 16–17	142	101	
112	DE 16–17	143	101	
113	DE 15–16	141	101	
114	DE 17–18	138	101	
115	ABC 15–16	138–158	112	
116	ABC 14–15	118–156	110, 112	(II)
117	HI 16–17	166	101	
118	DE 15–16	136–161	101	
119	CD 14–15	126–150	101, 110, 112	
120	ABC 17–18	160	101, 112	
121	J 16–17	182	101	
122	CD 16–17	135	101, 112	
123	HI 16–17	166	101	
124	JK 16–17	193	101	
125	KL 16–17	196	101	
126	KL 16–17	196	101	
127	IJ 16–17	182	101	
128	IJ 17–18	172	101	

No.	Square	Level	Locus	Stratum
129	HI 16–17	172	101	
130	HI 17–18	181	101	
131	DE 18–19	161	101	
132	EF 18–19	155	101	
133	DE 16–17	–	101	
134	ABC 13–14	182	110, 111	
135	DE 13–14	155	110	II (I)
136	GH 18–19	172	101	V—III
137	CD 14–15	155	110	
138	CD 13–14	150	110	II(I)
139	GH 15–16	147	101	
140	EF 10–11	131	107, 109	
141	DE 8–9	136	107	
142	DE 9–10	135	107	
143	HI 17–18	181	101	
144	GH 16–17	156	101	
145	GH 17–18	168	101	
146	EF 16–17	149–162	101	
147	DE 17–18	163	101	
148	IJ 16–17	177	101	
149	DE 14–15	–	110	
150	IJ 13–14	–	102	
151	JK 13–14	185	102	II
152	GH 15–16	–	101	
153	IJ 13–14	–	102	
154	CD 14–15	138	101	
155	CD 18–19	138	101	
156	EF 15–16	162	101	
157	JK 13–14	–	102	
158	CD 10–11	110–134	101	
159	EF 17–18	149	101	
160	EF 16–17	162	101	
161	HI 14–15	–	109	
162	GH 16–17	–	101	
163	DE 15–16	143	101	
164	JK 12–13	–	101	
165	JK 16–17	–	101	
166	HI 13–14	–	109	
167	EF 16–17	142	101	
168	CD 9–10	108	107	
169	ABC 18–19	–	101	
170	IJ 16–17	178	101	
171	GH 15–16	147	101	
172	GH 16–17	156	101	
173	EF 15–16	161	101	II
174	CD 17–18	134	101, 112	
175	IJ 16–17	173	101	
176	JK 13–14	–	102	II
177	CD 16–17	–	101, 112	
178	AB 14–15	–	110, 112	
201	DE 15–16	161–168	101	II
202	J 16	204–204	101	
203	H 15	153–178	101	
204	E 13–14	157–180	110	II
205	BC 13–14	125–154	110, 111	
206	BCD 13–14	150–157	110, 111	II
207	IJ 12–13	190–207	102	II
208	IJ 14–15	190–207	102	II
209	G 15–18		101	
	G 15	106		
	G 16	169		
	G 18	182		
210	AB 13–14	158–183	110	
211	G 15–17		101	
	G 15–16	164–178		
	G 17	178–181		
212	D 15–16–17	165–187	101	
213	EF 6	148–149	106	II
214	EF 7–8	138–183	106, 107	V
215	AB 13–14	170–183	110, 111	V
216	H 15–16	178–180	101	
217	DE 17	163–172	101	
218	CD 17	164–171	101, 112	II
219	E 15–16	168–178	101	

No.	Square	Level	Locus	Stratum
220	GH 12–13–14	163–186	109	III–II
221	EF 7–8	180–186	106, 107	II
222	C 9–10	128–138	107	
223	H 17	173–177	101	
224	G 15–18	180–204	101	III
225	CD 9–11	130–158	107	(III) (I)
226	DEF 2–5		104	II–I
	DE 2	217–240		
	DE 3	183–210		
227	I 17	179–195	101	
228	BCD 13–14	154–161	110	II
229	E 16	164	101	
230	ED 18	162–175	101	
	+C 18	150–156	101	
231	H 15–17	179–185	101	
232	I 15–17	201–211	101	V
233	IJ 18	189–209	101	
234	H 15–18	185–201	101	III
235	G 12–14	174–190	109	III, II
236	EF 9–10–11	157–177	107, 109	III
237	BCD 13–14	160–163	110	III
238	BC 11	130–155	111	II
239	C 10–11	115–140	107	(III)
	CD 10–11	155	107	
240	GH 12–13–14	178–186	109	II
241	JK 15	189–214	101	V-III
242	CD 10–11	150–154	107, 110	(III)
243	J 16–17	203–210	101	V–III
244	H 6–10	140–180	106, 108	II
245	AB 8–10	126–153	107	III (II)
246	H 18	175–186	101	
247	GH 12–14	187–197	109	III
248	HI 4–5		104	III
	H 4	203		
	I 5	174		
249	F 4–5, F 5	158		I
250	K 16–18	186–213	101	V–III
251	CD 9–11	150–165	107, 110	III
252	CD 5–7	94–150	104, 105, 106	
253	JK 14	193–215	102	III
254	GHI 15–18	187–218	101	V
255	CD 7–8	148–170	106, 107	
256	G 12	190–228	109	III
257	IJK 13	190–225	102	V—III
258	H 6–7	153–184	106	II
259	H 6–11	185	106, 108, 109	
260	IJK 10		102	II
	I 12	180–207		
	K 12	206–213		
261	JK 10		103	III
	J 10	178–221		
	K/10	208–221		
262	JK 9		103	
	J 9	180–227		
	K 9	210–227		
263	JK 8		103	
	J 8	190–227		
	K 8	220–229		
264	JK 7		103	
	J 7	190–238		
	K 7	220–238		
265	JK 6		103	
	J 6	200–249		
	K 6	230–249		
266	F 3–5		104	III
	F 3	217–234		
	F 5	168–234		
267	HI 2–3		104	
	H 2–3	208		
	I 2–3	240		
268	DE 15–17		101	V
	D 15–17	183–196		
	E 15–17	172–196		
269	AB 8–10	153–178	107	III (II)

No.	Square	Level	Locus	Stratum
270	IJK 12		102	(V)–III
	I 12	207–233		
	K 12	213–233		
271	H 5–6	157–180	106	
272	HI 4–5		104	
	H 4–5	190–237		
	I 4–5	233–237		
273	EF 6	146–180	106	
274	C 2–5	190–200	105	(III)
275	DE 2–5	140–203	104	III
276	AB 11	146–184	111	+
277	CD 10–11	155–176	107, 110	(III)
278	EF 10–11	157–184	107, 109, 110	III
279	EF 7–9	180–196	106, 107	II
280	CDE 6–7	100–159, 174–215	106	(V) II
281	B 13–14	158	110	III
282	AB 13–15	170–195	110	
283	D 7–9	165–185	106, 107	III–II
284	I 11	166–189	102	
285	B 16–18	170	101, 112	(II)
286	CD 2–4	200–228	104, 105	
287	L 12–13	205–213, 223	102	
288	CDE 18	156–175	101	
289	L 14–15	205–225	101, 102	
290	K 18	200–215	101	V
291		Surface	Wall No. 2	
292		Surface	Wall No. 1	
293		Surface	Wall No. 3	
294	F 15–16	127–147, 152	101	
295	F 17–18	147–167, 170	101	
296	L 16–18	200–217	101	V
297	E 19	160–169	101	V–III
298	G 12–14	188	109	
299	G 12–14	188–200	109	III
300	G 12–14	200–223	109	V
301	F 13–14	100–171	109	II
302	F 17–18	170–191	101	III
303	F 15–16	152–167	101	II
304	G 6–11	135–184	106, 108, 109	I
305	AB 6–7	80–156	106	II
306	B 20–21	118–138	101	
307	CD 20–21	124–140	101	V–III
308	D 20–24	138–170	101	V–III
309	GH 19	167–188	101	
310	AB 4–6	80–144	105	(III)
311	F 15–16	167–180	101	III
312	G 20–24	159–174	101	V–III
313	IJK 10	220	103	
314	F 13–14	175–185	109	III
315	F 15–16	183–203	101	V
316	AB 12	100	111	+
317	CD 12	110–137	110	I
318	E 12	120–154	109	I
319	G 7–12	184–187	106, 108	II
320	AB 1–5		105	(III)
	A 1	180		
	A 5	160		
321	IK 10	224–249	103	V
322	F 15–16	177	101	V
323	AB 6–7	144–165	105, 106	(V) II
324	AB 6–7	144–139	105, 106	III
325	CD 12	137–164	110	II
326	GH 11	142–167	109	
327	C 6	138–165	106	
328		138–158	Wall No. 3	
329	AB 6–7	147	105, 106	
330	B 6–7	149–174	105, 106	II
331	G 11	167–186	109	I
332	E 12	154–174	109	II
333	G 4–5	195–225	104	III (I)
334	F 12	124–174	109	III–II
335	C 15–16	168–180	112	V
336	C 16	180–198	112	V
337	IJKL 11	213–180, 231–219	102	III–I
338	G 11	186–201	109	III

No.	Square	Level	Locus	Stratum
339	HI 11	100–224	109	(III–II)
340	BC 9–10	–	107	
341	BC 10–11	–	107, 110, 111	
342	C 6	169	106	II
343	F 12	174–184	109	III
344	BC 11–12	167	110, 111	III
345	DEFG 6–9	206	106, 107, 108	
346	GH 6	–	106	
347	GH 9	200	108	III
348	DE 10	161–183	107	
349	G 11	187	109	II
350	DE 11–12	164	110	
351	E 7	–	106	
352	L 6–9	234–239	103	
353	BC 11	172–174	110, 111	(III)
354	AB 11	170–185	111	III
355	A 6–7	161	105, 106	
356	A 6	147–168	105	
357	BC 6	150–215	105, 106	II
358	BCD 12	190–204	110, 111	
359	D 12	170–204	110	
360	DEFG 7	190	106	
361	BCD 11	166–176	110, 111	
362	GH 6–10	209–218	106, 108	(IV)
363	GH 6–9	218–227	106, 108	V (IV)
364	DE 13	–	110	
365	CD 12–13	170–192	110	
366	F 12–13–14	174–185	109	III
367	F 12–13–14	199–209	109	III
368	G 12–15 TT–6	225	109	V (IV)
369	E 6	–	106	II
370	BCD 11–14 TT–11	164–183	110	(IV)
371	BD 11–14	183	110	(IV)
372	B 11	171	111	III
373	AB 11–12	185	111	(IV)
374	HI 6	180–227	103, 106	III
375	C 13–14	179–186	110	V
376	AB 12 TT–13	185–218	111	(IV)
377	EFG 7	186–205	106	
378	G 6 TT–16	180–222	106	(V–III)
379	DF 7–11	180–196	106, 107, 109, 110	
400				*400–470 are all III*
401	C 7		106	
402			106	
403			106	
404	B 10		107	
405			107	
406			106	
407	A–B 9–10		107	
408			106	
409		Surface	106	
410			106	
411			107	
412			107	
413			107	
414			107	
415	B 10		107	
416	D 8		107	
417	C 11		110	
418	B 10		107	
419	B 10		107	
420	C 8		107	
421	G 7		106	
422	D 7		106	
423	D 15		101 (W.1)	
424	A 10		107	
425	C 11		110	
426			106	
427	I 6		106	
428	C 11		107	
429	C 11		107	
430			107	
431			107	
432	C 10		107	
433	F 7		106	

No.	Square	Level	Locus	Stratum	No.	Square	Level	Locus	Stratum
434			107		537	TT 55	216–250	106, 107	
435	D-E 10		107		538	TT 51	229–235	109	
436			107		539	TT 52	222–239	108	
437	–		–		540	TT 59	176–185	106	
438			106		541	TT 60		107	(IV)
439			107		542	TT 58 (AB 13)	200–215	111	
440	C 10		107		543	TT 59	195	106	
441			107		544	TT 58	195–228	111	
442			107			(B 12.5–13.5)			
443	–		–		545	TT 58	195–228	111	V (IV)
444	C 10		107			(B 12.5–13.5)			
445	B 7		107		546	TT 60	185–196	107	
446			107		547	TT 55	237–257	107	
447			106 (W.2)		548	TT 58	200	111	(IV)
448			107		549	TT 58 (B 12)	190–206	111	V
449	–		–		550	TT 58	Surface	111	
450	B 12		111		551	TT 58	190–194	111	(IV)
451	D 9		107		552	TT 58	194–204	111	(IV)
452	A-B 7		106		553	TT 58	100	111	
453	C 7		106		554	TT 58 (B 12)	190–206	111	
454	A 9		107		555	TT 58	205	111	(IV)
455	E 16		101		556	TT 51	235–236	109	
456	B 9–10		107		557	TT 51	230–235	109	
457			107		558	TT 60	196–206	107	
458	–		–		559	TT 59	176–195	106	
459			107		560	TT 61a	Surface	109	
460	C 6		106		561	TT 63	204–216	109	IV
461	D 9		107		562	TT 63	216–224	109	V–IV
462	–		–		563	TT 61a	179–190	109	III
463	B 10		107		564	TT 61a	190–192	109	(V)
464	A 10		107		565	TT 60	206–226	107	
465	C-D 13		110		566	TT 62	170	107	
466			109		567	TT 62	211	107	(V–IV)
467			107		567A	TT 62	200–211	107	
468			106		568	TT 59	–	106	
469	E-F 7		106		568A	TT 59	–	106	
470	C 8–9		107		569	TT 63	215	109	V–IV
					570	TT 61a	192–203	110	(V)
500		Surface	110	+	571	TT 61a	206–216	110	V
501	TT 53	164–187	110	III	572	TT 61b	184–189	110	
502	TT 53	186–192	110		573	TT 61b	184–189	110	
503	TT 61 b	186–194	109	(IV)	574	TT 59	(1) –	106	
504	TT 51	231	109		575	TT 61b	(2) 192	110	(IV)–III
505	TT 53	163–186	110	(IV)	576	TT 61b	(3) 194	110	V
506	TT 51	175(?)–215	109	III	577	TT 61b	(4)	110	V
507	TT 64 and 54	185–207	110	(IV)	578	TT 61b	(5) 201	110	V
508	TT 53	186–202	110		579	TT 61b	(5) –	110	V
509	TT 51	215–223	109	(IV)	580	TT 59	165–234	105, 106	
510	TT 52	195–203	108, 109	III	581	TT 61b	(6) 199–206	110	
511	TT 52	203–209	108, 109	(III)	582	TT 61b	(5)	110	
512	TT 51	223–232	109		583	TT 61b	(7)	110	
513	TT 52	209–222	108		584	TT 61c	200–215	107	(III)
514	C 12	Unstratified	110		585	TT 61c	221–215	107	(III)
515	TT 52	209–222	108, 109	V	586	TT 61c	220	107	V
516	TT 50	202–239	110		587	TT 61c	–	107	
517	TT 56	201–223	109	V–IV	588	TT 61d	212–220	107	(IV)
518	TT 63	224	109	V–IV	589	TT 61d	212–220	107	(IV)
519	TT 55	203–219	107		590	TT 60a	211	107	(V–IV)
520	TT 58	Surface	111		591		No Box		
521	TT 62	Unstratified	107		592	TT 64	182–205	110	V
522	TT 52	222–239	108		593	TT 65	Surface	109	
523	TT 62	Unstratified	107		594	W 5	Unstratified	110	
524	TT 51	239–245	109		595	TT 65	216–230	109	III
525	TT 51	239–245	109		596	TT 65	230–251	109	(III)
526	TT 55	219–238	107	(IV)	597	TT 65	227–255	109	
527	TT 51	215–223	109	(IV)	598	TT 66	195–225	108	(III–II)
528	TT 57	–	106, 108		599	TT 66	195–225	108	(III–II)
529	TT 52	–	108		600	TT 51	–	109	
530	TT 51	223–232	109	V (IV)	601	EF 6–7	200	106	
531	TT 58	204–207	111		602	Cleaning W.1	–	109	
532	TT 55	238	107	(IV)	603	EF 6–7	220	106	
533	TT 57	226–255	108		604	Cleaning W.3	–	106	
534	TT 52	220–239	108		605	Cleaning W.1	–	109	
535	TT 58	Unstratified	111		606	Cleaning	–	109	
536	TT 57	235–250	106, 108			W.2 (N)			

1. ( ) = uncertain stratigraphy, perhaps stray finds; + = disturbed layer; 0–10 = present surface layer

Captions to the Colour Plates

Colour Plate 1: Egyptian Pottery

No.	Sample No.	Field No.		
1	S.N. 1529 I	28	(Malkata Group II)	Cream-coated body sherd
2	S.N. 1529 II	28	(Malkata Group II)	Cream-coated body sherd
3	S.N. 1501	228	(Transitional between Malkata Group II & III)	Burnished body sherd
4	S.N. 1502	502	(Nilotic affinity)	Base fragment of a cone-shaped jar
5	S.N. 1536	374	(Nilotic affinity)	Neck fragment
6	S.N. 1503	228	(Malkata Group II)	Body sherd
7	S.N. 1538	253	(Nilotic affinity)	Painted body sherd
8	S.N. 1502	502	(Malkata Group II)	Base fragment
9	S.N. 1558	303/646	(Malkata Group II)	Conical base fragment
10	S.N. 1570	234	(Malkata Group II)	Body sherd
11	S.N. 1563	269/78	(Nilotic affinity)	Painted neck fragment of a juglet
12	S.N. 1560	I-J 16–17	(Malkata Group II)	Handle fragment
13	S.N. 1564	232	(Nilotic affinity)	Painted handle fragment
14	S.N. 1506	505	(Malkata Group II)	Cream-slipped and burnished body sherd
15	S.N. 1505	594	(Malkata Group III)	Cream-slipped and burnished body sherd
16	S.N. 1499	241		Incised ceramic-like fragments

Colour Plate 2: Normal Pottery

No.	Sample No.	Field No.	
1	S.N. 1510	123	Cream-coated bottom fragment of a jar. Light inclusions of carbonate and gypsum fragments
2	S.N. 1514	253	Dark brown coated body sherd. Light inclusions are mainly gypsum
3	S.N. 1517	172	Cream-coated handle fragment of a large jar.
4	S.N. 1509	278	Cream-coated rim fragment. Rough elongate cavities resulting from burnt chaff
5	S.N. 1512	270	Body sherd with a highly developed cream surface and with rough elongate cavities resulting from chaff temper.
6	S.N. 1533	167	Handle fragment of a jar with a dark brown surface over a light buff body. Light inclusions are carbonate and gypsum fragments.
7	S.N. 1544	67	Shoulder fragment of a jar with a thin unhomogeneous cream surface. Wheel marks can be traced into the cream domains.
8	S.N. 1516	253	Cream coated body sherd with irregular turning marks.
9	S.N. 1531	119	Body sherd showing a rough texture due to abundant sand temper. Cavities caused by leaching of gypsum inclusions.
10	S.N. 1549	259	Black body sherd with white inclusions of carbonates and gypsum
11	S.N. 1547	GH 16–17	Handle fragment of a cooking pot showing a dark brown core with a high density of large cavities due to abundant shale fragments, and coarse rounded quartz grains, an outer red oxidised zone and a moderately developed light surface.
12	S.N. 1537	220	Inner surface of a cooking pot body sherd. Large cavities are due to extreme leaching of gypsum inclusions.
13	S.N. 1554	295	Body sherd with a well developed cream surface. Red inclusions are shale fragments.

Colour Plate 3: Rough Hand-made Pottery

No.	Sample No.	Field No.	
1	S.N. 1527	279	Bottom fragment of a rough hand-made pot with a rounded bottom (not typical of the Negev pottery repertoire). Dark inclusions are slag fragments intentionally added as temper.
2	S.N. 1522	258	Bottom fragment of a typical rough hand-mae Negev-ware cooking pot with flat bottom and vertical wall. Light inclusions are crushed fragments of Normal pottery. Elongated cavities are due to chaff temper.
3	S.N. 1524	323	Bottom fragment of a rough hand-made pot with a rounded bottom, not typical of the Negev-ware pottery repertoire.
4	S.N. 1523	237	Rim fragment of a Negev-ware cooking pot. Light inclusions are angular fragments of quartz and feldspars derived from a granitic sand.
5	S.N. 1521	325	Body sherd of a rough hand-made pot (Negev-ware cooking pot?). Dark inclusions are slag fragments. The greenish particle at the bottom right is a copper ore fragment.
6	S.N. 1520	325	Body sherd. Light inclusions are crushed fragments of Normal Pottery.

Colour Plate 4: Midianite Pottery

No.	Fig. No.	Field No.	Locus	Description
1	4:8	598/16	108	Bowl
2	4:10	271/1	106	Bowl
3	4:11	237/36	110	Bowl
4	5:5	295/206	101	Bowl
5	6:21	236/9	109	Goblet
6	6:19	67/1	101	Jug
7	–	–	–	Jug
8		similar to Fig. 4:13		Jug
9	–	241/67	101	Bowl
10	5:4	234/147	101	Bowl
11	5:1	241/67	101	Bowl
12	–	171/2	101	Bowl
13	5:16	234/146	101	Bowl
14	5:18	124/48	101	Bowl
15	5:15	287/32	102	Bowl

Colour Plate 5: Hathor votive mask (Eg. Cat. No. 25; Fig. 30:1)

Colour Plate 6: Amuletic wand (Eg. Cat. No. 175; Fig. 45:7)

Colour Plate 7: Egyptian glass (Fig. 85)

Colour Plate 8: Egyptian glass flask (Fig. 86:1)

Colour Plate 9: Egyptian glass (Fig. 86:3)

Colour Plate 10: Egyptian glass fragment

Colour Plate 11: Bronze serpent with gilded head (Met. Cat. No. 3; Fig. 53:3)

Colour Plate 12: As 11, close-up of head

Colour Plate 13: a. Iron ring, b. Gilded iron ring (Met. Cat. No. 22: Fig. 54:16)

Colour Plate 14: Bronze ram figurine (Met. Cat. No. 6: Fig. 53:6)

Colour Plate 15: Gold Jewellery (a – Fig. 84:130; b – Fig. 84:131)

Colour Plate 16: Midianite decorated jug (Fig. 7:4)

Colour Plate 17: Midianite sherd with human figure (Fig. 7:2)

Colour Plate 18: Midianite decorated sherds with unusual colours (Fig. 5:6)

Colour Plate 19: Beads and pendants (Figs. 78–80, 82, 84)

Colour Plate 20: Beads and pendants (Figs. 81, 83)

Colour Plate 21: Decorated faience sherd (Eg. Cat. No. 144; Fig. 37:34)

Colour Plate 22: Faience amulet (Eg. Cat. No. 206; Fig. 48:11)

Colour Plate 23: Faience amulet, the moon god Khons (Eg. Cat. No. 207; Fig. 48:12)

Colour Plate 24: Faience amulet, Harpocrates (Eg. Cat. No. 203; Fig. 48:5)

Colour Plate 25: Faience amulet, Pataikos (Eg. Cat. No. 197: Fig. 48:2)

Colour Plate 26: Faience amulet, Harpocrates (Eg. Cat. No. 204; Fig. 48:6)

Colour Plate 27 Faience amulet. Pataikos (Eg. Cat. No. 198; Fig. 48:1)

Colour Plate 28: Faience amulet, bearded Pataikos (Eg. Cat. No. 202; Fig. 48:9)

Colour Plate 29: Bronze male figurine (Met. Cat. No. 1; Fig. 53:1)

Detailed Captions to Selected Plates

Plate 107: Microphotographs of Normal and Midianite Pottery (Chapter III, 4)

No.	Sample No.	Field No.	
1	S.N. 1532	201	A typical example of a temper-poor (N.P.T II D) end member with a highly silty and dolomitic clay-rich groundmass. Elongate voids (white) are burnt out organic fragments. Grey inclusions with wavy outlines are decomposed carbonate fragments. (Normal light, ×35).
2	S.N. 1510	123	Vitrified groundmass of a temper-poor (N.P.T. II D) end member. The tiny white spots are quartz and dolomite. In the darker domains the dolomites and the other carbonates reacted with clay to form a yellowish isotropic groundmass and thus in these domains the white spots are the remaining refractory quartzes. (Polarised light, ×35).
3	S.N. 1535	322	Normal pottery with abundant carbonate inclusions. The voids (white) are carbonate inclusions which were either leached out or disintegrated during thin sectioning. The voids have complex reaction rims. The inner part of the rim is usually lighter in colour and built mainly of decomposed and recrystalised carbonate material, while the outer part of the rim is of a darker, isotropic material representing products of the reaction between the carbonate boundary and the clay groundmass. (Normal light, ×35).
4	S.N. 1534	259	Temper-rich Normal pottery. The major coarse non-plastic is a quartz sand (rounded white grains). The elongated dark grain with a peripheral cavity (white) and internal planar structure is a shale fragment. (Normal light, ×35).
5	S.N. 1514	253	Normal pottery with abundant shale fragments. As the shale fragments are highly argillaceous, peripheral shrinkage cavities and internal shrinkage cavities and internal shrinkage fractures occur abundantly resulting in a highly porous texture. (Normal light, ×35).
6	S.N. 1531	119	Coarsely tempered Normal pottery. In this sample the coarse non-plastic fraction is dominated by rounded quartz grains (white and grey). The clay-rich groundmass is anisotropic and contains a small volume proportion of fine non-plastics (mainly quartz). (Polarized light, ×35).
7	S.N. 1106	306/4	Midianite pottery with typical rectangular dark shale fragments. At the upper right edge of the photograph a light shale fragment can be seen. In this sample the fragments show only a few modifications. White patches are quartz grains. (Normal light, ×30).
8	S.N. 1214	287/31	Midianite pottery with dark shale fragments. The fragment at the bottom of the photograph shows an upper black portion with rounded white voids. This portion was formed by its partial melting, involving the development of black vitrified groundmass, gaseous bubbles and rounding of the shape. Tiny black spots are iron oxide concretions. (Normal light, ×30).
9	S.N. 1136	258/21	Midianite pottery with extreme development of partial melting in the dark shale fragments. The fragment at the centre and the gaseous voids are almost spherical in shape, and as expected, there are no shrinkage phenomena at this stage of modification.

Plate 108: Microphotographs of Egyptian and Rough Hand-made Pottery (Chapter III, 4)

No.	Sample No.	Field No.	
1	S.N. 1527	279/297	Rough Hand-made Pottery with crystalline slag fragments embedded in an anisotropic groundmass with zoned dolomite crystals. (Normal light, ×35).
2	S.N. 1125	304/2	Rough Hand-made Pottery tempered with a granitic sand composed of quartz, alkali feldspars (mainly perthitic), plagioclases and biotite embedded in an anisotropic groundmass. (Normal light, ×35).
3	S.N. 1518	303	Rough Hand-made Pottery with pegmatitic quartz fragments showing colour and extinction domains related to growth zoning, embedded in an anisotropic groundmass. (Polarised light, ×35).
4	S.N.1560	I-J 16–17	Egyptian pottery. Nile ware with a relatively coarse non-plastic fraction of mainly quartz grains and a silty fraction with various silicates including quartz, feldspars, hornblende, micas, pyroxene and epidote. (Normal light, ×35).
5	S.N. 1560	I-J 16–17	The same as No. 4, but under polarised light, ×35.
6	S.N. 1505	594	Egyptian pottery. Cream-slipped, a coarse non-plastic assemblage composed of rounded quartz grains and rounded calcareous fragments derived from fossiliferous carbonate rocks, embedded in a fine calcareous clay-rich groundmass with a few silty silicates of nilotic affinity (hornblende, mica, epidote and pyroxene). Analogous samples were found in Malkata Palace Group III and in Thebes, Dendera and Amarna. (Polarised light, ×35).
7	S.N. 1505	594	As No. 6, showing longitudinal shrinkage cracks characteristic for this petrographic type. (Normal light, ×35).
8	S.N. 1570	234	Egyptian pottery. Highly calcareous and non-plastic material. Calcareous components are mainly hornblende and micas. The dark elongate fragments in this photograph are weathered mica flakes. Analogous samples were found in Malkata Palace Group IV and in Thebes, Dendera and El Kab. (Normal light, ×35).
9	S.N. 1521	Surface	Rough Hand-made Pottery, tempered mainly with slag and shale fragments. A large black slag fragment is visible at the bottom of the photograph (the grey stripes in the slag fragment represent crystalline domains). A large shale fragment is visible at the centre of the photograph, its composition the same as that of the groundmass. (Normal light, ×35).

Plate 109: Microphotographs of pigments and coloured frits (Chapter III, 5)

1. Green copper frit on Egyptian ceramic. Flakes and scales of pigment particles embedded in partly molten frit. SEM; magnification ×500.
2. Same as 1 above; SEM; magnification ×2000.
3. Black decoration on Midianite pottery. SEM; magnification ×150.
4. Same as 3 above. SEM; magnification ×800.
5. Same as 3 above. SEM; magnification ×3000.

Plate 129: Textile impressions and their 'positives'

No.	Field No.	Scale
1–2	154/19	2:1

Plate 131: 1–4, Grape pips. 5–7 Olive kernals (1, 5–7 ×3; 2–4 ×10)

1	General view
2	Dorsal view
3	Ventral view
4	Side view
5	Field (Box) No. 211
6	Field (Box) No. 214
7	Field (Box) No. 294

Plate 132: Pistacia. 1–2 Ancient. 3–6 Recent (1 ×3; 2–5 ×30)

1	General view
2	The hilum
3	*P. atlantica* (Negev)
4	*P. khinjuk* (Iran)
5–6	*P. palaestina* (Gilead)

Plate 133: Textiles

No.	Field No.	Scale	Description
1	357/1	0.5:1	fibres
2	357a/1	3:1	wool fragment
3	324/34	2:1	fibres and wool
4	303/12.1	4:1	a rope
5	324/35	2:1	wood with fibres

Plate 134: Textiles

No.	Field No.	Scale	Description
1	303/6	2.3:1	weft faced
2	303/11a	4.2:1	weft faced
3	324/31	2.9:1	tabby
4	324/33	5.7:1	unidentified fibre

Plate 135: Wood (microphotographs to Chapter III, 20)

1	Item 252 I, identified as *Pinus* sp. Scanning electron micrograph of a cross section; resin ducts can be seen. (×58).
2,3	Item 289 II, identified as *Pinus nigra* (or *P. sylvestris*). Radial longitudinal sections showing part of a ray. (×530). No. 2. Fenestriform pits in cross-fields can be seen. No. 3. Dentate walls of ray tracheids can be seed.
4	Item 289, identified as *Pinus nigra* (or *P. sylvestris*). Tangential longitudinal section; a fusiform ray which included a radial resin duct (D) can be observed. (×126).
5	Item 270, identified as *Pinus halepensis*. Tangential longitudinal section showing two fusiform rays each containing a radial resin duct (D).
6	Item 296, identified as probably *Abies* sp. Cross section. (×42).

Plate 136: Wood (microphotographs to Chapter III, 20)

7,8	Item 296, identified as probably *Abies* sp. No. 7. Tangential longitudinal section. (×126). No. 8. Radial longitudinal section of a ray showing beaded tangential (end) walls of ray cells. (×530).
9,10	Item 319, identified as *Cupressus sempervirens*. The material had deteriorated to a large extent. (×126). No. 9. Radial longitudinal section showing part of a ray. No. 10. Tangential longitudinal section.
11–13	Item 289, identified as *Populus euphratica*. No. 11. Radial longitudinal section showing a vessel member (V) and part of a ray with cross-field pits (P). (×126). No. 12. Cross section. (×126). No. 13. Tangential longitudinal section. R – uniserate ray; V – vessel member. (×300).

Detailed Captions to Selected Plates

Plate 137: Wood (microphotographs to Chapter III, 20)

14,15	Item 315, identified as probably *Haloxylon persicum*: cross-sections.
	No. 14. Two pore clusters can be seen. (×42).
	No. 15. Enlargement of part of Fig. 14 showing thick-walled fibres and crystals in the cells. (×300).
16	Item 240, identified as Chenopodiaceae. Scanning electron micrograph of a cross-section. Most groups of included phloem (Ph) had disintegrated. (×199).
17,18	Item 211 I, identified probably as *Suaeda palaestina*. (×126).
	No. 17. Cross-section. C – conjunctive parenchyma; P – pore cluster.
	No. 18. Longitudinal section showing storied parenchyma. Some of the cells contain crystals.
19,20	Item 279, identified as *Tamarix* sp. (×126).
	No. 19. Cross-section.
	No. 20. Tangential longitudinal section.

Plate 138: Wood (microphotographs to Chapter III, 20)

21–23	Scanning electron micrographs of items identified as *Acacia* sp.
	No. 21. Item 319. RAdial longitudinal section Part of a ray can be seen. (×100).
	No. 22. Item 252 II. Tangential longitudinal section. (×100).
	No. 23. Item 319. Cross-section showing the border between two growth rings. (×220).
24	Cross-section of wood taken from a living tree of *Acacia gerrardii* var. *negevensis*, showing growth rings. (×126).
25	Scanning electron micrograph of item 252 II, cross-section showing alternating layers of fibres and parenchyma cells. (×100).

Plate 139: Wood (microphotographs to Chapter III, 20)

26–29	Scanning electrong micrographs of items which were identified as *Pistacia atlantica*.
	No. 26. Item 283. Cross-section showing the border between two growth rings. (×100).
	No. 27. Item 257. Longitudinal section of vessels showing spiral thickening on the secondary wall. (×1300).
	No. 28. Item 303. Tangential longitudinal section. Some ray cells which contained crystals (C) can be seen. (×110).
	No. 29. Item 283. Tangential longitudinal section showing two rays with resin ducts (D). (×200).

Plate 140: Wood (microphotographs to Chapter III, 21)

1	The comb, Field No. 211/1.
2	A microscopic photograph of a radial longitudinal section of a piece of the comb showing a vessel with a scalariform perforation plate. (×600).
3	Like No. 2 but of fresh wood of *Buxus sempervirens*. (×600).

Plate 141: Wood (microphotographs to Chapter III, 21)

4	A microscopic photograph of a cross-section of a disintegrating piece of wood, Field No. 166. (×50).
5	Like No. 4 but of fresh branch of *Tamarix tetragyna*. (×50).
6	Like No. 4 but a tangential section showing storied axial wood parenchyma similar to this in Fig. 7. (×150).
7	Like No. 5 but a tangential section. (×150).

Plate 142: Wood (microphotographs to Chapter III, 21)

8	A portion of No. 4 showing all cells to have thin walls. (×600).
9	A cross-section of microscopic photograph of *Tamarix aphylla*, showing the thickness of the cell walls when fresh. (×600).
10	A microscopic photograph of a cross-section of the charcoal Field No. 378. (×50).
11	Like No. 10, but of fresh wood of *Acacia* sp. (×50).

Plate 143: Sciaenidae

No.	Site	Fish	Bone	View	Scale
1	2	Argyrosomus regius (Meagre)	Vertebra truncalis 8	lateral left side	2:1
2	2	A. regius	V. truncalis 6–7	lateral left side	2:1
3	2	A. regius	V. 14–15	lateral left side	2:1
4	2	A. regius	V. 14–15	lateral left side	2:1
5	2	A. regius	V. 16	lateral left side	2:1
6	2	A. regius	V. 16–18	lateral left side	2:1
7	2 (Box 413)	A. regius	Left Otolith	inner side (medial)	4:1
7A	2 (Box 413)	A. regius	Left Otolith	outer side (lateral)	4:1
7B	2 (Box 413)	A. regius	Left Otolith	dorsal	4:1
8	200 (Box 319)	A. regius	Left Otolith	inner side (medial)	4:1
8A	200 (Box 319)	A. regius	Left Otolith	outer side (lateral)	4:1
8B	200 (Box 319)	A. regius	Left Otolith	dorsal	4:1
9	Recent	A. regius	Left Otolith	inner side (medial)	4:1
9A	Recent	A. regius	Left Otolith	outer side (lateral)	4:1
9B	Recent	A. regius	Left Otolith	dorsal	4:1

Plate 144: Sciaenidae

No.	Site	Fish	Bone	View	Scale
10	2 (Box 470)	Sciaenidae Species	Left Operculum	medial	2:1
10A	2 (Box 470)	Sciaenidae Species	Left Operculum	lateral	2:1
11	2 (Box 470)	Sciaenidae Species	Left Operculum		2:1
11A	2 (Box 470)	Sciaenidae Species	Left Operculum	lateral	2:1
12	Recent	A. regius	Left Operculum	medial	2:1
13	Recent	A. regius	Left Operculum	lateral	2:1

Plate 145: Sparidae (Seabreams)

No.	Site	Fish	Bone	View	Scale
32	2 (Box 423)	Sparus auratus or Rabdosargus haffara	Left Praemaxilla	lateral	2:1
32A	2 (Box 423)	Sp. auratus or R. haffara	Left Praemaxilla	occlusial surface	2:1
33	2 (Box 423)	Sp. auratus or R. haffara	Right Praemaxilla	lateral	2:1
33A	2 (Box 423)	Sp. auratus or R. haffara	Right Praemaxilla	occlusial	2:1
34	200 (Box 292)	Sp. auratus or R. haffara	Left Praemaxilla	oclusial	2:1
35	200 (Box 292)	Sp. auratus or R. haffara	Right Praemaxilla	occlusial	2:1
36	2	Sp. auratus or R. haffara	Left Maxilla	lateral	2:1
37	Recent	Sp. auratus–	Left Maxilla	lateral	2:1
38	2	Sp. auratus or R. haffara	Right Dentale	lateral	2:1
38A	2	Sp. auratus or R. haffara	Right Dentale	occlusial	2:1
39	2 (Box 423)	Sp. auratus or R. haffara	Left Dentale (with molar)	lateral	2:1
39A	2 (Box 423)	Sp. auratus or R. haffara	Left Dentale (with molar)	occlusial	2:1
40	200 (Box 324)	Sp. auratus or R. haffara	Left Dentale (with molar)	occlusial	2:1

Plate 146: Claridae (African Catfish)

No.	Site	Fish	Bone	View	Scale
41	30 (TT 62)	Clarias gariepinus (Nile Catfish)	Vertebra truncalis 11–13	lateral left side	4:1
42	30 (TT 62)	Clarias gariepinus	Vertebra caudalis 25–28	lateral left side	4:1
43	2 (Box 402)	Carcharinidae (Blue shark)	Vertebral centre Timna "shark bead")	surface caudal	4:1
44	2 (Box 402)	Carcharinidae (Blue shark)	Vertebral centre (Timna "shark bead")	surface dorsal	4:1
45	2 (Box 403)	Carcharinidae (Blue shark)	Vertebral centre (Timna "shark bead")	surface ventral	4:1
46	2 (Box 403)	Carcharinidae (Blue shark)	Vertebral centre (Timna "shark bead")	transverse section through middle of vertebral centre	6:1
47	2 (Box 403)	Carcharinidae (Blue shark)	Vertebral centre (Timna "shark bead")	transverse section near posterior end of vertebral centre	6:1

Plate 147: Mammalian remains

No.	Field No.	Animal	Sex	Age	Bone	View	Scale
1	310	Goat	male	adult	Metacarpus	dorsal	1:1
2	118	Sheep	female	juvenile	Metacarpus	dorsal	1:1
3	303	Sheep	female	juvenile	Metatarsus	dorsal	1:1
4	241	Goat	female	adult	Horncore	frontal	1:1
5	148	Goat	male	adult	Horncore	frontal	1:1
6	598	Sheep	male	adult	Horncore	frontal	1:1

Plate 148: Mammalian remains

No.	Field No.	Animal	Age	Bone	View	Scale
7	133	Goat or Sheep	infantile	Horncore	medial	2:1
8	291	Goat	juvenile	Horncore	frontal	2:1
9	311	Sheep	juvenile	Horncore	frontal	2:1
10	292	Goat or Sheep	infantile	Squama occipitalis	nuchal	1:1
11	203	Goat or Sheep	infantile	Squama occipitalis	nuchal	1:1
12	232	Goat or Sheep	juvenile	Squama occipitalis	nuchal	1:1
13	206	Goat or Sheep	juvenile	Squama occipitalis	nuchal	1:1
14	292	Sheep	adult	Squama occipitalis	nuchal	1:1

Plate 149: Mammalian remains

No.	Field No.	Animal	Age	Bone	View	Scale
15	112	Goat or Sheep	adult	Petro-Mastoid bone	from cerebral cavity	2:1
16	213	Goat or Sheep	adult	Petro-Mastoid bone	from cerebral cavity	2:1
17	292	Goat or Sheep	infantile	Mandibular branch: articular and muscular process	medial	1:1
18	303	Goat or Sheep	infantile	Mandibular branch: articular and muscular process	medial	1:1
19	282	Goat or Sheep	infantile	Mandibular branch: articular and muscular process	medial	1:1
20	293	Goat or Sheep	juvenile	Mandibular branch: articular and muscular process	medial	1:1
21	292	Goat or Sheep	juvenile	Mandibular branch: articular	medial and muscular process	1:1
22	252	Goat or Sheep	adult	Mandibular branch: articular and muscular process	medial	1:1

Plate 150: Mammalian remains

No.	Field No.	Animal	Age	Bone	View	Scale
23	232	Goat or Sheep	infantile	Mandibular corpus L	dorsal	1:1
24	292	Goat or Sheep	infantile	Mandibular corpus R	dorsal	1:1
25		Goat or Sheep	infantile	Mandibular corpus L	dorsal	1:1
26	123	Goat or Sheep	infantile	Mandibular corpus R	dorsal	1:1
27	143	Goat or Sheep	juvenile	Mandibular corpus L	dorsal	1:1
28	123	Goat or Sheep	juvenile	Mandibular corpus R	dorsal	1:1
29	319	Goat or Sheep	juvenile	Mandibular corpus L	dorsal	1:1
29A	319	Goat or Sheep	juvenile	Mandibular corpus L	buccal	1:1
30		Goat or Sheep	adult	Mandibular corpus L	dorsal	1:1
30A		Goat or Sheep	adult	Mandibular corpus L	buccal	1:1
31	252	Goat or Sheep	infantile	Femur, loose proximal epiphysis R	distal	1:1
32	310	Goat or Sheep	infantile	Femur, loose proximal epiphysis R	distal	1:1
33.1	293	Goat or Sheep	juvenile	Femur, proximal end without epiphysis R	distal	1:1
33.2	293	Goat or Sheep	juvenile	Femur, fitting loose proximal epiphysis R	proximal (surface of fusion)	1:1
34.1	280	Goat or Sheep	juvenile	Femur, proximal end without epiphysis R	distal	1:1
34.2	280	Goat or Sheep	juvenile	Femur, fitting loose proximal epiphysis	proximal (surface of fusion)	1:1
35		Goat or Sheep	juvenile	Femur, loose proximal proximal epiphysis R	proximal (surface of fusion)	1:1

Plate 152: Recent invertebrate, minerals and fossils

No.	Definition	Comments
1	Coscinarea monile	
2	Stylophora pistillata	Two examples
3	Platygyra lamellina	
4	Millepora dichotoma	
5	Pocillophora verrucosa	
6	Phyllacanthus imperialis	Sea urchin, plate fragments
7	Acropora	
8	Pinctada margaritifera	Ventral margin shows evidence of abrasion, possibly used as a scraper
9		Abrasion evident on the longest margin
10		Fragment shaped into an elongated 'plate'

Plate 153: Recent invertebrates

No.	Definition	Comments
1	Cypraea annulus	Left – ground down on side
	Cypraea turdus wickworthi	Right
2	Cypraea tigris or C. pantherina	Partially restored
3	Cypraea turdus wickworthi	With open dorsum
4	Lambis truncata sebae	Lip fragment
5	Strombus gibberulus albus	Two, holed
6	Strombus fasciatus	Holed
7	Terebra	
8	Rhinoclavis	
9	Corithium erythraeonense	Left – fresh
	C. erythraeonense	Right – worn and holed
10	Turbo opercula	Three items of two size groups
11	Conus catus	
12	Conus terebra	
13	Thais hippocastanus savigny	
14	Latirus polygonius	
15	Tridacna maxima	Two, complete

Plate 154: Recent invertebrates

No.	Definition	Comments
1	Tridacna squamosa	Two, fragmentary
2	Anadara	
3	Barbatia	
4	?Coenholectypus larteti	Two, Sea urchins
5	Heterodiadema libycum	Two, Sea urchins
6	Probably Neithea	Micritic limestone with pectinid fossil
7	Fusiform gastropods	Two
8	Trochiform gastropods	'Trochus' or 'Trochalis'
9	Exogyra-like Oyster	
10	Turriculate gastropods	Three (cf. Nerinea or Nerinella)

Detailed Captions to the Figures

Figure 1. Pottery from Stratum V. Macehead.

No.	Field No.	Locus	Description
1	564/4	109	Rim.
2	570/2	110	Rim.
3	596/2	109	Rim.
4	378/x	106	Rim with plastic rope decoration.
5	280/89	106	Macehead.

Figure 2. Flint Implements.

No.	Field No.	Locus	Description
1	–	–	Backed bladelet.
2	362	106	End-scraper.
3	–	–	Borer.
4	257	101	End-scraper – core.
5	297	101	Side-scraper.
6	331	109	End-scraper – core.
7	288	101	Borer.
8	257	102	Side-scraper.

Figure 3. Flint Implements.

No.	Field No.	Locus	Description
1	–	–	Bifacially retouched item.
2	570	110	End-scraper.
3	135	110	Core.

Figure 4. Midianite Pottery.

No.	Plate No.	Field No.	Locus	Description
1		250/25	101	Bowl. Very pale brown (10 YR 7/3), well fired. Small to large red grits.
2		277/106	110	Bowl. Very pale brown (10 YR 8/3), well fired. Small to large red grits.
3		234/144	101	Bowl. White (10 YR 8/2), well fired. Small red and black grits.
4		134/11	110, 111.	Bowl. Very pale brown (10 YR 8/3), well fired. Small to large red grits.
5		250/25	101	Bowl. Very pale brown (10 YR 7/3), well fired. Small black and small to large red grits.
6		296/7	101	Bowl. Very pale brown (10 YR 7/3), well fired. Small to medium red grits.
7		121/75	101	Bowl. Very pale brown (10 YR 8/3), well fired. Small to large red grits.
8	Col. Pl. 4:1	598/16	108	Bowl.
9		287/30	102	Bowl. White (lO YR 8/2), well fired. Small to medium red grits.
10	Col. Pl. 4:2	271/1	106	Bowl. Very pale brown (10 YR 7/3), well fired. Small to medium red grits.
11	Col. Pl. 4:3	237/36	110	Bowl. Light grey (10 YR 7/2), well fired. Small to medium red grits.
12		264/15	103	Bowl. Very pale brown (10 YR 7/3), well fired. Red grits.
13		241/68	101	Bowl. Very pale brown (10 YR 7/3). Red grits.

Figure 5. Midianite Pottery.

No.	Plate No.	Field No.	Locus	Description.
1	Col. Pl. 4:11	241/67	101	Bowl. Very pale brown (10 YR 7/3), well fired. Very small black and small to large red grits.
2		243/43	101	Bowl. Very pale brown (10 YR 8/3), well fired. Small to medium red grits.
3		233/13	101	Bowl. Very pale brown (10 YR 8/3), well fired. Small to medium red grits.
4	Col. Pl. 4:10	234/147	101	Bowl. Very pale brown (10 YR 7/3), well fired. Small to medium red grits.
5	Col. Pl. 4:4	295/206	101	Bowl. Grey (5 YR 5/1), poorly fired. Very small white grits.
6		211/2	101	Bowl. Very pale brown (10 YR 8/3), well fired. Very small red grits.
7		124/53	101	Bowl. Very pale brown (10 YR 7/3), well fired. Small to large reddish brown grits.
8		115/16A	112	Bowl. Pink (7.5 YR 8/4), well fired. Very small to small red grits.
9		241/69	101	Bowl. Very pale brown (10 YR 7/3), well fired. Medium to large red and large black grits.
10		319/299	106, 108	Bowl. Very pale brown (10 YR 8/3), well fired. Small to large red grits.
11		312/2	101	Bowl. Very pale brown (10 YR 7/3), well fired. Small to very large black grits.
12		254/4	101	Bowl. Very pale brown (10 YR 8/3), well fired. Medium to very large red grits.
13		241/70	101	Bowl. Very pale brown (10 YR 7/3), well fired. Very small to large red grits.
14		294/49	101	Bowl. Very pale brown (10 YR 7/3), well fired. Medium to very large red grits.
15	Col. Pl. 4:15	287/32	102	Bowl. Very pale brown (10 YR 7/4), well fired. Small to large red grits.
16	Col. Pl. 4:13	234/146	101	Bowl. Pale brown (10 YR 6/3), medium fired, dark grey core. Large to very large black, and small to large red grits; very small quartz inclusions.
17		269/76	107	Bowl. Very pale brown (10 YR 7/4), well fired. Small to large red grits.
18	Col. Pl. 4:14	124/48	101	Bowl. Everted rim. Very pale brown (10 YR 7/3), well fired. Small to large red grits.

Figure 6. Midianite Pottery.

No.	Plate No.	Field No.		Description.
1		90/3	101	Goblet(?). White (10 YR 8/1), well fired. Small to very large black grits.
2		337/118	102	Rim.
3		348/14	107	Flattened base. Pink (5 YR 8/3), well fired. Small to very large red grits.
4		170/8	101	Flat base. Very pale brown (10 YR 7/3), well fired. Small to large red grits.
5		257/284	102	Flat base. Very pale brown (10 YR 7/4), well fired. Very small to very large red grits.
6		363/5	106, 108	Flattened base. Very pale brown (10 YR 8/4), well fired. Very small to very large red grits.
7		257/285	102	Flattened base. Pale brown (10 YR 8/4), well fired. Small to very large red grits.
8		253/104	102	Flattened base. Very pale brown (10 YR 7/3), well fired. Small black and small to medium red grits.
9		277/107	107, 110	Flattened base. Very pale brown (10 YR 8/3), well fired. Medium to very large red grits.
10		289/21	101, 102	Flattened base, hand-made. Very pale brown (10 YR 8/3), well fired. Medium to very large red and brown grits.
11		598/15	108	Bowl.
12		234/145	101	Flattened base. Very pale brown (10 YR 8/3), well fired. Very small to large red grits.
13		50/7	101	Disc base. Very pale brown (10 YR 7/3), well fired. Small to large red grits.
14		295/203	101	Disc base. Very pale brown (10 YR 7/3), well fired. Small to very large red grits.
15		339/38	102	Disc base. Very pale brown (10 YR 7/3), well fired. Small to large red and brown grits.
16		241/71	101	Base, probably of small bowl.
17		337/34	102	Jug, flat base. Very pale brown (10 YR 8/3), well fired. Small to very large red and brown grits.
18		209/40	101	Jug.
19	Col. Pl. 4:6	67/3	101	Jug. Very pale brown (10 YR 8/3), well fired. Medium black and large red grits.
20		216/87	101	Goblet, flat base. Very pale brown (10 YR 7/4), well fired. Small to large red grits.
21	Col. Pl. 4:5	236/9	109	Goblet. White (10 YR 8/2), well fired. Small black and red grits.
22		345/16	107	Goblet, flat rim. Very pale brown (10 YR 7/3), well fired. Small to medium red and black grits.

Figure 7. Midianite Pottery.

No.	Field No.	Locus	Description
1	507/31	110	Jug(?).
2	245/6	107	Jug(?). Very pale brown (10 YR 8/3), well fired. Small red and black grits.
3	349/29	109	Juglet. Very pale brown (10 YR 8/3), well fired. Small to large red grits.
4	204/63	110	Juglet. Very pale brown (10 YR 8/3), well fired. Small to very large red grits.
5	250/27	101	Jug.
6	337/4	102	Juglet. Very pale brown (10 YR 7/3), well fired. Small to large black grits.

Figure 8.   Midianite Pottery.

No.	Field No.	Locus	Description
1	282/30	110	Juglet. Brown (7.5 YR 5/4), well fired. Medium to large red grits.
2	598/14	108	Juglet.

Figure 9.   Midianite Pottery.

No.	Field No.	Locus	Description
1	230/6	101	Juglet. Very pale brown (10 YR 7/3), well fired. Small red grits.
2	357/8	105, 106	Neck and handle, probably of a juglet. Very pale brown (10 YR 7/3), well fired. Small to medium red grits.
3	258/21	106	Juglet. Very pale brown (10 YR 7/3), well fired. Small to medium red grits.
4	233/14	101	Loop handle. Very pale brown (10 YR 7/4), well fired. Very small to large red grits.
5	CD/16–17/6	101	Juglet. White (10 YR 8/2), well fired. Very small to extra large red grits.
6	T–200/4	Surface	Juglet. Very pale brown (10 YR 7/3), well fired. Small to medium red grits.
7	248/5	104	Base.
8	224/134	101	Large bowl(?). Thickened, flat rim. Very pale brown (10 YR 7/4), well fired. Small to very large red grits.

Figure 10.   Midianite Pottery.

No.	Plate No.	Field No.	Locus	Description
1	106:1	241/6	101	Mug. Very pale brown (10 YR 7/3), well fired. Small red grits.

Figures 11–13.   Midianite Decorative Motifs.

Figure 14.   Negev-ware Pottery; 10-13, Rough Hand-made Pottery.

No.	Field No.	Locus	Description
1	200/8	Surface	Bowl, round rim, hand-made. Light brown (7.5 YR 6/4), well fired, but friable. Small to large white grits and small to medium white quartz inclusions.
2	135/52	110	Bowl, flat rim, hand-made. Dark grey (7.5 YR N4/), poorly fired, but friable. Small to large white, and very small to small red grits.
3	299/1	109	Bowl, rounded rim, hand-made. Brown (7.5 YR 5/2), poorly fired, black core. Small black and small to medium white grits, large quartz inclusions.
4	115/16B	112	Rounded rim, hand-made. Dark grey (7.5 YR N4/), poorly fired. Small to large white grits.
5	274/4	105	Rounded rim, probably of holemouth jar, hand-made. Light brownish grey (10 YR 6/2), poorly fired, black core. Medium to large white quartz inclusions.
6	T–200/3	Surface	Small bowl with rounded rim, and flat base, hand-made. Red (10 YR 5/6), poorly fired and friable, grey core. Medium black and medium to very large red grits, and white (5 YR 8/1) slip on exterior and interior.
7	308/4	101	Small protrusion at joint of base of side wall. Disc base, hand-made. Strong brown (7.5 YR 5/6), poorly fired, thick grey core. Medium to very large black grits, organic temper.
8	269/77	107	Ring base, hand-made. Brown (10 YR 4/3), medium fired, grey core. Medium black grits.
9	120/4	101, 112	Rounded bowl with rounded rim and convex base, hand-made. Brown (7.5 YR 4/4), well fired. Small to large black grits.
10	288/4	101	Rounded bowl with sharp rim and convex base, hand-made. Red (10 R 5/8), well fired. Medium white and grey grits. Medium quartz inclusions.
11	310/16	105	Bowl with rounded rim and flat base, hand-made. Light red (10 R 6/8), well fired. Small to large black grits. Very small and large quartz inclusions, organic temper.
12	298/2	109	Straight sloping-sided large bowl with rounded rim and disc base, hand-made. Reddish yellow (5 YR 6/8), poorly fired, thick light grey core. Medium to large black and medium red grits. Some inclusions of copper ore fragments.
13	339/120	102 (109)	Large bowl, thickened everted rim, hand-made. Brown (7.5 YR 5/4), poorly fired, very thick black core. Medium black grits.

Figure 15.   Rough Hand-made Pottery.

Figure No.	Plate No.	Field No.	Locus	Description
1	106:5	278/138	107	Goblet, rounded rim and rounded base, hand-made. Dark reddish grey (10 R 4/1), poorly fired. Small black and very large red grits.
2	106:4	249/1	104	Juglet, rounded base, hand-made. Reddish yellow (5 YR 6/6), medium fired, light grey core. Very small black, white and small red grits, organic temper. Reddish brown (5 YR 5/3) slip.
3	106:2	116/23	110(112)	Juglet, sharp rim and rounded base, hand-made. Red (10 R 5/6), poorly fired, grey core. Medium to large black grits. Base has holes punched from outside in.
4	–	253/105	102	Juglet, with spouted rounded rim, hand-made. Dark grey (7.5 R N4/), poorly fired, black core. Small to very large white grits, organic temper.
5	–	223/90	101	Pilgrim flask. Light brown (7.5 YR 6/4), well fired. Very small to small black and small white grits, organic temper.
6	–	289/23	101, 102	Juglet, ring base, hand-made. Light olive grey (5 Y 6/2), well fired. Very small quartz inclusions. Very pale brown (10 YR 8/3) slip.
7	–	319/302	106, 108	Pilgrim flask. Light red (2.5 YR 6/6), well fired. Small black, small to large white and reddish brown grits. Small quartz inclusions.
8	106.6	287/33	102	Holemouth jar, hand-made.

Figure 16.   1–3, Rough Hand-made Pottery; 4, 5, 6–8 Local Wheel-made Pottery.

No.	Field No.	Locus	Description
1	221/14	106, 107	Three handles (?), hand-made.
2	230/1	101	Appliqué of snake's body intertwined around two handles, hand-made.
3	85/3	110, 111	Appliqué of snake's head, hand-made. Yellowish red (5 YR 5/6), medium fired, thin light grey core. Small black, small to medium white and large red grits.
4	124/49	101	Jar, thickened rounded rim, hand-made. Light red (10 R 6/6), medium fired, grey core. Small black grits, organic temper.
5	224	101	Jar, thickened rounded rim.
6	262/13	103	Cooking pot, thickened rounded rim. Light red (10 R 6/6), medium-poorly fired, thick grey core. Large black grits.
7	343/102	102	Cooking pot, rounded rim. Red (2.5 YR 4/6), medium fired, thick grey core. Small white and large to very large black grits, very small quartz inclusions.
8	257/290	102	Carinated cooking pot, flat rim. Light red (2.5 YR 6/6), poorly fired, thin white core. Medium black, large brown, very large white and large red grits, organic temper.

Figure 17.   Local Wheel-made Pottery.

Figure No.	Plate No.	Field No.	Description
1	343/103	109	Krater, thickened rounded rim. Red (2.5 YR 5/6), well fired. Small to medium black grits, small rounded quartz inclusions, organic temper.
2	303/647	101	Krater, thickened rounded rim. Reddish yellow (5 YR 6/6), medium fired, thick grey core. Very small white grits, small rose quartz inclusions, organic temper.
3	209/41	101	Carinated cooking pot.
4	211/273	101	Carinated bowl with rounded rim. Light reddish brown (5 YR 6/4), medium fired, thin grey core. Small to large white and red grits, small rounded quartz inclusions, organic temper.
5	232/20	101	Krater, thickened rim sloping inwards. Light reddish brown (2.5 YR 6/4), poorly fired, very thick dark grey core. Medium to large white grits, organic temper.
6	232/22	101	Storage jar, thickened rim sloping inwards. Brown (7.5 YR 5/4), medium fired, grey core. Small to large black grits, small quartz inclusions, organic temper.

Figure 18. Local Wheel-made Pottery.

No.	Plate No.	Field No.	Locus	Description
1	–	257/287	102	Krater, thickened rounded rim with handle. Light red (2.5 YR 6/8), poorly fired, thick grey core. Small white and large black grits, organic temper.
2	–	257/293	102	Krater, thickened rim sloping outwards. Light red (2.5 YR 6/6), medium fired, thick dark grey core. Large black and white grits.
3	–	286/6	104, 105	Holemouth jar, flat rim. Red (2.5 YR 5/6), medium fired, grey core. Very small quartz inclusions.
4	106:3	224/136	101	Goblet, rounded rim, high ring-base. Red (10 R 5/6), well fired. Large black and medium to large red grits, small quartz inclusions.
5	–	80/12	102	Jug with pinched mouth. Yellowish red (5 YR 5/8), poorly fired, thick grey core. Small to large red and clear quartz inclusions, organic temper.
6	–	221/41	107	Chalice or pedestalled incense burner. Red (2.5 YR 5/6), medium fired, thin grey core. Medium to large black grits, organic temper.

Figure 19. Local Wheel-made Pottery.

No.	Field No.	Locus	Description
1	135	102	Oval (section) loop handle, probably of a jar.
2	224/135	101	Storage jar handle. Yellowish red (5 YR 5/6), medium fired, grey core. Small white grits, organic temper. Very pale brown (10 YR 7/3) slip on exterior.
3	257/288B	102	Storage jar handle. Reddish yellow (5 YR 6/6), poorly fired, thick grey core. Medium black and medium to very large white grits, organic temper. Very pale brown (10 YR 7/3) slip on exterior.
4	257/294	102	Storage jar handle. Red (2.5 YR 5/6), poorly fired, thick grey core. Medium black and white grits, organic temper. Pale brown (10 YR 5/6) slip on exterior.
5	252/32	104	Storage jar handle. Brown (7.5 YR 5/4), medium fired, grey core. Small red grits, rose quartz inclusions, organic temper. Very pale brown (10 YR 8/3) slip on exterior.
6	117/12	101	Cone-shaped storage jar base. Light red (2.5 YR 6/6), medium fired, yellowish grey core. Organic temper. Very pale brown (10 YR 7/4) slip on exterior.
7	295/205	101	Cone-shaped storage jar base. Reddish yellow (5 YR 6/8), poorly fired, thick grey core. Small rounded quartz inclusions. Original colour of slip unidentifiable.
8	209/42	101	Cone-shaped storage jar base. Light red (10 R 6/8), poorly fired, thick dark grey core. Small white and medium to large black grits, organic temper. Reddish brown (5 YR 5/3) slip on exterior and interior.
9	257/292	102	Rounded storage jar base. Reddish yellow (5 YR 6/6), poorly fired, thick grey core. Medium to very large black and white grits, organic temper.

Figure 20. 1–8, Local Wheel-made Pottery; 9–12, Egyptian Pottery.

No.	Field No.	Locus	Description
1	253/109	102	Jar or krater ring base. Light brown (7.5 YR 6/4), medium fired, thick black core. Medium quartz inclusions, organic temper.
2	206/59	110	Jar or krater. Reddish yellow (5 YR 6/6), poorly fired, thick light grey core. Medium to large black and red grits, medium quartz inclusions.
3	289/23	101	Base ring, perhaps of jug.
4	253/108	102	Storage jar, flat base. Reddish yellow (5 YR 6/6), medium fired, grey core. Small white and small to medium black grits, organic temper.
5	241/75	101	Storage jar, flat base. Reddish yellow (5 YR 6/6), medium fired, dark grey core. Large red grits, very small quartz inclusions, organic temper.
6	294/51	101	Storage jar, flat base. Yellowish red (5 YR 5/6), medium fired, thick grey core. Medium quartz inclusions, organic temper.
7	316/16	111	Storage jar, concave ring base. Yellowish red (5 YR 5/6), medium fired, grey core. Small white grits, organic temper.
8	257/289	102	Storage jar, disc base. Light red (2.5 YR 6/6), poorly fired, thick black core. Medium to large white grits, small quartz inclusions.
9	279/291	106	Jug, thickened everted rim. Dark reddish brown (2.5 YR 3/4), well fired. Small quartz inclusions.
10	217/3	101	Bowl, rounded rim. Red (2.5 YR 5/6), well fired, very thin light grey core. Medium black and medium to large white grits, small to medium rose quartz inclusions, organic temper.
11	167/5	101	Krater, thickened rim sloping outwards. Light red (2.5 YR 6/6), medium fired, thick grey core. Medium black and white grits, very small quartz inclusions.
12	113/4	101	Krater, thickened everted rim. Light red (10 R 6/8), poorly fired, very thick grey core. Medium black grits, small rounded red quartz inclusions, traces of organic temper.

Figure 21. 1–3, Egyptian Pottery; 4–10, Egyptian Painted Pottery; 11–14, Local Painted Pottery.

No.	Field No.	Locus	Description
1	153/3	102	Storage jar rim. Red (2.5 YR 5/6), poorly fired, thick grey core. Small to very large black grits, small to medium quartz inclusions.
2	502	110	Flat base, probably of jar or krater.
3	303/640	101	Juglet.
4	241/78	101	Body sherd.
5	235/31	109	Body sherd.
6	599	108	Body sherd.
7	–	–	Body sherd.
8	241/76	101	Juglet handle. Red (10 YR 5/8), well fired. Very small to small rounded rose quartz inclusions.
9	371/11	110	Juglet handle. Reddish yellow (5 YR 6/6), well fired. Small black and white grits.
10	239/19	107	Bowl.
11	304/15	101	Rim.
12	I-J/16–17	101	Rim, perhaps of jar.
13	232/23	101	Part of a holemouth jar. Thickened rim with ledge. Yellowish red (5 YR 5/6), medium fired, grey core. Small rounded quartz inclusions.
14	599	108	Body sherd.

Figure 22. Egyptian Alabasters, Inscribed Fragments.

No.	Eg. Cat. No.	Plate No.	Object	Material	Field No.	Locus	Square
1	234	–	Cup(?)	Alabaster	376/10	111	A-B 12
2	15	–	Statuette	Alabaster	125/1	101	K-L 16–17
3	233	116:3	Vase	Alabaster	328/1	104	E 5
4	235	–	Bowl	Alabaster	258/2	106	H 6–7
5	10	–	Fragment of inscription	White sandstone	276/1	111	A-B 11
6	11	–	Fragment of inscription	White sandstone	234/2	101	H 15–18
7	259	–	Offering table (?)	White sandstone	237/8	110	B-C-D 3–4
8	8	–	Fragment of inscription	White sandstone	227/1	101	I 17
9	9	–	Fragment of inscription	White sandstone	245/4	107	A-B 8–10
10	1	50;110:3	Fragment with inscription	White sandstone	464/1	107	A 10

Figure 23. Stone Pillars and Basins.

No.	Eg. Cat. No.	Plate No.	Object	Field No.	Locus	Square
1	3	116:1	Pillar	434/1	Unstratified stray find	
2	2	37–40	Pillar	433/1	106	F 7
3	4	–	Pillar	457/1	107	–
4	5	113:4	Inscribed Pillar	470/1	107	C 8–9
5	229	–	Basin	466/1	109	–
6	230	–	Pestle(?)	469/1	106	E-F 7
7	231	–	Basin	467/1	107	–
8	232	–	Basin	468/1	106	–

Figure 24. Cornices and Offering Table.

No.	Eg. Cat. No.	Plate No.	Object	Material	Field No.	Locus	Square
1	245	–	Offering table	White sandstone	429/1	107	–
2	244	–	Offering table	White sandstone	435/1	107	D-E 10
3	243	–	Offering table	White sandstone	463/1	107	B 10
4	7	110:2	Cornice (?)	White sandstone	454/1	107	A 9
5	6	110:1	Cornice	White sandstone	424/1	107	A 10

Figure 25.   Statuette and Sphinx.

No.	Eg. Cat. No.	Plate No.	Object	Field No.	Locus	Square
1	14	117:1	Statuette	245/3	107	B 9
2	12	114:1	Sphinx	380/1	107	B 8–10

Figure 26.   Sphinx.

No.	Eg. Cat. No.	Plate No.	Object	Field No.	Locus	Square
1	13	115:1, 2	Sphinx	305/1	106	A-B 6–7

Figure 27.   Sistra and Decorated Sherds.

No.	Eg. Cat. No.	Plate No.	Object	Field No.	Locus	Square
1	16	118:3	Sistrum	303/1	101	F 15–16
2	17	–	Sistrum	278/1	107, 109	E-F 10–11
3	18	–	Sistrum	289/1	101, 102	L 14–15
4	22	–	Sistrum	276/18B	111	A-B 11
5	163	–	Decorated sherd	332/7	109	E 12
6	164	–	Decorated sherd	257/281A	102	I-J-K 13
7	121	–	Decorated sherd	257/281B	102	I-J-K 13
8	149	–	Decorated sherd (jug)	263/25, 258	103, 106	–
9	145	–	Decorated sherd	78/21, 257/301	101, 102	–
10	146	–	Decorated sherd	337/2, 237, 276	102, 110, 111	–

Figure 28.   Sistrum, Ushabti and Bowl.

No.	Eg. Cat. No.	Plate No.	Object	Field No.	Locus	Square
1	19	119:1	Sistrum	241/4, 225/1	101	J-K 15
2	84	119:2	Ushabti	257/2	102	I-J-K 13
3	102	120:1	Bowl	337/1	102	I-J-K-L 11

Figure 29.   Votive Objects.

No.	Eg. Cat. No.	Plate No.	Object	Field No.	Locus	Square
1	220	–	Handle(?)	151/1	102	J-K 13–14
2	221	–	Handle(?)	135/3	110	D-E 13–14
3	222	120:4	Handle(?)	280/2	106	C-D-E 6–7
4	20	120:3	Sistrum	277/4	107	C-D 10–11
5	21	120:2	Sistrum	277/2	107	C-D 10–11
6	31	–	Menat counterpoise (?)	270/105	102	I-J-K 12

Figure 30.   Votive Objects.

No.	Eg. Cat. No.	Plate No.	Object	Field No.	Locus	Square
1	25	Col. Pl. 5	Hathor votive mask	242/1, 251/1	107	C-D 10–11
2	24	118:2	Sistrum	F-G 11–12	109	F-G 11–12
3	–	140:1	Comb	211/1	101	G 15–17
4	–	–	Ring	328/15	106,105	–
5	–	–	Ring	–	101	I-J 16–17B–52
6	–	–	Phallus (?)	120/7	112,101	A-B-C 17–18
7	–	–	Figurine (?)	263/5	103	J-K 8

Figure 31.   Votive Objects.

No.	Eg. Cat. No.	Plate No.	Object	Field No.	Locus	Square
1	34	120:5	Menat counterpoise	228/1	110	B-C-D 10–11
2	28	120:6	Menat counterpoise	206/1	110	B-C-D 13–14
3	26	121:1	Menat counterpoise	115/1	112	A-B-C 15–16
4	97	–	Jar stand	279/1	107	E-F 7–9
5	96	119:3	Jar stand	353/1, 370/2	110,111	–
6	98	121:2	Jar stand	245/1,	107	A-B 8–10
7	83a	125:3	Bracelet, Seti I	–	107	–
8	83b	125:2	Bracelet, Hathor	–	106	–

Figure 32.   Votive Objects.

No.	Eg. Cat. No.	Plate No.	Object	Field No.	Locus	Square
1	29	–	Menat counterpoise	258/20	106	H 6–7
2	33	–	Menat counterpoise	241/3	101	J-K 15
3	36	121:7	Menat counterpoise	65/1, 253/1 261/6,370/1	102,103, 110	–
4	32	121:5	Menat counterpoise	302/1, 301/1	101,109	–
5	27	121:6	Menat counterpoise	241/5, 248/1,118/1	101,104	–
6	30	121:8	Menat counterpoise	283/203, 231/1,263/4	106,101, 103	–

Figure 33.   Votive Objects.

No.	Eg. Cat. No.	Plate No.	Object	Field No.	Locus	Square
1	90	118:7	Feline figurine	95/1	110,112	B-C 14–15
2	91	–	Feline figurine	278/2	107,109	E-F 10–11
3	99	–	Feline figurine	282/1	110	A-B 13–15
4	93	–	Feline figurine	263/2	103	J-K 8
5	94	–	Feline figurine	257/1	102	I-J-K 13
6	35	–	Menat counterpoise	206/2	110	B-C-D 13–14
7	37	118:8	Menat counterpoise	269/2	107	A-B 8–10
8	38	–	Menat counterpoise	T200/2	Surface	

Figure 34.   Faience Objects.

No.	Eg. Cat. No.	Plate No.	Object	Field No.	Locus	Square
1	42	121:3	Bracelet	206/3	110,111	B-C-D 13–14
2	57	–	Bracelet	327/3	106	C 6
3	41	121:4	Bracelet	207/1	102	I-J 12–13
4	105	–	Bowl	253/103	102	J-K 14

Figure 35.   Bracelets.

No.	Eg. Cat. No.	Plate No.	Object	Field No.	Locus	Square
1	46	122:7	Bracelet	275/19	104	D-E 2–5
2	45	–	Bracelet	130/1	101	H-I 17–18
3	50	122:1	Bracelet	248/2	104	H-I 4–5
4	55	122:2	Bracelet	242/2	110	C-D 10–11
5	56	122:3	Bracelet	176/1	102	J-K 13–14
6	43	122:5	Bracelet	374/1	106	H 6
7	47	122:6	Bracelet	253/2	102	J-K 14
8	49	122:4	Bracelet	295/1	101	F 17–18

Figure 36.   Bracelets.

No.	Eg. Cat. No.	Plate No.	Object	Field No.	Locus	Square
1	51	122:11	Bracelet	277/3, 323/1, 371/1	106, 107, 110	
2	52	–	Bracelet	325/4	110	C-D 12
3	44	122:8	Bracelet	245/2	107	A-B 8–10
4	53	–	Bracelet	325/3	110	C-D 12
5	54	–	Bracelet	253/3	102	J-K 14
6	48	122:10	Bracelet	279/2	106, 107	E-F 7–9

Figure 37.   Bracelets and Faience Sherds.

No.	Eg. Cat. No.	Object	Field No.	Locus	Square
1	58	Bracelet	301/7	109	F 13–14
2	59	Bracelet	63/5	110	C-D 11–12
3	60	Bracelet	370/18	110	B-C-D 11–14
4	61	Bracelet	245/67, 339	107, 109	–
5	62	Bracelet	201/29, 277	107, 110	–
6	63	Bracelet	327/33	106	C 6
7	64	Bracelet	270/103	102	I-J-K 12
8	65	Bracelet	343/101	109	F 12
9	66	Bracelet	353/10	110	B-C 11
10	67	Bracelet	134/13	110, 111	A-B-C 13–14
11	68	Bracelet	279/293	101	E-F 7–9
12	69	Bracelet	237/37	110	B-C-D 13–14
13	70	Bracelet	133/6	101	D-E 16–17
14	71	Bracelet	208/45	102	I-J 14–15
15	72	Bracelet	343/105	109	F 12
16	73	Bracelet	343/104	109	F 12
17	74	Bracelet	216/88	101	H 125–16
18	75	Bracelet	I-J 16–17/54	101	–
19	76	Bracelet	350/3	110	D-E 11–12
20	77	Bracelet	316/17	111	A-B 12
21	78	Bracelet	T200/22	Surface	–
22	79	Bracelet	231/151	101	H 15–17
23	80	Bracelet	316/18	111	A-B 12
24	81	Bracelet	I-J 16–17/53	101	–
25	82	Bracelet	302/11	101	F 17–18
26	148	Decorated sherd (jug)	141/2	107	D-E 8–9
27	150	Decorated sherd (jug)	8/4	112	B-C 16–17
28	174	Juglet handle	219/2	101	E 15–16
29	173	Juglet handle	319/304	106 (108)	G 7–10
30	151	Decorated sherd (jug)	289/24	101, 102	L 14–15
31	147	Decorated sherd (jug)	43/1	109	E-F 13–14
32	223	Faience tile(?)	154/1, 236/3	107, 109, 110	–
33	218	Lid (?)	310/19	105	A-B 4–6
34	144 (*Col. Pl.* 21)	Decorated sherd (jar)	242/41, 325	107, 110	C-D 10–11

Figure 38.   Faience Objects.

No.	Eg. Cat. No.	Plate No.	Object	Field No.	Locus	Square
1	85	118:6	Feline figurine	156/1, 327/1	101, 106	–
2	86	–	Feline figurine	126/1	101	K-L 16–17
3	87	–	Feline figurine	241/1	101	J-K 15
4	88	–	Feline figurine	236/1	107	E-F 10–11
						E-F 9
5	89	–	Feline (?)	349/1	109	G 11
6	122	122:13	Decorated sherd	173/1, 245	101, 107	–

Figure 39.   Inscribed Faience Objects.

No.	Eg. Cat. No.	Object	Field No.	Locus	Square
1	225	Inscribed sherd	277/11	107, 110	C-D 10–11
2	107	Inscribed sherd	225/3	107	C-D 9–11
3	180	Cartouches	313	103	J-K 10
4	99	Ring stand	–	101	I-J 16–17
5	226	Inscribed sherd	294/52	101	F 15–16
6	100	Ring stand	257/300	102	I-J-K 13
7	101	Ring stand	253/110	102, 101	–

Figure 40.   Inscribed Objects.

No.	Eg. Cat. No.	Plate No.	Object	Field No.	Locus	Square
1	217	–	Inscribed sherd	348/16	107	D-E 10
2	108	–	Inscribed sherd	278/143	107	E-F 10–11
3	110	–	Inscribed sherd	232/25	101	I 15–17
4	224	–	Inscribed sherd	311/189	101	F 15–16
5	216	122:9	Inscribed sherd	337/3	102	I-J-K-L 11
6	104	122:12	Inscribed sherd	339/2, 323	109, 106	–
7	103	–	Cup	269/3, 260/1, 231, 242	101, 102, 107, 110	–
8	111	–	Dish	257/282B, 325, 332, 339	102, 110, 109	–

Figure 41.   Faience Objects.

No.	Eg. Cat. No.	Object	Field No.	Locus	Square
1	168	Sherd (jar rim)	239/18		C 10–11
2	169	Sherds (jug)	337/32, 111, 212, 206	102, 101, 110, 111	–
3	170	Sherds (bowl rim)	337/31	102	I-J-K-L 11
4	171	Cup	327/2	106	C 6
5	106	Decorated wine bowl sherd	70/2	109	H-I 13–14
6	159	Bowl	See Note 15		
7	160	Bowl or Cup	260/190, 93, 245, 257, 316	102, 107, 110, 111	–
8	112	Wine Bowl	213/8, 328	105, 106	–
9	123	Wine bowl sherds	241/63, 331, 232, 82	101, 102, 109	–

Figure 42.   Fish Bowl.

No.	Eg. Cat. No.	Object	Field No.	Locus	Square
1	113	Bowl with fish decoration	See note 11	–	–

Figure 43.   Decorated Faience.

No.	Eg. Cat. No.	Object	Field No.	Locus	Square
1	154	Decorated sherd Bowl rim	277/109	107, 110	C-D 10–11
2	126	Decorated sherd Bowl rim	166/3	109	H-I 13–14
3	153	Decorated sherd Bowl rim	310/15	105	A-B 4–6
4	127	Decorated sherd Bowl rim	289/20, 124	101, 102	–
5	116	Decorated sherd Bowl rim	256/15, 237, 332	109, 110	–
6	162	Decorated sherds Bowl rim	263/1,319 339	103, 106 108, 109	–
7	142	Decorated sherd Bowl rim	257/282A	102	I-J-K 13
8	16	Decorated sherds Bowl rim	344/12, 225,245	110, 111, 107	–
9	143	Decorated sherd Bowl rim	279/290A	106, 107	E-F 7–9
10	120	Decorated sherds Bowl rim	See Note No. 12	–	–
11	134	Decorated sherds Bowl base	339/121A	109	H-I 11
12	135	Decorated sherd Bowl base	339/121B	109	H-I 11
13	114	Decorated sherd Bowl base	278/142	107, 109 110	–
14	152	Decorated sherd Bowl base	234/151	101	H 15–18
15	130	Decorated sherd Bowl base	345/19	107	D-E-F-G 6–9

Figure 44.   Decorated Faience.

No.	Eg. Cat. No.	Object	Field No.	Locus	Square
1	124	Decorated sherds	See Note No. 14		
2	138	Decorated wine bowl sherds	137/21	110	C-D 14–15
3	118	Decorated wine bowl sherd	234/150	101	H 15–18
4	125	Decorated sherds	279/290B	107	E-F 7–9
5	129	Decorated sherds	302/12, 206	101, 110, 111	–
6	119	Decorated wine bowl sherds	241/64,	101, 106, 280, 232	– 110
7	117	Decorated sherds	374/13, 362	103, 106, 108	–
8	158	Decorated sherd (jug)	245/70	107	A-B 8–10
9	128	Decorated sherd	297/10	101	E 19
10	133	Decorated sherd	247/11	108	G-H 9
11	139	Decorated wine bowl sherds	337/36B	102	I-J-K-L 11

Figure 45.   Miscellaneous Faience Objects.

No.	Eg. Cat. No.	Plate No.	Object	Field No.	Locus	Square
1	219	118:5	Gaming piece(?)	223/1	101	H 17
2	178	–	Wand handle (?)	362/17, 204,233	101, 106, 108, 110	
3	172	–	Faience fragment	303/2	101	F 15–16
4	179	–	Wand handle	45/1	110, 111	B-C 11–12
5	176	–	Wand	225/1,	107, 101	I-J 16–17 B/1
6	177	Col. Pl. 6	Wand	278/4 337/2	102, 107, 109, 110	C-D 9–11
7	175	–	Wand	283/2	106, 107	D 7–9
8	–	–	Loom weight or gaming piece	279/287	107	H 5–16
9	–	–	Loom weight or gaming piece	234/140	101	H 15–18
10	–	–	Loom weight or gaming piece	323/18	105, 106	A-B 6–7

Figure 46.   Scarabs, Seals and Plaques.

No.	Eg. Cat. No.	Plate No.	Object	Material	Field No.	Locus	Square
1	18	–	Scarab	Faience	319/2	106,108	G 7–10
2	182	–	Scarab	Steatite	278/3	107,109,110	E-F 10–11
3	183	123:3	Scarab	Steatite	366/1	109	F 12–13–14
4	192	–	Seal (?)	Faience	283/1	107	D 7–9
5	190	–	Scaraboid	Faience	319/1	106,108	G 7–10
6	191	123:8	Scaraboid	Faience	97/1	101,112	CD 15–16
7	193	–	Scarab	Faience	598/13	108	
8	194	–	Scarab	Faience	503/6	109	E 12
9	189	123:6	Seal	Faience	340/1	107	B-C 9–10
10	184	123:5	Plaque	Faience	E 14/8	110	E 14
11	185	123:4	Plaque	Steatite	269/1	107	A-B 8–10
12	186	123:2	Plaque	Steatite	258/1	106	H 6–7
13	187	123:1	Plaque	Steatite	135/1	110	D-E 13–14
14	188	123:7	Plaque	Steatite	277/1	107	C-D 10–11

Figure 47.   Amulets.

No.	Eg. Cat. No.	Object	Material	Field No.	Locus	Square
1	210	Amulet	Faience	145/1	101	G-H 17–18
2	211	Amulet	Faience	137/1	110	C-D 14–15
3	a: 212 b: 213	Amulet	Faience	135/4	110	D-E 13–14
4	209	Amulet	Faience	347/1	108	G-H 9
5	215	Amulet	Faience	319/5	106, 108	G 7–10
6	201	Amulet	Faience	285/1	101, 112	B 16–18
7	214	Amulet	Faience	319/4	106, 108	G 7–10
8	195	Amulet	Faience	311/1	101	F 15–16
9		Amulet	Bronze	160/7	101	E-F 16–17
10		Amulet	Copper	295/197	101	F 17–18
11		Amulet	Bronze	309/1	101	G-H 19
12		Amulet	Bronze	319/27	106, 108	G 7–12
13		Amulet	Bronze	339/9	109	H-I 11

Figure 48.   Amulets.

No.	Eg. Cat. No.	Plate No.	Object	Material	Field No.	Locus	Square
1	198	–	Amulet	Faience	127/1	101	I-J 16–17
2	197	Col. Pl. 25	Amulet	Faience	127/2	101	I-J 16–17
3	196	–	Amulet	Faience	135/2	110	D-E 13–14
4	200	–	Amulet	Faience	339/3	109	H-I 11
5	203	Col. Pl. 24	Amulet	Frit	127/3	101	I-J 16–17
6	204	Col. Pl. 26	Amulet	Faience	127/4	101	I-J 16–17
7	208	–	Amulet	Faience	221/2	106, 107	F 7
8	205	118:1	Amulet	Gypsum	344/1	110, 111	B-C 11–12
9	202	Col. Pl. 28	Amulet	Faience	234/1	101	H 15–18
10	199	–	Amulet	Faience	338/1	109	G 11
11	206	Col. Pl. 22	Amulet	Faience	310/1	105	A-B 4–6
12	207	Col. Pl. 23	Amulet	Faience	129/1	101	H-I 16–17

Figure 49.   Stone Jar Lids.

No.	Eg. Cat. No.	Object	Material	Field No.	Locus	Square
1	236	Jar lid	Limestone	212/1, 238/1	101, 111	D 15–17
2	237	Jar lid	Limestone	269/4	107	A-B 8–10
3	241	Jar lid	Red sandstone	276/19	111	A-B 11
4	239	Jar lid	Limestone	276/2	111	A-B 11
5	240	Jar lid	White sandstone	309/2	101	G-H 19

Figure 50.   Stone Querns and Jar Lids.

No.	Eg. Cat. No.	Object	Material	Field No.	Locus	Square
1	–	Saddle-backed quern	Red sandstone (NSP)	66/1	101	I-J 15–16
2	–	Saddle-backed quern	NSP	254/3	101	G-I 15–18
3	–	Saddle-backed quern	NSP	266/4	104	F 3–5
4	–	Saddle-backed quern	NSP	295/193	101	F 17–18
5	–	Saddle-backed quern	NSP	275/15	104	J-K 17–18
6	238	Jar lid	Limestone	370/3, 316/1	110	B-D 11–14
7	242	Jar lid	Red sandstone	304/17	106, 108, 109	G 6–11

Figure 51.   Stone Altar Stands.

No.	Eg. Cat. No.	Plate No.	Object	Material	Field No.	Locus	Square
1	251	114:7	Stand	White sandstone	427/1	106	I 6
2	247	–	Stand	White sandstone	446/1	107	
3	248	124:2	Stand	White sandstone	452/1	106	A 7
4	249	124:1	Stand	White sandstone	421/1	106	G 7
5	250	–	Stand	White sandstone	423/1	101	D 15(W.1)
6	251	124:6	Stand	White sandstone	422/1	106	D 7
7	252	124:8	Stand	White sandstone	401/1	106	C 7
8	253	124:3	Stand	White sandstone	447/1	106	Wall 2
9	254	–	Stand	White sandstone	417/1	110	C 11
10	255	–	Stand	White sandstone	455/1	101	E 16
11	256	124:4	Stand	White sandstone	416/1	107	D 8
12	257	–	Stand	White sandstone	460/1	106	C 6
13	258	124:5	Stand	White sandstone	432/1	107	C 10

Figure 52.   Rock Stela, Eg. Cat 260; *Pl.* 105.

Figure 53.   Metal Figurines (*see* p. 147).

Figure 54.   Iron Objects (*see* pp. 147–8).

Figure 55.   Metal Ear-rings and Chain Links (*see* pp. 148–50).

Figure 56.   Metal Wires, Pins, Kohl Sticks, Sistrum Parts (*see* pp. 150–51).

Figure 57.   Metal Wires, Sistrum Fragments (*see* p. 151).

Figure 58.   Metal Wires, Pins, Kohl Stick (*see* pp. 151–3).

Figure 59.   Metal Wires, Rods, Tin and Lead Objects, Needles, Rivets (*see* pp. 153–4).

Figure 60.   Metal Rods and Wires (*see* pp. 154–6).

Figure 61.   Crook-shaped Wires (*see* pp. 156–7).

Figure 62.   Crook-shaped Wires (*see* pp. 157–8).

Figure 63.   Decorated Rings (*see* pp. 158–9).

Figure 64.   Metal Rings (*see* pp. 159–60).

Figure 65.   Metal Rings (*see* pp. 160–61).

Figure 66.   Metal Rings (*see* p. 161).

Figure 67.   Metal Rings (*see* pp. 161–2).

Figure 68.   Metal Rings (*see* pp. 162–3).

Figure 69.   Metal Rings (*see* pp. 163–4).

Figure 70.   Metal Rings (*see* p. 164).

Figure 71.   Metal Rings (*see* pp. 164–5).

Figure 72.   Metal Rings (*see* pp. 165–6).

Figure 73.   Metal Rings (*see* p. 166).

Figure 74.   Metal Spatulas and Bracelets (*see* pp. 166–7).

Figure 75.   Metal Spatulas, Projectile Point (*see* pp. 167–8).

Figure 76.   Metal Rods and Ferrules (*see* p. 168).

Figure 77.   Metal Spirals, Hooks, Punches and Balance Beam (*see* pp. 168–9).

Figure 78.   Beads.

No.	Col. Pl.	Object	Material	Field No.	Locus	Square
1	–	Disc bead	Conus Solander Quercinus	51/66	101	E-F 15–16
2	19.3	Disc bead	Quercinus	304/7	106,108,109	G 6–11
3	–	Disc bead	Quercinus	143/22A	101	H-I 17–18
4	–	Disc bead	Quercinus	203/25	101	H 15
5	19:5a	Disc bead	Mother of pearl	253/44	102	J-K 14
6	–	Disc bead	Flint	223/40	101	H 17
7	–	Disc bead	Flint	270/47	102	I-J-K 12
8	–	Disc bead	Flint	227/10	101	I 17
9	–	Disc bead	Mica	90/2	101	G-H 17–18
10	–	Disc bead	Chalcedony	74/10	101	I-J 16–17
11	19:6	Disc bead	Calcite	263/15	103	J-K 8
12	19:2	Disc bead	Calcite	319/268	106,108	G 7–10
13	19:1	Disc bead	Calcite	253/117	102	J-K 14
14	–	Disc bead	Calcite	277/20	107,110	C-D 10–11
15	–	Disc bead	Flint	259/1	106,108,109	H 6–11
16	19.8	Disc bead	Carnelian	206/46	10	B-C-D 13–14
17	19:5b	Disc bead	Carnelian	204/13	109	E 13–14
18	19:5d	Disc bead	Carnelian	154/17	101,109,110	D-E 14–15
19	–	Disc bead	Carnelian	292/2	101,110,112	(Wall l)
20	19:15c	Disc bead	Carnelian	343/19	109	F 12
21	–	Disc bead	Carnelian	13/7	101,112	C-D 16–17

Figure 79.   Beads.

No.	Col. Pl.	Object	Material	Field No.	Locus	Square
22	–	Disc bead	Mica	339/87	102, 109	H-I 11
23	–	Disc bead	Copper	270/92	102	I-J-K 12
23A	–	Disc bead	Copper	270/92A	102	I-J-K 12
23B	–	Disc bead	Copper	270/92B	102	I-J-K 12
24	–	Disc bead	Faience	107/1	101	G-H 15–16
25	–	Disc bead	Faience	44/1	106	F-G 6–7
26	–	Disc bead	Faience	260/12	102	I-J-K 12
27	–	Disc bead	Faience	226/3	104	D-E-F 2–5
28	–	Disc bead	Faience	241/53	101	J-K 15
29	–	Disc bead	Faience	213/1	106	E-F 6
30	–	Disc bead	Faience	223/46	101	H 17
31	–	Disc bead	Faience	121/25	101	J 16–17
32	19:10	Gadrooned disc bead	Faience	303/11B	101	F 15–16
33	19:11	Gadrooned disc bead	Faience	303/39	101	F 15–16
34	19:16	Gadrooned disc bead	Faience	366/2	109	F 12–14
35	–	Gadrooned disc bead	Faience	260/161	102	I-J-K 12
36	–	Short bead	Onxy	124/14	101	J-K 16–17
37	–	Short bead	Onxy	255/5	106, 107	C-D 7–8
38	–	Short bead	Carnelian	303/26	101	F 15–16
39	–	Short bead	Carnelian	135/8	109	D-E 13–14
40	–	Short bead	Carnelian	122/1	101	C-D 16–17
41	–	Short bead	Carnelian	340/5	107	B-C 9–10
42	–	Short bead	Faience	117/9	101	H-I 16–17
43	–	Short bead	Glass	208/34	101, 102	I-J 14–15
44	19:20	Short bead	Glass	303/18	101	F 15–16
45	–	Short bead	Glass	124/46	101	J-K 16–17
46	–	Standard bead	Calcite	277/82	107	C-D 10–11
47	–	Standard bead	Calcite	144/23	101	G-H 16–17
48	–	Standard bead	Carnelian	135/10	109, 110	D-E 13–14
49	–	Standard bead	Carnelian	51/67	101	E-F 15–16
50	–	Standard bead	Carnelian	44/2	106	F-G 6–7

Figure 80.   Beads.

No.	Col. Pl.	Object	Material	Field No.	Locus	Square
51	19:25	Standard bead	Faience	74/16	101	I-J 16–17
52	19:22	Standard bead	Glass	266/1	104	F 3–5
53	–	Standard bead	Glass	252/6	104	F 3–5
54	19:13	Standard bead	Glass	280/76	106	C-D-E 6–7
55	19:15	Standard bead	Glass	289/9	101, 102	L 14–15
56	–	Standard bead	Glass	303/18	101	F 15–16
57	19:14	Standard bead	Glass	283/40	106, 107	D 7–9
58	–	Standard bead	Glass	231/122	101	H 15–17
59	–	Standard bead	Glass	303/345	101	F 15–16
60	19:21	Standard bead	Glass	293/2	101, 111	Wall 3
61	–	Standard bead	Glass	224/11	101	G 15–18
62	–	Long bead	Flint	287/5	102	L 12–13
63	19:12	Long bead	Calcite	243/31	101	J 16–17
64	19:31	Long bead	Carnelian	338/24	109	G 11
65	19:30	Long bead	Faience	216/45	107	H 15–16
66	19:23	Long bead	Faience	121/6	101	J 16–17
67	–	Long bead	Faience	124/44	101	J-K 16–17
68	19:24	Long bead	Faience	216/54	101	H 15–16

Figure 81. Beads.

No.	Col. Pl.	Object	Material	Field No.	Locus	Square
69	20:11	Scallop bead	Glass	51/101	101	E-F 15–16
70	20:3	Scallop bead	Glass	51/102	101	E-F 15–16
71A	–	Long bead	Glass	330/2	105, 106	B 6–7
71B	–	Long bead	Glass	330/1	105, 106	B 6–7
72	–	Melon bead	Faience	211/252	101	G 15–17
73	–	Melon bead	Faience	241/52	101	J-K 15
74	–	Lenticular bead	Faience	229/8	101	E 16
75	–	Collared bead	Faience	331/1	109	G 11
76	–	Irregular bead	Glass	201/6	101	D-E 15–16
77	–	Irregular bead	Faience	266/2	104	F 3–5
78	–	Irregular bead	Faience	234/130	101	H 15–18

Figure 82. Beads.

No.	Col. Pl.	Object	Material	Field No.	Locus	Square
79	19:35	Shell bead	Cerithium Erythraconense	216/74	101	H 15–16
80	–	Shell bead	Strombus Gibberuius	234/135	101	H 15–18
81	–	Shell bead	Cymatium Pileare	260/16	102	I-J-K 12
82	–	Shell bead	Conus Quercinus	178/3	110, 112	A-B 14–15
83	–	Shell bead	Conus Quercinus	277/94	107	C-D 10–11
84	19:33	Shell bead	Drupa Tuberculata	141/1	107	D-E 8–9
85	–	Shell bead	Terebra Cerithina	115/6	112	A-B-C 8–10
86	19:34	Shell bead	Clanculus Pharaeonis	269/65	107	A-B 8–10
87	19:38	Shell bead	Cypraea Carneola	119/8	101, 110, 112	C-D 14–15
88	19:37	Shell bead	Cypraea Carneola	349/14	109	G 11
89	19:36	Shell bead	Engina Mendicaria	295/33	101	F 17–18
90	19:32	Shell bead	Nerita Polita	225/25	104, 106	C-D 9–11
91	–	Short bead	Gold	303/4	101	F 15–16
92	–	Disc bead	Gold	233/5	101	I-J 18
93	–	Short bead	Gold	234/4	101	H 15-18
94	–	Disc bead	Gold	311/2	101	F 15–16
95	–	Ring bead	Gold	337/5	102	I-J-K-L 11
96	–	Ring bead	Gold	270/28	107, 109	I/J/K 12
97	–	Ring bead	Gold	3214/1	109	F 13–14

Figure 83. Spacers and Pendants.

No.	Col. Pl.	Object	Material	Field No.	Locus	Square
98	–	Spacer	Faience	137/2	110, 112	C-D 14–15
99	20:2	Spacer	Bone	343/82	109	F 12
100	–	Spacer	Bone	278/92	107, 109	E-F 10–11
101	20:8	Spacer	Carnelian	213/1	106	E-F 6
102	–	Spacer	Carnelian	243/68	101	J 16–17
103	20:1	Spacer	Bone	278/10	109	E-F 10–11
104	20:9	Pendant	Hematite	280/7	106	C-D-E 6–7
105	–	Pendant	Hematite	277/23	107, 110	C-D 10–11
106	20:7	Pendant	Faience	337/18	102	I-J-K-L 11
107	–	Pendant	Bone	213/5	106	E-F 6
108	20:15,17	Pendants (2)	Mica	330/17–19	105, 106	B 6–7
109	20:16	Pendant	Mica	101/4	101	H-I 15–16
110	–	Pendant	Carnelian	265/10	103	J-K 6
111	20:12	Pendant	Rose quartz	–	–	-
112	19:29	Pendant	Carnelian	231/40	101	H 15–17
113	20:4	Pendant	Bone	303/5	101	F 15–16
114	–	Pendant	Bone	339/15	102, 109	H-I 11
115	–	Pendant	Bone	303/649	101	F 15–16
116	20:5	Pendant	Bone	241/7	101	J-K 15
117	–	Pendant	Bone	303/3	101	F 15–16

Figure 84. Pendants, etc.

No.	Col. Pl.	Object	Material	Field No.	Locus	Square
118	–	Pendant	Ceramic	85/10	110, 111	A-B-C 13–14
119	19:27	Pendant	Faience	118/72	101	D-E 15–16
120	19:26	Pendant	Faience	282/25	110, 112	A-B 13–15
121	19:28	Pendant	Faience	–	101	I-J 16–17B/2
122	–	Pendant	Faience	201/6	101	–
123	–	Pendant	Faience	303/48	101	F 15–16
124	–	Pendant	Gold	211/8	101	G 15–17
125	–	Pendant	Gold	283/30	106, 107	D 7–9
126	–	Pendant	Gold	254/4	215	
127	–	Pendant (3 pieces)	Gold	279/29	106, 107	E-F 7–9
128	–	Pendant	Gold	253/4	102	J-K 14
129	–	Pendant	Gold	260/2	102	
130	–	Pendant	Gold	319/7	594/14	G 7–10
131	–	Amulet	Gold	275/1	104	D-E 2–5
132	–	Head band(?)	Gold	89/1	102	I-J 13–14
133	–	Pendant	Bronze	278/131	107, 109	E-F 10–11
134	–	Pendant	Copper	339/9	102, 109	H-I 11
135	–	Pendant	Copper	295/197	101	F 17–18
136	–	Pendant	Copper	337/26	102	I-J-K-L 11
137	–	Pendant	Copper	319/27	106, 108	G 7–10

Figure 85. Egyptian Glass Fragments.

No.	Egypt Gl. Cat. No.	Col. Pl.	Object	Field No.	Locus	Square
1	1	7:4	Bowl	289/17	101,102	L 14–15
2	3	7:2	Krateriskoi	203/34	101	H 15
3	4	–	Krateriskoi	269/73	107	A-B 8–10
4	8	7:7	Krateriskoi	245/39	107	A-B 8–10
5	10	7:5	Krateriskoi	227/49	101	I 17
6	9	7:6	Krateriskoi	227/51	101	I 17
7	12	7:10	Krateriskoi	241/59	101	J-K 15

Figure 86. Egyptian and Roman Glass Fragments.

No.	Egypt Gl. Cat. No.	Col. Pl.	Object	Field No.	Locus	Square
Egyptian						
1	14	8	Flask	278/110–112	107,109,110	E-F 10–11
2	20	–	Amphoriskoi	245/40	107	A-B 8–10
3	30	9	Body fragment	273/1	106	E-F 6
4	18	–	Amphoriskoi	316/13–15	111	A-B 12
5	17	–	Amphoriskoi	245/41–43	107	A-B 8–10
6	26	–	Neck fragment	51/97	101	E-F 15–16
7	22	–	Pomegranate bottle	80/11	102	J-K 14–15
Roman						
8	–	–	Rim and neck fragment	317/1	110	C-D 12
9	–	–	Handle	344/7	110, 111	B-C 11–12

Figure 87.   Roman Pottery.

No.	Pl. No.	Rom. Pot. Cat. No.	Object	Field No.	Locus	Square
1	–	1	Bowl	552/3	111	TT 58
2	151:3	2	Bowl	226/13	104	D-E-F 2–5
3	–	3	Bowl or cup	120/7	101, 102	A-B-C 17–8
4	–	4	Cooking pot	325/23	110	C-D 12
5	–	5	Jar or jug	233/13	101	I-J 18
6	–	6	Jug	325/24	110	C-D 12
7	151:5	7	Jug	337/41	102	I-J-K-L 11
8	151:1	8	Bottle	325/25	110	C-D 12
9	–	9	Bottle or juglet	236/10	107, 109	E-F 9–11
10	–	10	Bottle or juglet	325/26	110	C-D 12
11	151:2	11	Goblet	337/140	102	I-J-K-L 11
12	–	12	Jug	325/21	110	C-D 12
13	–	13	Jug or jar	337/42	102	I-J-K-L 11

Figure 88.   Roman Pottery Oil Lamps.

No.	Pl. No.	Rom. Pot. Cat. No.	Object	Field No.	Locus	Square
1	–	14	Oil lamp	226/3	104	D-E-F 2–5
2	151:6	15	Oil lamp	245/5	107	A-B 8–10
3	–	16	Oil lamp	337/37	102	I-J-K-L 11
4	151:4	17	Oil lamp	325/1	110	C-D 12
5	151:7	18	Oil lamp	238/2	111	B-C 11

Figure 89.   Saddle-backed Querns.

No.	Object	Field No.	Locus	Material	Description
1	Saddle-backed quern	270/100	102	Red sandstone	L. 8.8, W. 6.3, Th. 3.9 mm
2	Saddle-backed quern	214/2	106, 107	Red sandstone	L. 83, W. 72.4, Th. 45mm.
3	Saddle-backed quern	362/16	106, 108	Red sandstone	L. 94.3, W. 81.5, Th. 45.5mm.
4	Saddle-backed quern	295/194	101	Red sandstone	L. 128, W. 11.3, Th. 61mm.
5	Saddle-backed quern	280/90	106	White sandstone	L. 118.2, W. 107.8, Th. 40.2mm.
6	Saddle-backed quern	203/36	101	Red sandstone	L. 122, W. 190, Th. 42.4mm.
7	Saddle-backed quern	311/193	101	Red sandstone	L. 132, W. 85, Th. 74mm.
8	Saddle-backed quern	297/8	101	Red sandstone	L. 151.4, W. 105, Th. 57.3mm.
9	Saddle-backed quern	314/46	109	Red sandstone	L. 155, W. 110, Th. 158.2mm.
10	Saddle-backed quern	314/47	109	Red sandstone	L. 157, W. 131.4, Th. 66.2mm.
11	Saddle-backed quern	334/6	109	White sandstone	L. 155, W. 143, Th. 50mm.
12	Saddle-backed quern	266/4	104	Red sandstone	L. 156, W. 149.7, Th. 61 mm.

Figure 90.   Bowls, Basins and Grinding Stone.

No.	Object	Field No.	Locus	Material	Description
1	Grinding bowl	409/1	106/Surface	White sandstone	Diam. 18, Th. 11cm.
2	Grinding bowl	–	–	Red Sandstone	–
3	Grinding bowl	226/12	104	Limestone	Diam. 258, Th. 5cm.
4	Shallow oval basin	248/4	103, 104	Limestone	L. 20.5, Th. 8.3cm.
5	Shallow bowl	275/16	104	Red sandstone	Diam. 38cm.
6	Rotary grinding stone	400/1	Surface	Granite	Diam. 36cm., with hole in centre

Figure 91.   Hammerstones, Anvil Stone and small Mortars.

No.	Object	Field No.	Locus	Material	Description
1	Crushing stone (anvil)	226/13	104	Red sandstone	L. 82, W. 78.5, Th. 72mm.
2	Hammerstone	368/4	109	Granite	L. 76.5, W. 71.8, Th. 46.6mm.
3	Hammerstone	108/1	109, 110	Limestone	L. 105.8, W. 101.9, Th. 73.6mm.
4	Small mortar	257/269	102	White sandstone	L. 151.3, W. 140.7, Th. 62.6mm.
5	Small mortar	258/19	106	White sandstone	L. 151.7, W. 134.4, Th. 65.1mm.

Figure 92.   Pestles, Anvils and Hammerstones.

No.	Object	Field No.	Locus	Material	Description
1	Pestle	366/41	109	–	L. 12.2, W. 8.3, Th. 7.4cm.
2	Crushing stone (anvil)	226/11	104	Limestone	L. 12, W. 7.7, Th. 6.3cm.
3	Hammerstone or pestle	–	–	–	–
4	Crushing stone (anvil)	330/25	105, 106	–	L. 9.2, W. 8.5, Th. 9.5cm.
5	Phallus	234	101	Stone	–
6	Hammerstone or pestle	254/2	101	Granite	Diam. 14 cm.

1

2

3   4

Plates 1, Egyptian pottery; 2, Normal pottery; 3, Rough hand-made pottery; 4, Midianite pottery. See p. 291–2 for details.

6

5

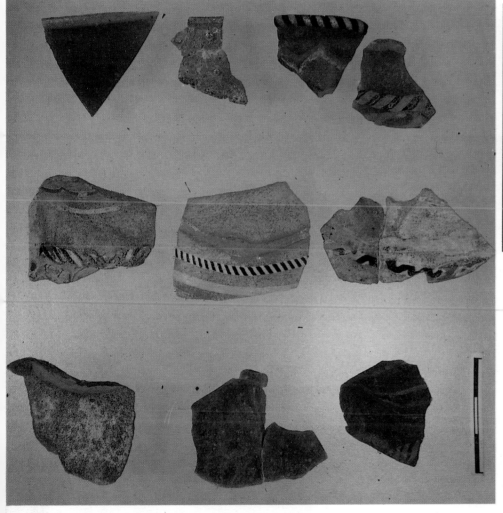

8

9

7

Plates 5, Hathor votive mask (Eg. Cat. No. 25, *Fig.* 30:1); 6, Amuletic wand (Eg. Cat. No. 175, *Fig.* 45:7); 7, Egyptian glass (*Fig.* 85); 8, Egyptian glass flask (*Fig.* 86:1); 9, Egyptian glass (*Fig.* 86:3); 10, Egyptian glass fragment. Scale 1:1.

10

11

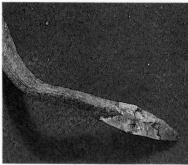

12

13

14

15

16

17

18

Plates 11, Bronze serpent with gilded head (Met. Cat. No. 3, *Fig.* 53;3); 12, Detail of head of bronze serpent; 13, a. Tin droplet (Met. Cat. No. 163), b. Gilded iron ring (Met. Cat. No. 22, *Fig.* 54:16); 14, Bronze ram figurine (Met. Cat. No. 6, *Fig.* 53:6); 15, Gold jewellery (a – *Fig.* 84:130, b – *Fig.* 84:131); 16, Midianite decorated jug (*Fig.* 7.4); 17, Midianite sherd with human figure (*Fig.* 7:2); 18, Midianite decoration with unusual colours (*Fig.* 5:6). Scale 1:1.

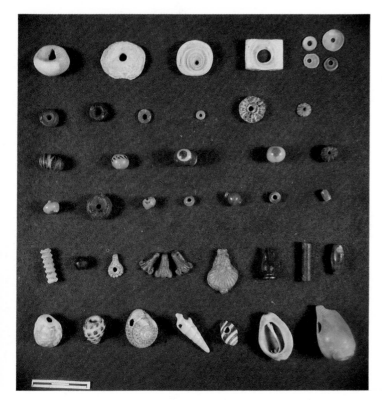

19

21

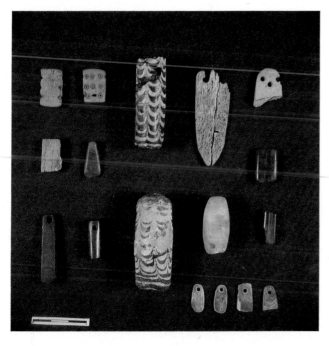

20

22

23

24   25   26   27

28   29

Plates 19, Beads and pendants (*Fig. 78–80, 82, 84*); 20, Beads and pendants (*Fig. 81, 83*); 21, Decorated faience sherd (Eg. Cat. No. 144, *Fig.* 37:34); 22, Faience amulet (Eg. Cat. No. 206, *Fig.* 48:11); 23, Faience amulet, the moon god Khons (Eg. Cat. No. 207, *Fig.* 48:12); 24, Faience amulet, Harpocrates (Eg. Cat. No. 203, *Fig.* 48:5); 25, Faience amulet, Pataikos (Eg. Cat. No. 197, *Fig.* 48:2); 26, Faience amulet, Harpocrates (Eg. Cat. No. 204, *Fig.* 48:6); 27, Faience amulet, Pataikos (Eg. Cat. No. 198, *Fig.* 48:1); 28, Faience amulet, bearded Pataikos (Eg. Cat. No. 202, *Fig.* 48:9); 29, Male figurine, bronze (Met. Cat. No. 1, *Fig.* 53:1). Scale 1:1.

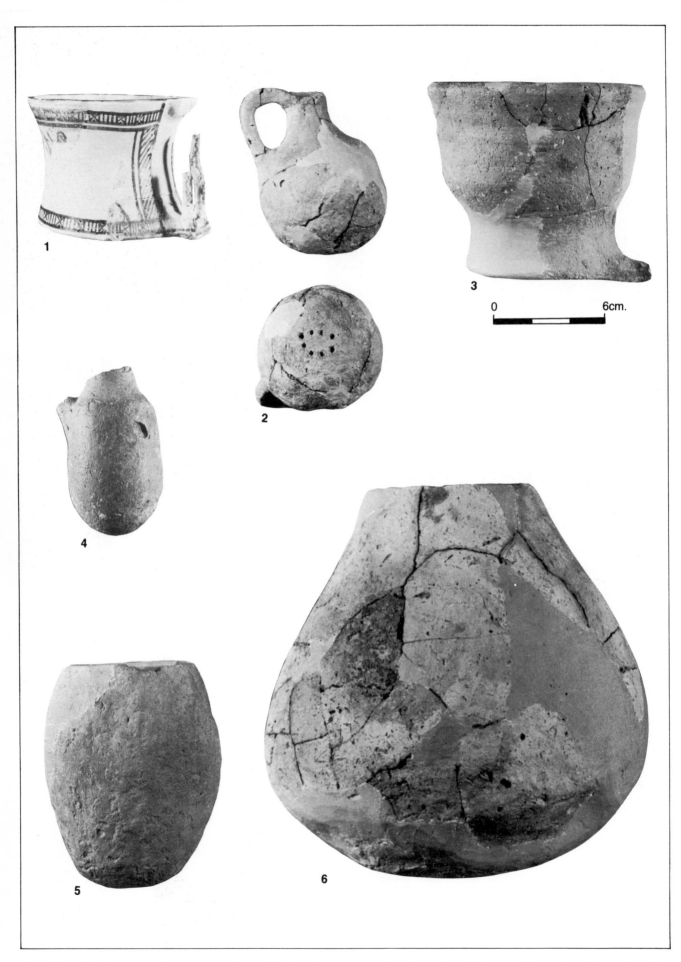

Plate 106.   Midianite, Rough Hand-made and Local Wheel-made pottery (1 – *Fig*. 10; 2 – *Fig*. 15:3; 3 – *Fig*. 18:4; 4 – *Fig*. 15:2; 5 – *Fig*. 15:1; 6 – *Fig*. 15:8).

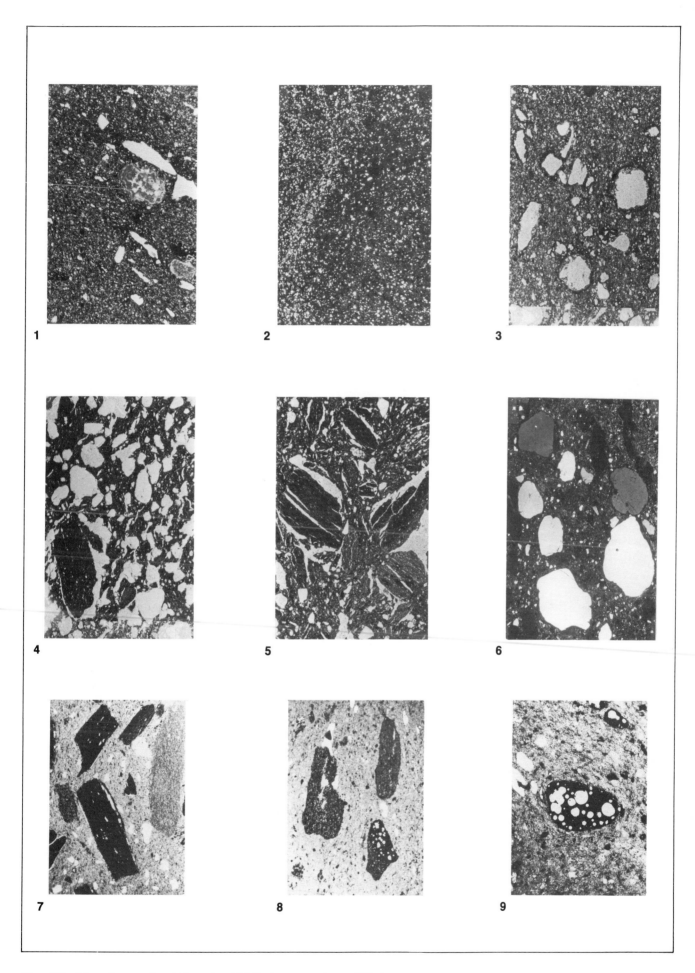

Plate 107. Microphotographs of Normal and Midianite pottery (*see* p. 293).

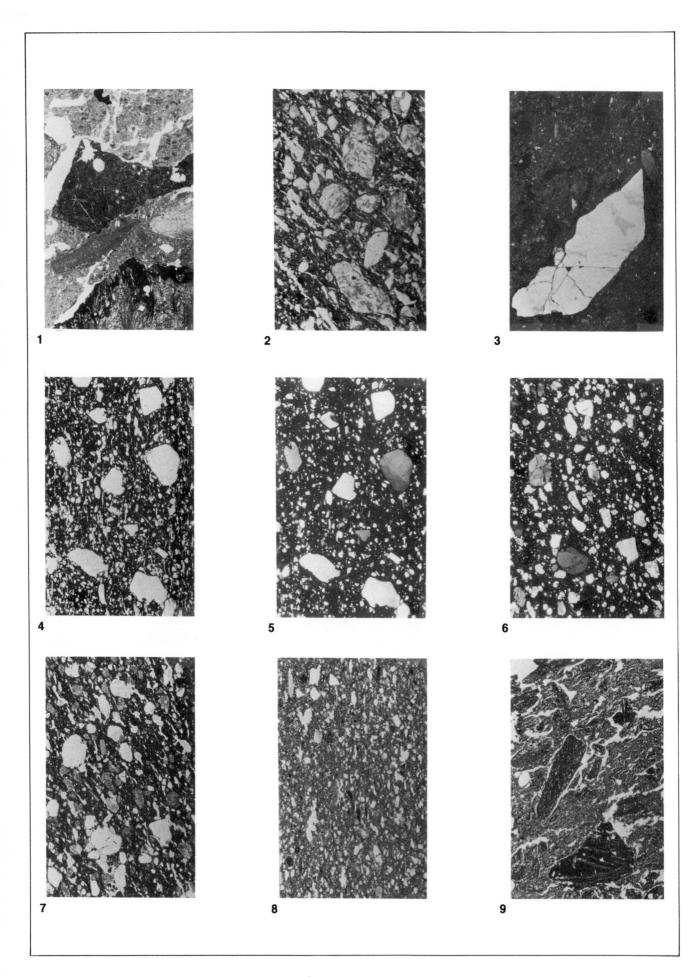

Plate 108. Microphotographs of Egyptian and Rough Hand-made pottery (*see* p. 293).

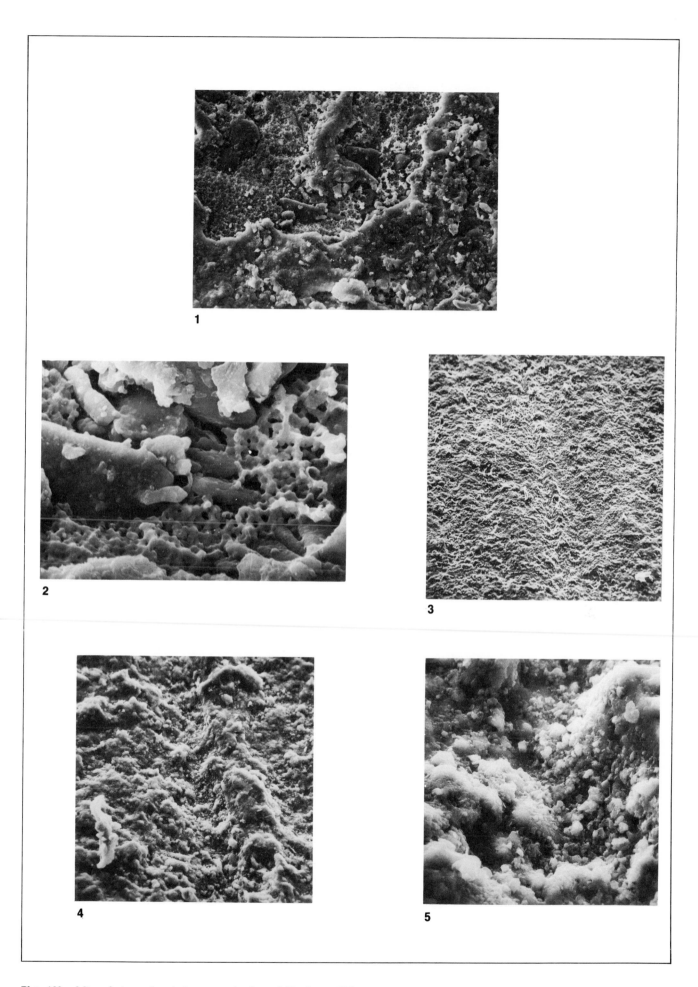

Plate 109. Microphotographs of pigment and coloured frits (*see* p. 294).

Plate 110.   Architectural elements from the *naos*. (1– *Fig*. 24:5; 2– *Fig*. 24:4; 3– *Fig*. 22:10).

Plate 111.   Fragments of Square Pillars (1 – *Fig.* 23:2 – side view; 2–4 no figure, *see* p. 268).

Plate 113. Architectural details from the *naos* and Square Pillar fragments (1, 2, 3, 5, 6, no figures; 4 – *Fig.* 23:4).

Plate 114. Fragments of white sandstone sculptures (1 – *Fig.* 25:2; 2–4 no figure, *see* p. 268).

0 3cm.

Plate 115.   White sandstone statuette (*Fig.* 26:1–2).

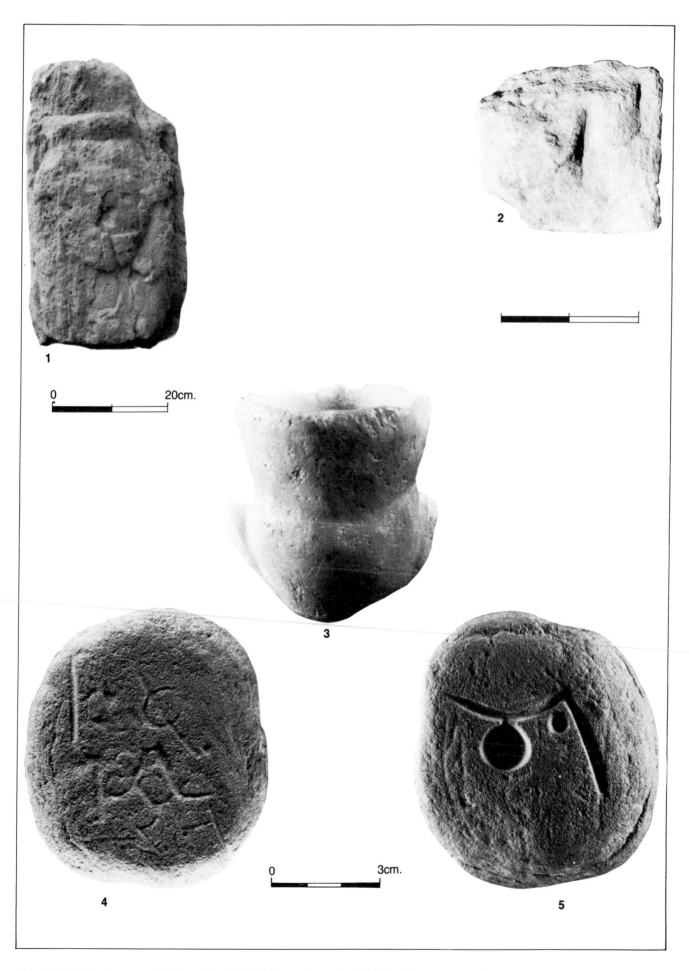

0   20cm.

0   3cm.

Plate 116.   Finds of stone and alabaster (1 – *Fig*. 23:1; 2 – no figure; 3 – *Fig*. 22:3; 4–5 (reverse) – no figure; *see* p. 268).

Plate 117. White sandstone statuette (Fig. 25:1).

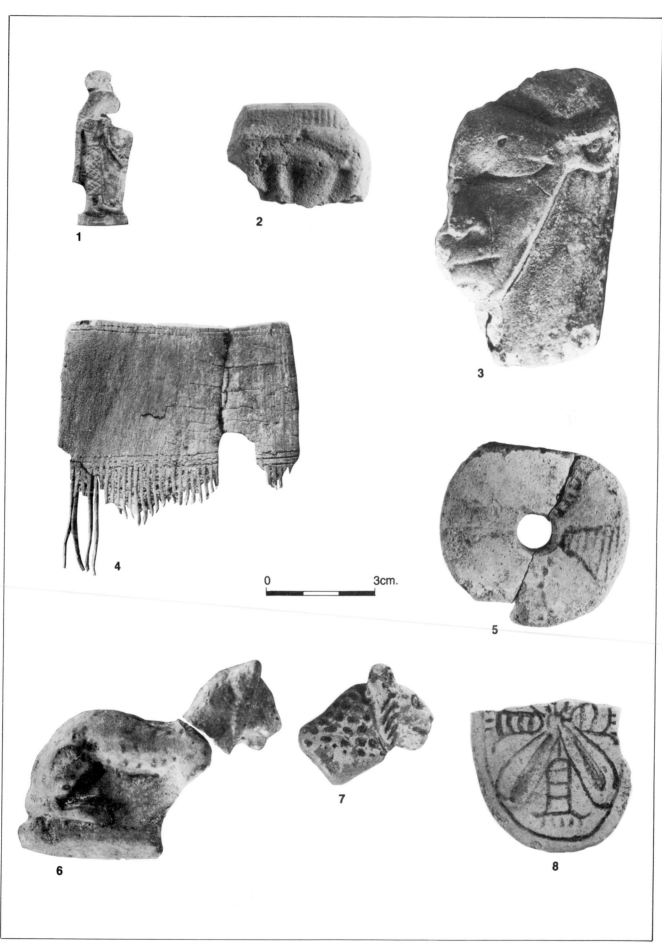

Plate 118.   Egyptian objects of faience, bone and wood (1 – *Fig.* 25:1; 2 – *Fig.* 30:2; 3 – *Fig.* 27:1; 4 – *Fig.* 30:3; 5 – *Fig.* 45:1; 6 – *Fig.* 38:1; 7 – *Fig.* 33:1; 8 – *Fig.* 33:7).

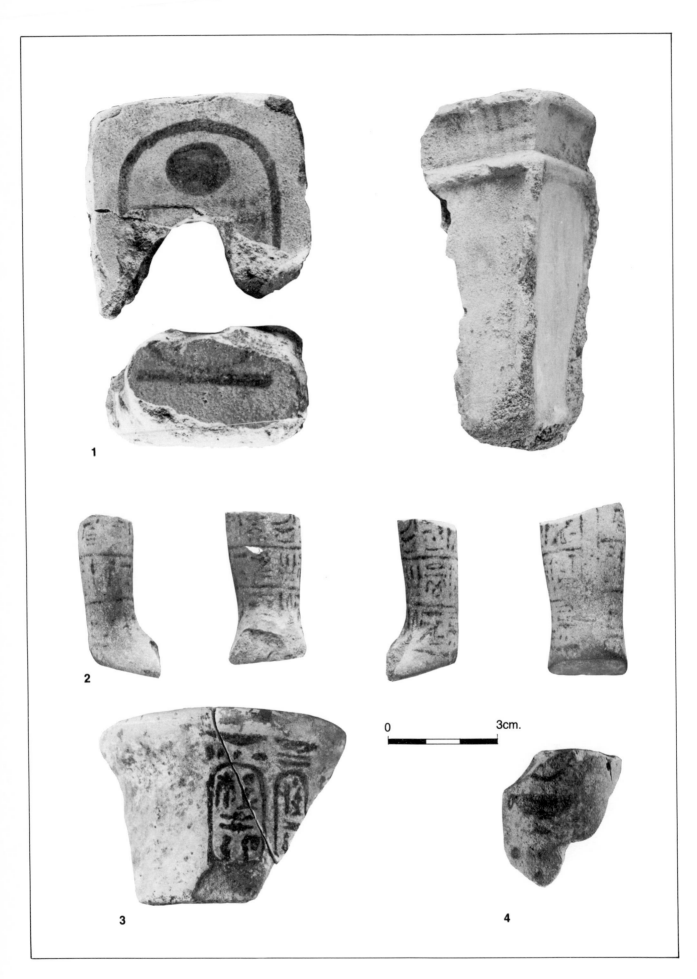

Plate 119. Egyptian faience objects (1 – *Fig.* 28:1; 2 – *Fig.* 28:2; 3 – *Fig.* 31:5; 4 – *Fig.* 29:1).

Plate 120.   Inscribed faience objects (1 – *Fig.* 28:3; 2 – *Fig.* 29:5; 3 – *Fig.* 29:4; 4 – *Fig.* 34:3; 5 – *Fig.* 31:1; 6 – *Fig.* 31:2).

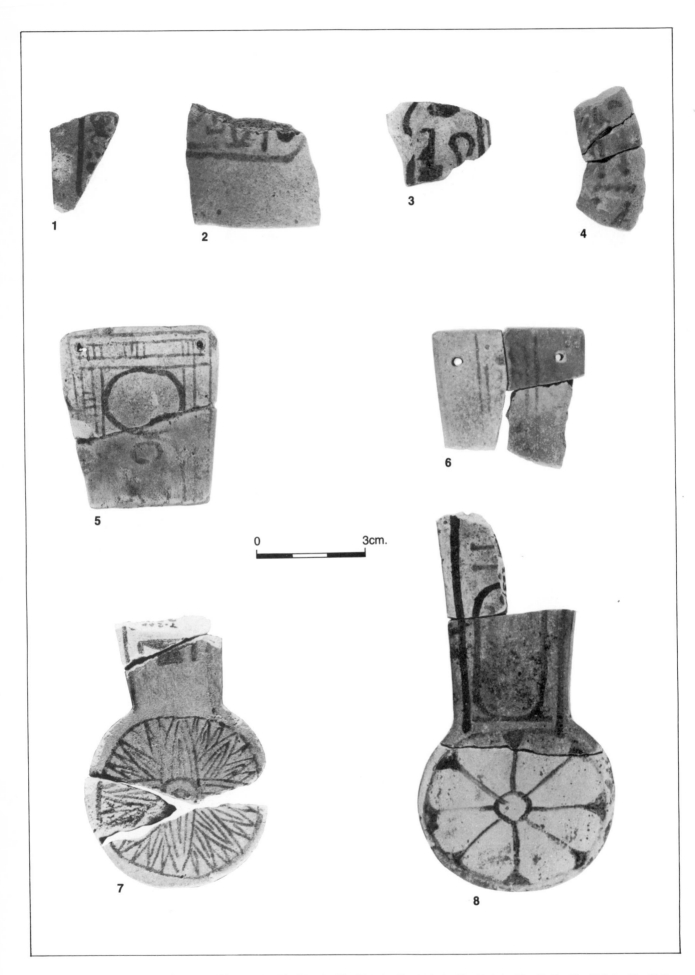

Plate 121.   Inscribed faience objects (1 – *Fig.* 31:3; 2 – *Fig.* 31:6; 3 – *Fig.* 34:1; 4 – *Fig.* 34:3; 5 – *Fig.* 32:4; 6 – *Fig.* 32; 7 – *Fig.* 32:3; 8 – *Fig.* 32:6).

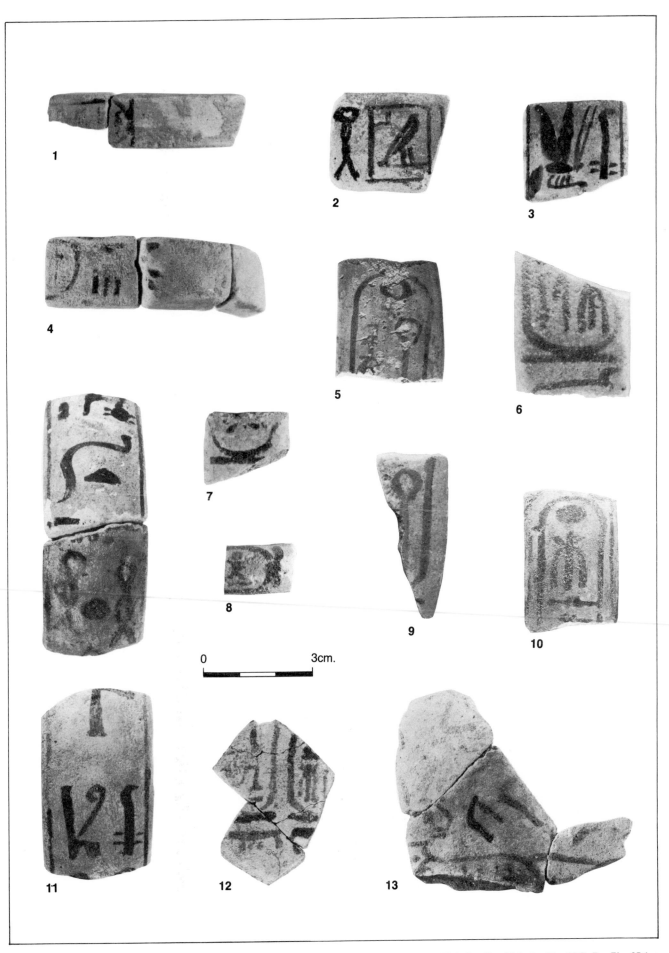

Plate 122. Inscribed faience objects (1 – Fig. 35:3; 2 – Fig. 35:4; 3 – Fig. Fig. 35:5; 4 – Fig. 35:8; 5 – Fig. 35:6; 6 – Fig. 35:7; 7 – Fig. 35:1; 8 – Fig. 36:3; 9 – Fig. 40:5; 10 – Fig. 36:6; 11 – Fig. 36:1; 12 – Fig. 40:6; 13 – Fig. 38:60).

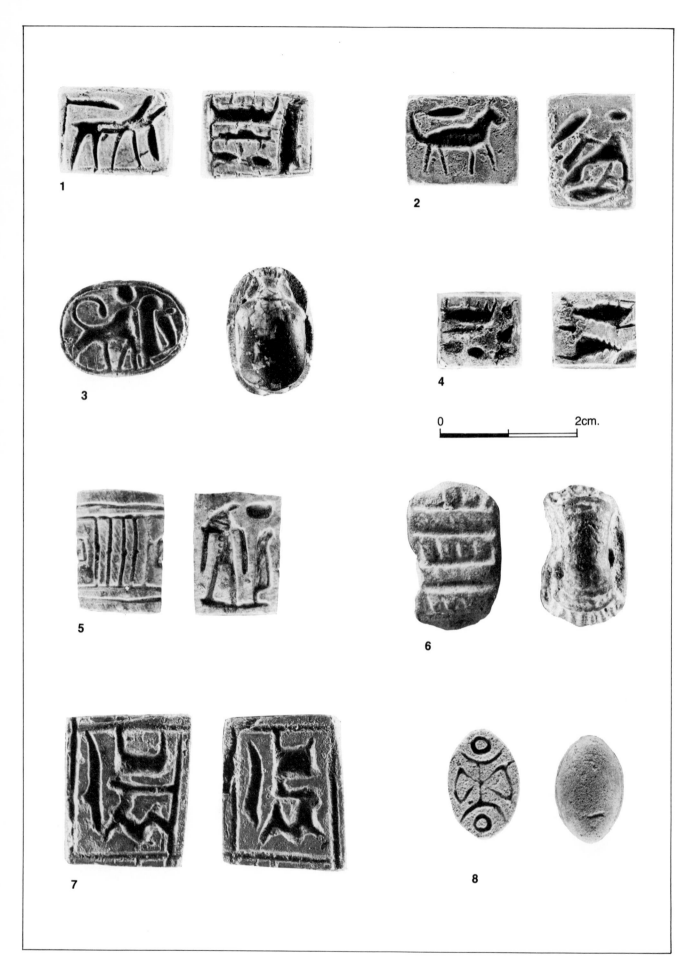

Plate 123.   Scarabs, seals and plaques (Fig. 46).

Plate 124.   Altar Stands (Fig. 51).

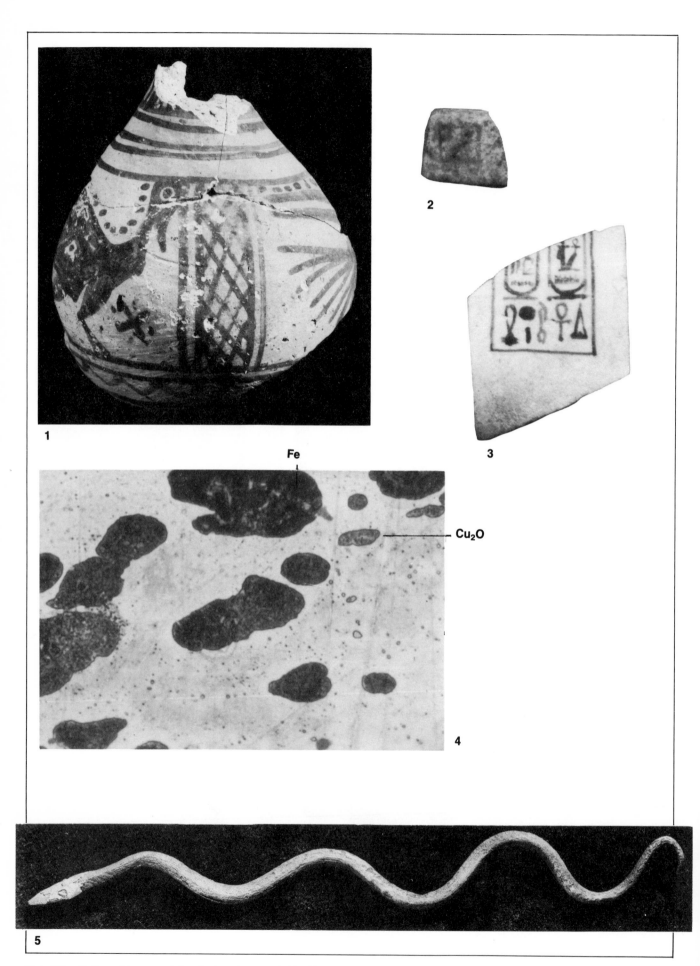

Fe

Cu₂O

Plate 125.   Pottery, faience, metal and microphotograph (1 – *Fig.* 7:4; 2 – *Fig.* 31:8; 3 – *Fig.* 31:8; 3 – *Fig.* 31:7; 4 – Chapter III.11: copper with globules of iron and some cuprous oxide; x 1260; 5 – *Fig.* 53:3).

Plate 126.   Metal objects (1 – *Fig*. 53:4; 2 – *Fig*. 53:5; 3 – *Fig*. 63:6; 4 – *Fig*. 55:16; 5 – *Fig*. 75:9; 6 – *Fig*. 53:2).

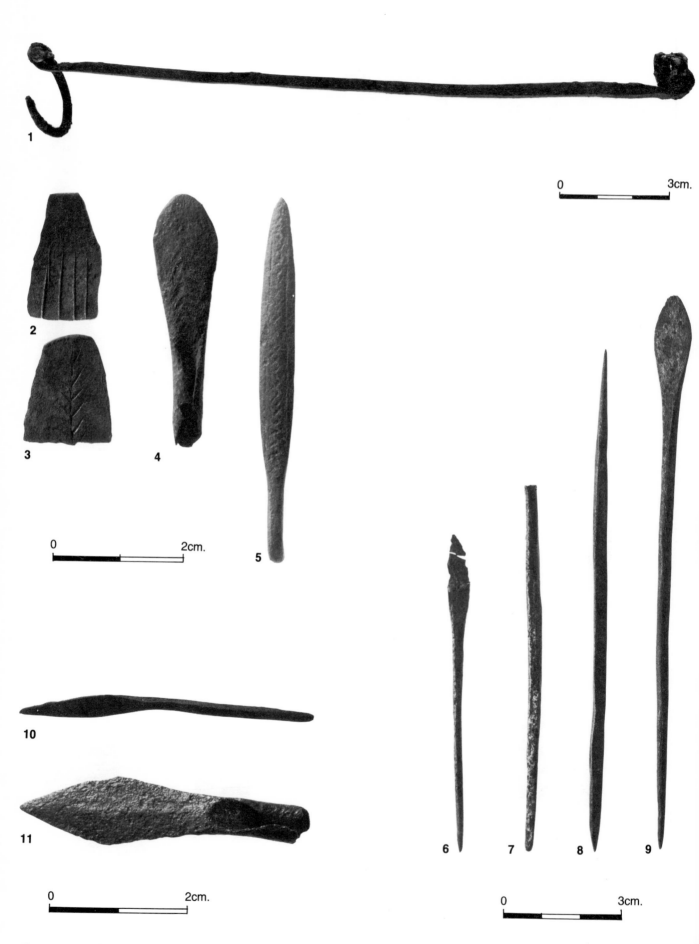

Plate 127. Metal objects (1 – *Fig.* 77:17; 2 – *Fig.* 47:10; 3 – *Fig.* 84:137; 4 – *Fig.* 74:1; 5 – *Fig.* 74:2; 6 – *Fig.* 74:5; 7 – *Fig.* 56:4; 8 – *Fig.* 56:5; 9 – *Fig.* 74:11; 10 – not identified; 11 – *Fig.* 75:15).

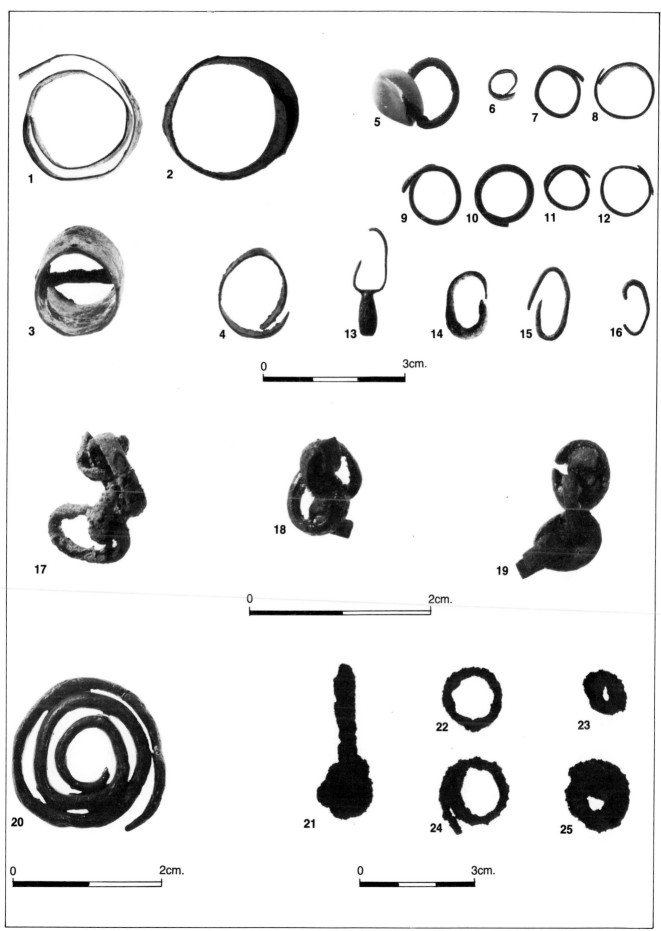

Plate 128. Iron and copper-base objects (1 – *Fig.* 76:9; 2 – *Fig.* 76:7; 3 – *Fig.* 76:11; 4 – not identified; 5 – *Fig.* 65:6; 6 – *Fig.* 65:5; 7 – *Fig.* 65:10; 8 – not identified; 9 – *Fig.* 69:7; 10 – *Fig.* 65:7; 11 – *Fig.* 68:6; 12 – *Fig.* 80:8; 13 – *Fig.* 55:13; 14 – *Fig.* 55:4; 15 – *Fig.* 55:7; 16 – *Fig.* 55:6; 17 – *Fig.* 55:33 part; 18 – *Fig.* 55:30; 19 – *Fig.* 55:34; 20 – *Fig.* 77:7; 21 – not identified; 22 – *Fig.* 54:1; 23 – *Fig.* 54:11; 24 – *Fig.* 54:2; 25 – *Fig.* 54:10).

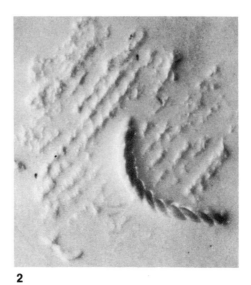

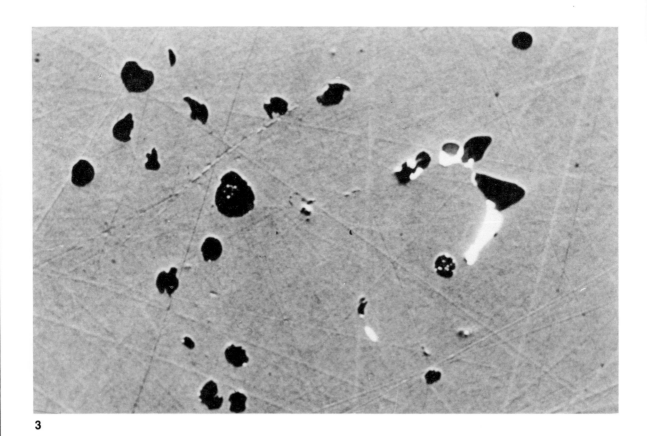

Plate 129.   1–2 Textile impressions on sherds to Chapter III.19 (Tex. Imp. Cat. No. 4, *see* p. 294 for details);
3 Microphotograph to Chapter III.8 (Met. Cat. No. 235).

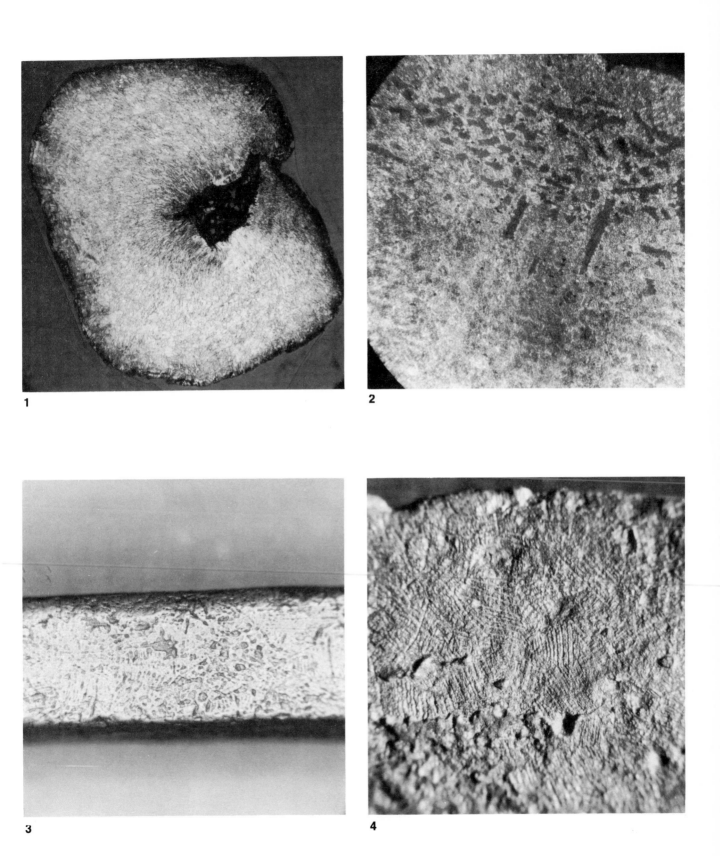

Plate 130.   Microphotographs to Chapter III.9 (1 – Met. Cat. No. 482; 2 – Met. Cat. No. 62; 3 – Met. Cat. No. 62; to Chapter III.10 (4-wire made by folding).

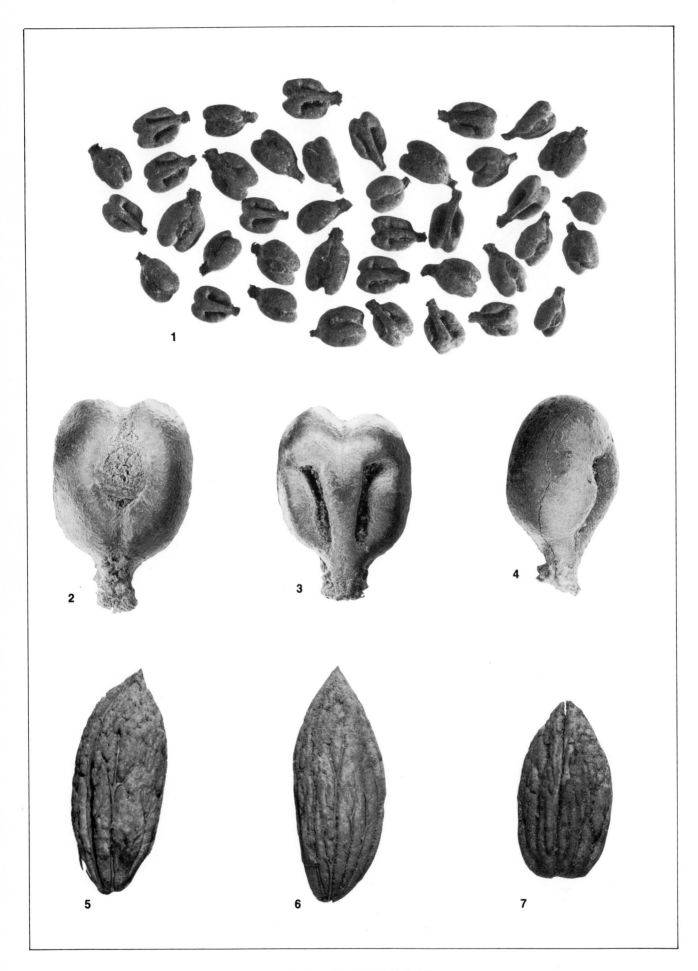

Plate 131. To Chapter III.22. 1–4 – Grape pips, 5–7 – Olive kernel (1, 5–7 ×3; 2–4 ×10).

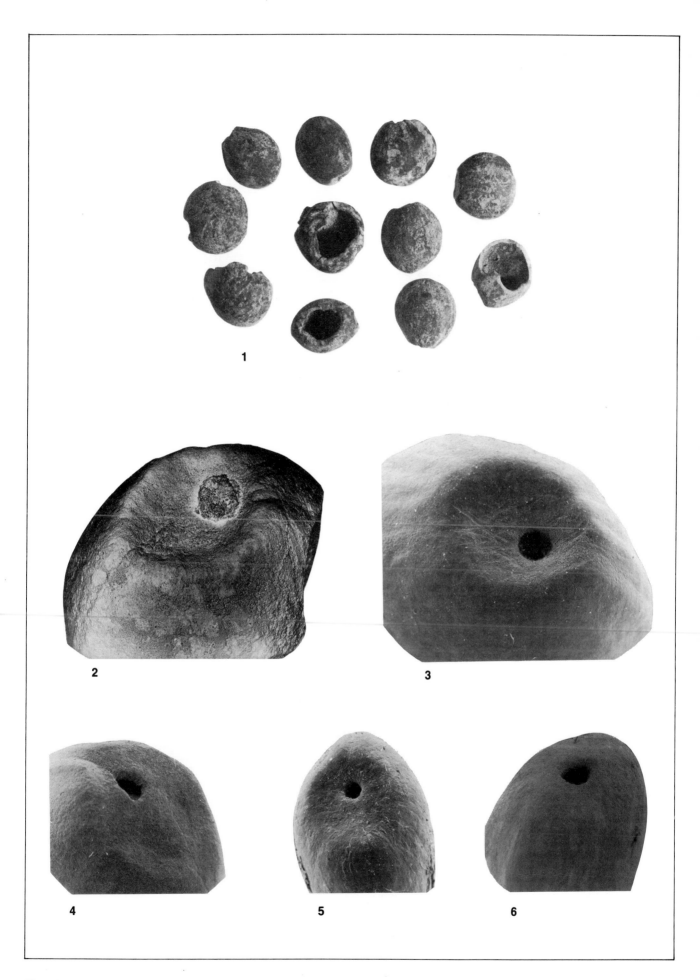

Plate 132. To Chapter III.22. Pistacia. 1–2 – ancient; 3–6 – recent (1 ×3; 2–5 ×30).

Plate 133.   To Chapter III.19. Textiles (*see* p. 294).

Plate 134.   To Chapter III.19. Textiles (*see* p. 294).

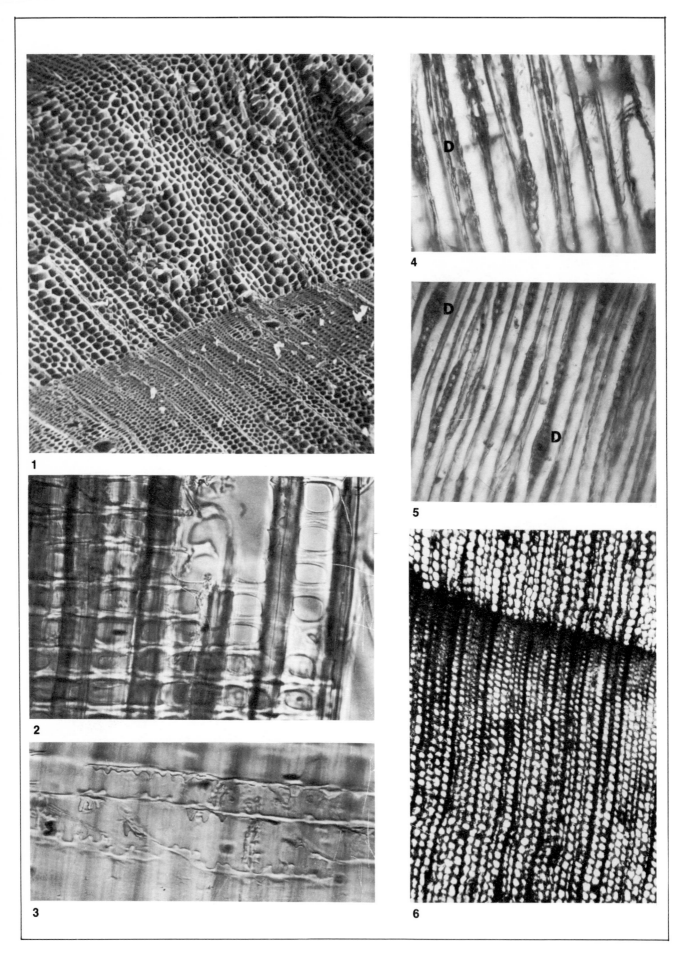

Plate 135.   Microphotographs to Chapter III.20 (*see* p. 294).

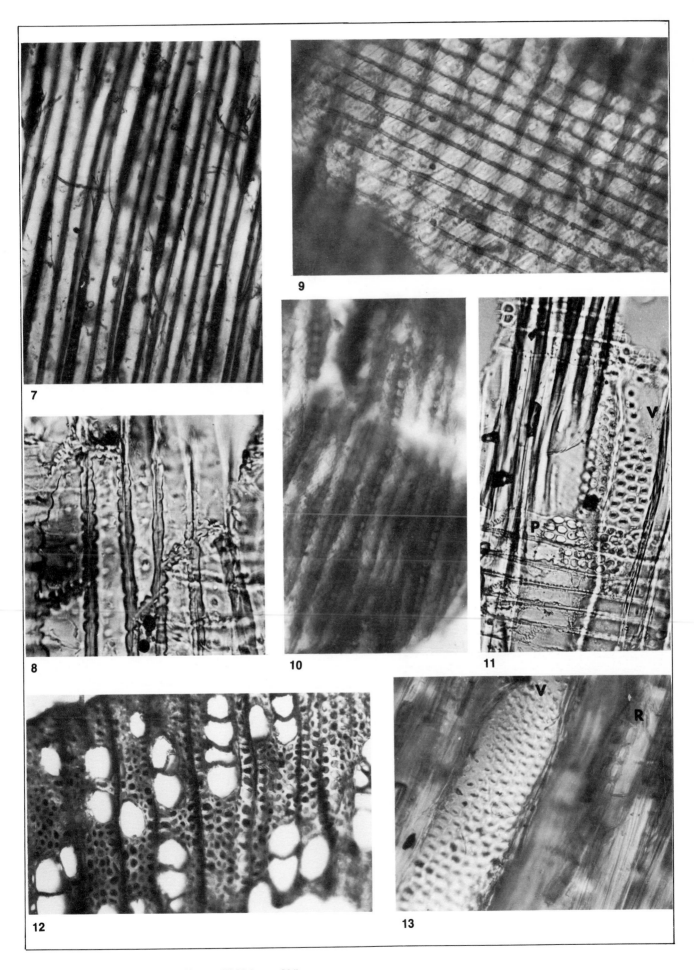

Plate 136.   Microphotographs to Chapter III.20 (*see* p. 294).

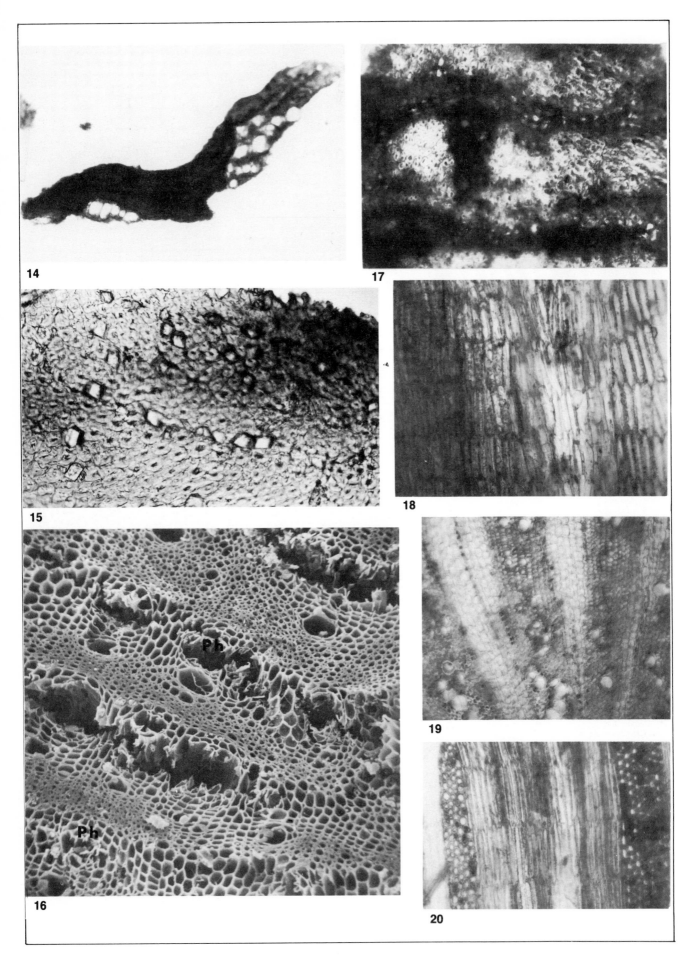

Plate 137. Microphotographs to Chapter III.20 (*see* p. 295).

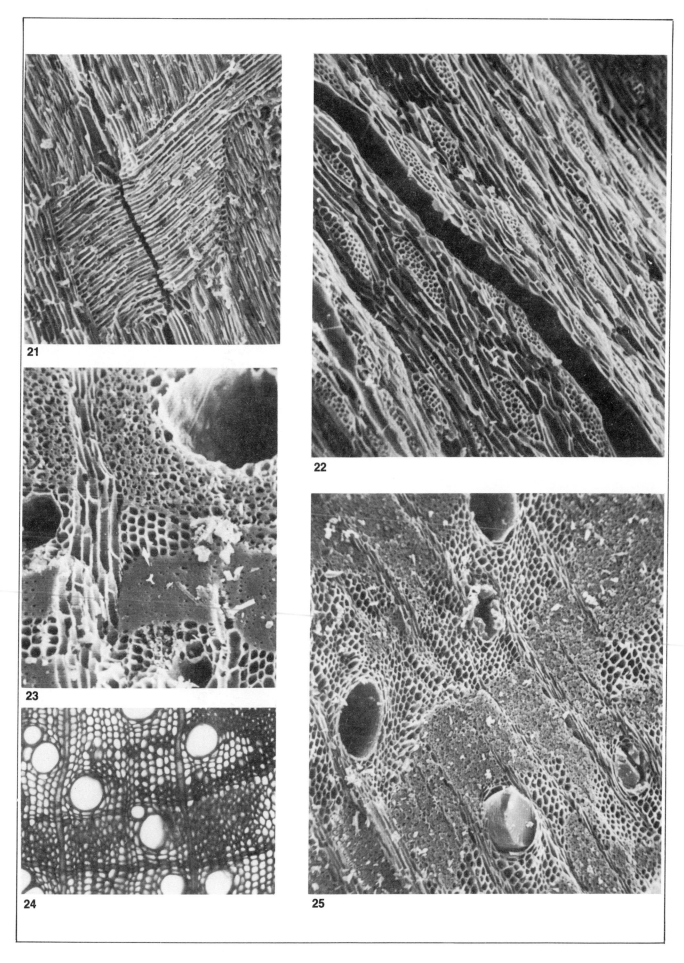

21

22

23

24

25

Plate 138. Microphotographs to Chapter III.20 (*see* p. 295).

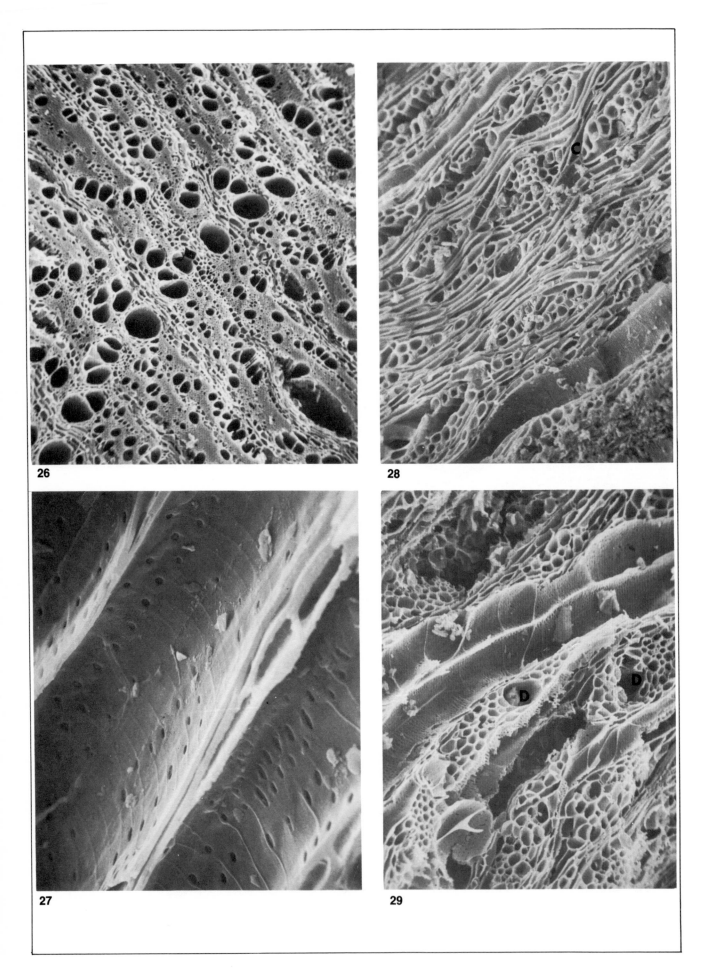

Plate 139.   Microphotographs to Chapter III.20 (*see* p. 295).

Plate 140. Wooden comb. 1 – photograph of comb; 2–3 microphotograph (to Chapter III.21) (*see* p. 295).

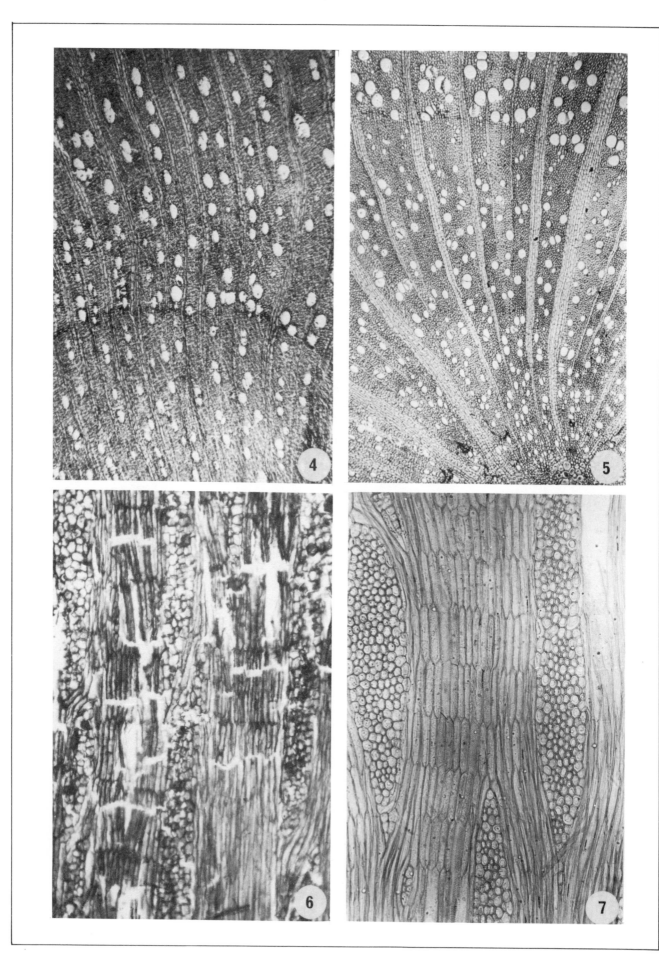

Plate 141.   Microphotographs to Chapter III.21 (*see* p. 295).

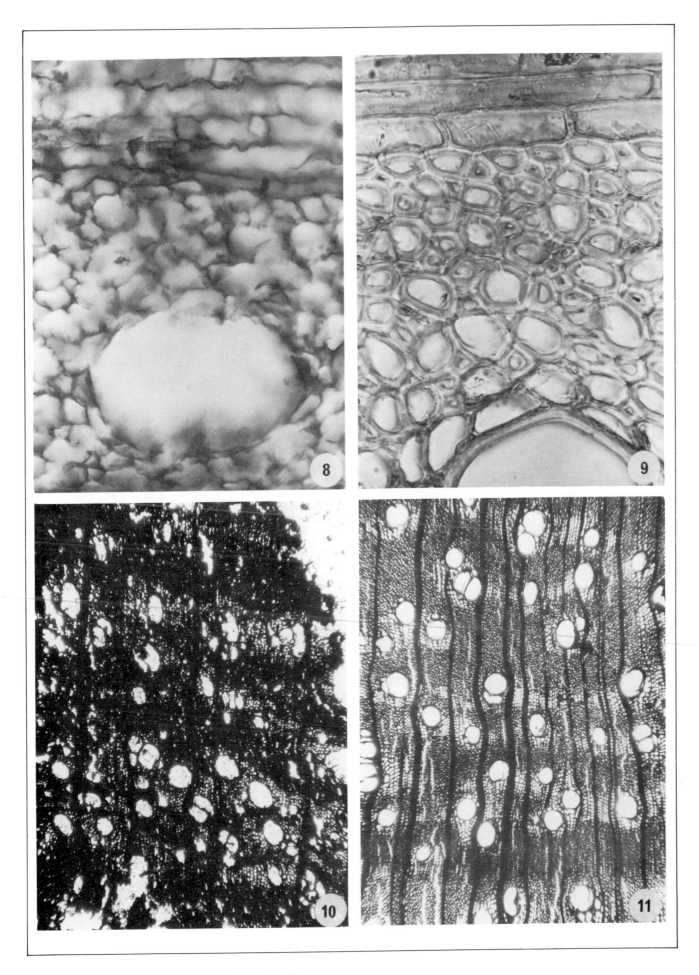

Plate 142.   Microphotographs to Chapter III.21 (*see* p. 295).

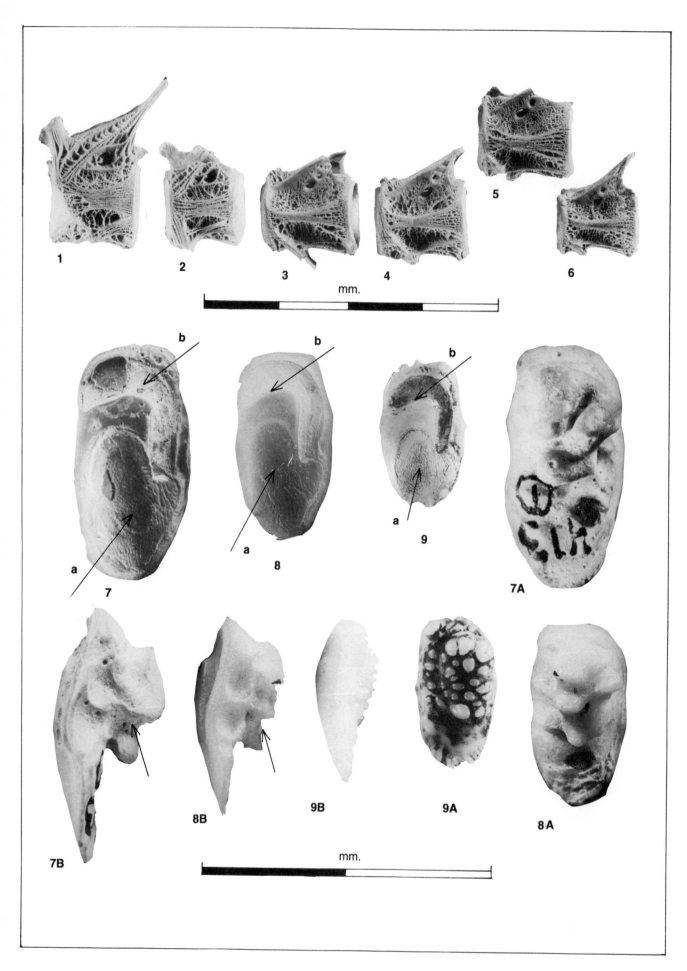

mm.

mm.

Plate 143.   Fish remains: Sciaenidae (*see* p. 296).

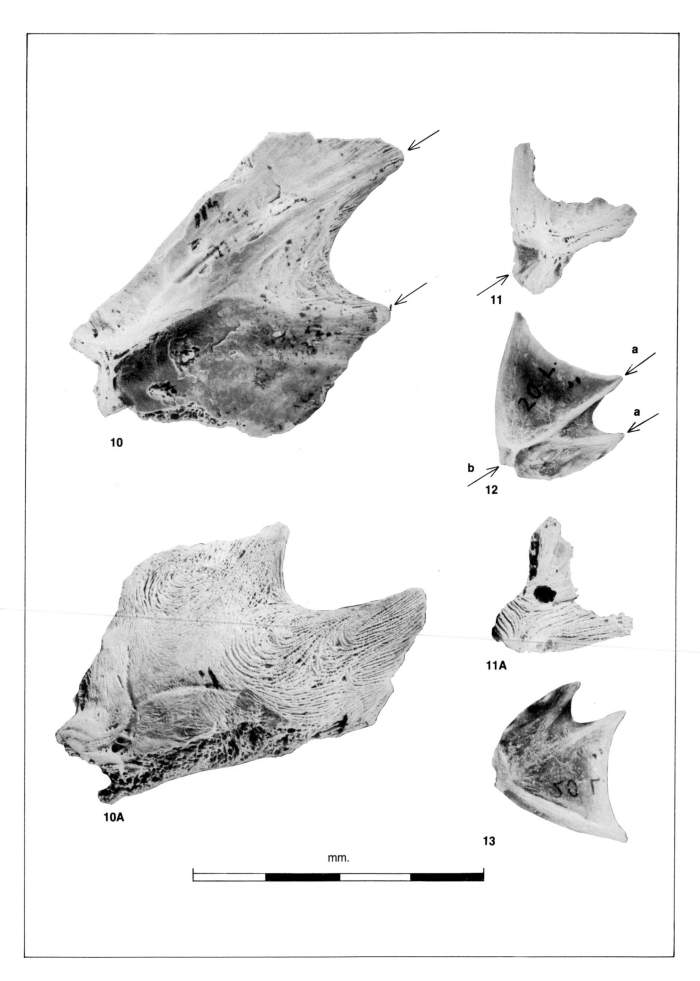

Plate 144.   Fish remains: Sciaenidae (*see* p. 296).

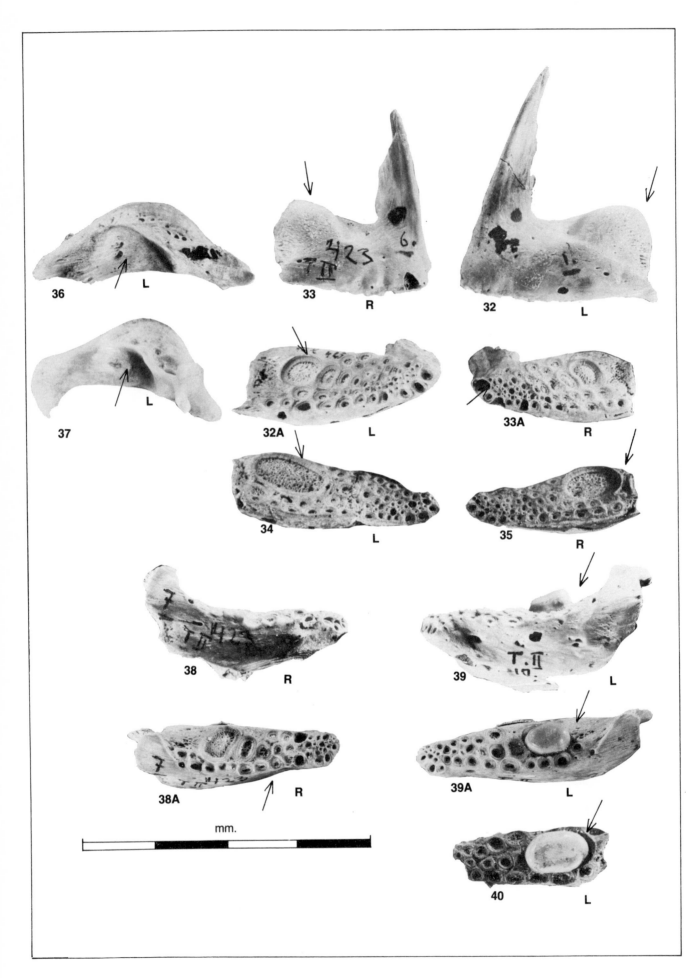

Plate 145.   Fish remains: Sparidae (*see* p. 296).

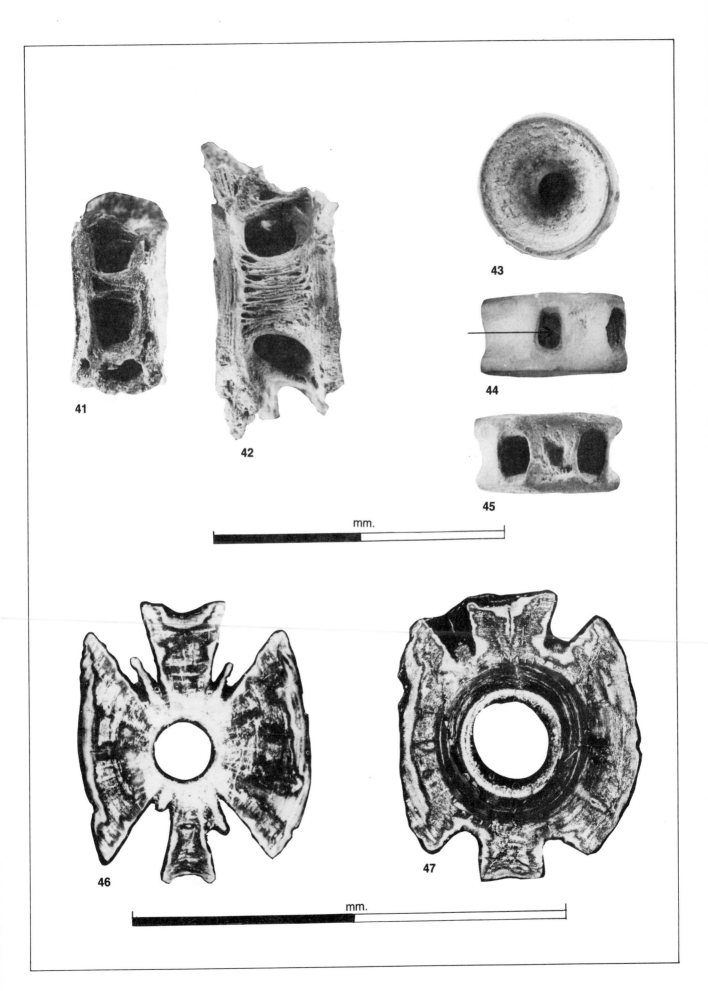

Plate 146.   Fish remains: Claridae (*see* p. 296).

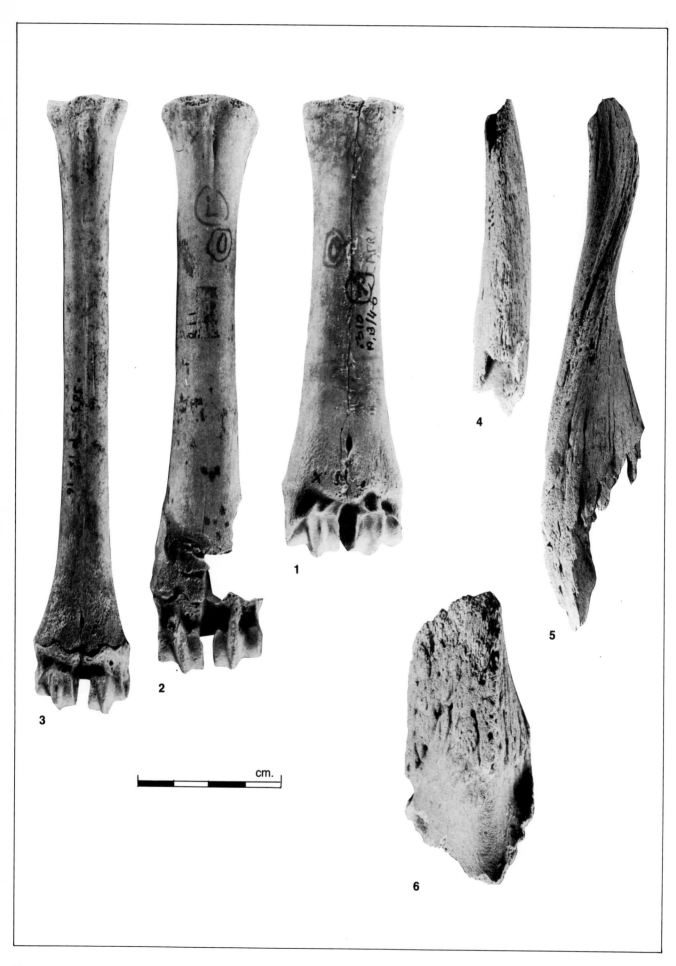

Plate 147. Mammalian remains (*see* p. 297).

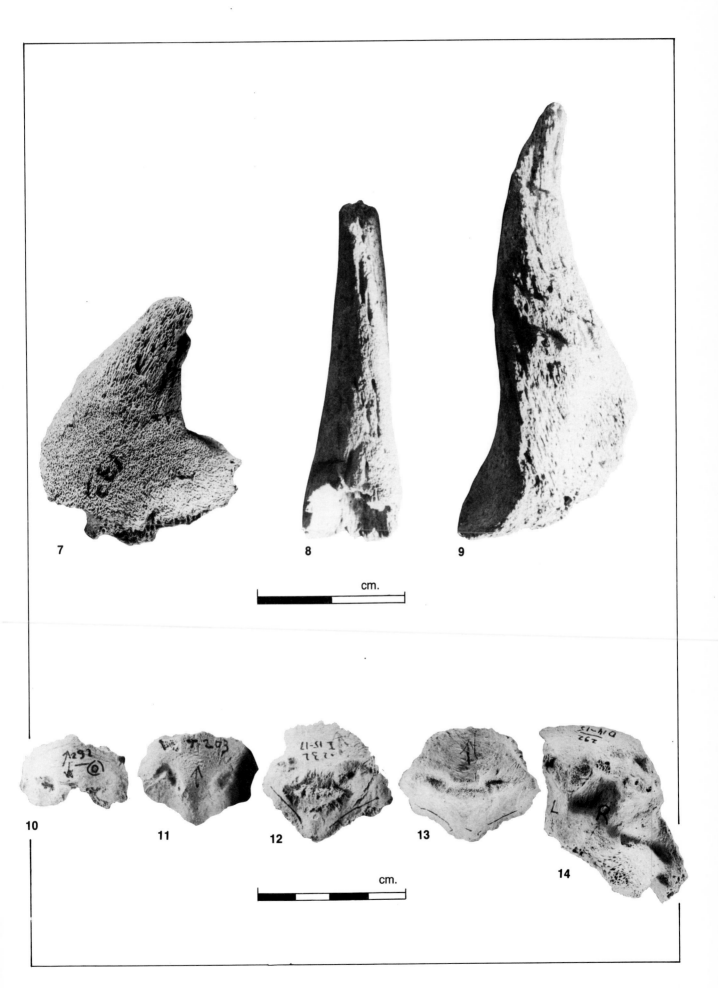

Plate 148. Mammalian remains (*see* p. 297).

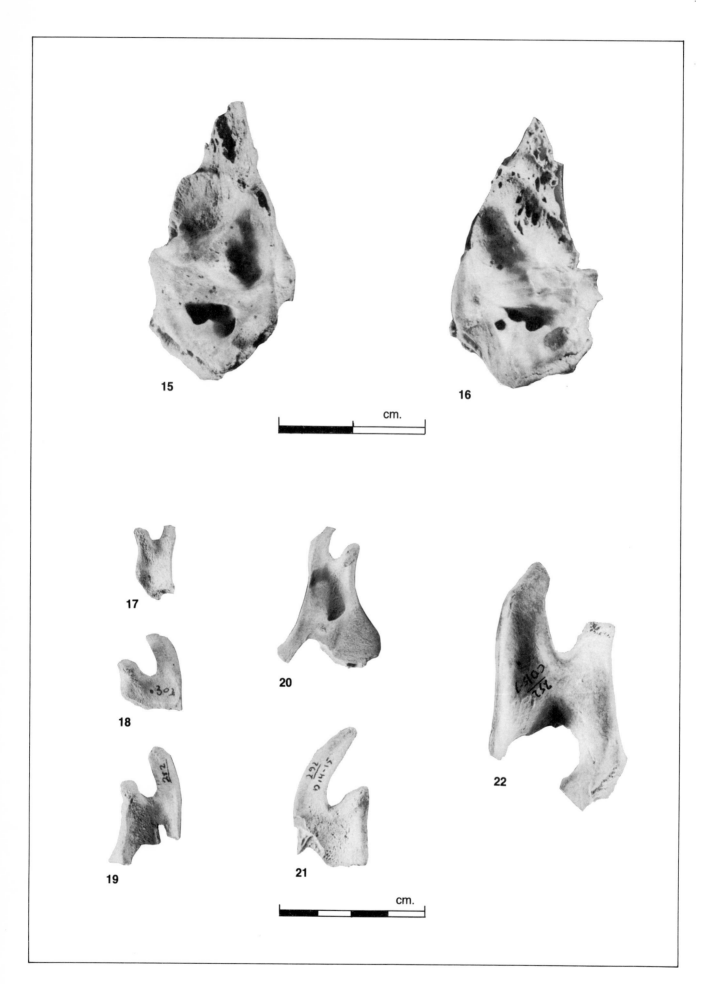

Plate 149.   Mammalian remains (*see* p. 297).

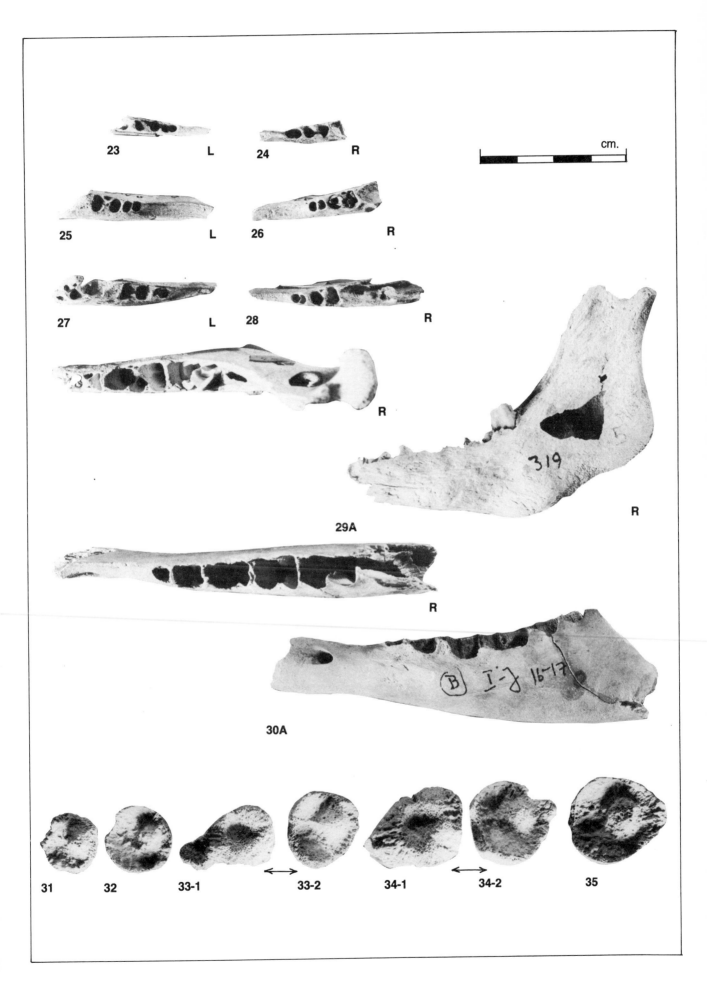

Plate 150. Mammalian remains (*see* p. 297).

Plate 151. Roman pottery (1 – *Fig.* 87:8; 2 – *Fig.* 87:11; 3 – *Fig.* 87:2; 4 – *Fig.* 88:4; 5 – *Fig.* 87:7; 6 – *Fig.* 88:2; 7 – *Fig.* 88:5).

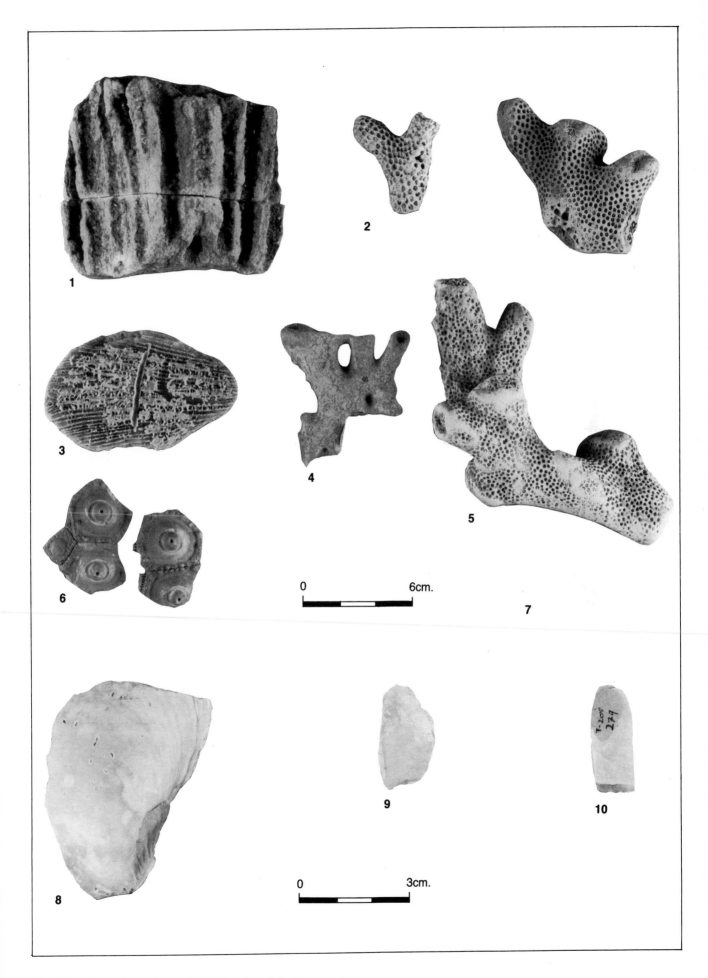

Plate 152.   Recent invertebrates; 8–10 Minerals and fossils (*see* p. 298).

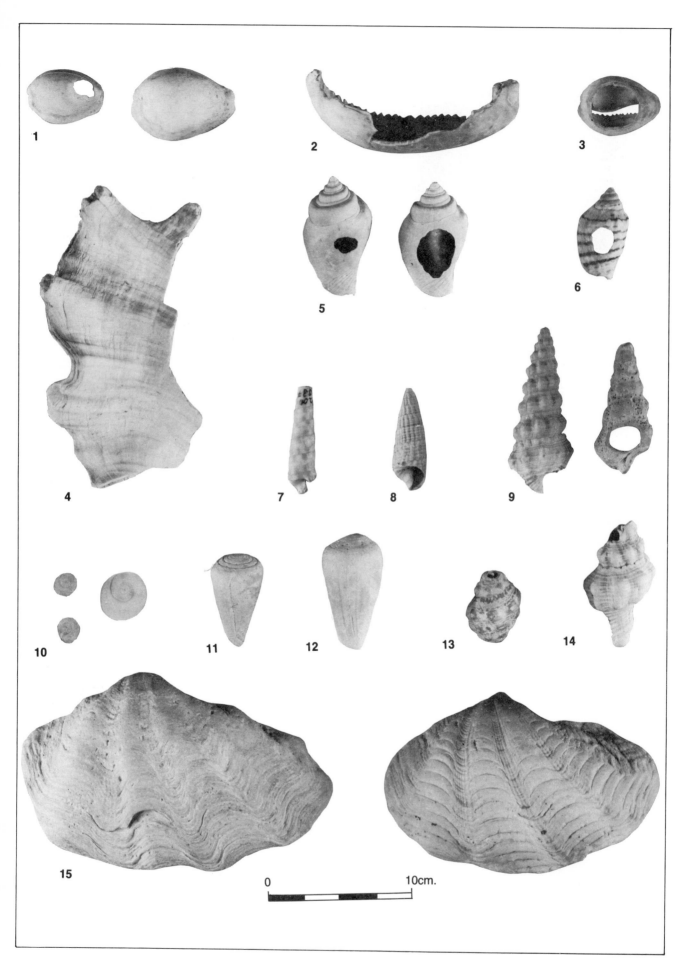

Plate 153.   Recent invertebrates (*see* p. 298).

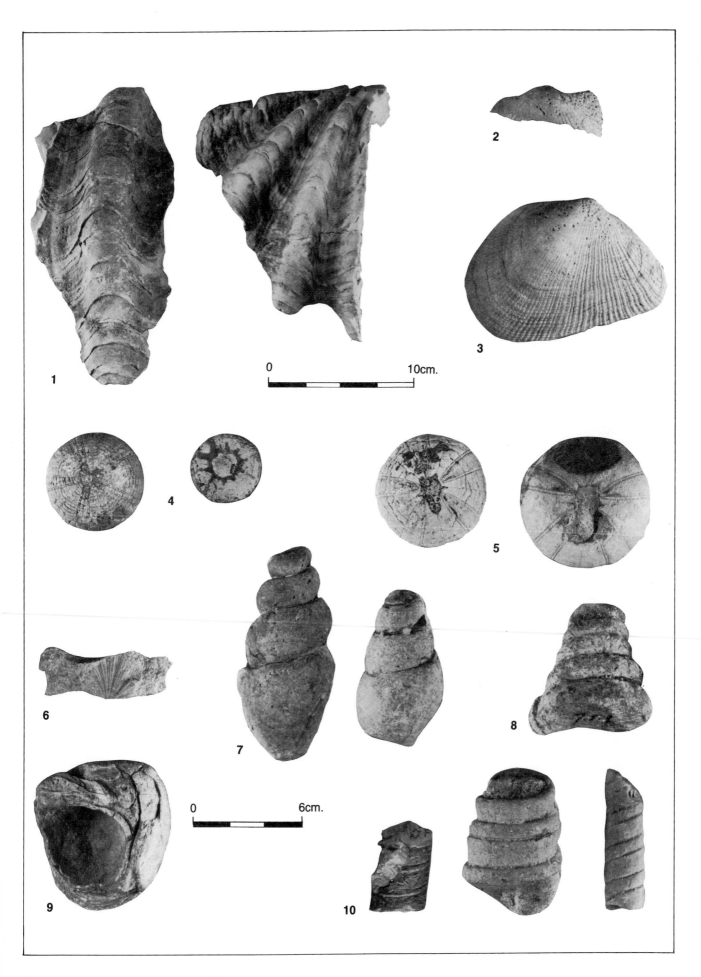

0   10cm.

0   6cm.

Plate 154.   Recent invertebrates (*see* p. 298).

Plate 155. Mineral votive gifts.

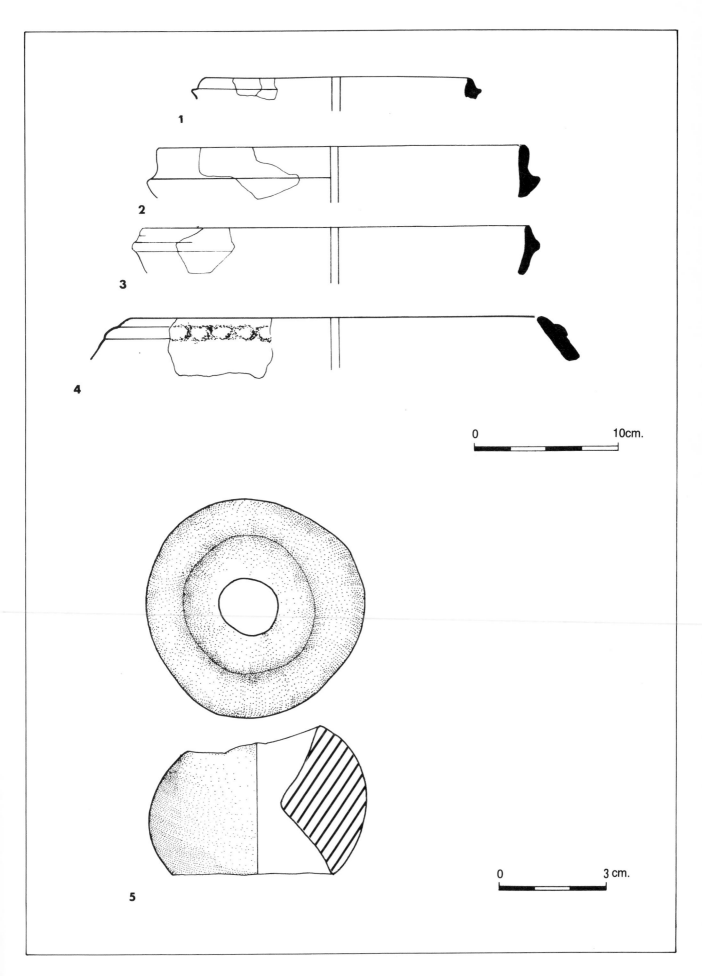

Fig. 1.   Pottery from Stratum IV, macehead (*see* p. 299).

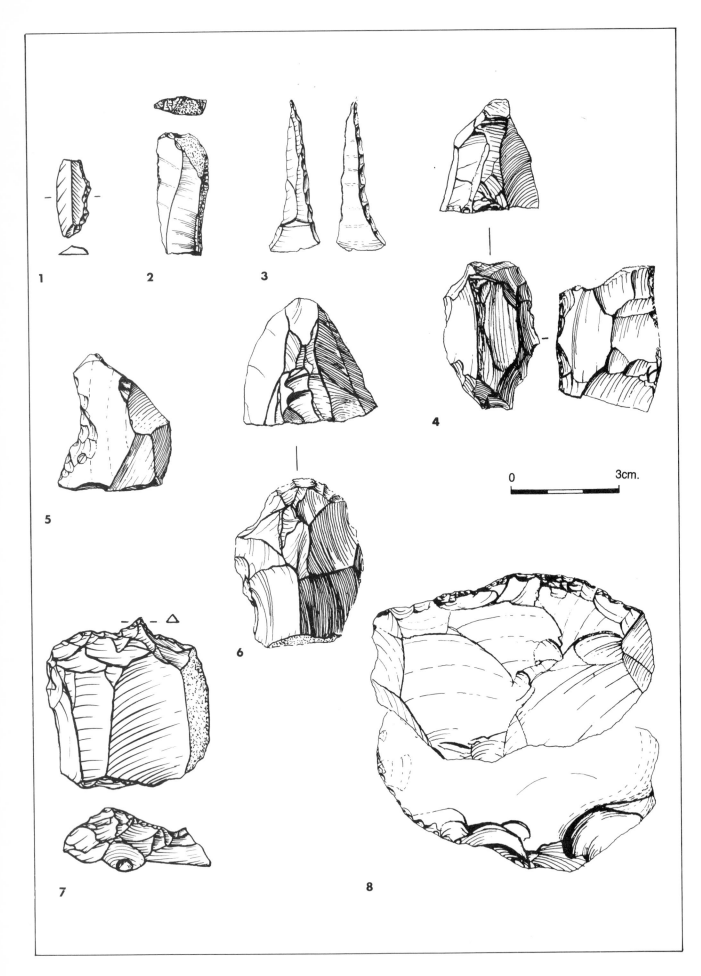

Fig. 2.   Flint implements (*see* p. 299).

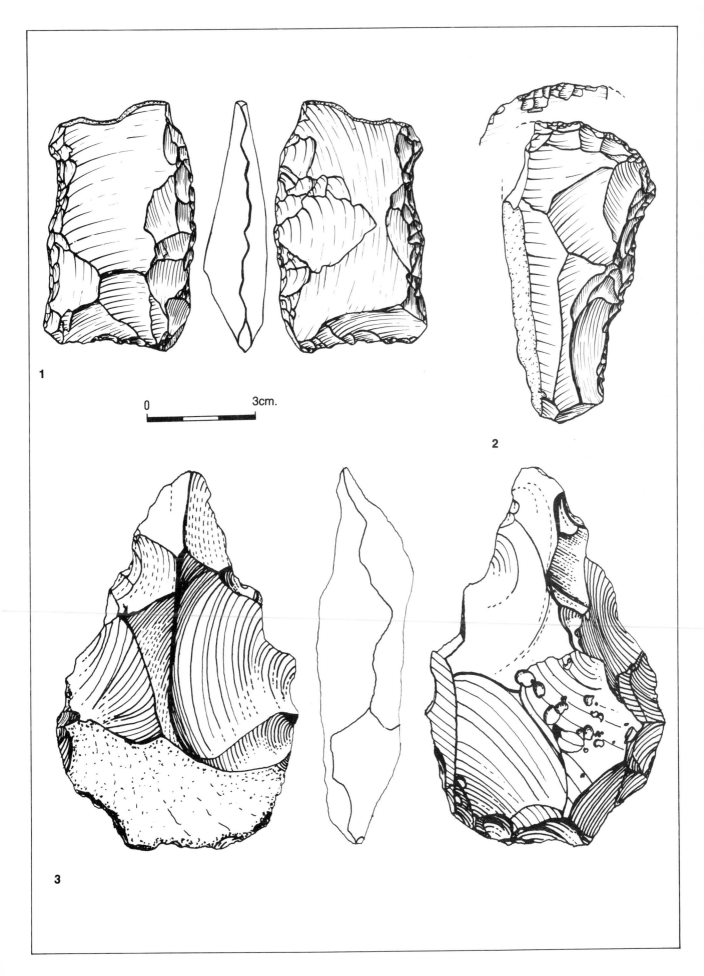

0 — 3cm.

1

2

3

Fig. 3. Flint implements (*see* p. 299).

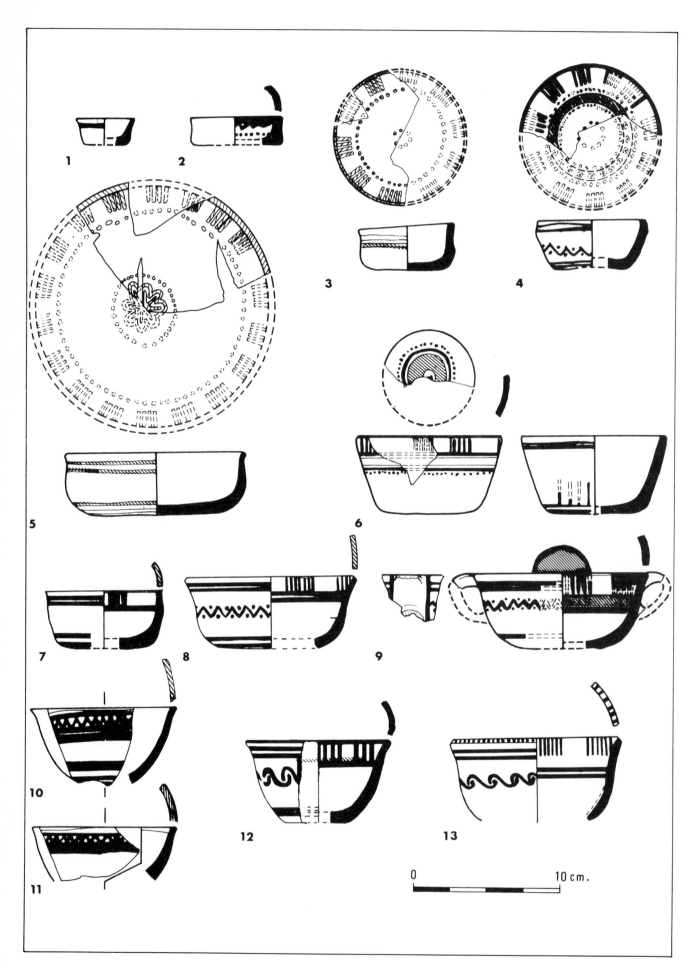

Fig. 4. Midianite pottery (*see* p. 299).

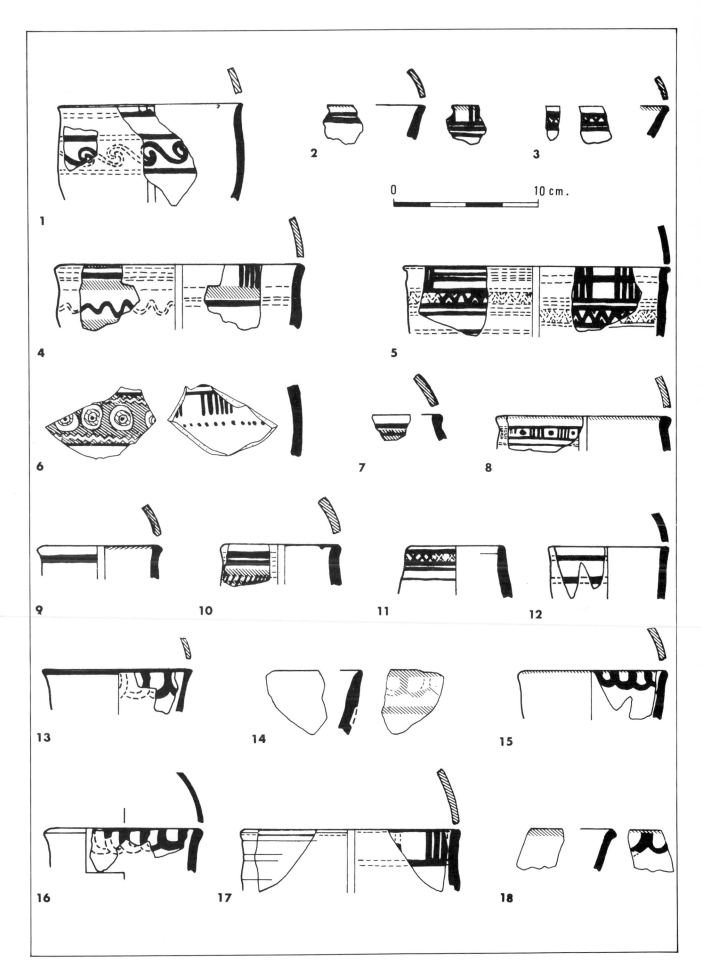

Fig. 5.   Midianite pottery (*see* p. 300).

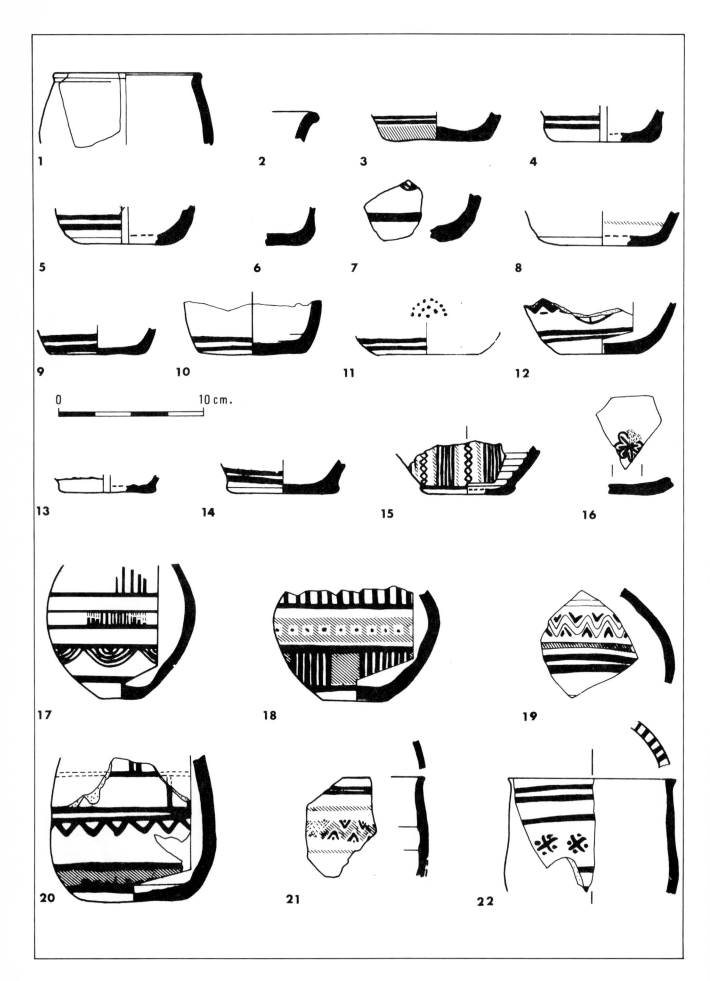

Fig. 6. Midianite pottery (*see* p. 300).

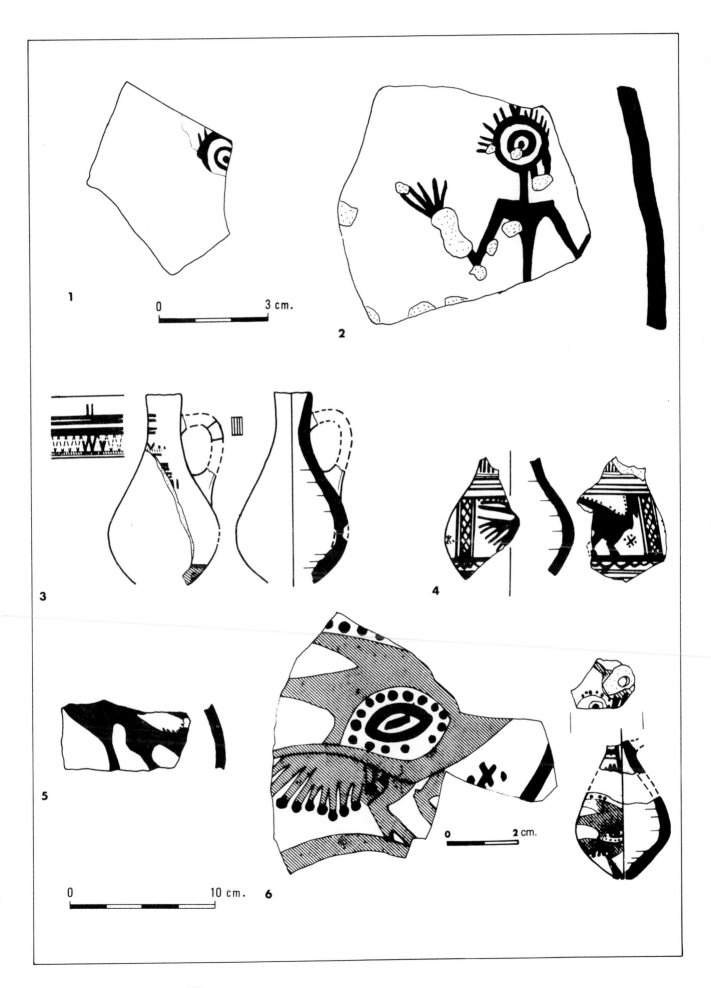

Fig. 7.   Midianite pottery (*see* p. 300).

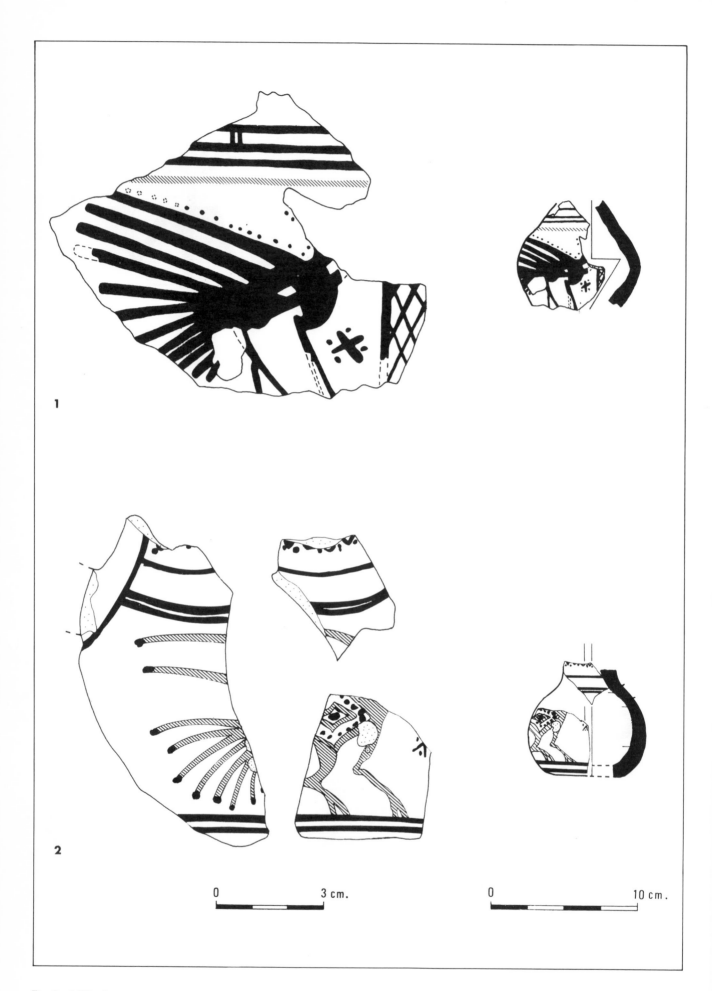

Fig. 8. Midianite pottery (*see* p. 301).

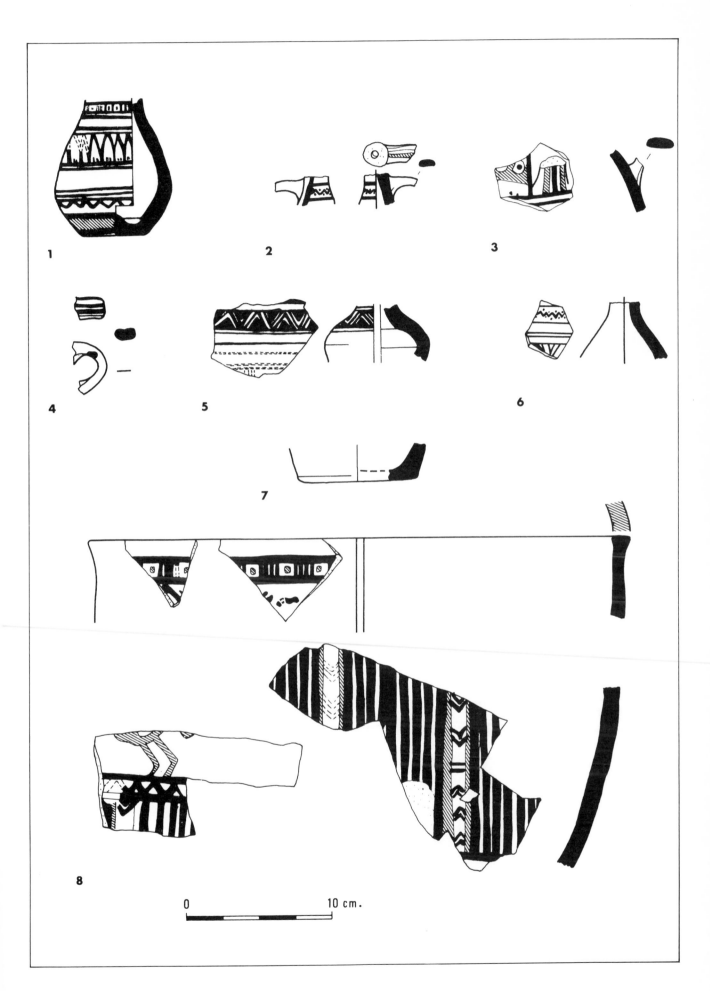

Fig. 9.   Midianite pottery (*see* p. 301).

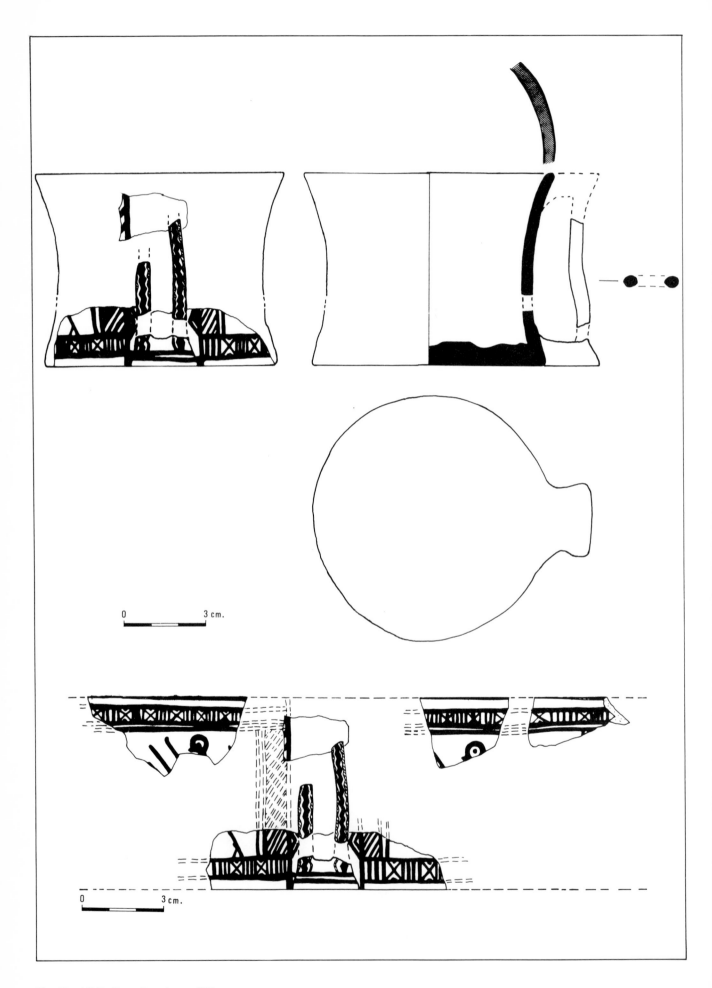

Fig. 10.   Midianite pottery (*see* p. 301).

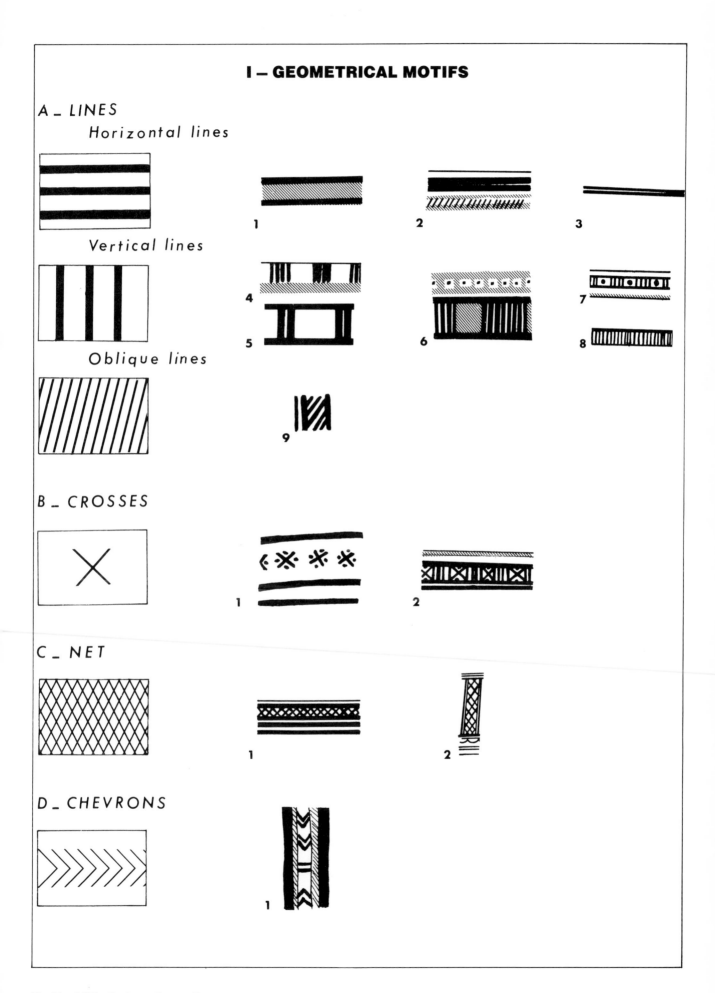

Fig. 11.   Midianite decorative motifs.

E _ TRIANGLES

1

F _ LOZENGES

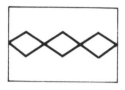

1

G _ ZIGZAG

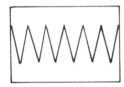

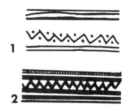

1

2

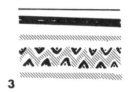

3

4

5

H _ ARCHES

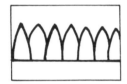

1

I _ JOINING SEMICIRCLES

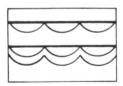

1

2

J _ WAVY LINES

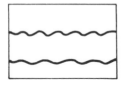

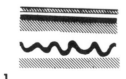

1

Fig. 12.   Midianite decorative motifs.

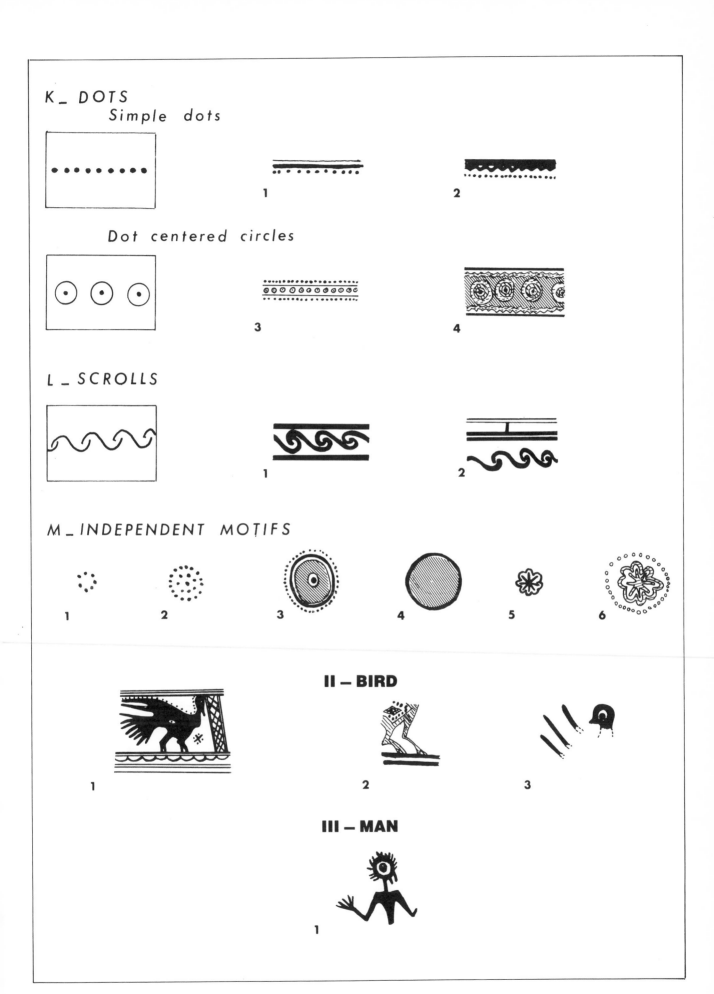

Fig. 13. Midianite decorative motifs.

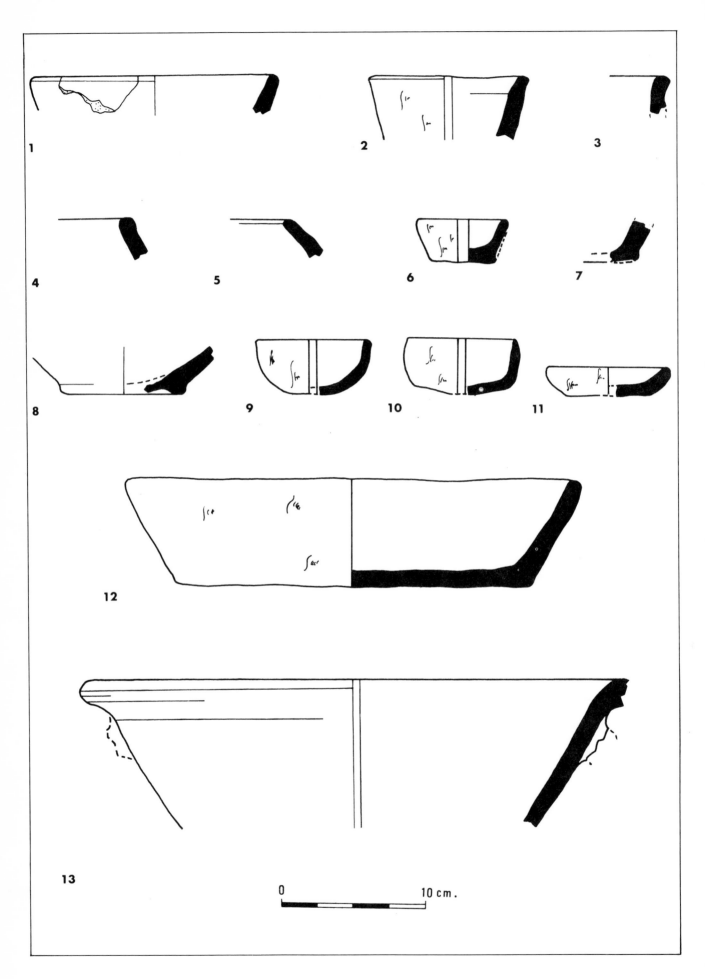

Fig. 14. 1–9 Negev-ware pottery; 10–13 Rough hand-made pottery (*see* p. 301).

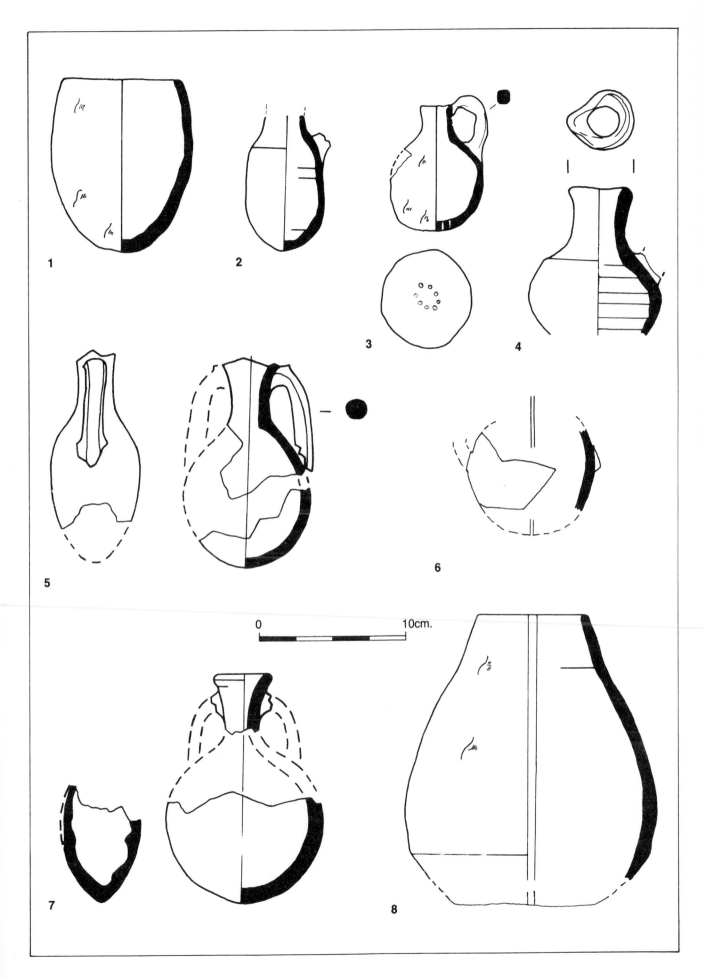

Fig. 15. Rough hand-made pottery (*see* p. 302).

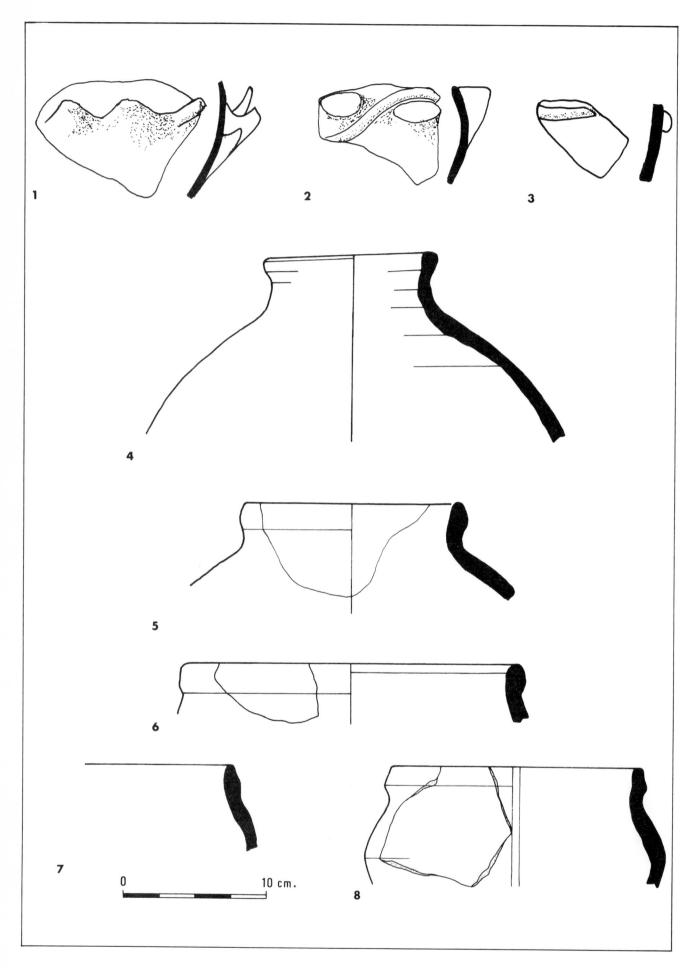

Fig. 16.   1–3, Rough hand-made pottery, 4, 5, 6–8 Local wheel-made pottery (*see* p. 302).

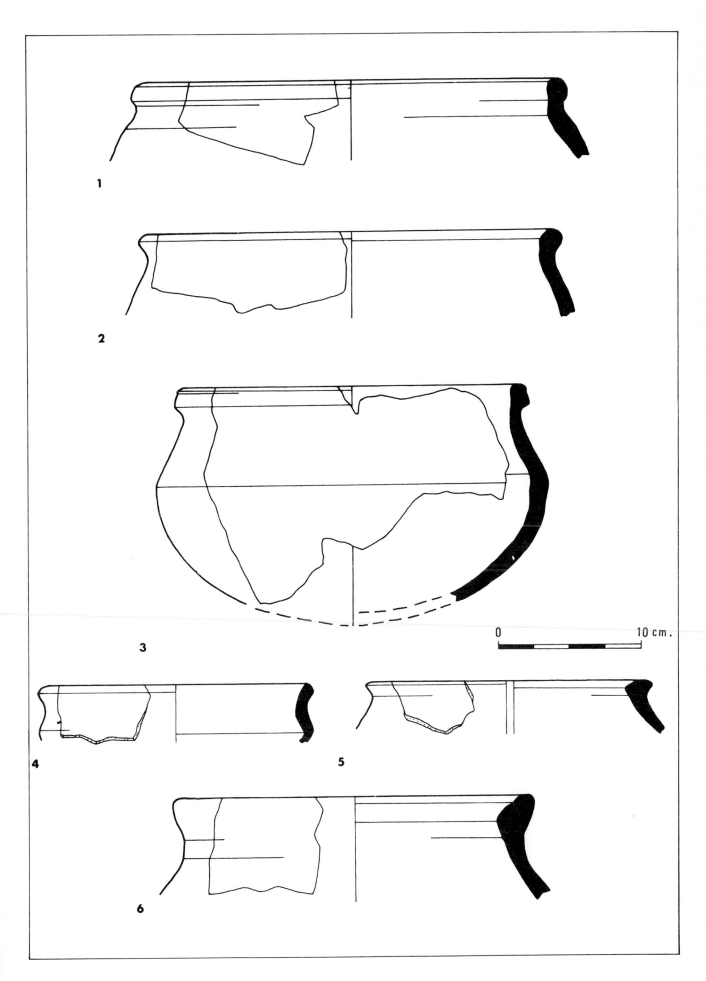

Fig. 17. Local wheel-made pottery (*see* p. 302).

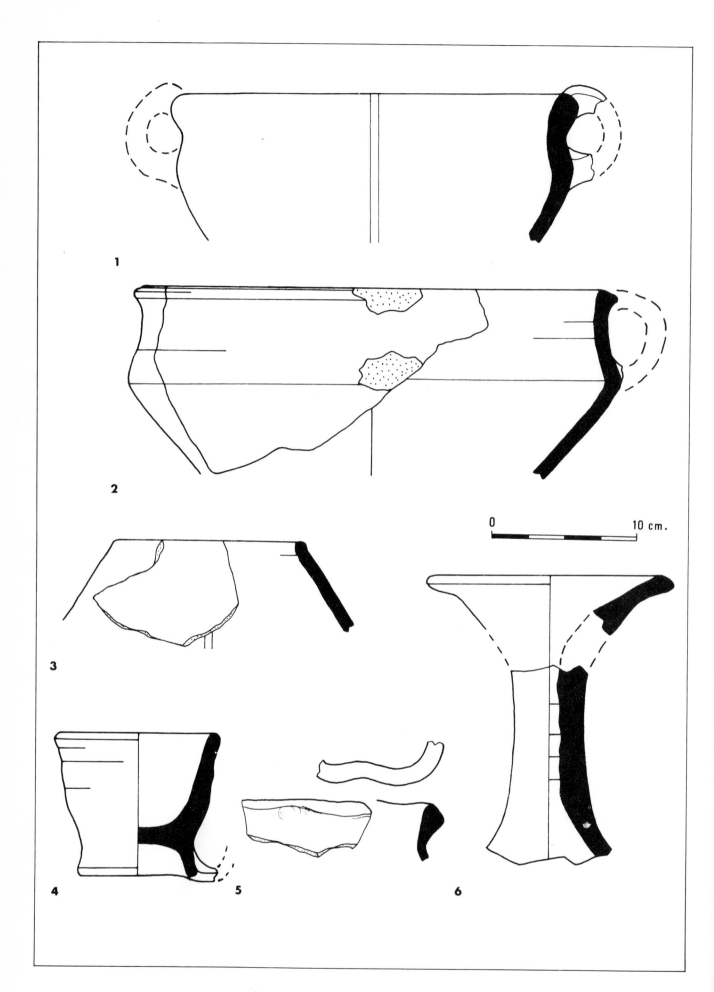

Fig. 18.   Local wheel-made pottery (*see* p. 303).

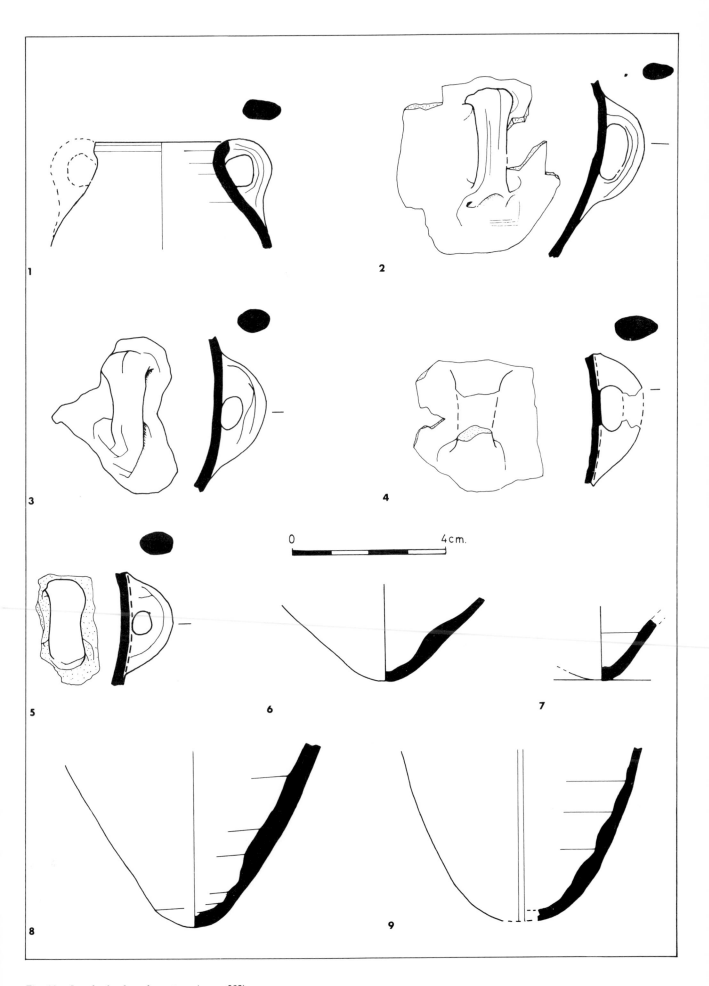

Fig. 19. Local wheel-made pottery (*see* p. 303).

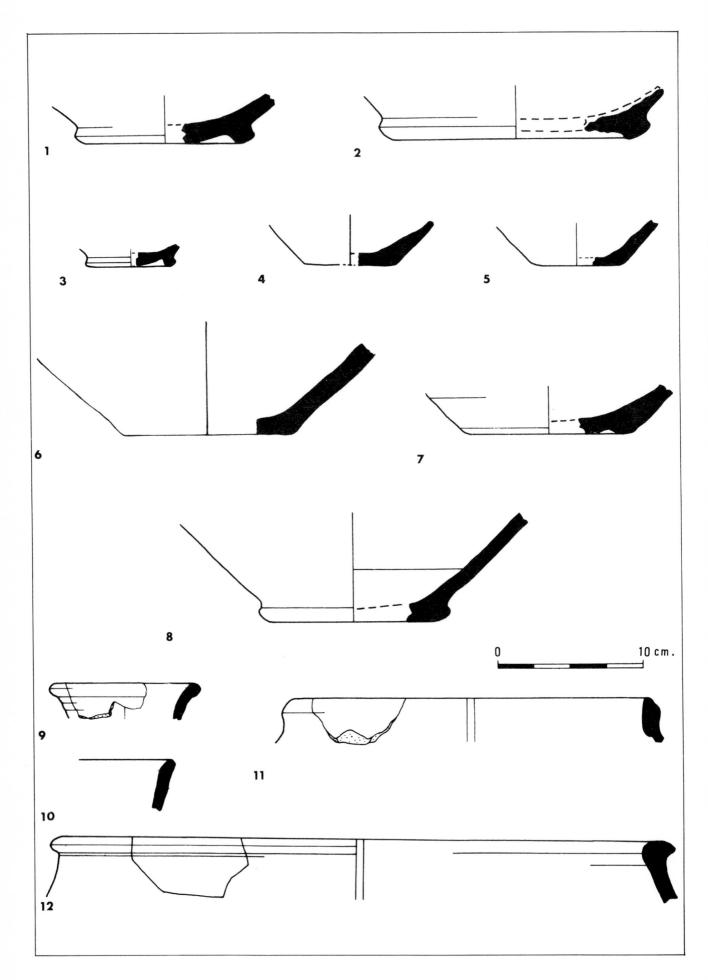

Fig. 20.   1–8, Local wheel-made pottery; 9–12 Egyptian pottery (*see* p. 303).

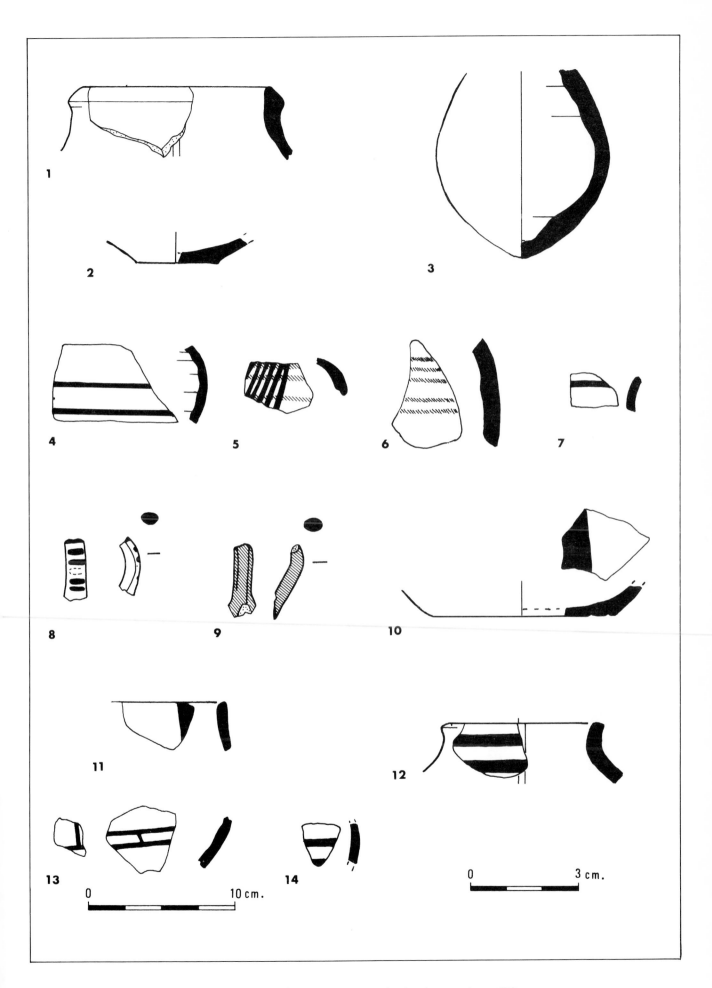

Fig. 21.   1–3, Egyptian pottery; 4–10, Egyptian painted pottery; 11–14, Local painted pottery (*see* p. 304).

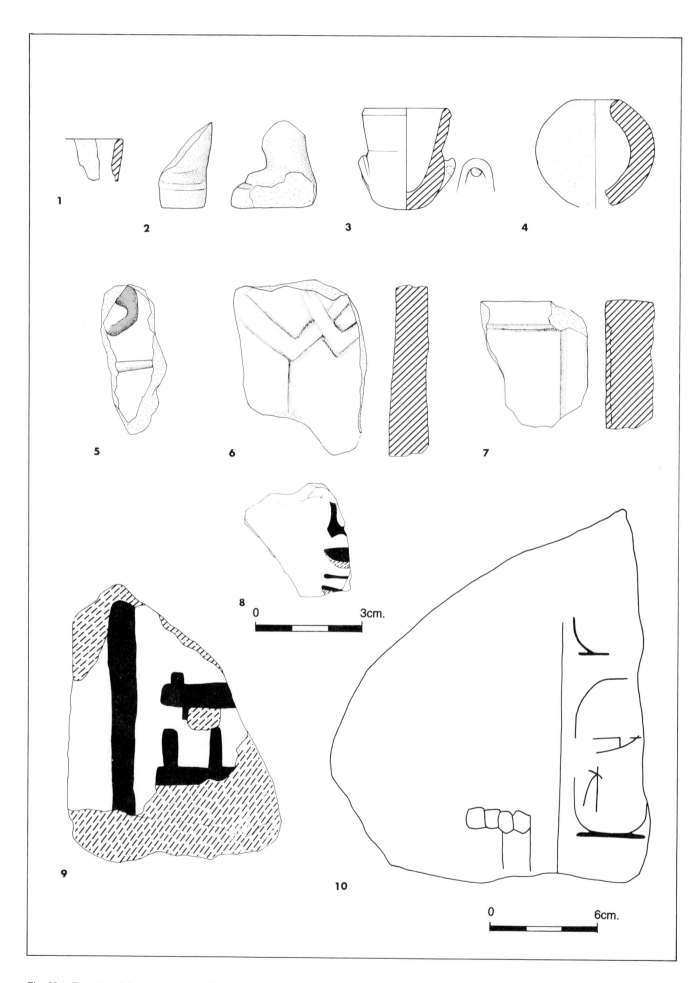

Fig. 22. Egyptian alabasters, inscribed fragments (*see* p. 304).

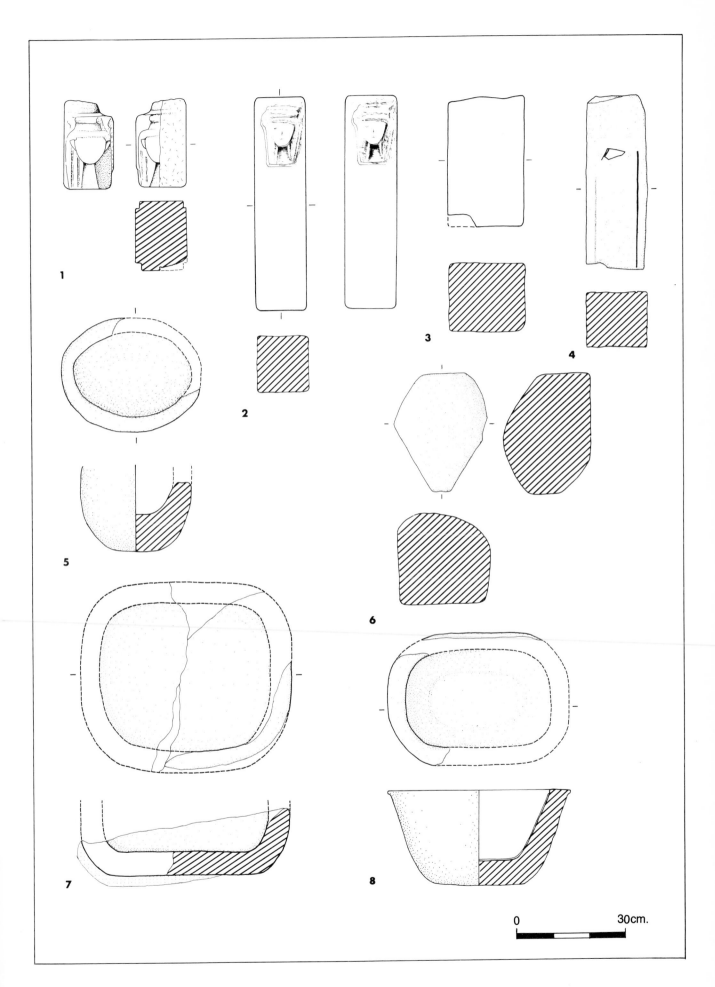

Fig. 23.   Stone pillars and basins (*see* p. 304).

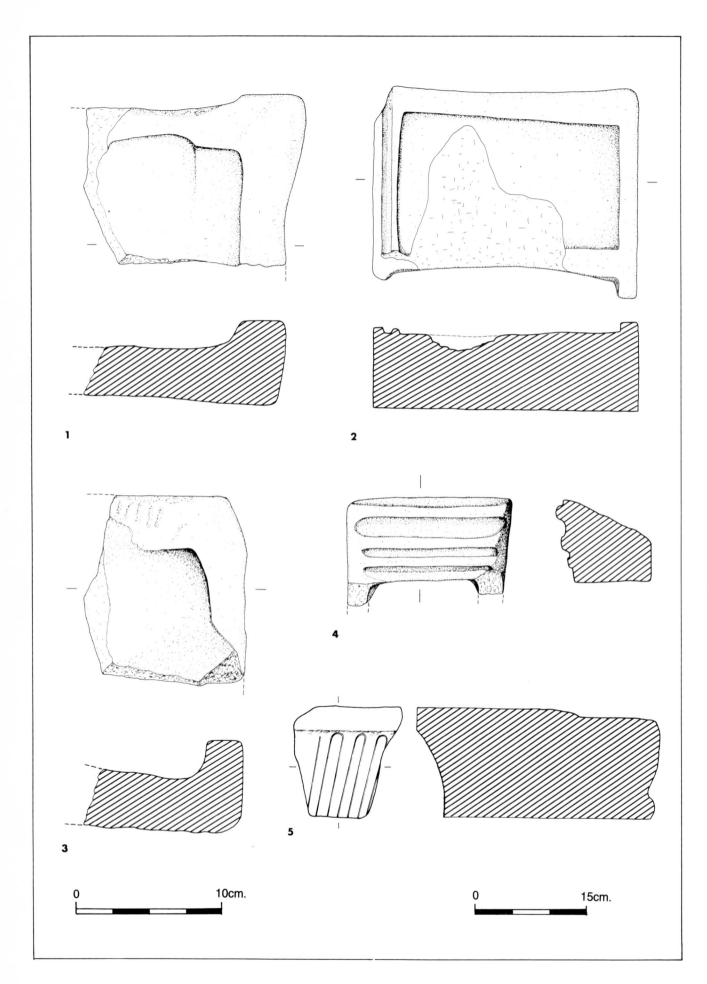

Fig. 24.   Cornices and offering tables (*see* p. 304).

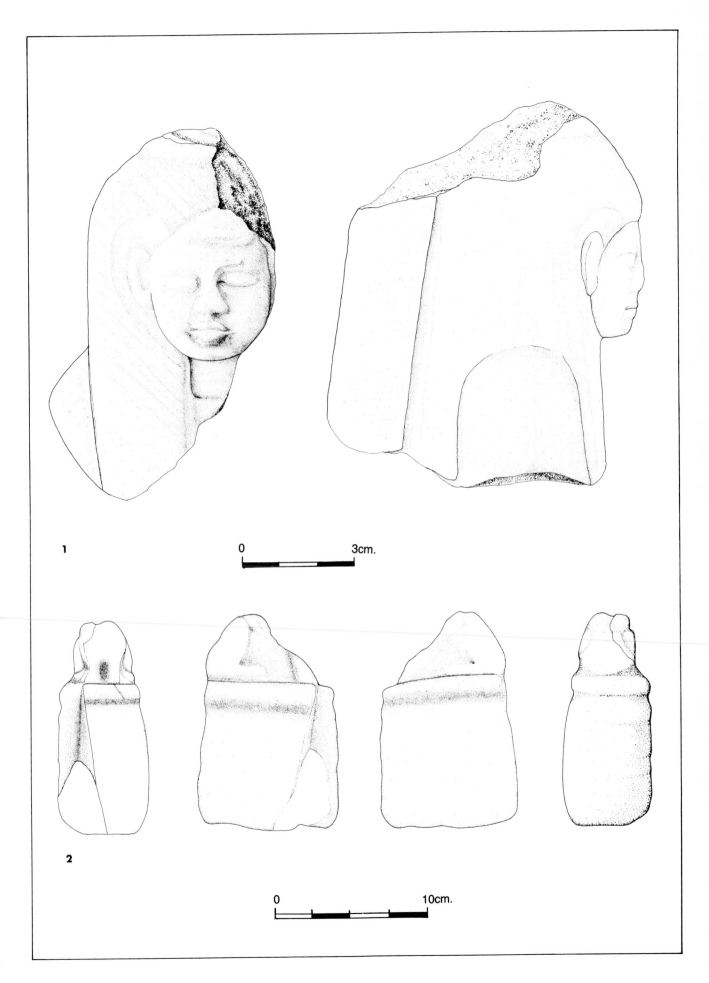

1

0   3cm.

2

0   10cm.

Fig. 25.   Statuette and sphinx (*see* p. 305).

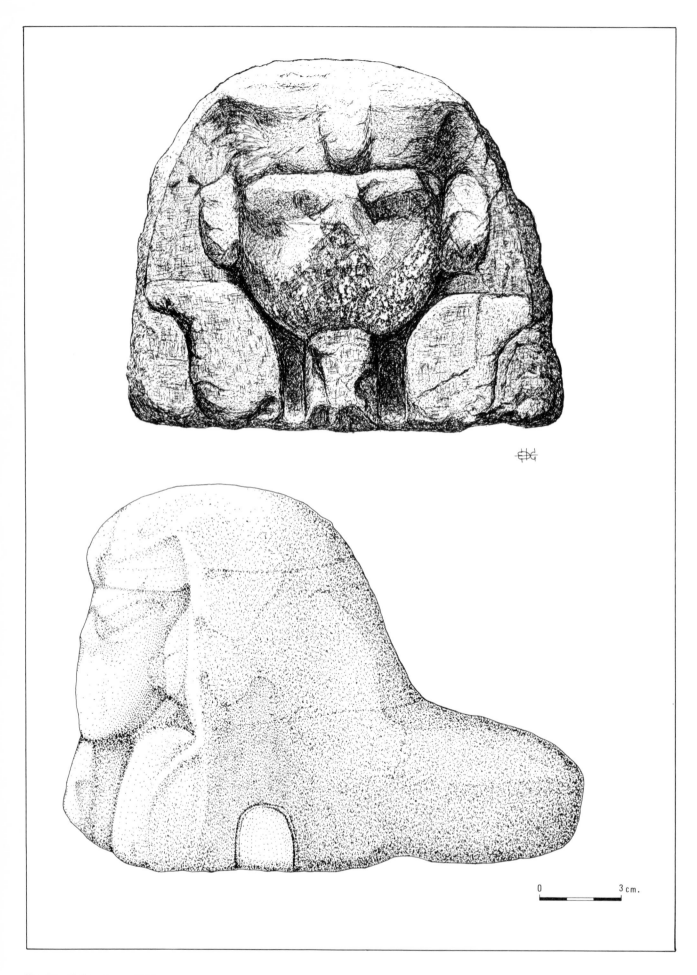

Fig. 26.   Sphinx (*see* p. 305).

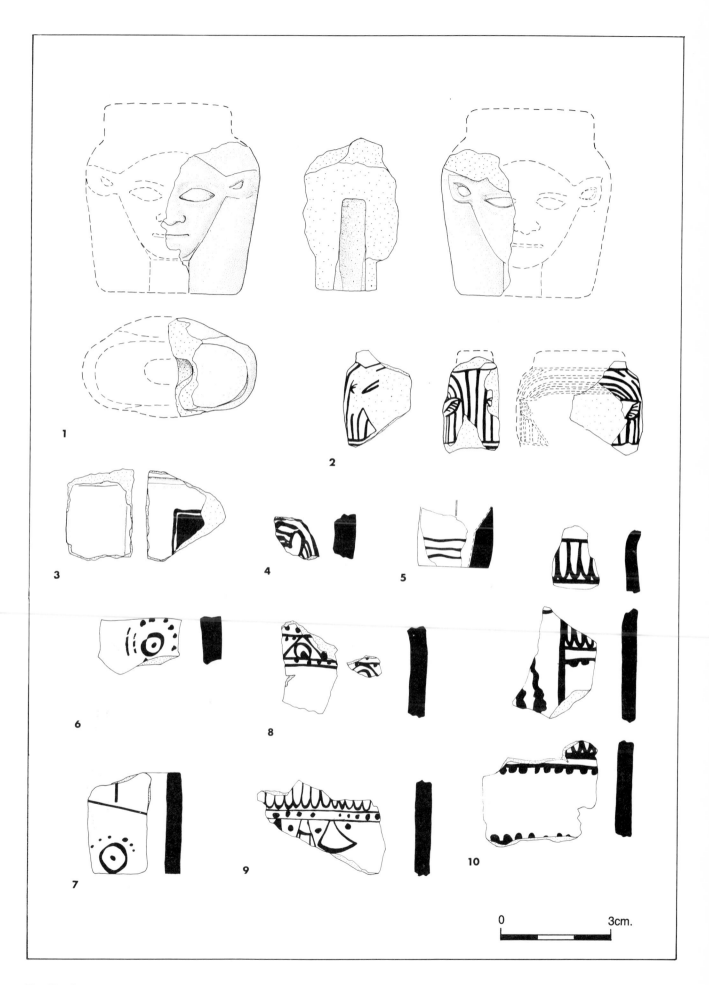

Fig. 27. Sistra and decorated sherds (*see* p. 305).

Fig. 28.   Sistrum, ushabti and bowl (*see* p. 305).

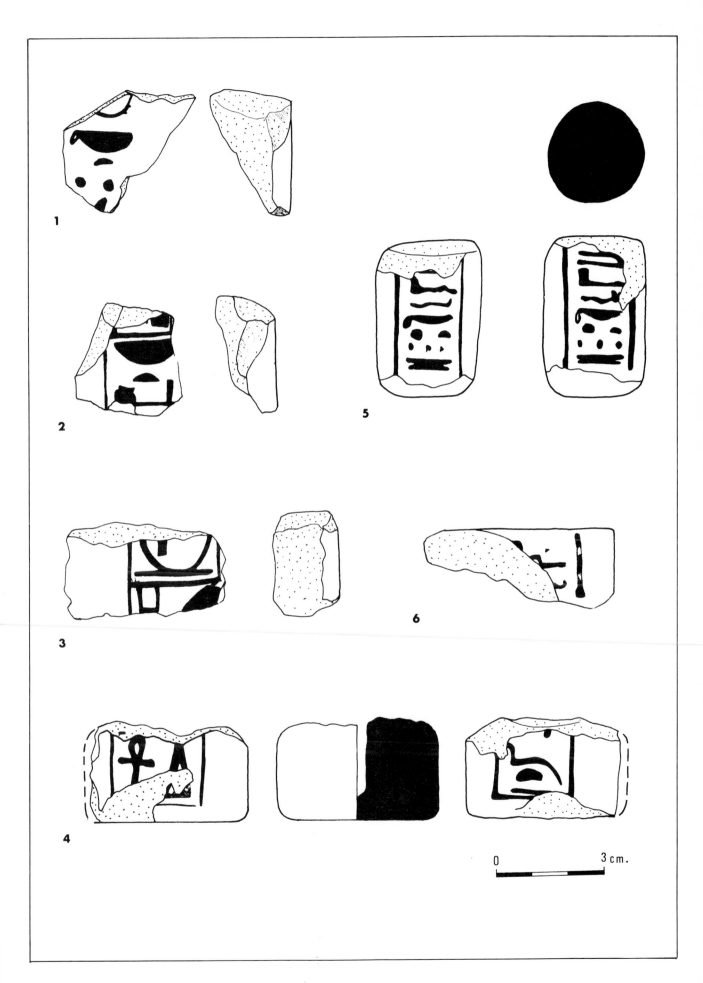

Fig. 29.   Votive objects (*see* p. 305).

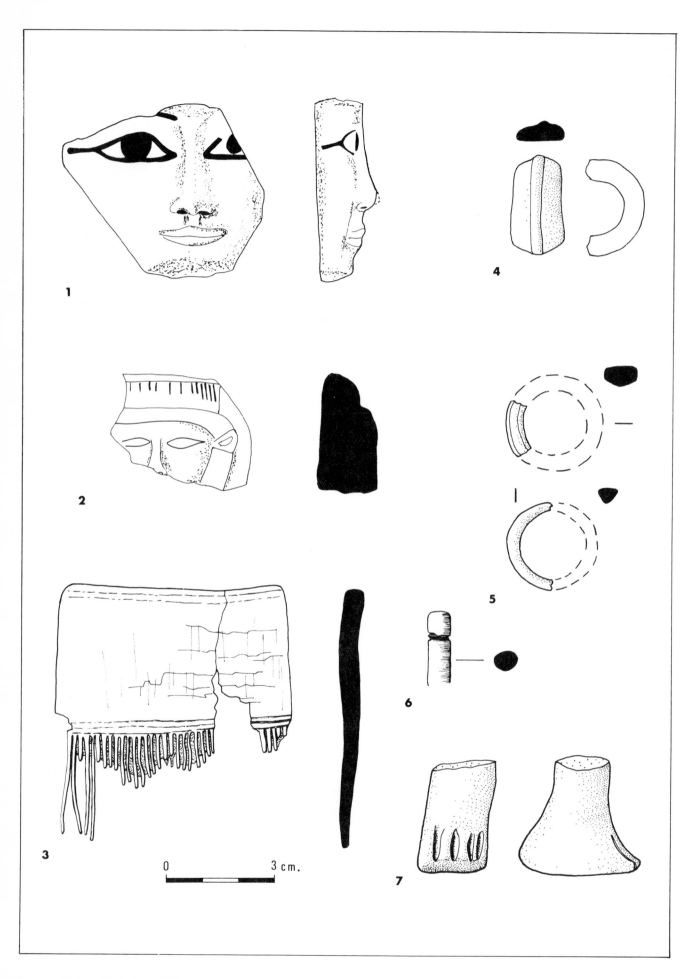

Fig. 30. Votive objects (*see* p. 306).

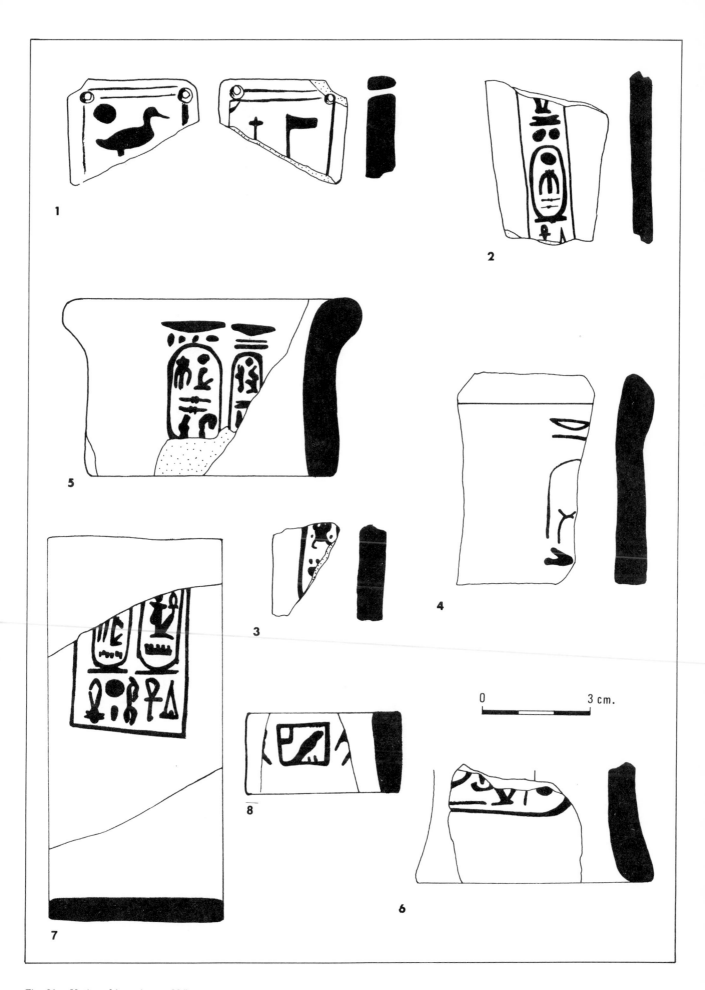

Fig. 31. Votive objects (*see* p. 306).

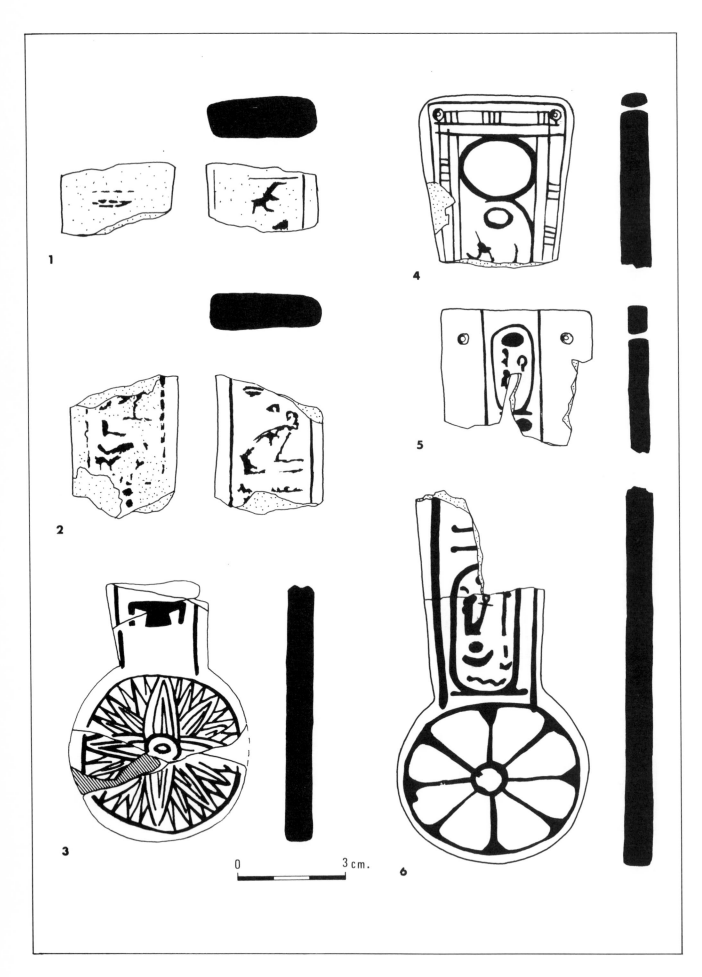

Fig. 32.   Votive objects (*see* p. 306).

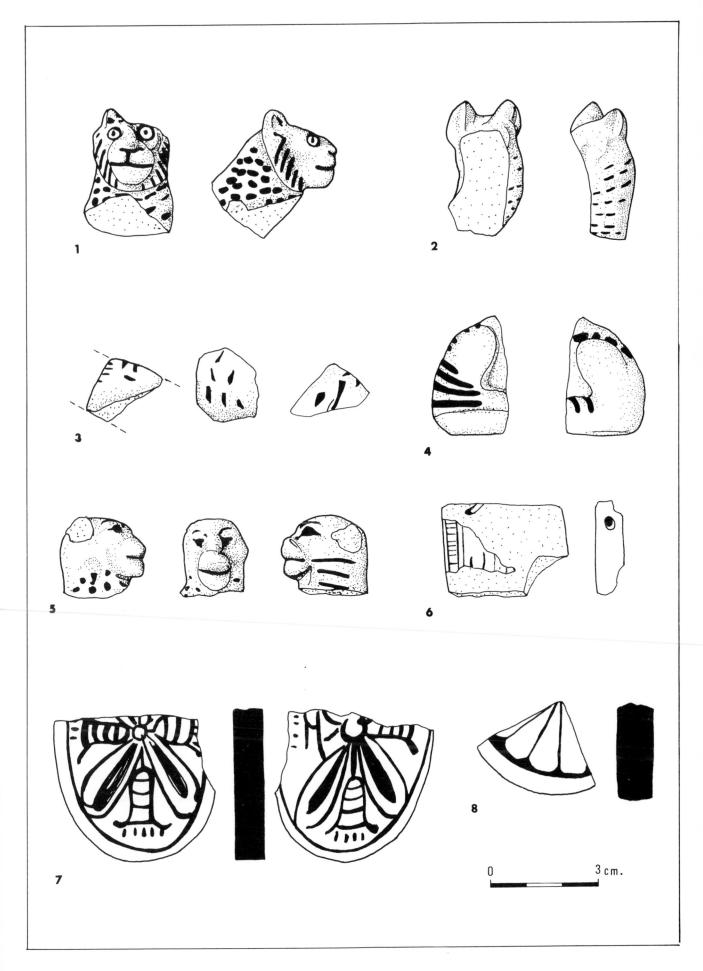

Fig. 33.   Votive objects (*see* p. 306).

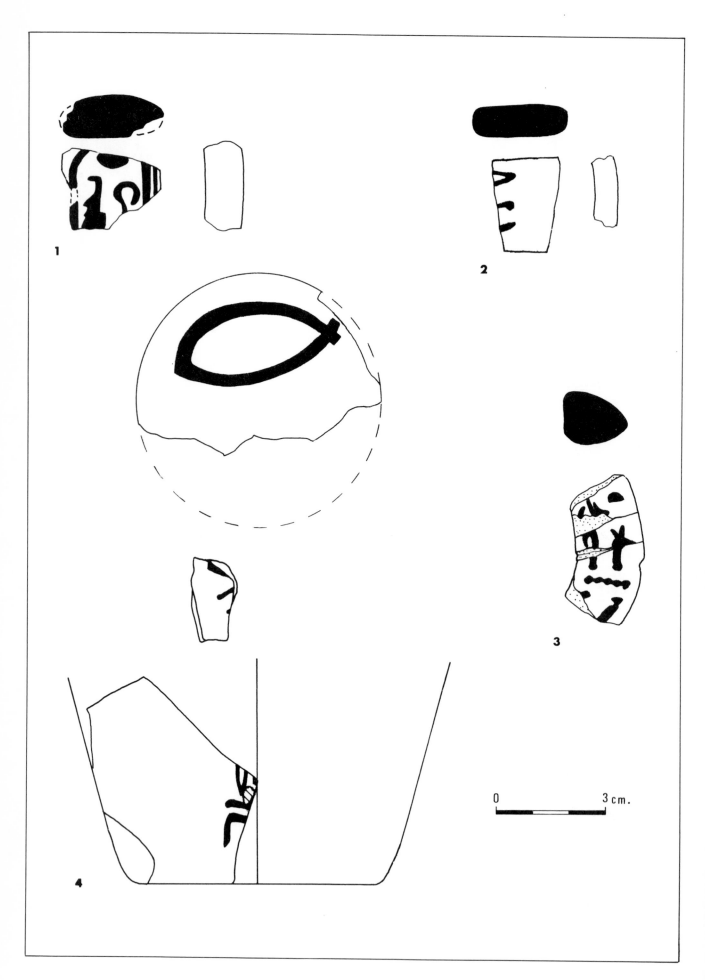

Fig. 34.   Faience objects (*see* p. 306).

Fig. 35. Bracelets (*see* p. 307).

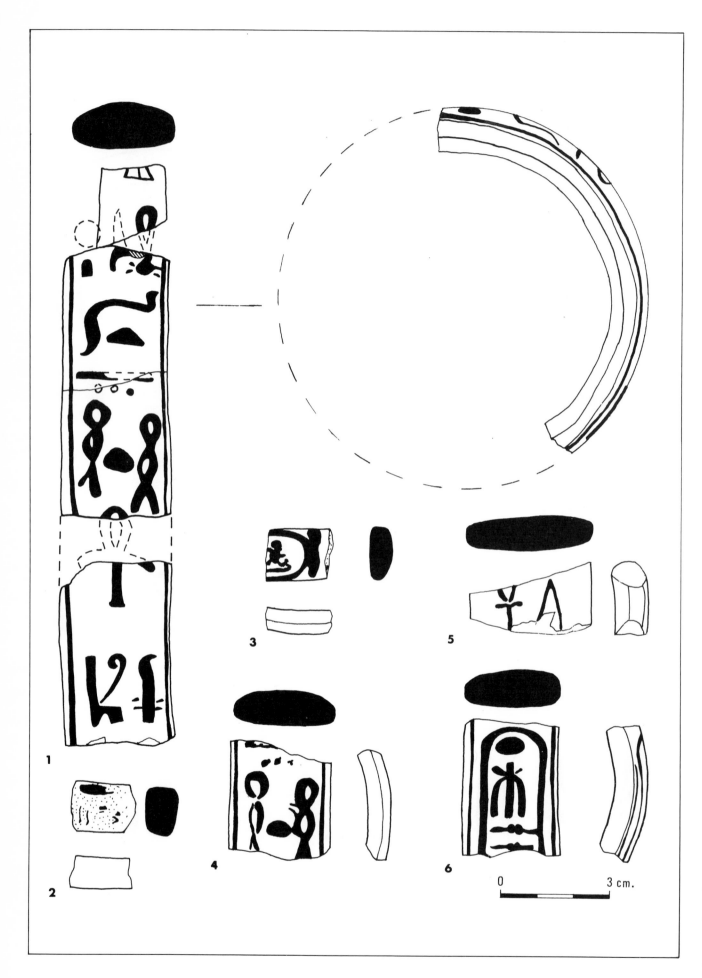

Fig. 36.   Bracelets (*see* p. 307).

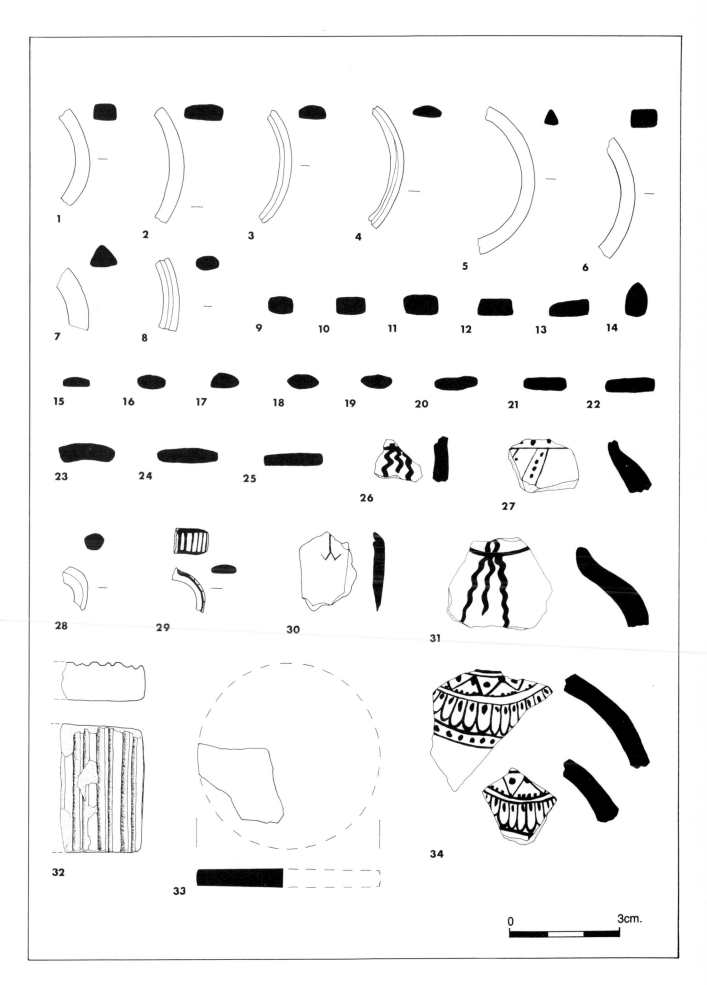

Fig. 37.   Bracelets and faience sherds (*see* p. 307).

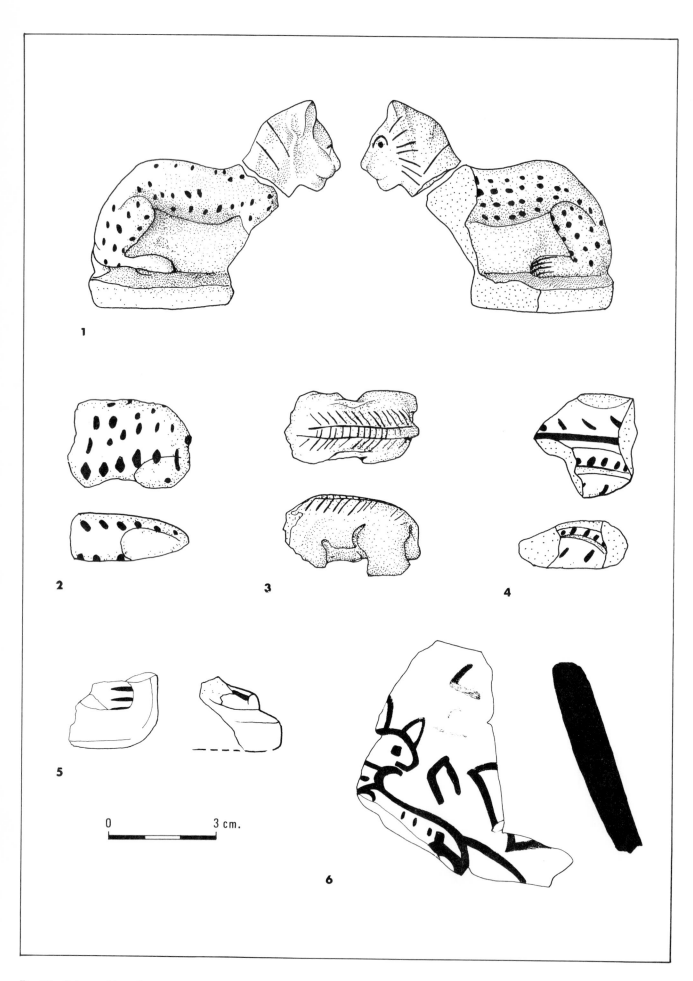

Fig. 38. Faience objects (*see* p. 308).

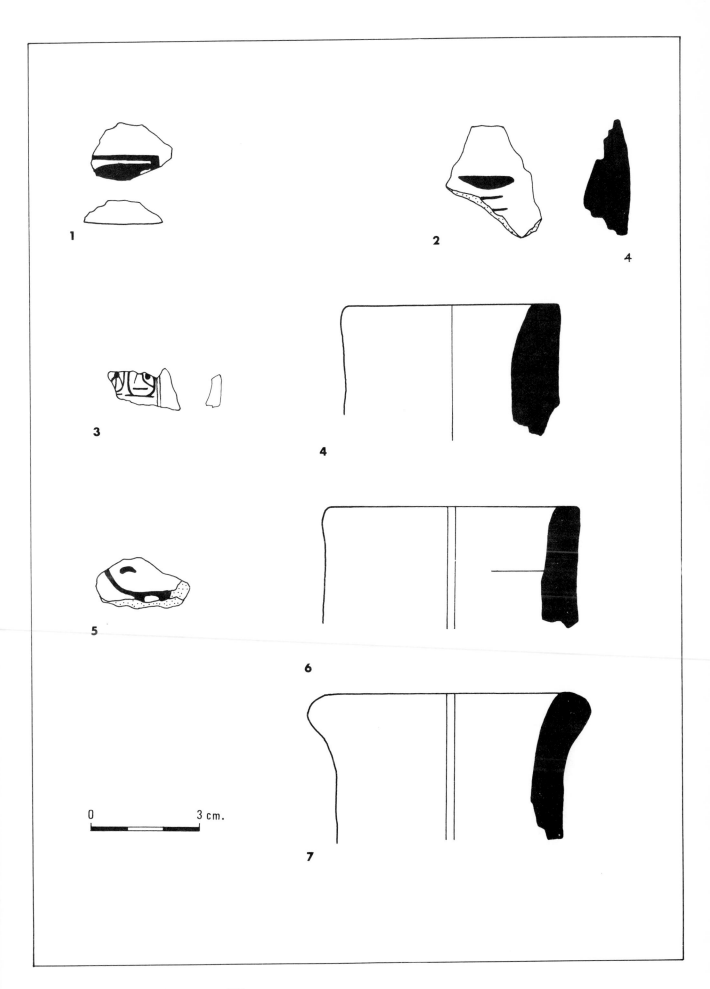

Fig. 39. Inscribed faience objects (*see* p. 308).

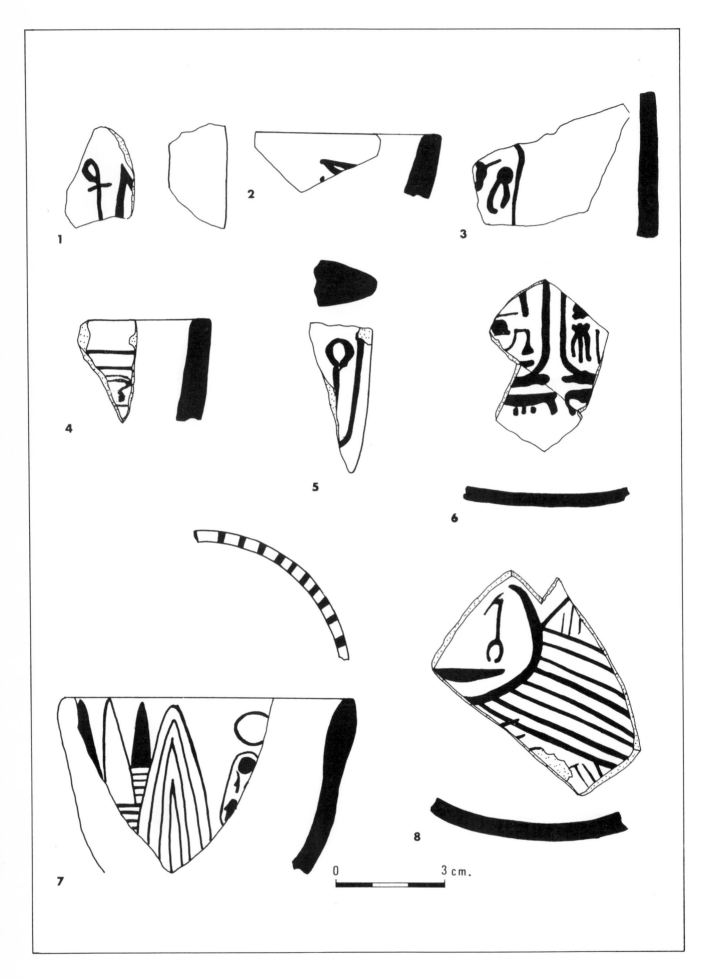

Fig. 40. Inscribed objects (*see* p. 308).

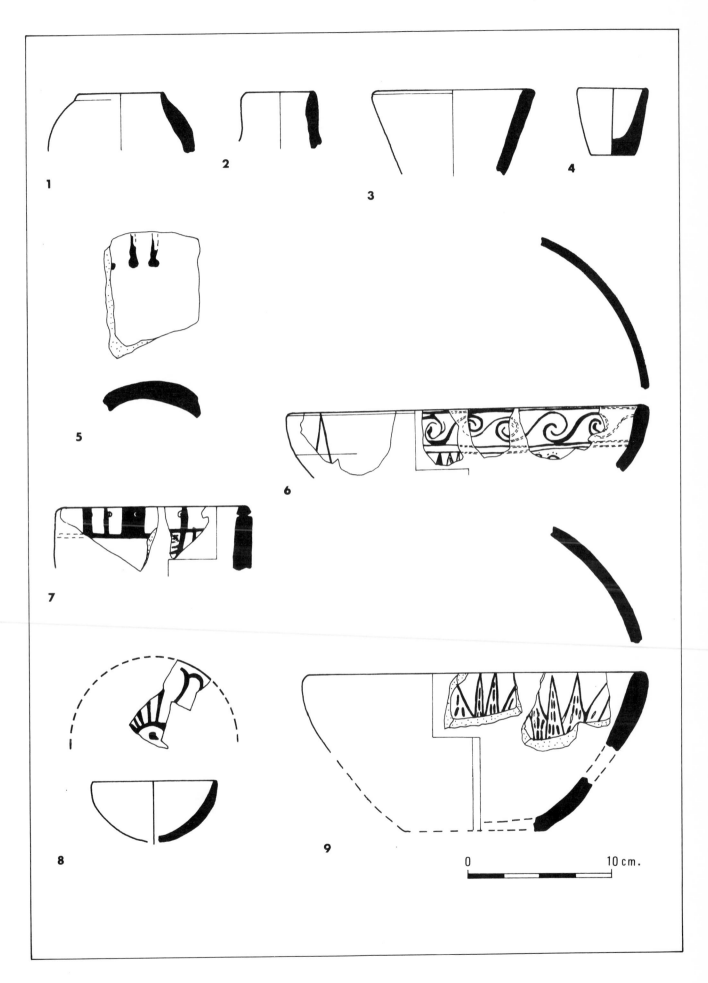

Fig. 41. Faience objects (*see* p. 308).

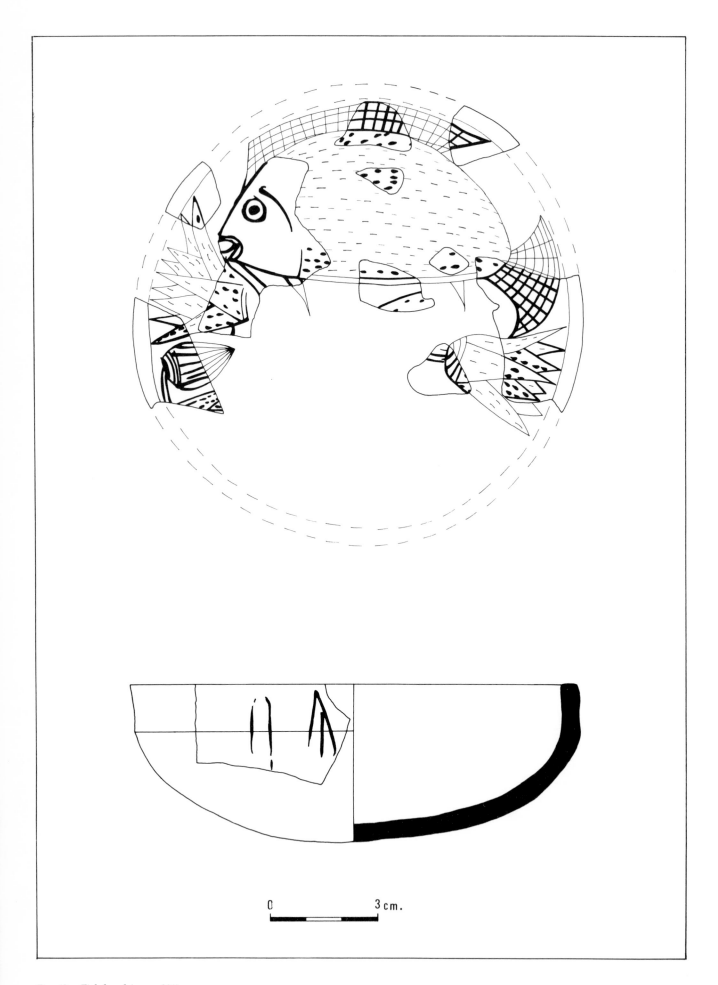

Fig. 42.   Fish bowl (*see* p. 308).

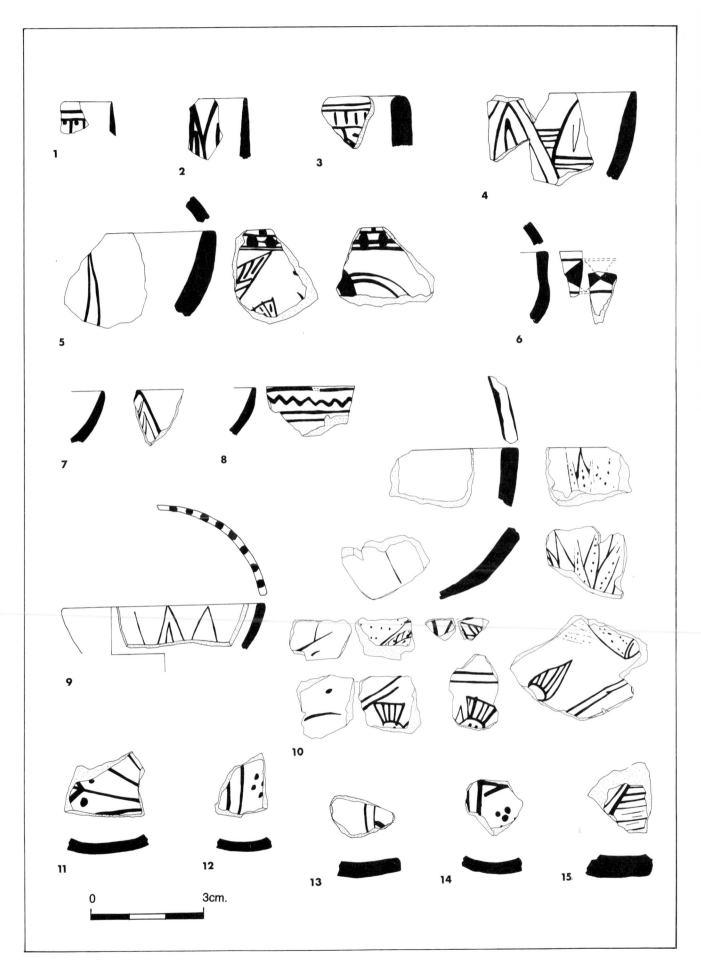

Fig. 43.   Decorated faience (*see* p. 309).

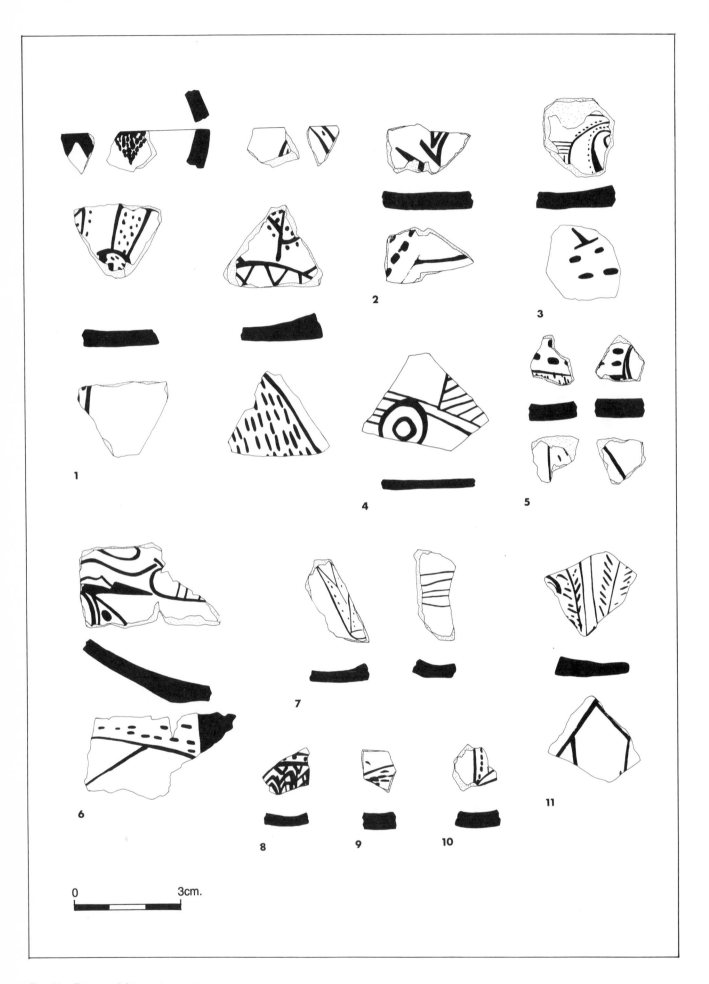

Fig. 44.   Decorated faience (*see* p. 309).

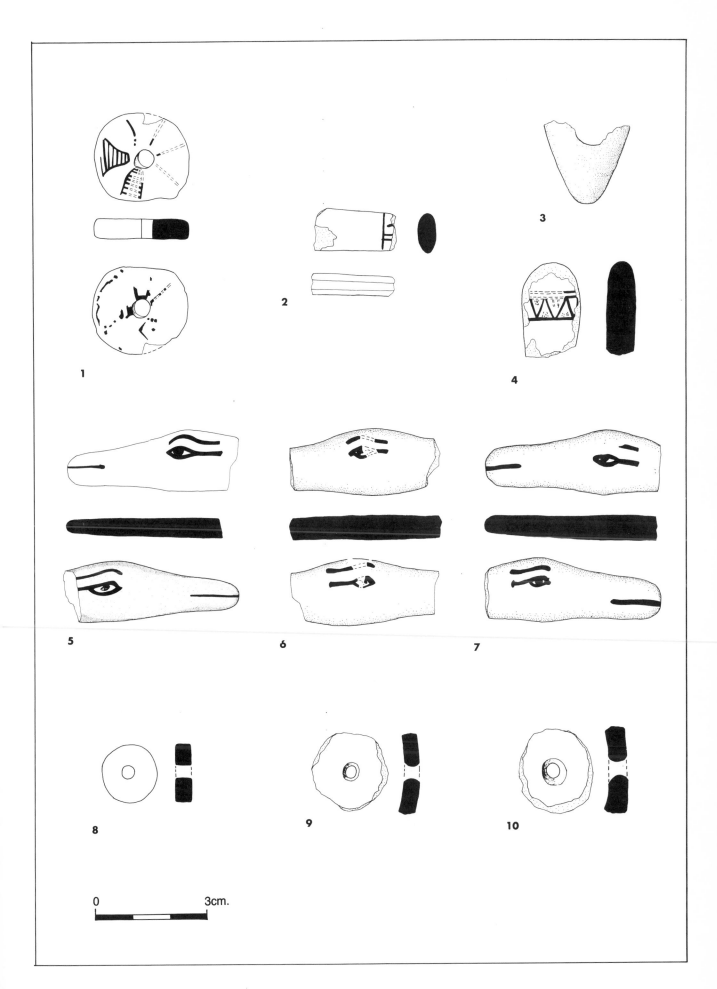

Fig. 45. Miscellaneous faience objects (*see* p. 309).

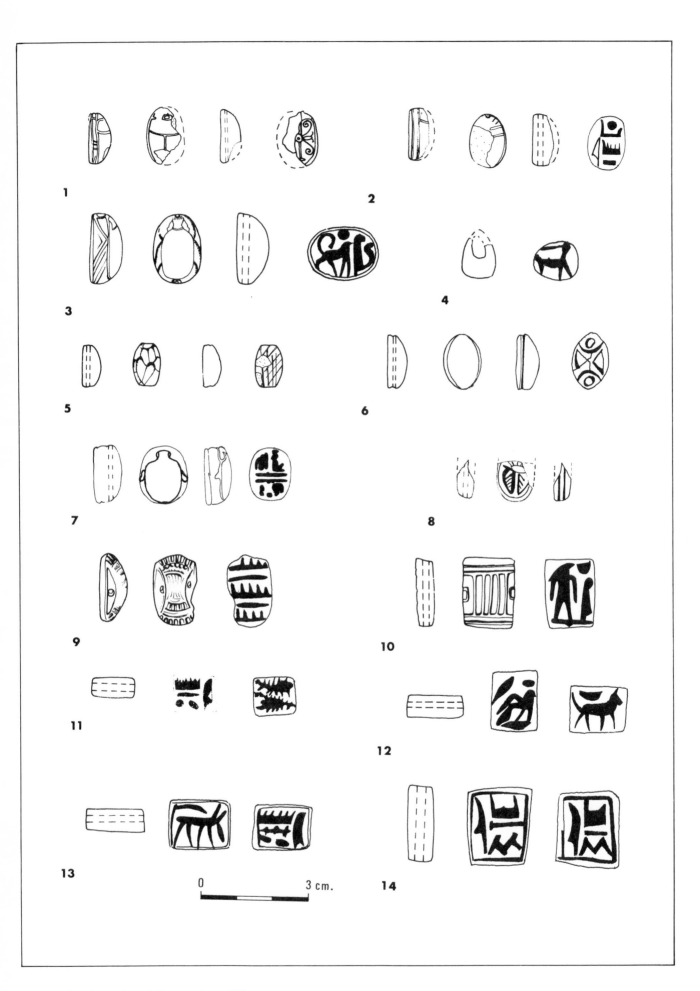

Fig. 46. Scarabs, seals and placques (*see* p. 310).

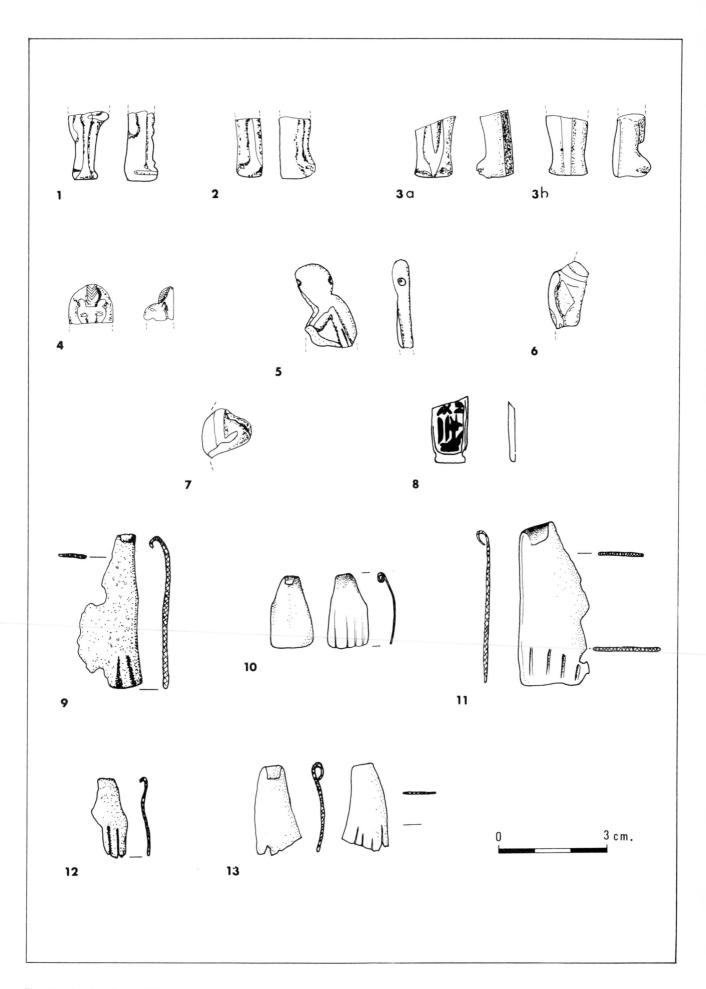

Fig. 47. Amulets (*see* p. 310).

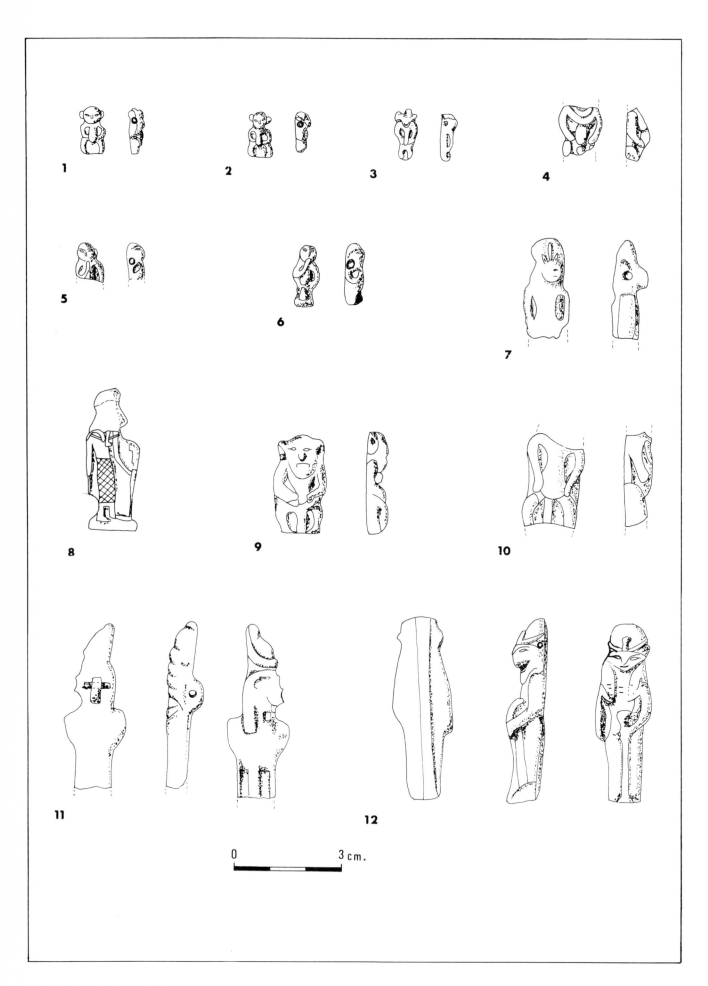

Fig. 48.   Amulets (*see* p. 310).

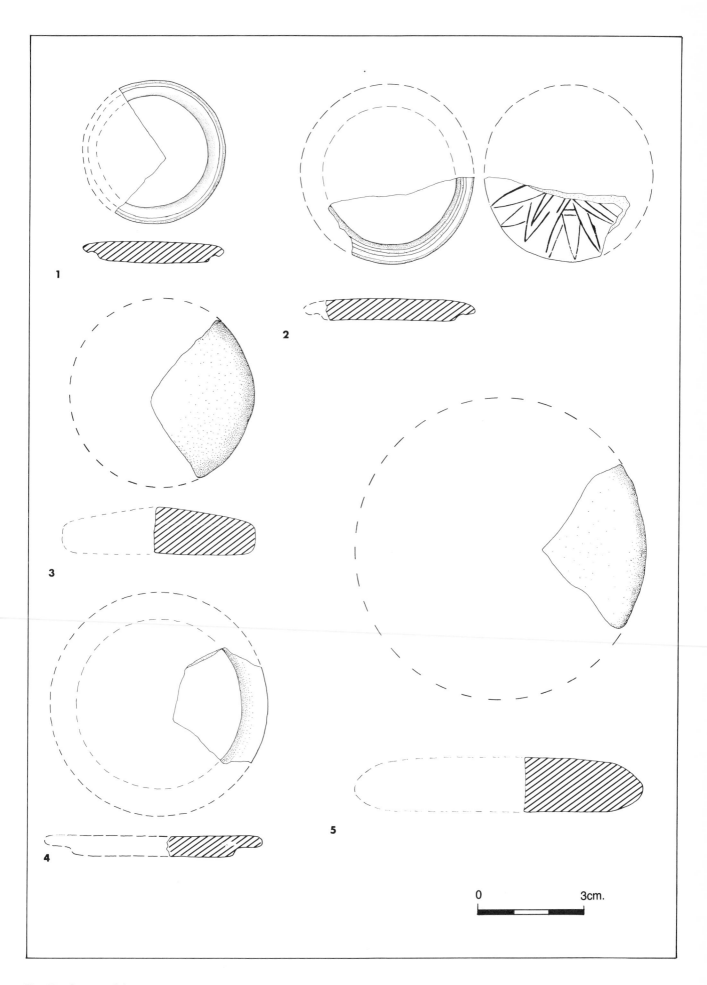

Fig. 49.   Stone jar lids (*see* p. 310).

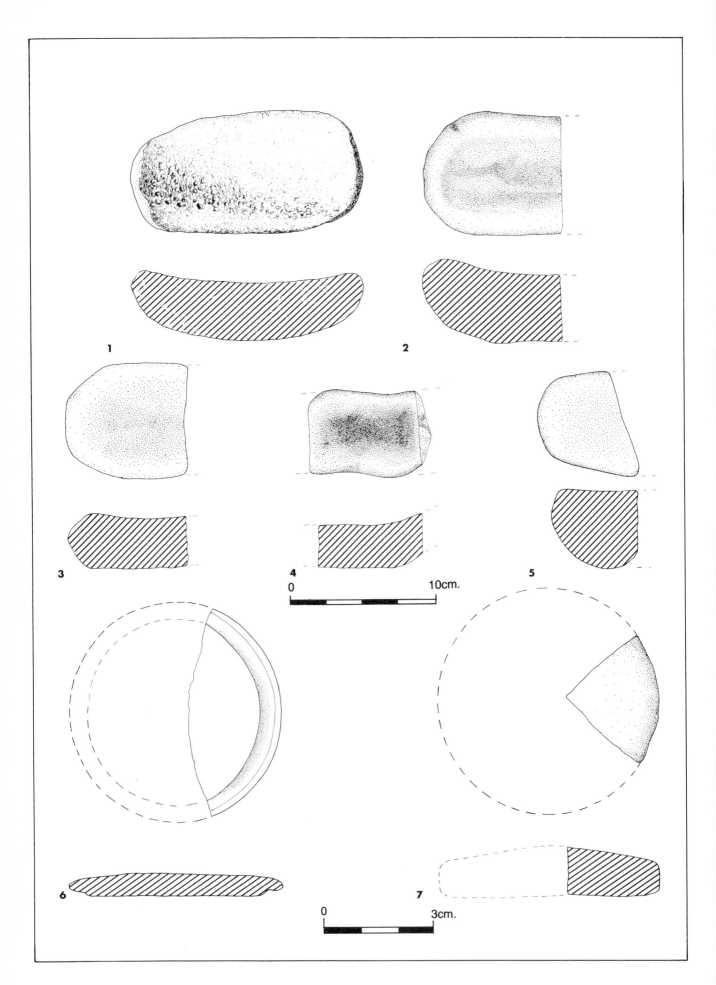

Fig. 50. Stone querns and jar lids (*see* p. 311).

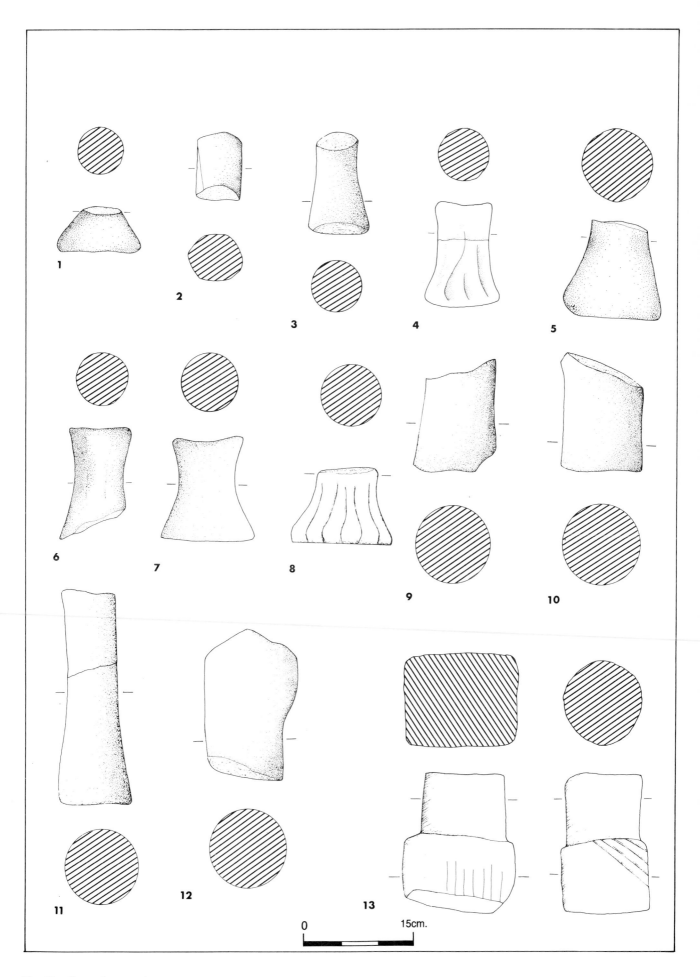

Fig. 51.   Stone altar stands (*see* p. 311).

Fig. 52.   Rock stela. Eg. Cat. No. 260, *Pl.* 105.

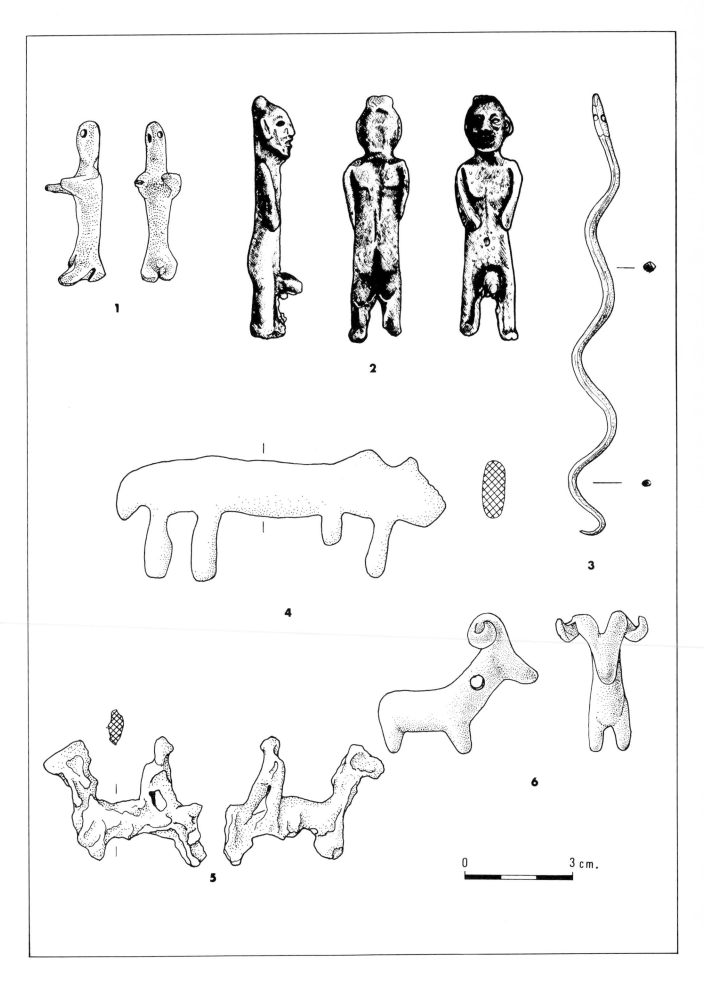

Fig. 53.   Metal figurines (*see* p. 147).

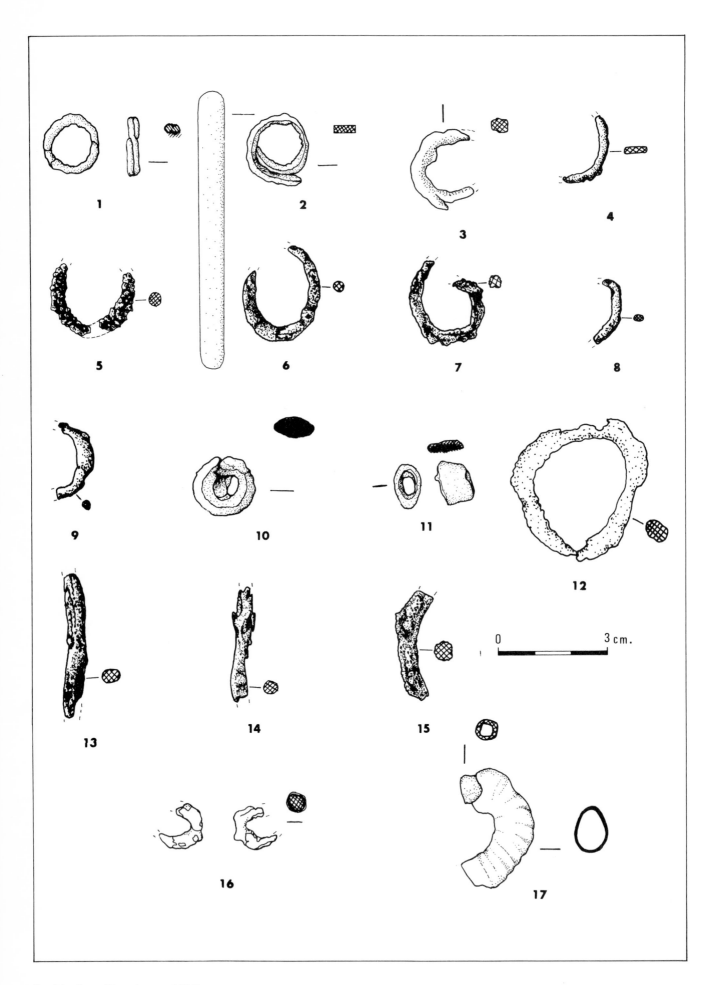

Fig. 54. Iron objects (*see* pp. 147–8).

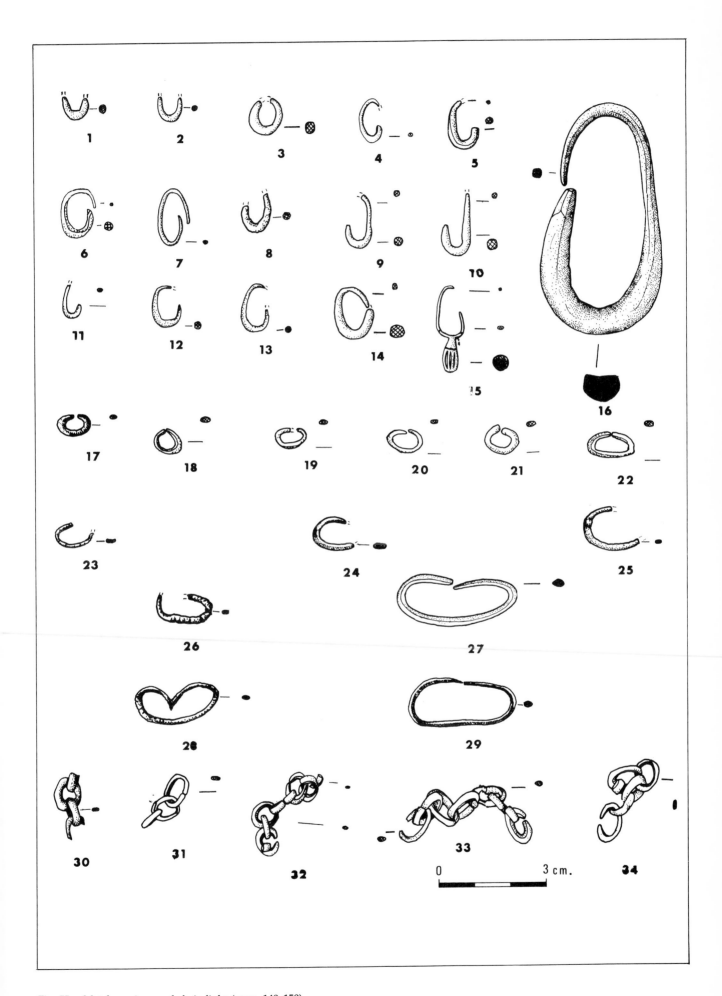

Fig. 55.   Metal ear-rings and chain links (*see* p. 148–150).

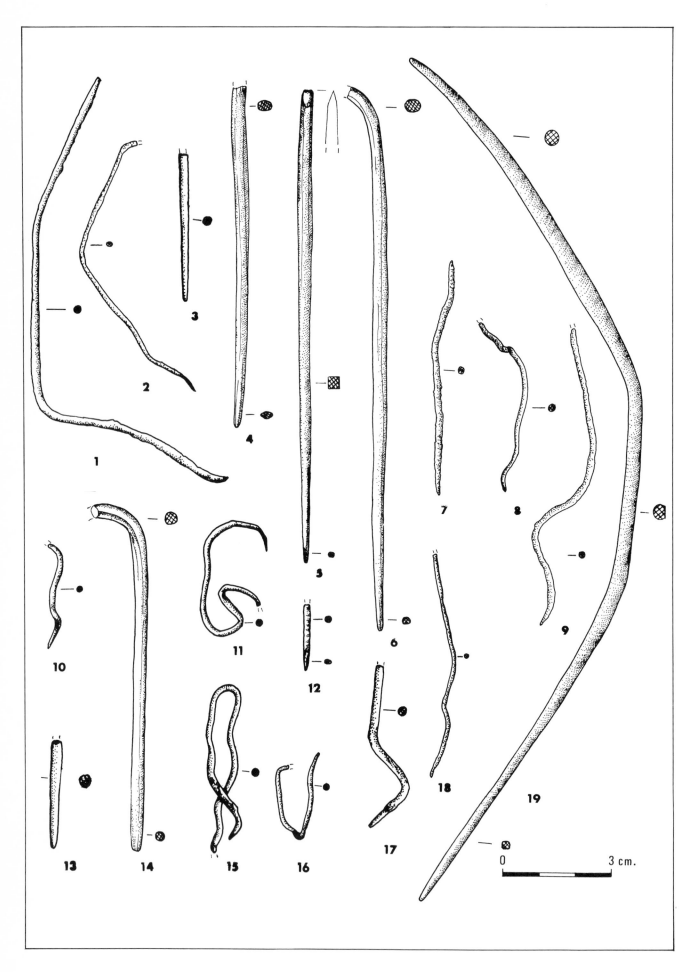

Fig. 56. Metal wires, pins, kohl sticks, sistrum parts (*see* pp. 150–51).

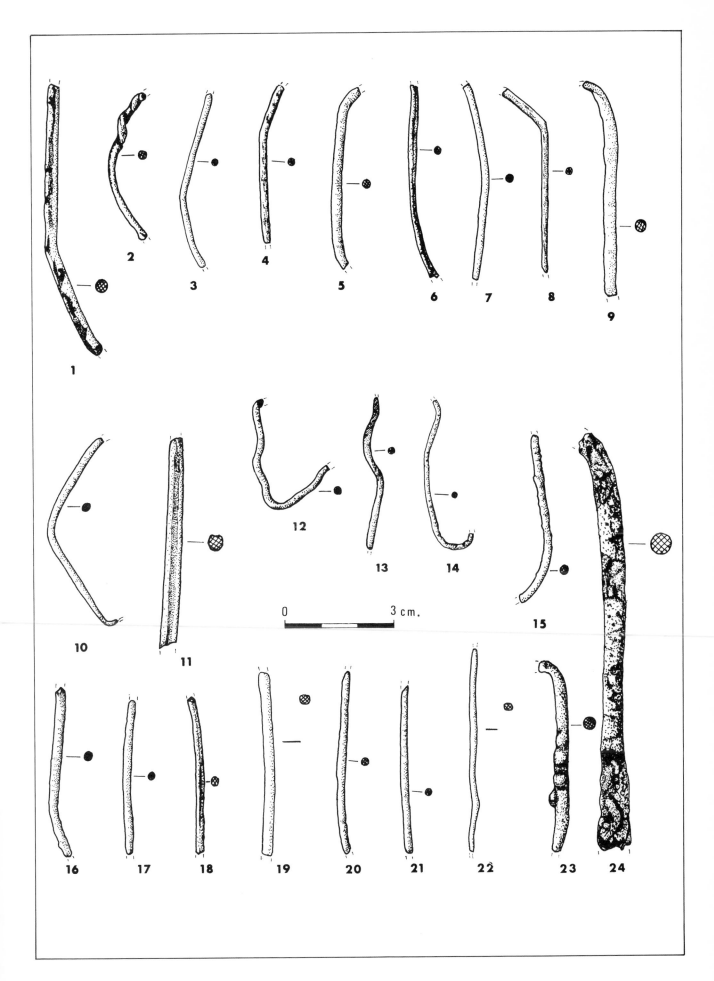

Fig. 57.   Metal wires, sistrum fragments (*see* p. 151).

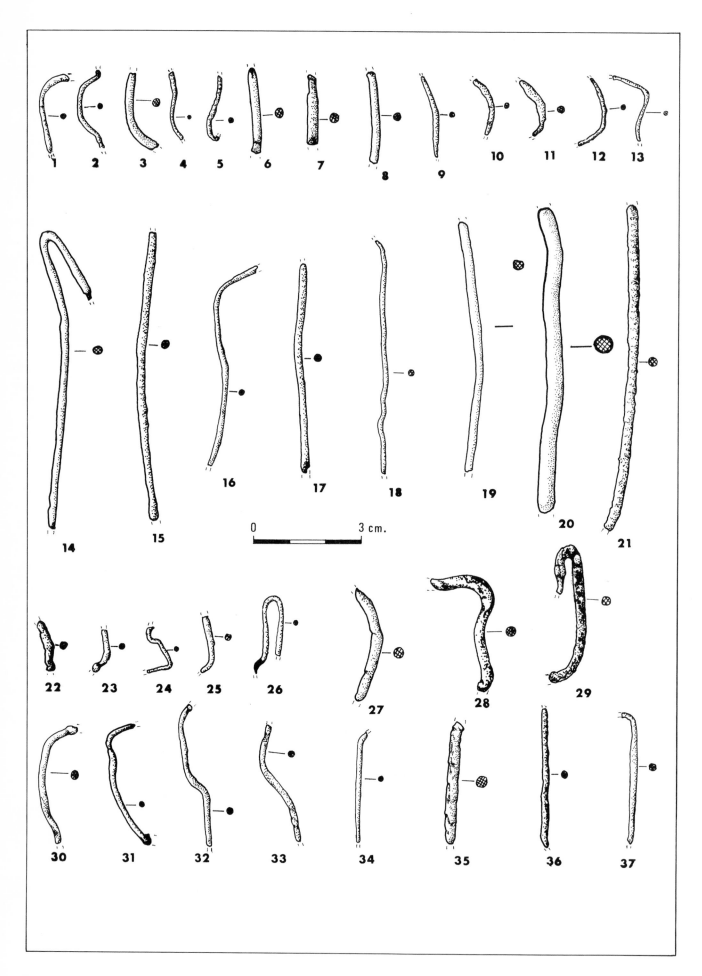

Fig. 58.   Metal wires, pins, kohl stick (*see* pp.151–3).

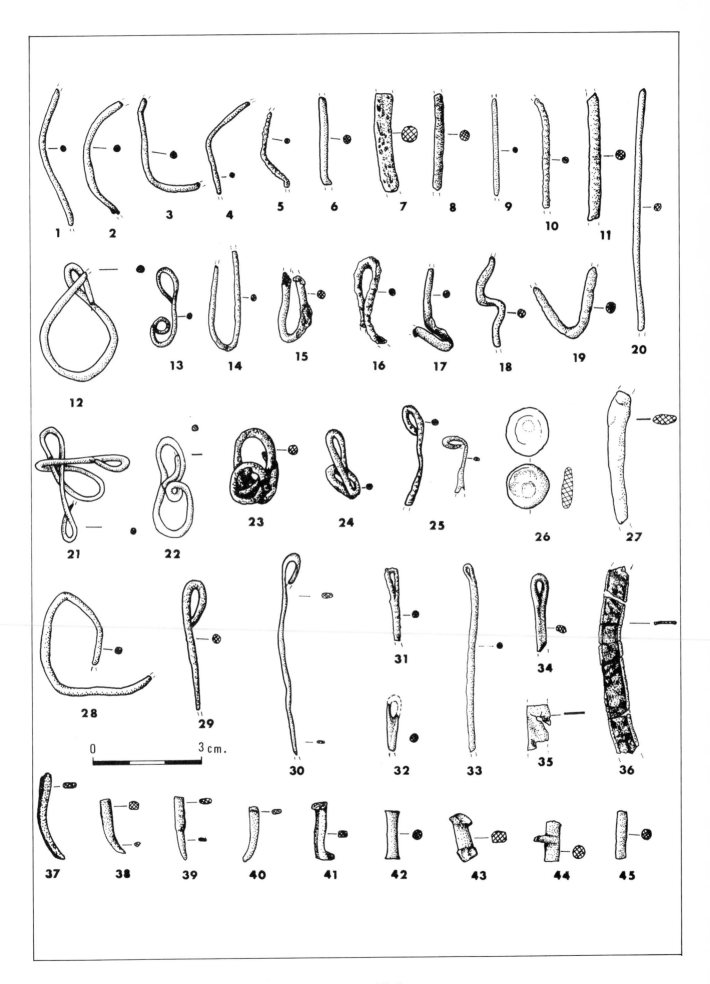

Fig. 59. Metal wires, rods, tin and lead objects, needles, rivets (*see* pp. 153–4).

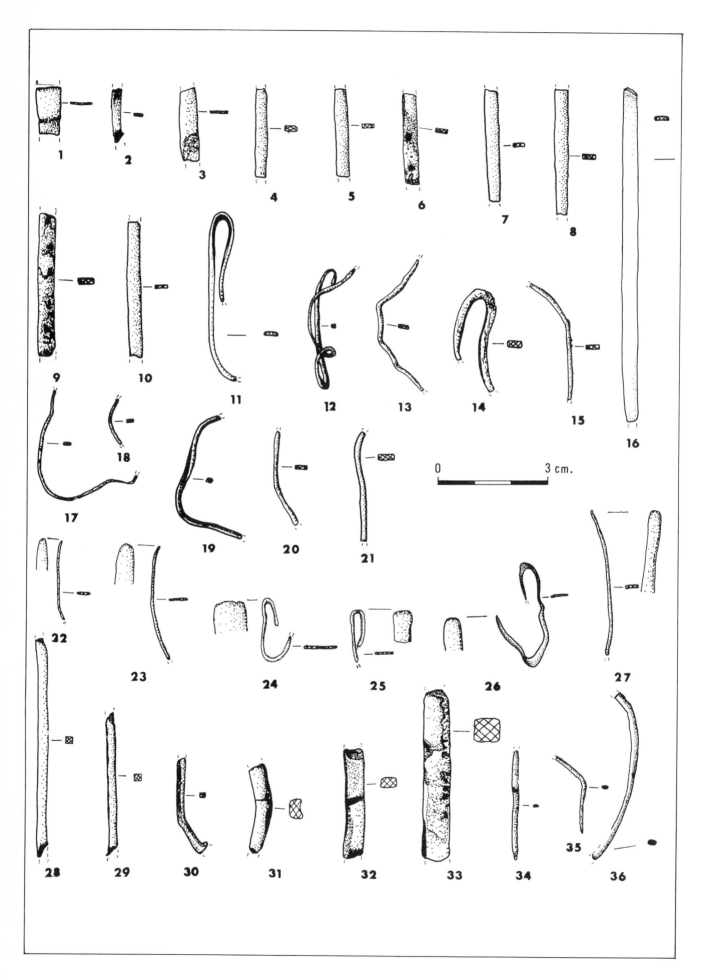

Fig. 60. Metal rods and wires (*see* pp. 154–6).

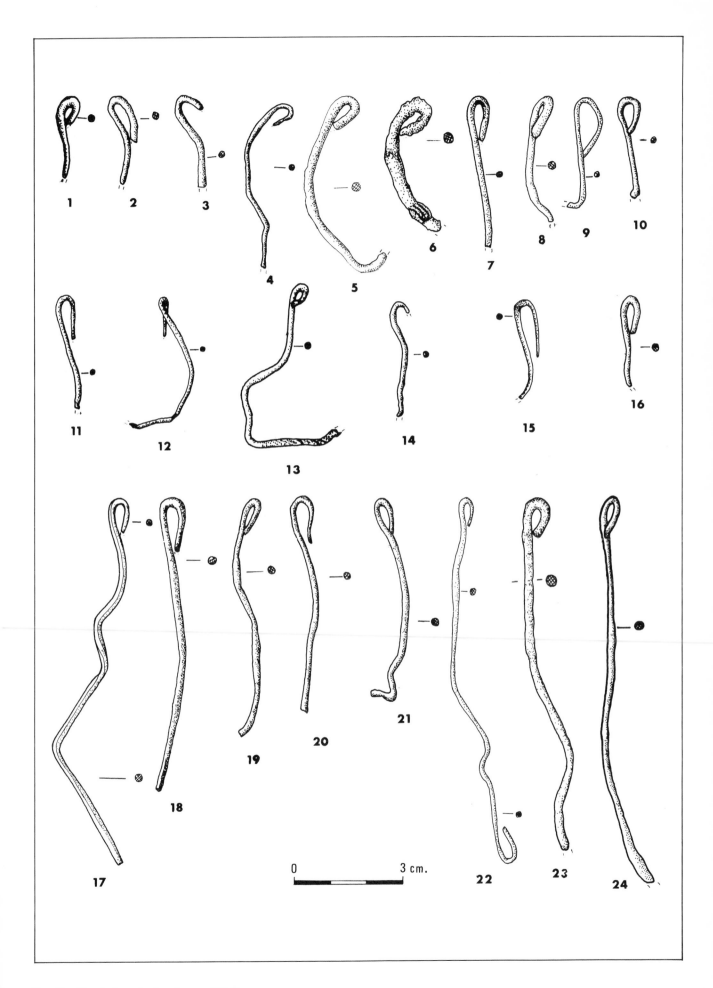

Fig. 61. Crook-shaped wires (*see* pp. 156–7).

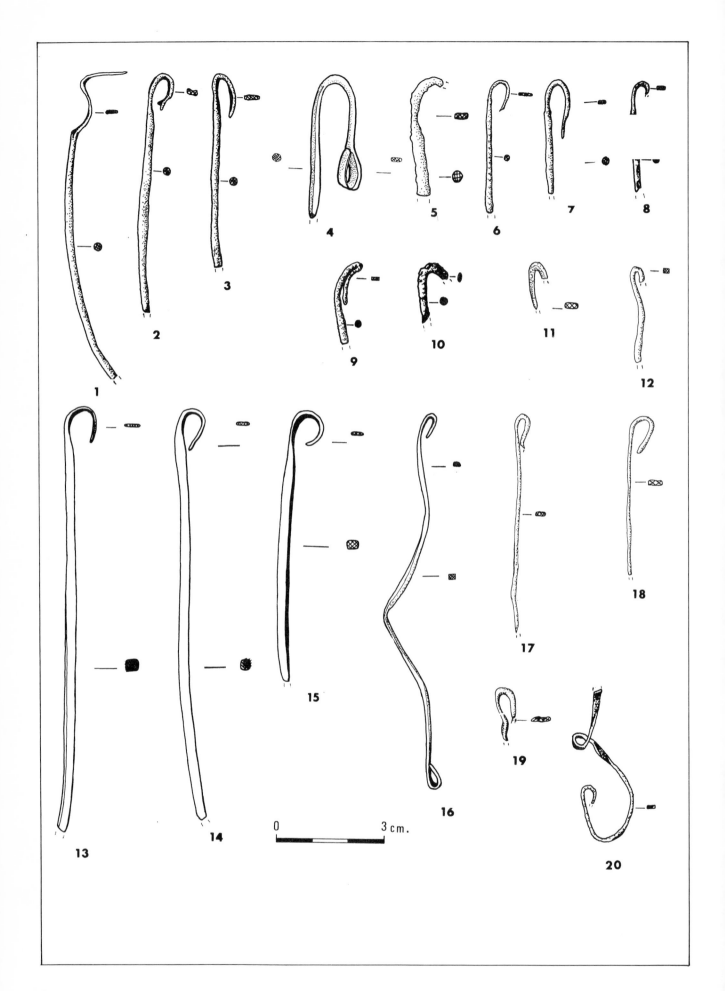

Fig. 62. Crook-shaped wires (*see* pp. 157–8).

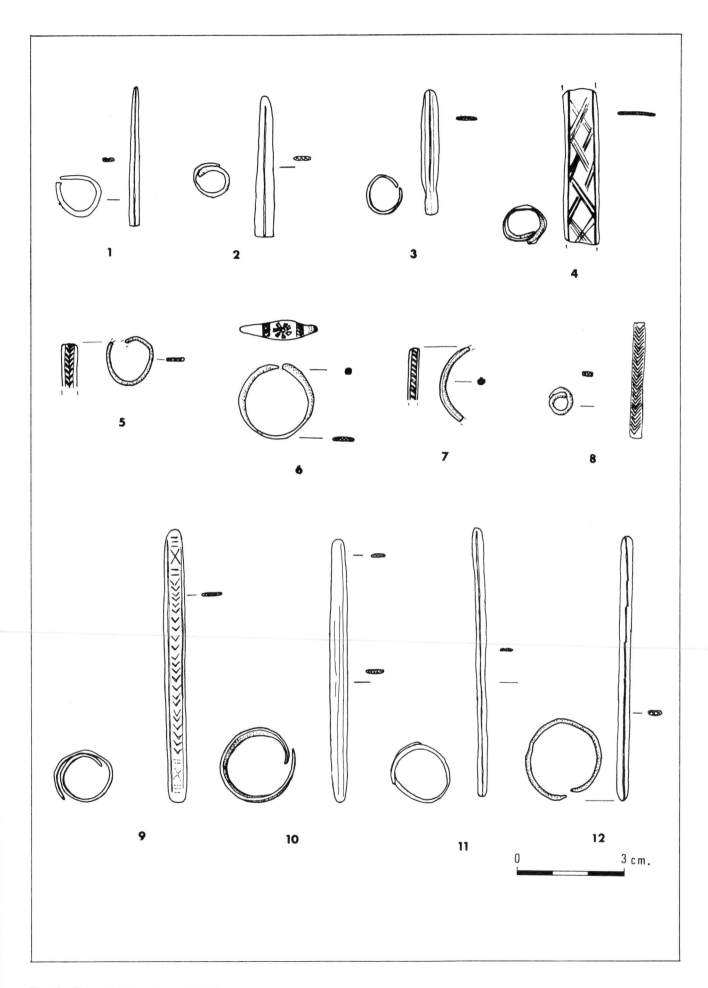

Fig. 63.   Decorated rings (*see* pp. 158–9).

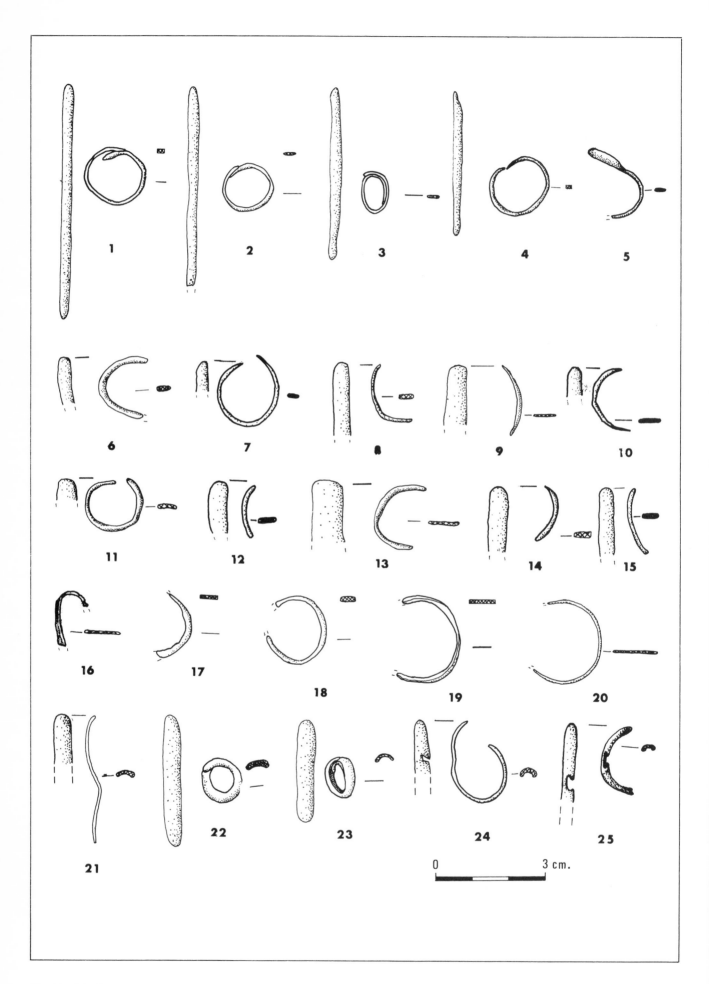

Fig. 64. Metal rings (*see* pp. 159–60).

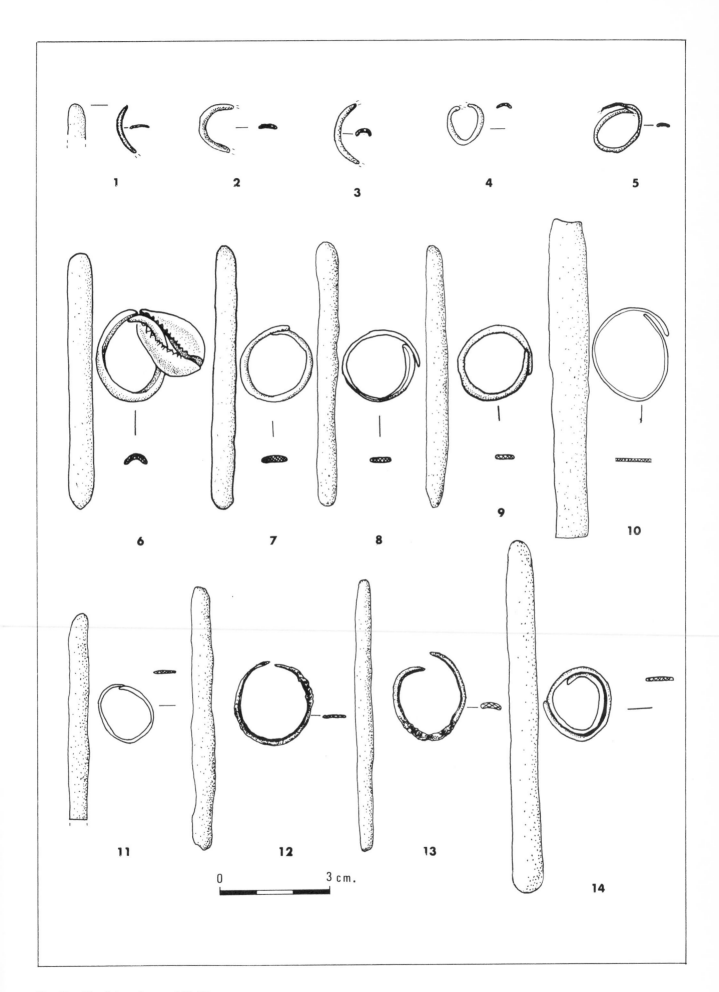

Fig. 65. Metal rings (*see* pp. 160–61).

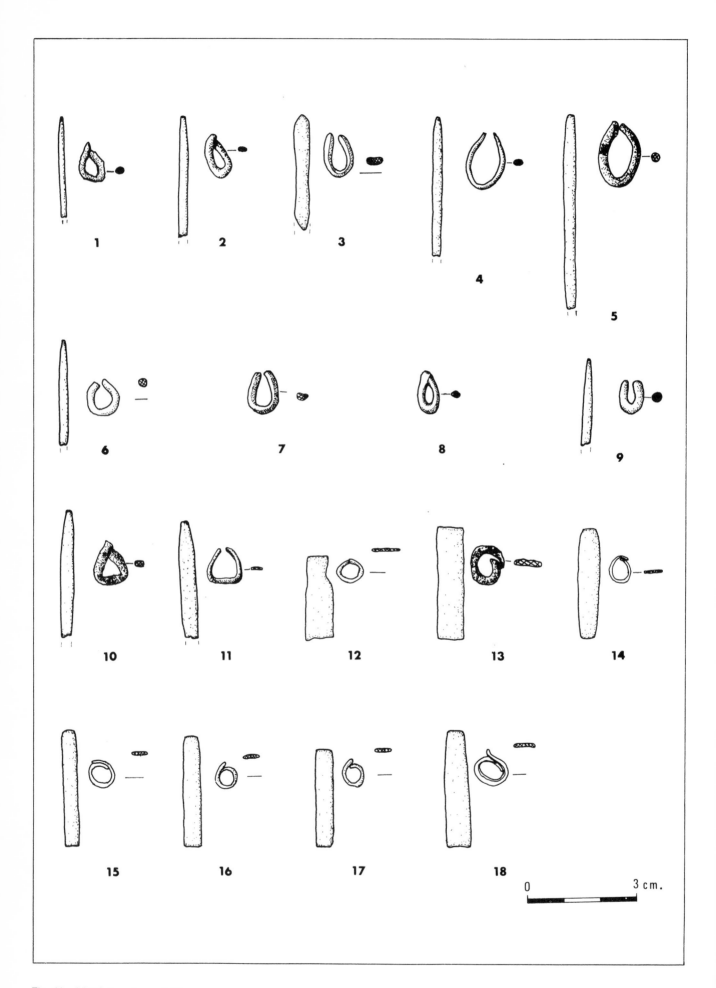

Fig. 66.   Metal rings (*see* p. 161).

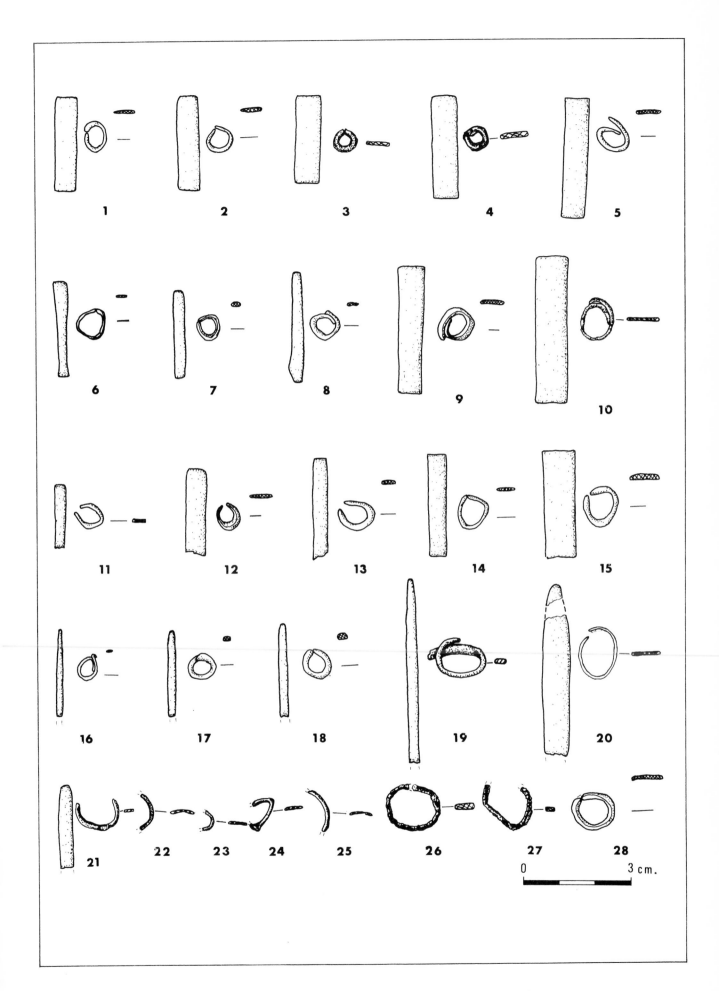

Fig. 67. Metal rings (*see* pp. 161–2).

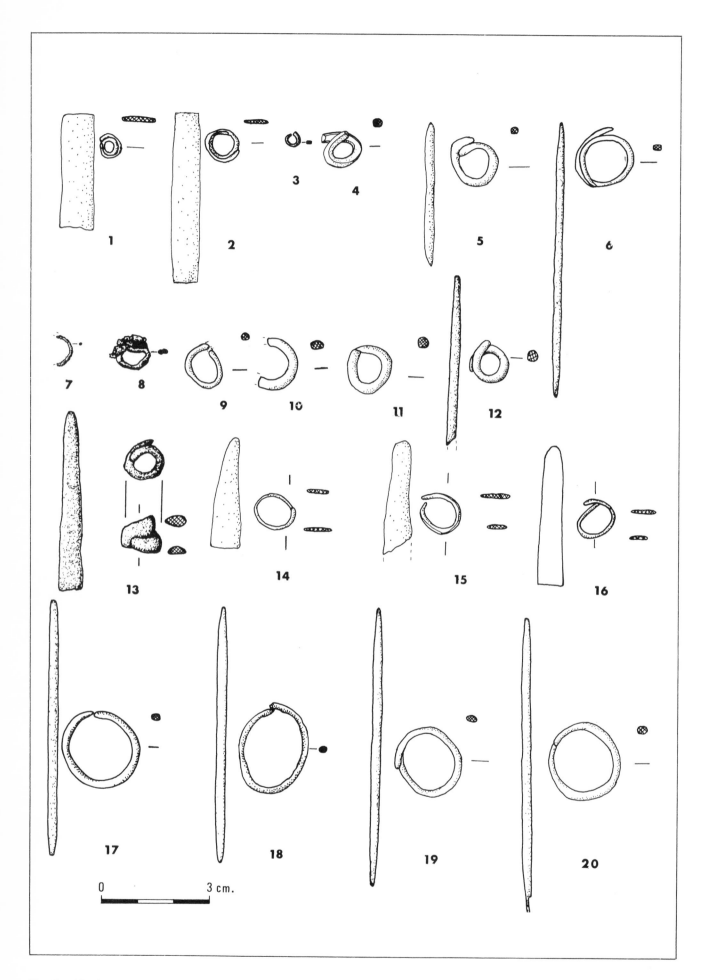

Fig. 68.   Metal rings (*see* pp. 162–3).

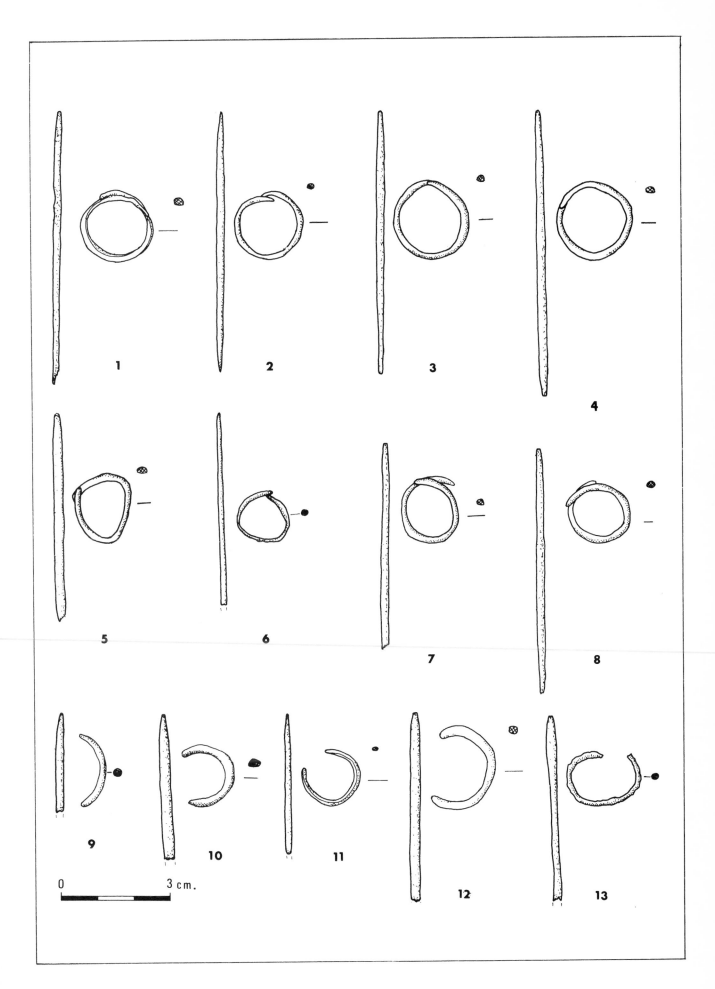

Fig. 69.   Metal rings (*see* pp. 163–4).

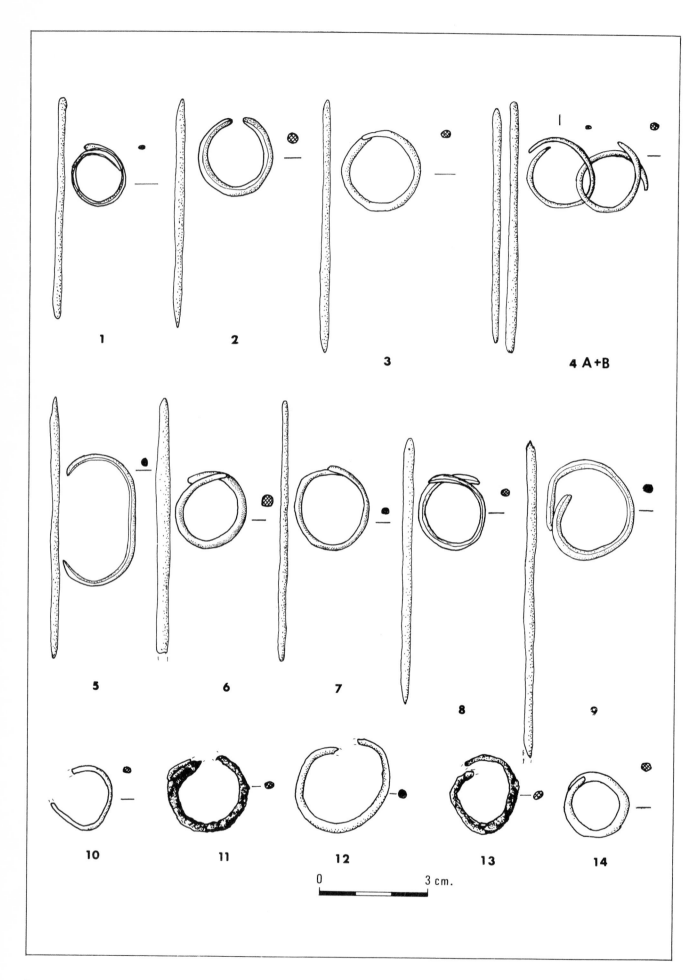

Fig. 70.   Metal rings (*see* p. 164).

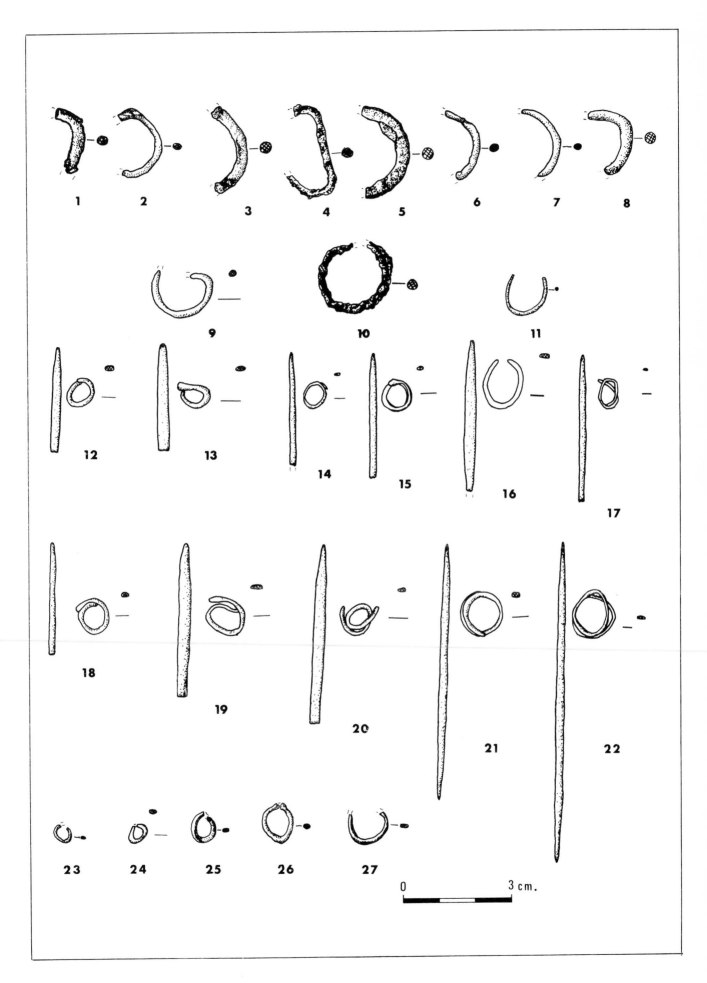

Fig. 71. Metal rings (*see* pp. 164–5).

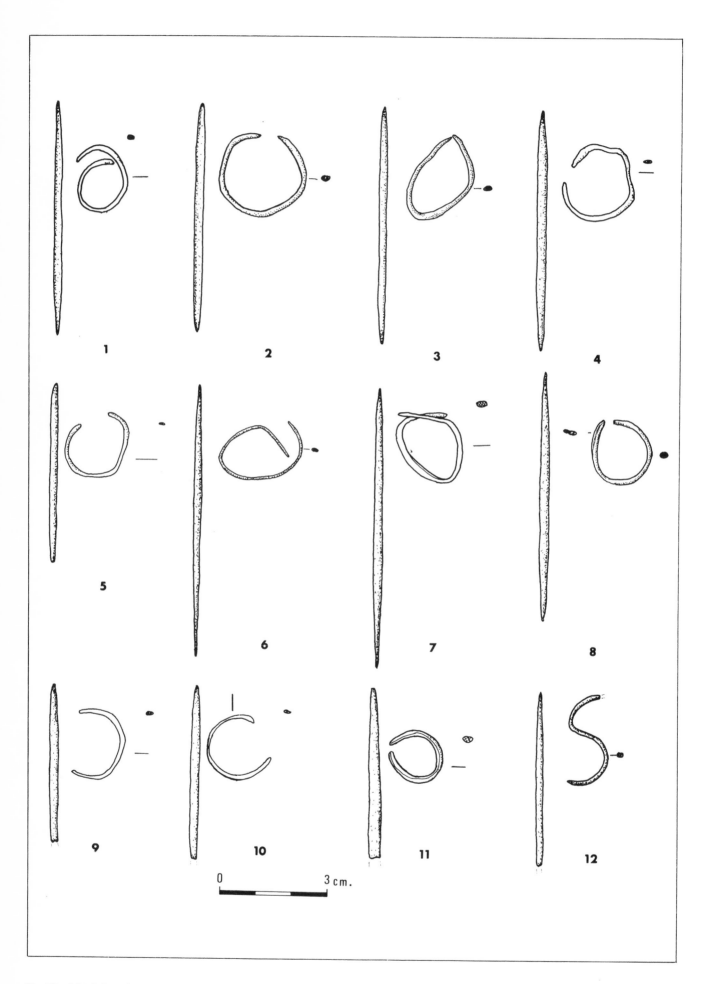

Fig. 72.   Metal rings (*see* pp. 165–6).

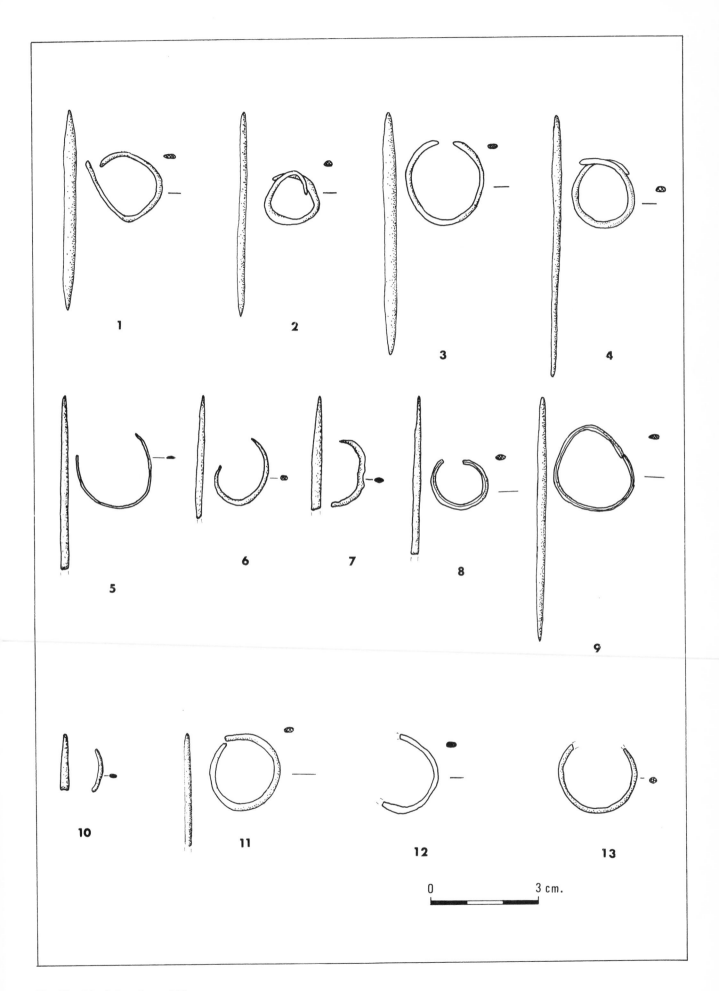

Fig. 73.   Metal rings (*see* p. 166).

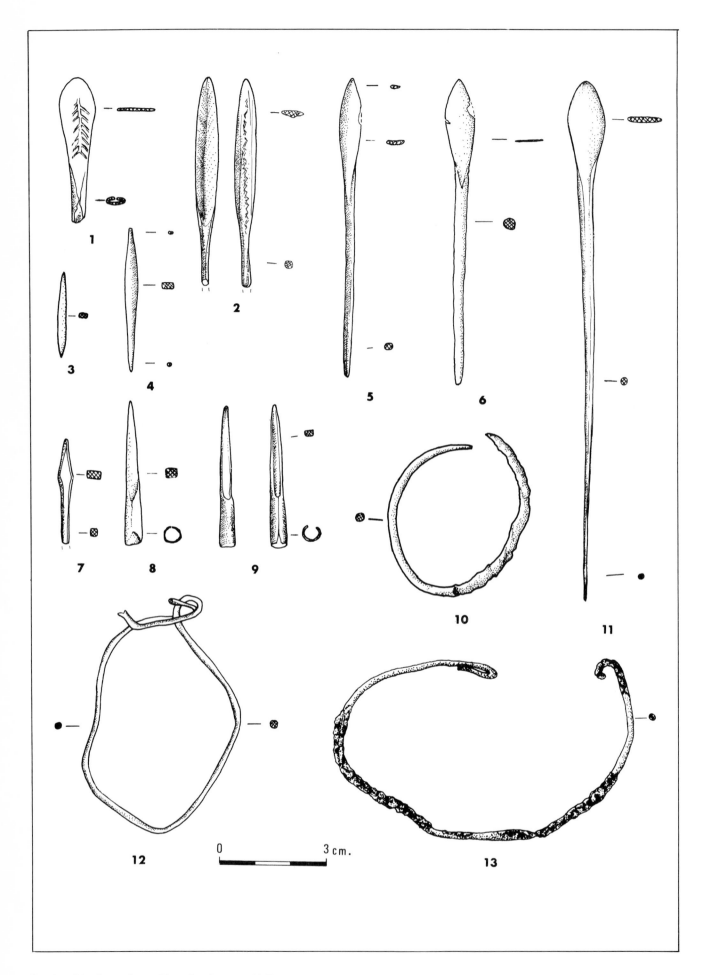

Fig. 74.   Metal spatulas and bracelets (*see* pp. 166–7).

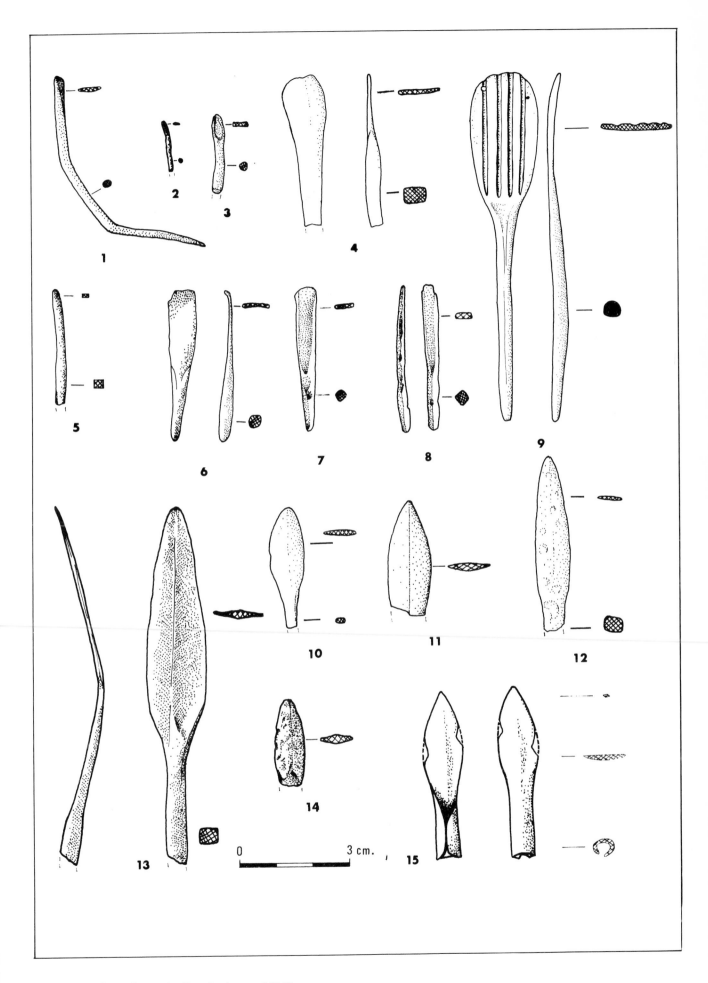

Fig. 75.   Metal spatulas, projectile point (*see* pp. 167–8).

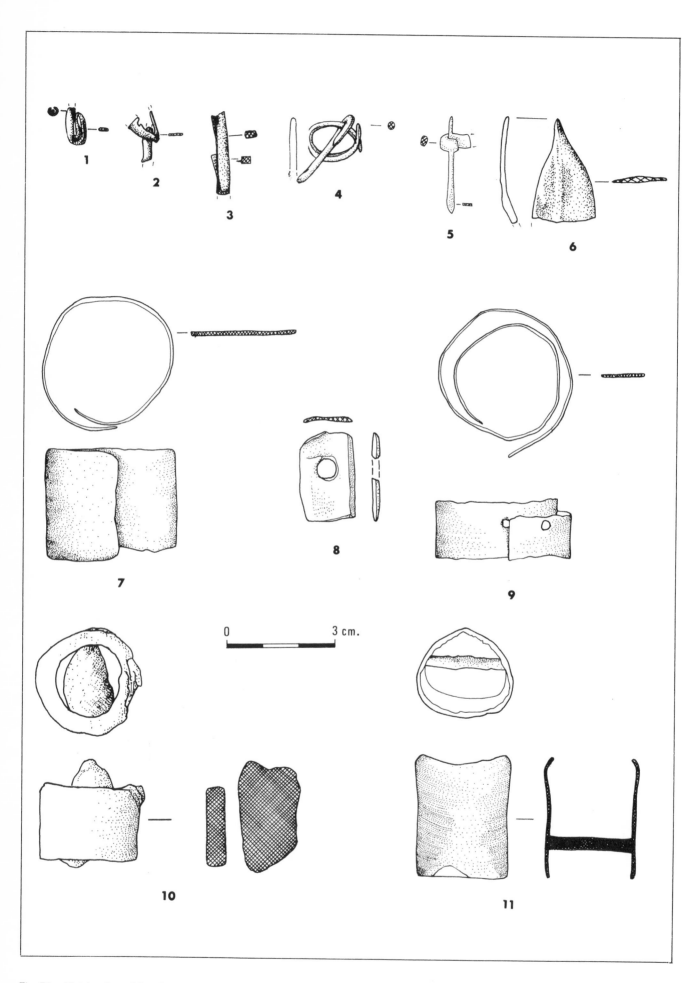

Fig. 76. Metal rods and ferrules (*see* p. 168).

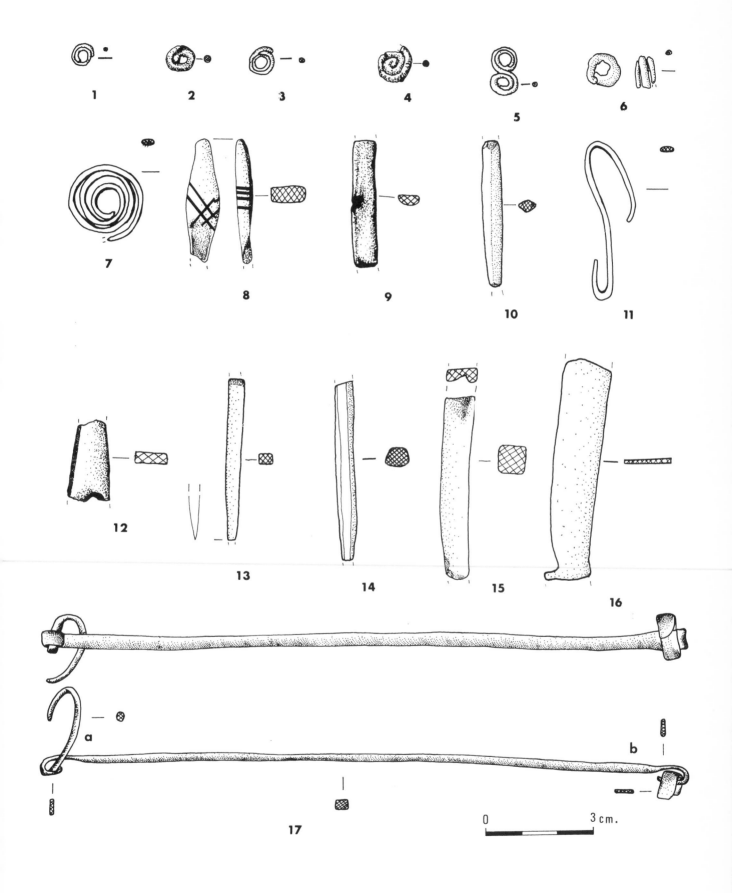

Fig. 77.   Metal spirals, hooks, punches and balance beam (*see* pp. 168–9).

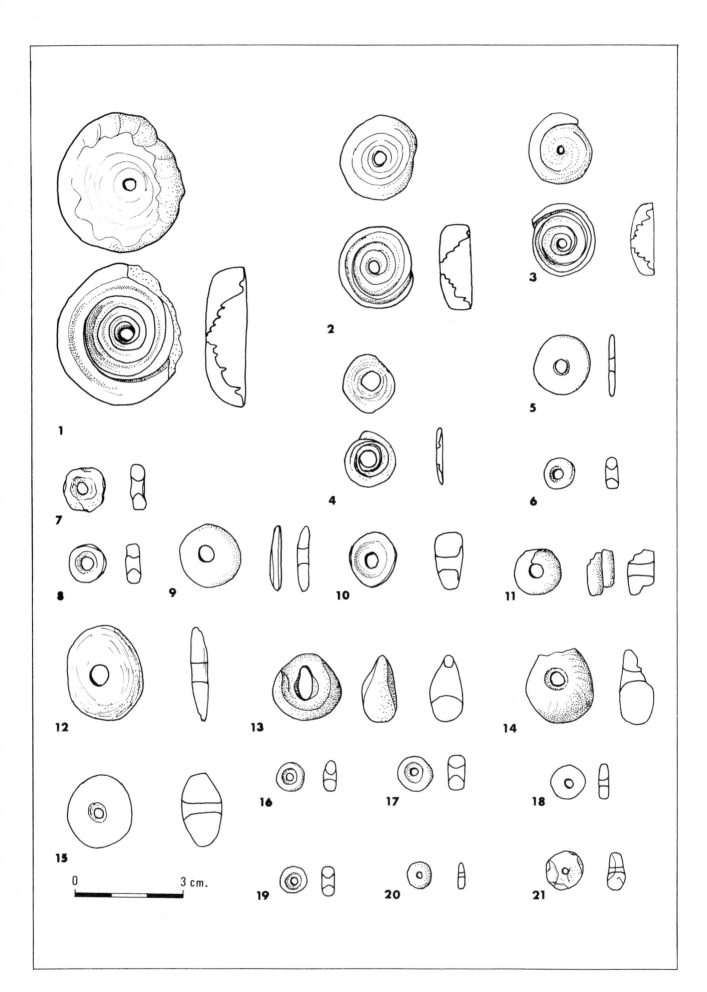

Fig. 78. Beads (*see* p. 312).

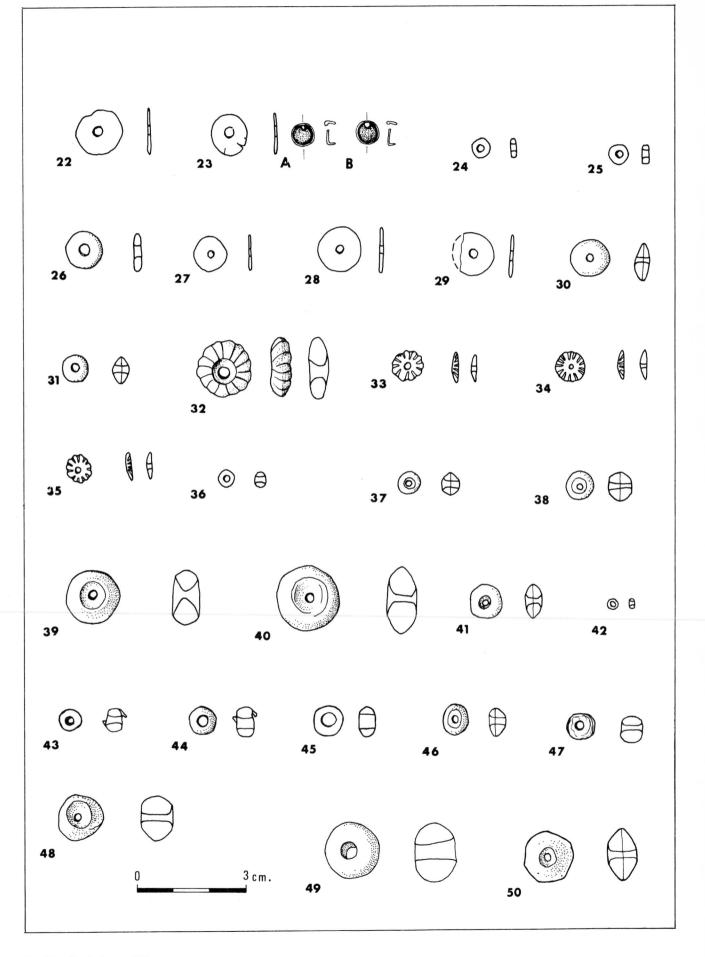

Fig. 79. Beads (*see* p. 313).

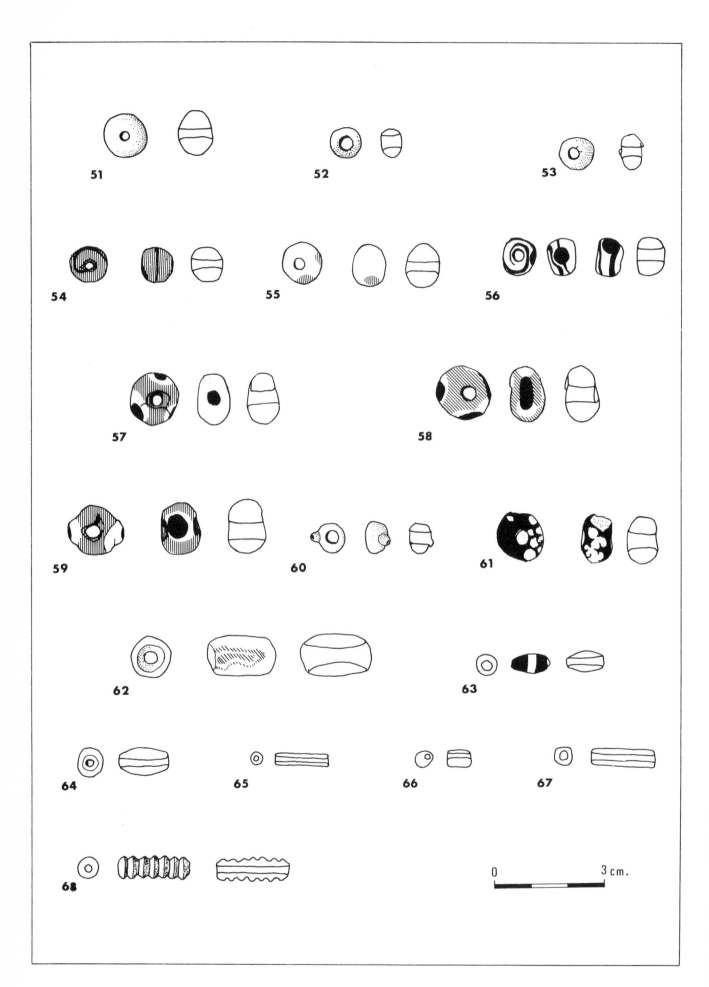

Fig. 80. Beads (*see* p. 313).

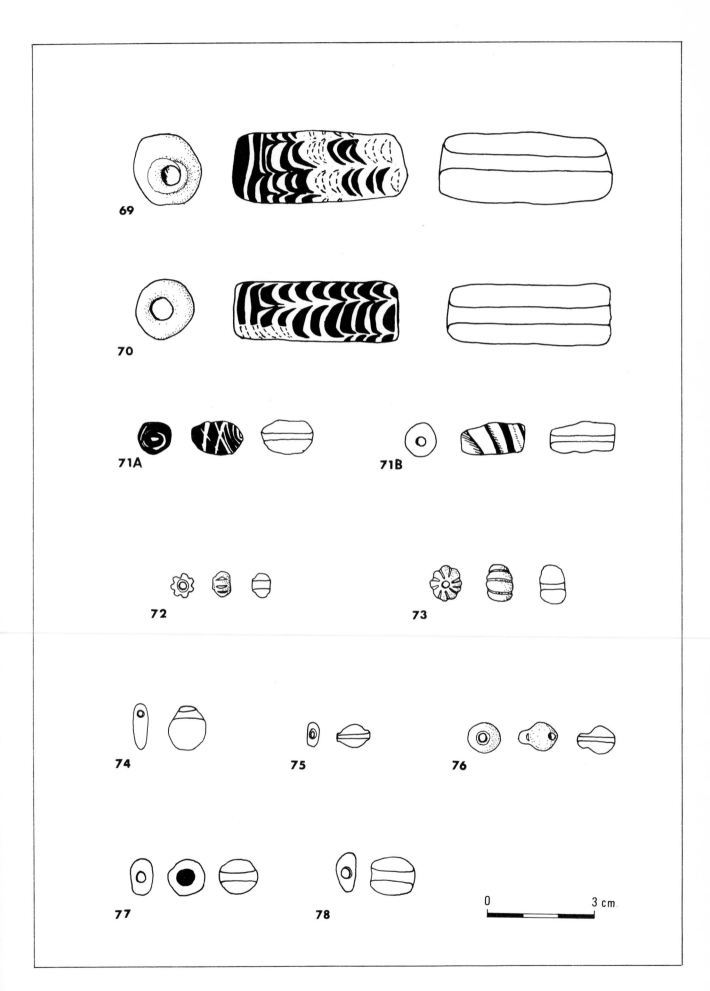

Fig. 81. Beads (*see* p. 314).

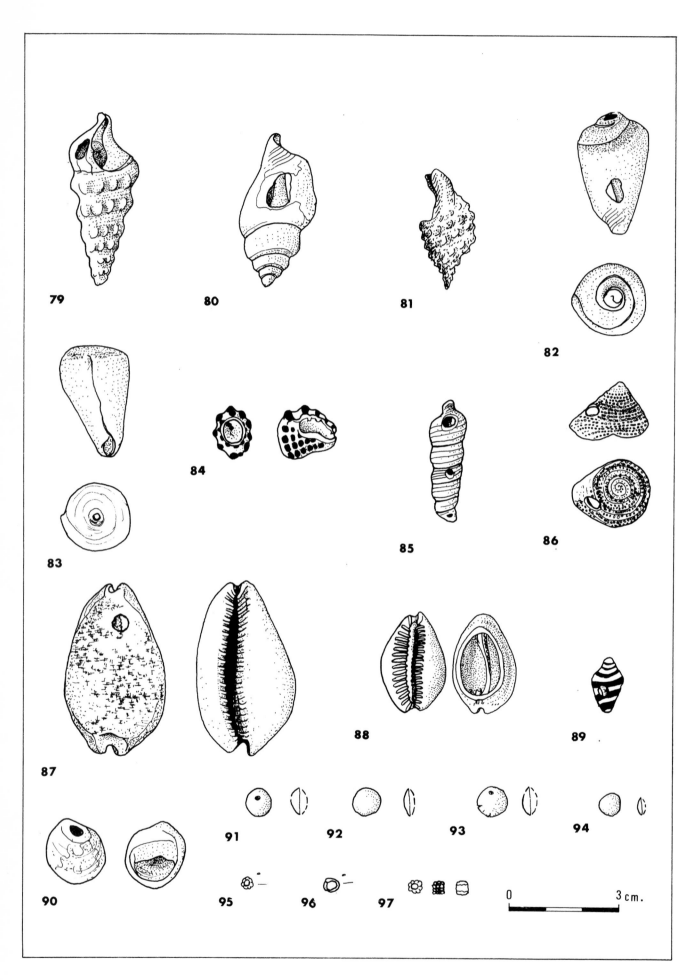

Fig. 82.   Beads (*see* p. 314).

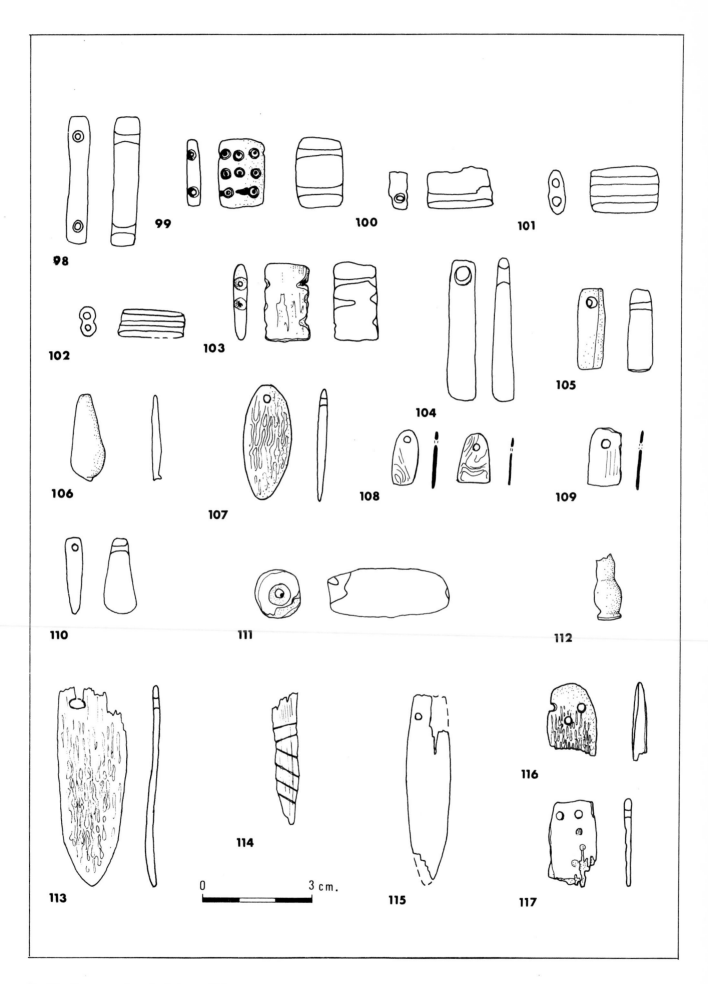

Fig. 83. Spacers and pendants (*see* p. 314).

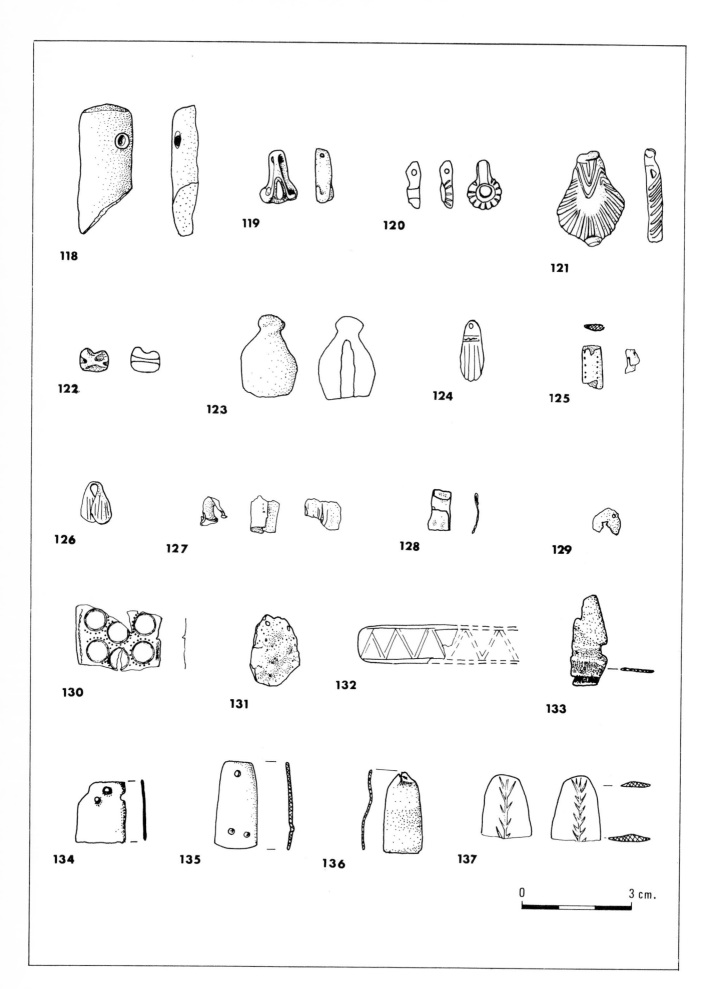

Fig. 84. Pendants (*see* p. 315).

Fig. 85. Egyptian glass fragments (*see* p. 315).

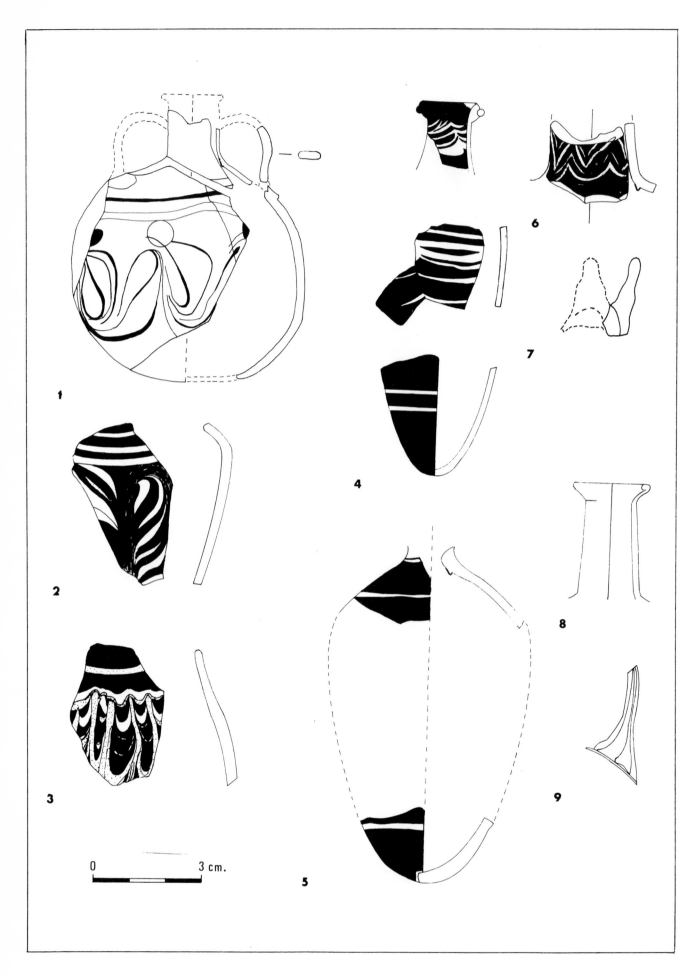

Fig. 86. Egyptian and Roman glass fragments (*see* p. 315).

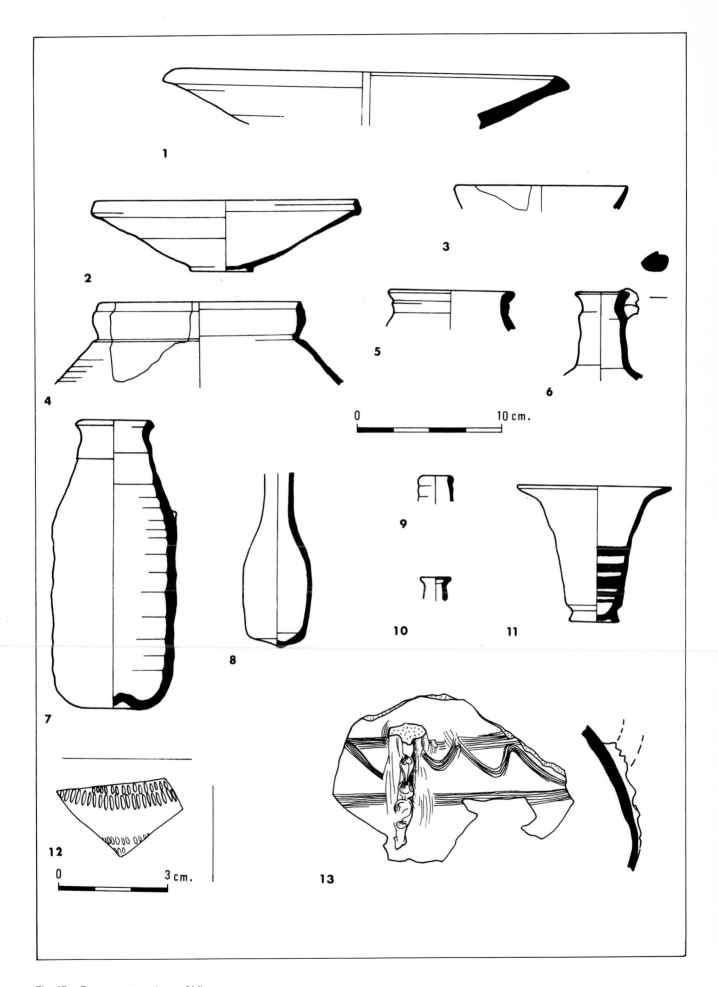

Fig. 87.   Roman pottery (*see* p. 316).

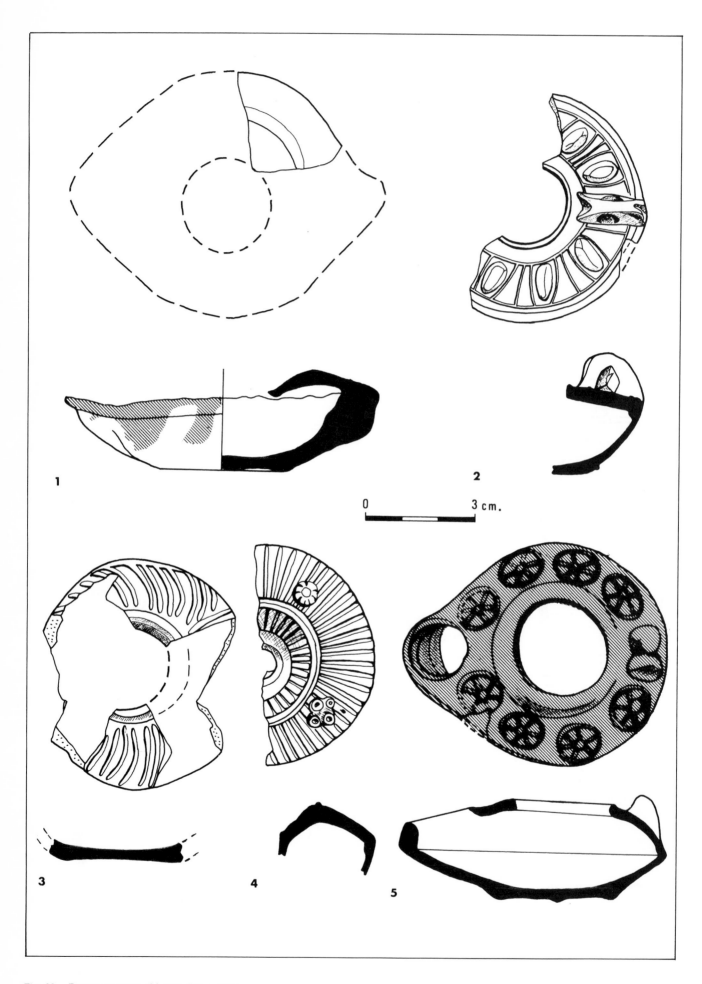

0   3 cm.

Fig. 88.   Roman pottery oil lamps (*see* p. 316).

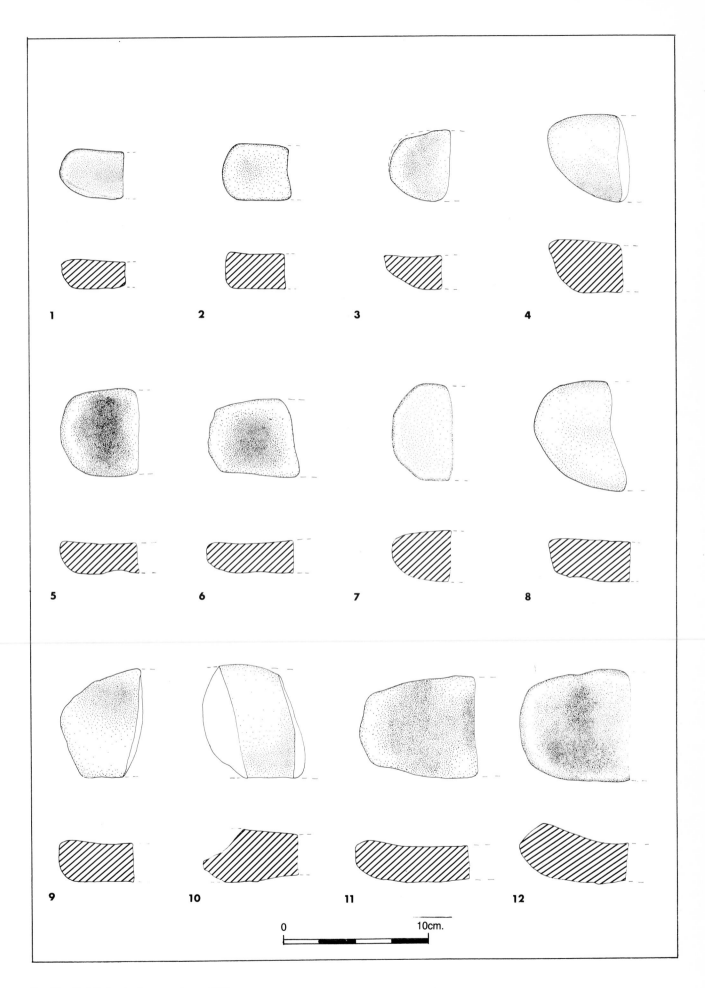

Fig. 89. Saddle-backed querns (*see* p. 316).

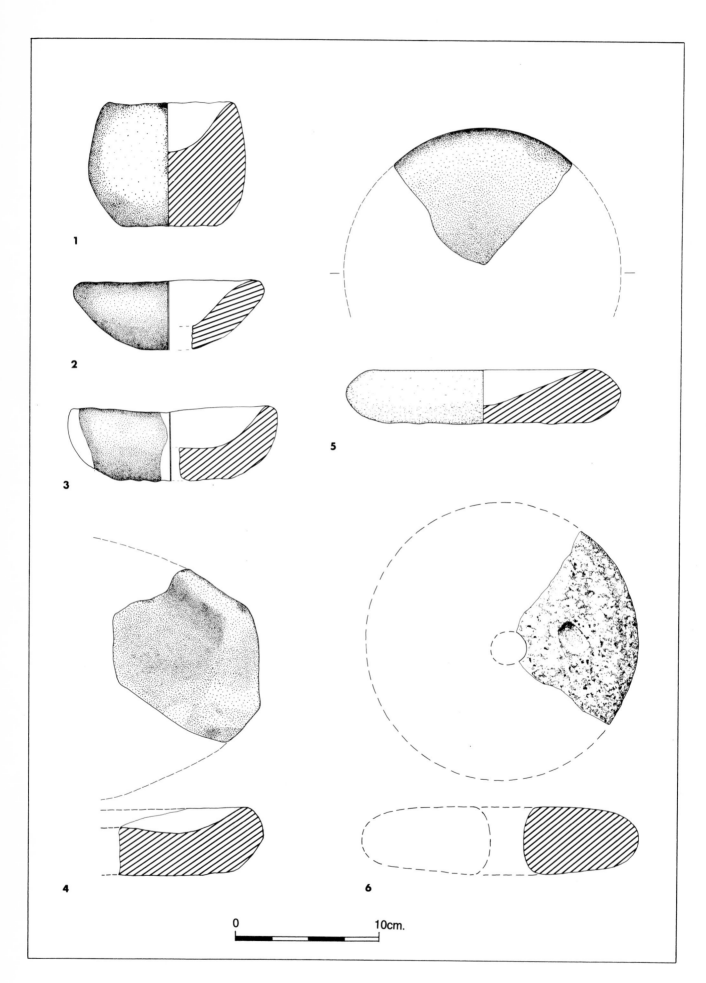

Fig. 90. Bowls, basins and grinding stone (*see* p. 316).

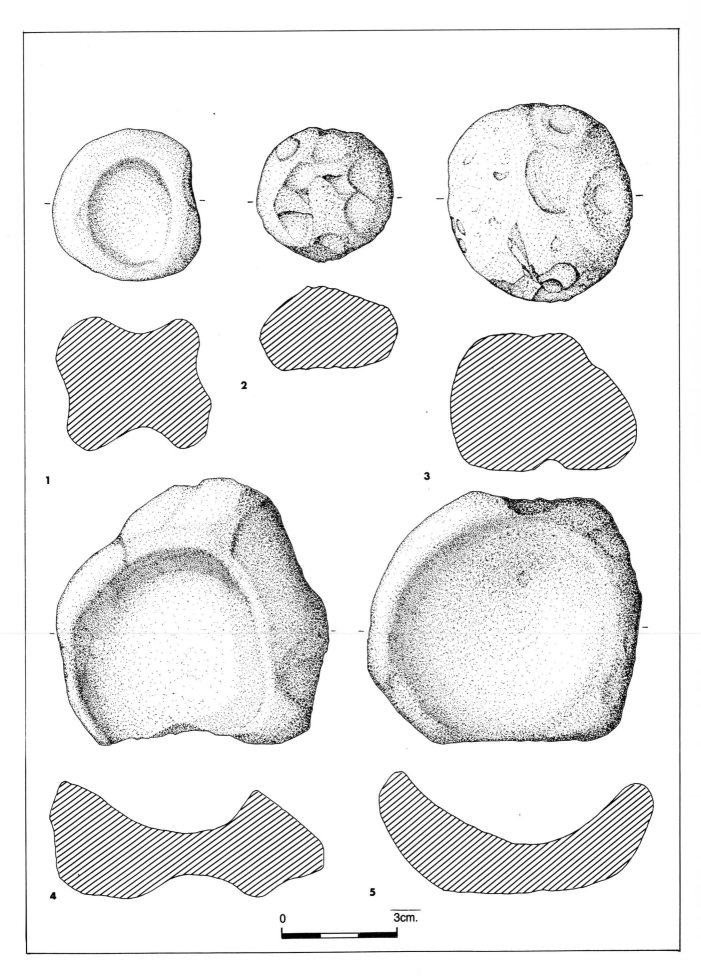

Fig. 91.   Hammerstones, anvil stone and small mortars (*see* p. 317).

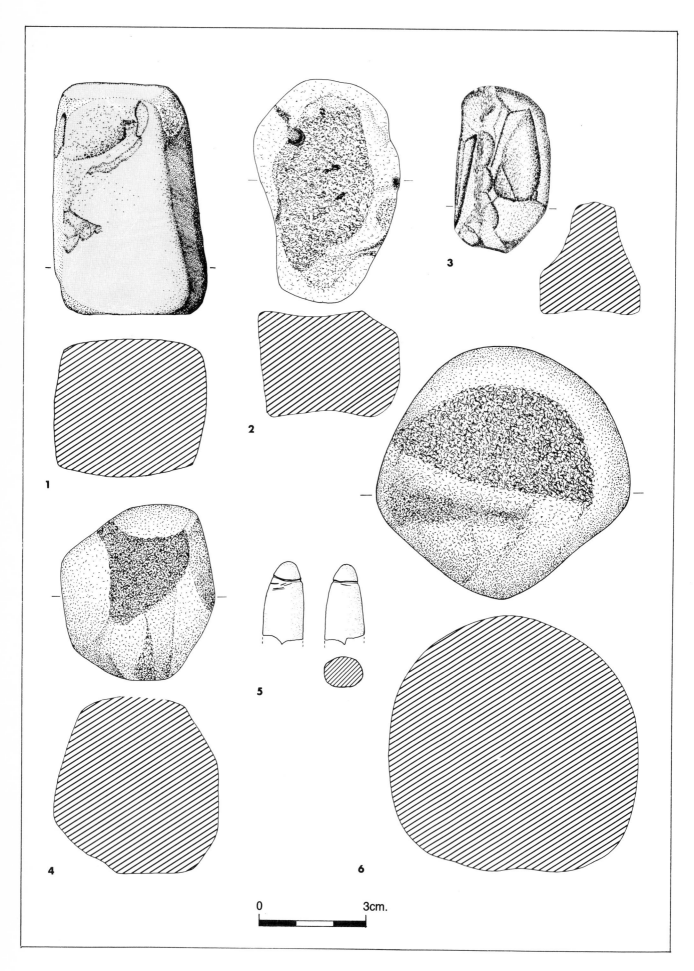

Fig. 92. Pestles, anvils and hammerstones (*see* p. 317).